COMPRESSED AIR AND GAS HANDBOOK

Sixth Edition

Compressed Air and Gas Institute

David M. McCulloch
Mac Consulting

www.cagi.org

Compressed Air and Gas Institute
1300 Sumner Avenue
Cleveland, Ohio 44115
216-241-7333
Email: cagi@cagi.org

Library of Congress pre-assigned control number: 2003091923

Editorial/production supervision: Leslie Schraff

ISBN 0-9740400-0-2

Publisher: Compressed Air and Gas Institute
1300 Sumner Avenue
Cleveland, OH 44115
Phone: 216-241-7333
Fax: 216-241-0105
Email: cagi@cagi.org
www.cagi.org

ISBN 0-9740400-0-2

Printed in the United States of America.

CONTENTS

FOREWORD

The Compressed Air & Gas Institute (CAGI) is a trade association with a long history of service to manufacturers and users of compressed air systems. Formed in 1915, CAGI membership comprises manufacturers of air and gas compressors, blowers, air dryers, filters, air powered tools, rock drills, rotary drills and other pneumatic devices used widely in manufacturing, construction, mining and process industries. CAGI strives to make available to educational establishments and potential users a wealth of knowledge accumulated over a considerable period of time.

The 6th edition of the Compressed Air & Gas Handbook has been reorganized into separate sections dealing with Applications, Production, Treatment and Distribution of compressed air. A further section deals with gases other than air. A completely new addition is a section covering positive displacement type blowers.

The individual sections also take into account the considerable changes that have taken place since the 5th edition was published in 1988. The dominant type of air compressor in manufacturing plants has changed from reciprocating to rotary. Controls have changed from simple electro-mechanical devices to solid state and microprocessor controls providing a much greater degree of flexibility and selection from multiple languages. Improvements in technology have increased compressor efficiency, drying and filtration techniques, allowing a much higher quality of the compressed air supplied to the point of use.

Much work also has been done on standards, not only for compressors, but for all related equipment and the quality of the air supply itself. Standards have been developed by the Compressed Air & Gas Institute in liaison with PNEUROP, its European equivalent. Some of these have become domestic standards through

ANSI, and some have become International standards through the International Organization for Standardization (ISO). Appropriate standards have been identified in this 6th edition of the handbook.

The Compressed Air & Gas Institute is a founding sponsor of The Compressed Air Challenge, initiated by the U.S. Department of Energy to help American Industry produce and use compressed air more efficiently and effectively. Institute member companies provide individuals who serve on the Advisory Board and the Project Development Committee, assist in preparation of printed materials and in training of end users, plant engineers, system designers and consultants.

The Compressed Air & Gas Institute has worked to provide the best engineering information and to make available an authoritative source for selection, installation, operation and maintenance of such equipment.

1

Compressed Air Uses

INTRODUCTION

The manufacture of compressors and of devices and machines to apply compressed air and utilize compressed gases is a large and essential industry that contributes greatly to the economy of all nations. Compressed air provides power for a multitude of manufacturing operations, some of which are listed in Figure 1.1. Carbon dioxide, compressed and added to the stage condensers of multi-stage flash evaporators, like that in Figure 1.2, helps prevent scale buildup and improves economy of operation. This is a process application, a type discussed at length in Chapter 7.

There are many different types of compressors, and their applications are numerous. By grouping them according to design objectives and differences, the reader may obtain a certain perception of the equipment and its applications.

The 80-125 psig air compressor is the first group. Perhaps this group is responsible for a slight technical error in the title of the book. Air is a gas. But air compressed to 100 psig is essentially a utility, and its compressors are so large a class that air compressors are considered separately from compressors for other gases. Most plant air systems operate at or near this pressure, and many air tools may be plugged into the system using quick-disconnect fittings, in much the same way the householder plugs in a toaster or food processor.

In the text of this book, every application of 100 psig air cannot possibly be taken up, since they are so numerous. As an example, consider the synthetic fibers that are now the main ingredient of clothing and rugs. When this material was first introduced into the marketplace, it had a smooth silky texture. The clothing items were considered uncomfortable, especially to men. To overcome this objection, the smooth synthetic strands were cut in lengths and woven into a combined thread of

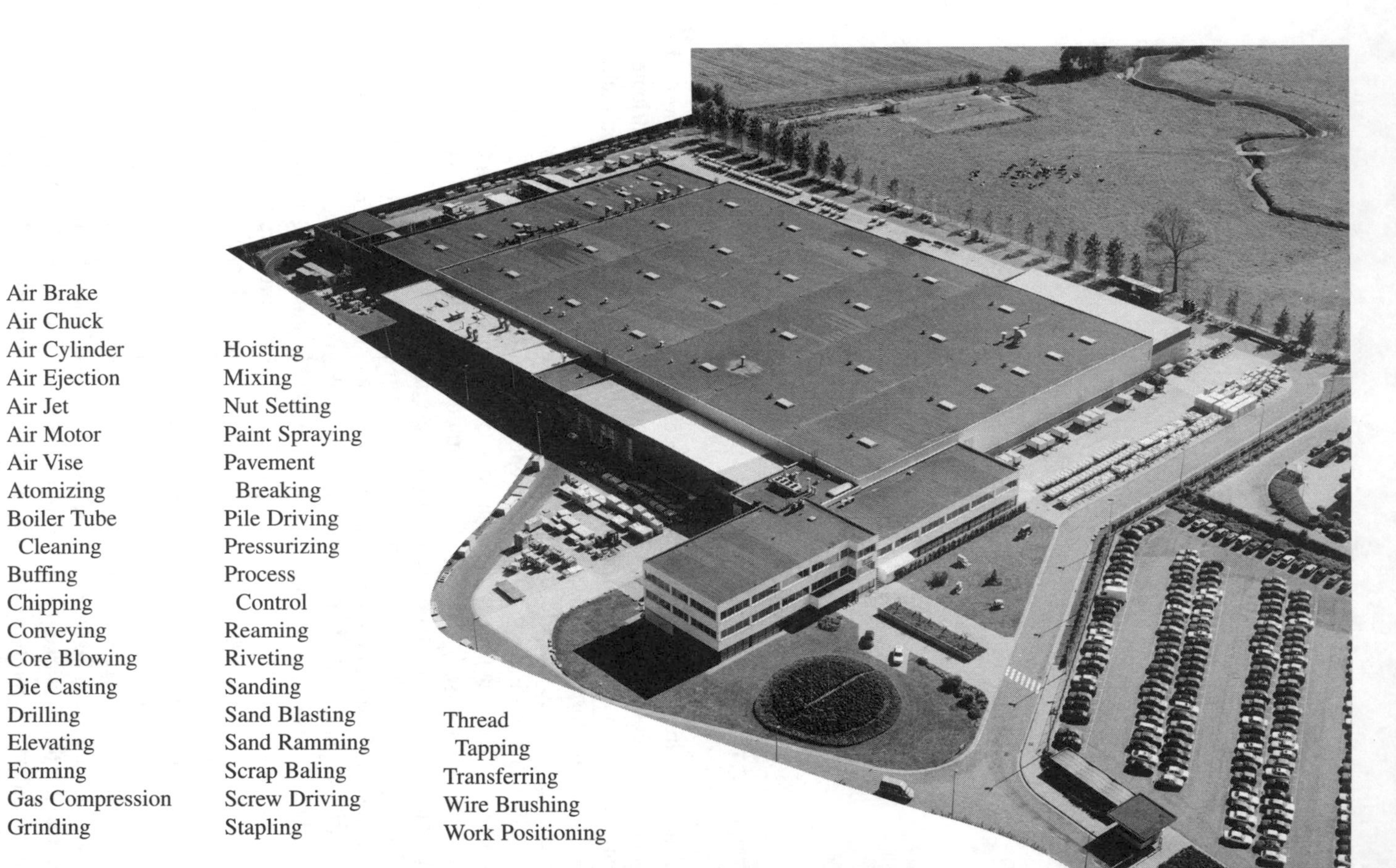

Figure 1.1 Typical major applications of compressed air in manufacturing.

synthetic and natural fiber, such as wool or cotton. The process produced a final product having the texture of the natural fiber component. However, the synthetic components would curl and form balls of the separated synthetic. The final solution was to tangle and loop the synthetic fiber by using nonlubricated air in a turbulent flow of air and fiber passing through a venturi.

Figure 1.2 Desalination plant extracting fresh water from sea water can have a profound effect on arid and semi-arid areas. Carbon dioxide, compressed and added to the stage condensers of such multi-stage flash evaporators, helps prevent scale buildup and improves economy of operation.

The design of the 100 psig air compressor permits a wide range of standard models, from the small home air compressor to the large-flow-capacity centrifugal compressor. The machine is self-contained in that all components required in the compressor design are furnished in a neat, compact package. The air compressor package should require the minimum amount of power per unit of flow. It may have one or two stages, or more, depending on the type of compressor used.

The second group is made up of gas compressors. In this group, special compressors are designed to meet a chemical process requirement on pressure or flow, and are capable of handling whatever gas or gas mixture is to be used. The gas compressor is the conveyor of the raw materials needed by the process. Within the process, the gas is chemically combined to produce the desired product. Several pressure levels are often required by the process. While the gas compressor may be similar in appearance to the 100 psig air compressor, much more engineering is generally involved. Such compressors will be discussed in Chapter 7.

The third division is made up of oil and gas field compressors. These are separated from the gas compressor group for several reasons. First, they are units that can be moved from one oil field to another by truck. They are usually designed to operate at 100 rpm to match the gas-engine-drive rotation speed. The compressor and driver are on a combined support skid. This requirement imposes restrictions on the size of the unit and necessitates more compact design. The use of the compressor in any application may be of short duration, and in many instances the user will rent the compressor-driver package. The oil and gas compressor is utilized to pressurize oil wells with natural gas to force the remaining oil out of the formation. The user may have many small wells positioned in various locations. These compact machines are also ideally suited for offshore platforms.

Larger reservoirs of oil, such as those located offshore, may require large-flow-capacity machines, and these are also considered oil and gas field compressors. The large-capacity units are often centrifugal compressors driven by a gas turbine. The manufacturer assembles the centrifugal compressor and its driver on steel modules that conform to the large offshore platforms. As noted earlier, these compressors will be discussed in Chapter 7.

The use of carbon dioxide and nitrogen to force oil to the surface is replacing the use of natural gas, reserving the natural gas for energy use. The nitrogen is produced by an air-separation process. The air is fed to the process using a standard 100 psig centrifugal air compressor. The nitrogen is at approximately 40 to 50 psig pressure from the air-separation process. The nitrogen is then compressed by a reciprocating compressor to a final pressure ranging from 1000 to 4500 psig, depending on the depth of the oil formation.

From the standpoint of applications, compressed air and gas may be divided into power, process, and control. Power service includes those applications in which air is used either to produce motion or to exert a force, or both. Examples are linear actuators, pneumatic tools, clamping devices, air lifts, and pneumatic conveyors. Process service is defined as any application in which air or other gas enters into a process itself. Examples are combustion, liquefaction and separation of gas mixtures into components, hydrogenation of oils, refrigeration, aeration to support biological processes, and dehydration of foods. Control applications are those in which air or gas triggers, starts, stops, modulates, or otherwise directs machines or processes.

Control applications occur throughout power and process use. Some steady-flow process plants are virtually automatic, and Detroit-style, batch-type manufacturing may be highly automatic, too. Pneumatic controls have special attributes that make them ideal for many situations, such as control of pneumatic machines (Figure 1.3) or control with explosion-proof requirements.

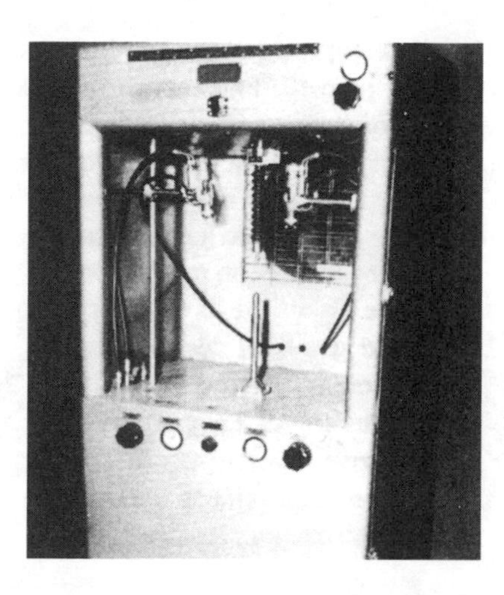

Figure 1.3 Automatic spray painting machine with two reciprocating air guns and automatic cycling and paint spray adjustment.

In some industries, compressed air may be limited to power service alone, as in quarrying and road building (Figure 1.4). Others may use air only in process service. But, in many cases, compressed air is used in both power and process services with pneumatic control within a single plant.

Figure 1.4 Crawler mounted blast hole drill sinking presplitting holes at desired angles for finished cut.

The United States Pavilion at Expo 70, Osaka, Japan, seen in Figure 1.5, was designed by Davis, Brady, Chermayeff, Giesmer, de Harak Associates under the direction of the U.S. Information Agency. It is thought to be the first compressed air-supported cable-roof structure ever built. Four air compressors were needed to support the 270 by 462 ft. elliptical roof. The translucent fiber-glass skin permits sunlight to pass through into the interior of the pavilion below.

Figure 1.5 United States Pavilion at Expo 70, just after the compressed-air-supported roof was completed, which was the first compressed air support roof ever built.

In addition to industrial plant applications, small portable air compressors for the home and job site have become very popular. These provide consumers and contractors with a productive alternative energy source to perform many jobs faster and easier. Do-it-yourself and professional applications abound for small portable air compressors and pneumatic tools.

Primary home site applications include inflating, cleaning, painting, wrenching, sanding, caulking, and grinding. Standard air chucks and other attachments are readily available to inflate tires, shocks, air beds, etc. Air operated spray guns, typically connected to a small portable air compressor, can be used to apply finishes or paint to furniture, automobiles, shutters, trim, and lawn equipment. Numerous woodworking related projects can be accomplished with grinders, sanders, and drills which are smaller and lighter than their electric counterparts. These small air compressors can be driven by an electric motor or gasoline engine where electrical supply is not convenient or close to water. These can drive impact wrenches and ratchets for loosening or tightening stubborn bolts and lug nuts.

Job site applications for electric or gasoline engine driven small portable air compressors usually include nailing and fastening. Some of these are small and light enough to be hand carried and include up to a 4 gallon air receiver tank. With the growing popularity of these air compressors, conventional hammers and nails have been replaced by pneumatic framing, roofing, and finish nailers. Typically, a framing and roofing nailer requires a 1 hp air compressor, while finish nailers for molding and trim work require only about 2 hp. In addition to nailers, small portable air compressors have been designed to operate staplers which have replaced older manual staple guns. Roof shingles, molding, trim, flooring, and industrial board are installed quickly and easily using the proper combination of small portable air compressor and pneumatic stapler combination.

These introductory examples are given to provide the reader with an insight into the major divisions of the compressed air and gas industry. In this chapter, only air compressors and compressed-air applications will be discussed.

SOME GENERAL USES OF COMPRESSED AIR

Compressed air is helping us to cope with some of the problems of our complex society. For example, in the production of food, orchards are sprayed by means of compressed air. Fish farming, which is a rapidly developing industry with trout and other fish already supplied in considerable quantity, depends on compressors to aerate the pools to keep the water fresh. In water supplies, deep reservoirs and lakes are similarly aerated to improve the water.

In underwater geological exploration for minerals and oil, a sudden release of compressed air produces acoustic waves by which the sea bottom is probed without the damage to marine life that resulted when explosive charges were used.

In sewage treatment plants, large volumes of air are used to help purify the water so that it may be discharged directly back into streams where fish may live.

Compressed air has found applications in virtually all fields of commerce and industry. It is used in the primary products industries supplying semifinished raw materials to manufacturing industries and in the manufacture of heavy goods. It is important in the light goods or consumer products industries, as well as in the processing and packaging of consumer goods. It plays an important part in transportation, building, and construction, and in service operations for maintenance of all industry.

Compressed air is used in virtually every phase of manufacturing. In one medium-sized plant, there may be a hundred different uses of air. Air from the same compressed air system may actuate stamping presses, air wrenches, aerating equipment and pneumatic controls and, at the same time, be part of a chemical or manufacturing process. In addition to the well-known applications for air, individual manufacturers find many special uses tailored in their own techniques.

Some applications of compressed air are so widespread that to discuss them in detail under each industry where they apply would be unnecessarily repetitive. Several of these are taken up in the following sections without reference to specific industries.

Industrial Plant Maintenance

Although there are many ingenious special uses of compressed air in individual plants, practically all industries find the same applications in at least the common problems of maintenance of buildings and machines.

Air tools such as paving breakers, described more fully under the construction classifications, are used for repairing concrete floors, opening masonry walls for various service lines, and similar work (Figure 1.6). Smaller air hammers are used for caulking and chipping. Plants with well-placed air outlets use air-operated drills, screwdrivers, and wrenches for other maintenance work. Portable paint-spraying outfits enable smaller crews to carry out this important maintenance job. Sprinkler systems, especially in unheated portions of a plant where freezing is a hazard, are controlled by air pressure, which prevents water from entering the pipes until heat breaks the seal and releases the pressure. The cleaning of machines, floors, remote ceiling areas, and overhead pipes progresses faster with the aid of air jets. Boiler tubes are quickly and thoroughly cleaned with air pressure. A familiar use of compressed air is tuck pointing of masonry walls. Metalizing of worn parts is done with compressed air.

Figure 1.6 A compressed air operated paving breaker for road construction/repair.

On the Production Line

Pneumatic tools have many advantages in industrial production. They have a low ratio of weight-to-power, which is especially important whenever human fatigue is a problem, and they may be used steadily for long periods without overheating and with low maintenance costs (Figure 1.7).

Among the air tools that find general application are chipping and scaling hammers used in railroads, oil refineries, chemical refineries, shipyards, and many other industries. These tools are used in the foundry for cleaning large castings. In other industries, they remove weld scale, rust, and paint. They are also useful in cutting and sculpturing stone.

Pneumatic drills are of special value in the machine shop for all classes of reaming, tapping, and drilling whenever the work cannot be conveniently carried to the drill press and for all classes of breast drill work. These machines are also frequently employed for operating special boring bar and, in emergency cases, for independent drive of a machine tool where the horsepower required is within their capacity.

Pneumatic drills, like other portable air tools, achieve great time and cost savings over the corresponding hand tools for reaming, tapping, and drilling, especially in the automotive, aircraft, rail car, locomotive, and other heavy machinery industries.

Figure 1.7 Pneumatic screwdrivers are used on an assembly line.

Grinding, sanding, wire brushing, buffing, and polishing are also speeded up and facilitated in these same industries and many others. Finishing surfaces and preparing them for finishing operations are the prime objectives.

Two regular assembly operations, driving screws and turning up nuts, are greatly speeded up by pneumatic screwdrivers and nut runners. In some instances, magazine feed and attached locating fixtures further speed the operations.

Air-operated riveting hammers of either the percussion or compression type produce reliable, inexpensive joints. A later paragraph includes discussion of riveting as it is extensively applied in aircraft manufacture.

Integral-gear-type centrifugal air compressors are generally used in central plant air applications requiring volumes ranging from 1000 to 30,000 cfm and discharge pressures in the 100 to 125 psig pressure range. Central air supplies are utilized extensively in manufacturing for pneumatic tool operation and robotics, as in the automotive industry, and for instrument and control air purposes in many industries. Figure 1.1 provides a list of typical applications of air compressors in manufacturing. Such units have also found utilization in process industries to provide an oxidation air source in chemical and petrochemical plants and to aid bacterial growth in the pharmaceutical industry. Figure 1.8 shows a typical installation of a centrifugal compressor for plant air.

Figure 1.8 Three-stage integral-gear-type centrifugal compressor for plant air service.

Air Motors, Vacuum and Other Auxiliary Devices

Air motors are widely used as a source of power in operations that involve lammable or explosive liquids, vapors, or dust. They may be operated in hot, cor-osive, or wet atmospheres without damage.

Air-motor speeds may be varied easily. Such motors may be started and topped quickly and are not harmed by stalling and overloading.

In general, air motors may be classified as either the rotary sliding vane type r the piston type. Either type may be supplied with a gear-type speed reducer. otary sliding vane motors generally deliver lower torques at higher speeds, while iston motors, whether axial or radial, deliver higher torques at lower speeds.

A great many hand-held air tools are driven by rotary sliding vane motors, ften with a speed-reducing gear train. Air hoists may be powered by either type of notor, the piston-type being especially suited for accurate positioning of loads. The afety aspects of both types led to many applications in underground tunnels and nines, and in industrial areas where there are solvents or other flammable sub-tances. They drive many pumps used in construction and many positioning evices, such as indexing tables, used in manufacturing.

Vacuum has many applications in production. A vacuum pump is a compres-or in which the intake vacuum rather than the pressurized air is the desired effect. or vacuum chucking, as an example of vacuum applications in production, the ump holds a vacuum in a tank located near the machine, while bleeder holes under he part to be machined are opened to hold the part in place. An intake filter on the nlet line to the vacuum pump cleans the air of any foreign material that could oth-rwise be picked up from the machining process.

The extensive use of pneumatic auxiliary production equipment should be oted as well. Clamps, positioners, presses, feeders, air chucks, and many other evices actuated by air cylinders are found to be effective, economic speeders of roduction. After the cost of accurate welding and millwrighting is taken into ccount, it is usually found that pneumatic cylinders, plus ratchets or stops, provide eciprocating or rotating interrupted motions much more economically than can be rovided by many traditional mechanical devices. Vacuum devices perform similar unctions for smaller or lighter parts. Air hoists and winches greatly facilitate the andling of heavier products in shipping, as well as in production, construction, nd mining.

In areas of finishing and packaging, pneumatic devices find many important pplications. For example, certain dry powders may be fluidized by compressed air nd applied by electrostatic deposition. Preheating the surface can cause polymer owders to fuse on contact, producing a continuous plastic coating. Pneumatic sta-lers facilitate the closure of many packing cartons. Pneumatic sanders help pro-ide a smooth finish on many appliances.

Blast cleaning has been successfully applied to such jobs as putting a satin fin-sh on completed work, removing scale, paint, and rust from surfaces, preparing etal surfaces for painting, enameling, tinting, sheradizing, or galvanizing, and

cleaning and finishing castings. Cleaning castings and building exteriors is proba
bly the most extensive application. In addition, it is used for cleaning pottery anc
crockery, bottle molds, forgings and steel plates, and sheets. Blast nozzles are usec
alone or are incorporated into varying designs for blast equipment. The commo
term sandblasting is a misnomer since many abrasives besides sand are used, result
ing in a wider range of applications.

Air Separation

A large number of integral-gear-type units have been utilized since 1960 i
cryogenic air-separation plants for both main air and nitrogen booster recycle serv
ice. Figure 1.9 features an installation of a 2700 hp, motor-driven integral gear uni
for main air compressed service in a 150-ton/day oxygen plant. In the same plant
a special high-pressure version of the integral gear unit (Figure 1.10) is utilized fo
nitrogen recycle service.

Figure 1.9 The main air compressor of an air-separation plant. It has a capacity c
14,600 cfm compressing from atmospheric pressure to 97 psia and requires 2700 bhp.

Figure 1.10 A large nitrogen compressor used in air separation. It has a capacity of 5900 cfm, with inlet pressure of 85 psia and discharge pressure of 515 psia with a 6200 bhp driver.

This unit has rated inlet and discharge pressures of 85 psia and 515 psia, respectively. Because of the high density of the high-pressure nitrogen being compressed, this unit of relatively small volume, 5900 cfm, requires a 6200 hp motor drive for three stages of compression.

The previous applications point out the versatility of the integral gear centrifugal compressor designed for air and nitrogen service in a large number of industries at varying volume and pressure ratings, and in a considerable number of configurations of stages and intercooling arrangements. See also nitrogen.

Automation

Where production quantities warrant, production lines may be automated profitably. There have been many exciting trends in the field of automation by pneumatics. For example, air circuitry and pneumatic controls permit the integration of conventional and special air tools and auxiliary devices into single automatic machines (Figure 1.11). One system uses a building-block approach (Figures 1.11 and 1.12) with a high degree of interchangeability of pneumatic tools and controls. Fluidics offers simple devices for pneumatic control at lower pressures and with virtually no moving parts. Pneumatic positioners, capable of positioning parts to an accuracy of 1/1000 inch without the use of mechanical stops, have been developed. Pneumatic punch card and tape readers offer all the advantages of digital controls and programming to pneumatic systems.

Figure 1.11 Drilling and other machining operations are carried out automatically on this pneumatically controlled machine.

Figure 1.12 Precision machine parts are produced by this machine using compressed air not only for control but also for milling, drilling, and indexing.

Other important developments in automation with compressed air include pneumatic handling of materials. Many substances in granular, chip, pelleted, or powdered form are very successfully handled in this way. The different methods utilized in pneumatic transfer of materials are discussed in a later section of this chapter under the heading Pneumatic-type Conveyors.

The range of solids transferred extends from cement to pelleted rubber. In cement production and in the unloading of grain ships, to mention only two examples, industry is highly dependent on pneumatic conveying; such applications are large consumers of compressed air. The reader is referred to the specific topics in this chapter on these industries, as well as lumbering and woodworking, the food industries, and rubber manufacturing, for examples of this important materials-handling technique.

Painting is often automated (Figure 1.13), using air circuitry and pneumatic controls in robot machines similar to that shown. Masks are automatically cleaned in a solvent and dried by air jet.

Figure 1.13 Large recreational vehicles are painted on an automated robotic paint line.

Compressed air is used widely in automatic packaging machinery for sealing, locating the work, and actuating arms that fold paper to wrap the work. Pneumatic packaging is most commonly applied to small, mass-produced articles such as dry cells, candy bars, and writing tablets, but its use on other products is rapidly increasing. Vacuum devices find many similar applications, such as picking up and transferring sheet metal and sheet-metal parts.

Automated Assembly Stations

Compressed air plays a vital role in the design and operation of automated assembly systems. Automated assembly stations (Figure 1.14) are speeding up assembly operations on high-volume production lines in the automotive, appliance, electronics, communications, and business machines industries.

Air power is especially suitable for automated systems because it is safe and clean, and the work it produces is easily controllable over a broad operating range by means of simple, low-cost control devices. Air power can produce either reciprocating or rotating motions, and the tool or feed mechanism being powered can be installed without injury to the system.

Typical air-power applications in automatic machines are the following:

1. To tighten threaded fasteners such as screws, nuts, and bolts to specified torque.
2. To drive plugs, pins, and rivets with air. There is the option of either pressing or hammering.
3. To feed fasteners or parts.
4. To actuate positioning cylinders, slides, or work heads.
5. To operate indicator lights showing such conditions as satisfactory completion of work cycle, reject, or the possible shutdown of feed, drive, or positioning components.
6. To transmit signals to recording computers.

Figure 1.14 Semi-automatic and fully automatic assembly stations utilize air power to feed parts and position them, tighten or drive fasteners to specified torque, and inspect the completed assemblies.

Acid Manufacture, Agitating Liquids

In one process for making nitric acid, compressed air and ammonia are passed through a catalytic converter and the resulting gas, with additional air, is then passed through an absorption tower where nitric acid is formed. The process is included here because it utilizes air compressors. Waste gas for the process is put through an expander to provide power for the compressor and to improve the process efficiency. A power recovery unit of this kind in a nitric acid plant is seen in Figure 1.15.

Compressed air has important applications in agitating, elevating, and transferring acids and acid solutions. Agitating is usually accomplished by means of an air pipe run along the bottom of the tank. Air issues from the openings along the pipe and bubbles through the liquid to provide the desired turbulence. Corrosion problems in the acid industry require the proper selection of materials, and they are often expensive. The simplicity of the equipment needed for such air agitation thus gives it an economical advantage over mechanical equipment for the same purpose. Air or gas may also be a part of the process itself and may be very simply introduced during agitation.

Figure 1.15 A power-recovery turbine providing power for an air compressor in a nitric acid plant.

Aeration Blending

Modern aeration blending uses a system that can blend dry, pulverized materials efficiently on an industrial scale. This method has been applied successfully to materials as varied as cement, raw materials, and polyvinyl chloride. By contrast, mechanical blending does not easily produce as uniform a mixture as the aeration method. Most dry, pulverized materials that can be fluidized by aeration can also be blended by this method. Blending aeration commonly reduces deviations from plus or minus 3 percent in the unblended materials to plus or minus 0.1 percent in the final product. Composition is homogeneous enough to permit discontinuing regular sampling and testing in many plants. Improved quadrant blending and the use of pulsating air are responsible for this progress.

The aeration blending system has demonstrated these advantages: thorough homogenization even when the blended materials differ in bulk density, fineness, and specific gravity, high tonnage and hourly capacity with rapid mixing and continuous discharge of blended material following an initial mixing period, and excellent economy in electrical power per ton of blended product.

The quadrant blending method incorporates a round, flat-bottomed silo that is filled with material to a level about equal to the silo diameter. The silo bottom is covered with aeration units closely spaced to give a uniform distribution of air into the material throughout the entire dispersion region. Each aeration unit is faced with a porous refractory block to release air into the material in innumerable fine jets.

To effect blending, enough compressed air is let into one-quarter of the silo bottom to expand and fluidize the material above that section. As a result, this material rises above the level of the adjacent quadrants and flows rapidly out across them, running off at a slope of 2 to 5 degrees. At the bottoms of the temporarily inactive quadrants, a small amount of compressed air is meanwhile released, with just enough air to make the material mushy but not enough to produce much change in bulk density.

The rising column of expanded material in the actively aerated quadrant is thus partly surrounded with heavier material, which tends to slump over into the active quadrant, where it subsequently becomes aerated, expands, rises, and flows outward in its turn. A strong, continuous roll-over motion in a vertical circuit is obtained, which results in a very thorough mixing of the silo contents from top to bottom and throughout the active quadrant. The intensive aeration is automatically switched to a previously inactive adjacent quadrant under variable, timed control and then successively to each of the other quadrants in rotation, blending all the material in one cycle, although two circuits are often used to assure virtually perfect blending.

Following the blending of the initial charge, the process becomes continuous, because the incoming stream with variable analysis is rapidly mixed in the large mass of blended material in the silo, and a stream having uniform analysis can be withdrawn continuously from the bottom of the silo.

The time required may be from 1 to 2 hours following the establishment of a sufficient head of unblended material to initiate the fluid motion, depending on the character of the mix, the size of the silo, and the length of the period of air admission to each quadrant section. For fast blending, these periods may be from 5 to 15 minutes per quadrant. Results are improved when the flow to the active quadrant is pulsating flow.

Applications include blending of crystalline or other finely divided materials showing finenesses in the general range of 100% through 150 mesh and 85% or more passing 200 mesh, and especially those powdery materials that have been found to flow well in dense-stream conveyors.

In addition to the porous refractory blocks already described, there are several other important parts. Air-distribution pipes and headers are installed inside the silo for selective control of the air supply to the quarter-sections of the silo bottom.

Aeration, Air Screens, Agitation, and Bubbling De-Icing

For many years, the principle of releasing compressed air beneath the surface of a liquid and allowing the resulting bubbles to rise to the surface has been put to a multitude of uses. One is the agitation of the liquid itself, to stir it without mechanical equipment, either to mix several liquids of different viscosities or to mix solids with a liquid. This same principle of agitation may be used to distribute heat throughout the liquid, the same as is done by mechanical stirring.

A second application is the release of a compressed gas into a liquid for the purpose of mixing the gas with the liquid to speed up a chemical reaction. Frequently, air is bubbled through a liquid in order to use the oxygen content of the air to oxidize the liquid or material being carried by liquid. This is the principle of bubbling sewage to oxidize it more rapidly and, at the same time, keep the entire mass mixed sufficiently to be handled by pumping equipment, that is, to prevent setting of the solids. An aerated lagoon with air bubbling from a number of sources is shown in Figure 1.16. Also, water purification by oxidation of reservoirs (Figure 1.17) is quite common. The reduction of algae that results means that much less chlorination is required, with resulting better taste.

Figure 1.16 A lagoon being aerated by sources at or near the bottom.

Another application is the formation of a screen of air bubbles beneath the surface of the water that can act as a shock absorber to protect a subsurface such as a dam from explosive forces, such as blasting under water in the lake immediately behind the dam. Properly engineered air screens can absorb the explosive force since the air is compressible while the water is not, and thus prevent this extreme hydraulic impact from being imparted to the underwater structure or device that is to be protected.

Figure 1.17 Bubbles may be seen rising to the surface of a lake. The air supply at the lake bottom not only helps to oxidize impurities but also brings stagnant water to the surface from depths at which thermal currents may not be caused by the heat of the sun. Both actions contribute subtantially to the quality of the water.

Air screens have been used successfully to divert schools of fish in commercial fishing. Also, air bubble screens have been tried for shark protection on bathing beaches.

For many years, the principle of releasing air bubbles to control wave action, the air thus serving as a breakwater, has been successfully used.

With the rapid growth of the boating industry in recent years, the increased number of marinas has brought another bubbling principle into common use (Figure 1.18). In those areas where freezing occurs during winter months, the formation of heavy ice around piles and docks can cause extreme damage. In the past most boats were removed from the water to protect them in winter months when freezing could occur. The dry-land storage problem now has become a serious one because of space limitations and, at the same time, it is better to keep most boats in the water to prevent their drying out and to avoid the expensive recaulking that would then be necessary.

Air bubbling can take advantage of the natural inversion that occurs as water cools down to near the freezing point. At approximately 39°F, the warmer water no longer stays on the surface of a pond, lake, or stream; rather, it settles to the bottom. As the temperature continues to drop until it reaches the freezing point of 32°F, the water will then finally freeze on the surface. If the natural inversion did not take place, water would freeze from the bottom up.

Figure 1.18 Compressed air released below the docks brings warmer water to the surface to keep the dock area ice-free.

The release of air bubbles beneath the surface of a body of water will bring some of this warm water to the surface where the air and water mixture will rise above the surface of the water and tend to flow outward, thus spreading the warmer water over a larger area, thereby preventing freezing. This procedure can even melt ice that has previously formed. This principle is effective with any body of water having sufficient depth to maintain a temperature above freezing at its bottom. Large areas need not be included. Bubbler tubing along docks or around boats will maintain a sufficiently open area to afford protection against ice damage.

This same principle is used to keep ponds and lakes open for wildlife, such as birds that depend on open water to survive and fish located in ponds that might normally freeze over solid and possibly suffocate the fish by lack of oxygen in the water.

Finally, there are applications making use of the principle that rising bubbles will cause the water to reach a height above the surrounding water. One of these is the containment of oil spillage, and another is creation of a barrier against saltwater intrusion into freshwater streams because of tidal flow. In both of these applications, the rapidly rising bubbles actually make the water level higher at the point at which the bubbles break through the surface. Thus, the water at this higher point tends to flow back upon itself, causing, in effect, a barrier.

Agriculture

Applications of compressed air in agriculture are so numerous and varied that no attempt is made here to list them all. Only a few specific uses are mentioned.

Compressed air is widely used in farm equipment and for farm operations in most of the primary and auxiliary agricultural activities, such as erosion control, land drainage and irrigation, tilling, planting, insect and weed control, pruning, harvesting, and threshing. Compressed air performs useful services in connection with livestock raising and dairying. There are many compressed air applications around farm buildings, too, such as pumping, material handling, and primary processing, as well as maintenance and repair of farm machinery.

Some more recent applications of compressed air in agriculture include spraying trees, dusting insecticides and fungicides, feeding livestock in transit, disinfecting poultry houses, handling rice hulls, changing tractor tires, cleaning eggs, picking raw cotton, and seeding and fertilizing with compressed air guns.

Vacuum also finds many uses in agriculture, including milking machines. Vacuum lines can transport milk directly to tank trucks for improved sanitation. Vacuum egg lifters are also used to lift individual eggs gently for packaging. Vacuum seed planters deposit single seeds and eliminate later thinning. Vacuum is also used to impregnate eggs with medicinal solutions to reduce mortality rates of chicks.

An interesting compressed-air development is the use of foam to protect delicate crops from frost. Early tests indicate that the method is promising.

Aircraft

Several hundred thousand rivets go into the manufacture of an airplane or helicopter. Powerful, lightweight, air-operated drills, wrenches, and riveting hammers are used in the assembly of airplane fuselages, wings, and other components. The importance of such air tools is seen in Figure 1.19, in which an air drill is being used in the construction of a small plane.

Figure 1.19 An air drill in use in the manufacture of an airplane.

Airplanes provide many applications for compressed air in addition to those used in the manufacturing stages. Compressed air stored in the air springs of the landing gear softens the shock when the airplane is landing or taxiing. Cabins are pressurized and air conditioned for high-altitude flights. Compressed air is used for de-icing plane wings, for heating the engines, for various actuating and control functions, and for operating the refueling equipment. The modern safety devices with which particularly the overseas airplanes are equipped, such as life belts, rafts, and emergency chutes, are inflated quickly and reliably, when needed, by compressed air or gas stored in high-pressure bottles.

Compressed air is used to start jet engines and to provide cabin service for the comfort of passengers while the plane is at the terminal. A compressor specifically adapted to this purpose is seen in Figure 1.20.

Figure 1.20 A compressor unit specifically designed for aircraft support while the plane is at the terminal.

Some airlines at certain airports are changing over to stationary versions of these units, electrically driven, thus reducing overall energy costs. The compressor in each case is a regular single-stage unit, but the control system is fairly critical due to the nature of the application.

Automobiles

From manufacture to maintenance, compressed air plays an important role in the automobile industry. Air-operated drills, nutsetters, grinders, buffers, pneumatic hammers, and impact wrenches are among the hand tools commonly used in both factory and garage. A pneumatic angle wrench is used (Figure 1.21) in automobile assembly.

Air chucks, air-operated tailstocks for safe and quick travel, and other air devices are used on machine tools where a high rate of production is a factor. Air hoists to take heavy pieces to and from a machine are fast-acting and easy to control. They permit assembly-line techniques to be used with parts and assemblies such as engines or transmissions that would otherwise be too heavy to handle.

The casual visitor to an automobile assembly plant may not be aware of the extent to which compressed air facilitates automation. An electric welding machine, for example, has air lines to supply the pneumatic clamps that hold the work in position during welding, and these are easily mistaken for electric cables. Few general visitors are aware that the machine for pressing recesses or dimples into automobile firewalls to strengthen them is pneumatically operated.

As production line operations become more and more automated, whether it be drilling holes or welding frames or bodies, the smooth power characteristics and compactnes … that they will remain essential parts of the equip … produced. Air circuitry and controls facilitate th … sembly machines.

Figure 1.21 Utilizing computer control, lug nuts are tightened precisely to the vehicle manufacturer's torque requirement.

Servicing the automobile provides many more applications for compressed air such as painting. It is used in equipment for removing and inflating tires, in air lifts, air jacks, pneumatic grease guns, and air jets, in blowing out clogged gasoline lines, and in cleaning out car interiors. It is also used in air guns for spraying paint or antirust coatings and for oiling springs. Compressed air is also employed in retreading and regrooving tires, sandblasting and cleaning pistons, and sandblasting spark plugs. Agitating solutions for cleaning metal parts is still another use.

Automobile manufacture is a field that has already made widespread use of industrial robots. Among their many applications, these devices clamp and hold frame members while they are welded automatically. While robot control is usually electrical, pneumatic manipulators are in widespread use on robots because of the well-regulated movement of which they are capable.

Beverages

Both soft-drink bottlers and brewers depend on compressed air for a number of bottling operations, including capping bottles and kegging beer (Figures 1.22A and 1.22B). The immediate, sensitive response of compressed air makes it the choice of power on many types of controls in production processes. Cooperage departments of breweries use compressed air to coat kegs with pitch by forcing it through a length of pipe leading into the kegs. Pipe orifices spread the pitch to all the interior surfaces. Transferring liquids from cask to cask, unloading grain from cars and hoisting it to storage bins, and testing kegs for leaks are other applications for compressed air. Automatic bottling machines have many applications for compressed air in control and in actuating some of the necessary reciprocating and intermittent motions. See Distilleries.

Figure 1.22A Plastic bottles, produced with compressed air, have replaced much heavier glass bottles for many beverages.

Figure 1.22B A variety of plastic beverage bottles molded with the use of compressed air.

Blast Cleaning

Blast cleaning, also known as abrasive blasting, is a process in which abrasive material entrained in a jet is directed onto the surface to be cleaned and abraded. The particles may be natural sand, man-made mineral, or steel abrasive granules, shot, and the like. Blasting may be used for cleaning, controlling surface roughness, such as required when preparing for subsequent surface treatments or application of coatings, or carving a design upon a surface. Sandblasting to renovate a stairway is shown in Figure 1.23.

Formerly, the removal of mold sand from the surface of castings was a tedious job, and some sand often remained to dull the tools in subsequent machining operations. Such cleaning is now commonly performed by sandblasting with excellent results.

Similar applications are found in removing scale formed on steel during rolling, forging, or heat treatment. Most of the surface coating processes, like painting, plating, and enameling, are much more effective if the surface to be treated has been properly prepared by blasting. Cleaning of building walls and carving of monuments are two other well-known applications.

Sandblasting machines are usually designed to operate on air at a pressure of 90-150 psig available from the shop or construction lines. Basically, the size of the nozzle aperture determines the air flow needed to operate the machine. For example, the air flow is 23.5 cfm for 1/8-in. nozzle and about 210 cfm for a 3/8-in. nozzle, assuming an air pressure of 90 psig. The larger air stream raises the productivity of the operation, but it is uneconomical if only small-sized parts have to be treated or if the equipment is operated only occasionally.

Wet sandblast-type machines are mainly designed for outdoor work where they are widely used because of their many advantages. The main advantages are freedom from operating dust, independence from weather conditions, and eliminating the need for drying the sand. In some applications, the possibility of introducing liquid corrosion inhibitors into the sand mixture is another valuable feature of this system.

Figure 1.23 Blast cleaning a stairway as part of a renovation project.

This system of sandblasting has many applications in removing mill scale, marine scale, and fouling growth from marine hulls. It is also used in cleaning large tanks, pipe lines, bridges, and many other structures.

Some blast-cleaning machines use soft absorbent materials such as ground corn cobs, nutshells, and sawdust instead of sand. The soft action of the blast does not injure wiring and other delicate components, and because there is usually no need to dismantle the assembly, this type of cleaning is more convenient for many applications than the conventional method of washing individual parts in a solvent bath. Most types of blast-cleaning machines are manufactured either as stationary or portable units. Air blast cabinets designed for handling small-sized parts are usually stationary. There are also various types of rotary table blast cleaning machines used for continuous operations.

Breathable Air for Contaminated Environments

Workers in environments contaminated by toxic wastes, paint vapors, paint spray, or other potentially harmful substances are provided with a supply of clean breathable air by a system like that shown in Figure 1.24, or by similar, larger systems. The air is taken from a clean environment so that no cartridge replacement is required, and the compressor is an oil-free, nonlubricated type. Besides painting, another wide field of application is the removal of asbestos from the walls and ceilings of old buildings.

Figure 1.24 Breathing-air equipment for a worker in a contaminated environment. The compressor is placed in a clean environment outside the contaminated region.

Carpet Industry Application

Compressed air is used extensively in the yarn extrusion and yarn entanglement processes. The resulting texture of air entangled products has made them very popular in recent years and compressed air is the main ingredient. The yarn can be of several types (nylon, polyester, polypropylene, etc.), colors, and deniers (fineness or thickness) that are entangled through an air jet. Most air entangled products use air pressure up to 175 psig. Two stage rotary screw compressors are the preferred choice to meet energy efficiency and high pressure requirements.

Cement Production and Products

The Portland cement industry is one of the largest consumers of compressed air, and probably few industries surpass it in diversity of application, as seen in Figure 1.25. The largest portion of the compressed air required in a cement plant is utilized in conveying. All but a negligible portion of the Portland cement manufactured in the United States is transported from grinding mills to storage silos by compressed air pumps. Many mills also utilize the compressed-air pumping method for conveying cement from silos to packer bins, for loading and unloading cars, and for unloading and loading ships and barges. Kiln flue dust, packer spill, and pulverized coal are frequently handled in this manner.

Figure 1.25 Compressed air is a vital part in the production of cement. Big air users include bag houses, control air, and conveying systems.

In many dry process plants, the raw materials are both conveyed and blended for precise chemical control of the composition by compressed-air pumping and aeration.

In wet process plants, compressed air is utilized to mix and blend the slurries and to maintain the individual mineral particles in intimate mixture and suspension. To decrease fuel consumption in burning, many mills dewater the slurry by filters served by vacuum pumps. To ensure free flow and discharge of dry pulverized materials, aeration of bins is a universal practice throughout the industry.

Rock drills are essential in cement-plant quarries, and compressed-air rock hoists and car dumpers are used by most crushing departments. Air-operated grinders and other tools are commonly used in the large maintenance shops that cement plants require, and compressed-air and vacuum lines are essential in plant laboratories.

Large volumes of air at relatively low pressures are required in cement manufacturing. Blowers and fans supply the fuel or primary combustion air stream to kilns, whether fired by pulverized coal or oil. The use of air-swept unit mills for pulverizing coal is rapidly increasing. Air, in large volumes and at fan pressure, quenches the hot clinker, reducing its temperature abruptly from 2500°F to about 150°F for the purposes of improving cement quality, recovering heat, and reducing the clinker to a temperature suitable for grinding. Most modern plants employ air-swept pulverizers in closed grinding circuits to control cement fineness and to economize on power. Similar circuits are used in dry process plants for the preparation of raw materials. Fans also serve as dust collectors in almost every department of a modern cement plant.

Chemical Plants

A two-stage, integral-gear-type centrifugal compressor to provide process air in an amino acid plant is shown in Figure 1.26. This unit is discharging at only 35 psig to match process requirements.

Figure 1.26 A low pressure two-stage Centrifugal Air Compressor utilized in a chemical plant.

Construction

Construction on roads, buildings, dams, bridges, and tunnels probably accounts for more of our gross national product than any other single phase of industrial activity (Figures 1.27 and 1.28). In most construction operations, compressed air provides the power from the time ground is broken until the job is completed.

Figure 1.27 Largest tunnel under the Elbe River in Germany.

Figure 1.28 For well digging, pneumatic drills consume 400 cfm at 125 psig.

The paving breaker (Figure 1.29) was among the first pneumatic tools to find widespread use. One person with this tool can do the work of 15 working with hand sledges and chisels in cutting asphalt or concrete pavements or in demolishing concrete foundations, retaining walls, floors, partitions, and other structures.

Figure 1.29 The paving breaker is one of the most widely used construction tools. It facilitates removal of old paving so that new construction may proceed.

Similar tools equipped with spades, diggers, root cutters, drivers, and tampers are very widely used for more sophisticated tasks.

The reciprocating mass used in the paving breaker with the added feature of a slow, revolving motion is used in the rock drill, one of the most important adjuncts of modern construction. Pneumatic rock drills are available in a wide range of sizes from comparatively light portable tools to heavy, propelled machines. These tools are capable of drilling blast holes at any angle, even drilling multiple holes into hard rock (Figure 1.30) to depths of many feet and at speeds that one or two decades ago would have been considered unbelievable. Drill points are cooled and the drill cuttings removed from the holes by means of an air stream directed at the spot where needed. Compressed air is also used to actuate the auxiliary motions of these units and even to propel these heavy-duty rock drilling machines. Thousands of these tools are now being used in huge road construction programs in many parts of the world.

Figure 1.30 A 35-lb class sinker drill producing an 8-ft. exploratory hole to determine the structure of the rock below.

Steelwork, essential in all modern construction, is joined by rivets or bolts and nuts, millions of which are used. Whatever fastener is selected, pneumatic tools are available to do the field drilling and reaming, riveting, or assembly of bolts and nuts.

Pneumatic drills, reamers, riveters and holder-ons, nut runners, and highly advanced designs of impact wrenches are used because no other kinds of power tools are capable of exerting as much power per pound of tool or are so rugged and dependable in maintenance as pneumatic tools. Finally, the operating safety of pneumatic tools should be stressed, a feature of utmost importance when one considers the hazardous locations in which power tools are commonly used on steel construction jobs.

Compressed air is also used for many auxiliary operations on construction since it is a versatile and readily available source of power. Some uses are drilling, hoisting, pumping, riveting, forging, and there are many others. Air tools have sufficient power to enable them to perform difficult tasks without excessive weight for the operator to handle. There is no overheating from constant use, and tool maintenance is low. Backfilling excavations, tamping dirt and concrete, testing and caulking pipe lines, operating drainage pumps, power brushing of pipes to remove rust, driving metal road markers, brushing concrete surfaces, driving sheet pile, constructing caissons, casting concrete piles, sharpening drills, and cutting metals under water are all jobs for which air-operated tools and equipment are used by contractors.

Air guns to apply concrete on new construction and on repairs are a common application. Air guns for spray painting save time and money on innumerable projects. On tunnel projects, air-operated rock drills, clay diggers, and other pneumatic tools speed up many otherwise slow operations. Pneumatic placers are used to line tunnel interiors. Lightweight, air-operated pumps are used by contractors for pumping out sumps, trenches, manholes, caissons, cofferdams, tanks, and bilges.

Compressed air is used in the drilling of sand drains for stabilizing soft, wet soils such as saturated clays and silts. An annular pipe mandrel is driven into the ground by an air hammer or vibrator. Air is introduced just above the driving shoe and blows the soil to the surface through the center. Self-destructive drilling mud is fed through the outer, annular space as the mandrel is withdrawn. When the mud disintegrates, it leaves a sand-filled column through which water may drain when the soil is placed under load.

Liquid nitrogen has been used successfully in freezing earth for underground construction. Injected into the ground in a series of holes, it can create an impervious front for the retention of ground water during construction.

Seeding and fertilizing after backfill operations can be done quickly by an air gun similar to a sandblast gun. Seed and granular fertilizer are picked up from bags carried in a small truck and scattered by air from a portable compressor towed by the truck. Even steep banks along a highway may be seeded in this manner, the truck and compressor moving along at grade level. Seeding of highway banks at 10 acres per hour is reported.

In construction work, there are important applications for compressed air other than those given here, and most of the pneumatic tools already referred to have other uses besides those specifically mentioned. The typical applications listed for compressed air can only suggest the variety of work that this versatile power can successfully perform.

Computers and Business Machines

The development of small computers has led to greatly increased computer applications throughout industry for machine and process control and for many businesses, hospitals, and countless other applications. Manufacture of chips and microchips on which these computers are based is done in clean rooms where compressors are part of the equipment that maintains a controlled, virtually dust-free atmosphere. Compressed air and vacuum are used in molding of plastics from which computer cases are made. Many air tools find applications in computer manufacture similar to those in other industries.

Pneumatic-type Conveyors

Pneumatic conveyors occupy a position of great importance in the materials-handling field. Materials may be transferred by air under pressure or partial vacuum depending on how the materials may best be handled in any given situation. In the pressure system, the compressor precedes the system, while in the vacuum system, it follows the system. Pneumatic conveying is especially suitable for many dry,

TABLE 1.1 Some Materials Handled by Pneumatic Conveyors

Alum	Iron oxide
Alumina (calcined)	Lime (hydrated)
Ammonium sulfate	Lime (pebble)
Arsenic (trioxide)	Limestone (pulverized)
Asbestos dust	Magnesium chloride
Barium sulfate	Magnesium oxide
Barley	Malt
Beef cracklings	Meat scraps (dried)
Bentonite	Middlings
Blood (dried)	Oats
Bone char	Phenolic resin
Borax	Polyethylene (powdered)
Bran	Polystyrene beads
Calcium carbonate	Polyvinyl chloride
Carbon (black)	Resin (synthetic)
Catalysts	Rice
Cellulose acetate	Rubber pellets
Cement	Rye
Clay (dried)	Salt
Clay (air-floated)	Salt cake
Cocoa beans	Sawdust
Coffee beans (green)	Silex
Copra	Soap chips
Corn	Soap flakes
Corn flakes (brewers')	Soda ash
Cottonseed hull bran	Sodium tetraphosphate
Cyanamid (pulverized)	Soybean meal
Dolomite (crushed)	Starch
Feldspar (pulverized)	Steel chips
Ferrous sulfate	Stucco (hydrocol)
Flax seed	Sugar
Flour	Titanium dioxide
Fly ash	Vinylite
Foundry sand (green and core)	Volcanic ash (pulverized)
Fuller's earth	Water-conditioning chemicals
Grains (dry-spent)	Wood chips
Grain dust	Wood flour
Grits (corn)	Zinc sulfide
Gypsum (raw, pulverized)	
Gypsum (calcined)	
Hops	

free-flowing bulk materials that are handled by pneumatic pipeline conveyors (Figure 1.31) and for parts that may be placed in carriers of specific size and shape, the carriers to be transmitted through pneumatic-tube conveyors.

Figure 1.31 Bulk PVC product being pneumatically conveyed to storage silo.

The air speed needed to keep material moving in conventional pneumatic systems depends on particle size, shape and weight, buoyancy, friction, turbulence, and other factors and is difficult to determine theoretically. If the air speed is great enough to convey material horizontally, it is generally more than ample for elevating. For grain and normal mill stocks, horizontal conveying speeds range from 3000 to 5000 fpm with conventional pneumatic systems.

Materials such as cement or flour depend for their successful handling on having enough air entrained throughout the material to give the mixture fluid properties. Such solids are said to be fluidized. The material remains generally in suspension in the air which carries it through the system, there being pressure and turbulence enough to reaerate any of the material that may tend to settle out. A great many powder, chip, granular, and pelletized materials are successfully handled in this way (see Table 1.1).

Solids pumps, air locks, blow tanks, and many other units are available for feeding and aerating the solids handled in the various systems with pressures adapted to the requirements of the solids handled, the conveying distance, and other factors. Solids pumps are shown in Figure 1.32. An air-activated gravity conveyor, like that seen in Figure 1.33, depends on air to fluidize the solid and keep it fluidized so that it moves along the slide under only the force of gravity.

Figure 1.32 Two solid pumps installed in a cement plant, one to convey dry raw materials, the other dry, finished cement. Conveying distances here are 600 and 500 ft. respectively.

Figure 1.33 A slide channel system for bulk loading in a cement plant. Dry, aerated cement is handled pneumatically, without moving equipment parts.

Air Versus Mechanical Systems

It is difficult to make a single comparison of pneumatic and mechanical systems. For short conveying distances, the mechanical conveyor is usually cheaper, but as length increases, pneumatic systems become relatively more favorable. Power consumption is also higher for pneumatic systems. These costs, however, often become less important than the savings in production costs that may result from pneumatic conveying. Dependability, low maintenance cost, and elimination of spillage are factors in this connection. The main advantage of pneumatic conveying is its flexibility. Materials may be taken around corners, through walls, or literally anywhere a pipe may be located, avoiding obstructions that would be serious obstacles in mechanical conveying. Safety and cleanliness are also important in certain cases. With air supply of sufficient capacity, most air conveyors are self-cleaning.

Air conveyors and mechanical conveyors both have uses that will remain exclusive to each. There are, however, overlapping areas where users may well benefit from a careful study of both types.

Dentistry

High-speed pneumatic drills (Figure 1.34) have greatly reduced the pain associated with dental operations. These drills are powered by an air turbine and rotate at speeds sometimes exceeding 200,000 rpm and permit faster cutting. To remove heat, the sensitive area is sprayed with an air-water mist. To keep the mirror clean, it is rotated so fast by a small air motor that moisture is thrown off by centrifugal force. Vacuum also removes accumulated moisture.

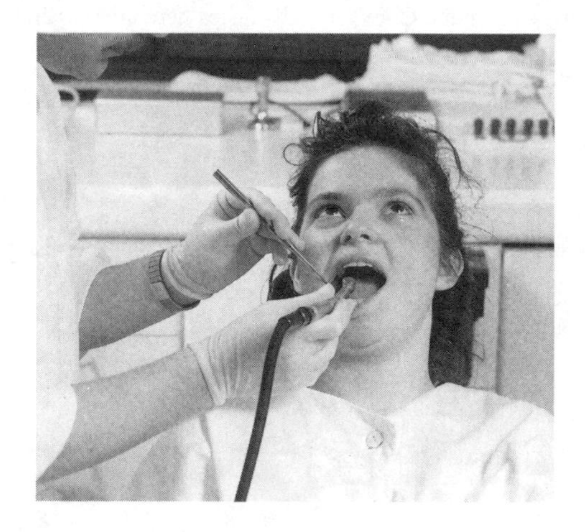

Figure 1.34 High rotative speeds achieved by air-turbine-driven dentists' drills permit faster cutting and alleviate the pain often associated with certain dentistry.

Die Casting

Injection of molten metal into die cavities is generally done hydraulically, but many auxiliaries are operated by compressed air. Small air grinders and impact wrenches are used to repair and assemble dies. Many trimming fixtures are air operated or have pneumatic clutches. Pneumatic clamps are used to hold castings during finishing operations, and in these operations, grinding and chipping hammers are extensively applied. Pneumatic tools are also utilized by maintenance and shipping departments.

Distilleries

Pneumatic power is used extensively in distilleries for conveying grain and malt. The entire system of handling grain can be dependent on pneumatic power from the boxcar to the grain elevator to storage in the distillery and, finally, to the production processes themselves.

Dry Cleaning Plants

Compressed air is one of the principal forms of power in dry cleaning plants. Collar, shirt, and garment presses use compressed air to press clothing against the steam-heated chest. Spotters spray cleaning solutions with small air guns, and dry cleaners use air-spray guns for mothproofing rugs and storage cabinets. The positive pressure of compressed air is considered by many rug and upholstery cleaners to be the best means of removing loose dirt. Furs that have been cleaned are given an electric look by a brief jet of compressed air against the fur. Filters for recovery of cleaning solvents employ compressed air to drive out gas and to cake sludge so that it can be handled conveniently. Cotton lint, which is a fire hazard in laundries, is easily removed by periodic cleaning of the building with air lines that reach to the ceiling and under machines.

Electric Products

Actually, this is not a single industry but a group of industries extending from tiny instruments through the wide range of communication equipment and giant electric generators. It would be extremely difficult to assess the hundreds of applications in which compressed air facilitates the manufacture of electric products, in process service as well as in power service. Vacuum is indispensable for the production of electric bulbs and vacuum tubes. At the same time, it is also by air power that the various motions of automatic equipment used for manufacturing bulbs are actuated and controlled. Some electric machinery is protected against the corroding effects of the atmosphere in which they have to operate by impregnation using a sequence of vacuum and compression.

Pneumatic screwdrivers are often used in assembly operations, as are specialized tools such as wire wrapping tools and wire cutting and stripping machines. The latter have been automated using pneumatic controls so that a circuit board may be programmed and wired automatically. Fusion soldering machines are also powered by compressed air.

Farm Machinery

The makers of farm machinery find a great variety of production applications among those discussed in the earlier parts of this chapter, including the use of rapid-action air chucks, drills, grinders, and polishing machines, as well as air-blast equipment, and in the manufacture of farm equipment such as pruners, posthole augers, and the like.

Food Industries

Packers, bakers, millers, refiners, and many others in food industries use compressed air to expedite many processes. Transferring liquids and granular materials (Figure 1.35) from trucks and railroad cars is a common application. A pneumatic conveyor can unload a carload of grain in three hours.

Sugar refineries and bakeries use compressed air for transferring syrups. Vegetable fats and other liquids are transferred in the same manner. The agitating of certain liquid or liquid-immersed foods such as pickles is accomplished by bubbling compressed air through them. Pressure filtering also uses the same source of power. Canneries use compressed air in can-filling machines and for cooking and sterilizing.

Figure 1.35 Sugar being conveyed vertically by pneumatic means from truck to candy plant.

Nitrogen-enriched atmospheres in storage areas retard the spoilage of many foods, extending the storage life of apples, for example, by many months. The reduced level of oxygen is largely responsible for this, but carbon dioxide and other gases are added in small quantities to control enzyme reactions and retard fungus or mold growth. One system for charging rail cars with inert gases utilizes a supply truck (Figure 1.36).

Figure 1.36 Mobile trailers carry nitrogen and other special atmospheres to rail sidings where these gases are injected into rail cars to help preserve foods and other perishables in transit.

Lettuce is chilled and crisped by subjecting it to a vacuum that evaporates free moisture from between the leaves, removing their latent heat. Nitrogen-enriched atmospheres are also used in rail and highway carriers especially adapted to the insulation and isolation requirements to bring lettuce and other perishable foods fresh to market.

Liquid nitrogen is also used as a refrigerant for trucks and rail cars transporting perishable foods. The liquid is stored in specially insulated containers and released to the cargo area as refrigeration is required.

Foods can be frozen much faster using liquid nitrogen than they can in a conventional refrigerated-air freezer.

Many foods are packaged in inert gases to exclude oxygen, to which most spoilage is directly related. Gases most commonly used are nitrogen and carbon dioxide, with nitrogen preferred in most cases since it is highly inert. Carbon dioxide is soluble in water so that, in some cases, it can impart an acid taste. Nitrogen is preferred for foods with high aroma. Carbon dioxide appears to retain the color of cured meats. Where vacuum packing may cause slices of food to crush or to adhere and be difficult to separate, gas packaging is preferred (Figure 1.37). Gas packaging is also especially suitable for fragile, freeze-dried foods.

Figure 1.37 Machine for automatic packaging of foods in individual plastic bags. Vacuum or inert gases retard spoilage.

In bakeries, air is used for cleaning biscuit dies, spraying butter in pans, and pressing out dough in measured amounts from automatic roll machines. Air jets are the best method of cleaning crumbs from bread-slicing machines without having to stop the machines. Air hoists and lifts are also commonly used in the bakery and at the loading dock. Compressed air is also used to spray insecticides in bakery storage rooms.

One of the more recent but rapidly expanding and important applications of compressed air in bakeries is that of flour handling by means of pneumatic conveyors. Such conveyors do the complete flour handling automatically, starting with the unloading of freight cars or trucks. The whole flour-handling process is performed economically and hygienically in a tightly closed pipe system with no direct handling by warehouse personnel. The reader is also referred to a later paragraph on packing houses.

Forging

Forge shops find extensive use for air hoists, grinders, chipping hammers, clamps, sandblast, and so on. Many plants have installed air-operated hammers or are converting steam-operated hammers to air operation. A study of the economy should be made comparing the cost of electric power to operate the compressors to the cost of generating steam. No standby power costs are involved with air during idle shifts or weekends, as is usual when steam is used.

Foundry

Most of the cast iron for our foundries is melted in the cupola furnace in which burning coke provides the necessary heat. The combustion of the coke requires a large volume of air, which is introduced into the cupola through the tuyeres. About 30,000 ft^3 of air is required to melt one ton of metal.

Foundries where most of the work was formerly done very inefficiently by hand are now being mechanized to a very large extent. Most of the equipment of the mechanized foundry relies on compressed air for its motive power because of its flexibility and the sturdiness of air-powered equipment. A pneumatic cone-wheel grinder is shown in Figure 1.38 employed on a foundry application.

Figure 1.38 Foundries depend extensively on compressed air. Here recesses in a heavy casting are being finished by a cone-wheel grinder turning at 8500 rpm.

In one large foundry, prepared coremaking sand is handled at a rate of up to 960 tons per two-shift day by the blow-tank type of pneumatic conveyor, which delivers sand to any of 38 stations selected by the master control. Air also transports molding sand from delivery points to storage bins, supplies molding machines, and recovers used mold and core sand.

One of the most important pneumatic devices in the foundry is, in fact, the molding machine. Large-diameter cylinders provide the force necessary for squeezing sand into the mold, with squeeze pressure being regulated by a valve in the air supply line to suit conditions. The high-frequency jolting action needed to pack the sand evenly is produced by air cylinders with quick, reciprocating action. In several modern molding machines, particularly for use on higher floor levels or for producing very clear-cut impressions, jolting is replaced by vibration, also produced by compressed air. In heavy-type rollover molding machines, the rollover and draw operations are also air powered, working in combination with hydraulic booster devices. Sand is introduced into the molds with the air, and compressed air is used for auxiliary operations such as blowing and lifting.

Compressed air has long been used in foundries, and the trend toward mechanization and automatic control has only increased its use. Figure 1.39 shows a fully automatic shell molding machine operated by compressed air.

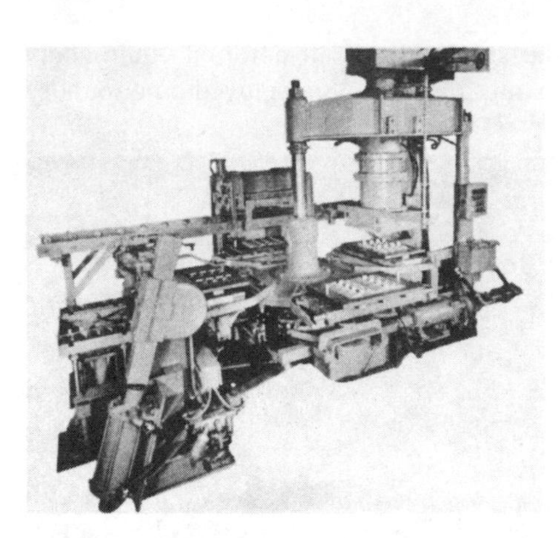

Figure 1.39 An integrated machine for shell molding with all the operations fully automated by means of compressed air.

Cores are also made by blowing the sand into the core box instead of jolting or ramming. Core boxes with vents for the escape of air are used, and the core blowing machine, operated by compressed air, carries the sand, which is suspended in the air stream, into the core box from which the air escapes, leaving the sand. The sand is compacted enough for most purposes, and the process takes only a few seconds. A pneumatic core blowing machine is shown in Figure 1.40.

Figure 1.40 An air-operated core blowing machine, with pneumatically operated clamping and drawing action.

Furniture

Better production is obtained from many machines used in a furniture factory because of compressed air devices. A good example is a front-skirting operation for chairs. Two air pistons in the center of the machine raise and lower the saws. At either end of the sawing, boring, and framing machines are air pistons that furnish the drive for boring tools. Compressed air is used for these operations because its resilient pressure adjusts itself to varying stocks, thus avoiding breakage.

Air jets clean out sanding machines, exhaust and separate sawdust and shavings automatically from mortisers and boring machines, and blow off pumice after rubbing. Air clamps are used to hold work in such machines as automatic shapers and to straighten pieces in presses. Pneumatic screwdrivers, drills, and wrenches are used in assembly, and other pneumatic devices are used in buffing and carving wood. Painting, varnishing, and enameling are done with air spray guns.

Pneumatic sanders, buffers, polishing machines, discs, and belt grinders are particularly useful in furniture manufacturing. These tools are insensitive to the dust developed in woodworking because of the self-protecting properties of compressed air-driven machines and, in addition, are spark-free, thus avoiding fire hazards. Air tools in furniture production are shown in Figures 1.41 and 1.42.

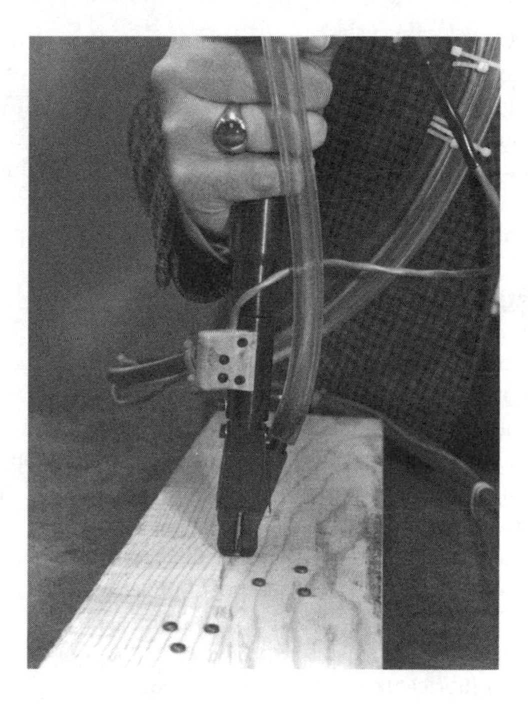

Figure 1.41 A pneumatic screwdriver fitted with an automatic screw feed device. Such tools are especially useful in furniture manufacture.

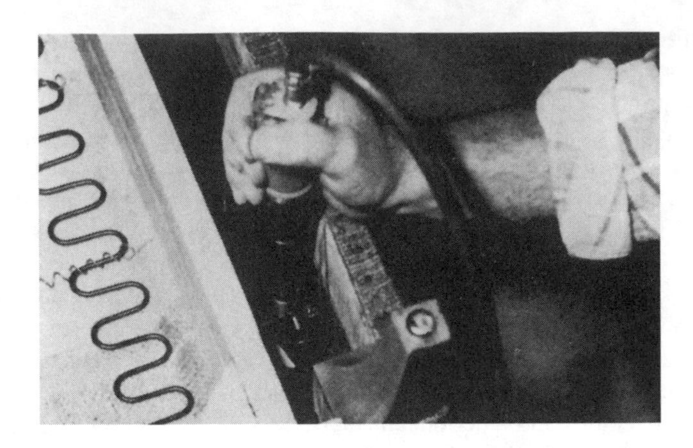

Figure 1.42 Modern furniture production depends on equipment such as pneumatic staplers and the angle nut-setter shown, as well as specialty tools such as the hog-ringer used to fasten upholstery material to springs.

Garages

Virtually every garage has a compressor at least for inflating tires. But further use is being made of compressed air to help with the more difficult jobs, such as changing tires where impact wrenches, tire changers and spreaders, lifters, and the like, greatly facilitate the work. Many of these same pneumatic devices are used in tire recapping plants along with air-operated molds. Also see the section on automobiles.

Air pressure is used for cleaning engine parts. Pneumatic sanders, grinders, polishers, and paint spray equipment are useful in auto-body repair shops.

Gas Bearings

There has been extensive research and development in gas bearings, and a considerable volume of literature has been generated on the subject. Gas bearings have configurations greatly similar to those of other hydrostatic and hydrodynamic bearings. In these bearings, however, air or another gas replaces the customary liquid lubricant, producing the film necessary to carry the bearing load and to separate the bearing surfaces for minimum friction under load.

A principal advantage of gas bearings is their extremely low friction torque. This torque tends to be very nearly constant as long as full film separation is maintained, since the viscosity of gases varies very little over a considerable range of temperature. Gas bearings may also be applied at temperatures where the viscosity characteristics of oil lubricants would be unsuitable.

There are two basic types of air or gas bearings, the externally pressurized and the self-acting types. Externally pressurized bearings receive gas under pressure from an outside source, such as a compressor or an accumulator, from which it is introduced under the journal so as to float the load and separate the bearing surfaces without dependence on relative motion to maintain the pressure in the film. The self-acting type does not depend on external pressurization but instead produces its own pressure in the same way that pressure is produced in an oil-lubricated bearing. The journal, slightly smaller than the bearing, has a varying clearance due to its eccentricity. The air or other gas is carried into the converging space because of its viscosity and is thus compressed, providing enough pressure to support the load.

Glassworks

The ancient art of glass blowing has been transferred to automatic machines that produce far greater quantities of glassware than were once possible through individual efforts. In this field, compressed air is an essential factor.

Typical applications are the handling of glass sand, blowing glass, operating molds and presses, operating sandblasts, supplying oil burners, etching glass, and molding and frosting lamp bulbs.

A machine for forming glass tumblers employs compressed air in several important ways, and each year millions of bottles are formed by compressed air in blow molding machines.

Compressed air and gas are mixed in glass-plant burners for high combustion rates. High-pressure air is applied to drive cleaning cloths through glass tubes and to blast holes in glass with abrasives. Designs are etched in glass by sandblast, which cuts the exposed surfaces not protected by a masking stencil. Some types of sheet glass are heated to high temperatures and then cooled by compressed air to chill the outside surface. This process gives shatterproof qualities and greater tensile strength. Glass sheets are lifted by vacuum cups, a method that has reduced the risk of breakage.

As the glass industry has continued to increase the variety of its products, it has greatly extended its need for compressed air.

Golf Courses

Golf courses and similar areas can be seeded and fertilized by using a portable compressor and a gun similar to a sandblast gun. A suction hose inserted into a bag of seed or granular fertilizer carries the material to the gun, where it is ejected and distributed by a jet of air. It is reported that 500 acres per week have been seeded and fertilized in this manner using two guns and one portable compressor and truck.

Compressed air may be used to blow out sprinkler systems on golf courses to prevent frost damage. This usually permits lines to be installed in trenches of uniform depth without much regard to gravity drainage. A portable compressor rented by the day makes this an economical method. Care should be taken not to exceed the safe pressure for the system. In the absence of more exact information, this may be taken as equal to the water pressure to which the system is subjected.

Heat-recovery System

Heat removed from a compressor during the compression process to improve the volumetric and compression efficiency may be used to meet part of the building heat load. The installation may be indoors or outdoors; an example of the latter is shown in Figure 1.43.

Figure 1.43 A 300-hp unit mounted outdoors in a northern climate. A heat-recovery system is used that heats a major portion of a warehouse.

High Energy Rate Forming (HERF)

Compressed air has been used to replace dynamite or other charges in the explosive forming process, which is used for the manufacture of items like metal kegs with double curvature and formed rings. Releasing compressed air against the large piston in a HERF machine and applying the resulting force to a small piston or on water in the forming chamber produces more safely the same kind of irregular shapes for which explosive charges were originally used. The same degree of care must be taken, however, as in handling other high-pressure gases.

Hospitals

Hyperbaric chambers, in which the patient breathes air or oxygen under pressure, have been constructed in several hospitals (Figure 1.44). Usually used where a higher than normal concentration of oxygen is desired in the patient's tissues or blood, hyperbaric treatment has been effective against carbon monoxide poisoning and respiratory disorders and shows promise in conjunction with certain surgical operations. There appear to be many other possible applications. Decompression techniques are used that are similar to those in deep-sea diving.

For inhalation therapy, clean, dry air is provided at positive pressures of a few inches of water to breathing apparatus in unpressurized treatment rooms. Oil-free compressors providing air for hospital use are shown in Figure 1.45.

Figure 1.44 The hyperbaric chamber is utilized in a varicty of medical procedures including deep sea diving decompression.

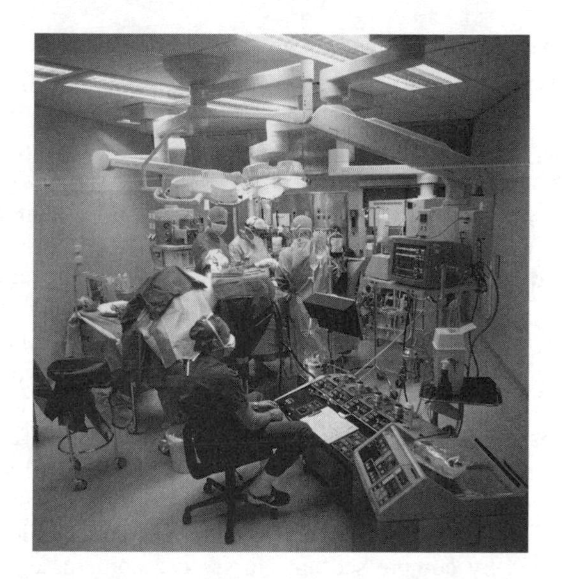

Figure 1.45 Oil-free compressors supply medically clean air for a hospital.

In an alternative system for supplying breathing air, which is convenient for the patient to use either in the hospital or at home, oxygen-enriched air is provided by a molecular sieve bed, which removes nitrogen from atmospheric air. Such oxygen concentrators achieve oxygen concentrations of 90 to 95%. A unit of this kind is seen in Figure 1.46.

Figure 1.46 An oxygen concentrator with cover removed to show the molecular sieve containers and other internal parts.

Many gases, especially anesthetics and oxygen, are supplied in pressurized cylinders. These are tested periodically at hydraulic pressures considerably above their service pressure. Color coding is used to avoid confusion among gases, and a distinctive valve and thread are used for each gas.

A high-speed, turbine-driven pneumatic surgical drill has exciting capabilities. In craneotomies, it permits removal of the top of the skull in as little as two minutes, compared with the half-hour needed by surgeons using hand tools. It has been used to carve transplanted chest cartilage to form a realistic artificial ear and to shape a grafted bone to rebuild a badly broken nose. Weighing less than 1 lb, the drill brings increased speed and power to bone cutting, drilling, and shaping. The compressed air or nitrogen that provides its power is exhausted at the drill tip to cool the area of the cut. Airtight seals permit ready sterilization of this surgical tool.

Instruments to control hospital air conditioning and humidity in operating rooms are operated by compressed air. Nurses use compressed air for cleaning catheters and other tubing, and for spraying medication. Vacuums are used to draw off blood and secretions during operations when there is no way of clamping off and sponging to give the surgeon a clear operative field. Operating rooms are pressurized to keep out dust. Air-operated doors in surgery are manipulated by foot pedals. Hospital laboratory, laundry, and maintenance personnel also use compressed air for many purposes.

Industrial Applications of Portable Compressors

Portable compressors have often provided plant air during the interim when, for one reason or another, the regularly used stationary compressor was out of service. Such an emergency situation caused the silenced portable compressor in Figure 1.47 to be used as the plant air supply at a plastic container plant.

Figure 1.47 A portable, oil-free screw air compressor is shown outside a plastic container plant. When two in-plant compressors failed, this rental, oil-free unit solved the emergency problem that the failures created.

Another silenced compressor is shown in Figure 1.48 providing compressed air during remodeling of a plant. Although there is a clear trend toward stationary compressors at ski areas, as reported in a separate section on that topic later in this chapter, rented portable compressors have long provided air for snow making.

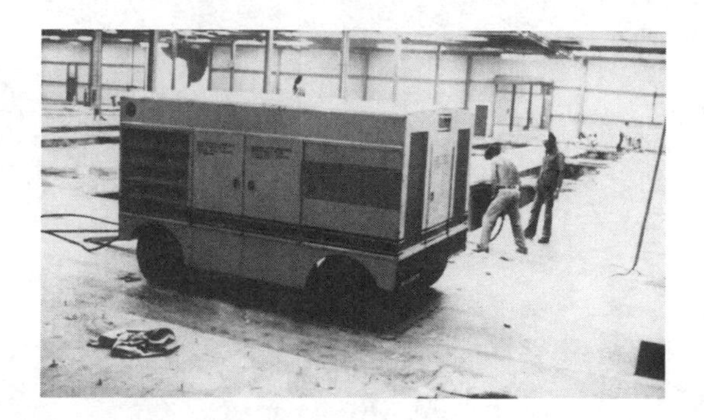

Figure 1.48 A 425 cfm, 125 psig oil-injected portable screw compressor supplying air for remodeling a manufacturing plant.

Household Appliances

Manufacturers of refrigerators, stoves, vacuum cleaners, radios, television sets, kitchen sinks, cabinets, toasters, mixing machines, washing machines, fans, clocks, and other home appliances utilize compressed air for so many purposes that only a few will be named here. These are press clutches in sheet metal fabrication, air cylinders for process machinery, hand tools in appliance assembly, spray guns in appliance finishing, blow offs to purge liquids and solids from metal parts, air ejection of press products, air agitation in plating, pressure lead testing in refrigerator assembly, servopilots for hydraulic machine control, and instrument pilots to control pressure, temperature, and flow.

Iron and Steel

One of the largest process uses for air is in the manufacture of iron and steel. It takes about 100,000 ft^3 of compressed air to make one ton of steel. Iron has been used for several thousand years in weapons, tools, and implements; early production depended on the use of hand-operated bellows. It was not until the advent of steam engines to drive large blowers that sufficient air could be provided for large-scale production.

The real evolution in steel making started with the invention of the Bessemer converter, which uses large quantities of compressed air to eliminate the foreign elements from the pig iron produced in the blast furnace. Air compressed to a pressure of 25 to 32 psig was forced through the molten metal of the converter for a period of 12 to 18 minutes to remove these impurities. Open-hearth furnaces are likewise very large users of compressed air. A modern compressor installation supplying air for a blast furnace is shown in Figure 1.49.

Figure 1.49 An axial flow compressor for blast furnace duty in a steel mill.

Oxygen enrichment substantially raises the productive capacity of iron and steel manufacturing facilities. The oxygen, used in very large quantities for the process, is obtained by means of high-pressure compression and liquefaction of air (Figure 1.50).

Figure 1.50 Two centrifugal compressors in an oxygen plant supplying tonnage oxygen for a steel mill.

Vacuum melting, which makes use of vacuum pumps, is a relatively new but very important method of producing the high-purity steel, alloys, and nonferrous metals required in certain special applications.

Compressed air is used in the iron and steel industry for many other purposes besides the basic process, particularly as motive power for functional equipment and for tools. Such applications include chipping and grinding billets, furnace notching and tapping, ladle dumping, operating hoists and lifts, agitating solutions, sandblasting, and caulking tanks. It also includes operating tamping machines, furnace door hoists, ore bin doors, steel coil lifting and traverse mechanisms, instrumentation and process controls, balance for lifting mill tables, billet turners and billet ganging tables, pinch rolls on processing lines, blast furnace bell operation, coke gates, lubrication systems, air clutches and brakes, mixing of materials, descaling, air clamps, blow downs, and more.

Lumbering and Woodworking

Wood is a material that serves a thousand purposes such as to build homes, panel walls, construct furniture, and make newsprint. Sawmills, woodworking shops, and lumber mills are dependent on compressed air in power service to do much of the heavy work. Unloading, splitting, cutting, sawing, and trimming logs are a few typical applications. Spark-free air hoists handle heavy loads easily and quickly without any hazard of fire in the presence of combustible materials. Compressed air conveys chips and sawdust from the mills and is used for cleaning rafters, timbers, and framework, as well as machines.

Air is required for the drying kilns of veneer plants. Pneumatic sanders, drills, and screwdrivers are used extensively in woodworking and finishing.

In process service, railroad ties, bridge timbers, poles, and other wood members that will be exposed to dampness are treated with creosote and other preservatives infused into the porous surface by means of compressed air.

Machinery Production

Pneumatic tools and auxiliary equipment have found countless applications in the production of machines of all types. It is customary to provide an air outlet at each work point in a machine shop so that the machinist or operator may make easy use of the many air tools usually available.

Pneumatic controls are frequently applied to machine production processes. Actuating mechanisms may be interlocked for safety, and hazards such as explosions and electrical shock are eliminated by the use of air. Devices such as pneumatic positioners, air circuitry, fluidics, and pneumatic logic controls have put compressed air even more into the manufacture of machines. Material feed as well as machining operations are subject to pneumatic control.

On regular production lines, air tools also accelerate manufacturing and assembly processes, and wide use is made of multiple or gang tools to improve personnel utilization. Gang drills and multiple nut runners and screwdrivers are examples. Dozens of profitable applications of air tools are discussed in Chapter 5. The simplicity and lightness of air tools, such as the impact wrench shown in Figure 1.51, together with the fact that they do not burn out or overheat under constant heavy use, make them highly adaptable to the manufacture of machines.

Figure 1.51 An impact tool with a hand grip being used to tighten nuts on a machine part.

Auxiliary equipment such as feeders, reciprocating motions, indexing devices, revolving tables, positioners, and many other devices actuated by pneumatic cylinders also find many ingenious applications. Air hoists (Figure 1.52) greatly facilitate the handling of heavy machine parts.

For a discussion of materials handling by pneumatic means and the use of pneumatic tools in metalworking processes, the reader is referred to other paragraphs under the appropriate headings in this chapter and Chapter 5.

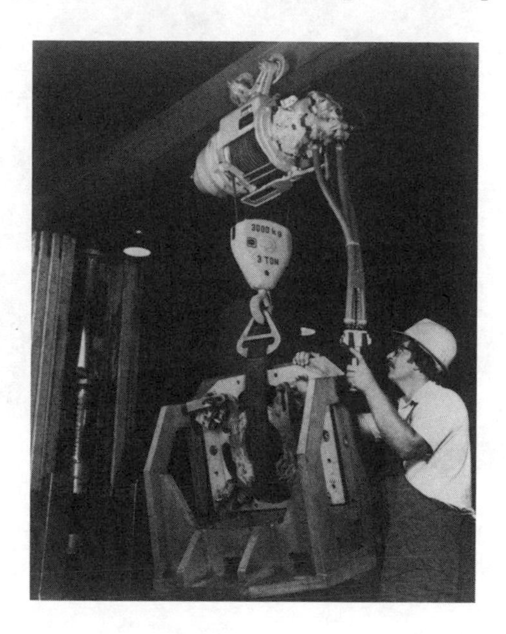

Figure 1.52 A 3-ton air hoist greatly facilitates moving a heavy machine part. Many such hoists have an air motor drive along the I-beam that serves as its track.

In pressworking of metals, many different uses are found for compressed-air devices, including pneumatic clutches, air-actuated stock feeders, pneumatic hold-downs, vacuum suction cups, and spray lubricators for metal strips preceding deep drawing operations. Presses may be powered or controlled by compressed air, although hydraulic actuation is usual.

The spraying of lubricants and coolants by means of compressed air is also common in many other areas of metalworking production. It provides an economical means to apply lubricant and avoids the accumulation of dirt on the machines and the spoilage of products. Points in obscure locations can be reliably lubricated by this method. Coolant application by means of air spraying is of particular importance whenever the conventional method of coolant flooding cannot be used. In many cases, where liquid coolants are not needed or are for some reason not admissible, effective cooling and chip removal are attained by means of an air stream directed at the edge of the cutting tool.

Many machine parts and frames are fabricated by welding. Pneumatic scaling hammers are used for weld flux scaling (Figure 1.53).

Figure 1.53 A weld flux scaling operation using a lightweight scaling hammer equipped with an angle chisel operating at 5700 blows per minute.

Mining

Compressed-air tools in all types of mines have lightened the burden of the miner, made the job safer, and greatly increased output (Figure 1.54). Pneumatic tools like percussion and auger rock drills, coal drills, drifters, stopers, as well as air-actuated, powerfeeding appliances for most of these tools, are in general use as standard equipment for the miner. Unwatering by the air-lift system, unloading cars, running direct-acting pumps, loading ore, filling cracks and seams with cement, conveying, ventilating remote areas, pile driving for shaft work, operating coal punches, chain machines and radial-ax coal cutters, spreading stone dust to prevent dust explosions, removing methane from mine shafts for safety (this gas can then be sold to gas pipe lines), and operating pick and drill sharpeners are typical applications of compressed air in mining.

Figure 1.54 A pneumatic jackleg drill in operation with an air-driven lamp in the foreground.

Mobile Homes

There are many uses for air tools in the manufacture of mobile homes. In general, the applications of drills, riveters, nut runners, nibblers, screwdrivers, and the like, are similar to their uses in aircraft production. Figure 1.55 shows a pneumatic screwdriver being used in an assembly operation.

Figure 1.55 A pistol-grip pneumatic screwdriver is being used to install a trim strip in a mobile home.

Monuments and Cut Stone

It would be quite difficult to find any company in the stone-cutting field that does not rely on compressed air for its work (Figure 1.56). The task of cutting raised or carved lettering has been enormously reduced by the use of sandblasting to remove the part of the stone that is not protected by a stencil. Artisans carve figures and special letters in granite and marble with air chisels in one-third the time required by hand chisels. Air-operated surfacing machines are used to line up granite blocks, and air drills are used to prepare the blocks for plugs and feathers.

Figure 1.56 Pneumatically operated engraving tools are used for wording of monuments and headstones.

Nitrogen

The cost of producing an acceptable purity level of nitrogen has been greatly reduced by the introduction of membrane technology. Atmospheric air is composed primarily of nitrogen (78%) and oxygen (21%), with the remaining 1% containing several constituents in various amounts. Compressed air at around 175 psig is passed to the outside of a tightly grouped pack of small bore hollow fibers, made of a semi-permeable membrane (Figure 1.57B). Oxygen, carbon dioxide and water vapor preferentially permeate the fiber wall more readily than nitrogen, allowing relatively high purity nitrogen to be delivered. The reduced cost has opened up a wide variety of applications, including apple storage and food packaging, where the elimination of oxygen prevents oxidation and spoilage. Heat treatment also benefits from a nitrogen blanket to prevent oxidation in the product.

Figure 1.57A Membrane type nitrogen generator with air supplied by a compressor at approximately 175 psig.

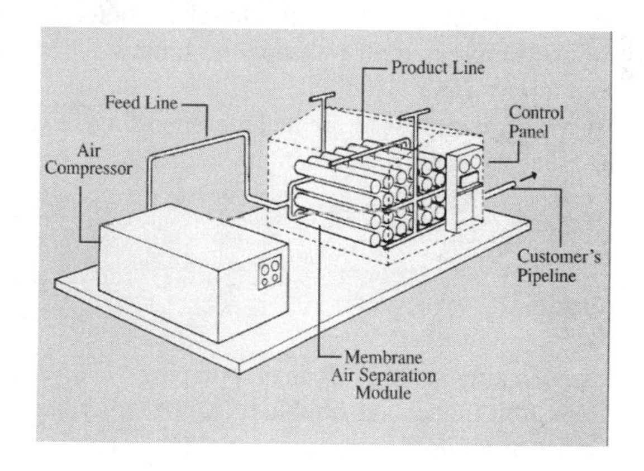

Figure 1.57B A diagram of a typical air separation installation.

Nonferrous Metals

The metallurgical processing of nonferrous metals from the ore to the metal offers varied applications for compressed air. Typical of the industries in this field is the extraction of aluminum.

In the raw material state, bauxite ore is refined to produce aluminum oxide. A major application of compressed air is the unloading of ore from oceangoing ships or from gondola cars. Another important use is the agitation of the material in the precipitators. The agitation is accomplished by bubbling air through the material in a liquid suspension. Further applications air conveyor systems, pneumatic instruments, and general maintenance equipment.

The aluminum oxide is next converted into aluminum metal by the electrolytic reduction process. In the reduction cell rooms, the anodes are lifted by means of pneumatic jacks. Air is used for metal-tapping siphons and for various service applications. In carbon plants where anodes and cathodes are produced for use in aluminum pots, air conveyors, pneumatic gates and controls, air hoists, and cranes are used, in addition to various service equipment including dust collectors, air-blast cleaners, tampers, vibrators, sanders, and wrenches.

Besides these specific applications, there is wide use of air power in supporting facilities such as shops, warehouses, administrative areas, boiler plants, and water treatment plants.

Office Buildings, Hotels, Stores, and Institutions

Compressed air has found many uses in building services. For example, elevator doors and other automatic doors are operated by compressed-air cylinders.

Pneumatic-tube systems for conveying cash and messages speed up office and commercial functions. Sprinkler systems in churches, hotels, stores, and many other buildings require compressed air to keep lines empty until fusible plugs melt to release the air and admit water.

Other typical uses of compressed air include operating sewage and drainage ejectors, removing ashes by electropneumatic systems, operating certain dentists' and physicians' services, cleaning electric motors by air jets, and operating hotel and institutional laundry machines.

Packing Houses

Stuffing sausages, testing sausage casings, pumping water, and operating loin presses and shoulder-cutting machines are but a few of the typical applications for compressed air in meat packing. Compressed air aids combustion in smokehouses, in burning hairs from hog snouts, and so on. Air hoists are used to lift calves from conveyor lines and for other lifting where speed and ease of operation are important.

Air entrained in meat during processing is removed by vacuum before the meat is canned. Inert gases, especially nitrogen and carbon dioxide, are also used in packing meat.

Paint Factories

Filling cans is one of the ways compressed air is used by paint manufacturers. Stamping cans and transferring liquids are other uses. Air hoists are especially chosen in the presence of flammable varnish to handle varnish filters because of the absence of electric sparks.

Paper Mills

Manufacturing of newsprint and molding, drying, and sterilizing pulp-paper containers are typical applications for compressed air in the manufacture of paper. The air cylinder, widely used throughout industry, is frequently used here to provide power for baling and packaging paper. Heavy bundles of paper are easily moved on machine tables of paper shears when there is air-float equipment creating a film of compressed air between load and table.

Petroleum

Many uses for compressors in the petroleum industry are for process compressors, which are taken up in later chapters of this book. However, there are important uses for air compressors both in the oil fields and in petroleum product manufacturing.

Three 3000-cfm, three-stage, package centrifugal compressors for centralized plant and instrument air control purposes are shown in Figure 1.58. Operation flexibility and power outage backup for reliability are provided by selection of two steam-turbine-driven units and one motor-driven unit operating in parallel.

An interesting use, referred to as in situ combustion, is made of compressed air in the secondary recovery of high-viscosity oil. Compressed air is pumped into the ground in oil-bearing sand or strata and the oil is ignited. Thc resulting combustion is supported and regulated by the compressed air. Pressures up to 1000 psig are developed, and the heat generated reduces the viscosity of the oil so that it flows to the pumping location.

Figure 1.58 Three 3000-cfm, integrally geared, packaged centrifugal air compressors installed in a refinery. Two of the compressors are driven by steam turbines. The third compressor is driven by an induction motor.

Pharmaceuticals

Antibiotics, the miracle drugs that have almost eliminated some diseases, require large volumes of air for the fermentation process by which they are produced. Centrifugal compressors in one plant supply 52,000 ft^3 of air per minute at 25 psig to provide oxygen for the microorganisms essential to the production of these drugs.

Air is also used as a transporting and drying vehicle in spray dryers. The liquid concentrate is sprayed in a uniformly atomized sheet into a stream of heated air where moisture is evaporated, leaving a solid powder that is removed in a collector.

Air is used in packaging. One drug manufacturer uses air to blow headache powder into tissue envelopes. Many use pneumatics to assist in filling and sealing plastic containers of many kinds.

Plastics

The starting materials from which plastics are manufactured are usually gases such as acetylene, ethylene, propylene, methane, and natural gas. Large compressors are often required for air, as well as for the hydrocarbon gases. Only air applications will be discussed here.

Molding presses are sometimes operated by compressed air cylinders or air hydraulic cylinders. Parts are ejected and swept from the molds pneumatically, and many finishing operations, such as drilling, buffing, grinding, and painting, are done with compressed air equipment (Figure 1.59).

Figure 1.59 Watercooled heat exchangers are compact and allow compressors to be installed in applications where cooling air may not be available.

Plastic molding and forming have become a very large and important industry. Air or vacuum exerting a uniform, controllable force against a softened plastic sheet gently molds it to the contours desired. Usually, this is not a simple process. Various techniques have evolved to achieve uniform thickness in the parts and to perform other functions, such as those illustrated in Figures 1.60A and B. Sometimes air acts as the mold itself, as in the manufacture of helicopter bubbles and other similar optical plastic parts where contact with a solid mold could reduce or destroy the transparency of the plastic.

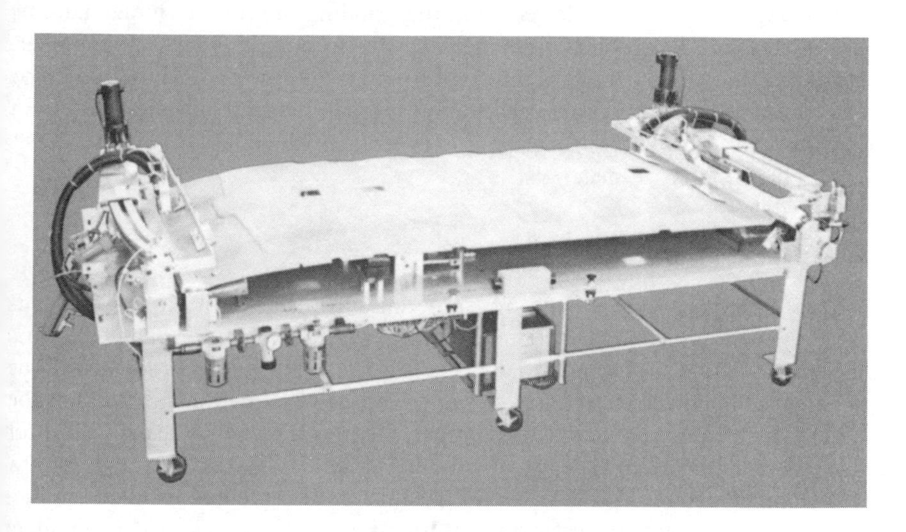

Figure 1.60A A die used in custom molding of plastics has an air-powered insert that helps make holes in the part during the molding process.

Figure 1.60B Ejection of a molded piece is accomplished by compressed air. Five ejector pins can be seen on the die half.

Pneumatic clamps hold sheets in place, and flashing is removed from molds by air tools. The plastic extrusion process uses compressed air to support thin-wall tubing, to size plastic tubing after it leaves the die, and to cool strips of flat plastic materials as they come out of the die.

Potteries and China Works

Sandblasting smoothes away blemishes from bisque chinaware before glazing. Liquid glaze material is transferred from the grinding mill to the storage tank by compressed air, and air motors drive machines for forming clay bats on the molds.

Pottery dust and other dirt is removed from electric motors and machinery by air jets. Air-operated rattlers are used to recover material from containers used for firing. One maker of flower pots uses air to operate molds and for drying. Air hoists facilitate the handling of materials.

Power and Light Plants

Gas-turbine-driven generators are sometimes available to meet peak demands and emergencies. Such units can deliver full power in a matter of minutes from a cold start. To start a gas turbine, compressed air is usually used to bring the turbine up to a speed at which its own power can take over when fuel is injected into the combustion chamber and ignited. A relatively small compressor is usually used for charging the air receivers that supply the starting air.

Automatic controls, predominantly pneumatic, measure and regulate the fuel fed to the boilers and the necessary draft for complete combustion. They regulate the flow of acid and caustic in the right ratio to the makeup water in a closed system. In many stations, the use of compressed air for soot blowing is also very important to assure good heat transfer from flame to water or steam.

A unique combination of separate three-stage and two-stage centrifugal air compressors operating in series and tandem is shown in Figure 1.61. This arrangement of five stages with four intercoolers located below the mezzanine floor supplies air soot blowing for a coal-fired boiler in an electric utility. Typical discharge pressures range from 300 to 350 psig in this application.

Figure 1.61 Two- and three-stage compressors for a soot-blowing application in an electric utility plant.

Air tools and air hoists are extensively used for erecting and caulking boilers and for maintenance. Condensers, for example, may be opened up by pneumatic impact wrenches and tubes cleaned by air-operated plugs.

In nuclear generating stations, the uses of air are much the same as those in fuel-burning stations, except those related directly to the fuel. However, there is one special need for air. The reactor is enclosed in a shield, and the shield, in turn, is enclosed in an airtight, steel-plate sphere. To be certain that there is no leakage to the atmosphere from within this sphere, air pressure of about 2 psig is maintained inside, and any drop in pressure is sensed so that corrective measures may be taken immediately. The use of portable compressors in leak testing in a nuclear power plant is shown in Figure 1.62. It must be remembered, however, that safety considerations dictate that pressure vessels when tested for strength must be tested hydraulically.

Figure 1.62 Eleven portable, dry screw, oil-free air compressors arranged in tandem to provide air for a 9-day leak tightness and structural integrity test of domed containment structures at a nuclear power station.

Utilities have many maintenance jobs for which air tools are especially useful in keeping trucks and other equipment in good operating condition.

Compressed air and air tools are also needed along distribution lines. A maintenance truck for utility service is shown in Figure 1.63. Air tools are also kept on the track so that many repairs can be made at the site of the trouble. Utility companies use portable compressors for emergency work (Figure 1.64) as well as new construction.

An experimental plant intended to reduce the cost of petroleum fuels used for power generation shows great promise. Base load electricity operates compressors during off-peak hours to store compressed air in underground reservoirs. During periods of peak load requirements, the air is taken from storage, mixed with fuel, burned, and expanded through gas turbines to generate the extra electrical power that is needed.

Figure 1.63 A utility maintenance truck that takes a compressor and assorted air tools right to the source of the trouble for speedy, efficient repairs.

Figure 1.64 A portable air compressor used by utility employees on an emergency repair job. Such compressors are also used on scheduled maintenance and new construction.

Pneumatic Tools

There are many applications using a wide variety of pneumatic tools. These are covered in Chapter 5.

Printing and Newspaper Plants

Compressed air is used for cleaning machinery, hoisting stereotype plates, pumping water, and operating steam tables and monotype machines. It is used to prevent wet labels from smudging and for powder dusting of freshly printed sheets. In the rotogravure pressroom, it is used in the tension system for brakes, imprinters, and folders, as well as for automatic strapping of the finished printed material. Pneumatic clamps facilitate wrapping and tying. Suction cups are used in feeding presses, and a gentle air blast riffles paper to assist in separating sheets. Pneumatic extrusion pumps supply printing ink in some types of presses (Figure 1.65).

Figure 1.65 Air cylinders operate the ram-mounted extrusion pumps in a lithograph plan supplying printing ink to the ink fountains of four-color offset presses.

Quarries

Compressed air is the principal form of power for most quarrying operation (Figure 1.66). In power service, stone channelers, rock drills, and plug drills are al air operated. Air hoists move heavy stone slabs. In the quarry blacksmith shop, ham mers, sharpening tools, and other cutters are operated by compressed air. Many quarries are now using compressed air to operate steam pumps for improved efficiency.

Figure 1.66 A portable compressor in a stone quarry.

Railroads

Diesel and electric railroads have found that hundreds of operations can be performed best with compressed-air-operated tools and equipment. Along the tracks and in carshops, maintenance facilities, and railroad yards, this vital industry is doing essential work with compressed air.

A few of the many railroad applications for pneumatic tools include lubrication of locomotives and general maintenance operations on rolling stock such as drilling, wheel alignment, reaming, and surface grinding. There are also many applications in the maintenance of buildings and other fixed facilities. An extensive air-distribution system is a regular feature of the railroad yard.

A mechanized track crew uses air-operated tools for driving spikes, digging, grinding rails, tightening bolts, tamping, pulling spikes, and other jobs, which can thus be done much faster than by manual methods. Large pneumatic tamping machines utilize a number of tampers in simultaneous operation for tamping ballast along the track.

In connection with railways, perhaps the most universally known and appreciated application of compressed air is that of the air brake. The original basic system has been retained with modern refinements and additions and is still used on most railways throughout the world. The air brake is typical of many compressed-air applications in that air is used to perform several functions. Compressed air actuates the brakes so that all are applied simultaneously and controls operation so that braking can be regulated sensitively over a wide range of braking force. The remote control by means of air becomes automatic if the brake hose running from car to car along the train should rupture at any point, an outstanding safety feature. Air brakes are relatively simple, and their maintenance is easy and inexpensive.

Favorable experience with air brakes on railways contributed to their use in trucks and buses. Most heavy vehicles of this type are equipped with air brakes.

Refrigeration Plants and Ice Plants

Pneumatic applications here include hoisting ice tanks and loading cars, pumping and aerating water by compressed air, descaling condenser coils, and cleaning boiler flues with compressed air. Standard types of compressors with special seals are essential to mechanical refrigeration systems.

Rubber

Compressors are essential in the process by which synthetic rubber is produced. Air blasts clean out the rubber molds. Small air drills clean out the vent hole in the wall of the mold to release gases from the heated rubber during vulcanizing. Air pistons press unvulcanized casings into the general contour of tires. They also are used to ram crude rubber into mixer rolls.

Temperatures in rubber mold presses and vulcanizers are subject to pneumatic control.

Sewage Plants

The activated sludge system of sewage treatment depends on compressed air to convey materials and to agitate sewage gently in preparation tanks so that particles of grease and oil will be coagulated. In deep aeration tanks, compressed air is used for violent agitation to supply oxygen to bacteria. Typical applications for pneumatic tools are for caulking pipe lines and for driving sheet pile. Paving breakers and rock drills are used in maintenance. One type of blower utilized in the treatment of sewage and industrial wastes is shown in Figures 1.67A and B.

Figure 1.67A Two-impeller rotary blower installation for aeration and treatment of industrial wastes.

Figure 1.67B Two-impeller rotary blowers for sewage aeration.

Ships

Shipboard installations include compressors for air separation plants producing liquid oxygen and nitrogen for use on board. Other compressors supply air for ship maintenance, for vacuum-blast cleaning of salt from the skin of aircraft carriers, and cleaning the holds of cargo ships (Figure 1.68).

Figure 1.68 The hold of a cargo ship is being cleaned with a high-pressure spray from a piston pump mounted directly on a drum of cleaning solution.

Pneumatic controls are now used extensively on ships. As a typical example, one such system controls the diesel propulsion on a harbor tug with two propellers and two propulsion engines. For greater operating flexibility, the tug has a separate and independent set of these controls for each engine. With these controls, the tug can be maneuvered and steered from any of three positions, from either side of the pilot house, or from the afterdeck. This arrangement lets the tug operator handle the vessel with increased speed and facility from the point of best visibility.

There are, of course, many other applications for compressed air as the actuating or control medium on various naval vessels. Examples of the latter group of uses is the operation of torpedoes by compressed air motors and torpedo launching through pneumatic tubes.

Shipyards

Compressed air has long been essential in building and repairing ships. Pneumatic tools have eliminated many slow, tedious steps in construction.

Air hammers are used in driving nails and spikes in keel laying, for grooving plates prior to welding, for chipping weld scale, for riveting, and for caulking. Air grinders have innumerable applications in smoothing surfaces. Spraying speeds up the painting phase of shipbuilding to a rate 20 times faster than that achieved by the hand-brushing method. Air-operated reamers and drills are important for shipbuilders.

Since much of their work is done outdoors, shipbuilders find compressed air especially convenient and useful in blowing off snow, rain, or dirt from their work areas. Testing welded seams is made easier by blowing away water accumulated along the seams.

Air motors, hoists, and lifts are used in a number of shipbuilding operations. Compressed air is used to start diesel engines and for various specific applications, too, such as operating dry docks, salvaging sunken wrecks, and various underwater operations.

Ski Areas

Uncertainties of the weather made operating a ski area rather a risky business until the advent of snow-making equipment. Portable compressors in capacities of 600-1900 cfm are well adapted to the high-volume air requirements for making snow. Stationary, electric-motor-driven compressors are also often used in snow-making systems. Some systems have heads consisting of a fan, nose cone, and nucleator mounted on a sled or carriage along with its own compressor (Figures 1.69A and 1.69B). Such equipment enables operators to lengthen the skiing season and to compensate for periods with too little snow.

Figure 1.69A An integrated snow-making system with its own compressor projects snow over a 150-ft. radius on slopes as steep as 45 degrees.

Figure 1.69B Large ski resorts have substantial centrifugal air compressor installations for snow making when atmospheric conditions require it.

Soap and Detergents

The manufacture of soap is a progressive industry whose research activities have substantially eased some of the more laborious household chores. Modern production methods include the extensive application of compressed air for operating instruments, blowing out stock lines, agitating liquids, operating pneumatic cylinders and tools, aerating stock, general cleanup, ejecting cans, pressurizing enclosures and fluidizing granular solids.

Steel Mills

A somewhat different intercooler containment vessel design is shown in Figures 1.70A and 1.70B in two views of an 1880-hp, motor-driven, four-stage integral gear unit in a European steel works compressed-air station.

Figure 1.70A

Figure 1.70B Four-stage package geared centrifugal compressor in the compressed air station of a steelworks. The unit is rated at 6700 cfm at 128 psia requiring an 1880-hp motor.

Tanneries

Handling tannic acid and pumping water are typical uses of compressed air in this industry. Air hoists and lifts are production aids.

Telephone Companies

An annoyance sometimes experienced during a telephone conversation is cross-talk from another conversation. This problem is caused largely by moisture in the lines. Compressed air dryers supply compressed dry air to the communications cables to prevent this condition and also to help locate potential cracks or leaks in the cables, thus reducing the amount of digging needed to gain access to such trouble spots. Such a system is seen in Figure 1.71.

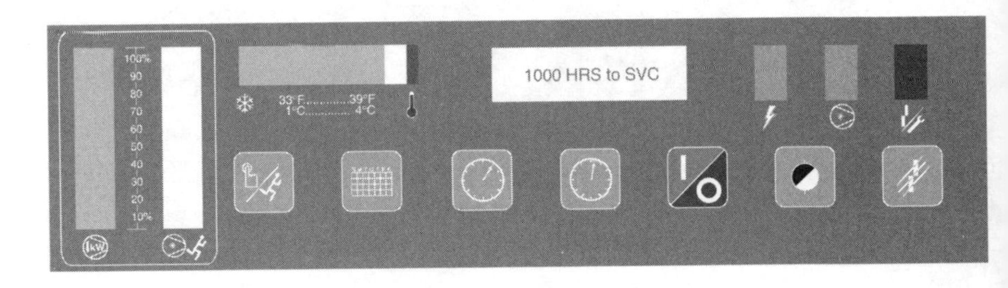

Figure 1.71 A refrigerated air dryer control panel.

Textiles

Compressed air is especially suited to cleaning looms, spindles, and other machinery in cotton and woolen mills. Cotton-finishing works use air for cleaning such machinery as presses and slashers and for operating baling presses. Humidifying systems employ compressed air in their operation. Other textile plant uses for compressed air are agitating, elevating, and transferring of dies or other solutions, pumping water, moistening goods, automatic control of steam and water lines, starting engines, and operating pressure accumulators. The tufting of rugs has improved from 45% to 95% by using compressed air to hold the thread in a hollow needle.

In hat factories, jets are used for raising nap on velour hats and for cleaning machines. Presses, like most laundry presses, are air-operated because of ease of control and comfort for the operator.

A lingerie manufacturer uses air to operate cutting tools and, in a unique application, for inflatable bras and inflatable hip pads.

A two-stage version of the integral gear type, rated 4000 scfm at 100 psig, is shown in central station service in the power house of a southern textile plant for air jet weaving in the textile industry (Figure 1.72). A larger volume, 4500 scfm, four-stage unit is shown in Figure 1.73 in another textile plant installation. Note that in both cases the intercoolers are housed within the outboard end cast housing.

Figure 1.72 Three stage integrally geared centrifugal type air compressor. Electronic digital controls include communication ports for host computer, phone modem, printer, internet monitoring.

Figure 1.73 Centrifugal air compressors provide plant air pressure in a textile manufacturing facility.

Theaters and Amusement Parks

Permanently installed air jets and push-button valves permit the projectionist to blow dirt from TV movie projector gates without stopping the film. Amusement parks provide for one of the fastest growing uses of compressed air. Thrilling rides at the amusement park make use of air brakes based on the same air-release principle used in brakes on railway trains (Figure 1.74).

Figure 1.74 Many amusement park ride designers utilize compressed air to power their latest designs.

Underwater Exploration

Experiments continue to explore the conditions under which people can live for extended periods beneath the surface of the sea (Figure 1.75). Continental shelves are potentially very important sources of minerals and food.

Air under pressure has an increased partial pressure of its oxygen. If this partial pressure becomes too great, the nervous system is affected and convulsions or blacking out may result. Nitrogen under pressure can also have serious effects. Nitrogen narcosis is accompanied by a numbing of the senses and a feeling of drunkenness. Air for breathing must, in any case, be free from oil and other contaminants.

Synthetic atmospheres are being studied in which the oxygen partial pressure is controlled by reducing the percentage of oxygen below that found in atmospheric air and substituting other inert gases, notably helium. While the helium atmosphere increases the pitch of the human voice and makes voice communication difficult, it does eliminate the intoxication associated with nitrogen and diffuses well from the blood during decompression.

Figure 1.75 Divers need good, clean air when they are exploring the wonders of nature under the sea.

Compressed air is used in offshore seismographic exploration for oil and other minerals. An air gun automatically emits several pulses per second of air at about 2000 psig. These pulses, reflected from the strata below, supply useful information to the geologist conducting the exploration. This method, as formerly conducted with rapid explosions of dynamite, was dangerous and not acceptable to fish and game commissions. The compressed-air method is also reported to be more effective.

Waterworks

Compressed air jets have proved especially effective in several important cleaning and maintenance operations in waterworks. Washwater tanks and beam supports are quickly cleaned with air jets and corrosion is thus checked. Rust is cleaned from chlorinator machines and ammonia feeders with compressed air.

Water with a heavy iron content is treated by aerating it with compressed air to remove carbon dioxide and to cause the iron to separate when no other treatment is needed. Compressed air is also used for pumping water, for cleaning boiler flues, and for operating tools for caulking, sheet pile driving, and trenching.

Well Drilling

An increasingly vital role is being played by compressed air and gas in the recovery of oil and water. In oil-well drilling, compressed air is used increasingly in dry formations to replace mud for removal of cuttings. Compressed gas is used to increase the volume of oil to the surface by gas-lift techniques. Compressed-air oil-field drilling rigs are shown in Figure 1.76. The offshore drilling platform in Figure 1.77 uses a similar skid-mounted compressor.

Figure 1.76 Three 850 cfm, 200 psig oil field drilling rigs supplying the air for air drilling of gas and oil. The compressors are single-stage, helical-screw type.

Figure 1.77 A 425 cfm, 125 psig, skidmounted, silenced compressor is used for drilling air on this offshore rig.

In water drilling, a common method of sand and gravel well development is to release surges of compressed air intermittently from the storage tank while pumping water by the air-lift method. This creates a uniform condition of gravel surrounding the screen. Since air lifts for pumping water have no moving parts in the well, pump maintenance cost is minimized. The power plant can be located any place where natural sources of power may be utilized. The purity of the water is also improved by aeration.

Large volumes of air are required for removing chips and cooling the bits used in drilling large-diameter holes such as are used for atomic tests. These have diameters up to 90 in. and depths as great at 6000 ft. Atomic bombs exploded at the bottom of such wells produce vibrations that are sensed at distant points.

Wind Tunnels

Wind tunnels, especially those for research and development in supersonic aircraft, require very large volumes of air in pressures ranging up to 2500 psig. Other laboratory and research applications also often depend heavily on compressed air and gas.

2

Compressed Air Production (Compressors)

OIL INJECTED ROTARY SCREW COMPRESSORS

The oil injected rotary screw compressor, with electric motor driver, has become a dominant type for a wide variety of industrial and mining applications. The engine driven portable version also enjoys great popularity for mining, construction and energy exploration applications. Natural gas engines also are used.

The stationary version is characterized by low vibration, requiring only a simple load bearing foundation and providing long life with minimal maintenance in broad ranges of capacity and pressure. They also are used widely in vacuum service.

The oil injected rotary screw compressor is a positive displacement type, which means that a given quantity of air or gas is trapped in a compression chamber and the space that it occupies is mechanically reduced, causing a corresponding rise in pressure prior to discharge.

Compression Principle

The oil injected rotary screw compressor consists of two intermeshing rotors in a stator housing having an inlet port at one end and a discharge port at the other. The ***male*** rotor has lobes formed helically along its length while the ***female*** rotor has corresponding helical grooves or flutes. The number of helical lobes and grooves may vary in otherwise similar designs.

Air flowing in through the ***inlet port*** fills the spaces between the lobes on each rotor. Rotation then causes the air to be trapped between the lobes and the stator as the inter-lobe spaces pass beyond the inlet port. As rotation continues, a lobe on one rotor rolls into a groove on the other rotor and the point of intermeshing moves progressively along the axial length of the rotors, reducing the space occupied by the air, resulting in increased pressure. Compression continues until the inter-lobe spaces are exposed to the ***discharge port*** when the compressed air is discharged. This cycle is illustrated in Figure 2.1.

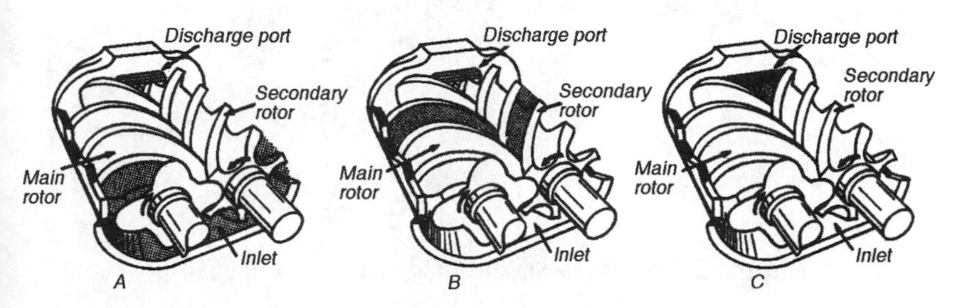

Figure 2.1 Helical Screw-Type Compressor; Compression Cycle, Single Stage

Oil is injected into the compression chamber during compression and serves three basic functions:

1) It ***lubricates*** the intermeshing rotors and associated bearings.
2) It ***takes away*** most of the ***heat*** caused by compression.
3) It ***acts as a seal*** in the clearances between the meshing rotors and between rotors and stator.

Lubrication

The generic term ***oil*** may be a hydrocarbon product but most compressors now use cleaner and longer-life synthetic lubricants, including diesters, polyglycols, polyalphaolefins, polyol esters and silicon based fluids. These newer products are suitable for a wider range of temperatures and have higher flash points. The lubricant chosen should be compatible with the compressor gaskets and seals.

A mixture of compressed air and injected oil leaves the air end and is passed to a sump/separator where most of the oil is removed from the compressed air. Directional and velocity changes are used to separate most of the liquid. The remaining aerosols in the compressed air then are separated by means of a coalescing filter, resulting in only a few parts per million of oil carry-over (usually in the range 2 - 5). A minimum pressure device, often combined with a discharge check valve, prevents excessive velocities through the separator element until a normal system pressure is achieved at start-up. Most oil injected rotary screw compressor packages use the air pressure in the oil sump/separator, after the discharge of the air end, to circulate the oil through a filter and cooler prior to re-injection to the compression chamber. Some designs may use an oil pump.

Bearings at each end of each rotor are designed to carry the radial and axial thrust loads generated. See Figure 2.2A. These bearings are lubricated directly with the same filtered oil as is injected into the compression chamber. A similar arrangement with built-in Spiral or Turn Valve for capacity control, is shown in Figure 2.2B.

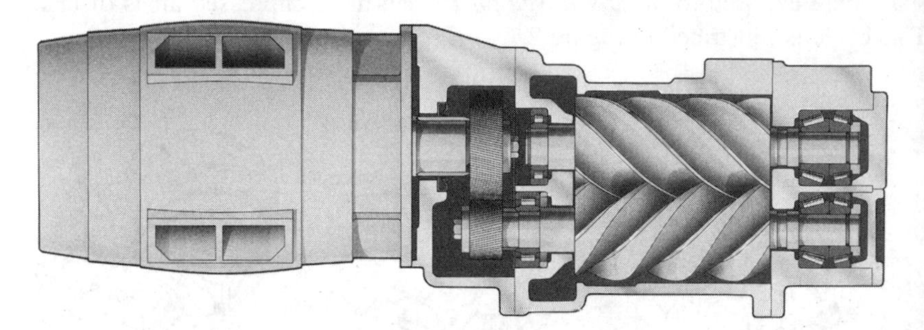

Figure 2.2A Single Stage Oil Injected Screw Compressor

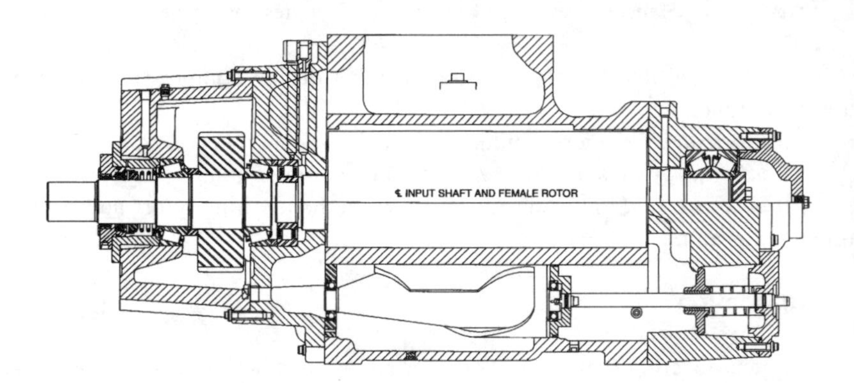

Figure 2.2B Single Stage Oil Injected Screw Compressor with Integral Variable Displacement Control Valve

Multi-stage compressors may have the individual stages mounted side by side, either in separate stators or within a common multi-bore stator housing. See Figure 2.2C. Alternatively, the stages may be mounted in tandem with the second stage driven directly from the rear of the first stage. See Figure 2.2D. Multiple stages are used either for improved efficiency at a given pressure or to achieve higher pressures.

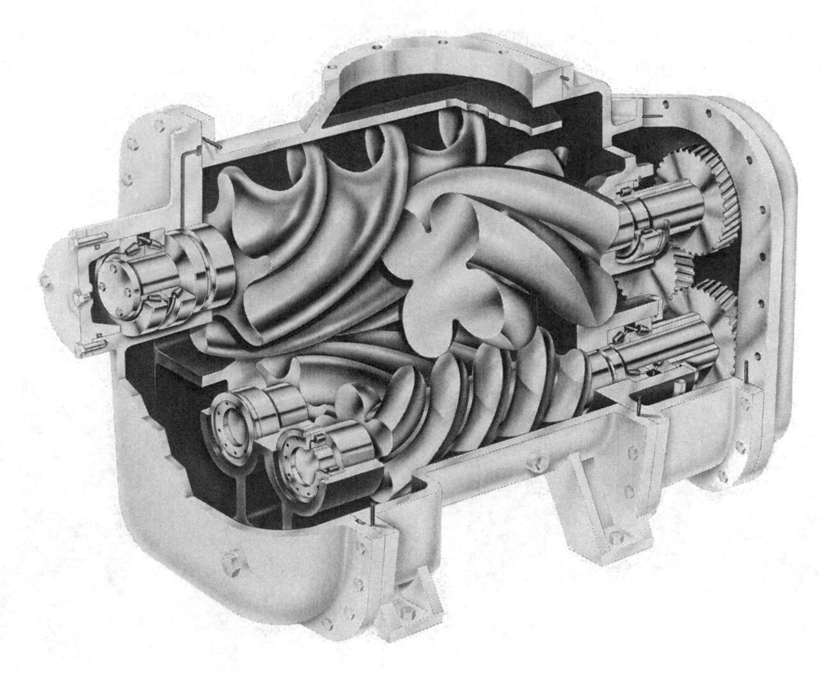

Figure 2.2C Two Stage "Over/Under" Design of Oil Injected Rotary Screw Compressor

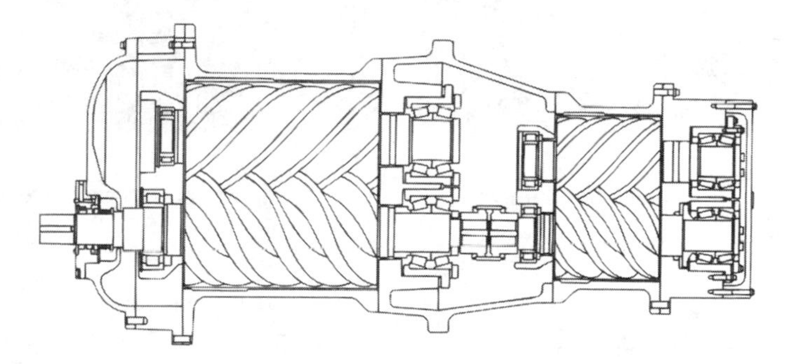

Figure 2.2D Two Stage "Tandem" Oil Injected Rotary Screw Compressor

Cooling

The temperature of the oil injected into the compression chamber generally is controlled to somewhere in the area of 140°F, either directly or indirectly by controlling the discharge temperature. The discharge temperature must remain above the pressure dewpoint to avoid condensation of moisture that would mix with the oil. A thermostatic bypass valve allows some or all of the oil being circulated to bypass the oil cooler to maintain the desired temperature over a wide range of ambient temperatures.

Generally, a suitable temperature and viscosity of the oil are required for proper lubrication, sealing, and to avoid condensation in the oil sump. It also is necessary to avoid excessive temperatures that could result in breakdown of the oil and reduced life.

In addition to oil cooling, an air aftercooler is used to cool the discharged air and to remove excess moisture. In the majority of applications, radiator type oil and air coolers are employed and provide the opportunity of heat recovery from the compression process for facility heating (Figure 2.3). Water cooled heat exchangers, with water control valves, also are available on most rotary screw compressor packages.

Figure 2.3 Installation of oil injected rotary screw compressors with heat recovery ducting.

In multi-stage designs, oil may be removed and the air cooled between the stages in an intercooler, or the air/oil mixture may pass through a curtain of oil as it enters the next stage.

Single stage oil injected rotary screw compressor packages are available from 3 - 700 hp, or 8 - 4000 cubic feet per minute, with discharge pressures from 50 - 250 psig. Two stage versions can improve specific power by approximately 12 - 15% and some can achieve discharge pressures up to 250 psig. Oil injected rotary screw vacuum pumps also are available from 80 - 3000 inlet cfm and vacuum to 29.7 in. Hg.

Capacity Control

Demand for compressed air seldom matches exactly the output from the compressor so some form of capacity control is essential. The type of capacity control is based on both the type and size of rotary air compressors, the application, and the number of compressors in the system. (See also Chapter 4, Compressed Air Distribution Systems). Typical capacity control systems for stationary air compressors are as follows:

Start/Stop Control is the simplest form of control, in which a pressure switch, sensing system pressure at the discharge of the compressor, sends a signal to the main motor starter to stop the compressor when a pre-set pressure is reached. When pressure falls to another pre-set pressure, the pressure switch sends a signal for the compressor to be restarted. The pressure switch will have an adjustable upper pressure setting and a fixed or adjustable differential between the upper and lower pressure settings. An air receiver is essential to prevent too frequent starting and stopping, which affects life of motor insulation due to high inrush current at each start. This type of control normally is limited to compressors in the 30 hp and under range. Its advantage is that power is used only while the compressor is running but this is offset by having to compress to a higher receiver pressure to allow air to be drawn from the receiver while the compressor is stopped.

Constant Speed or Continuous Run Control allows the compressor to continue to run, even when there is reduced or no demand for compressed air. This term may be used with **Load/Unload Control** and/or with **Inlet Valve Modulation.**

Load/Unload Control. In this type of control when the upper pressure setting is reached, the pressure switch sends a signal to close a valve at the inlet of the compressor (but maintaining a calibrated low flow), reducing the mass flow through the compressor. Simultaneously a blow-down valve, installed in a line coming from the compressor discharge but prior to a discharge check valve, is opened. When the blow-down valve is opened, the compressor air end discharge pressure is lowered gradually and the discharge check valve prevents back flow from the system or receiver.

Closing the inlet valve reduces the inlet pressure, increasing the pressure ratio across the air end. This reduces as the air end discharge pressure is reduced, resulting in reduced power requirements. In the case of oil injected rotary compressors, the rate of blow-down must be limited to prevent foaming of the oil in the sump/separator. An adequate receiver/system volume is required to allow fully unloaded operation for a sufficient period of time. It may take from 30 to 100 seconds for the sump pressure to be fully reduced, during which time the compressor bhp will reduce from 70% to about 20 - 25%. The average power consumption for a given reduced flow rate will be reduced as receiver/system volume is increased.

In oil free rotary screw compressors and oil free lobe type rotary compressors, this is the most common type of control, with both stages being unloaded simultaneously but without the blow-down time associated with oil injected types, allowing fully unloaded power almost immediately.

Inlet Valve Modulation allows compressor capacity to be adjusted to match demand. A regulating valve senses system or discharge pressure over a prescribed range (usually about 10 psi) and sends a proportional pressure to operate the inlet valve. Closing (or throttling) the inlet valve causes a pressure drop across it, reducing the inlet pressure at the compressor and, hence, the mass flow of air. Since the pressure at the compressor inlet is reduced while discharge pressure is rising slightly, the compression ratios are increased so that energy savings are somewhat limited. Inlet valve modulation normally is limited to the range from 100% to about 40% of rated capacity, at which point the discharge pressure will have reached full load pressure plus about 10 psi and it is assumed that demand is insufficient to require continued air discharge to the system. At this point the compressor will be unloaded as previously described in Load/Unload Control.

Dual Control is a term used to describe a rotary air compressor with a selector switch to enable selection of either **Modulation** (in some cases **Start/Stop**) or **Load/Unload** capacity control. This arrangement is suitable for locations where different shifts have substantially different compressed air requirements.

Automatic Dual or Auto Dual Control is a further refinement to each of the above systems. When a compressor is unloaded a timer is started. If compressed air demand does not lower system pressure to the point where the compressor is required to be re-loaded before the pre-set time has expired, the compressor is stopped. The compressor will re-start automatically when system pressure falls to the predetermined setting.

Variable Displacement (Slide, Spiral or Turn Valve) is a device built into the compressor casing to control output to match demand. Rising discharge pressure causes the valve to be repositioned progressively. This reduces the effective length of the rotors by allowing some bypass at inlet and delaying the start of compression. The inlet pressure and compression ratio remain constant so part load power requirements are substantially less than for inlet valve modulation. The normal capacity range is from 100% to 40-50%, below which inlet valve modulation may be used down to 20-40%, after which the compressor is unloaded. Curve shows Average Power v Percent Capacity with Variable Displacement Capacity Control (Slide/Spiral/Turn Valve) from 100% to 50% capacity followed by Inlet Valve Modulation to 40% capacity, then unloading. With this type of control, the inlet pressure to the air end does not change, hence, the pressure ratio remains essentially constant. The effective length of the rotors is reduced.

Geometric Lift or Poppet Valves may be used to have a similar effect to Slide, Spiral or Turn Valves but with discreet steps of percent capacity rather than infinitely variable positioning. The normal range is from 100% down to 50% capacity and normally with four valves. An inlet modulation valve may be added for capacities below 50%.

Variable Speed may be achieved by variable frequency AC drive, or by switched reluctance DC drive. Each of these has its specific electrical characteristics, including inverter and other losses. Air end displacement is directly proportional to rotor speed but air end efficiency depends upon male rotor tip speed. Most variable speed drive (VSD) package designs involve full capacity operation above the optimum rotor tip speed, at reduced air end efficiency and increased input power, when compared with a constant speed compressor of the same capacity, operating at or near its optimum rotor tip speed. While energy savings can be realized at all reduced capacities, the best energy savings are realized in applications where the running hours are long, with a high proportion in the mid to low capacity range.

Some designs stop the compressor when a lower speed of around 20% is reached, while others may unload at 40-50%, with an unloaded power of 10-15%. The appropriate amount of storage volume should be considered for each of these scenarios.

Field conversion of an existing compressor to variable speed drive must consider the electric motor, the proposed male rotor tip speed at 100% capacity and the reduction of air end efficiency at reduced speeds and capacity. (Figure 2.4).

Steam turbines and engines also can be used as variable speed drivers.

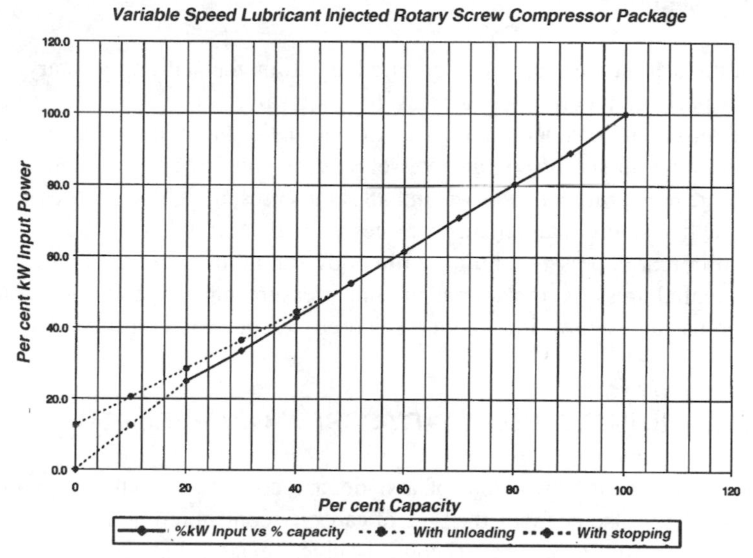

Figure 2.4 Variable Speed Curve

Multiple Compressor Sequencing is desirable in larger installations so that only a sufficient number of compressors will be in operation to meet current demand. There are two types of sequencing systems available:

1. **Cascading System:** In this system, the pressure settings in the compressors installed are overlapped so that as the pressure increases or decreases, the appropriate compressor is started or stopped, loaded or unloaded. This requires a large overall system pressure band.

2. **Rate of Change or Target Pressure System:** In this system, the individual pressure settings in each compressor are over-ridden by the master sequencer. This measures the change in pressure over time, and calculates the system flow requirement. As a result the overall system pressure band is reduced with resulting energy savings over the cascading system.

Compressors are started and stopped, loaded and unloaded, as required to maintain current system requirements. It is desirable to have only one compressor in the system at any given time to be in a reduced capacity mode of operation. This optimizes energy requirements. Sequencing can be arranged to equalize running hours of each compressor or to operate the compressors in a specified sequence, particularly where there is a mix of larger and smaller compressors. The sequence can be changed manually or automatically. Most modern compressors have microprocessor controls, which facilitate appropriate programming. Some microprocessor controls do not require a master sequencer in this type of system.

Safety Systems

In addition to pressure and temperature indicators, the stationary rotary screw compressor package incorporates a pressure relief valve for relief of excess pressure in the air/oil sump/separator vessel. An automatic, blow-down valve relieves pressure from the oil sump/separator vessel on each shutdown. Also included is a high discharge temperature shutdown and, in some cases, a high discharge pressure shutdown. Other safety/maintenance devices typically provided include high air inlet filter differential pressure, high oil filter differential pressure, high air/oil separator differential pressure, low unloaded sump pressure and motor overloads. Most compressor packages now incorporate microprocessors for controls and safety devices.

PORTABLE OIL INJECTED ROTARY SCREW COMPRESSORS

The basic design is similar to that of a stationary compressor but employing an engine driver. This also changes the type of capacity control required and the safety features. Normally, a pneumatic control valve, sensing compressor discharge pressure, is used to progressively close the inlet modulating valve as system pressure rises with decreased air demand. Simultaneously, engine speed, and hence compressor speed, is decreased.

As air demand increases, the system pressure falls, the engine speeds up and the inlet valve opens. Usually, the pressure range from fully closed inlet and idling engine to fully open inlet valve and engine at full speed is about 25 psig or, e.g., an operating pressure range of 100 to 125 psig.

Portable oil injected rotary screw compressors are available in a similar range of capacity and discharge pressure as for the stationary industrial type.

As for stationary packages, a safety relief valve is incorporated to relieve excess pressure at the compressor discharge. An automatic blow-down valve, actuated and held closed by engine oil pressure, relieves pressure in the oil sump/separator vessel when the engine is shut down. The engine fuel supply also is cut off, stopping the engine in the event of high compressor discharge temperature. The engine itself is protected from low oil pressure, high cooling water temperature and low coolant level by being shut down if either of these malfunctions occurs.

OIL INJECTED SLIDING VANE COMPRESSORS

The oil injected sliding vane rotary compressor was introduced as an engine driven portable air compressor in 1950 and a few years later was applied as a stationary industrial air compressor with electric motor driver.

The oil injected rotary sliding vane compressor is a positive displacement type, which means that a given quantity of air or gas is trapped in a compression chamber and the space that it occupies is mechanically reduced, causing a corresponding pressure rise prior to discharge.

The basic design consists of a circular stator in which is housed a cylindrical rotor, smaller than the stator bore and supported eccentrically in it. The rotor has radial (sometimes off-set) slots in which vanes, or blades, slide. Rotation of the rotor exerts centrifugal force on the vanes, causing them to slide out to contact the bore of the stator, forming "cells" bounded by the rotor, adjacent vanes and the stator bore. Some designs have means of restraining the vanes so that a minimal clearance is maintained between the vanes and the stator bore.

An inlet port is positioned to allow air to flow into each cell exposed to the port, filling each cell by the time it reaches its maximum volume. After passing the inlet port, the size of the cell is reduced as rotation continues, as each vane is pushed back into its slot in the rotor. Compression continues until the discharge port is reached, when the compressed air is discharged. See Figure 2.5.

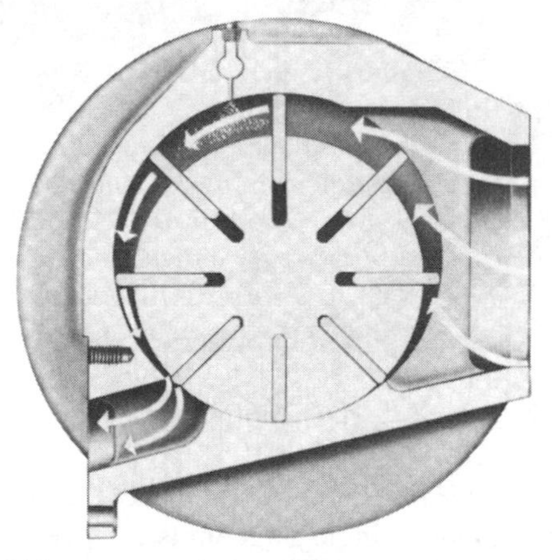

Figure 2.5 An Oil-Injected, Sliding Vane Rotary Compressor

Similar to the oil injected rotary screw compressor, oil is injected into the compression chamber to act as a lubricant, as a seal, and to remove the heat of compression. Single and two-stage versions are available with either in-line or over-under arrangement of the stages.

The oil injected sliding vane compressor normally is sold as a completely pre-engineered package in the range from 10 to 200 hp, with capacities from 40 to 800 acfm and discharge pressures from 80 to 125 psig.

Packaging, oil injection and separation, lubrication, cooling, capacity control and safety features essentially are similar to those for the oil-injected rotary screw compressor, either stationary or portable.

OIL FREE ROTARY SCREW COMPRESSORS

The oil free rotary screw compressor also is a positive displacement type of compressor. The principle of compression is similar to that of the oil injected rotary screw compressor but without oil being introduced into the compression chamber. Two distinct types are available - the ***dry type*** and the ***water injected type.***

In the ***dry type***, the intermeshing rotors are not allowed to touch and their relative positions are maintained by means of lubricated timing gears external to the compression chamber. See Figure 2.6. Since there is no injected fluid to remove the heat of compression, most designs use two stages of compression with an intercooler between the stages and an aftercooler after the second stage. The lack of a sealing fluid also requires higher rotative speeds than for the oil injected type.

Figure 2.6 Lube Free Screw Air End

Dry type oil-free rotary screw compressors have a range from 20 - 900 hp or 80 - 4,000 cubic feet per minute. Single stage units can operate up to 50 psig while two-stage generally can achieve 150 psig.

In the ***water injected type***, similar timing gear construction is used but water is injected into the compression chamber to act as a seal in internal clearances and to remove the heat of compression. This is shown in Figure 2.7. This allows pressures in the 100 - 150 psig range to be accomplished with only one stage. The injected water, together with condensed moisture from the atmosphere, is removed from the discharged compressed air by a conventional moisture separation device.

Similar to the oil-injected type, oil-free rotary screw compressors generally are packaged with all necessary accessories.

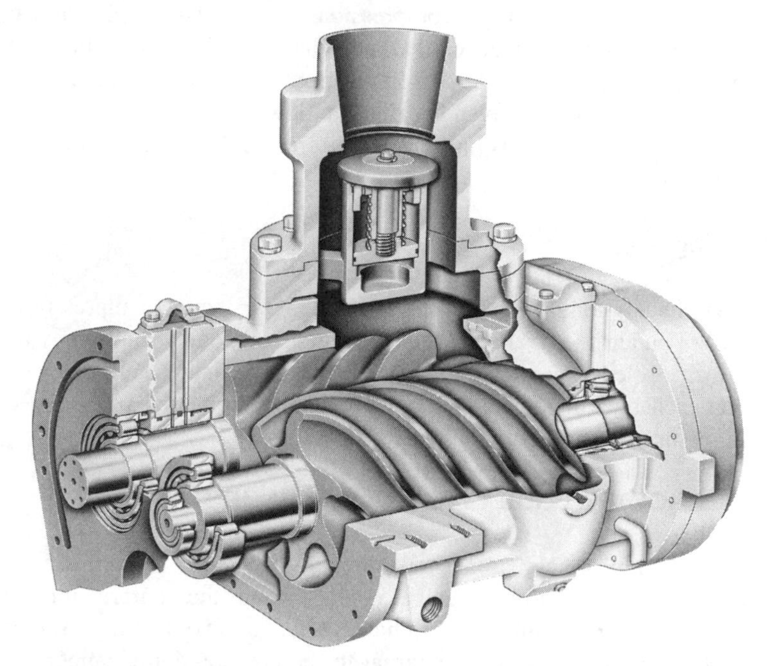

Figure 2.7 Oil Free, Water Injected Rotary Screw Compressor

Capacity Control

Capacity control for the ***dry type*** oil-free rotary screw compressor normally is the simple load/unload as sometimes employed for oil-injected rotary screw. Inlet valve modulation is avoided as this would result in a greatly increased pressure ratio and discharge temperature rise in the second stage. The inlet valve is totally closed except for a small, calibrated flow for cooling purposes. Simultaneously the unloading valve is opened, releasing any back pressure to atmosphere. Unloaded power of approximately 15 – 20% is normal.

Automatic start/stop control also may be used where demand is infrequent or sporadic. Dual control, combining automatic start/stop with load/unload is useful for long periods of high demand followed by long periods of low demand (e.g. day shift vs. night shift).

Capacity control for the water-injected oil-free rotary screw compressor normally utilizes inlet valve modulation as for the oil-injected type, the water taking away the heat of compression at increased pressure ratios combined with reduced mass flows.

Lubrication

Oil-free rotary screw compressors utilize oil for lubrication of bearings and gears, which are isolated from the compression chamber. The oil also may be used for element jacket cooling on air-cooled units due to the lack of cooling water.

Typically, an oil pump is directly driven from a shaft in the gearbox, assuring oil flow immediately at start-up and during run-down in the event of power failure. An oil filter, typically with 10 micron rating, protects bearings, gears and the oil pump from damage.

Cooling

The cooling system for the dry type oil-free rotary screw compressor normally consists of an air cooler after each stage and an oil cooler. These may be water cooled or air cooled radiator type. Some two-stage designs also employ an additional heat exchanger to cool a small portion of the compressed air for recycling to the compressor inlet during the unloaded period.

OIL-FREE ROTARY LOBE TYPE COMPRESSORS

The rotary lobe type compressor is a positive displacement, non-contact, or clearance type, design. With no mechanical contact within the compression chamber, lubrication within this chamber is eliminated. Lubrication of bearings, timing gears and speed increasing gears is all external to the compression chamber and shaft seals prevent any migration of oil to the compression chamber. This ensures oil-free compression and air delivery. A cross-sectional view of the intermeshing rotors in their stator is shown in Figure 2.8. The compression principle is illustrated in Figure 2.9. Two intermeshing rotors have lobe profiles that intermesh during rotation. Air flows into the compression chamber from the two inlet ports while the discharge ports are sealed by the rotor lobes. Rotation continues until the two discharge ports are exposed to the compression chamber, when the air is discharged. The dual inlet/outlet ports eliminate any axial thrust loads.

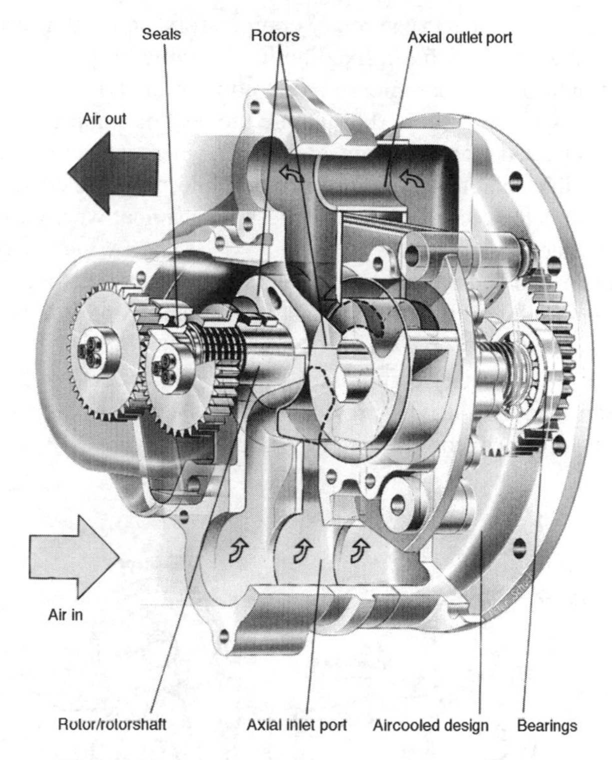

Figure 2.8 Oil-Free Rotary Lobe Compressor

Working principle

1. Suction: air at inlet pressure enters the compression chamber. Outlet ports sealed off by female rotor.

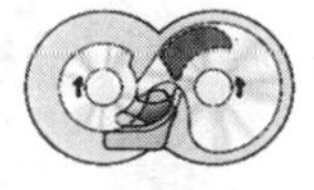

2. Compression starts: in- and outlet ports closed off. Volume is reduced, pressure increses.

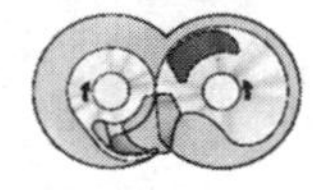

3. End of compression: entrapped air is compressed to its maximum. Suction starts again as inlet ports are opened.

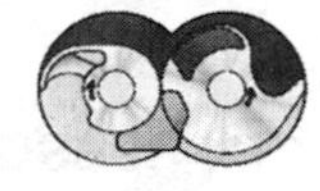

4. Delivery: recess in female rotor uncovers outlet ports and compressed air flows out.

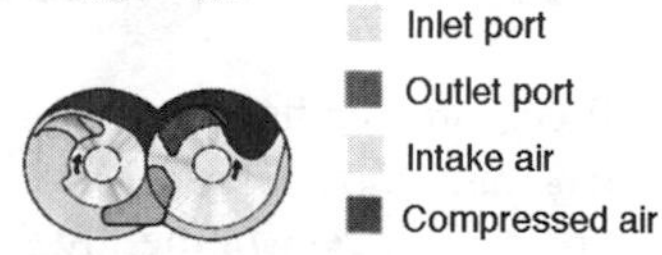

Figure 2.9 Rotary Lobe Compression Principle

As in other displacement type rotary compressors, no inlet or discharge valves are incorporated. Each stage has a fixed built-in volume (or pressure) ratio.

For most industrial air compressors operating in the 80 - 125 psig range, two stages of compression are required to handle the heat of compression. Air leaving the first stage is passed through an air intercooler, where its temperature is reduced as close as possible to atmospheric temperature and resulting condensate drained off, before it enters the second stage compression chamber where it is compressed to the desired system pressure.

A typical air end consists of a cast iron stator housing that may have passage for air or water cooling. Rotational speeds are chosen to optimize volumetric efficiency for a given profile.

Like other oil-free rotary compressors, the normal method of capacity control is the common load/unload type previously described. Both stages are unloaded simultaneously. A typical control system is shown in Figure 2.10.

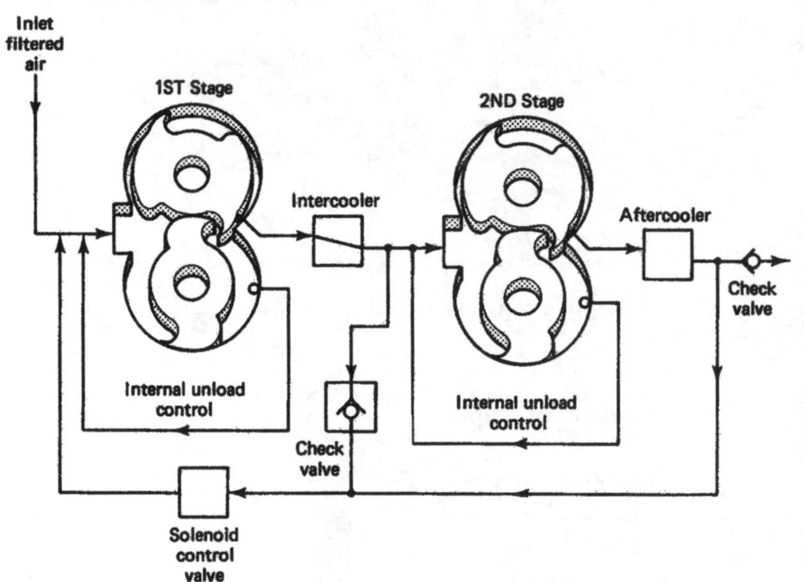

Figure 2.10 Method of Unloading the Rotary, Oil-Free, Lobe-Type Compressor

Both air cooled and water cooled packages are available from 25 - 75 hp with capacities from 85 to 315 cfm.

ROTARY SCROLL TYPE COMPRESSORS

The rotary scroll compressor has become a popular compressor as a domestic air conditioning refrigerant compressor. More recently it has been introduced to the standard air compressor market in the lower end of the horsepower range of rotary air compressors.

The operating compression principle is accomplished by means of two intermeshing spirals or scrolls, one scroll being stationary and the other orbiting in relation to the stationary scroll. See Figure 2.11. The stationary scroll is shown in black and the orbiting scroll in white. Air entering through the suction port in the stationary scroll, fills the suction chamber consisting of the outer labyrinth of the stationary scroll and on the outside edge of the orbiting scroll, as shown in the illustration at position 1. At this position, the portion of the compression chamber at an intermediate pressure is sealed by adjacent portions of the two scrolls.

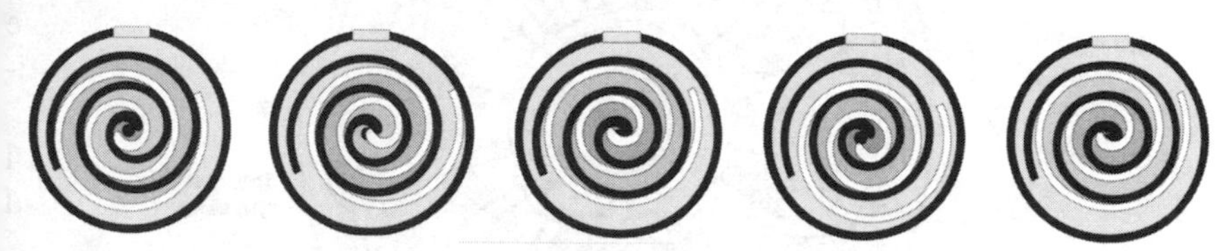

Figure 2.11 Operating Principle for a Scroll Compressor

As orbiting continues, the space occupied by the air becomes progressively reduced as shown in steps 2 through 5 and moves progressively toward the discharge port in the center of the stationary scroll.

It should be noted that the flow through the Suction Port and through the Discharge Port is continuous, providing pulsation free delivery of compressed air to the system. There is no metal to metal contact between the scrolls, eliminating the need for lubrication in the compression chamber and ensuring oil free air delivery from the scroll compressor. However, without the removal of the heat of compression, the efficiency is less than comparable oil injected air compressors.

Current models are air cooled and range from approximate 6 to 14 acfm, 2 through 5 hp, with discharge pressures up to 145 psig. This size range is expected to steadily increase. Noise levels with a sound attenuating canopy are extremely low, in the range 52 - 59 dBA at 1 meter, in accordance with the CAGI/Pneurop test code.

LIQUID RING ROTARY COMPRESSORS

The liquid ring (or liquid piston) rotary compressor also is a positive displacement type compressor. The mode of compression is similar to that of the sliding vane rotary compressor but the vanes (or blades) are fixed on the rotor. See Figure 2.12. The stator bore may be circular with the rotor eccentric to it, or elliptical with the rotor concentric to it. The former provides one compression per revolution while the latter provides two.

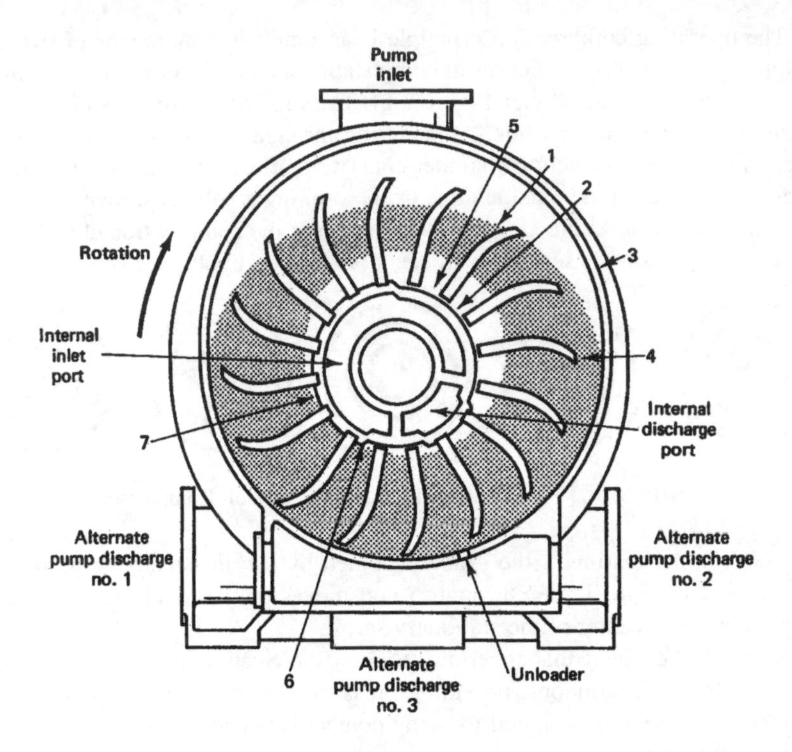

Figure 2.12 Cross Section Diagram of a Liquid Piston Rotary Compressor

A ring of liquid is swirled around the bore of the stator as the rotor turns. The depth of penetration of each vane, or blade, into the liquid, varies as rotation occurs. The space between the rotor hub and the liquid ring therefore varies. Axial inlet porting allows air to fill the space between adjacent vanes until its maximum volume. Further rotation then causes the space to be reduced and compression to occur until the discharge port is reached when the compressed air is discharged. The liquid ring also serves to remove the heat of compression and is discharged into gas-liquid separator, which removes the liquid from the gas. Because of the intimate contact of gas and liquid, the final discharge temperature can be held close to the supply temperature of the liquid, eliminating the need for an aftercooler.

Vapors are condensed in a liquid ring compressor when the liquid is cooler than the saturation temperature of the air/vapor or gas/vapor mixture in the compressor. The liquid ring, therefore, is acting as a condenser. The condensate becomes part of the sealant liquid during the compression cycle and is removed from the gas-liquid separator with the sealing fluid.

The liquid also scrubs the gas, removing solid particles from significant down to micron sizes, without damage to the unit unless the particles are abrasive.

The lubricated bearings are external to the compression chamber and isolated from it, so oil-free compression is obtained.

Capacities range from 2 to 16,000 acfm with a discharge pressure up to 35 psig in a single stage and 125 psig in a two stage version. As a compressor it is much less efficient than other rotary positive displacement types due to the energy required to swirl the liquid in the stator.

This type of design is used most commonly as a vacuum pump up to 26 in. Hg., single stage. Two stage units can achieve higher vacuum levels. This type of vacuum pump is used widely in the pulp and paper industry.

Compression of gases other than air are possible and a liquid is chosen, which is compatible with the gas being compressed. Rotor and stator materials may also require to be changed.

PORTABLE AIR COMPRESSORS

History

The history of the portable compressor follows closely the history of the construction and mining fields. These industries required compressors that could be easily moved with the work. The portable compressor, therefore, is a complete air compressor plant, sufficiently light in weight, yet strong enough to withstand the severe service encountered in construction and mining work.

Stationary air compressors were well established in industry by the turn of the century. They were heavy, large machines that required bulky, solid foundations. Often, it was necessary to use long lengths of air hose or pipe to reach the work, and this resulted in losses due to friction and leakage.

Around 1900, portable air compressors were introduced. These compressors were little more than stationary compressors on wheels and were limited in their application by the drills of that period, which weighed up to 500 lb and did less work than today's lightest drills while using a great deal more air.

In 1910, the portable compressor in most common use had one large, single-stage compression cylinder driven horizontally by a steam or oil engine. Probably the greatest single factor that stimulated portable compressor development was the advent of the lightweight air drill.

In 1933, the first two-stage, air-cooled portable compressor was manufactured. Shortly afterward, compressor manufacturers established standard sizes and rated portable compressors on actual free air delivery. In 1938 and 1939, portable compressors became more modernized, with pneumatic tires and streamlined enclosures. By 1939, the first multispeed regulation system had been introduced. This allowed the compressors and engines to idle when little or no air was demanded. In the late 1940s, regulation was improved by providing variable engine speed and controlled air intake flow throughout the air requirement range. By the late 1950s, selective loading of cylinders and simultaneous variable-speed features were added to reciprocating portable compressors.

In the 1950s, the oil-injected, sliding-vane rotary compressor was introduced. Higher-speed, overhead-valve engines made possible considerable reduction in the size and weight of the portable compressor. Oil injected into the rotating compressor acts as a coolant, lubricant, and sealant. The oil is then separated from the air, cooled, filtered, and reinjected.

In 1961, the first oil-injected, rotary-screw compressors were manufactured in this country. Since then, improvements in oil separation and cooling systems for these rotaries have resulted in lightweight units. Now, materials have also made the reciprocating units lighter and more maneuverable. Oil-free, rotary-screw portable compressors, which do not require oil in the compression chamber, also became available for primarily industrial applications.

In 1968, the first quiet or silenced portable compressors were introduced into the construction market. They were sold as hospital and residential noise attenuated machines (Fig. 2.13) in and around the Eastern megalopolis. Since 1978, all portable compressors sold in U.S. commerce that are 75 cfm and larger have been required by law to comply with the sound levels stipulated by thc EPA in the Noise Control Act of 1972.

Figure 2.13 Portable compressor with a noise limiting housing suitable for use in residential areas, even near hospitals.

In 1982, a new type of two-stage portable air compressor was introduced. It uses a diesel-exhaust-driven turbocharger to drive centrifugal compressors as the first stage, and the same diesel mechanically drives the reciprocating piston compressor as the second stage.

Description

A portable air compressor is usually defined as a self-contained unit mounted on wheels. The unit consists of an air compressor, prime mover with silencing, cooling, control, air induction, lubrication, fuel tank, exhaust, and starting systems.

Additionally, reciprocating compressors and non-oil-injected compressors are equipped with air receivers requiring no oil separators. Oil-injected rotary units are equipped with oil-air separator systems, which also act as air receivers. Since skid-mounted units are generally obtained by removing the running gear from a portable unit, they are usually classified with the portable unit. Such a unit is seen in Fig. 2.14. Mounting designs vary, depending on weight and application of the unit. Lighter-weight units with low air capacities (below 125 cfm on 210 cmh) are usually mounted on two wheels and are predominantly gasoline engine driven, but compressors as high as 600 cfm may now be mounted on two wheels. Manufacturers now offer both gasoline- and diesel-driven units up to 250 cfm, and they are generally two-wheel units. Larger units, 250 cfm and up are normally two- or four-wheel units and diesel driven.

Figure 2.14 Three portable air compressors provide high pressure compressed air for pile driving operations on a U.S. interstate highway project. Each compressor delivers 1600 cfm of air at 150 psig.

A less common but highly useful type of portable compressor is the electric-motor-driven machine (Fig. 2.15). In external appearance, this compressor looks similar to engine-driven portable compressors, but the prime mover is an electric motor having the proper electrical and mechanical characteristics for the compressor and its application. The electric motor starter is usually an integral part of such a portable compressor package.

Large manufacturing, process industries, petrochemical, and refinery facilities find the electric-driven portable compressor useful. Furthermore, in some cases the engine-driven compressor is undesirable for reasons of safety in some of the aforementioned facilities.

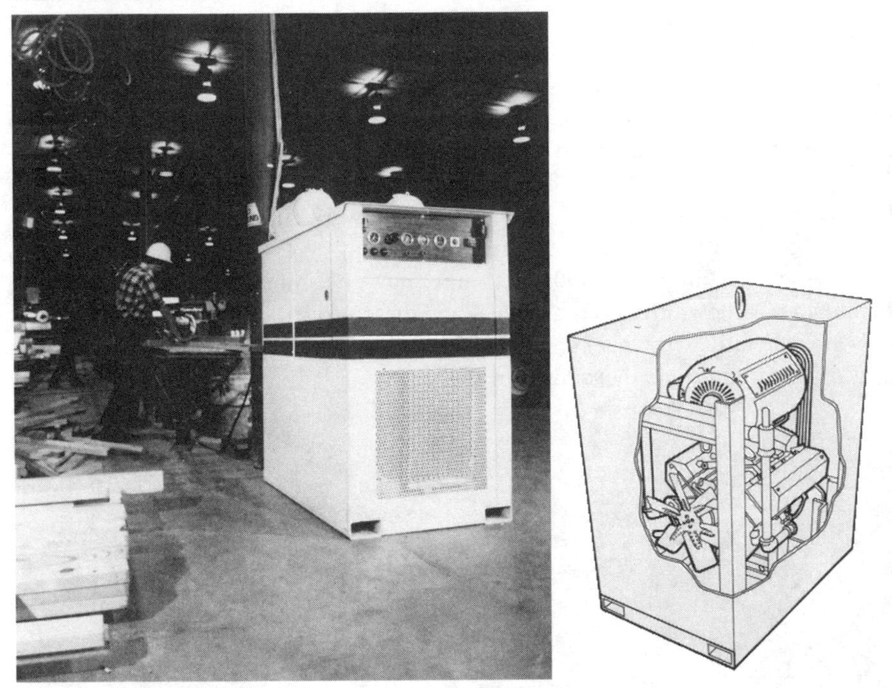

Figure 2.15 Sixty horsepower electric motor driven reciprocating compressor, located near work-stations without hazard or annoyance to workers. Units are virtually vibration free and require no foundation or bolting to the floor.

Muck-mounted units are used by utility companies because they require a highly mobile, relatively small compressor of usually 160 cfm or less for their particular type of small, short job. The same truck with its air compressor may move through congested city streets and be used on as many as three or four different jobs in one day. A truck-mounted booster unit is seen in Figure 2.16.

Figure 2.16 Portable booster compressors are now compact enough to fit into the bed of a pickup truck. This unit boosts 750-900 cfm from 250-350 psig to 900 psig.

One type of truck-mounted unit is the utility skid. This is a skid-mounted unit, normally not larger than 175 cfm without wheels. It is mounted across the frame behind the cab inside the utility body. A second type of truck-mounted unit is the power take-off driven type, which utilizes the truck engine for its prime mover. The extra cost for the vehicle and drive line maintenance are considerations in this configuration. Noise emissions are a major factor in these installations because of their use in populated areas.

Self-propelled compressors have the advantage that they can be moved without the need for a separate towing unit. Such compressors can be equipped with attachments to accomplish other work. Attachments include either rear- or front-mounted rock drills for highway drilling, front-end loaders, and backhoes.

Capabilities

Portable compressors are manufactured in sizes ranging from 20 to 5000 cfm, with delivery at 100 to 250 psig operating pressure. The model numbers normally designate the approximate air delivery, and size increments are such as to cover the full range available. Figures 2.17 and 2.18 are typical of the construction of a 175 and 742 cfm unit.

Figure 2.17 A silenced portable compressor in use in a residential area.

Figure 2.18 A 742 cfm portable compressor running several hand-held drills on a construction site.

Pressure

Since the most common application of the portable compressor is to operate air-powered tools (Chapter 5), the units are designed and rated for 100 psig or higher discharge pressure. The units normally have sufficient reserve for operating at high altitudes. Some manufacturers offer portable compressors and skid-mounted units with higher pressure ratings, up to 350 psig for the increasingly common higher-pressure applications, such as down-the-hole hammer, pipeline testing, oil- and gas-well servicing, sandblasting, rock drilling, and pile driving.

Air Receiver

A reciprocating portable compressor is normally equipped with an approved ASME pressure vessel. The ratings of temperature and pressure stated on the ASME data plate attached to the ASME-approved vessel must not be exceeded.

Oil-injected rotary compressors are equipped with an air-oil separator that is an approved ASME pressure vessel. The vessel contains the separator system and also serves as the compressor lubricant and compressed-air storage vessel.

Receivers are equipped with a safety valve for protection against excessive pressures and with a drain valve for removing moisture that will accumulate. The manufacturer's instruction manual should be consulted for the proper procedure for draining moisture from the receiver.

Fuel Storage Tank

Fuel tanks normally contain sufficient fuel for eight hours of operation. A fuel strainer is usually furnished and should be kept clean and in good condition. The fuel tank should be drained regularly to avoid problems from moisture condensation and accumulation of foreign particles in the tank.

Lubrication System

The lubricating system is a critical part of all designs of portable compressors. The manufacturer's recommendations should be followed in regard to lubricating oil specifications. This applies to oil-injected and dry rotary compressors, as well as reciprocating compressors.

Regulation

Since the portable compressor is most widely used to furnish air power for operating pneumatic tools and other devices designed to operate on constant pressure, the manufacturer furnishes, as standard equipment, a regulating system designed to hold the discharge pressure practically constant while the volume of air delivered to the tools varies with the demand. The manufacturer's instruction book contains detailed instructions for the care and maintenance of the particular design of regulating system, but one requirement common to all types of regulating devices is that they must not be tampered with unnecessarily nor by persons unfamiliar with their functioning and adjustment. Figure 2.19 shows a capacity curve for a typical portable control with proper adjustment.

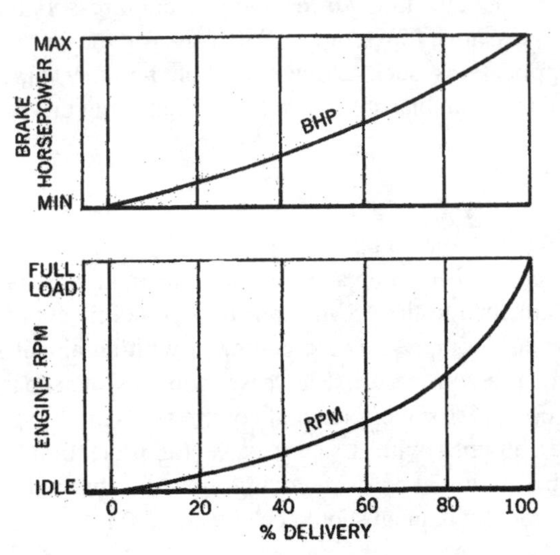

Figure 2.19 Capacity curves for a typical portable compressor capacity control.

Towing

Air compressors towed on highways are subject to state, local, and federal regulations. Department of Transportation federal regulations presently require brakes on equipment weighing 3000 lb gross weight or less if W3 is greater than 40% of the sum of W1 and W2 (Fig. 2.20).

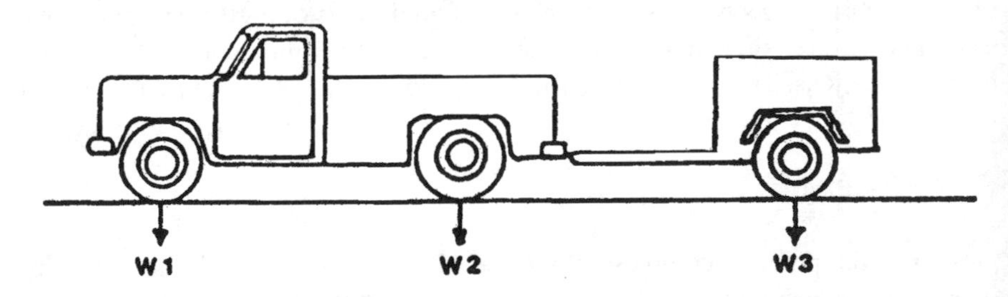

Figure 2.20 Federal regulations require brakes on towed compressors weighing 3000 pounds or less, depending upon the weight distribution W1, W2, and W3.

Applications

Portable compressors have wide and varied uses in both construction and industrial applications. Figures 2.21, 2.22, and 2.23 show some typical as well as some unusual applications. Other examples of applications are described elsewhere.

Figure 2.21 The drill on this air tool is working in rock six feet below ground level. The hole will accommodate the steel reinforced foundation for a highway sign. The compressor is a 100 cfm, 100 psig unit.

Figure 2.22 Two portable compressors being readied for the job of sandblasting a ship in dry dock to remove rust. The housing doors will be closed during the actual work.

Figure 2.23 A 265 cfm, 125 psig silenced portable air compressor on site at an oyster shell barge.

Typical Specifications

To aid prospective purchasers, the Compressed Air and Gas Institute offers the following as a set of typical specifications that may be used in the procurement of engine-driven portable compressors:

1. *General:* The portable air compressor shall consist of an air compressor and a gasoline or diesel engine rigidly connected together in permanent alignment and mounted on a common frame. The air compressor and engine shall be monoblock or direct connected through a heavy-duty industrial type clutch or coupling. The portable air compressor shall be provided with complete cooling, lubricating, regulating, starting, silencing and fuel systems, fuel tank, air receiver (if required), and other equipment to constitute a complete self-contained unit. The unit shall be new and unused, if a model currently manufactured.
2. *Compressors:* The compressor shall deliver its rated capacity of free air per minute when compressing to a discharge pressure of 100 psig.
3. *Engine:* The engine shall be of ample power to drive the compressor at full load. It shall be equipped with suitable air filter and muffler.
4. *Cooling system:* The cooling system shall be of suitable design and sufficient capacity for satisfactory operation in an ambient temperature of 125°F. A radiator guard shall be furnished.
5. *Regulation:* The unit shall be equipped with a regulator to vary automatically the volume of air delivered so as to meet the air demands when they are less than full capacity.

6. *Capacity ratings:* The capacity of the unit should be specified as the minimum requirement for the equipment to be operated, plus some allowance for normal decrease in efficiency. The manufacturer will then select the unit(s) nearest this capacity.
7. *Air receiver:* If one is required, the air receiver shall be made according to the ASME code and approved by the National Board, and the portable compressor shall be complete with pressure gage, safety valve, service manifold, and drain openings.
8. *Mountings:* The assembled portable air compressor shall be on a heavy frame equipped with one of the following:
   a. Two pneumatic tires with suspension springs or torsion bar suspension system
   b. Four pneumatic tires with suspension springs
   c. Skid mounts
9. *Housing:* The unit shall be enclosed by a complete housing with removable or hinged side doors that can be securely locked in place. Noise emission must meet current EPA specifications.
10. *Instrumentation:* The unit shall be equipped with the instruments required for the safe and efficient operation of the machine and should be enclosed in the above lockable housing or behind a lockable panel door.

RECIPROCATING AIR COMPRESSORS

Reciprocating compressors are positive displacement type in which a quantity of air or gas occupies a space that is mechanically reduced, resulting in a corresponding increase in pressure. A variety of such compressors is described in this chapter. Also included are vacuum pumps, which may be regarded as compressors having sub-atmospheric inlet pressure.

SINGLE-ACTING RECIPROCATING AIR COMPRESSORS

This type of compressor is characterized by its automotive type piston driven through a connecting rod from the crankshaft. Compression takes place on the top side of the piston on each revolution of the crankshaft, Figure 2.24. A design variation in small single stage oil-less compressors is a combined piston and connecting rod that tilts or rocks in the cylinder during its travel within the cylinder.

Figure 2.24 Two Cylinder Single-Acting Reciprocating Compressor

Single acting reciprocating air compressors may be air cooled, Figure 2.25 or liquid cooled, Figure 2.26, although the vast majority are air cooled. These may be single-stage, usually rated at discharge pressures from 25 to 125 psig, two-stage usually rated at discharge pressures from 125 psig to 175 psig, or multi-stage for pressures above 175 psig.

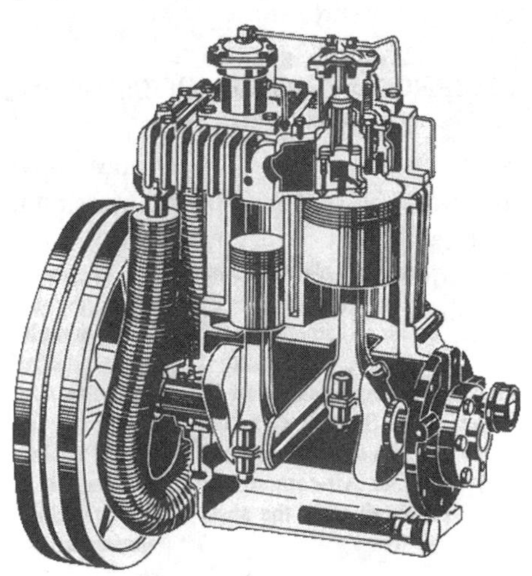

Figure 2.25 Air Cooled, Single-Acting, Two-Stage Reciprocating Compressor

Figure 2.26 A Single-Acting Liquid-Cooled Compressor

The most common air compressor in the fractional and single digit hp sizes is the air cooled reciprocating air compressor. In larger sizes, single-acting reciprocating compressors are available up to 150 hp, but above 25 hp are much less common.

Two-stage and multi-stage designs include interstage cooling to reduce discharge air temperatures for improved efficiency and durability. Coolers may be air cooled, Figure 2.27 or liquid cooled, Figure 2.28.

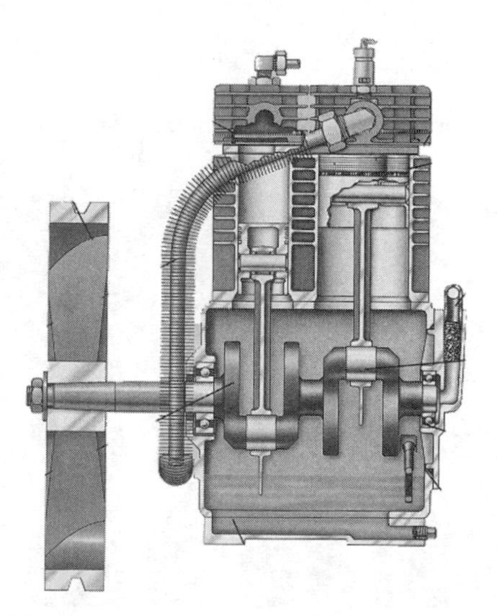

Figure 2.27 Air-Cooled, Single-Acting Reciprocating Compressor

Figure 2.28 Water-Cooled, Single-Acting Reciprocating Compressor

Pistons

Pistons used in single-acting compressors are of the automotive or full skirt design, the underside of the piston being exposed to the crankcase. Lubricated versions have a combination of compression and oil control piston rings which:

1. Seal the compression chamber.
2. Control the oil to the compression chamber.
3. Act (in some designs) as support for piston movement on the cylinder walls.

In lubricated units, compression rings generally are made of cast iron and oil control rings of either cast iron or steel, Figure 2.29.

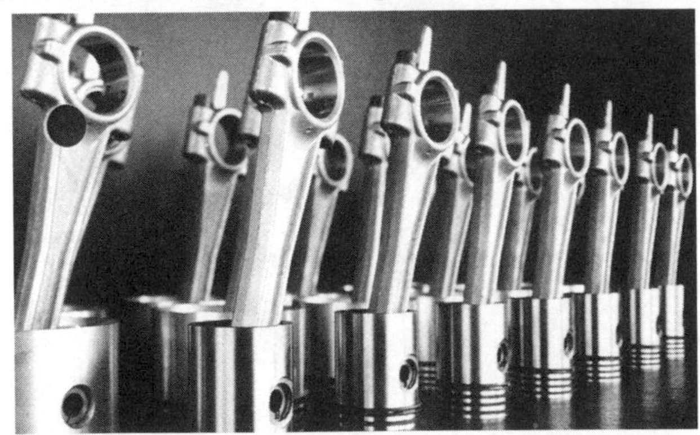

Figure 2.29 Single-Acting Compressor Cylinders and Crankshafts

Oil-free, or non-lube, designs do not allow oil in the compression chamber and use pistons of self-lubricating materials or use heat resistant, non-metallic guides and piston rings that are self-lubricating. These are shown in Figure 2.30. Some designs incorporate a distance piece or crosshead to isolate the crankcase from the compression chamber.

Figure 2.30 Balance-Opposed, Oil-Free, Single-Acting Reciprocating Compressor

Oil-less designs have piston arrangements similar to oil-free versions but do not have oil in the crankcase. Generally these have grease pre-packed crankshaft and connecting rod bearings, Figure 2.31.

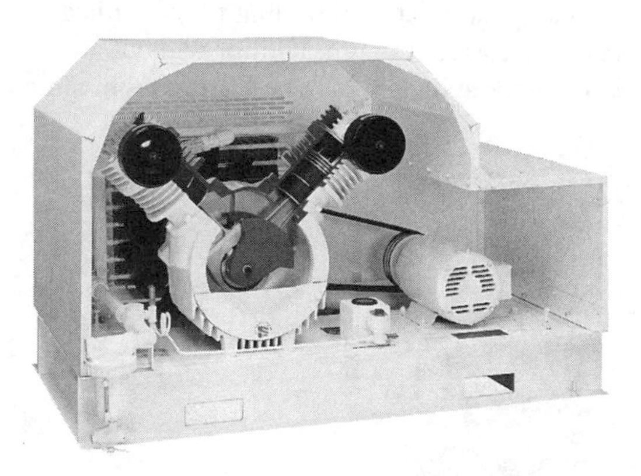

Figure 2.31 Oil-Free, Single-Acting Reciprocating Compressor

Cylinders

A variety of cylinder arrangements is used. These include:

1. A single vertical cylinder.
2. In-line or side by side vertical cylinders.
3. Horizontal, balance opposed cylinders.
4. V or Y configuration.
5. W configuration.

These are illustrated in Figure 2.32.

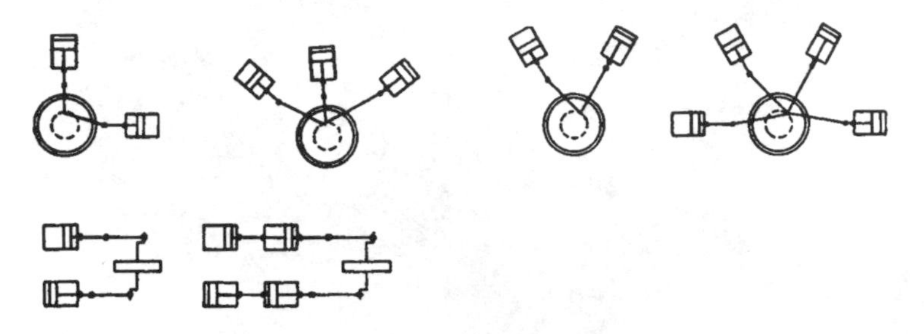

Figure 2.32 Various cylinder arrangements used in displacement compressors. Some are suitable for single-acting compressors, while others are double-acting, and require a cross-head and guide.

The number of cylinders is dependent on the capacity of air required and the number of stages. Cylinders may be separate castings, cast en-bloc, or a combination of the two, as shown in Figures 2.33A and B. Cylinders and heads for air cooled designs normally have external finning for better heat dissipation. Materials may be cast iron or die cast aluminum, with or without an iron or steel bore liner. Small oil-less compressor cylinders may be formed from aluminum tubing.

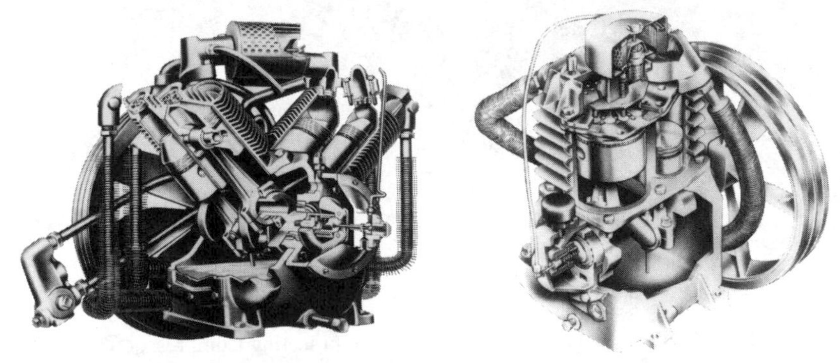

Figures 2.33A and 2.33B Cylinders may be individual castings (A), cast en-bloc, or a combination of the two (B).

Valves

In general, both inlet and discharge valves for single-acting compressors are of the automatic pressure type that open and close on a small differential pressure. To secure rapid action, the valve elements are made light in weight and are proportioned for low lift. As a valve opens, increasing spring pressure minimizes the impact forces on the valve elements and reduces noise levels. In some designs, cushioning pockets, which form as the valves approach the full open position, further minimize impact. There are three types of compressor valves. These are:

1. The reed type, Figure 2.34, has only one moving part, which flexes between its closed and open positions. It requires no lubrication, has the advantage of low clearance volume with high flow area and is easily cleaned or replaced. Reed valves may be designed in various configurations for a tandem, two-stage cylinder arrangement as shown in Figure 2.35.
2. The disc type, Figure 2.36, includes a flat disc, ring or plate, which seats on the edges of a slightly smaller opening in the valve seat. The disc is backed by a valve spring. Springs used for this type of valve are of various configurations, such as a coil spring or a spring shaped like the disc except for its wave deformity, which allows it to act as a wave spring. Some disc type valves have multiple springs. In some applications, several valves per cylinder are used.
3. The strip or channel type, Figure 2.37, which consists of a valve seat having a number of slots or ports and a corresponding number of valve strips or channels that cover and close off the slots. A bowed leaf spring for each channel, or the bowed valve strip itself, returns the valve strip or channel to its seat after the air has passed through. A stop plate limits the lift of the valve.

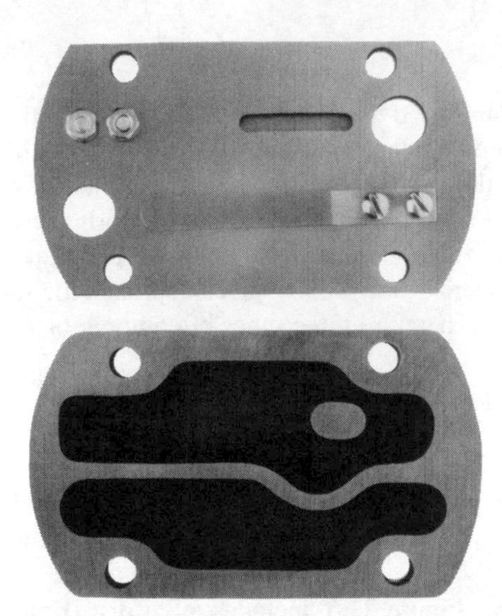

Figure 2.34 A reed valve showing the reed (upper view) and the air or gas passages (lower view).

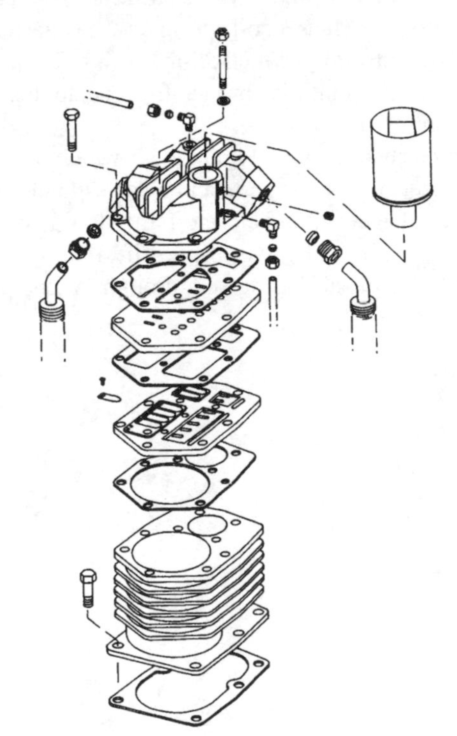

Figure 2.35 Cylinder, valve and head assembly, showing the reed valve for a tandem two-stage cylinder arrangement.

Figure 2.36 Disc valve, also known as a ring valve.

Figure 2.37 A channel-type valve. The bowed springs serve to return the valves to their seats.

Cooling

Single-acting air compressors have different arrangements for removing the heat of compression. Air cooled versions have external fanning for heat dissipation on the cylinder, cylinder head and, in some cases, the external heat exchanger. Air is drawn or blown across the fans and the compressor crankcase by a fan that may be the spokes of the drive pulley/flywheel. This is illustrated in Figure 2.38.

Figure 2.38 Air-Cooled Single-Acting Reciprocating Compressor

Liquid cooled compressors have jacketed cylinders, heads and heat exchangers, through which liquid coolant is circulated to dissipate the heat of compression. See Figure 2.39. Water, or an ethylene glycol mixture to prevent freezing, may be employed.

Figure 2.39 Water-Cooled Single-Acting Reciprocating Compressor

Bearings

The main crankshaft bearings usually are anti-friction ball or tapered roller bearings. Some designs may employ sleeve type main bearings. Crank pin and piston pin bearings normally are of the journal or sleeve type. Some connecting rod designs include precision bore bearings at the crank pin and needle roller bearings at the piston pin, or a plain piston pin. Figure 2.40 shows an exploded view of a typical design.

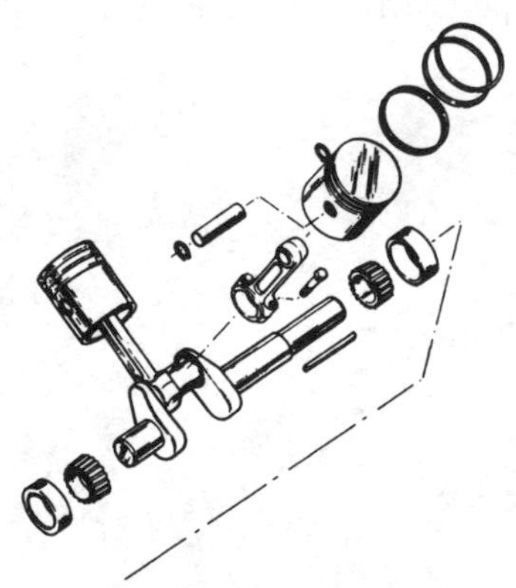

Figure 2.40 Blow-up of the working parts of a compressor showing tapered roller main bearings and insert-type connecting rod bearings.

Lubrication

Single-acting air compressors may utilize a splash or a full pressure lubrication system. A controlled splash lubrication system is shown in Figure 2.41. A dipper on the connecting rod dips into the oil reservoir in the crankcase each revolution and produces a splash of oil which lubricates the connecting rod, piston pin, main bearings and cylinders.

Figure 2.41 Splash-feed lubrication showing oil dipper and oil level float gage.

A variation, as shown in Figure 2.42, is the use of rings, running in a groove in the crankcase and dipping into the oil reservoir, to provide positive lubrication to critical wear areas.

A full pressure lubrication system is shown in Figure 2.43. In this design, a positive displacement oil pump draws oil from the reservoir in the crankcase and delivers it under pressure through rifle drilled passages in the crankshaft and connecting rod to the crank pin and the piston pin. This pressure system provides spray or drop lubrication to the anti-friction main bearings and cylinders.

Figure 2.42 Splash-feed lubrication using oil rings.

Figure 2.43 A full pressure lubricating system in a two-stage, air cooled, reciprocating air compressor.

Exposure of the piston and cylinder walls to the oil results in some oil carry-over into the air stream delivered from the compressor and carbon deposits on valves and pistons. The specified oil should be used and changed regularly as specified by the compressor manufacturer.

As previously stated, oil-less designs use self-lubricating piston and guide materials and pre-lubricated and sealed bearings with no liquid oil in the crankcase.

Balance

Rotative and reciprocating motion produce forces that must be counterbalanced for smooth operation. The weight of the reciprocating parts and counterbalanced crankshafts are used to achieve optimum running balance. Minimal vibration not only minimizes maintenance but is essential for compressors mounted on an air receiver or storage tank. These may be referred to as tank mounted compressors. This type of air receiver must be capable of accepting not only the static pressures to which it is subjected but also the dynamic loading caused by unbalanced reciprocating forces from the air compressor and its driver mounted on it.

Drives

The most common drive arrangement is belt drive from an electric motor, internal combustion engine or engine power take-off. The compressor sheave also acts as a flywheel to limit torque pulsations and its spokes often are used for cooling air circulation. Belt drive allows a great degree of flexibility in obtaining the desired speed of rotation.

Flange mounted or direct coupled motor drives provide compactness and minimum drive maintenance. Belts and coupling must be properly shielded for safety and to meet OSHA requirements in industrial plants. Fractional horsepower compressors normally are built as integral assemblies with the electric motor driver.

Unit Type and Packaged Compressors

A packaged compressor may be defined as an air compressor with its driver and associated components self contained and ready for installation and operation. These may include the compressor, driver, starter, intake filter, cooling system with aftercooler and all necessary mechanical and electrical controls. An air receiver and interconnecting piping also may be included. These require no foundation and simplify installation. Figures 2.44 A, B, C, D and E illustrate these compressors. These compressors may use vertical or horizontal air receivers. Smaller models are available for portable use with wheel mounting or very small hand carried tanks.

Unit type compressors are very suitable for installation close to a point of use where a dedicated air compressor is desirable, as discussed in the chapter on Compressed Air Distribution Systems.

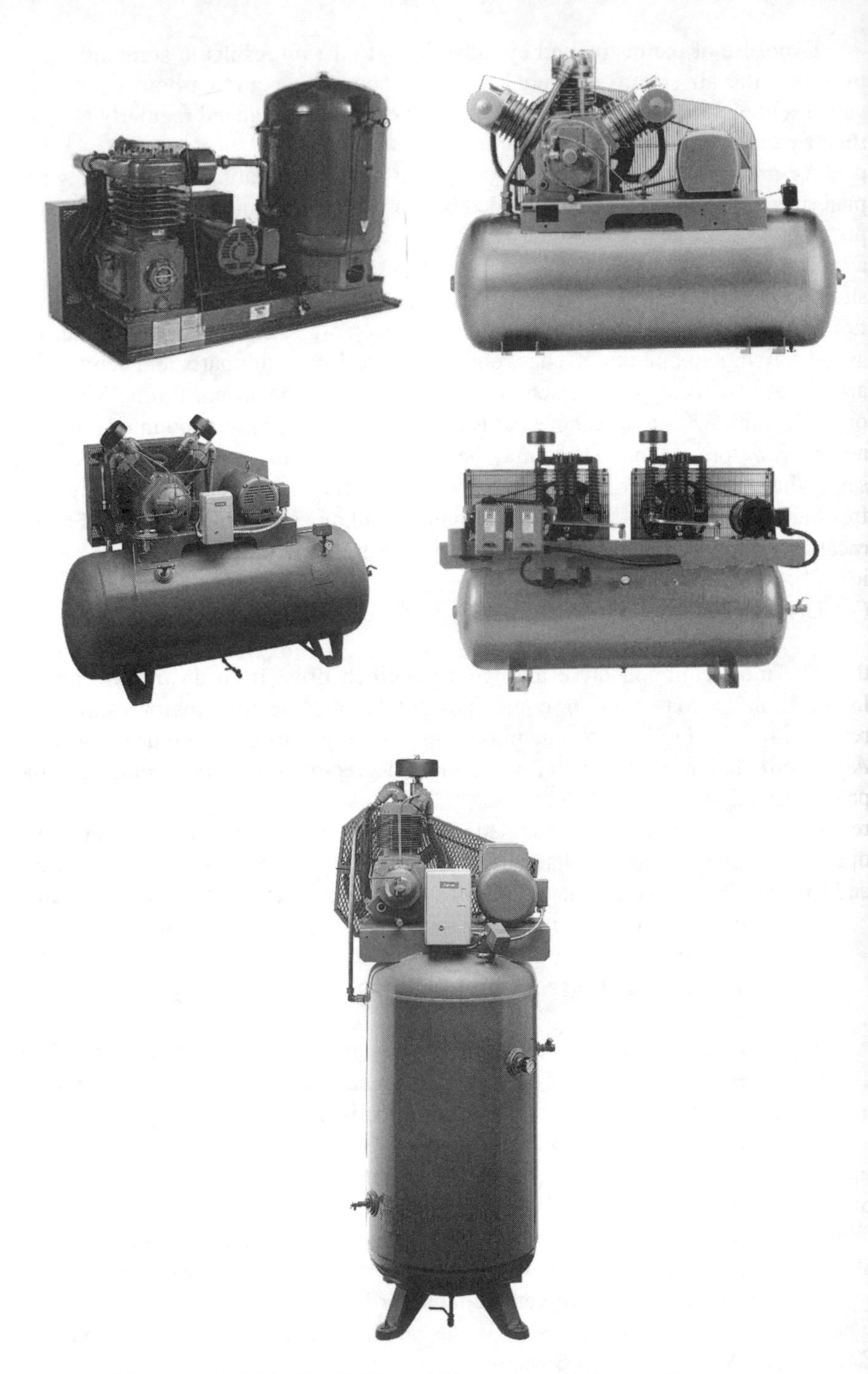

Figures 2.44A, B, C, D and E Various Unit-Type Compressors

Capacity Control

Start/Stop Control is the simplest form of control, in which a pressure switch, sensing system pressure, sends a signal to the main motor starter to stop the compressor when a pre-set pressure is reached. When pressure falls to another pre-set pressure, the pressure switch sends a signal for the compressor to be restarted. The pressure switch may have an adjustable upper pressure setting and a fixed or adjustable differential between the upper and lower pressure settings. Adjustments should be made only by qualified personnel and in accordance with the manufacturer's specifications.

An air receiver is essential to prevent too frequent starting and stopping, which affects the life of motor insulation due to high inrush current at each start. This type of control normally is limited to compressors in the 30 hp and under range. Its advantage is that power is used only while the compressor is running but this is offset by having to compress to a higher receiver pressure to allow air to be drawn from the receiver while the compressor is stopped. This type of control is best suited to light or intermittent duty cycles. Air cooled reciprocating air compressors typically are rated for duty cycles ranging from 50/50 (50% on and 50% off) to 75/25 (75% on and 25% off).

Constant Speed Control allows the compressor to continue to run, even when there is reduced or no demand for compressed air. This term may be used with Load/Unload Control. In this type of control when the upper pressure setting is reached, a pilot device sends a signal to actuate an inlet valve unloader, Figure 2.45. A common method holds the inlet valve(s) open so that air is drawn into and pushed back out of the cylinder without any compression taking place. This also requires an adequate air receiver since air delivery is either 100% or zero but may operate within a narrower pressure differential than stop/start control. Load/unload capacity control should be used where the duty cycle is heavy and continuous.

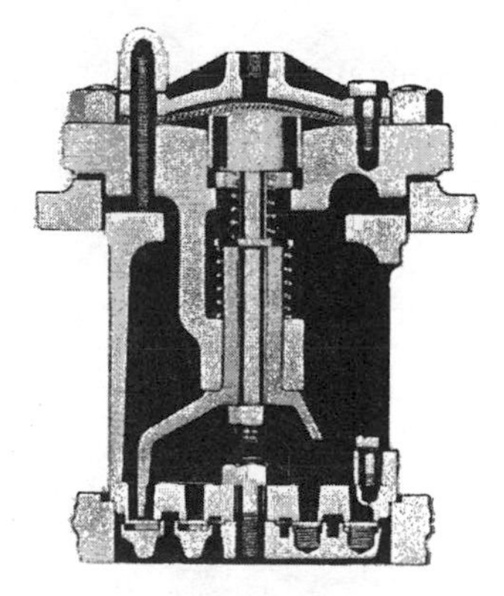

Figure 2.45 An intake valve unloader in which fingers hold the valve open until compressed air pressure is reduced.

Load/unload control also may be obtained using valves in the compressor discharge lines. When actuated by a pressure signal, the valve is held open so that air is released to atmosphere instead of being delivered to the receiver/tank. The air passes through the compressor discharge valves and possibly through the discharge line, before being released to atmosphere.

Internal combustion engine driven compressors normally use load/unload control to avoid the need to start and stop the engine. The same signal that operates the valve unloaders, can be used to operate an engine throttle control, so that the engine runs at idle speed with reduced power and noise during the unload cycle. The amount of power used by the compressors during the unload cycle is dependent on how well the design minimizes flow resistance into and out of the compression cylinder(s) and the magnitude of the mechanical friction.

When load conditions are changeable, special control systems are available that select start/stop or constant speed control to match the prevailing air demand cycle.

Operating Conditions

Stationary single-acting and unit type air compressors are adaptable to a wide range of conditions of temperature, altitude and humidity. High altitude or high humidity conditions may require a specially rated driver and cooling arrangements. Where oils are used, oils should be selected for the application and temperatures involved and in accordance with the manufacturer's recommendations.

Some applications of single-acting reciprocating air compressors are discussed in the chapter on Applications.

DOUBLE-ACTING RECIPROCATING AIR COMPRESSORS

This type of reciprocating air compressor uses both sides of the piston to compress the air. The piston is driven by a piston rod extending through a packing gland from a crosshead, which, in turn, is driven through a connecting rod from the main crankshaft. This is illustrated in Figure 2.46. The crosshead and its guide ensure that the piston and its rod operate in a straight line. Air is compressed as the piston moves in each direction of one complete stroke or one revolution of the crankshaft.

Figure 2.46 Double-acting compressor cylinder showing cooling water jackets around the cylinder and in the crank end cylinder head. Stud bolts shown are used to connect the cylinder to the distance piece between the cylinder and the crankcase.

The basic double-acting compressor has a single cylinder, single throw crankshaft, crosshead and connecting rod. The arrangement may be horizontal, as shown in Figure 2.47, or vertical, as shown in Figure 2.48.

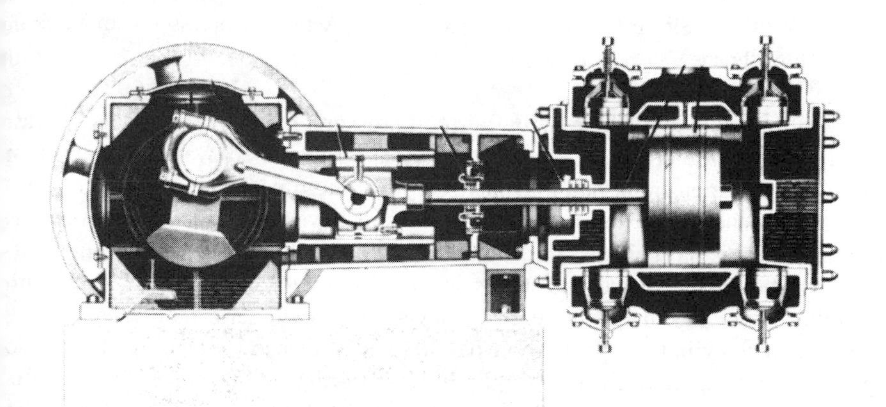

Figure 2.47 Single Cylinder, Single Throw, Horizontal, Double-Acting Compressor

Figure 2.48 A Single Cylinder, Oil Free, Double-Acting Compressor with Vertical, Single-Throw Frame

Multiple cylinder, double-acting air compressors, may have cylinders operating in parallel for increase flow rate, or in series for increased overall compression ratios. The common crankshaft may have single or multiple throws to drive the connecting rods and, hence, the pistons. Compressors have been built with as many as 10 crank throws on a single crankshaft. Figure 2.49 shows a two-stage air compressor with the first and second stages arranged at a right angle to each other, with a single throw crankshaft. Figure 2.50 shows a four cylinder arrangement having two first-stage cylinders operating in parallel and two second-stage cylinders also operating in parallel, also with a single throw crankshaft. Multi-stage compressors use water cooled heat exchangers after each stage to cool the air before entering the next stage, improving overall compression efficiency.

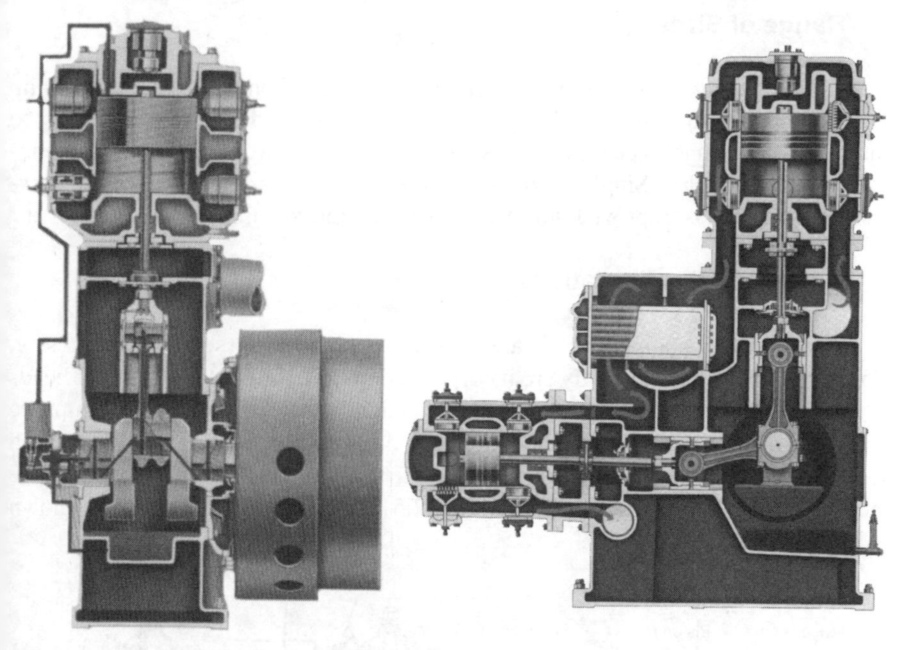

Figure 2.49 Two-Stage, Single Throw, Double-Acting Compressor with Vertical First Stage and Horizontal Second Stage Cylinder

Figure 2.50 Packaged Vertical Single-Cylinder Compressor Complete with Air Receiver

Range of Sizes

Double-acting reciprocating air compressors range from approximately 10 hp to 1,000 hp, although for standard plant air applications, they have given way to rotary and centrifugal type air compressors. Discharge pressures up to several thousand psig are possible. Single-stage air compressors are common for 100 psig service but efficiency is improved with two stages and intercooling.

Types and Configurations

This type of compressor is a heavy duty, continuous service compressor. Cooling water jackets normally are incorporated in the cylinders and cylinder heads to remove some of the heat of compression, maintain thermal stability and improve lubrication, reducing carbonization of valve parts. Water cooling jackets around valves and piston rod packing are essential due to localized heating. Valves may be located in the cylinders, as shown in Figure 2.51, or in the cylinder heads, as shown in Figure 2.52.

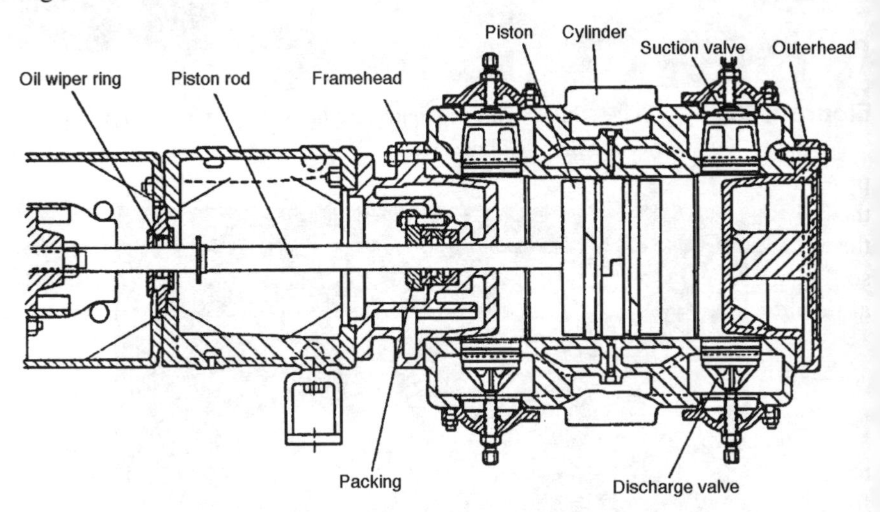

Figure 2.51 Horizontal Cylinder Arrangement with Suction and Discharge Valves around the Cylinder

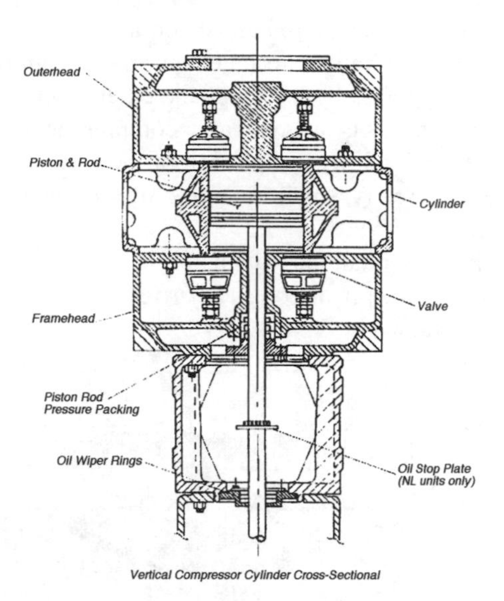

Figure 2.52 Cylinder arrangement with suction and discharge valves located in the cylinder heads.

A distance piece between the crankcase and the cylinder may incorporate the piston rod packing that prevents leakage of compressed air from the cylinder, along the rod and scraper rings that prevent migration of lubricant from the crankcase to the cylinder. An extended distance piece is used for oil-free, or non-lube compressors, to prevent any portion of the piston rod that enters the oil-free cylinder from also entering the lubricated crankcase.

Lubrication

Lubrication of the crankcase may be from a splash system, a forced feed system, or a combination of both. Cylinders are lubricated by means of a mechanical, force feed lubricator with one or more feeds to each cylinder. The type, size and application of the compressor will determine the method of lubrication and the type of lubricant. Modern synthetic lubricants now are common for cylinder lubrication.

Oil-Free, or Non-lube Compressors

These terms normally are used for air compressors that do not have any lubricant fed to the cylinder(s). Piston rings and rod packing usually are of PTFE-based materials, carbon, or other synthetic materials, which can operate without added lubrication.

In the majority of oil-free compressors, the piston rides in the cylinder bore on the synthetic or carbon wearing (or rider) ring or shoe, see Figure 2.53.

Alternatively, but less common, a tailrod and external crosshead may be added outboard of the cylinder head. With this arrangement, the weight of the piston is carried by the piston rod, supported by the main and external crossheads.

Where oil-free air delivery is not critical, a compressor with a standard length distance piece may be used, allowing a portion of the piston rod to enter both the cylinder and the crankcase. A small quantity of oil then migrates from the crankcase to the cylinder.

Where oil-free air is essential, the extended distance piece must be used, A baffle plate also is attached to the piston rod as a barrier to prevent the migration of oil along the piston rod, as shown in Figure 2.53.

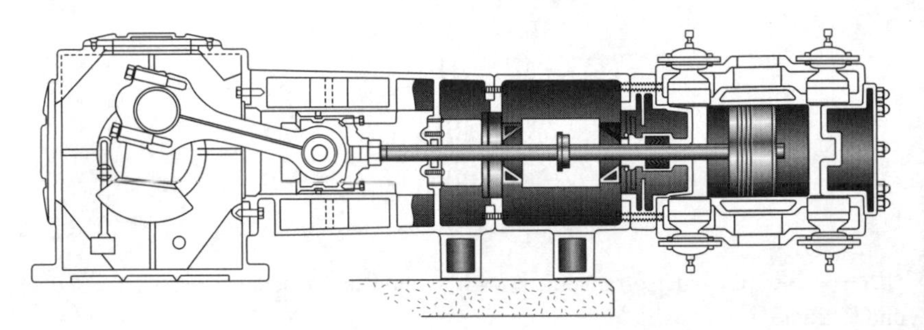

Figure 2.53 A non-lubricated compressor with piston riding on carbon wearing rings.

Capacity Controls for Double Acting Reciprocating Air Compressors

Reciprocating air compressors are positive displacement type, essentially having constant speed and capacity with variable pressure. The capacity can be varied to meet required demand by means of several types of capacity control.

These control systems are based upon maintaining the discharge pressure within prescribed limits. A pressure sensing element, or pilot, allows control air to operate the unloading mechanism. These are of three basic types:

1) Mechanically holding open the suction valves, allowing air drawn into th cylinder to escape without being compressed. See Figure 2.54.
2) The use of clearance pockets, which allows a predetermined portion of the compressed air to be diverted to the clearance pocket(s), then re-expand into the cylinder as the piston returns on its suction stroke. This reduces the amount compressed air delivered to the system and the amount of atmospheric air entering the cylinder.
3) Closing off the inlet air to the first stage cylinder of a two-stage compressor, then venting the second stage cylinder to atmosphere after a near vacuum has been reached within the compressor.

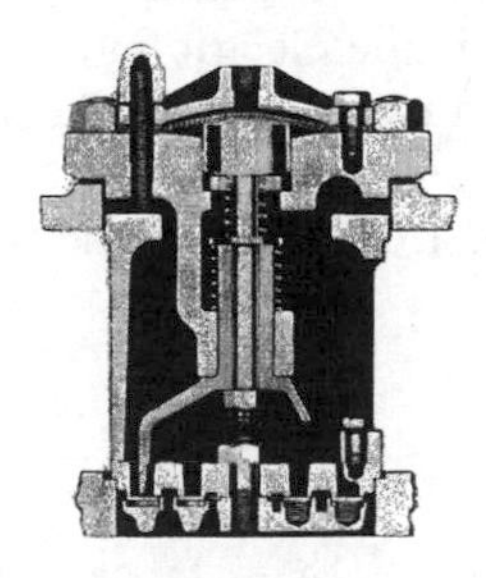

Figure 2.54 Suction valve unloader in which fingers hold the valve open until compressed air pressure is reduced.

Type 1) normally is used to provide compressor capacities of 0% or 50%, by holding open the suction valves on one end or on both ends of the cylinder. 100% capacity is obtained by allowing the suction valves to operate normally. This generally is called three step capacity control.

A combination of types 1) and 2) can provide 0%, 25%, 50%, 75% and 100% capacity and generally is known as five step capacity control. These are illustrated in Figure 2.55.

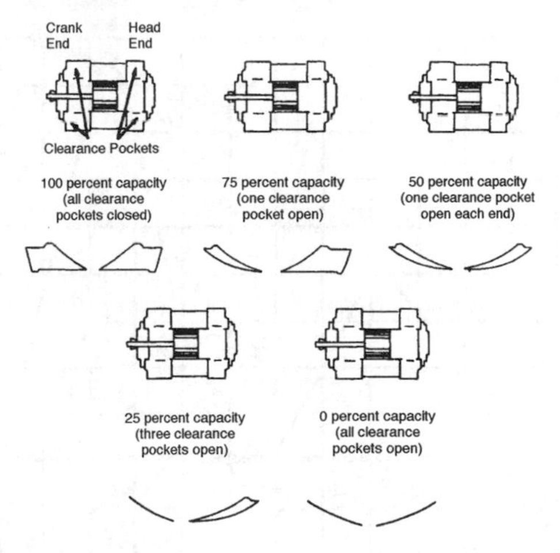

Figure 2.55 Capacity control steps by means of suction valve unloaders and clearance pockets.

Prime Movers for Double-Acting Reciprocating Air Compressors

Motive power for this type of compressor may be provided by one of the following principal drivers:

1. Electric motor.
2. Oil or gas engine.
3. Steam engine or turbine.

The most common driver is the electric motor, which is energy efficient and reliable. This may be induction type, synchronous type, wound rotor type or DC type, the first two types being the most common.

The compressor drive motor can vary by the type of connection between the motor and the compressor:

1. Motor with belt drive.
2. Flange mounted motor.
3. Direct connected motor.
4. Motor and flexible coupling.
5. Motor and speed reducing gearbox.

Belt drives generally are limited to about 150 hp with an 1800 rpm drive motor. The selection of other types depends on the specific characteristics of the compressor, including speed and torque. Arrangements generally are made for the motor to be started with the compressor unloaded. Flywheel mass may be necessary to keep current fluctuations within allowable standards. A typical compressor torque-effort diagram is shown in Figure 2.56. In all cases, NEMA Standards must be observed.

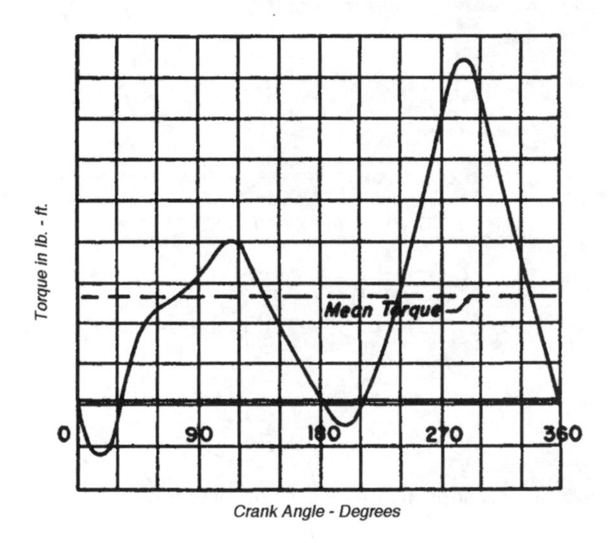

Figure 2.56 A Typical Torque-effort Diagram

Varying installation requirements will determine some of the required design characteristics of the drive motor. Typical enclosures include: Open Drip-Proof; Totally Enclosed and Explosion Proof (dependent upon type of risk in specific location). Non-sparking belt drive arrangements may also be necessary. Electrical controls also must meet the requirements of the location. Ambient temperature and altitude also can be major considerations.

Synchronous motors, requiring direct current field excitation, can provide a power factor of 1.0 and can have a leading power factor to offset lagging power factor of other equipment in the plant, reducing energy costs.

Full voltage starting normally is used for this type of compressor. Reduced voltage starting, where essential, must take into account the required starting torque.

Steam Turbine Driven Compressors

Steam engines have given way to steam turbines, where steam is used as an available and economical source of power. Generally, a speed reducing gear is required between a relatively high speed steam turbine and a relatively low speed double-acting reciprocating compressor. Suitable coupling and flywheel arrangements also are required. Steam turbine speed governors may be manual or automatic in response to changes compressed air system pressure and/or flow.

Performance Guarantees

Capacities are based upon prevailing ambient conditions, while power normally is stated in bhp at full load capacity and pressure. The capacity and bhp normally are guaranteed within 3 percent but not cumulative. This means that the specific power in bhp/cfm also is guaranteed within 3%, so that factors such as compression efficiency and/or mechanical efficiency are not significant for comparison of different compressors. The full load bhp and motor efficiency will allow specific power in kW/100 cfm for comparison with other types of packaged compressors. Table 2.1 is given as a typical example of compressor performance at full load, 75% and 50% capacity. ASME PTC-9 or ISO 1217 are common standards for performance tests. Members of the Compressed Air and Gas Institute now use a standardized Performance Data Sheet to show compressor performance based upon a standardized test.

Table 2.1 Data on Direct-connected, Motor-driven, Two-stage Compressor

Size of compressor cylinders, in	20 1/2 and 12 1/2 x 8 1/2		
Piston displacement, cfm (1st stage)	1662		
Rpm	514		
Discharge pressure, psig	100		
Altitude, ft above sea level	0		

	Full Capacity	3/4 Capacity	1/2 Capacity
Actual capacity, ft^3 free air per min.	1395	1046	697.5
Bhp at compressor shaft	260	200	143
Bhp per 100 cfm actual capacity	18.6	19.25	20.5
Motor efficiency, percent	93.2	92.3	90.2
Electrical hp input per 100 cfm actual capacity	19.95	20.85	22.77

INSTALLATION AND CARE OF DOUBLE-ACTING RECIPROCATING AIR COMPRESSORS

Manufacturers normally supply a manual for installation and operation of their compressors. It is recommended that in addition, ASME B19.1, Safety Standards for Air Compressor Systems, also be consulted.

Location

Proper location and installation will materially reduce maintenance and operating costs. The compressor should be located in a clean, well-lighted area of sufficient size to permit cleaning, ready inspection and any necessary dismantling, such as removal of pistons with rods, flywheels, belt sheaves, crankshafts, or intercooler tube bundles. The installation drawings furnished by the manufacturer show space required for major dismantling. The location should be such as to keep piping runs short, with a minimum of elbows to minimize pressure losses. Locate the compressor accessories, such as the aftercooler and air receiver, to permit short, straight runs of piping to minimize vibration caused by pressure pulsations from the compressor discharge.

In plants such as foundries and woodworking plants, where dusty conditions prevail, the compressor(s) should be located in a separate machinery room or dust-free room with provision for drawing clean air from outside the building to the compressor inlet. An air inlet filter should be installed and, for dusty atmospheres, select the heavy duty type. It should be remembered that this filter is to protect the compressor. Additional filtration downstream of the compressor may also be necessary to protect equipment at points of use.

Foundation

A properly designed and constructed compressor foundation performs two functions:

1. It maintains compressor alignment and at proper elevation;
2. It minimizes vibration and prevents its transmission to any building structures external to the foundation.

The foundation must satisfy these requirements:

1. The foundation base area must be properly distributed so that the pressure it imposes on the supporting soil will not at any point exceed the safe bearing capacity for the particular soil encountered. The foundation will sink if this requirement is not satisfied (Figure 2.57A).
2. The base area should be so disposed that there is not too great a difference in unit loading on different equal areas of the supporting soil. If the unit loading varies too greatly under different parts of the foundation, the block probably will tilt (Figure 2.57B).
3. The foundation must have such proportions that the net result of the total vertical load and horizontal unbalanced reciprocating forces always falls within the base area of the foundation (Figure 2.57C). If the resultant falls outside, the foundation and compressor probably will topple or stand at an angle.
4. There must be enough weight in the foundation to prevent it from sliding on the supporting soil because of unbalanced forces (Figure 2.57D).

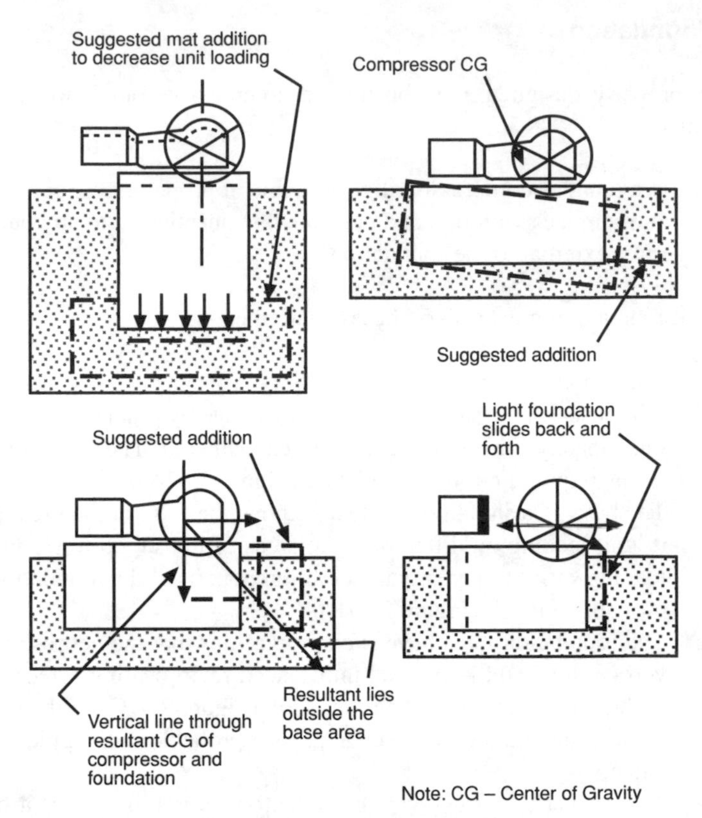

Figure 2.57A, B, C, D The foundation must distribute the weight evenly over an area sufficiently large that the machine does not settle.

To determine the required thickness of a concrete foundation, it is necessary to know:

1. Compressor weight and its center of gravity.
2. Magnitude, frequency, location, and direction of any unbalanced forces.
3. Quality of the subsoil.

Foundation Plans

Manufacturer's foundation plans, submitted with the compressor, are designed for good, hard, firm ground, such as well cemented sand and gravel or hard clay, always dry. The recommended foundation provides for enough thickness to support the compressor's weight plus disturbing forces, and for installation of sufficiently long foundation bolts to secure good anchorage. Carefully worded notations, appended to the plans, stress the importance of soil quality and urge investigation by a foundation expert if there is any question of the soil's ability to carry the load.

Subsoil Characteristics

To prevent objectionable vibrations and to eliminate very expensive correction after the compressor has been erected, it is highly important that the soil characteristics (static loading limits and elasticity) be known and understood. Soil quality may vary at different seasons. Soil may be wet in spring and early fall. These conditions must be considered carefully, because foundations have been known to move when the soil is wet, the movement completely disappearing after the soil has thoroughly dried. Soft clay, alluvial soils, loose sand and gravel, silt, and filled ground are poor supports for foundations of large reciprocating compressors. When such soils are found, the bearing area should be increased by placing the foundation on a mat, as shown in Figure 2.57A. With poor or wet soil, piling may be necessary to provide the necessary vertical support. It is usually advisable to also install batter piles (piles driven at an angle at the foundation ends) to absorb the horizontal unbalanced forces. Where possible, neighboring installations on similar soil should be observed to help determine the necessary precautions.

In many instances where substantial concrete depth is available in existing basement floors or where the footing is solid rock, the only requirement is to apply a surface covering of properly bonded concrete to support the compressor. There is little reason to excavate a perfectly sound structure to provide for a concrete block, such as may be indicated by the compressor manufacturer's foundation plan.

Regardless of past experience in a given area, it is recommended on any sizable installation that soil tests be made before proceeding with the foundation. Corrections on any foundation are extremely costly. The character of the subsoil should be ascertained by borings on the site of the foundation. Such borings should be at least four in number, one near each corner of the proposed foundation. It is highly recommended that the borings be made and judged by a competent foundation engineer, who should recommend suitable construction. Table 2.2 covers various soils encountered and gives the safe bearing capacity for static loads in tons per square foot. For dynamic loading, the allowable or design loads for a compressor foundation should not exceed one-quarter to one-sixth of the values given.

Table 2.2 Safe Static Load in Tons per Square Foot (For allowable dynamic loading use 1/4 - 1/6 of loadings shown.)

Type of Soil	Tons per sq ft
Solid ledge of hard rock, such as granite	25 to 100*
Sound shale and other medium rock, requiring blasting for removal	10 to 15
Hardpan, cemented sand and gravel, difficult to remove by picking	8 to 10
Soft rock in disintegrated ledge, or in natural ledge difficult to remove by picking	5 to 10
Compact sand and gravel, requiring picking for removal	5 to 6
Hard clay, requiring picking for removal	4 to 5
Gravel, coarse sand, in thick, natural beds	4 to 5
Loose medium or coarse sand; fine, dry sand	3 to 4
Medium clay, stiff but capable of being spaded	2 to 4
Fine, wet sand, confined2 to 3	
Soft clay	1

*It is ordinary not advisable to impose on any type of soil a unit load greater than 15 tons because the safe crushing strength of concrete (in the foundation) is usually taken at about 15 tons per sq. ft.

Subsoil Elasticity

For many years, design considerations for compressor foundations have dealt only with the static load problem. Recent experience has illustrated clearly that not enough emphasis has been placed on the elastic characteristics of sub-soils. In addition to the static load imposed on the subsoil by all structures, reciprocating machines also exert a dynamic loading on the foundation. To cope with the dynamic loading, installation and care of stationary reciprocating compressors it is necessary to consider the elastic characteristics of the ground on which the foundation rests. The following example illustrates this characteristic.

An oil storage tank was observed to settle when filled, and position when emptied. The test was made with a transit and repeated several times with the same results. There was a definite relation between load that is typical of elastic materials.

The foundation and the ground form an elastic system that will produce excessive vibrations if excited by periodic forces (the unbalanced forces in the reciprocating equipment). Having a frequency near the natural frequency of this bearing pressures keep the natural frequency of the foundation high, and also reduces the possibility of transmitted vibrations. A thorough treatment of this subject is beyond the scope of this book; therefore, it is recommended that a foundation engineer, familiar with the local area, be contacted. The compressor's unbalanced forces can be obtained from the manufacturer's drawings or by contacting the manufacturer.

Foundation Depth, Area, and Placement

Where possible, the foundation should be carried down to a firm footing. Where the foundation is exposed to freezing temperatures, its depth should extend below the frost line. Where the depth of the foundation is made greater than that shown on the manufacturer's drawing, the supporting area of the foundation should be proportionately increased by means of a mat. Such an increase in base area also is necessary whenever a compressor is raised above the floor line to an extent greater than that shown on the drawings. Where the sides of the foundation do not abut well-tamped soil, the base area must be increased. It also is advisable to isolate the foundation from any building footings, walls, or floors, to prevent any vibration being carried into the building structure. Where more than one compressor is being installed and they are relatively close to each other, it is advisable to cast their foundations en-bloc or separately on a common concrete mat.

Concrete Mix

It is recommended that a concrete engineer be consulted as to the proper mix for the location. In general, a foundation concrete mix of 3000 lb/in.^{2} compressive strength at 28 days of age and a slump not to exceed 6 in. should be suitable. Such concrete mix should be made with durable aggregates of such gradation that not over 40 gal total water, including moisture in sand and aggregates, be used per cubic yard of concrete. A cement dispersing agent may be used in the concrete mix. If aggregates are used that contain reactive silica, then cement must be used that contains less than 0.6% sodium and potassium alkalis.

The minimum amount of water described above should be adhered to. This should provide a ready placement and desired strength. No foundation should be poured when there is any possibility of freezing. For foundations, it is important that continuous pouring be used rather than section pouring.

The top surface of the foundation before the concrete takes its final set should be roughened by raking (preferably with a coarse rake) to provide an irregular surface for the bonding of the grout, which will be applied after the equipment is aligned and set on the foundation. The surfaces should be roughened to the extent that the aggregate is exposed and indentations in the surface are at least 2 in. deep and irregular. Figure 2.58 illustrates the proper surface texture.

The foundation should cure for at least a week before equipment is put on it. Curing can be accomplished by using a curing compound or by keeping the surface wet for seven days if normal temperatures exist or longer at lower-than-normal temperatures.

Figure 2.58 Proper surface to provide bonding of the grout.

Foundation Bolts

Foundation or anchor bolts serve two purposes: (1) they hold the compressor down firmly on the foundation, and (2) they prevent it sliding laterally on the foundation. The real function of the bolts then is to bind the compressor firmly so that, structurally, the compressor and foundation constitute a single mass.

Most manufacturers furnish a drawing that specifies anchor-bolt dimensions and their location in the foundation. Foundation bolts can be purchased from the manufacturer or can be made locally. In the event the bolts are obtained locally, the materials should not be of lower quality steel than AISI-C-1120, which has the following physical properties for hot-rolled material:

Tensile strength, 65,000 psi
Yield point, 38,000 psi
Elongation, 25%
Brinnel hardness, 117, approximate

Providing casings of ample length around the bolts permits lateral movement to compensate for slight inaccuracies in locating the bolts in the foundation and unavoidable shifts of the hole core during casting of the compressor frame and supports. Casings can be made of metal drain pipe, ordinary steel, or wrought-iron pipe. They should be at least 2 in. larger in diameter than the foundation bolts.

Foundation Template

The template (Figure 2.59) is a pattern or frame usually fabricated from wood strips. It supports the bolts and casings, while the foundation is poured around them. Engineering practice dictates using a template because it is more economical to build one than to endeavor to locate the bolts without it. It is best to build a template on the job. If the manufacturer furnished it, damage and warping during shipment might distort the framework and destroy its accuracy. Dimensions for building the template can be secured from the manufacturer's foundation drawing.

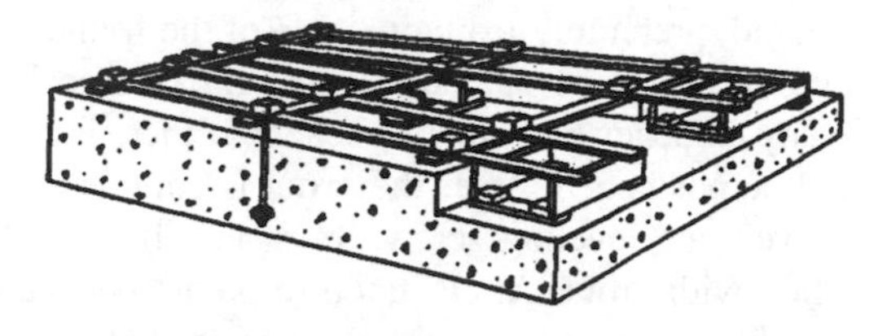

Figure 2.59 Foundation template fabricated from wood strips.

Alignment and Leveling

The compressor should be accurately aligned and leveled in accordance with the manufacturer's instructions. Before setting the compressor crankcase on the foundation, any glaze or loose scale from the exposed aggregate and any other particles not solidly bonded together must be removed with a star chisel. The surface must be swept and thoroughly cleaned to make sure that all loose particles are removed.

Leveling Wedges

Iron or steel is used in preference to wood for leveling wedges. This eliminates the possibility of swelling and thus disturbing the alignment. It is recommended that the wedges be removed after grouting because the grout may shrink away from contact with the bottom of the frame or base, thus allowing the wedges to support the whole load. When the wedges have been removed, the base must necessarily rest firmly on the grout.

Where the frame is provided with leveling set screws, wedges should not be used, but steel plates should be placed under the set screws. These plates can be left in place after the grouting has set, but the set screws must be backed off.

Grouting

Cement-type grout. The top of the foundation should be cleared, brushing and washing it off to remove any loose particles and checked to make sure that no tools or equipment used in leveling the compressor have been left.

A dam or form should be built around the foundation high enough to make a grout level inside the hollowed-out portion of the frame at least 1 in. above the bottom flange face. Forms should also be built around supports and motor sole plates or base plates. The foundation should be wetted thoroughly at least 30 minutes before the grout is poured and kept wet to ensure a good bond between the foundation and grout. The entire unit should then be rechecked for alignment and prepared for grouting, using a grout recommended by either a competent concrete engineer or by the compressor manufacturer.

The grout is poured, preferably from one side of the frame, to all points within the form. The grout should be spread evenly and worked beneath the frame flange to eliminate any tendency to form air pockets. If there is any danger of freezing, grouting should be postponed. After the grout is completely hard, all wedges and shims are removed and leveling screws retracted, if used. Voids left by the wedges are then filled with grout. At all juncture points between the compressor unit and grout, it is advisable to apply several coats of shellac, followed by paint, to prevent any seepage of oil into these points. In fact, the whole foundation should be similarly treated, as oil has a very harmful effect on the concrete or grout.

Plastic-type grout. The basic advantages of plastic-base, thermosetting epoxy resin and an inert filler type of grout, are oil resistance, higher tensile strength, very low shrinkage, fast hardening, and high bonding strength to metal and to rough concrete. It has chemical resistance and properties of impact strength.

The concrete surfaces must be dry and clean. Metal surfaces, where, a bond is desired, must be cleaned by sandblasting or with solvents, and all dust must be removed. Cleaned metal surfaces can be coated with a bonding film, as recommended by the manufacturers of plastic grouts.

Areas where bonding is not desired, such as forms or water hoses used for scaling around the oil pan, must be waxed. Specific instructions on the use of plastic grout must be requested from the equipment or plastic grout manufacturers or both. Because of its low viscosity, particular attention must be paid to scaling of forms and all possible seepage passages.

Air Intake

A clean, cool, dry air supply is essential to the satisfactory operation of a compressor. Wherever possible, the compressor inlet should be taken from the outside air. The open end of the intake pipe must be well hooded and screened to prevent rain and dirt or dust from entering. The filter should take air from at least 6 ft or more from the ground or roof and should be located several feet away from any wall to minimize the pulsating effect on the structure. The pulsations immediately surrounding the intake may rattle windows and disintegrate a weakly constructed building wall.

It is recommended that the compressor intake not be located in an enclosed courtyard. In such a yard, the air compressor intake could cause pressure pulsations that would cause building vibrations even in solidly constructed buildings.

The air intake must always be located far enough from steam, gas, or oil engine exhaust pipes to ensure that the air will he free from dust, dirt, moisture, and contamination by exhaust gases.

The intake piping should be as short and direct as possible with long-radius elbows where bends are necessary. It should be the full diameter of the intake opening of the compressor. If the intake pipe is extremely long, a larger size should be used.

If the air intake pipe is above the floor, aluminum, plastic galvanized pipe, standard steel pipe, or sheet-metal pipe can be used.

Glazed vitrified pipe and reinforced polyvinyl are convenient materials to use for underground air intakes, as long sweep elbows of the same material can be obtained. All joints should be cemented to make them watertight, as any water seeping into the intake is carried into the compressor, washing away the lubricant and causing the piston and cylinder to cut or wear. If a concrete duct is built for use as an air intake, it must have a smooth, hard interior surface, for if the concrete crumbles or disintegrates due to the air rushing through, the ingredients are carried into the compressor cylinder, causing rapid wear of the valve and piston, as well as possible scoring of the compressor cylinder. Glazed, vitrified pipe is preferred. Painting the interior of a concrete duct, if used, with a high-grade special waterproof paint or epoxy coating is advisable.

Discharge lines, steam lines, hot water lines, and the like, must never be put in the intake duct as such practice will raise the temperature of the intake air and cause considerable loss in the volume flow of the unit. A fact that must be remembered in this connection is that for every 5°F reduction of the temperature of the intake air there is a gain of approximately 1 percent in air weight. Table 2.3 shows the effect of intake or initial temperature on the delivery of air compressors.

Where a compressor is used in or adjacent to a chemical plant, the air drawn into the compressor may contain acid fumes that attack iron and steel, causing corrosion and wear of the valves, pistons, and cylinders. Similarly, the exhaust from other industries may contain contaminants that will be injurious to the compressor. If these conditions are known to exist, the manufacturer of the compressor should be informed when the machine is purchased so that the proper precautions may be taken. Every effort should be made to locate the intake away from such fumes or other contaminants.

The suction line to the compressor should be thoroughly cleaned before the machine is first started to remove accumulation of pipe scale and grit or other foreign objects inadvertently placed in the line during the installation, It is recommended that the piping be fabricated with a sufficient number of flanged joints so that it can be dismantled easily for cleaning and testing. It is far better to clean and test piping in sections before actual erection than after it is in place. The use of chill rings for butt welds in piping is recommended. This prevents welding beads getting into the pipe and being carried through, not only on the original startup, but later during operation.

Table 2.3 Effect of Initial or Intake Temperature on Delivery of Air Compressors Based on a Normal Intake Temperature of 60°F

Initial Temperatures			Initial Temperatures		
°F	°F abs.	Relative Delivery	°F	°F abs.	Relative Delivery
-20	440	1.18	70	530	0.980
-10	450	1.155	80	540	0.961
0	460	1.13	90	550	.0944
10	470	1.104	100	560	0.928
20	480	1.083	110	570	0.912
30	490	1.061	120	580	0.896
32	492	1.058	130	590	0.880
40	500	1.040	140	600	0.866
50	510	1.020	150	610	0.852
60	520	1.00	160	620	0.838

Depending on the material used and its condition, the cleaning can be accomplished by one of several methods or a combination of methods. These include wire brushing and blowing out, hammering and blowing out, sandblasting and blowing out, wiping with lintless cloth, washing or flushing, pickling and washing, or other methods practical at the site to ensure a clean suction line.

If it is impossible to mount the filter immediately adjacent to the compressor and if other than a short straight intake is to be used, it is recommended that a consulting engineer be contacted to make sure that the intake sizing and configuration will not introduce objectionable pulsation and excessive pressure drop.

Discharge Piping

The discharge pipe should be the full size of the compressor outlet or larger, and it should run directly to an aftercooler, if one is used. If no aftercooler is used, the discharge pipe should run directly to the receiver, the latter to be set outdoors if possible but kept as close to the compressor as practical. The discharge pipe should be as short and direct as possible, with a minimum of fittings and with long radius elbows where bends are necessary. Unnecessary pockets should be avoided. If a pocket is formed between the compressor and the aftercooler or receiver, it should be provided with a drain valve or automatic trap to avoid accumulation of oil and water moisture in the pipe itself.

The piping should be sloped away from the compressor with sufficient pitch to prevent either condensate or oil draining back into the compressor. A drop leg with a drain valve or automatic trap is a good idea at the compressor discharge to positively prevent liquids reaching the compressor.

The use of plug valves should be considered for the discharge line because these do not have pockets found in gate and globe valves. With outdoor installations in severe cold weather, this could eliminate freezing and breaking of valves.

The hot discharge line should not contact wood or other flammable materials. Any gaskets in the discharge piping should be of asbestos, if permitted, or other oil-proof, noncombustible material. If the discharge line is more than 100 ft. long, pipe of the next larger diameter should be used throughout.

Under certain conditions of installation and operation, pipeline surges or pulsations may be set up in intake or discharge lines; these pulsations not only may cause vibrations of the pipe if it is not well anchored and supported, but may also influence the performance of the compressor.

A frequent cause of pressure surge in the discharge line is a rather long line with a receiver located at considerable distance from the compressor. The surge may be avoided by installing as near to the compressor as possible a surge drum of suitable size, which will damp out the vibrations. To isolate compressor vibrations from the system, it may be desirable to make the connection to the system by means of a short length of flexible hose. When in doubt, a competent engineer should be consulted who is familiar with the handling and piping of compressed air and gases.

Pipelines through which hot air passes should be kept clean to avoid the danger of a fire starting in the accumulated dirt and oil. It is recommended that a removable portion of the discharge pipe be installed directly out of the compressor so that this section can be readily removed when necessary, inspected, and cleaned of any buildup of carbon. Piping should drain toward the aftercooler and receiver.

All piping connected to the compressor should be arranged with flange fittings or unions close to the compressor to permit removal of the cylinder at any time without disturbing the piping. All overhead piping must be well supported to relieve the compressor of any incidental strains.

Cautions

A globe or gate valve may be placed in the discharge line between the compressor and aftercooler or between the compressor and the receiver when more than one compressor discharges into a single aftercooler or receiver. When such a shutoff valve is used, a safety valve of proper size positively must be placed in the line between the compressor and the shutoff valve and must be checked periodically. This is very important, since the compressor may at some time be started with the stop valve closed, and if no safety valve is used, sufficient pressure may build up to burst the cylinder. Figure 2.60 shows the wrong way and the right way to do this, if a stop valve must be employed. The safety valve or valves should have a total capacity more than sufficient to handle the entire output of the compressor. The globe valve on the safety valve branch is to allow for manual relief of pressure in the cylinder before opening it for inspection or repair.

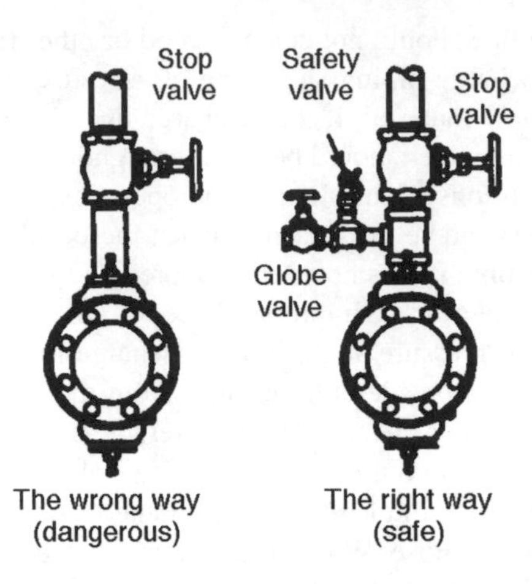

Figure 2.60 The right way and the wrong way to install a stop valve.

Figure 2.61 indicates in a general way how the discharge of a compressor should be connected to an aftercooler, if one is used, and to the receiver; the receiver inlet should be near the top of the tank and the discharge near the bottom. The arrangement of air piping for large compressors is, in general, the same as shown here.

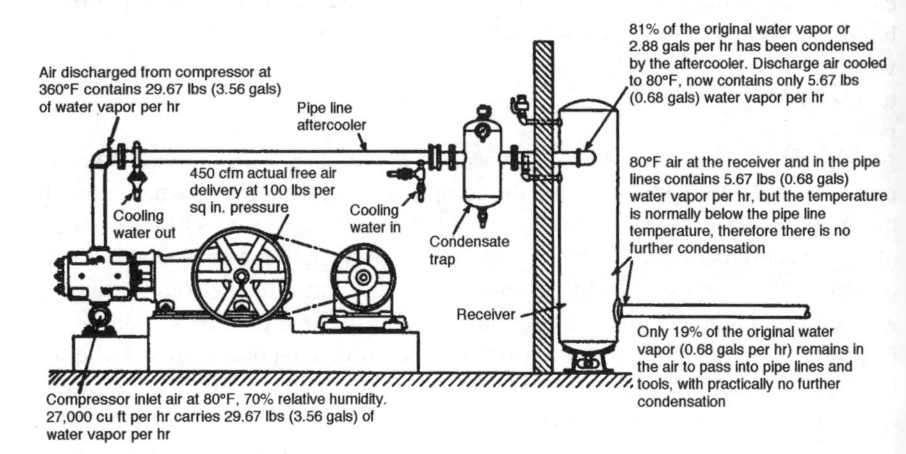

Figure 2.61 Compressor arrangement showing the piping to a receiver with an after-cooler.

A bypass valve or unloading valve should be provided in the discharge line to help in the starting of the compressor on occasions when receiver pressure is atmospheric or below the pressure necessary to permit the unloaders to function properly.

Circulating Water

A liberal supply of cooling water for cylinder jackets, cylinder heads, intercoolers, and aftercoolers must be provided. The operator should wait long enough after turning on the water before starting the compressor to ensure that water jackets are completely filled and the flow of water is established from the compressor. The use of dirty or scale-depositing water should be avoided, as it clogs the water passages and will reduce the cooling efficiency and result in considerable shutdown time to clean the jackets and maintain the overall efficiency of the compressor.

The inlet water connection should be located at the lowest point of the cylinder so that water can be easily drained from the cylinder when the compressor is shut down. The discharge connection should be at the highest point to ensure complete filling of the water jackets with no air pockets. The water piping should be provided with a valve for controlling the flow of water. The water flow control valve should be on the inlet water side to avoid water pressure on the cylinder when the unit is shut down.

When cooling water is very cold, condensation may form in the air inlet passage of the high pressure cylinder as the air enters from the intercooler, because the air may be much warmer than the water. Similarly, condensation may form in the first-stage cylinder of a multi-stage unit or in the cylinders of a single-stage unit handling saturated air if condensation forms. It will be carried into the cylinder and will destroy the lubricant, causing rapid cylinder and valve wear. To relieve this condition, it is advisable to pass the cold water through the intercooler first. This heats the water up considerably and allows a more normal relationship between water temperature and incoming gas temperature in the cylinder. The inlet water temperature to the cylinder jackets should never be less than the incoming gas temperature and, in general, should be 10 to 15°F above the incoming gas temperature. Except in very special cases, water temperature should never be higher than 160°F, with 120°F being a preferred maximum.

Under ordinary circumstances, the cooling water is piped first to the intercooler and aftercooler in parallel. The intercooler saves power by reducing the volume of air handled by the high pressure stage, and the aftercooler moisture separator removes moisture and prevents carryover into the lines. Usually it is advisable, therefore, to supply each of them with the coldest water. From the intercooler, the water is then taken through the low and high pressure cylinder jackets.

If the jacket water flow is regulated automatically by thermostatic valves, the valves should be equipped with a bypass arranged so that at no time will the flow through the cylinder be completely stopped. This bypass should be designed to provide for enough circulated water to eliminate the formation of air pockets and hot spots and to make sure that the jackets are full of water at all times.

All jacket and cooler drains must be opened in freezing weather when shutting down the compressor.

Outlet water should flow into open funnels, allowing frequent temperature readings to be made; an excessive rise in temperature indicates insufficient water, carbonized discharge valves, leaking piston rings, or broken valve parts.

When closed water systems are used, sight flow indicators should be put on the water discharge from each cylinder and intercooler to show positively that water is circulating, and the discharge pipe should be bypassed to an open funnel so that the cylinders and coolers can be tested frequently to detect leaks in the water jacket or in the intercooler tubes. If the water discharge is opened to the overflow funnel and the valve on the discharge line shut off, any leakage between the air and the water spaces will be revealed by air blowing out with the water. Figure 2.62 shows how the water piping should always be arranged when a closed system is used. A relief valve should always be installed on a closed water system and should be between the shut-off valve and the cylinder jackets so as to prevent excessive pressure from building up in the jackets. The compressor manufacturer should be consulted for the proper setting of the water jacket relief valve.

The thermosyphon cooling system normally is used only in the case of portable or temporary installations where a continuous water supply is not available.

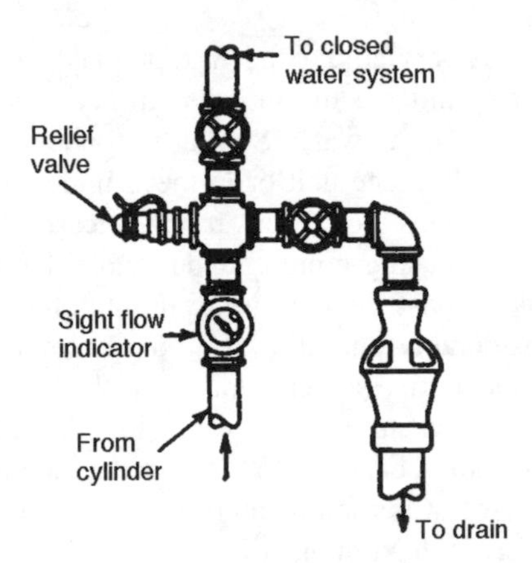

Figure 2.62 Arrangement of air-cylinder water piping when a closed system is used.

Table 2.4 shows the cooling water quantities recommended for the water coolers, cylinder jackets, and aftercoolers. It is expected that the temperature of the air leaving the intercooler or aftercooler will be within 10 to 15 °F, respectively, of the temperature of the water entering the cooler for ordinary working conditions.

Table 2.4 Summary of Industrial Plant Compressed Air Systems (including ratio of air consumption to area of plant).

Type of Plant	Manufacturing Area (ft²)	Compressor Capacity (scfm at psig)		Horsepower	CFM per 1000 ft²
Automotive hardware manufacturing	580,000	9,600	110	1800	17
Laminated glass manufacturing	1,200,000	10,000	100	-------	8.3
Automobile component manufacturing	580,000	6,024	100	-------	10.5
Electrical switchgear manufacturing	252,000	500	100	250	2
Electrical switchgear manufacturing	135,000	400	100	100	3
Electronic computer manufacturing	750,000	1,525	100	300	2
Standard large electric manufacturing company	Any size	-------	100	-------	7
Glass bottle manufacturing, using automatic glass-blowing equipment	129,000	5,720 330*	50 100	800 15	45 2.6
Glass stemware manufacturing, using automatic glass-blowing equipment	234,000	7,160 1,400	50 100	500 250	30 6
Automated foundry	74,000	2,400	125	450	33

* Boosted from 50-psi system.

DYNAMIC COMPRESSORS

Dynamic-type compressors are machines in which air is compressed by the mechanical action of rotating impellers imparting velocity and pressure to the air. The centrifugal type discussed in this chapter has a flow classified as being in the radial direction. For further information on dynamic compressors, including types, definitions, and applications, the reader is referred to Chapter 7.

Continued demand in the 1960s for more efficient compression with lower operating costs resulted in a transition to cooling between compressor stages. This was true for many centrifugal compressors used for plant air and other applications. The reason for this change is shown in Fig. 2.63. The figure shows the relative effect of additional coolers for a given compression ratio. The centrifugal compressor assumed is for a typical plant air, 100 psig application.

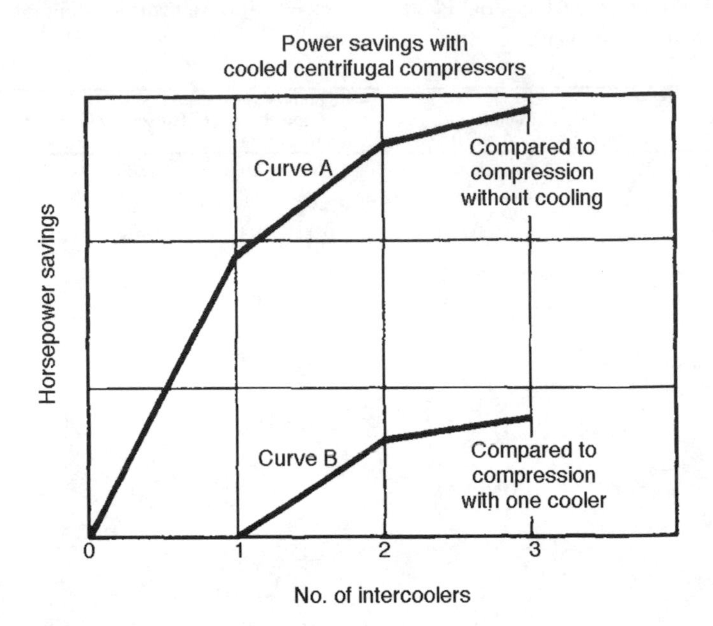

Figure 2.63 Relative power savings for a compressor with coolers compared with a compressor having no cooler and with one having one cooler.

The rising cost of energy in the 1970s resulted in a further drive to reduce the power requirements and led to a considerable number of arrangements, extending from one- to three-cooler compressor designs for plant air to three- or four-cooler compressor designs for 350 psig soot-blower applications in coal-fired boilers. Impeller and stage designs were improved aerodynamically to provide maximum air supply for minimum operating costs in standard product lines.

Typical of such intercooled compressors is the integral-gear-type centrifugal compressor seen in Fig. 2.64. It consists of a low-speed gear directly connected to the motor drive and two high-speed pinions having extended shafts that carry four centrifugal compressor impellers. Two shaft speeds are used to provide selection of more optimal impeller speeds and, consequently, improved efficiency. The use of axial entry to each centrifugal compressor stage and the adaptation of matching single-stage centrifugal compressor scrolls surrounding each impeller provided an ideal flow pattern and an efficient conversion of the velocity head leaving the impellers.

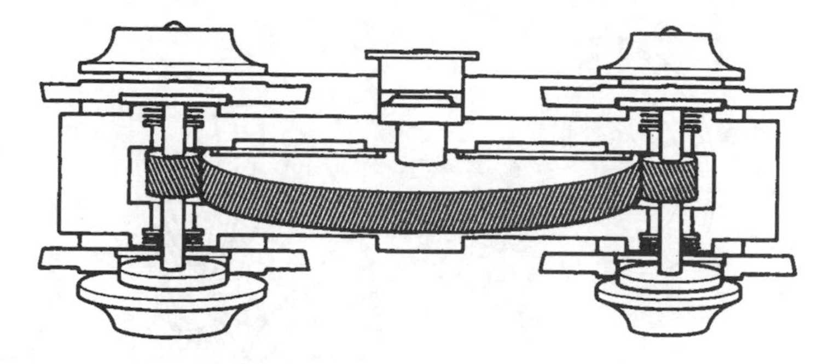

Figure 2.64 The input full gear of atypical integral-gear unit runs at motor speed. The low-speed pinion drives the first-stage impeller (upper right) and the second-stage impeller (lower right). The high-speed pinion drives the third-stage impeller (upper left) and the fourth-stage impeller at higher speed to match the reduced volume flow due to pressure increase and cooling.

Further input power reduction is obtained in the compression cycle by inter-cooling between the four stages of compression. Figure 2.65 shows the flow path and the cooling between stages. The open impeller design generally is used for air compressors up to and over 100,000 cfm. This impeller design is seen in Figure 2.66. Closed or shrouded impellers are used for very large flow volumes.

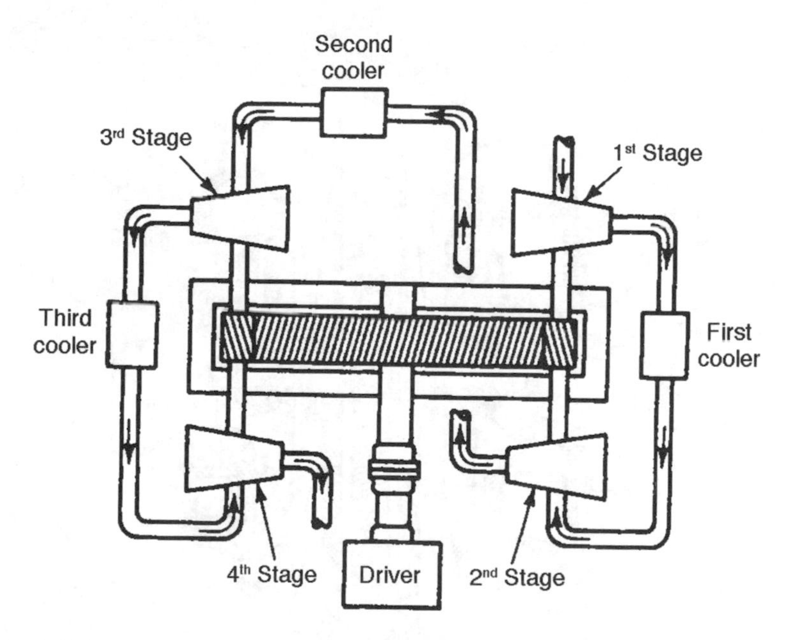

Figure 2.65 Flow diagram of an integral-gear-type compressor showing stages of compression and including the cooling arrangement.

Figure 2.66 Pinion of an integral-gear unit having open, backward-curve-bladed impellers.

Opposed impellers on each pinion shaft (Fig. 2.67) help to balance the aerodynamic thrust load. The unit is equipped with special radial load bearings, which provide stability for the lightweight, high-speed pinions. An integral thrust bearing or thrust transfer ring on the pinion shaft absorbs the remaining net thrust between impellers. The low-speed gear shaft has full sleeve bearings and an integral-type thrust bearing. The gears shown are precision single helical type.

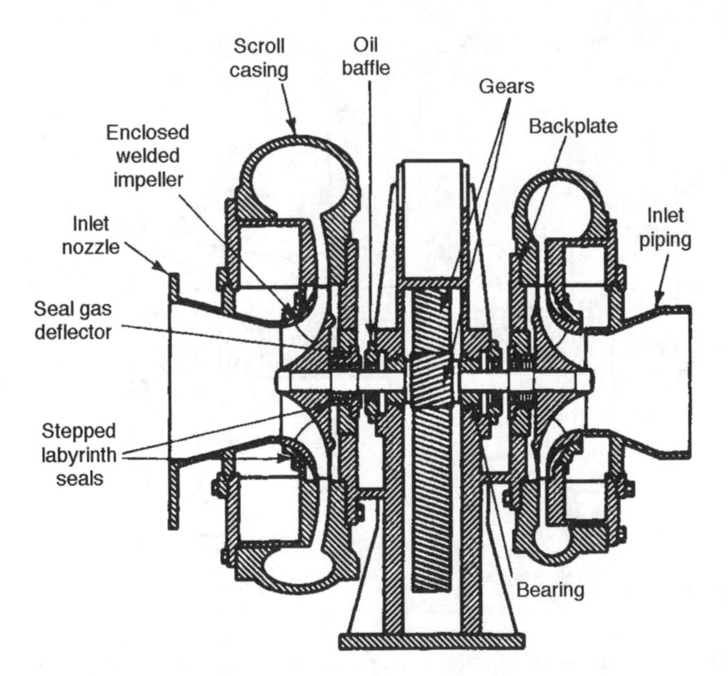

Figure 2.67 Opposed impellers help balance thrust on pinions.

COMPRESSED AIR USES

The low-speed gear shaft has full sleeve or anti-friction bearings and an integral-type thrust bearing. Principle uses are for the operation of hand tools, actuation of control devices, textile weaving, air separation, cleaning, and prevention of contamination and to power many motions required in automatic and semiautomatic production equipment. The design of the compressed-air system of a plant depends on the use to which the compressed air is to be put.

How Many Compressors? How Large?

Due consideration should be given to the need to have standby or future capacity when establishing the size and number of compressors. Equipment may be sized for less than the peak demand, provided the peak will be of short enough duration that the receiver and compressor together can carry the peak without compromising the pressure available at the points of use, and providing further that the valleys are deep and broad enough so that the receiver pressure can be restored before the next peak.

Where the size of compressors is great enough that such a feature is available, capacity modulation should be specified. For sizes at which capacity modulation is not available, the compressor capacity should be adequate to permit the unit to carry the anticipated eight-hour operating load with not over six hours of operation. For such installations, the capacity of the receiver should be such as to assure at least 2 minutes of operation of the compressor between the pressure control cut-in and cut-out pressures, with zero draw from the receiver. Table 2.4 shows typical data for the capacity required in cubic feet per minute per 1000 ft^2 of plant area for a variety of facilities.

Some attempt has been made to indicate in this table the general nature of the facility and the principal use made of compressed air therein. It will be noted that a relatively broad range of capacity is shown, even for closely similar plants. This indicates that the table should be used primarily as a check on the computed size or to establish approximate plant capacity if data are not available for more accurate computation.

Sizing Considerations

Centrifugal air compressors normally are specified on the basis of required air flow volume. However, there are several ways to calculate volume, and serious problems can result unless both user and manufacturer use the same method. At the very least, the user can have trouble comparing bids from competing manufacturers. At worst, he may choose the wrong compressor.

These problems can be avoided by specifying capacity in terms of the actual inlet conditions and by understanding how compressor capacity is affected by variable ambient conditions, such as inlet pressure, temperature, and relative humidity.

Also, factors such as cooling-water temperature and motor load must be considered before a compressor and its drive motor can be sized.

The accompanying data sheet shows tabulated factors that must be considered when choosing a centrifugal air compressor. Some factors are fixed; others vary fairly frequently, daily, or seasonally.

Compressed Air and Gas Institute
Data Sheets
for
Packaged, Integrally Geared Centrifugal
Compressors
Plant-Process-Instrument

1.0	Customer site considerations:		
1.0.1	Barometer (psia)		*______
1.0.2	Inlet pressure at first stage flange, psia		______
1.0.3	Inlet temperature, °F		*______
1.0.4	Relative humidity, %		*______
1.0.5	Cooling-water temperature, °F		*______
1.0.6	Cooling-water pressure, psig		*______
1.0.7	Source of cooling water (i.e., tower city, etc.)		*______
1.0.8	Outlet pressure from aftercooler psig/psia (if supplied)		*______
2.0	Vendor model designation		______
3.0	Number of stages		______
4.0	SCFM at 14.5 psia, 68°F, 0 percent RH (dry):		
4.0.1	Required by purchaser		*______
4.0.2	Vendor's rating		______
5.0	Weight flow:		
5.0.1	Inlet weight flow, lb/min dry, based on SCFM at 14.5 psia, 68°F, 0 percent RH (dry)	For 4.0.1 ______	For 4.0.2 ______
5.0.2	Specific humidity (vapor content) at site conditions, lb water vapor/lb dry air	______	______
5.0.3	Inlet total weight flow wet (5.0.1 x 5.0.2 + 5.0.1 = 5.0.3)	______	______
6.0	ICFM at inlet flange		______
7.0	Horsepower at compressor coupling:		
7.0.1	For 4.0.1		______
7.0.2	For 4.0.2		______
8.0	Cooling water required and pressure drop:	GPM	Pressure drop, psi
8.0.1	Intercoolers, total	______	______
8.0.2	Aftercooler	______	______

				Type
8.0.3	Oil cooler		______	______
9.0	Bearings:			
9.0.1	Bull gear bearings:			
9.0.1.1	Journal			______
9.0.1.2	Thrust			______
9.0.2	Pinion bearings:			
9.0.2.1	Journal			______
9.0.2.2	Thrust			______
10.0	Drive Coupling:			
10.0.1	Make			______
10.0.2	Type			______
11.0	Impeller shaft seal:			
11.0.1	Air seal			______
11.0.2	Oil seal			______
			Pressure	Flow
11.0.3	Air seal requirements		______	______
11.0.4	Oil seal requirements		______	______
12.0	Impellers:			
12.0.1	Type, backward leaning or radial blades			______
12.0.2	Material			______
13.0	Compressor casings:			
13.0.1	Type, horizontal or vertical split			______
14.0	Gear case:			
14.0.1	Type, horizontal or vertical split			______
14.0.2	Material			
15.0	Pinions:			
15.0.1	Material			______
15.0.2	AGMA quality number			______
16.0	Bull gear:			
16.0.1	Material			______
16.0.2	AGMA quality number			______
			Pressure	Flow
17.0	Oil reservoir retention time		______	______

The main item that must be specified is inlet air volume. Standard cubic feet per minute (scfm) is a common unit of measure for compressor capacity in the United States; however, several definitions of this unit exist. The definition of Standard Air adopted by the Compressed Air & Gas Institute, PNEUROP and ISO, is air at 14.5 psia, 68˚F and 0% relative humidity. The petrochemical industry commonly uses 14.7 psia, 60˚F and 0% relative humidity. A third (metric) definition specifies a standard(or normal) cubic meter of air at 1 atmosphere, 32˚F, and 0 percent relative humidity.

A packaged integrally geared, multistage, centrifugal compressor typically includes the following:

1. Multistage centrifugal compressor
2. Prime mover (motor or turbine)
3. Coupling and guard
4. Lube oil system including:
   a. Single oil cooler
   b. Single oil filter
   c. Auxiliary oil pump (full capacity)
5. Intercoolers
6. Aftercooler-moisture separator
7. Vibration monitoring system
8. Controls and instrumentation
9. Control panel
10. Inlet filter-silencer
11. Discharge check valve
12. Inlet valve
13. Blowoff silencer
14. Discharge blowoff valve-antisurge valve
15. Base plate

To avoid the confusion caused by these variable standards, some users have adopted a simpler unit that expresses inlet volume in terms of the actual inlet pressure, temperature, and humidity. This inlet cubic feet per minute (icfm) indicates the actual volume entering the first stage of a multistage compressor at the expected operating conditions. This volume, in turn, determines the impeller design, nozzle diameter, and casing size that provides the most efficient operation.

The relationship between icfm and equivalent scfm can be expressed as:

$$Q_s = Q_i \frac{T_s}{T_i} \frac{P_i - P_v}{P_s}$$

where:

Q_i = inlet air volume, icfm
Q_s = standard air volume, scfm
T_i = inlet temperature, degrees Rankin
T_s = standard temperature, °R
P_i = inlet pressure, psia
P_s = standard barometric pressure, psia
P_v = partial vapor pressure, psia

This last term is equivalent to the saturated steam pressure at temperature T_i multiplied by the relative humidity. This equation indicates how ambient air pressure, inlet temperature, and relative humidity affect capacity.

Inlet pressure is determined by taking the barometric pressure and subtracting a reasonable loss for the inlet air filter and piping. A typical value for filter and piping loss is 0.3 psig.

The need to determine inlet pressure at the compressor flange accurately is particularly critical in high-altitude installations. Because barometric pressure varies with altitude, a change in altitude of more than a few hundred feet can greatly reduce compressor capacity. Often, the lost capacity can be restored by using larger-diameter impellers, but occasionally a different-sized compressor must be used.

Other variables that influence volume flow include temperature and relative humidity of the inlet air. These must be considered over the range of conditions expected in service. Air volume is lowest at the highest expected operating temperature, and vice versa. Therefore, the impellers must be designed to deliver the required flow at the highest temperature expected. This guideline also applies to the temperature of the cooling water, which controls the temperature of the air delivered to the stage following an intercooler.

Relative humidity also affects the useful volume of air available at the compressor inlet. The higher the humidity, the less is the effective air volume available; thus the impellers must be sized for the highest humidity expected.

A typical multistage centrifugal compressor for plant air service compresses air in several stages, with intercooling between each pair of stages. The relationship of pressure versus volume flow of a typical compressor is such that the pressure decreases at an increasing rate as volume flow rate increases.

Compressed air is often used in some sort of pneumatic device or is involved in chemical-processing operations. When it is used in a machine to do work, the amount of work done depends on the mass flow of air passing through the device. Mass is also a common denominator in cases in which it is involved in a chemical reaction and becomes part of the product. For the chemical equation to balance, a specific mass of product requires a specific mass of air. Therefore, in the final analysis, the mass flow (weight flow) of air delivered by a compressor should be the fundamental factor in specifying its capacity.

Weight Flow of Air Delivered

The key word is delivered, that is, air available for use at the discharge flange of the compressor. The work that can be done is based on what comes out of the compressor. Given a properly defined specification, the manufacturer is responsible for making sure that the compressor takes in enough air to make up for seal losses and the like so that the required weight flow is available at the discharge.

Importance of Air Density and Volume Flow

Volume flow does not tell very much because the weight of air in each cubic foot depends on the temperature and pressure of the air. In other words, the weight flow is related to the density of the air as well as the volume flow. The following formula relates weight flow to volume flow:

$$W = Q\rho$$

where:

W = weight flow, lb/min
Q = volume flow at the given air density, cfm
ρ = weight density, lb/ft^3

Air density (weight of a cubic foot of air) is inversely proportional to its absolute temperature. Thus, the higher the temperature, the less weight flow in each cubic foot. The weight flow delivered in summer is less than in winter. Therefore, the specification for a compressor should provide for the required weight flow to be delivered on a hot summer day. A slightly larger compressor will be required if the air temperature is 90°F rather than 68°F. If the manufacturer's rating (based on air at 68°F) is accepted, the compressor will not deliver the same weight of air per minute at 90°F.

Another reason that the volume flow by itself must be qualified is that air density also depends directly on air pressure. Because atmospheric air pressure depends on altitude, a compressor installed at a higher elevation (above sea level) gets less weight of air in each cubic foot of intake air than the same compressor installed at sea level. This change in weight flow due to differences in barometric pressure can be significant. For example, because of the lower atmospheric pressure, a compressor in Kansas City will deliver nearly 5% less air than the same compressor installed in Miami.

For dry air, the relationship of density to temperature and pressure is:

$$\rho = \frac{144P}{RT}$$

where:

P = absolute air pressure, psia
R = gas constant of dry air
T = absolute air temperature, °R

However, barometric pressure is not the only factor that affects inlet pressure. The effects of air filter, inlet valve, and piping leading to the compressor should also be considered. Because these components can cause significant pressure drop, a cubic foot of air measured just ahead of the compressor flange will contain less air by weight than a cubic foot measured ahead of the filter. Then the density of air entering a compressor becomes:

$$\rho = \frac{144(p_b - \Delta p)}{R_m T}$$

where:

p_b = absolute barometric air pressure, psia
Δp = pressure loss in inlet air filter, piping, and inlet valves, psia
R_m = gas constant of air mixture (i.e. with water vapor)

Relative Humidity Is Important

Another variable that often causes confusion in sizing an air compressor is relative humidity. Atmospheric air always contains water vapor. As a result, the compressor takes in a mixture of air and water vapor. This affects compressor operation and performance because the higher the pressure of the air, the less water vapor it can hold. And what it can no longer hold condenses in the intercoolers and aftercooler and is drained as water. So, once again, the weight of the cubic foot in is not the same as the weight out. The compressor must be sized slightly larger to allow for the water vapor loss, which, although it is part of the inlet flow, is not part of the delivered weight flow.

A portion of the inlet volume is attributable to water vapor. This depends on the relative humidity of the intake air and can be calculated by:

$$E = \frac{P_b}{P_b - rh(p_{vs})} - 1$$

where:

E = factor representing the amount of increase in inlet volume due to water vapor
rh = relative humidity, percent
p_{vs} = vapor pressure at saturation at given air temperature, psia (from steam tables; selected data are given in Table 2.5)

Table 2.5 Pressure of Water Vapor at Saturation

Temperature (°F)	Pressure (psia)	Temperature (°F)	Pressure (psia)	Temperature (°F)	Pressure (psia)
32	0.08854	60	0.2563	86	0.6152
34	0.09603	62	0.2751	88	0.6556
36	0.10401	64	0.2951	90	0.6982
38	0.11256	66	0.3164	92	0.7432
40	0.12170	68	0.3390	94	0.7906
42	0.13150	70	0.3631	95	0.8153
44	0.14199	72	0.3886	96	0.8407
46	0.15323	74	0.4156	98	0.8935
48	0.16525	76	0.4443	100	0.9492
50	0.17811	78	0.4747	102	1.0078
52	0.19182	80	0.5069	104	1.0695
54	0.20642	82	0.5410	106	1.1345
56	0.2220	84	0.5771	108	1.2029
58	0.2386	85	0.5961	110	1.2748

These relationships provide all that is needed to relate the real need, that is, weight flow of dry air to a quantity (volume flow), which is generally used by compressor manufacturers for performance rating purposes.

Many relationships become apparent from the following formula:

$$W = 144Q\frac{P_b - \Delta P}{R_m T(1+E)}$$

where:

P = pounds per square foot
= $144p$

Specifying Ambient Conditions

Another point that can be inferred from the last equation for Q, is that in order to be sure to have enough air, the compressor buyer must be careful in specifying ambient conditions. These should tend toward the minimal conditions, that is, high air temperature, normal barometric pressure, and high humidity. This does not mean that the specification should be based on the maximum air temperature on record. The result would be an unnecessarily large compressor.

For example, an air-conditioning guide gives a 1 percent confidence limit on a summer temperature of 95°F in Chicago. This means there is only one chance in 100 that a 95°F temperature will be exceeded, and not much more chance that even 90°F will often be exceeded by a significant amount. On the other hand, specifying a lower air temperature such as 60°F will result in running short of air on the days when this temperature is exceeded.

A related item to be considered in selecting a compressor of the correct capacity is cooling-water temperature, since most air compressors are intercooled. Water temperature has much the same effect as air temperature. This is easy to understand because the water cools the air before it enters the next compression stage. The warmer the cooling water, the warmer the air, and the less dense it will be. Therefore, water temperature should be specified at the highest anticipated temperature; otherwise, the compressor will deliver less air than expected. When cooling-water temperature is lower than specified, water flow can be reduced.

Centrifugal Air Compressor Characteristics

Centrifugal air compressor performance can be represented by a characteristic curve of discharge pressure versus flow. This is a continuously rising curve from right to left (Fig. 2.68). The effect of environment on performance requires understanding of two phenomena associated with this curve: choke (stonewall) and surge.

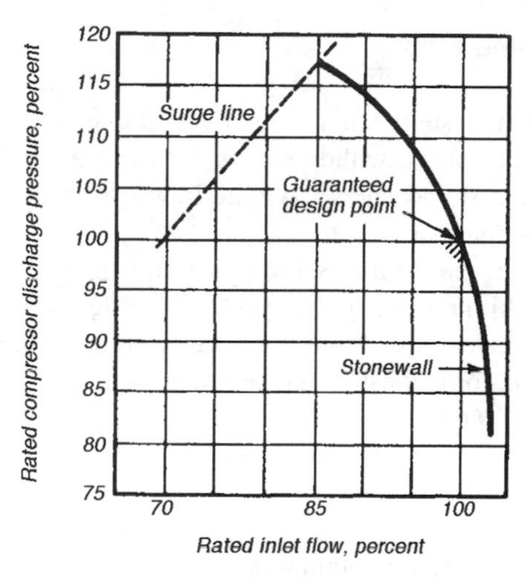

Figure 2.68 Typical Centrifugal Compressor Performance Curve

When the compressed-air system pressure decreases, a centrifugal compressor delivers an increased volume of air. As the system pressure continues to decrease, the air delivery from the compressor continues to increase until the air velocity somewhere in the compressor reaches the speed of sound. At this point, the flow is said to be choked because further reduction in system pressure does not result in additional air delivery by the compressor.

On the other hand, the maximum discharge pressure of the centrifugal compressor is a function of the intersection of the surge line and the sloping performance curve. When the compressed-air system pressure increases, the compressor furnishes less air as higher pressures are encountered until the system resistance is matched. This relationship may continue until the compressor is unable to maintain a steady flow of air into the system.

When the compressor cannot maintain a steady flow of air, backflow from the system through the compressor occurs until a momentary equilibrium is established between the compressor and the system. This backflow is commonly referred to as surge. This phenomenon is roughly equivalent to the stalled condition of an airfoil. Under this condition, compressor operation moves from surge to some point below the operating point shown on the performance curve. When the compressor continues to operate against sustained excessive system pressure, compressor operation moves up the curve and surge occurs again.

Neither of these conditions is desirable and both should be avoided. Control systems that allow the compressor to function without reaching the choked or surge condition must be based on prevailing environmental conditions. Therefore, it is helpful to examine individual environmental factors that can affect compressor performance. Further detailed information on compressor performance can be found in Chapter 7.

Weight or Volume Flow

The compressed-air system is in reality a vessel that stores energy in the compressed air, energy that can be withdrawn by instruments and air-powered tools. When a portion of this stored energy is withdrawn from the system, it must be replenished by the compressor.

The performance of the compressed-air system is measured by the pressure of the air in the system. Air pressure in the system for a steady air usage and relatively constant system temperature depends on the weight of the air in the system. Boyle's law states that, for a constant gas temperature,

$$p_1 v_1 = p_2 v_2$$

where:

v = specific volume, ft³/lb

Therefore, $$p_1 = \frac{V_1}{w_1} = P_2 \frac{V_2}{w_2}$$

where:

V_1 = total volume
w = total weight

The compressed-air system volume is constant, so

$$V_1 = V_2$$

Therefore, assuming no change in temperature or relative humidity,

$$\frac{P_1}{w_1}=\frac{P_2}{w_2}$$

or

$$p_2=p_1\frac{w_2}{w_1}$$

As air is withdrawn from the compressed-air system, the weight of the air, if not replaced, and the air pressure decrease. The performance of the compressed-air system therefore depends on the weight of the air delivered by the compressor. Because each centrifugal compressor has a fixed volume design capacity, in cubic feet per minute or cubic meters per hour, the weight flow capacity is determined by the pressure temperature and relative humidity of the air entering the compressor.

The compressor characteristics curve and work input are related to flow as a function determined by physical geometry, blade angle, speed of rotation, molecular weight of gas, and other factors to a minor degree. Once this characteristic has been established for a compressor, it can be affected by inlet air pressure, temperature, relative humidity and cooling water temperature.

Effect of Inlet Air Temperature

The head relationships discussed in Chapter 7 can be used to explore the effects of inlet air temperature. Aerodynamic work input to a centrifugal compressor is proportional to polytropic head and weight flow of air to which the head is imparted. Polytropic head is measured in foot-pounds (work) per pound of air or, more simply, as feet of head. Power is then obtained by multiplying head times total weight flow and considering mechanical losses and efficiency.

Polytropic head is obtained by the equation:

$$H_p = ZRT\left(\frac{n}{n-1}\right)\left[\left(\frac{P_2}{P_1}\right)^{(n-1)/n} - 1\right]$$

where:

H_p = polytropic head, ft-lb/lb
Z = supercompressibility factor for air; $Z = 1.0$ except at cryogenic temperatures
R = gas constant (1545/molecular weight)
T = inlet air temperature, °R (°R = °F + 460)

$$\frac{n}{n-1} = \eta x \frac{k}{k-1}$$

where:

n = polytropic exponent
η = polytropic efficiency, percent
k = ratio of specific heats
p_2 = discharge pressure, psia
p_1 = inlet pressure, psia

The head, H, required to raise the air from the inlet pressure p_1 to the discharge pressure p_2 is:

$$H = C_1 \times T \left[\left(\frac{p_2}{p_1} \right)^{c_2} - 1 \right]$$

where:

C_1 = constant = $ZR\frac{n}{n-1}$

C_2 = constant = $\frac{n-1}{n}$

For a fixed geometry and constant speed, air compressor head per stage is constant. The only variables are the inlet temperature and the pressure ratio. Therefore, if inlet pressure is constant and inlet temperature is increased, discharge pressure must necessarily drop to maintain the equality. Conversely, when inlet temperature decreases, discharge pressure must increase.

Inlet air temperature also affects the weight flow through all types of compressors:

$$W = \frac{QP}{RT} = \text{constant} \times \frac{P}{T} = \text{constant} \times \rho$$

where:

W = weight flow, lb/min
Q = volume flow rate, cfm
ρ = weight density, lb/ft^3

Weight flow through a centrifugal compressor is proportional to inlet volume and inlet pressure and indirectly proportional to inlet air temperature. Because P1/RT1 is weight density, another way of stating this relationship is that weight flow is proportional to density. As inlet temperature decreases, weight flow through the compressor increases, and vice versa, although volume flow remains constant.

Often the compressor manufacturer states the capacity for a standard air temperature, which may be as low as 60°F. Then, when the compressor operates with 90°F inlet air, for example, the weight *flow* is reduced by the ratio of the absolute temperatures:

$$\frac{60 + 460}{90 + 460} = 0.945$$

For a fixed air usage, the compressed-air system pressure is reduced by the ratio of the weight flows or 5.5 percent, other factors remaining unchanged. In SI units, capacity stated in terms of a standard temperature of 68°F with an actual inlet temperature of 86°F would mean a weight reduction of:

$$\frac{20 + 273}{30 + 273} = 0.967$$

Proper performance of the compressed-air system requires that the compressor rating be guaranteed for the summertime air inlet temperature or that a weight *flow* rating be guaranteed at the same conditions.

The effect of wintertime air temperature on air density must also be considered. A compressor rated for 90°F summertime inlet air temperature, for example, will have a 17% higher weight flow when operating with a 10°F wintertime inlet air temperature:

$$\frac{w_2}{w_1} = \frac{90 + 460}{10 + 460} = 1.17$$

or in S.I. units:

$$\frac{32 + 273}{-12 + 273} = 1.17$$

This increased weight flow will not impair the pressure performance of the compressed-air system because the resultant increase in pressure can be relieved through a relief valve or through more frequent cycling of the compressor. But this increased weight flow will add to the cost of compressing the air because of the increased power required:

$$P = C_1 \text{ x} W$$

where:

P is the power required. Therefore, for this example, the power required by the compressor will increase 17%. Alternative expressions for power are

$$P = C_2 \text{ x} \frac{p}{t}$$

and where C_1, P_2 and C_3 are constants:

$$P = C_3 \text{ x } \rho$$

Figure 2.69 shows the effect of inlet air temperature. Increasing temperature means decreasing flow and power requirement, and decreasing temperature means increasing flow and power requirement. The implications to the buyer are twofold. First, the compressor must be rated at a sufficiently high temperature so that the plant does not run short of air on a hot day-perhaps not the highest temperature of the year, but a mean temperature based on a reasonable confidence level. Second, controls must be provided to prevent the compressor from drawing excessive additional power when the air is cooler.

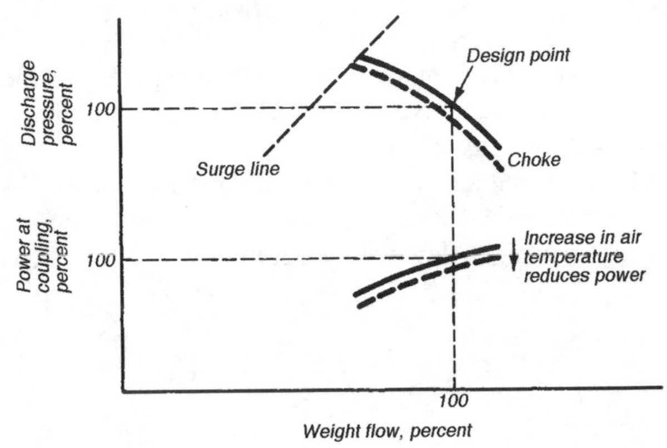

Figure 2.69 Inlet air temperature has an inverse relationship with flow and power in a centrifugal compressor. Decreasing temperature at the inlet increases flow, and more power is required to compress the denser air.

Effect of Inlet Air Pressure

A change in the inlet pressure does not affect the established pressure ratio, but the discharge pressure varies directly with changes in the inlet pressure (Fig. 2.70). Reducing the inlet pressure also reduces the weight flow through a compressor, but volume flow remains the same. Because weight flow is reduced, the power requirement is also lower.

If a given discharge pressure is required, a higher pressure ratio is required when the inlet pressure is lower, which, in turn, causes a higher work input. This factor should be considered when the compressor operates at high elevations.

Under normal operating conditions, the daily change in inlet air pressure is relatively small, except when the inlet air filter becomes dirty and needs cleaning.

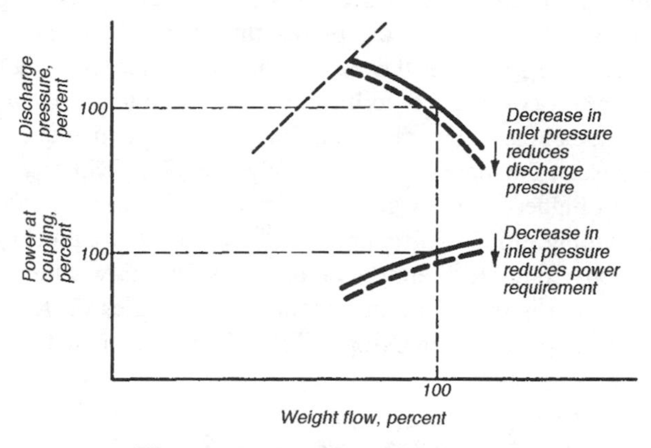

Figure 2.70 Inlet pressure effects.

Effect of Cooling-water Temperature

First-stage performance is not affected by cooling-water temperature. However, all successive stages undergo a change in performance similar to that related to air temperature (Fig. 2.71). Changes in cooling-water temperatures directly affect the temperature of the air entering the second and third or any later stages.

A reduction in cooling-water temperature increases the discharge pressure, weight flow, and the power consumption. Conversely, a higher cooling-water temperature decreases the discharge pressure, weight flow, and power consumption.

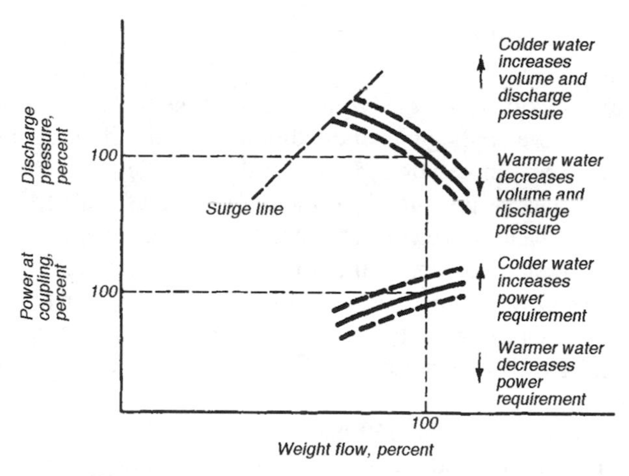

Figure 2.71 Cooling-water temperature effects.

Most significant are the combined effects of simultaneous changes in water and air temperatures, because in most plants they tend to increase or decrease together (Fig. 2.72). On a summer day, higher air and water temperatures are normal, whether the water comes from cooling towers, a public supply, a river, or the sea, although changes in air temperatures are more extreme than changes in water temperature because of the moderating effect of heat storage capacity in water. The combined effect of higher air and water temperatures is to depress the compressor characteristic, resulting in lower discharge pressure, lower weight flow, and lower power consumption. Colder temperatures increase weight flow and power requirements. This discussion is related to an uncontrolled compressor. It also describes what happens to an installed compressor selected for rated ambient conditions at a given site.

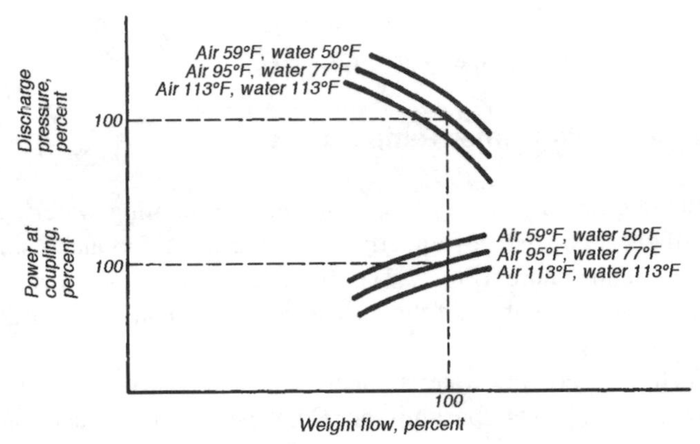

Figure 2.72 Combined effects of cooling water and inlet air temperatures.

Motors and Controls

Impellers are sized to deliver the required flow at the highest expected operating temperature. However, during winter conditions of colder air and cooling water, air density increases and more power is required to handle the increased weight flow at the same discharge pressure. The motor rating must therefore be chosen with the increased demand of the coldest expected air temperature in mind.

The extra cold-weather output required from a motor often can be covered by its service factor. Thus, for instance, a 2100 cfm compressor with a required horsepower close to 450 hp can be fitted with a motor rated at 450 hp and a 1.25 service factor. Under winter conditions, the service factor provides the extra horsepower required.

However, motors rated above 450 usually have a 1.15 service factor, which may not cover the extra horsepower requirement. One alternative in this situation is to use a motor with a horsepower rating higher than the compressor nominal rating, but this approach can be expensive and can require the use of controls to limit compressor output under normal operating conditions.

A more economical approach may be to use a motor load control so as to more closely match the motor and compressor. The cost of a load control, obviously, must be balanced against the cost of a larger motor.

Motor controls may also be necessary, regardless of cost, in applications that cannot accommodate the extra weight flow at the cold conditions. This is particularly true of process-control installations, where the process can accept only a certain weight flow of air. In addition, load controls may be needed in installations where the electrical system cannot supply the inrush current required by a large motor.

Typical load controls measure the current applied to the motor and close the compressor inlet throttle valve when the current exceeds a maximum allowable value. Other types of controls may be a control based on air flow or a device to adjust the inlet throttle valve in response to inlet air temperature, as seen in Figure 2.73. More detailed information on compressor controls is included in the Gas Section of Chapter 7.

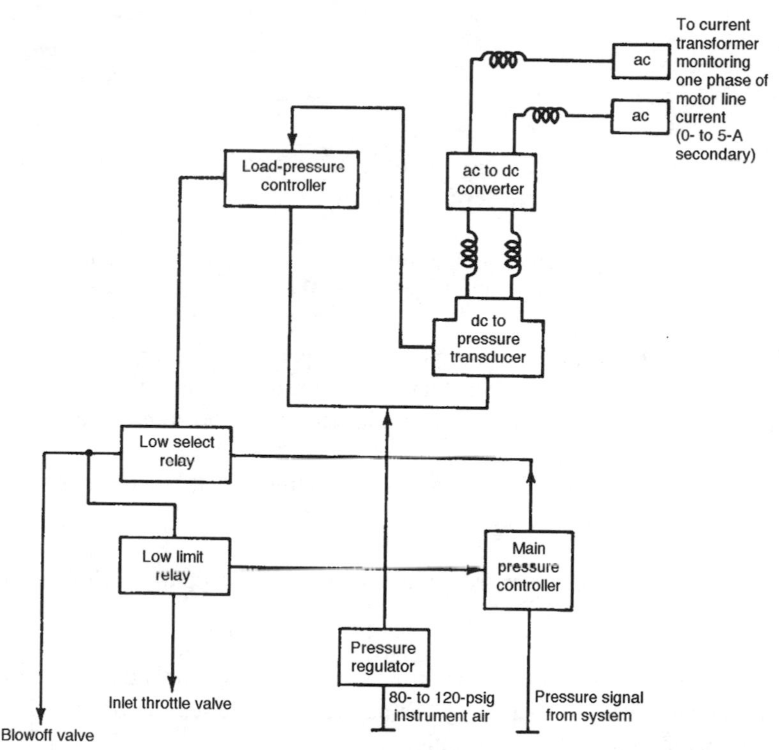

A typical motor load control limits motor horsepower by adjusting the opening of the compressor inlet throttle valve. The control measures motor line current with current transformers, and the transformer signal is led to a current-to-pressure transducer that converts the signal to a control pressure. This pressure signal is compared to the output of the main pressure controller. The lower of the two signals is transmitted to the inlet throttle valve, adjusting its setting to limit the motor current and horsepower.

Figure 2.73 Controlling motor load.

3

Compressed Air Treatment (Dryers and Filters)

Compressed air leaving an air compressor is not normally of a quality suitable for the intended use. This is due to several factors:

Atmospheric air, particularly in an industrial environment, contains pollutants, which include: particulate matter, moisture and hydrocarbons.

The air inlet filter on an air compressor is a particulate filter, designed to protect the air compressor rather than any downstream equipment and may not provide sufficient filtration for the equipment using the compressed air.

The air compressor itself will contribute contaminants in the form of wear particles and compressor oil carry-over.

The discharge temperature from the compressor may be too high for distribution and use.

Cooling after compression results in condensation of moisture and saturated air leaving the aftercooler. This will have a serious effect on pneumatic tools, spray painting and other applications.

APPLICATIONS REQUIRING CLEAN, DRY AIR

Plant Air

In almost every operation, clean, dry compressed air will result in lower operating costs. Dirt, water and oil entrained in the air will be deposited on the inner surfaces of pipes and fittings, causing an increase in pressure drop in the line. A loss of pressure is a loss of energy used to compress the air. A reduced pressure at the point-of-use results in a loss of performance efficiency.

Liquid water carry-over accelerates corrosion and shortens the useful life of equipment. Corrosion particles can plug valves, fittings and instrument control lines. When water freezes in these components, similar plugging will occur.

Valves and Cylinders

Deposits of sludge formed by dirty, wet and oily air act as a drag on pneumatic cylinders so that the seals and bearings need frequent maintenance. Operation is slowed down and eventually stopped. Moisture dilutes the oil required for the head and rod of an air cylinder, corrodes the walls and slows response. This results in loss of efficiency and production.

Moisture flowing to rubber diaphragms in valves can cause these parts to stiffen and rupture. Moisture also can cause spools and pistons to pit.

In high speed production, a sluggish or stuck cylinder could create costly downtime. A clean, dry air supply can prevent many of these potential problems.

Air Powered Tools

Pneumatic tools are designed to operate with clean, dry air at the required pressure. Dirty and wet air will result in sluggish operation, more frequent repair and replacement of parts due to sticking, jamming and rusting of wearing parts. Water also will wash out the required oils, resulting in excessive wear.

A decrease in pressure at the tool caused by restricted or plugged lines or parts will cause a reduction in the efficiency of the tool.

Clean, dry air at the required pressure will enable the production worker to start operating immediately at an efficient level, with no time lost to purge lines or drain filters and will help to maintain productivity and prolong tool life.

Instrument Air

Control air supplied to transmitters, relays, integrators, converters, recorders, indicators or gauges is required to be clean and dry. A small amount of moisture passing through an orifice can cause malfunction of the instrument and the process it controls. Moisture and resultant corrosion particles also can cause damage to instruments and plug their supply air lines.

Pneumatic thermostats, which control the heating and air conditioning cycles in large and small buildings, also require clean, dry air.

Instruments and pneumatic controllers in power plants, sewage treatment plants, chemical and petrochemical plants, textile mills and general manufacturing plants, all need clean, dry air for efficient operation.

The Instrument Society of America (ISA) has published ISA-S7.3 Quality Standard for Instrument Air.

Preservation of Products

When used to mix, stir, move or clean a product, air must be clean and dry. Oil and water in compressed air used to operate knitting machinery will cause the tiny latches on the knitting needles to stick. When used to blow lint and thread off finished fabrics, contaminants in the air may cause product spoilage.

If air is used to blow a container clean before packaging, entrained moisture and oil may contaminate the product. Moisture in control line air can cause the wrong mixture of ingredients in a bakery, the incorrect blend in liquor, waterlogged paint, or ruined food products.

In some printing operations, air is used to lift or position paper which will be affected by dirty, wet air and any water on the paper will prevent proper adhesion of the inks. In pneumatic conveying of a product such as paper cups or cement, dry air is essential.

Test Chambers

Supersonic wind tunnels are designed to simulate atmospheric conditions at high altitudes where moisture content is low. These chambers use large volumes of air, which must be dried to a very low dew point to prevent condensation in the tunnel air stream.

Breathing Air

The air coming from an air compressor, whether lubricated or oil-free type, is not suitable for breathing. Treatment of the air is required before the air can be considered suitable for breathing and certain health and safety standards must be met.

In industrial plants, air may be supplied to respirators, hoods and helmets and for applications such as sand blasting. Occupational Safety and Health Administration (OSHA) standard OSHA:1910:13d applies and requires drying, filtration and treatment to meet specific levels, including carbon monoxide, with an alarm system.

The Compressed Gas Association (CGA) Standard Commodity Specification G-7.1, Grade D, commonly is specified for plant breathing air systems. Medical air for hospitals must meet National Fire Protection Association (NFPA) Standard NFPA-99.

Air Quality Classes encompassing these pollutants have been established in an International Standard ISO 8573-1. These are as shown in Tables 3.1A, 3.1B and 3.1C.

Table 3.1A Maximum Particle Size and Concentration of Solid Contaminants

Class	Max particle size *	Max concentration **
	microns	mg/m^3
1	0.1	0.1
2	1	1
3	5	5
4	15	8
5	40	10

* Particle size based on a filtration ratio $\beta_\mu = 20$

** At 1 bar (14.5 psia), 20 °C (68 °F) and a relative vapor pressure of 0.6 (60%).

Table 3.1B Maximum Pressure Dew Point

Class	Class Max pressure dew point	
	ºC	ºF
1	-70	-94
2	-40	-40
3	-20	-4
4	+3	+37.4
5	+7	+44.6
6	+10	+50
7	not specified	

Table 3.1C Maximum Oil Content

Class	Max concentration ***
	mg/m^3****
1	0.01
2	0.1
3	1
4	5
5	25

*** At 1 bar(14.5 psia), 20°C(68°F) and a relative vapor pressure of 0.6 (60%).

**** 1 mg/m^3 is a weight of oil in a volume of air and is approximately equal to 0.83 ppm by weight.

COMPRESSED AIR TREATMENT

In discussing the treatment of compressed air, the normal sequence of treatment of compressed air after the compressor will be followed:

Temperature

High temperatures could cause unsafe surface temperatures and thermal expansion in piping systems and could adversely affect gaskets, seals and other downstream components. An aftercooler normally is installed after the discharge from an air compressor and often as an integral part of the compressor package. The aftercooler may use atmospheric air for cooling and provide a final compressed air temperature within 15 to 30°F of the ambient temperature. A water cooled aftercooler also can provide a final compressed air temperature in the range of 5 to 15°F of the cooling water temperature. The lower the temperature leaving the aftercooler, the more moisture will be removed from the air as condensate.

Moisture

Moisture in compressed air systems causes problems of various types. Corrosion (rust) will reduce the effective life of piping and vessels and result in scale particles being carried downstream, which can block orifices and instrument tubing. Moisture also may freeze in piping and control lines with serious effects. The principal problems may be summarized as follows:

1. Washing away required oils.
2. Rust and scale formation within pipelines and vessels.
3. Increased wear and maintenance of pneumatic devices.
4. Sluggish and inconsistent operation of air valves and cylinders.
5. Malfunction and high maintenance of control instruments and air logic devices.
6. Product spoilage in paint and other types of spraying.
7. Rusting of parts after sandblasting.
8. Freezing in exposed pipelines during cold weather.
9. Further condensation and possible freezing of moisture in mufflers whenever air devices are rapidly exhausted.

In the last case, some rock drills exhibit a 70ºF drop in temperature from inlet to exhaust. Most portable pneumatic tools have a considerably smaller temperature drop but the problem can exist. The increased use of automatic machinery and controls has made these problems more serious and has caused increased awareness of the need for better quality compressed air. This has led to standards such as ISO 8573-1 mentioned above.

Moisture Content of Air

All atmospheric air contains water vapor. A rule of thumb is that for every 20°F of temperature increase of air, its potential for holding moisture doubles. Table 3.2 shows the water content in grains per cubic foot of saturated air at various temperatures. Table 3.3 shows similar data in gallons per 1,000 cubic feet at various relative humidities. Table 3.4 shows the water content of 1,000 cubic feet of dry air at 14.7 psia and 60°F when heated to the temperature shown and moisture added until saturation level then compressed at constant temperature to the pressure shown.

Determination of the Water Content of Compressed Air Systems

A question of concern to the user of compressed air is, "How much water or condensate must be removed from the compressed air system?" Using Tables 3.3 and 3.4 permits the simple determination of the amount of condensate to be found in a compressed air system under a variety of operating conditions of pressure, temperature and relative humidity. The data presented, water content of saturated air at various temperatures and pressures, represent the worst possible condition. There is no guarantee that the water vapor content of compressed air will be any less than saturation at any given operating temperature and pressure; therefore, the vapor content at saturation should be used in all calculations.

The following example will illustrate the calculation of water content in a compressed air system:

Example:

On a warm, humid day, the outdoor temperature is 90°F and the relative humidity is 70%. This air is drawn into an air compressor intake and compressed to 100 psig. From the compressor, the air flows through an aftercooler and into an air receiver. The hot compressed air is cooled to 100°F by the time it leaves the receiver. As it flows through the compressed air piping within the plant, it is further cooled to 70°F. Suppose this compressed air system is in a medium-sized manufacturing plant utilizing a total of 50 air tools, each rated at 20 scfm. Also assume that the tools are utilized 50% of the time. This makes an average air flow requirement of 20 x 50 x 0.5, or 500 scfm. Determine the amount of water collecting in the receiver and in the downstream piping.

Table 3.2 Moisture Content of Saturated Air at Various Temperatures

°F	Grains per cu. ft.	°F	Grains per cu. ft.	°F	Grains per cu. ft.	°F	Grains per cu. ft.
-60	.01470	-15	.218	+30	1.935	+75	9.356
-59	.01573	-14	.234	+31	2.023	+76	9.749
-58	.01677	-13	.243	+32	2.113	+77	9.962
-57	.01795	-12	.257	+33	2.194	+78	10.38
-56	.01914	-11	.270	+34	2.279	+79	10.601
-55	.02047	-10	.285	+35	2.366	+80	11.04
-54	.02184	-9	.300	+36	2.457	+81	11.27
-53	.02320	-8	.316	+37	2.550	+82	11.75
-52	.02485	-7	.332	+38	2.646	+83	11.98
-51	.02650	-6	.350	+39	2.746	+84	12.49
-50	.02826	-5	.370	+40	2.849	+85	12.73
-49	.03004	-4	.389	+41	2.955	+86	13.27
-48	.03207	-3	.411	+42	3.064	+87	13.53
-47	.03412	-2	.434	+43	3.177	+88	14.08
-46	.03622	-1	.457	+44	3.294	+89	14.36
-45	.03865	0	.481	+45	3.414	+90	14.94
-44	.04111	+1	.505	+46	3.539	+91	15.23
-43	.04375	+2	.529	+47	3.667	+92	15.84
-42	.04650	+3	.554	+48	3.800	+93	16.15
-41	.04947	+4	.582	+49	3.936	+94	16.79
-40	.05249	+5	.640	+50	4.076	+95	17.12
-39	.05583	+6	.639	+51	4.222	+96	17.80
-38	.05922	+7	.671	+52	4.372	+97	18.44
-37	.06292	+8	.704	+53	4.526	+98	18.85
-36	.06677	+9	.739	+54	4.685	+99	19.24
-35	.07085	+10	.776	+55	4.849	+100	19.95
-34	.07517	+11	.816	+56	5.016	+101	20.33
-33	.07962	+12	.856	+57	5.191	+102	21.11
-32	.08447	+13	.898	+58	5.370	+103	21.54
-31	.08942	+14	.941	+59	5.555	+104	22.32
-30	.09449	+15	.986	+60	5.745	+105	22.75
-29	.09982	+16	1.032	+61	5.941	+106	23.60
-28	.10616	+17	1.080	+62	6.142	+107	24.26
-27	.11258	+18	1.128	+63	6.349	+108	24.93
-26	.11914	+19	1.181	+64	6.563	+109	25.41
-25	.12611	+20	1.235	+65	6.782	+110	26.34
-24	.13334	+21	1.294	+66	7.069	+111	27.07
-23	.14113	+22	1.355	+67	7.241	+112	27.81
-22	.14901	+23	1.418	+68	7.480	+113	28.57
-21	.15739	+24	1.483	+69	7.726	+114	29.34
-20	.166	+25	1.551	+70	7.980	+115	30.14
-19	.174	+26	1.623	+71	8.240	+116	30.95
-18	.184	+27	1.697	+72	8.508	+117	31.79
-17	.196	+28	1.773	+73	8.782	+118	32.63
-16	.207	+29	1.853	+74	9.153	+119	33.51

7000 grains moisture = 1 pound

Table 3.3 Water Content of Air in Gallons per 1000 Cubic Feet

	Temperature, °F									
%RH	35	40	50	60	70	80	90	100	110	120
5	.0019	.0024	.0035	.0050	.0071	.0099	.0136	.0186	.0250	.0332
10	.0039	.0047	.0069	.0100	.0142	.0198	.0273	.0372	.0501	.0668
15	.0058	.0071	.0104	.0150	.0213	.0298	.0411	.0561	.0755	.1007
20	.0078	.0095	.0139	.0200	.0284	.0398	.0549	.0750	.1012	.1351
25	.0098	.0119	.0174	.0251	.0356	.0498	.0689	.0940	.1270	.1699
30	.0117	.0143	.0209	.0301	.0427	.0599	.0828	.1132	.1531	.2051
35	.0137	.0166	.0244	.0351	.0499	.0700	.0969	.1325	.1794	.2407
40	.0156	.0190	.0279	.0402	.0571	.0801	.1110	.1519	.2060	.2768
45	.0176	.0214	.0314	.0453	.0644	.0903	.1251	.1715	.2328	.3133
50	.0195	.0238	.0349	.0503	.0716	.1005	.1394	.1912	.2598	.3502
55	.0215	.0262	.0384	.0554	.0789	.1107	.1537	.2110	.2871	.3876
60	.0235	.0286	.0419	.0605	.0861	.1210	.1681	.2310	.3146	.4254
65	.0254	.0310	.0454	.0656	.0934	.1313	.1825	.2511	.3424	.4637
70	.0274	.0334	.0490	.0707	.1007	.1417	.1970	.2713	.3705	.5025
75	.0294	.0358	.0525	.0758	.1081	.1521	.2116	.2917	.3988	.5418
80	.0313	.0382	.0560	.0810	.1154	.1625	.2263	.3122	.4273	.5816
85	.0333	.0406	.0596	.0861	.1228	.1730	.2410	.3328	.4562	.6219
90	.0353	.0430	.0631	.0913	.1302	.1835	.2559	.3536	.4853	.6627
95	.0372	.0454	.0666	.0964	.1376	.1940	.2707	.3745	.5147	.7041
100	.0392	.0478	.0702	.1016	.1450	.2046	.2857	.3956	.5443	.7460

Table 3.4 Water Content of 1000 Cubic Feet of Dry Air at 14.7 psia and 60°F When Heated to the Temperature Shown and Moisture Added until Saturation Level and then Compressed at Constant Temperature to the Pressure Shown

	Temperature, °F									
PSIG	35	40	50	60	70	80	90	100	110	120
0	.0392	.0479	.0702	.1016	.1450	.2046	.2857	.3956	.5443	.7460
10	.0233	.0283	.0416	.0600	.0854	.1200	.1667	.2290	.3119	.4217
20	.0165	.0201	.0295	.0426	.0605	.0849	.1176	.1612	.2186	.2939
30	.0128	.0156	.0229	.0330	.0469	.0657	.0909	.1243	.1682	.2256
40	.0105	.0128	.0187	.0269	.0383	.0536	.0741	.1012	.1367	.1830
50	.0089	.0108	.0158	.0228	.0323	.0452	.0625	.0853	.1152	.1540
60	.0077	.0093	.0137	.0197	.0280	.0391	.0540	.0737	.0995	.1329
70	.0068	.0082	.0121	.0174	.0246	.0345	.0476	.0649	.0876	.1169
80	.0060	.0074	.0108	.0155	.0220	.0308	.0425	.0580	.0782	.1043
90	.0055	.0067	.0098	.0140	.0199	.0279	.0385	.0524	.0706	.0942
100	.0050	.0061	.0089	.0128	.0182	.0254	.0351	.0478	.0644	.0858
110	.0046	.0056	.0082	.0118	.0167	.0234	.0323	.0439	.0592	.0789
120	.0043	.0052	.0076	.0109	.0155	.0216	.0298	.0407	.0548	.0729
130	.0040	.0048	.0071	.0102	.0144	.0201	.0278	.0378	.0509	.0678
140	.0037	.0045	.0066	.0095	.0135	.0188	.0260	.0354	.0476	.0634
150	.0035	.0042	.0062	.0089	.0126	.0177	.0244	.0332	.0447	.0595
160	.0033	.0040	.0058	.0084	.0119	.0167	.0230	.0313	.0421	.0561
170	.0031	.0038	.0055	.0080	.0113	.0158	.0217	.0296	.0398	.0530
180	.0029	.0036	.0052	.0075	.0107	.0149	.0206	.0281	.0378	.0503
190	.0028	.0034	.0050	.0072	.0102	.0142	.0196	.0267	.0359	.0478
200	.0027	.0032	.0048	.0068	.0097	.0136	.0187	.0254	.0342	.0455

From Table 3.3, the amount of water per 1,000 cubic feet of air at 90°F and a relative humidity of 70% is 0.1970 gallons. From Table 3.4, the quantity of water per 1,000 cubic feet of air at 100ºF and 100 psig is 0.0478 gallons. The difference is 0.1970 - 0.0478 or 0.1492 gallons per 1,000 standard cubic feet of air. This is the amount of water which would condense in the receiver for every 1,000 cubic feet of air used. Every hour, 30,000 cubic feet (60 minutes x 500 scfm) of air will flow through the receiver. An excess of 4.476 gallons of water will collect in the receiver each hour. In an 8 hour shift, this amounts to 35.8 gallons.

In the compressed air piping, the temperature dropped to 70ºF. At that temperature and 100 psig, the compressed air can hold only 0.0182 gallons per 1,000 scfm of air. The air leaving the receiver has a moisture content of 0.0478 gallons per 1,000 scfm of air, therefore 0.0296 gallons per 1,000 cubic feet of air will condense in the piping system. Again, with 30,000 cubic feet of air each hour, the excess water will be 0.888 gallons. In an 8 hour shift the quantity will be 7.1 gallons which, if not removed, will be flowing through air tools, cylinders, paint guns and any other device using this air. Adding this amount to the 35.8 gallons collected in the air receiver makes a total of 42.9 gallons of water which will have condensed at some point in the compressed air system during an eight hour shift.

Relative Humidity

Relative humidity (or, more correctly, relative vapor pressure) is defined as the ratio of the actual water vapor partial pressure to its saturation pressure at the same temperature. This term usually is considered in connection with atmospheric air. It is dimensionless and normally expressed as a percentage. Relative humidity often is incorrectly defined as the amount of moisture content of the air at a given temperature, divided by the amount of moisture which would be present if the air was saturated. The error incurred by this method is more significant in atmospheric, or meteorological, applications than at higher pressure applications (above 50 psig). Therefore, in this discussion, both methods will be considered close enough to be synonymous.

The moisture in the air will begin to condense into liquid water when the air is cooled to its saturation point (dew point). When air is saturated, it is said to have 100% relative humidity. When air contains less moisture than it could contain at that temperature, the amount is expressed as a percentage.

(Example: Air containing half of the potential amount is said to have 50% relative humidity).

When air is compressed, the space occupied by the air is reduced. For example, when atmospheric air has been compressed to 103 psig and cooled to its original temperature, the volume has been reduced to one eighth of its atmospheric volume. This is a compression ratio of 8:1 [(103 psig + 14.7 psia) ÷ 14.7 psia]. If the final volume is 1 cubic foot, the original volume was 8 cubic feet. It is highly improbable that the final volume of air can contain the original amount of moisture in vapor form, so condensation will occur.

Figure 3.1 illustrates the ability of air to hold water vapor at various temperatures while maintaining constant pressure. It will be noted that as the temperature decreases, the quantity of water vapor that can be held also decreases but the relative humidity remains constant. When the volume is reheated, its ability to hold moisture is increased but since the excess moisture, in the form of condensate, has been drained off, no additional moisture is available, The relative humidity therefore decreases. The relative humidity can be approximated as follows: The 0.0858 gallons/1000 cu. ft. moisture content for saturated air can be determined from Table 3.4 at the intersection of the 120°F column and the 100 psig row. The moisture content for 80°F and 40°F also can be determined in this same manner. As the temperature is increased from 40°F to 120°F it again has the ability to hold 0.0858 gallons/1000 cu. ft. However, since the excess moisture has been drained off as condensate at the lower temperature, only 0.0061 gallons/1000 cu. ft. of moisture remain. The relative humidity, therefore, is reduced to 0.0061/0.0858 or 7%.

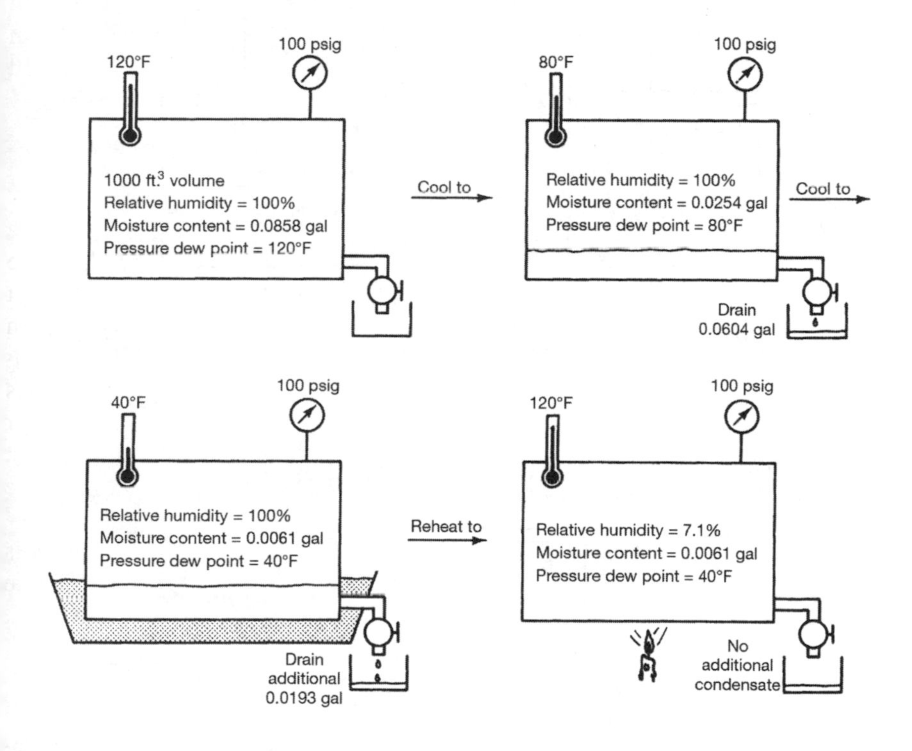

Figure 3.1 How temperature influences the capacity of air to hold water vapor, the pressure remaining constant.

An illustration of pressure effect on vapor content and relative humidity is shown in Figures 3.2A, 3.2B, 3.2C, 3.2D and 3.2E. The amount of water vapor in 10 cubic feet of air at 80% relative humidity and 80°F was determined from Table 3.3: 0.1625 x 10/1000 = 0.0016 gallons. See Figure 3.2.

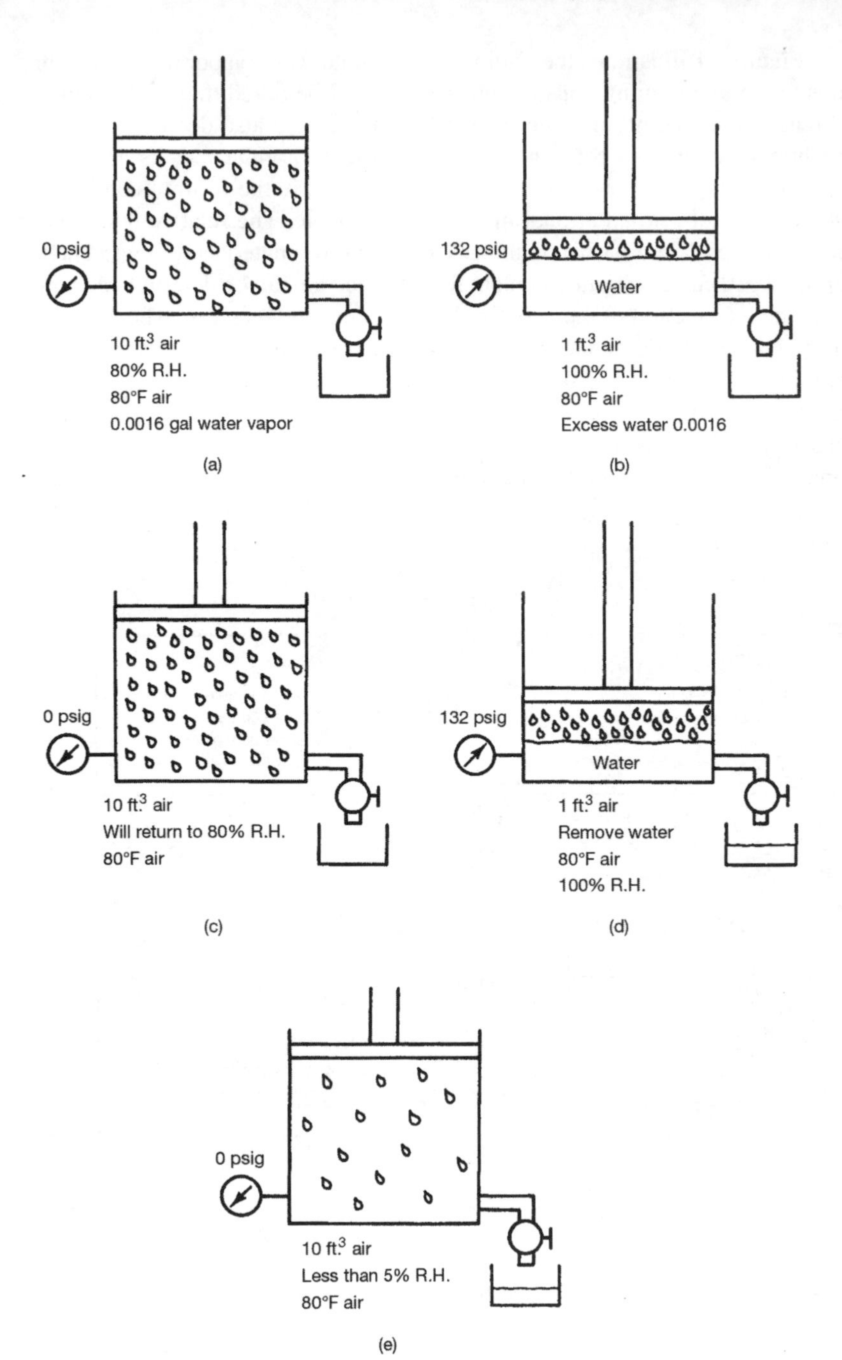

Figures 3.2A, 3.2B, 3.2C, 3.2D and 3.2E Reflect how change in pressure affects moisture content of air at constant pressure.

Now, if air is compressed from 10 cubic feet to 1 cubic foot, the pressure is increased from atmospheric to 132 psig.[(132 psig + 14.7 psia) ÷ 14.7 psia]. At 132 psig and 80°F, the 1 cubic foot volume can hold only 0.00002 gallons of moisture. Since there were 0.0016 gallons in the 10 cubic feet of atmospheric air but as 1 cubic foot it now can hold only 0.00002 gallons, the excess moisture will condense. If the excess moisture is not removed (see Figure 3.2) and the pressure is reduced to atmospheric, the excess water will gradually evaporate back into the air until an equilibrium is established. This will happen because the air under this condition can again hold 0.0016 gallons of water vapor. If the condensed water had been removed at pressure, as shown in Figure 3.2D and the pressure again reduced as shown in Figure 3.2E, the excess water is no longer available to evaporate back into the air. The water vapor content of the 10 cubic feet then will be 0.00002 gallons, which is the maximum vapor content that 1 cubic foot of air can hold at 80°F and 132 psig. The 0.00002 gallons per 10 cubic feet thus determined is 0.002 gallons per 1,000 cubic feet. Referring to Table 3.3 it will be seen that 0.002 gallons per 1,000 cubic feet is less than any quantity listed for 80°F. Therefore, the relative humidity is less than 5%.

Dew Point

A more useful term than relative humidity for indicating the condition of water vapor in a compressed air system is dew point. The dew point is the temperature at which condensate will begin to form if the air is cooled at constant pressure. At this point the relative humidity is 100%. In Figure 3.3 it will be noted that the dew point is equal to the saturated air temperature and follows the air temperature until it reaches its lowest point of 40°F. At this point, the air temperature was increased but the dew point remained at 40°F since the excess moisture had been removed. No further condensation would occur unless the temperature dropped below 40°F. This is further illustrated in Figure 3.4. If a refrigerant type air dryer set at +35°F were installed at point A, Figure 3.3 it would cool the air to +35°F and the moisture in the air would condense and be removed at this point. No further condensation would occur unless the temperature somewhere downstream was dropped below +35°F.

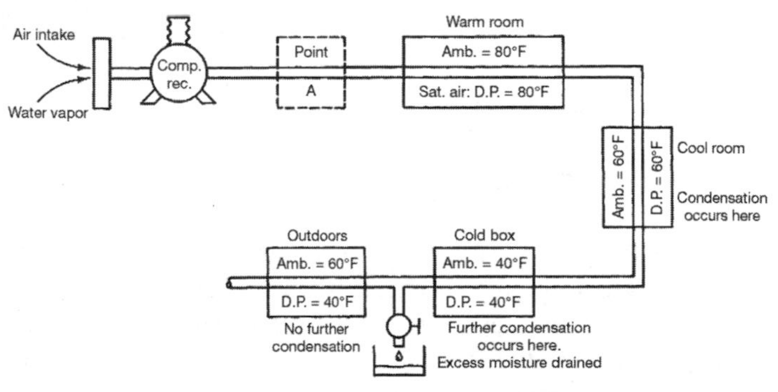

Figure 3.3 Effect of dew point on condensation.

To obtain the dew point temperature expected if the gas were expanded to a lower pressure proceed as follows:

1. Using "dew point at pressure," locate this temperature on scale at right hand side of chart.
2. Read horizontally to intersection of curve corresponding to the operating pressure at which the gas was dried.
3. From that point read vertically downward to curve corresponding to the expanded lower pressure.
4. From that point read horizontally to scale on right hand side of chart to obtain dew point temperature at the expanded lower pressure.
5. If dew point temperatures of atmospheric pressure are desired, after step 2, above read vertically downward to scale at bottom of chart which gives "Dew Point at Atmospheric Pressure."

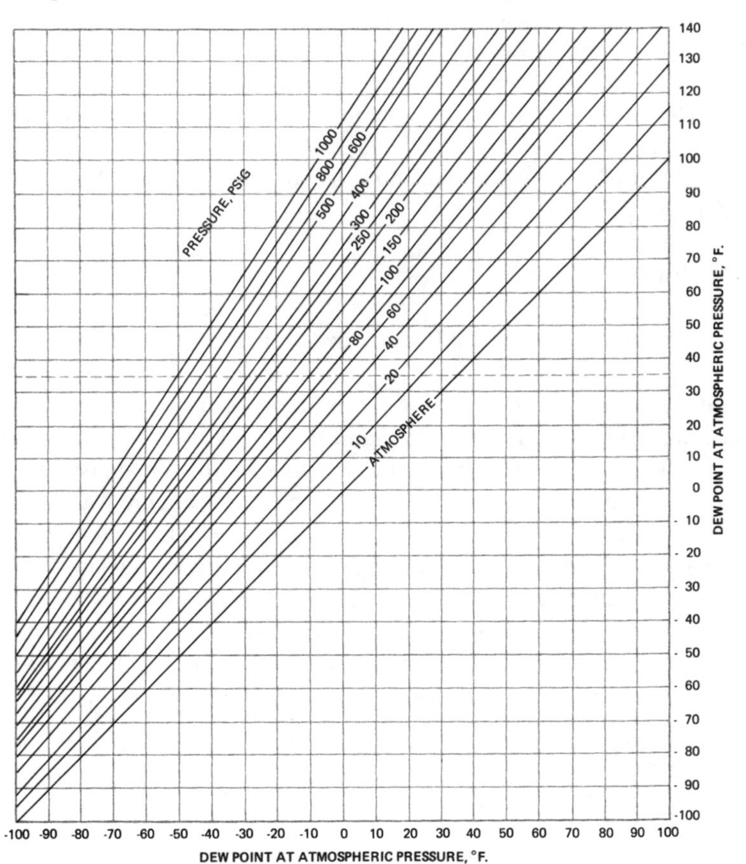

Figure 3.4 Dew Point Conversion

It should be noted that as air leaves a compressor it is under both an elevated pressure and an elevated temperature. A delicate balance exists under this condition since the air under pressure has less capacity for water vapor, whereas air at an elevated temperature has a greater capacity for water vapor. The air leaving the compressor generally is cooled in an aftercooler causing water to condense and allowing saturated air to flow downstream. Any further cooling in the piping will result in additional condensation in the piping system.

Pressure Dew Point

A distinction must be made between the dew point at atmospheric conditions and the dew point under operating conditions at elevated pressure. It is desirable to know the temperature at which water vapor will begin to condense into liquid water within the pressurized system. The relationship between the atmospheric dew point and the pressure dew point is shown in Figure 3.4. To convert the pressure dew point at 100 psig of 35°F to an atmospheric dew point, draw a horizontal line at +35°F from the scale on the right until it intersects the 100 psig pressure line, then draw a line vertically downward to the scale at the bottom where it shows approximately -10°F. This demonstrates quite clearly that when water is removed from air under pressure, the resulting atmospheric dew point will be substantially lower. This is one reason why dryers normally are located after the air compressor. Another reason is the reduced volume which has to pass through the dryer for the same mass flow of air.

Moisture Separators

A moisture separator is a mechanical device designed to remove liquid condensate from an air stream. Such a device normally is installed after an air cooler such as an aftercooler following an air compressor. It should be noted that such devices are not 100% efficient and that some condensate will pass through with the air stream and that any further cooling downstream will result in additional condensate in the piping system. Moisture separators also require a trap or similar device to allow condensate to be drained automatically from the pressurized system during operation.

COMPRESSED AIR DRYERS

Dryer Types

Different methods can be used to remove the moisture content of compressed air. Current dryer types include the following:

Refrigerant type	- Cycling
	- Non-cycling
Regenerative desiccant type	- Heatless (no internal or external heaters)
	- Heated (internal or external heaters)
	- Heat of Compression
	- Non-regenerative single tower
Deliquescent	
Membrane	

Each of these dryer types will be discussed in some detail.

Refrigerant Type Dryers

Although it does not offer as low a dew point as can be obtained with other types, the refrigerant type dryer has been the most popular, as the dew point obtained is acceptable in general industrial plant air applications. The principle of operation is similar to a domestic refrigerator or home air conditioning system. The compressed air is cooled in the air to air heat exchanger and then is cooled further in the refrigerant to air heat exchanger to about 35°F, at which point the condensed moisture is drained off. The air is then reheated in the air to air heat exchanger by means of the incoming air which also is pre-cooled before entering the air to refrigerant heat exchanger. This means that the compressed air leaving the dryer has a pressure dew point of 35 to 40°F. A lower dew point is not feasible in this type of dryer as the condensate would freeze at 32°F or lower.

In a non-cycling refrigerant dryer, the refrigerant circulates continuously through the system. Since the flow of compressed air will vary and ambient temperatures also vary, a hot gas bypass valve often is used to regulate the flow of the refrigerant and maintain stable operating conditions within the refrigerant system. In most designs, the refrigerant evaporates within the air to refrigerant heat exchanger (evaporator) and is condensed after compression by an air or water to refrigerant heat exchanger (condenser). A typical schematic diagram is shown in Figure 3.5.

This design provides rapid response to changes in operating loads.

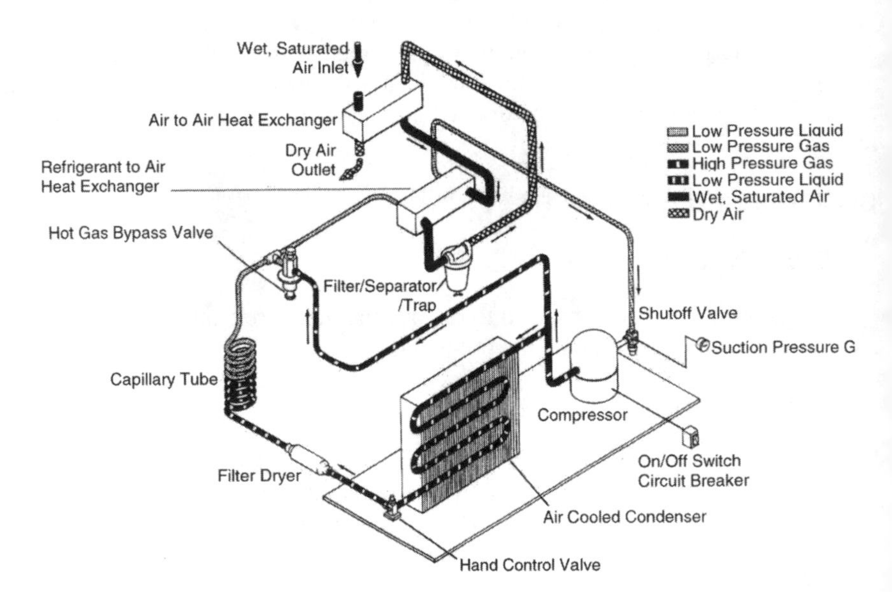

Figure 3.5 Typical Refrigerant Type Dryer with Hot Gas By-pass Valve

Figure 3.6 A non-cycling, direct expansion refrigerant dryer.

Figure 3.7 A large capacity refrigerant air dryer.

While older refrigerant air dryers have used refrigerants such as a CFC and R-12, newer designs are in compliance with the Montreal Protocol and use either an HCFC such as R-22 or an HFC such as R-134A or R-404A.

Cycling type refrigerant dryers chill a thermal mass surrounding the air passage in the evaporator. This mass may be a liquid such as glycol or a metal such as aluminum block, beads or related substance, which acts as a reservoir of coldness. The compressed air is cooled by the thermal mass which has its temperature controlled by a thermostat and shuts off the refrigerant compressor during reduced loads, providing savings in operating costs, but at higher initial capital cost.

Advantages of refrigerant type air dryers include:

- Low initial capital cost.
- Relatively low operating cost.
- Low maintenance costs.
- Not damaged by oil in the air stream.

Disadvantages include:

- Limited dew point capability.

Advantages of Non-Cycling Control:

- Low and precise dew point.
- Refrigerant compressor runs continuously.

Disadvantages of Non-Cycling Control:

- No energy savings at partial and zero air flow unless.
- Refrigrant compressor can be unloaded.

Advantages of Cycling Control:

- Energy savings at partial and zero air flow.

Disadvantages of Cycling Control:

- Dew point swings.
- Reduced refrigerant compressor life due to on/off cycling.
- Increased weight and size to accommodate heat sink mass.
- Increased initial capital cost.

Regenerative Desiccant Type Dryers

These dryers use a desiccant, which adsorbs the water vapor in the air stream. A distinction needs to be made between adsorb and absorb. Adsorb means that the moisture adheres to the desiccant, collecting in the thousands of small pores within each desiccant bead. The composition of the desiccant is not changed and the moisture can be driven off in a regeneration process by applying dry purge air, by the application of heat, or a combination of both. Absorb means that the material which attracts the moisture is dissolved in and used up by the moisture as in the deliquescent desiccant type dryer.

Regenerative desiccant type dryers normally are of twin tower construction. One tower dries the air from the compressor while the desiccant in the other tower is being regenerated after the pressure in the tower has been reduced to atmospheric pressure. Regeneration can be accomplished using a time cycle, or on demand by measuring the temperature or humidity in the desiccant towers, or by measuring the dew point of the air leaving the on-line tower. In the heatless regenerative desiccant type, no internal or external heaters are used. Purge air requirement can range from 10 to 18% of the total air flow. The typical regenerative desiccant dryer at 100 psig has a pressure dew point rating of -40°F but a dew point down to -100°F can be obtained. A typical schematic diagram is shown in Figure 3.8.

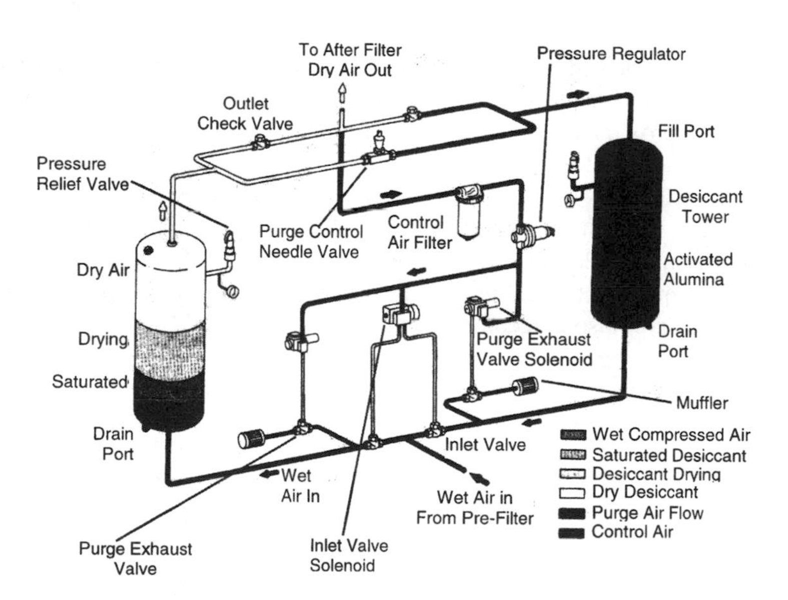

Figure 3.8 Typical Twin Tower Regenerative Desiccant Type Dryer

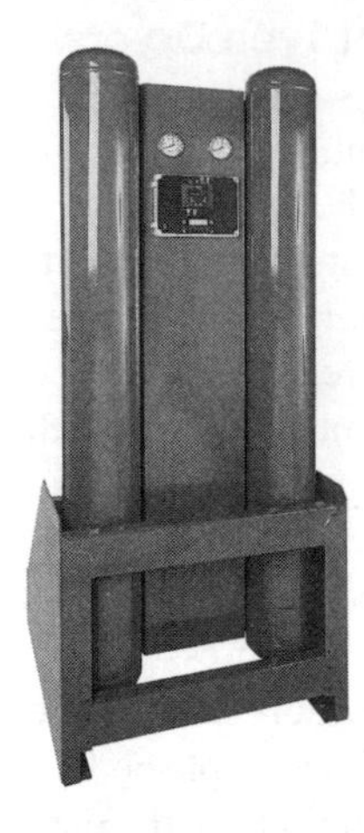

Figure 3.9 A heatless regenerative dryer with automatic operation. Dry purge air is used to remove moisture from the desiccant in the bed being regenerated. The towers are filled with an activated alumina desiccant.

Heat reactivated regenerative desiccant dryers may have internal or external heat applied by heaters. In the internal type, steam or electricity may be used in heaters embedded in the desiccant bed. This reduces the amount of purge air required for regeneration to about 5%. The purge air plus normal radiation is used to cool the desiccant bed after regeneration to prevent elevated air temperatures going downstream.

In externally heated regenerative desiccant dryers, the purge air is heated to a suitable temperature and then passes through the desiccant bed. The amount of purge air is approximately 0 to 8% of the air flow through the dryer. The purge air can be eliminated if a blower is used for circulation of atmospheric air through the desiccant bed.

To protect the desiccant bed from contamination from oil carry over from the air compressor, a coalescing filter is required upstream of the dryer. To protect downstream equipment from desiccant dust or "fines," a particulate filter downstream of the dryer also is recommended.

Figure 3.10 An externally heated blower purge heat reactivated dryer.

Figure 3.11 An externally heat reactivated dryer.

Advantages of regenerative desiccant type dryers include:

- Very low dew points can be achieved without potential freeze-up.
- Moderate cost of operation for the dew points achieved.
- Heatless type can be designed to operate pneumatically for remote, mobile or hazardous locations.

Disadvantages of regenerative desiccant type dryers include:

- Relatively high initial capital cost.
- Periodic replacement of the desiccant bed (typically 3-5 years).
- Oil aerosols can coat the desiccant material, rendering it useless if adequate pre-filtering is not maintained.
- Purge air usually is required.

Heat of Compression Type Dryers

Heat of compression type dryers are Regenerative Desiccant Dryers which use the heat generated during compression to accomplish desiccant regeneration, so they can be considered as heat reactivated. There are two types, the Single Vessel Type and the Twin Tower Type.

The Single Vessel Heat of Compression Type Dryer provides continuous drying with no cycling or switching of towers. This is accomplished with a rotating desiccant drum in a single pressure vessel divided into two separate air streams. One air stream is a portion of the hot air taken directly from the air compressor at its discharge, prior to the aftercooler, and is the source of heated purge air for regeneration of the desiccant bed. The second air stream is the remainder of the air discharged from the air compressor after it passes through the air aftercooler. This air passes through the drying section of the dryer rotating desiccant bed where it is dried. The hot air, after being used for regeneration, passes through a regeneration cooler before being combined with the main air stream by means of an ejector nozzle before entering the dryer.

The Twin Tower Heat of Compression Type Dryer operation is similar to other Twin Tower Heat Activated Regenerative Desiccant Dryers. The difference is that the desiccant in the saturated tower is regenerated by means of the heat of compression in all of the hot air leaving the discharge of the air compressor. The total air flow then passes through the air aftercooler before entering the drying tower. Towers are cycled as for other Regenerative Desiccant Type Dryers.

Advantages of Heat of Compression Type Dryers include:

- Low electrical installation cost.
- Low power costs.
- Minimum floor space.
- No loss of purge air.

Disadvantages of Heat of Compression Type Dryers include:

- Applicable only to oil free compressors.
- Applicable only to compressors having a continuously high discharge temperature.
- Inconsistent dew point.
- Susceptible to changing ambient and inlet air temperatures.
- High pressure drop and inefficient ejector nozzle on single vessel type.
- Booster heater required for low load (heat) conditions.

Single Tower Deliquescent Type Dryers

The Deliquescent Desiccant Type Dryer uses a hygroscopic desiccant material having a high affinity for water. The desiccant absorbs the water vapor and is dissolved in the liquid formed. These hygroscopic materials are blended with ingredients to control the pH of the effluent and to prevent corrosion, caking and channeling. The desiccant is consumed only when moist air is passing through the dryer. On average, desiccant must be added two or three times per year to maintain a proper desiccant bed level.

The Single Tower Deliquescent Desiccant Type Dryer has no moving parts and requires no power supply. This simplicity leads to lower installation costs. Dew point suppression of 15 to 50 F degrees is advertised. This type of dryer actually dries the air to a specific relative humidity rather than to a specific dew point.

Advantages of Single Tower Deliquescent Desiccant Type Dryers include:

- Low initial capital and installation cost.
- Low pressure drop.
- No moving parts.
- Requires no electrical power.
- Can be installed outdoors.
- Can be used in hazardous, dirty or corrosive environments.

Disadvantages of Single Tower Deliquescent Type Dryers include:

- Limited suppression of dew point.
- Desiccant bed must be refilled periodically.
- Regular periodic maintenance.
- Desiccant material can carry over into downstream piping if dryer is not drained regularly and certain desiccant materials may have a damaging effect on downstream piping and equipment. Some desiccant materials may melt or fuse together at temperatures above 80°F.

Membrane Type Dryers

Membrane technology has advanced considerably in recent years. Membranes commonly are used for gas separation such as in nitrogen production for food storage and other applications. The structure of the membrane is such that molecules of certain gases (such as oxygen) are able to pass through (permeate) a semi-permeable membrane faster than others (such as nitrogen) leaving a concentration of the desired gas (nitrogen) at the outlet of the generator.

When used as a dryer in a compressed air system, specially designed membranes allow water vapor (a gas) to pass through the membrane pores faster than the other gases (air) reducing the amount of water vapor in the air stream at the outlet of the membrane dryer, suppressing the dew point. The dew point achieved normally is 40ºF but lower dew points to -40ºF can be achieved at the expense of additional purge air loss.

Advantages of Membrane Type Dryers include:

- Low installation cost.
- Low operating cost.
- Can be installed outdoors.
- Can be used in hazardous atmospheres.
- No moving parts.

Disadvantages of the Membrane Type Dryers include:

- Limited to low capacity systems.
- High purge air loss (15 to 20%) to achieve required pressure dew points.
- Membrane may be fouled by oil or other contaminants.

DRYER RATINGS

When a compressed air system is started (for example, at the beginning of a shift, day or week), the lack of system pressure can result in excessive air velocities through moisture separators, dryers and filters with resultant moisture carry-over.

This problem can be eliminated by having a "minimum pressure valve" located downstream of the primary treatment equipment, prior to distribution to the plant.

Oil injected rotary compressors have a similar valve to prevent excessive velocities through the air/oil separator but this does not protect treatment equipment downstream of the compressor.

The standard conditions for the capacity rating in scfm of compressed air dryers, are contained in The Compressed Air & Gas Institute (CAGI) Refrigerated Air Dryer Standard ADF100. These commonly are called the three 100s. That is, a dryer inlet pressure of 100 psig, an inlet temperature of 100°F and an ambient temperature of 100°F. If the plant compressed air system has different operating conditions, this will affect the dryer rating and must be discussed with the supplier to ensure compatibility.

SELECTING A COMPRESSED AIR DRYER

Before a compressed air dryer is selected, a number of factors must be evaluated. The dew point required for the point of use application(s) must be known. The compressed air flow rate should be determined. This normally matches the maximum flow rate of the compressor. Variations in flow also should be considered. The compressed air pressure and temperature entering the dryer and the ambient temperature must be known. Available utilities such as electricity also should be known.

Know the Specific Use of the Compressed Air

It is important to establish the pressure dew point required for each application. In some cases this may be specified by the manufacturer of the equipment using the compressed air. Air considered dry for one application may not be sufficiently dry for another. Dryness is relative. Even air in the desert has some moisture content. In a compressed air system there will still be an amount of moisture present even after passing through a dryer, since the pressure dew point will vary according to the dryer type. An adequate pressure dew point for the anticipated range of ambient conditions is essential to prevent system problems.

For example, an accumulation of moisture in the air supply to instrumentation can cause faulty indication or operation. Bulk pneumatic conveying of cement cannot tolerate the presence of water. Similarly, insufficiently dried air for pneumatic unloading of railroad tank cars carrying liquid chlorine would result in the formation of hydrochloric acid resulting in serious corrosion. Droplets of moisture in wind tunnel air at high velocities may have the effect of machine-gun bullets, damaging test models. Low temperature applications are particularly vulnerable due to the potential of water freezing.

An industrial plant air system may supply a variety of applications within the plant and each application may require a different degree of dryness. This may require a dryer for each application rather than drying all of the plant air to a dew point much lower than needed for some applications.

Know the Required Dew Point

Specifying a dew point lower than is required for an application can add to the capital and operational costs and is not good engineering practice. On the other hand, a dew point which is marginal may not meet requirements in varying operating conditions and could result in costly shutdown of a process and/or damage to the product or equipment.

It is recommended that the supplier of the end use equipment be asked to specify the dew point required for their equipment. One of the classes established in ISO 8573-1 should be selected. This will allow a dryer meeting that class to be chosen. The experience of the dryer manufacturer also should be drawn upon in this decision.

Typical pressure dew points from dryers are as follows:

Refrigerant Types:	33 - 39°F.
Regenerative Desiccant Types:	-40°F to -100°F
Single Tower Deliquescent Types:	65 to 80°F (20 to 50°F below inlet temperature)
Membrane Types:	40°F standard and lower

It is desirable to know the amount of moisture remaining in saturated air at various operating pressures. This can be seen in Figure 3.12.

Know the Flow Capacity

The flow capacity of a dryer normally is stated in standard cubic feet per minute (scfm). The Compressed Air & Gas Institute and PNEUROP now define standard air as measured at 14.5 psia (1 bar), 68°F (20°C) and 0% relative vapor pressure (0% relative humidity). The capacity rating also is based upon inlet conditions to the dryer of saturated compressed air at 100 psig and 100°F and an ambient temperature of 100°F. An increase in inlet pressure raises the capacity of the dryer while an increase in inlet temperature or ambient temperature lowers it.

Generally a dryer is selected having the same capacity as the air compressor which it follows. If the dryer is placed between the air compressor and an air storage vessel (air receiver), the flow through the dryer cannot exceed the output from the air compressor, even though air flow from the air receiver to the plant air system may vary according to demand. If the dryer is placed downstream of the air receiver, the dryer could see a surge in demand which exceeds the capacity of the dryer. However, the dryer then benefits from some radiant cooling of the air in the receiver and is shielded from potentially harmful pressure pulsations if the air compressor is the reciprocating piston type.

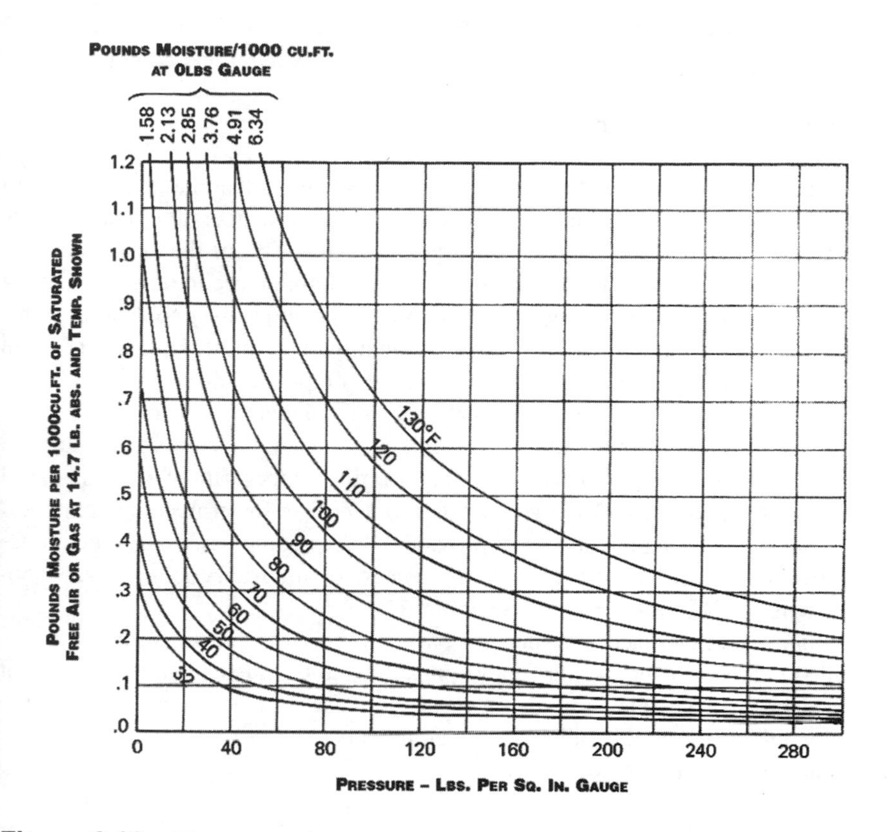

Figure 3.12 Illustrates moisture remaining in saturated air when compressed isothermally to the pressure shown.

In some cases a dryer is sized to meet the output of two air compressors operating in parallel. In such cases it should be recognized that if an air compressor is out of operation, the dryer will be operating at half capacity and if the dryer is out of service, all of the output from two air compressors will have to bypass the dryers.

In some cases it is desirable to have a dryer close to the point of use to meet the specific requirements of an application. In this case, the dryer capacity should be sufficient to meet the maximum requirement of the application.

Know the Operating Pressure

The higher the inlet pressure to the dryer, the lower the moisture content at saturated conditions and the lower the load on the dryer. A higher operating pressure can result in a smaller, more efficient and more economical dryer but will require higher operating costs of the air compressor to produce the higher pressure. A rule of thumb is an additional ½% of energy costs for each additional 1 psi over 100 psig. The dryer design also must be capable of the maximum operating pressure.

Know the Operating Temperature

Knowing the operating temperature requires knowing the inlet temperature to the dryer, the variations in ambient temperature and the temperature requirements at the points of use. Normally, the air dryer is downstream of the air compressor aftercooler and moisture separator and has a temperature close to that of the aftercooler outlet. It should be recognized that 15 to 20 °F above inlet coolant temperature is the norm for aftercoolers. This means that with the most popular air cooled radiator types using ambient air for cooling, the compressed air temperature leaving the aftercooler will be 15 to 20 degrees above ambient. This also means that if a maximum ambient temperature of 100°F is anticipated, the compressed air inlet temperature to the dryer will exceed the dryer rating temperature of 100°F and will affect the dryer capacity. Water cooled aftercoolers with lower than ambient cooling water supply temperatures may provide a lower compressed air temperature from the aftercooler but the temperature of water coming from cooling towers generally will be above ambient air temperature.

Often the aftercooler is an integral part of the air compressor package and is installed indoors at a temperature above prevailing ambient temperature but the piping which distributes the compressed air to the plant system may pass outside the building and be exposed to ambient temperatures below freezing. The pressure dew point of the air then must be suitable for the lowest anticipated ambient temperature.

Know the Utility Requirements

Plant utilities generally available may include electric power, natural gas, cooling water, steam and compressed air. The type of location (e.g. hazardous or remote), the availability at the proposed location of each utility and its relative cost, will influence the selection of the type of dryer. Some dryer types require electric power while others do not.

HOW TO SPECIFY

The following sample specifications will provide minimum data to the dryer manufacturer:

Refrigerant Type

1. Inlet Conditions
   Flow - scfm
   Inlet Pressure - psig
   Inlet Air Temperature (saturated) - °F
2. Ambient Conditions
   Ambient temperature range - °F
3. Performance
   Pressure Dew Point at Dryer Outlet - °F
   Pressure Drop Across Dryer(max allowable) - psid
4. Design
   Pressure - psig
   Temperature - °F
   Ambient Temperature °F
      Maximum
      Minimum
   Air cooled or water cooled
5. Utilities
   Electric: ____Volts_____Phase_____Hz_____Amps
   Cooling water temperature, pressure, flow rate (if applicable)
   Power Consumption_____kW
   Electrical Enclosure - NEMA _____
      NEMA type 1 General Purpose
      4 Weatherproof (water & dust)
      7 Explosion Proof
      12 Dust Tight

Regenerative Desiccant Type

1. Inlet Conditions
   Flow - scfm
   Operating Pressure - psig
   Inlet Air Temperature (saturated) - °F
2. Ambient conditions
   Ambient temperature range - °F
3. Performance
   Outlet Flow (Inlet Flow less Purge Air Requirement) - scfm
   Pressure Dew Point at Dryer Outlet - °F
   Pressure Drop Across Dryer - psid

4. Design
   Pressure - psig
   Ambient Temperature - °F
      Maximum
      Minimum
   Cycle time - hours/minutes
   Adsorption Period
   Regeneration Period
   Heating (if applicable)
   Cooling (if applicable and other than ambient)
5. Utilities
   Electric:_____Volts_____Phase_____Hz_____Amps
   Power Consumption_____kW
   Electrical Enclosure - NEMA _____
      NEMA type 1 General Purpose
      4 Weatherproof(water & dust)
      7 Explosion Proof
      12 Dust Tight

   Steam_____psig_____°F (if applicable)
   Steam Consumption - lb/hr
   Water_____psig_____°F
   Water Consumption_____gph

Single Tower Deliquescent Desiccant Type

1. Inlet Conditions
   Flow - scfm
   Operating Pressure - psig
   Inlet Air Temperature °F
2. Ambient Conditions
   Ambient Temperature range - °F
3. Performance
   Outlet Dew Point - °F
   Pressure Drop Across Dryer (maximum allowable) - psid
4. Design
   Pressure - psig
   Temperature - °F
   Ambient Temperature - °F
      Maximum
      Minimum

Membrane Type

1. Inlet conditions
    Flow - scfm
    Operating Pressure - psig
    Inlet Air Temperature (saturated) - °F
2. Performance
    Outlet Flow (Inlet Flow less Sweep/Purge Air Requirement) - scfm
    Pressure Dew Point at Dryer Outlet - °F
    Pressure Drop Across Dryer - psid
3. Design
    Pressure - psig
    Temperature - °F
    Ambient Temperature - °F - Maximum_____ Minimum_____

COMPRESSED AIR FILTERS

Particulate matter normally refers to solid particles in the air and it has been estimated that there are as many as 4 million particles in one cubic foot of atmospheric air. When compressed to 103 psig the concentration becomes over 30 million. Over 80% of these are below 2 microns.

One micron = One millionth of a meter
or = 0.04 thousandths of an inch

The inlet filter of a typical air compressor has a rating of about 10 microns and is designed for the protection of the air compressor and not any downstream equipment. In addition, wear particles from reciprocating compressors and deposits from degradation of oils exposed to the heat of compression can add to downstream contamination.

The main mechanisms of mechanical filtration are Direct Interception, Inertial Impaction and Diffusion. These also may be enhanced by Electrostatic Attraction.

Direct Interception occurs when a particle collides with a fiber of the filter medium without deviating out of the streamline flow. Usually this occurs on the surface of the filter element and affecting mainly the larger sized particles (usually over 1 micron).

Inertial Impaction occurs when a particle, traveling in the air stream through the maze of fibers in the filter element, is unable to stay in the streamline flow and collides with a fiber and adheres to it. Usually this occurs with particles in the range from 0.3 to 1.0 microns.

Diffusion (or Brownian Movement) occurs with the smallest particles, below 0.3 microns. These tend to wander through the filter element within the air stream, with increased probability of colliding with a filter fiber and adhering to it.

Particulate Filters

Particulate filter designs are such that some overlap occurs with the different mechanisms and the desired degree of contaminant removal. A higher degree of contaminant removal than is necessary will result in a higher pressure drop across the filter, requiring a higher pressure from the air compressor and additional energy costs. Particulate filters have a pressure drop rating when new but the actual pressure drop will increase with continued use until the pressure drop reaches the level at which an element change is required. This results in a reduced downstream pressure and/or the need to increase the compressor discharge pressure to maintain the required pressure at end-use applications.

A particulate filter is recommended downstream of the air dryer, before any operational equipment or process.

Coalescing Filters

Small droplets of moisture or oil adhere to the filter medium and coalesce into larger liquid droplets. Flow through the filter element is from the inside to the outside where the larger diameter allows a lower exit velocity. An anti re-entrainment barrier normally is provided to prevent droplets from being re-introduced to the air stream. The cellular structure allows the coalesced liquid to run down by gravity to the bottom of the filter bowl from which it can be drained, usually by means of an automatic drain. The liquid may contain both oil and water.

The coalescing action should not result in any increase in pressure drop over the life of the filter. Pressure drop increase normally is due to the accumulation of particulate matter if the coalescing filter is not preceded by an adequate particulate filter. The normal pressure drop should be the "wet" pressure drop after the element by design has become saturated. The "dry" pressure drop before the element is properly wetted will be lower.

A coalescing filter is recommended before any dryer whose drying medium may be damaged by oil. The term oil includes petroleum based and synthetic hydrocarbons plus other synthetic oils such as di-esters which can affect materials such as acrylics in downstream equipment or processes.

Adsorption Type Filters

Particulate and coalescing type filters are capable of removing extremely small solid or liquid particles down to 0.01 microns but not oil vapors or odors. Adsorption is the attraction and adhesion of gaseous and liquid molecules to the surface of a solid. Normally the filter elements contain activated carbon granules which have an extremely high surface area and dwell time. The activated carbon medium is for the adsorption of vapors only. An adsorption filter must be protected by an upstream coalescing type filter to prevent gross contamination by liquid oil.

With the combination of all three types of filters downstream of a dryer, it is possible to obtain an air quality better than the atmospheric air entering the air compressor.

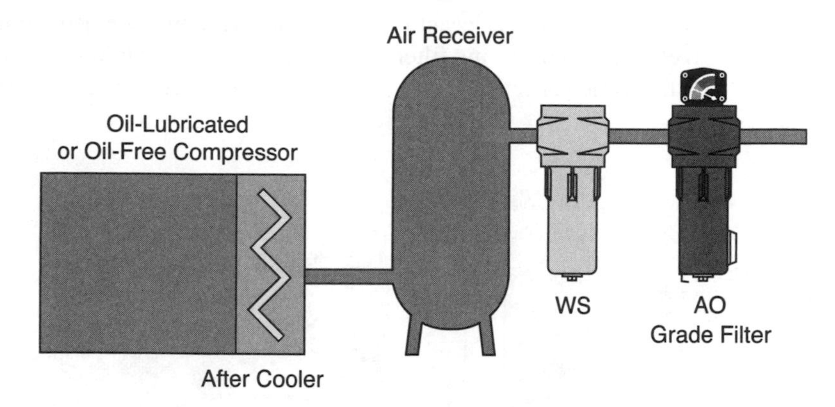

Figure 3.13

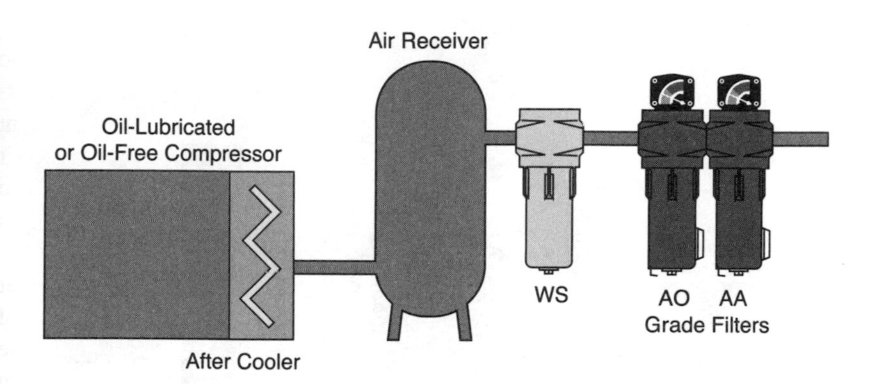

Figure 3.14

Dryer and Filter Arrangements

In the Air Quality Classes of ISO 8573-1, (see Tables 3.1A, B and C) the first class concerns particulate content, the second moisture content and the third hydrocarbons.

A general purpose coalescing filter capable of removing particles down to 1 micron, and liquids down to 0.5 ppm (rated at 70°F), placed after an air aftercooler and moisture separator (Fig.3.13), will meet the requirement of Classes 1.-.3. That is Class 1 particulate, no rating for moisture and Class 3 for hydrocarbons.

The same general purpose coalescing filter used as a pre-filter followed by a high efficiency coalescing filter (Fig. 3.14) to remove liquid particles down to 0.01 microns, would meet Classes 1.-.2.

The same type of filter normally used in conjunction with a refrigerant type dryer (Fig.3.15) will meet Classes 1.4.1.

In the case of a regenerative desiccant type dryer having a pressure dew point of -40°F, the high efficiency coalescing filter placed before the dryer to protect the desiccant bed and the same particulate filter placed after the dryer (Fig. 3.16) will meet Classes 1.2.2.

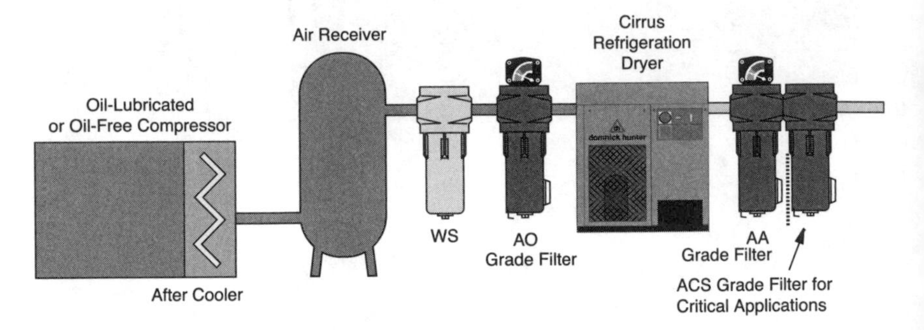

Figure 3.15

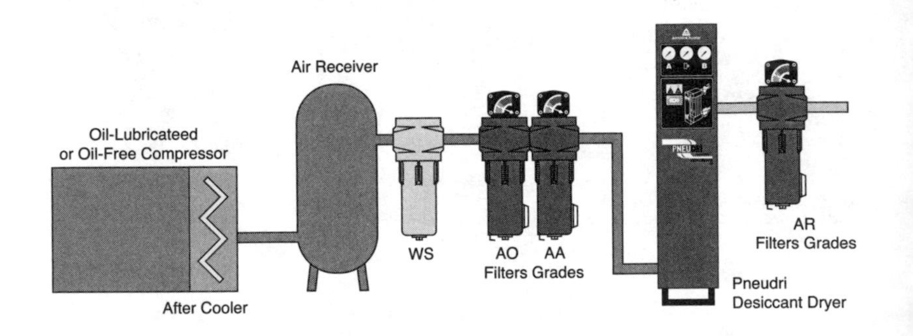

Figure 3.16

The combination of this latest arrangement followed by an activated carbon filter capable of removing oil content to 0.003 ppm, will then meet Classes 1.2.1 which may be claimed as "oil free," having less hydrocarbon content than a normal industrial atmosphere. If the regenerative desiccant type dryer had a pressure dew point rating of -100°F, the combination then would meet Classes 1.1.1. Substituting a refrigerant type dryer would still meet the "oil free" condition but with a higher (35 to 38°F) pressure dew point, or Classes 1.4.1.

Oil-Free Air

Atmospheric air, particularly in industrial environments, contains condensible hydrocarbons from incompletely burned fuels exhausted by engines, heaters and other sources. It has been estimated that these can range from 0.05 to 0.25 ppm. Aerosols also atomize down to 0.8 to 0.01 microns. An oil-free air compressor does not introduce oil into the compression chamber but the atmospheric air entering the compressor contains these atmospheric pollutants to a lesser or greater extent. For this reason, oil-free compressors also require adequate drying and filtration after the compression process to meet Class 1.2.1 or 1.1.1 air quality. Additional treatment is required to meet Breathing or Medical Air requirements as listed under Breathing Air.

4

Compressed Air Distribution (Systems)

COMPRESSED AIR DISTRIBUTION SYSTEMS

When a compressed air distribution system is properly designed, installed, operated and maintained, it is a major source of industrial power, possessing many inherent advantages. Compressed air is safe, economical, adaptable and easily transmitted and provides labor saving power. The cost of a complete compressed air system and pneumatic tools is relatively small in comparison with the savings effected by their use.

Object of the Compressed Air Distribution System

The primary object of a compressed air distribution system is to transport the compressed air from its point of production (compressors) to its points of use (applications) in sufficient quantity and quality and at adequate pressure for efficient operation of air tools and other pneumatic devices. However, many other considerations come into the design of the system to ensure the efficiency and safety of the total system. These will be discussed in this chapter. These include:

- Air volume flow rate
- Air pressure requirements
- Type(s) and number of compressors
- Air quality
- Air system efficiency
- Air system safety
- Air system layout
- Air volume flow rate requirements

Air Volume Flow Rate Requirements

The proper capacity to install is a vital and basic question and often misunderstood. The capacity rating of air compressors generally is published in terms of "free air," which is at atmospheric conditions of pressure, temperature and relative humidity and not at the pressure, temperature and relative humidity required at the air tool or pneumatic device to be operated.

The Applications chapter of this book contains many illustrations of current uses of compressed air power. The air tools chapter also provides much useful information on applications of air powered tools and other pneumatic devices.

A study of air-operated devices in a typical manufacturing plant will show that some of these devices operate almost constantly while others operate infrequently but may require a relatively large volume of air while in use. It also will be found that the amount of air actually used by the individual devices will vary considerably in different applications. The total air requirement therefore should not be the total of the individual maximum requirement but the sum of the average air consumption of each. Sufficient controlled storage capacity of compressed air also is essential to meet short-term high volume demands.

Recommendations for efficient components for the compressed air system have been discussed in earlier chapters. This chapter deals with the compressed air distribution system which feeds the production operation. Proper design of the distribution system is essential to avoid energy waste and to ensure proper use of all pneumatic devices.

Determination of the average air consumption is facilitated by the use of the concept of load factor. Pneumatic devices generally are operated only intermittently and often are operated at less than full load capacity. The ratio of actual air consumption to the maximum continuous full load air consumption, each measured in cubic feet per minute of free air, is known as the load factor. It is essential that the best possible determination or estimate of load factor be used in arriving at the plant capacity needed.

Two items are involved in the load factor. The first is the time factor, which is the percentage of work time during which a device actually is in use. The second is the work factor, which is the percentage of the air required for maximum possible output of work per minute that is required for the work actually being performed by the device. For example, the air consumption of a grinder with full open throttle varies considerably, depending on how hard the operator applies the grinding wheel against the work piece. The work factor also is affected by the system operating pressure. For example, a system pressure of 125 psig will provide a work factor 22% higher than a system pressure of 100 psig. (See Table 4.10). The work factor therefore is the ratio (expressed as a percentage) of the air consumption under actual conditions of operation, to the air consumption when the tool is fully loaded. The load factor is the product of the time factor and the work factor. In one plant studied, the air actually consumed by 434 portable air tools on production work was only 15% of the total rated full time air consumption of all the tools.

In designing an entirely new compressed air distribution system, it is highly desirable to utilize experience with a similar plant. The established load factor can be used as the basis of a good estimate for the new system. A log of pressures throughout an existing facility will reveal trends, including peaks and lulls in demand and potential irregularities to be avoided in the new system. Another source of this type of information is the manufacturer of the air tools and pneumatic devices involved.

Table 4.1, shows the maximum air requirements of various tools and can be used for preliminary estimates. These figures are approximate and individual tools from different manufacturers may vary by more than 10% from the figures given. Since load factor may vary considerably from one plant to another, any general figures should be used with caution. For example, one manufacturer states that the compressor capacity should be about one third of the requirement of all the pneumatic tools. See Table 4.2. It is recommended that the manufacturer of each air tool, device or machine, be consulted as to recommended requirements. Table 4.1 should not be used for constant demand applications, including sandblasting requirements shown in Table 4.2.

Table 4.1 Air Requirements of Various Tools

Tool	Free Air, cfm at 90 psig, 100% Load Factor
Grinders, 6" and 8" wheels	50
Grinders, 2" and 2 1/2" wheels	14-20
File and burr machines	18
Rotary sanders, 9" pads	53
Rotary sanders, 7" pads	30
Sand rammers and tampers,	
1" x 4" cylinder	25
1 1/4" x 5" cylinder	28
1 1/2" x 6" cylinder	39
Chipping hammers, weighing 10-13 lb	28-30
Heavy	39
Weighing 2-4 lb	12
Nut setters to 5/16" weighing 8 lb	20
Nut setters 1/2" to 3/4" weighing 18 lb	30
Sump pumps, 145 gal (a 50-ft head)	70
Paint spray, average	7
Varies from	2-20
Bushing tools (monument)	15-25
Carving tools (monument)	10-15
Plug drills	40-50
Riveters, 3/32"-1" rivets	12
Larger weighing 18-22 lb	35
Rivet busters	35-39
Wood borers to 1" diameter weighing 4 lb	40
2" diameter weighing 26 lb	80
Steel drills, rotary motors	
Capacity up to 1/4" weighing 1 1/4-4 lb	18-20
Capacity 1/4" to 3/8" weighing 6-8 lb	20-40
Capacity 1/2" to 3/4" weighing 9-14 lb	70
Capacity 7/8" to 1" weighing 25 lb	80
Capacity 1 1/4" weighing 30 lb	95
Steel drills, piston type	
Capacity 1/2" to 3/4" weighing 13-15 lb	45
Capacity 7/8" to 1 1/4" weighing 25-30 lb	75-80
Capacity 1 1/4" to 2" weighing 40-50 lb	80-90
Capacity 2" to 3" weighing 55-75 lb	100-110

Table 4.2 Cubic Feet of Air Per Minute Required By Sandblast

	Compressed Air Gage Pressure (psig)			
Nozzle Diameter	60	70	80	100
1/16"	4	5	5.5	6.5
3/32"	9	11	12	15
1/8"	17	19	21	26
3/16"	38	43	47	58
1/4"	67	76	85	103
5/16"	105	119	133	161
3/8"	151	171	191	232
1/2"	268	304	340	412

For tools used regularly on one operation, a study of active and inactive times may be made. Judgement may be exercised at this time as to the work factor to be applied if other than unity. If air requirements of a manufacturing process are evaluated on the basis of unit production in cubic feet of free air per piece produced, they may then be combined on the basis of total production to arrive at the average volume rate of air required.

Many pieces of production equipment are actuated by pneumatic cylinders. These include automatic feed devices, chucks, vises, clamps, presses, intermittent motion devices, both reciprocating and rotary, door openers and many other devices. Such devices usually have low air consumption and are themselves inexpensive. They find increasing use in automated production processes. Air consumption for such cylinders is shown in Table 4.3. This table shows only the theoretical volume swept out by the piston during one full stroke, which must be converted into a flow rate of free air. Many cylinders contain air cushioning chambers which increase the volume somewhat over the tabled figures. In addition, in actual use the air pressure to the cylinder may be throttled to a pressure considerably below the system line pressure. If a limit switch cuts off the air supply when a certain force is exerted by the cylinder, the corresponding pressure should be calculated and used rather than full line pressure in converting the tabled figures to free air conditions. In many applications the full available piston stroke is not needed. In fact, a reduced length of stroke may be an advantage in reducing operating time. The air consumption for such cases is calculated using only the actual stroke.

Table 4.3 Volume of Compressed Air in Cubic Feet Required per Stroke to Operate Air Cylinder

Piston Diameter in Inches	Length of Stroke in Inches*											
	1	2	3	4	5	6	7	8	9	10	11	12
1 1/4	.00139	.00278	.00416	.00555	.00694	.00832	.00972	.0111	.0125	.0139	.0153	.01665
1 7/8	.00158	.00316	.00474	.00632	.0079	.00948	.01105	.01262	.0142	.0158	.0174	.01895
2	.00182	.00364	.00545	.00727	.0091	.0109	.0127	.0145	.01636	.0182	.020	.0218
2 1/8	.00205	.0041	.00615	.0082	.0103	.0123	.0144	.0164	.0185	.0205	.0226	.0244
2 1/4	.0023	.0046	.0069	.0092	.0115	.0138	.0161	.0184	.0207	.0230	.0253	.0276
2 3/8	.00256	.00512	.00768	.01025	.0128	.01535	.01792	.02044	.0230	.0256	.0282	.0308
2 1/2	.00284	.00568	.00852	.01137	.0142	.0171	.0199	.0228	.0256	.0284	.0312	.0343
2 5/8	.00313	.00626	.0094	.01254	.01568	.0188	.0219	.0251	.0282	.0313	.0345	.0376
2 3/4	.00343	.00686	.0106	.0137	.0171	.0206	.0240	.0272	.0308	.0343	.0378	.0412
2 7/8	.00376	.00752	.0113	.01503	.01877	.0226	.0263	.0301	.0338	.0376	.0413	.045
3	.00409	.00818	.0123	.0164	.0204	.0246	.0286	.0327	.0368	.0409	.0450	.049
3 1/8	.00443	.00886	.0133	.0177	.0222	.0266	.0310	.0354	.0399	.0443	.0488	.0532
3 1/4	.0048	.0096	.0144	.0192	.024	.0288	.0336	.0384	.0432	.0480	.0529	.0575
3 3/8	.00518	.01036	.0155	.0207	.0259	.031	.0362	.0415	.0465	.0518	.037	.062
3 1/2	.00555	.0112	.0167	.0222	.0278	.0333	.0389	.0445	.050	.0556	.061	.0644
3 5/8	.00595	.0119	.0179	.0238	.0298	.0357	.0416	.0477	.0536	.0595	.0655	.0715
3 3/4	.0064	.0128	.0192	.0256	.032	.0384	.0447	.0512	.0575	.064	.0702	.0766
3 7/8	.0068	.01362	.0205	.0273	.0341	.041	.0477	.0545	.0614	.068	.075	.082
4	.00725	.0145	.0218	.029	.0363	.0435	.0508	.058	.0653	.0725	.0798	.087
4 1/8	.00773	.01547	.0232	.0309	.0386	.0464	.0541	.0618	.0695	.0773	.0851	.092
4 1/4	.0082	.0164	.0246	.0328	.041	.0492	.0574	.0655	.0738	.082	.0903	.0985
4 3/8	.0087	.0174	.0261	.0348	.0435	.0522	.0608	.0694	.0782	.087	.0958	.1042
4 1/2	.0092	.0184	.0276	.0368	.046	.0552	.0643	.0735	.0828	.092	.101	.1105
4 5/8	.0097	.0194	.0291	.0388	.0485	.0582	.0679	.0775	.0873	.097	.1068	.1163
4 3/4	.01025	.0205	.0308	.041	.0512	.0615	.0717	.0818	.0922	.1025	.1125	.123
4 7/8	.0108	.0216	.0324	.0431	.054	.0647	.0755	.0862	.097	.108	.1185	.1295
5	.0114	.0228	.0341	.0455	.0568	.0681	.0795	.091	.1023	.114	.125	.136
5 1/8	.01193	.0239	.0358	.0479	.0598	.0716	.0837	.0955	.1073	.1193	.1315	.1435
5 1/4	.0125	.0251	.0376	.0502	.0627	.0753	.0878	.100	.1128	.125	.138	.151
5 3/8	.0131	.0263	.0394	.0525	.0656	.0788	.092	.105	.118	.131	.144	.158
5 1/2	.01375	.0275	.0412	.055	.0687	.0825	.0962	.110	.1235	.1375	.151	.165
5 5/8	.0144	.0288	.0432	.0575	.072	.0865	.101	.115	.1295	.144	.1585	.173
5 3/4	.015	.030	.045	.060	.075	.090	.105	.120	.135	.150	.165	.180
5 7/8	.0157	.0314	.047	.0628	.0785	.094	.110	.1254	.142	.157	.1725	.188
6	.0164	.032	.0492	.0655	.082	.0983	.1145	.131	.147	.164	.180	.197

Air turbines may be used for starting gas turbines and for other purposes. Air consumption of turbines may be calculated by the usual methods of thermodynamics. For single-stage impulse turbines with converging nozzles, the air consumption may be found by applying Fliegner's equation to the nozzles. The air turbine manufacturer can supply the needed data.

Other devices have an air flow condition approximating simple throttling. A steady jet used for blowing chips from a tool would fall within this classification. Another device with approximately the same flow characteristics is the vibrator actuated by a steel ball propelled around a closed circular track by means of an air jet. Table 4.4 may be used for estimating air flow through such devices. These data are not intended for use in air measurements and should be used only for estimating system air requirements.

Table 4.4 How to Determine Compressor Size Required

				CFM required*		
		Number of Tools	Load Factor (per cent of time tools actually operated)	Per Tool When Operating	Total if All Tools Operated Simultaneously	Total Actually Used (A x B x C ÷ 100)
Type of Tool	Location	(A)	(B)	(C)	(D)	(E)
Blowguns, chucks and vises	Machine Shop	4	25	25	100	25
8-in. grinders	Cleaning	10	50	50	500	250
Chippers	Cleaning	10	50	30	300	150
Hoists	Cleaning	2	10	35	70	7
Small screwdrivers	Assembly	20	25	12	240	60
Large nutsetters	Assembly	2	25	30	60	15
Woodborer	Shipping	1	25	30	30	7 1/2
Screwdriver	Shipping	1	20	30	30	6
Hoist	Shipping	1	20	40	40	8
Total		47			1270	528 1/2

*Cfm is cubic feet of free air per minute.
Note: Total of column (E) determines required compressor sizes.

Regenerative desiccant type compressed air dryers require purge air which may be as much as 15% of the rated dryer capacity and this must be added to the estimate of air required at points of use.

An often neglected consideration is system leaks. Theoretically, a new system should have no leaks but experience shows that most systems have varying amounts of leakage sources.

Electronic leak detectors are available and should be used on a regularly scheduled basis. It also can be useful to determine how long an air compressor runs to maintain system pressure during a shutdown period when there is no actual usage of compressed air.

Air Pressure Requirements

This is one of the more critical factors in the design of an efficient compressed air distribution system. One problem is that the variety of points of application may require a variety of operating pressure requirements. Equipment manufacturers should be consulted to determine the pressure requirement at the machine, air tool or pneumatic device. If these operating pressure requirements vary by more than 20%, consideration should be given to separate systems. In a typical plant with an air distribution system operating at a nominal 100 psig, an increase of one half per cent in the air compressor energy costs is required for each additional 1 psi in system pressure. Operating the complete system at 20% higher pressure to accommodate one point of use, would result in the air compressor(s) using 10% more energy and an increase in work factor as previously noted. This, obviously, is to be avoided.

Allowance also must be made for pressure drops through compressed air treatment equipment, including air dryers and filters.

An inadequately sized piping distribution system will cause excessive pressure drops between the air compressors and the points of use, requiring the compressor to operate at a much higher pressure than at the points of use. This also requires additional energy. For example, if the distribution piping size is only half of the ideal, the cross-sectional area is only one fourth, resulting in velocities four times the ideal and sixteen times the pressure drop. In an air distribution system where a given pipe diameter piping may be sufficient, it should be remembered that the installation labor cost will be the same for double the pipe diameter and only the material cost will increase. The savings in energy costs from reduced pressure drop will repay the difference in material costs in a very short time. and could provide for future capacity.

Air velocity through the distribution piping should not exceed 1800 ft. per minute (30 ft. per sec.). One recommendation, to avoid moisture being carried beyond drainage drop legs in compressor room header upstream of dryer(s), is that the velocity should not exceed 1200 ft. per minute (20 ft. per sec.). Branch lines having an air velocity over 2000 ft. per minute, should not exceed 50 ft. in length. The system should be designed so that the operating pressure drop between the air compressor and the point(s) of use should not exceed 10% of the compressor discharge pressure.

Pressure loss in piping due to friction at various operational pressures is tabulated in Tables 4.5, 4.6, 4.7, 4.8 and 4.9 can be used to determine pipe sizes required for the system being designed. These tables are based upon non-pulsating flow in a clean, smooth pipe.

Table 4.5 Loss of Air Pressure Due to Friction

Cu ft Free Air Per Min	Equivalent Cu ft Compressed Air Per Min	Nominal Diameter, In.											
		1/2	3/4	1	1 1/4	1 1/2	2	3	4	6	8	10	12
10	1.96	10.0	1.53	0.43	0.10								
20	3.94	39.7	5.99	1.71	0.39	0.18							
30	5.89	....	13.85	3.86	0.88	0.40							
40	7.86	....	24.7	6.85	1.59	0.71	0.19						
50	9.84	....	38.6	10.7	2.48	1.10	0.30						
60	11.81	....	55.5	15.4	3.58	1.57	0.43						
70	13.75	....	....	21.0	4.87	2.15	0.57						
80	15.72	....	....	27.4	6.37	2.82	0.75						
90	17.65	....	....	34.7	8.05	3.57	0.57	0.37					
100	19.60	....	....	42.8	9.95	4.40	1.18						
125	19.4	....	....	46.2	12.4	6.90	1.83	0.14					
150	29.45	....	....	....	22.4	9.90	2.64	0.32					
175	34.44	....	....	....	30.8	13.40	3.64	0.43					
200	39.40	....	....	....	39.7	17.60	4.71	0.57					
250	49.20	....	....	....	....	27.5	7.37	0.89	0.21				
300	58.90	....	....	....	....	39.6	10.55	1.30	0.31				
350	68.8	....	....	....	....	54.0	14.4	1.76	0.42				
400	78.8	....	....	....	....	....	18.6	2.30	0.53				
450	88.4	....	....	....	....	....	23.7	2.90	0.70				
500	98.4	....	....	....	....	....	29.7	3.60	0.85				
600	118.1	....	....	....	....	....	42.3	5.17	1.22				
700	137.5	....	....	....	....	....	57.8	7.00	1.67				
800	157.2	....	....	....	....	....	....	9.16	2.18				
900	176.5	....	....	....	....	....	....	11.6	2.76				
1,000	196.0	....	....	....	....	....	....	14.3	3.40				
1,500	294.5	....	....	....	....	....	....	32.3	7.6	0.87	0.29		
2,000	394.0	....	....	....	....	....	....	57.5	13.6	1.53	0.36		
2,500	492	....	....	....	....	....	....	....	21.3	2.42	0.57	0.17	
3,000	589	....	....	....	....	....	....	....	30.7	3.48	0.81	0.24	
3,500	688	....	....	....	....	....	....	....	41.7	4.68	1.07	0.33	
4,000	788	....	....	....	....	....	....	....	54.5	6.17	1.44	0.44	
4,500	884	....	....	....	....	....	....	....	....	7.8	1.83	0.55	0.21
5,000	984	....	....	....	....	....	....	....	....	9.7	2.26	0.67	0.27
6,000	1,181	....	....	....	....	....	....	....	....	13.9	3.25	0.98	0.38
7,000	1,375	....	....	....	....	....	....	....	....	18.7	4.43	1.34	0.51
8,000	1,572	....	....	....	....	....	....	....	....	24.7	5.80	1.73	0.71
9,000	1,765	....	....	....	....	....	....	....	....	31.3	7.33	2.20	0.87
10,000	1,960	....	....	....	....	....	....	....	....	38.6	9.05	2.72	1.06
11,000	2,165	....	....	....	....	....	....	....	....	46.7	10.9	3.29	1.28
12,000	2,362	....	....	....	....	....	....	....	....	55.5	13.0	3.90	1.51
13,000	2,560	....	....	....	....	....	....	....	....	....	15.2	4.58	1.77
14,000	2,750	....	....	....	....	....	....	....	....	....	17.7	5.32	2.07
15,000	2,945	....	....	....	....	....	....	....	....	....	20.3	6.10	2.36
16,000	3,144	....	....	....	....	....	....	....	....	....	23.1	6.95	2.70
18,000	3,530	....	....	....	....	....	....	....	....	....	29.2	8.80	3.42
20,000	3,940	....	....	....	....	....	....	....	....	....	36.2	10.8	4.22
22,000	4,330	....	....	....	....	....	....	....	....	....	43.7	13.2	5.12
24,000	4,724	....	....	....	....	....	....	....	....	....	51.9	15.6	5.92
26,000	5,120	....	....	....	....	....	....	....	....	....	....	18.3	7.15
28,000	5,500	....	....	....	....	....	....	....	....	....	....	21.3	8.3
30,000	5,890	....	....	....	....	....	....	....	....	....	....	24.4	9.5

In psi in 1000 ft of pipe, 60 lb gage initial pressure. For longer or shorter lengths of pipe the friction loss is proportional to the length, i.e., for 500 ft, one-half of the above; for 4,000 ft, four times the above, etc.

Table 4.6 Loss of Air Pressure Due to Friction

Cu ft Free Air Per Min	Equivalent Cu ft Compressed Air Per Min	Nominal Diameter, In.											
		1/2	3/4	1	1 1/4	1 1/2	2	3	4	6	8	10	12
10	1.55	7.90	1.21	0.34									
20	3.10	31.4	4.72	1.35	0.31								
30	4.65	70.8	10.9	3.04	0.69	0.31							
40	6.20	....	19.5	5.40	1.25	0.56							
50	7.74	....	30.5	8.45	1.96	0.87							
60	9.29	....	43.8	12.16	2.82	1.24	0.34						
70	10.82	....	59.8	16.6	3.84	1.70	0.45						
80	12.40	....	78.2	21.6	5.03	2.22	0.59						
90	13.95	....	....	27.4	6.35	2.82	0.75						
100	15.5	....	....	33.8	7.85	3.74	0.93						
125	19.4	....	....	46.2	12.4	5.45	1.44						
150	23.2	....	....	76.2	17.7	7.82	2.08						
175	27.2	....	....	....	24.8	10.6	2.87						
200	31.0	....	....	....	31.4	13.9	3.72	0.45					
250	38.7	....	....	....	49.0	21.7	5.82	0.70					
300	46.5	....	....	....	70.6	31.2	8.35	1.03					
350	54.2	....	....	....	....	42.5	11.4	1.39	0.33				
400	62.0	....	....	....	....	55.5	14.7	1.82	0.42				
450	69.7	....	....	....	....	....	18.7	2.29	0.55				
500	77.4	....	....	....	....	....	23.3	2.84	0.67				
600	92.9	....	....	....	....	....	33.4	4.08	0.96				
700	108.2	....	....	....	....	....	45.7	5.52	1.32				
800	124.0	....	....	....	....	....	59.3	7.15	1.72				
900	139.5	....	....	....	....	....	....	9.17	2.18				
1,000	155	....	....	....	....	....	....	11.3	2.68				
1,500	232	....	....	....	....	....	....	25.5	6.0	0.69			
2,000	310	....	....	....	....	....	....	45.3	10.7	1.21	0.29		
2,500	387	....	....	....	....	....	....	70.9	16.8	1.91	0.45		
3,000	465	....	....	....	....	....	....	....	24.2	2.74	0.64	0.19	
3,500	542	....	....	....	....	....	....	....	32.8	3.70	0.85	0.26	
4,000	620	....	....	....	....	....	....	....	43.0	4.87	1.14	0.34	
4,500	697	....	....	....	....	....	....	....	54.8	6.15	1.44	0.43	
5,000	774	....	....	....	....	....	....	....	67.4	7.65	1.78	0.53	0.21
6,000	929	....	....	....	....	....	....	....	....	11.0	2.57	0.77	0.29
7,000	1,082	....	....	....	....	....	....	....	....	14.8	3.40	1.06	0.40
8,000	1,240	....	....	....	....	....	....	....	....	19.5	4.57	1.36	0.54
9,000	1,395	....	....	....	....	....	....	....	....	24.7	5.78	1.74	0.69
10,000	1,550	....	....	....	....	....	....	....	....	30.5	7.15	2.14	0.84
11,000	1,710	....	....	....	....	....	....	....	....	36.8	8.61	2.60	1.01
12,000	1,860	....	....	....	....	....	....	....	....	43.8	10.3	3.08	1.19
13,000	2,020	....	....	....	....	....	....	....	....	51.7	12.0	3.62	1.40
14,000	2,170	....	....	....	....	....	....	....	....	60.2	14.0	4.20	1.63
15,000	2,320	....	....	....	....	....	....	....	....	68.5	16.0	4.82	1.84
16,000	2,480	....	....	....	....	....	....	....	....	78.2	18.2	5.48	2.13
18,000	2,790	....	....	....	....	....	....	....	....	....	23.0	6.95	2.70
20,000	3,100	....	....	....	....	....	....	....	....	....	28.6	8.55	3.33
22,000	3,410	....	....	....	....	....	....	....	....	....	34.5	10.4	4.04
24,000	3,720	....	....	....	....	....	....	....	....	....	41.0	12.3	4.69
26,000	4,030	....	....	....	....	....	....	....	....	....	48.2	14.4	5.6
28,000	4,350	....	....	....	....	....	....	....	....	....	55.9	16.8	6.3
30,000	4,650	....	....	....	....	....	....	....	....	....	64.2	19.3	7.5

In psi in 1000 ft of pipe, 80 lb gage initial pressure. For longer or shorter lengths of pipe the friction loss is proportional to the length, i.e., for 500 ft, one-half of the above; for 4,000 ft, four times the above, etc.

Table 4.7 Loss of Air Pressure Due to Friction

Cu ft Free Air Per Min	Equivalent Cu ft Compressed Air Per Min	Nominal Diameter, In.											
		1/2	3/4	1	1 1/4	1 1/2	2	3	4	6	8	10	12
10	1.28	6.50	.99	0.28									
20	2.56	25.9	3.90	1.11	0.25	0.11							
30	3.84	58.5	9.01	2.51	0.57	0.26							
40	5.12	....	16.0	4.45	1.03	0.46							
50	6.41	....	25.1	9.96	1.61	0.71	0.19						
60	7.68	....	36.2	10.0	2.32	1.02	0.28						
70	8.96	....	49.3	13.7	3.16	1.40	0.37						
80	10.24	....	64.5.	17.8	4.14	1.83	0.49						
90	11.52	....	82.8	22.6	5.23	2.32	0.62						
100	12.81	....	....	27.9	6.47	2.86	0.77						
125	15.82	....	....	48.6	10.2	4.49	1.19						
150	19.23	....	....	62.8	14.6	6.43	1.72	0.21					
175	22.40	....	....	....	19.8	8.72	2.36	0.28					
200	25.62	....	....	....	25.9	11.4	3.06	0.37					
250	31.64	....	....	....	40.4	17.9	4.78	0.58					
300	38.44	....	....	....	58.2	25.8	6.85	0.84	0.20				
350	44.80	....	....	....	....	35.1	9.36	1.14	0.27				
400	51.24	....	....	....	....	45.8	12.1	1.50	0.35				
450	57.65	....	....	....	....	58.0	15.4	1.89	0.46				
500	63.28	....	....	....	....	71.6	19.2	2.34	0.55				
600	76.88	....	....	....	....	....	27.6	3.36	0.79				
700	89.60	....	....	....	....	....	37.7	4.55	1.09				
800	102.5	....	....	....	....	....	49.0	5.89	1.42				
900	115.3	....	....	....	....	....	62.3	7.6	1.80				
1,000	128.1	....	....	....	....	....	76.9	9.3	2.21				
1,500	192.3	....	....	....	....	....	....	21.0	4.9	0.57			
2,000	256.2	....	....	....	....	....	....	37.4	8.8	0.99	0.24		
2,500	316.4	....	....	....	....	....	....	58.4	13.8	1.57	0.37		
3,000	384.6	....	....	....	....	....	....	84.1	20.0	2.26	0.53		
3,500	447.8	....	....	....	....	....	....	....	27.2	3.04	0.70	0.22	
4,000	512.4	....	....	....	....	....	....	....	35.5	4.01	0.94	0.28	
4,500	576.5	....	....	....	....	....	....	....	45.0	5.10	1.19	0.36	
5,000	632.8	....	....	....	....	....	....	....	55.6	6.3	1.47	0.44	0.17
6,000	768.8	....	....	....	....	....	....	....	80.0	9.1	2.11	0.64	0.24
7,000	896.0	....	....	....	....	....	....	....	....	12.2	2.88	0.87	0.33
8,000	1,025	....	....	....	....	....	....	....	....	16.1	3.77	1.12	0.46
9,000	1,153	....	....	....	....	....	....	....	....	20.4	4.77	1.43	0.57
10,000	1,280	....	....	....	....	....	....	....	....	25.1	5.88	1.77	0.69
11,000	1,410	....	....	....	....	....	....	....	....	30.4	7.10	2.14	0.83
12,000	1,540	....	....	....	....	....	....	....	....	36.2	8.5	2.54	0.98
13,000	1,668	....	....	....	....	....	....	....	....	42.6	9.8	2.98	1.15
14,000	1,795	....	....	....	....	....	....	....	....	49.2	11.5	3.46	1.35
15,000	1,923	....	....	....	....	....	....	....	....	56.6	13.2	3.97	1.53
16,000	2,050	....	....	....	....	....	....	....	....	64.5	15.0	4.52	1.75
18,000	2,310	....	....	....	....	....	....	....	....	81.5	19.0	5.72	2.22
20,000	2,560	....	....	....	....	....	....	....	....	....	23.6	7.0	2.74
22,000	2,820	....	....	....	....	....	....	....	....	....	28.5	8.5	3.33
24,000	3,080	....	....	....	....	....	....	....	....	....	33.8	10.0	3.85
26,000	3,338	....	....	....	....	....	....	....	....	....	39.7	11.9	4.65
28,000	3,590	....	....	....	....	....	....	....	....	....	46.2	13.8	5.40
30,000	3,850	....	....	....	....	....	....	....	....	....	53.0	15.9	6.17

In psi in 1000 ft of pipe, 100 lb gage initial pressure. For longer or shorter lengths of pipe the friction loss is proportional to the length, i.e., for 500 ft. one-half of the above; for 4,000 ft, four times the above, etc.

Table 4.8 Loss of Air Pressure Due to Friction

Cu ft Free Air Per Min	Equivalent Cu ft Compressed Air Per Min	Nominal Diameter, In.											
		1/2	3/4	1	1 1/4	1 1/2	2	3	4	6	8	10	12
10	1.05	5.35	0.82	0.23									
20	2.11	21.3	3.21	0.92	0.21								
30	3.16	48.0	7.42	2.07	0.47	0.21							
40	4.21	....	13.2	3.67	0.85	0.38							
50	5.26	....	20.6	5.72	1.33	0.59							
60	6.32	....	29.7	8.25	1.86	0.84	0.23						
70	7.38	....	40.5	11.2	2.61	1.15	0.31						
80	8.42	....	53.0	14.7	3.41	1.51	0.40						
90	9.47	....	68.0	18.6	4.30	1.91	0.51						
100	10.50	....	....	22.9	5.32	2.36	0.63						
125	13.15	....	....	39.9	8.4	3.70	0.98						
150	15.79	....	....	51.6	12.0	5.30	1.41	0.17					
175	18.41	....	....	....	16.3	7.2	1.95	0.24					
200	21.05	....	....	....	21.3	9.4	2.52	0.31					
250	26.30	....	....	....	33.2	14.7	3.94	0.48					
300	31.60	....	....	....	47.3	21.2	5.62	0.70					
350	36.80	....	....	....	....	28.8	7.7	0.94	0.22				
400	42.10	....	....	....	....	37.6	10.0	1.23	0.28				
450	47.30	....	....	....	....	47.7	12.7	1.55	0.37				
500	52.60	....	....	....	....	58.8	15.7	1.93	0.46				
600	63.20	....	....	....	....	....	22.6	2.76	0.65				
700	73.80	....	....	....	....	....	30.0	3.74	0.89				
800	84.20	....	....	....	....	....	40.2	4.85	1.17				
900	94.70	....	....	....	....	....	51.2	6.2	1.48				
1,000	105.1	....	....	....	....	....	63.2	7.7	1.82				
1,500	157.9	....	....	....	....	....	....	17.2	4.1	0.47			
2,000	210.5	....	....	....	....	....	....	30.7	7.3	0.82	0.19		
2,500	263.0	....	....	....	....	....	....	48.0	11.4	1.30	0.31		
3,000	316	....	....	....	....	....	....	69.2	16.4	1.86	0.43		
3,500	368	....	....	....	....	....	....	....	22.3	2.51	0.57	0.18	
4,000	421	....	....	....	....	....	....	....	29.2	3.30	0.77	0.23	
4,500	473	....	....	....	....	....	....	....	37.0	4.2	0.98	0.24	
5,000	526	....	....	....	....	....	....	....	45.7	5.2	1.21	0.36	
6,000	632	....	....	....	....	....	....	....	65.7	7.5	1.74	0.52	0.20
7,000	738	....	....	....	....	....	....	....	....	10.0	2.37	0.72	0.27
8,000	842	....	....	....	....	....	....	....	....	13.2	3.10	0.93	0.38
9,000	947	....	....	....	....	....	....	....	....	16.7	3.93	1.18	0.47
10,000	1,051	....	....	....	....	....	....	....	....	20.6	4.85	1.46	0.57
11,000	1,156	....	....	....	....	....	....	....	....	25.0	5.8	1.76	0.68
12,000	1,262	....	....	....	....	....	....	....	....	29.7	7.0	2.09	0.81
13,000	1,368	....	....	....	....	....	....	....	....	35.0	8.1	2.44	0.95
14,000	1,473	....	....	....	....	....	....	....	....	40.3	9.7	2.85	1.11
15,000	1,579	....	....	....	....	....	....	....	....	46.5	10.9	3.26	1.26
16,000	1,683	....	....	....	....	....	....	....	....	53.0	12.4	3.72	1.45
18,000	1,893	....	....	....	....	....	....	....	....	66.9	15.6	4.71	1.83
20,000	2,150	....	....	....	....	....	....	....	....	....	19.4	5.8	2.20
22,000	2,315	....	....	....	....	....	....	....	....	....	23.4	7.1	2.74
24,000	2,525	....	....	....	....	....	....	....	....	....	27.8	8.4	3.17
26,000	2,735	....	....	....	....	....	....	....	....	....	32.8	9.8	3.83
28,000	2,946	....	....	....	....	....	....	....	....	....	37.9	16.4	4.4
30,000	3,158	....	....	....	....	....	....	....	....	....	43.5	13.1	5.1

In psi in 1000 ft of pipe, 125 lb gage initial pressure. For longer or shorter lengths of pipe the friction loss is proportional to the length, i.e., for 500 ft, one-half of the above; for 4,000 ft, four times the above, etc.

Table 4.9 Loss of Air Pressure Due to Friction

Cu ft Free Air Per Min	Nominal Diameter, In.												
	1/2	3/4	1	1 1/4	1 1/2	1 3/4	2	3	4	6	8	10	12
5	12.7	1.2	0.5										
10	50.7	7.8	2.2	0.5									
15	114.1	17.6	4.9	1.1									
20	202	30.4	8.7	2.0	0.9								
25	316	50.0	13.6	3.2	1.4	0.7							
30	456	70.4	19.6	4.5	2.0	1.1							
35	811	95.9	26.2	6.2	2.7	1.4							
40	....	125.3	34.8	8.1	3.6	1.9							
45	....	159	44.0	10.2	4.5	2.4	1.2						
50	....	196	54.4	12.6	5.6	2.9	1.4						
60	....	282	78.3	18.2	8.0	4.2	2.2						
70	....	385	106.6	24.7	10.9	5.7	2.9						
80	....	503	139.2	32.3	14.3	7.5	3.8						
90	....	646	176.2	40.9	18.1	9.5	4.8						
100	....	785	217.4	50.5	22.3	11.7	6.0						
110	....	950	263	61.2	27.0	14.1	7.2						
120	....	....	318	72.7	32.2	16.8	8.6						
130	....	....	369	85.3	37.8	19.7	10.1	1.2					
140	....	....	426	98.9	43.8	22.9	11.7	1.4					
150	....	....	490	113.6	50.3	26.3	13.4	1.6					
160	....	....	570	129.3	57.2	29.9	15.3	1.9					
170	....	....	628	145.8	64.6	33.7	17.6	2.1					
180	....	....	705	163.3	72.6	37.9	19.4	2.4					
190	....	....	785	177	80.7	42.2	21.5	2.6					
200	....	....	870	202	89.4	46.7	23.9	2.9					
220	....	....	....	244	108.2	56.5	28.9	3.5					
240	....	....	....	291	128.7	67.3	34.4	4.2					
260	....	....	....	341	151	79.0	40.3	4.9					
280	....	....	....	395	175	91.6	46.8	5.7					
300	....	....	....	454	201	105.1	53.7	6.6					
320	....	....	....	....	....	....	61.1	7.5					
340	....	....	....	....	....	....	69.0	8.4	2.0				
360	....	....	....	....	....	....	77.3	9.5	2.2				
380	....	....	....	....	....	....	86.1	10.5	2.5				
400	....	....	....	....	....	....	94.7	11.7	2.7				
420	....	....	....	....	....	....	105.2	12.9	3.1				
440	....	....	....	....	....	....	115.5	14.1	3.4				
460	....	....	....	....	....	....	125.6	15.4	3.7				
480	....	....	....	....	....	....	137.6	16.8	4.0				
500	....	....	....	....	....	....	150.0	18.3	4.3				
525	....	....	....	....	....	....	165.0	20.2	4.8				
550	....	....	....	....	....	....	181.5	22.1	5.2				
575	....	....	....	....	....	....	197	24.2	5.7				
600	....	....	....	....	....	....	215	26.3	6.2				
625	....	....	....	....	....	....	233	28.5	6.8				
650	....	....	....	....	....	....	253	30.9	7.3				
675	....	....	....	....	....	....	272	33.3	7.9				
700	....	....	....	....	....	....	294	35.8	8.5				
750	....	....	....	....	....	....	337	41.4	9.7				
800	....	....	....	....	....	....	382	46.7	11.1				
850	....	....	....	....	....	....	433	52.8	12.5				
900	....	....	....	....	....	....	468	59.1	14.0				
950	....	....	....	....	....	....	541	65.9	15.7				
1,000	....	....	....	....	....	....	600	73.0	17.3	1.9			
1,050	....	....	....	....	....	....	658	80.5	19.1	2.1			

Table 4.9 Loss of Air Pressure Due to Friction (continued)

Cu ft Free Air Per Min	Nominal Diameter, In.												
	1/2	3/4	1	1 1/4	1 1/2	1 3/4	2	3	4	6	8	10	12
1,100	....	....	....	....	....	....	723	88.4	21.0	2.4			
1,200	....	....	....	....	....	....	850	105.2	25.0	2.8			
1,300	....	....	....	....	....	....	....	123.4	29.3	3.3			
1,400	....	....	....	....	....	....	....	....	33.9	3.8			
1,500	....	....	....	....	....	....	....	....	39.0	4.4			
1,600	....	....	....	....	....	....	....	....	44.3	5.1			
1,700	....	....	....	....	....	....	....	....	50.1	5.7			
1,800	....	....	....	....	....	....	....	....	56.1	6.4			
1,900	....	....	....	....	....	....	....	....	62.7	7.1	1.6		
2,000	....	....	....	....	....	....	....	....	69.3	7.8	1.8		
2,100	....	....	....	....	....	....	....	....	76.4	8.7	2.0		
2,200	....	....	....	....	....	....	....	....	83.6	9.5	2.2		
2,300	....	....	....	....	....	....	....	....	91.6	10.4	2.4		
2,400	....	....	....	....	....	....	....	....	99.8	11.3	2.6		
2,500	....	....	....	....	....	....	....	....	108.2	12.3	2.9		
2,600	....	....	....	....	....	....	....	....	117.2	13.3	3.1		
2,700	....	....	....	....	....	....	....	....	126	14.3	3.3		
2,800	....	....	....	....	....	....	....	....	136	15.4	3.6		
2,900	....	....	....	....	....	....	....	....	146	16.5	3.9		
3,000	....	....	....	....	....	....	....	....	156	17.7	4.1		
3,200	....	....	....	....	....	....	....	....	177	20.1	4.7		
3,400	....	....	....	....	....	....	....	....	200	22.7	5.3		
3,600	....	....	....	....	....	....	....	....	224	25.4	5.6	1.8	
3,800	....	....	....	....	....	....	....	....	250	28.4	6.6	2.0	
4,000	....	....	....	....	....	....	....	....	277	31.4	7.3	2.2	
4,200	....	....	....	....	....	....	....	....	305	34.6	8.1	2.4	
4,400	....	....	....	....	....	....	....	....	335	38.1	8.9	2.7	
4,600	....	....	....	....	....	....	....	....	366	41.5	9.7	2.9	
4,800	....	....	....	....	....	....	....	....	399	45.2	10.5	3.2	
5,000	....	....	....	....	....	....	....	....	433	49.1	11.5	3.4	
5,250	....	....	....	....	....	....	....	....	477	54.1	12.6	3.4	
5,500	....	....	....	....	....	....	....	....	524	59.4	13.9	4.2	1.6
5,750	....	....	....	....	....	....	....	....	....	64.9	15.2	4.6	1.8
6,000	....	....	....	....	....	....	....	....	....	70.7	16.5	5.0	1.9
6,500	....	....	....	....	....	....	....	....	....	82.9	19.8	5.9	2.3
7,000	....	....	....	....	....	....	....	....	....	96.2	22.5	6.8	2.6
7,500	....	....	....	....	....	....	....	....	....	110.5	25.8	7.8	3.0
8,000	....	....	....	....	....	....	....	....	....	125.7	29.4	8.8	3.6
9,000	....	....	....	....	....	....	....	....	....	159	37.2	10.2	4.4
10,000	....	....	....	....	....	....	....	....	....	196	45.9	13.8	5.4
11,000	....	....	....	....	....	....	....	....	....	237	55.5	16.7	6.5
12,000	....	....	....	....	....	....	....	....	....	282	66.1	19.8	7.7
13,000	....	....	....	....	....	....	....	....	....	332	77.5	23.3	9.0
14,000	....	....	....	....	....	....	....	....	....	387	89.9	27.0	10.5
15,000	....	....	....	....	....	....	....	....	....	442	103.2	31.0	12.0
16,000	....	....	....	....	....	....	....	....	....	503	117.7	35.3	13.7
18,000	....	....	....	....	....	....	....	....	....	636	148.7	44.6	17.4
20,000	....	....	....	....	....	....	....	....	....	....	184	55.0	21.4
22,000	....	....	....	....	....	....	....	....	....	....	222	66.9	26.0
24,000	....	....	....	....	....	....	....	....	....	....	264	79.3	30.1
26,000	....	....	....	....	....	....	....	....	....	....	310	93.3	36.3
28,000	....	....	....	....	....	....	....	....	....	....	360	108.0	42.1
30,000	....	....	....	....	....	....	....	....	....	....	413	123.9	48.2

*To determine the pressure drop in psi, the factor listed in the table for a given capacity and pipe diameter should be divided by the ratio of compression (from free air) at entrance of pipe, multiplied by the actual length of the pipe in feet, and divided by 1000.

Piping from the header to points of use should connect to the top or side of the header to avoid being filled with condensate for which drainage drop legs from the bottom of the header should be installed. Properly located and maintained compressed air dryers should prevent condensate in headers. Headers and piping also should have an ample number of tapped connections to allow evaluation of air pressure at points throughout the system.

Air tools generally are rated at 90 psig. They can operate at lower or higher pressure but at the expense of efficiency. Torque wrenches will vary in torque output depending on the air pressure at the tool affecting the quality of the work piece. Similarly, paint spray may be too sparse or too dense if the air pressure at the paint gun fluctuates significantly.

Ideally, the pressure throughout an air distribution system should remain in steady state. This is not possible due to the variations in air flow requirements, pressure losses in the system and the types of controls used. This will be discussed later under air system efficiency.

The air pressure at a point of use will be the air pressure at the compressor discharge less the pressure drop due to friction from the flow rate between these two points, and the pressure drop through equipment such as compressed air dryers and filters. In addition, the use of flexible hose and quick disconnect fittings between piping and the tool may cause a significant pressure drop which must be accounted for. A point of use requiring compressed air at 100 psig requires an air compressor rated at 110 psig or higher, depending on the distribution system, controlled storage capacity and the type of treatment used for the air quality required at the point of use. Remember that the higher the pressure at the air compressor, the higher the operational energy costs. It follows then, that when procuring new production equipment requiring compressed air power, it should be specified to have the lowest possible efficient operating pressure.

Consideration also should be given to the potential addition of equipment creating additional air demand on the system, which could result in a fall in the system pressure, particularly if there is marginal air compressor capacity.

Artificial Demand in a Compressed Air System

When a compressed air system operates at a pressure higher than required, not only is more energy consumed in compressing the air, but end uses consume more air and leakage rates also increase. This increase may be referred to as Artificial Demand.

A compressed air system should be operated with compressor and system controls set to achieve the lowest practical pressure.

Intermittent Demand in a Compressed Air System

Compressed Air Systems are dynamic, meaning that, conditions of flow rate and pressures throughout the system are not static but constantly changing.

The steadiest conditions usually occur in process type applications, where the demand for compressed air is relatively constant and/or changes are gradual. This simplifies the necessary controls for air flow and pressure.

In many industrial plants, demand can vary widely as a variety of tools are used and as isolated demand events occur. Often, a demand event occurs at some considerable distance from the compressor(s) supplying the compressed air. This often is aggravated by an initial distribution, sized for a given flow rate and distance, having been extended due to plant expansion. The original distribution pipe size has not been increased but the length and the flow rate have been increased. Pressure drop throughout the extended distribution system can vary erratically. This is because the compressor controls are sensing discharge pressure, an increase being interpreted as a reduction in demand and a decrease being interpreted as an increase in demand.

In some cases, a specific piece of machinery is installed, requiring a relatively large amount of compressed air but only for a relatively short period of time. If the total demand is measured over an hour or a day, the average flow rate in cubic feet per minute is well within the capacity capability of the compressor(s). However, during thc time when the demand event occurs, the flow rate may exceed the capacity of the compressor(s), dryer(s) and filter(s).

Air Receiver Capacity

Theoretically, if the distribution system volume (air receivers plus distribution piping) was large enough, the air compressors would see a constant discharge pressure and there would be no artificial demand. Obviously a grossly oversized system volume is not practical but it demonstrates that many problems can be eliminated with adequately sized and located air receivers and sufficiently large diameter distribution piping. The basic problem is in getting compressed air from Point A to Point B, at the required flow rate and pressure. The air pressure, the length of the distribution piping and the air velocity due to its diameter, will determine the pressure drop. Excessive length and too small a pipe diameter can create significant problems. Hysteresis also will cause delays in response time of controls, further aggravating the problems.

An adequately sized secondary receiver, located close to points of high and/or intermittent demand, can provide the required flow rate(s) without significant pressure drop in the system and the air compressor(s) has adequate time to replenish the pressure in the secondary air receiver. A restriction orifice at the inlet to a secondary air receiver can limit the rate of flow to replenish the receiver in the available time, without depleting air supply needed at other points in the system. In some cases, a check valve prior to the inlet to the receiver will ensure the availability of the required intermittent demand flow rate and pressure, as the secondary receiver will not then be supplying air to other demand events which may occur prior to the receiver. This is particularly so for situations of high intermittent demand.

The size of an air receiver can be calculated as follows:

$$V = T x \frac{C x P_a}{P_1 - P_2}$$

where:

- V = Receiver volume, ft^3
- T = Time allowed (minutes) for pressure drop $P_1 - P_2$ to occur.
- C = Air demand, cfm of free air
- P_a = Absolute atmospheric pressure, psia
- P_1 = Initial receiver pressure, psig
- P_2 = Final receiver pressure, psig

The formula assumes the receiver volume to be at ambient temperature and that no air is being supplied to the air receiver by the compressor(s). If the compressor(s) is running while air is being drawn from the receiver, the formula should be modified so that C is replaced by $C - S$, where S is the compressor capacity, cfm of free air. The initial formula also can be used with a known receiver size, to determine the time to restore the air receiver pressure. In this case, C is replaced by S, which is the compressor capacity, cfm of free air.

In a compressed air distribution system, the Supply Side consists of the air compressors, air aftercoolers, dryers and associated filters, and a primary air receiver. The primary air receiver provides storage volume of compressed air and tends to isolate the compressors from the dynamics of the system. The compressor capacity controls respond to the pressure seen at the discharge connection of the compressor package. Multiple compressors should be discharging into a common header, which will be at slightly higher pressure than the primary air receiver pressure, due to pressure drop through the dryers and filters (presuming the dryers and filters are between the compressors and the primary air receiver).

In the past, mainly with reciprocating compressors, rules of thumb for sizing a primary air receiver have been from 1 gallon per cfm to 3 gallons per cfm of compressor capacity. This is no longer regarded as good practice and the recommended primary receiver size will vary with the type of compressor(s), its type of capacity control, anticipated or actual supply side and demand side events.

Many oil injected rotary screw compressors are equipped with capacity control by inlet valve modulation designed to match the output from the air compressor with the demand from the points of use. On this basis, it has been stated that an air receiver is not needed. At best, this is misleading. An air receiver near the discharge of a rotary screw compressor will shield the compressor control system from pressure fluctuations from the demand side, downstream of the receiver and can allow the compressor to be unloaded for a longer period of time, during periods of light demand. The addition of an over-run timer can stop the compressor if it runs unloaded for a pre-set time, saving additional energy.

Some oil injected rotary screw compressors are sold with load/unload capacity control, which is claimed to be the most efficient. This also can be misleading, since an adequate receiver volume is essential to obtain any real savings in energy.

At the moment an oil injected rotary screw compressor unloads, the discharge pressure of the air end is that in the sump/separator vessel, the Unload Pressure Set Point. At the same time, the inlet valve closes but that is as inlet valve modulation to zero capacity. At this point, the power is approximately 70% of full load power. It is not until the pressure in the sump separator vessel has been bled down to the fully unloaded pressure that the fully unloaded power of approximately 25% is reached.

Rapid bleed down of pressure would cause excessive foaming of the oil and increased oil carry-over downstream. To prevent this, from 30 – 100 seconds bleed-down time is normal and, in some cases, the compressor will re-load before the fully unloaded condition is reached. For this reason, adequate primary storage volume is essential to benefit from Load/Unload Capacity Control.

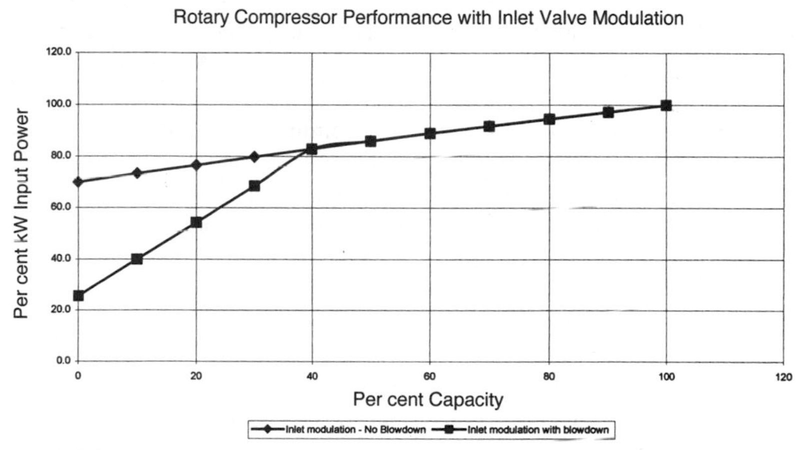

Figure 4.1A Shows Inlet Valve Modulation from 100% to 40% capacity and unloading at that point.

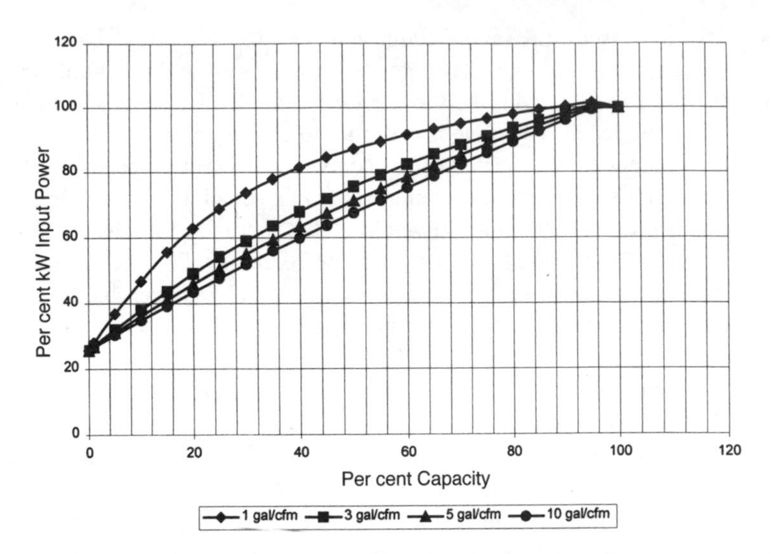

Figure 4.1B Shows Average Power *v* Percent Capacity with various sizes of primary receiver.

One solution, sometimes proposed, is to eliminate modulation and have the compressors operate in a load/unload mode. Certain factors must be recognized before making such a change. The standard full capacity, full load pressure, often has the compressor running at around 110% of motor nameplate rating, or using 10% of the available 15% continuous overload service factor. The remaining 5% is meant to cover tolerances and items such as increased pressure drop through the air/oil separator before it is required to be changed.

If the discharge pressure is allowed to rise by an additional 10 psi without the capacity being reduced by inlet valve modulation, the bhp will increase by 5% and the motor could be overloaded. A reduction in discharge pressure may be necessary to operate in this mode.

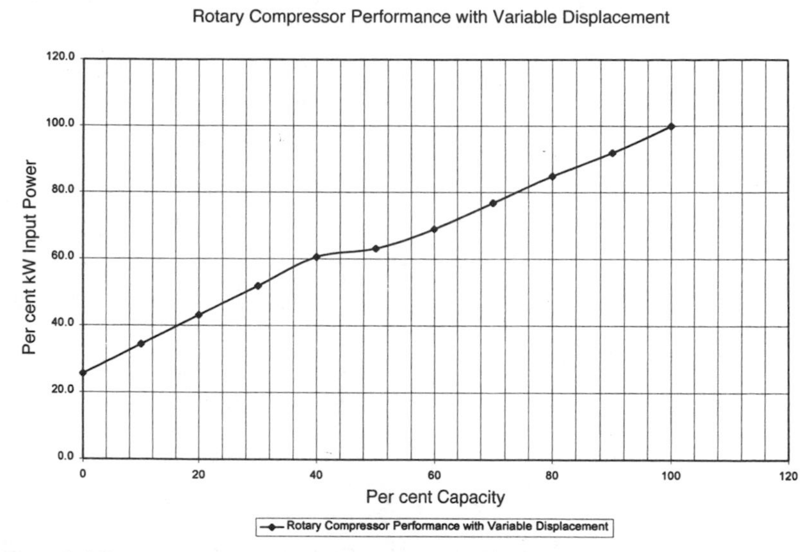

Figure 4.1C Shows Average Power *v* Percent Capacity with Variable Displacement Capacity Control (Slide/Spiral/Turn Valve) from 100% to 50% capacity followed by Inlet Valve Modulation to 40% capacity, then unloading. With this type of control, the inlet pressure to the air end does not change, hence the pressure ratio remains essentially constant. The effective length of the rotors is reduced.

In each of the above types of capacity control, the compressors are essentially the same, running at constant speed. Only the method of control changes.

Variable Speed may be achieved by variable frequency AC drive, or by switched reluctance DC drive. Each of these has its specific electrical characteristics, including inverter and other losses.

Air end displacement is directly proportional to rotor speed but air end efficiency depends upon male rotor tip speed. Most variable speed drive (VSD) package designs involve full capacity operation above the optimum rotor tip speed, at reduced air end efficiency and increased input power, when compared with a constant speed compressor of the same capacity, operating at or near its optimum rotor tip speed. While energy savings can be realized at all reduced capacities, the best energy savings are realized in applications where the running hours are long, with a high proportion in the mid to low capacity range.

Some designs stop the compressor when a lower speed of around 20% is reached, while others may unload at 40-50%, with an unloaded power of 10-15%. The appropriate amount of storage volume should be considered for each of these scenarios.

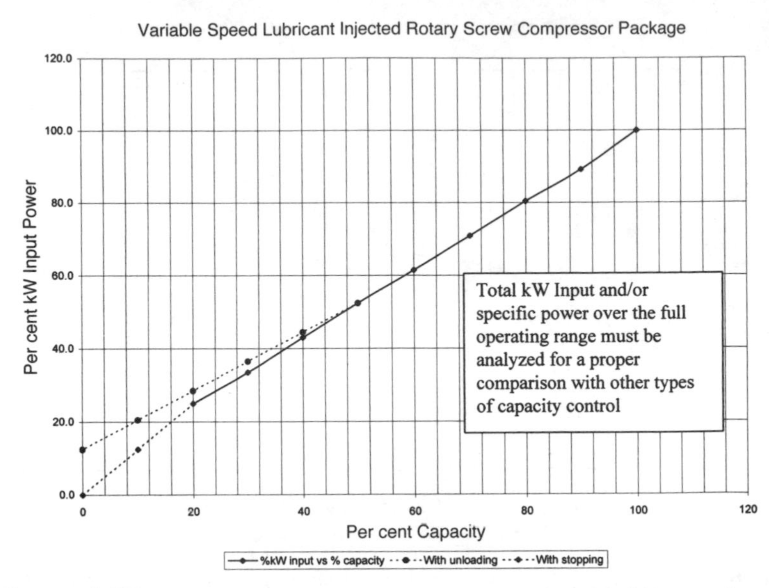

Figure 4.1D Shows Average Power *v* Percent Capacity with this type of control.

The control mode chosen should take into account the receiver/system volume relative to compressor capacity, the range of flow rate normally experienced, and the mean flow rate during a 24 hour period.

It should be noted that in systems with multiple compressors and sequencing controls, it is possible to have most of the compressors running fully loaded on base load with only one compressor on "trim" or part load, providing the most efficient mode for the system. It also is not necessary to have the air receiver/system storage capacity based upon the total capacity of all the compressors, provided they are not all on the same load and unload pressure settings. In such cases only the capacity of the "trim" compressor needs to be considered, provided it is the same as, or close to, the capacity of a compressor that may be fully unloaded or stopped while the "trim" compressor continues to operate.

A primary air receiver allows the compressor(s) to operate in a given discharge pressure range (usually 10 psi) from load to unload. Multiple compressors also can be sequenced as needed and with all but one operating in the most efficient, fully loaded mode. The capacity of the one compressor is modulated to match system demand.

Another option to minimize the effects of artificial demand, is the use of a Pressure/Flow Controller. This normally is located downstream of the primary air receiver and is a sophisticated form of pressure regulator. It is designed to allow flow at the required rate of demand, to maintain a stable downstream pressure, often within +/-1 psi. The stable downstream pressure can be set at the lowest practicable level for satisfactory operation of the pneumatic equipment, reducing the rate of any leakage from the system and allowing improved quality control from pneumatic processes, tools and devices. While this may reduce the flow rate experienced, the compressors may still operate at a higher than required discharge pressure. The primary receiver then will provide a certain amount of storage volume but compressor controls also must be addressed to reduce the compressor discharge pressure, if optimum energy use is to be achieved.

Selecting the Air Compressor Type(s) and Number

Air compressors vary in design characteristics and, although there is some overlap, each has its optimum range of capacity and/or pressure. The design and operational characteristics of each type is discussed in Chapter 2.

Generally, air cooled reciprocating compressors are best suited to a capacity requirement of 40 acfm (approx. 10 hp) or less, although sizes up through 150 hp are available. Standard pressure ratings of 100 psig and 175 psig are common. The 175 psig rating is common in automotive repair facilities for tire changers, hoists, etc., but seldom required for typical industrial applications. Most have cylinder lubrication but lubricant free and lubricant less designs also are available. Smaller sizes generally run on a start/stop type of control requiring an air receiver storage tank with a significant pressure difference between the start and stop settings. Larger sizes may have continuous running with load/unload controlled by pressure settings. In some cases, a specific point of use may benefit from one of these compressors dedicated to it rather than drawing from the main distribution system.

Double-acting reciprocating air compressors are efficient in operation but require relatively large installation spacc and foundations. These once were the work horses of plant air systems but have been largely displaced by less costly packaged rotary air compressors.

Rotary air compressors are available up through 3,000 acfm with pressures up to 200 psig although most operate around 100-125 psig. These are available both lubricant injected and lubricant free and have a variety of control types available as described in the chapter on rotary air compressors.

Centrifugal air compressors are best suited to relatively high volume, base load conditions and are considered more economical above 1,500 acfm (approx. 300 hp). Although pressures up to 10,000 psig are possible, most industrial centrifugal air compressors operate in the 100-125 psig range. Capacity control may be by inlet throttling, inlet guide vanes and/or discharge bypass. These are described in the chapter on dynamic air compressors.

Depending on the total system requirements, more than one type of compressor may be the best choice. For example, a large volume automotive plant may benefit from centrifugal compressor(s) capable of handling the base load demand and rotary or reciprocating air compressor(s) to function as trim compressor(s) for fluctuating loads. Standby air compressors also would be required to allow for maintenance and any unscheduled down time. A plant shutdown can be much more costly than an additional air compressor. Consideration also should be given to potential plant expansion.

Some plants operate only one shift per day so air demand does not fluctuate as much as in a plant with three shifts of operation and only one shift at full production rate. In plants having three shifts with widely differing requirements, the base load compressor should be capable of handling the demands of the least loaded shift and an additional compressor or compressors running only for the other shift(s). Maintenance needs also must be taken into consideration and a number of identical air compressors can minimize replacement parts considerations.

A centralized air compressor room can facilitate installation and maintenance considerations and minimize the number of standby compressors required and, if required, the number of operators. On the other hand, the distance from the compressor room to the furthest point of use must be considered as extensive lengths of piping cause increased pressure drop, potential leaks and, when run outside, potential line freezing problems. The centralized room also must allow room for maintenance and for future plant expansion and additional air requirements. Depending on compressor type(s), a centralized compressor room keeps the need for noise attenuation at one location, away from the work place.

Comparison of standard air compressor performance generally is made at a discharge pressure of 100 psig. The nameplate hp rating may not correspond with the total package kW required by the compressor so it is important to have the actual kW at the specified operating conditions of air flow rate and pressure when comparing air compressor types and manufacturers. A useful comparison at 100 psig is the total package input kW per 100 acfm of free air delivered. These are at compressor full load operation. Consideration also must be given to efficiency at part load and no load operation. Power costs vary widely throughout the United States and compressor efficiency considerations will be more serious in some areas. For samples of the data sheets, visit the CAGI website: www.cagi.org.

Lubricant free air may be required for all or for some specific point of use applications. If only one point of use or a few points require lubricant free air, there are two ways of dealing with this. One is to draw air from the plant air distribution system and treat it immediately prior to the point of use with the appropriate filtration (see the chapter on Air Treatment). The other is to have a separate lubricant free compressor for the point(s) of use. In general, lubricant free compressors have a higher initial cost and higher maintenance costs than their lubricated or lubricant injected counterparts but have the advantage of condensate uncontaminated by lubricant.

A significantly higher (or lower) pressure may be required for a specific point or points of use. A higher pressure requirement generally would be better served by a dedicated higher pressure air compressor or by a booster compressor drawing from the main air distribution system and boosting it to the pressure required at the specific point(s) of use. This prevents the total air distribution from operating at a higher pressure and absorbing more energy. A lower pressure at a specific point of use with significant air demand may justify a separate air compressor. The alternative is drawing from the main air distribution system through a pressure regulator which will maintain the required lower pressure at the point of use.

The location of the compressors chosen must take into account the type of cooling required. The vast majority of air compressors under 100 hp and about 50% of air compressors 200 hp and larger are air cooled. This eliminates the need for cooling water and its drainage considerations, or cooling water systems with higher than ambient water temperatures and the possible need for water treatment and associated costs. However, large radiator type coolers located outdoors in Northern climates also can present problems of lubricant temperature and viscosity at start-up when the compressor is idle overnight. Heated air from radiator type coolers located indoors can be used for space heating in plants in winter months and vented outside when heated air is not required. Adequate ventilation in the compressor room also must be considered.

Another consideration, often overlooked, is the source of inlet air to the compressor. Drawing air from the compressor room may be using air which has been air conditioned and either cooled or heated at cost and a higher than outside ambient temperature also results in a reduced mass flow of air through the compressor. On the other hand, air drawn from outside should be from a location where contaminants such as industrial gases will not be a problem. It also should be remembered that the air intake filter on a standard air compressor package is designed to protect the air compressor and not necessarily the equipment downstream of the compressor.

Available electrical power to the compressor room also must be considered, including voltage and kW capability.

Air Quality

The applications at the points of use will determine the quality of the air required at each point. Considerations include the content of particulate matter, condensate and lubricant. In the chapter on Compressed Air Treatment, reference is made to Air Quality Classes for these contaminants as published in International Standard ISO 8573-1. The chapter also describes the equipment available to meet these classes, including various air dryer types and various filter types. The manufacturer of process machinery and other pneumatic devices should be consulted to determine the air quality required.

When the air intake filter for the compressor(s) is mounted remotely from the compressor(s), the inlet air piping from the air intake filter to the compressor inlet must be clean and, being at atmospheric pressure, may be of plastic material. It should be remembered that the air intake filter is for the protection of the air compressor and does not necessarily protect the compressed air distribution system or equipment installed downstream. Downstream filtration is recommended as discussed under Compressed Air Treatment.

The compressed air distribution system itself may contribute to the contaminant problem, particularly if standard steel piping and air receivers are used. Stainless steel or copper piping is essential to some processes but may be considered too expensive in most industrial plants. Galvanized piping is one alternative and internal epoxy coating of air receivers another. The introduction of plastics for compressed air distribution systems has potential risks. These pipes generally are pressure rated at around 80°F and their pressure capability falls rapidly as temperature is increased. Such piping located near the roof of a building may see relatively high temperatures from the surrounding atmosphere as well as from the compressed air it contains. The plastic material and the joint compounds also may not be compatible with the type of air compressor lubricant used, particularly some of the popular synthetic lubricants. Plastic piping also is much more vulnerable to damage from fork lift trucks, etc., when located at or near floor level. For these reasons, plastic piping is not recommended for compressed air systems.

Fick's Law

This issue relates in general to very low pressure dew points, typically minus 100°F and below. Any leaks in supply and distribution piping downstream of the dryer(s) will allow back diffusion of atmospheric water vapor to enter the compressed air line even if it is at full line pressure.

Dalton's Law states that the total pressure of a mixture of ideal gases is equal to the sum of the partial pressures of the constituent gases. The partial pressure is defined as the pressure that each gas would exert if it alone occupied the volume of the mixture at the mixture temperature.

Fick's law of diffusion applies here. It states that "The rate of diffusion in a given direction is proportional to the negative of the concentration gradient." The concentration gradient relates to the partial pressure of water vapor and is very high at low pressure dew points.

The gradient difference of thousands of ppm of water vapor is the driving force; rather like an osmosis effect across a membrane. There are certain materials recommended to prevent this effect at low dew points and any dew point below minus 100°F should be sampled with nickel, PTFE, or preferably stainless steel tubing. Materials such as PVC or rubber should be totally avoided in these cases.

Also, experience has shown that to purge the piping downstream from the dryer takes an extraordinary amount of dry air to purge out the residual moisture. In tests to obtain minus 94°F dew point, if you have, for example, 1 $ft.^3$ of piping downstream from the dryer, then you will need about 1,000,000 $ft.^3$ of dry air to purge it down to that dew point. This could have considerable impact in an industrial compressed air system, if a very low pressure dew point is required. It is essential that the proper piping material be used and any leaks in piping or joints be rectified to prevent this problem and to save the waste of compressed air.

Air System Efficiency

Efficient operation of the compressed air system requires that pressure fluctuations be minimized. Air compressors have an on load and an off load pressure setting. The differential between these two settings should not be allowed to exceed

10% of the maximum pressure setting. Normally this will require adequate storage volume in the form of an air receiver.

Generally, air receivers are sized in gallons up through 200 gallons, whereas larger sizes are sized in cubic feet. There are different old rules of thumb for the minimum size of an air receiver. One, based upon gallons, is: 1 gal per cfm of compressor capacity.

A similar rule, based upon cubic feet, is: 1 ft^3 per 10 cfm of compressor capacity. (100 psig gives a compression ratio at sea level conditions of 114.7/14.7=7.8 which is close to the conversion factor of 7.481 gallons per cubic foot).

Smaller air compressors (usually mounted on a receiver/tank) and operating with start/stop capacity control, generally have a receiver size, in cubic feet, approximating the compressor free air capacity in cfm divided by three.

The time taken for the pressure in an air receiver to drop from one pressure to another is:

$$T = V \frac{P_1 - P_2}{CP_a}$$

where:

T = time, minutes
P_1 = Initial receiver pressure, psig
P_2 = final receiver pressure, psig
P_a = atmospheric pressure, psi abs.
C = air requirement, cubic feet per minute
V = Receiver volume, cubic feet

The equation assumes the receiver to be at a constant atmospheric temperature and that no air is supplied to the receiver during the time interval. If air is being supplied constantly to the receiver at S cubic feet per minute, then C should be replaced by $(C - S)$.

It should be noted that a receiver sized on the latter basis, with the compressor running at full load, will allow a drop in pressure of 2 psi in 4.08 seconds if there is a demand 20% above the compressor capacity. The pressure will fall by 10 psi in 20.4 seconds under the same conditions.

It is essential that such transient loads be taken into account when sizing an air receiver. A receiver located near a point of intermittent high flow rate can help isolate the air compressor(s) and the remainder of the system from severe pressure fluctuations.

The various types of capacity control for each compressor type, and their benefits, are described in the chapter dealing with air compressors. For very small compressor systems, it is common for the air compressor to be mounted on the air receiver. This means that compressed air drying and filtration equipment must be downstream of the air receiver. The receiver may provide some cooling of the air and a means of some of the condensate being collected but a sudden demand for air exceeding the dryer capacity rating could result in air to the point of use at a higher dew point.

When the air receiver is located downstream of the compressed air dryer, it stores air already dried and, since the capacity rating of the dryer normally matches the capacity rating of the air compressor, the dryer is shielded by the air receiver from a sudden high demand for air, so that the point(s) of use always receive(s) properly dried air.

Additional air receiver capacity close to the point of use can be very beneficial where the point of use may require a large volume of air for a relatively short period of time. The rest of the system then is shielded from a potential sharp drop in pressure due to the sudden high demand.

The minimum receiver capacity may be calculated but experience and judgment also are important. It has been claimed that unlike reciprocating compressors, which only load/unload or step type capacity control, rotary compressors with modulation of capacity by inlet throttling or similar means, do not need a receiver. This generally is not true and the installation of an air receiver is a sound investment, minimizing fluctuations in compressor operating conditions.

The following example illustrates the calculation of receiver capacity for a certain application.

Example:

$$V = \frac{TCP_a}{P_1 - P_2}$$

$$= \frac{(3 \text{ x } 15)(90)(14.7)}{(60)(100\text{-}80)}$$

$$= 49.6 \text{ ft}^3$$

The next larger standard air receiver from Table 4.10 should be used. Air receivers should meet the Code for Unfired Pressure Vessels published by The American Society of Mechanical Engineers (ASME). This normally is a requirement of insurance companies. All federal, state and local codes and/or laws also must be satisfied. ASME coded vessels also have approved safety valves, pressure gauge, drain valve and hand holes for manhole covers. Vertical receivers also have a skirt base. The location of the air receiver is important and regular draining is essential. The potential of freezing must be considered and provision made for heated drains where necessary. Provision for connecting a portable compressor for emergency conditions is also recommended.

Table 4.10 Data for Selection of Receiver

Diameter, in.	Length, ft	Actual Compressor Capacity*	Volume, ft^3
14	4	40	4 1/2
18	6	110	11
24	6	190	19
30	7	340	34
36	8	570	57
42	10	960	96
48	12	2115	151
54	14	3120	223
60	16	4400	314
66	18	6000	428

*Cubic feet of free air per minute at 40 to 125 psig for constant-speed regulation. For automatic start-and-stop service, the receivers are suitable only for capacities one-half of the actual compressor capacities listed here-to avoid starting too frequently.

Energy management systems, with adequate and strategically placed air receivers, intermediate controls and pressure regulation, have been developed to suit specific plant systems, maintaining a very narrow band of system pressures, minimizing run time of compressors and improving quality of work performed by pneumatic tools and equipment.

Machines can be equipped to detect an idle threshold and stop air flow to an idle machine. A simple solenoid valve, arranged to close when the machine is idle, can reduce air consumption significantly.

When an air distribution system is operated at a higher than required pressure, this requires approximately one percent additional power for each 2 psi of operating pressure. In addition, leakage increases as pressure increases. This can be seen from Table 4.11 which shows that a 1/4 inch diameter orifice with 100 psig on one side and atmospheric pressure on the other will have a flow rate of 104 acfm of free air. At 110 psig the flow rate increases to 113 acfm. The combined effect of an increase of 5% in power for the 10 psi pressure increase, combined with the 8.6% increase in leakage flow rate, results in an overall power increase of approximately 14%.

Table 4.11 Discharge of Air through an Orifice

Gage Pressure before Orifice, psi	Nominal Diameter, In.										
	1/64	1/32	1/16	1/8	1/4	3/8	1/2	5/8	3/4	7/8	1
	Discharge, Cu. ft. Free Air Per Min.										
1	.028	0.112	0.450	1.80	7.18	16.2	28.7	45.0	64.7	88.1	115
2	.040	0.158	0.633	2.53	10.1	22.8	40.5	63.3	91.2	124	162
3	.048	0.194	0.775	3.10	12.4	27.8	49.5	77.5	111	152	198
4	.056	0.223	0.892	3.56	14.3	32.1	57.0	89.2	128	175	228
5	.062	0.248	0.993	3.97	15.9	35.7	63.5	99.3	143	195	254
6	.068	0.272	1.09	4.34	17.4	39.1	69.5	109	156	213	278
7	.073	0.293	1.17	4.68	18.7	42.2	75.0	117	168	230	300
9	.083	0.331	1.32	5.30	21.1	47.7	84.7	132	191	260	339
12	.095	0.379	1.52	6.07	24.3	54.6	97.0	152	218	297	388
15	.105	0.420	1.68	6.72	26.9	60.5	108	168	242	329	430
20	.123	0.491	1.96	7.86	31.4	70.7	126	196	283	385	503
25	.140	0.562	2.25	8.98	35.9	80.9	144	225	323	440	575
30	.158	0.633	2.53	10.1	40.5	91.1	162	253	365	496	648
35	.176	0.703	2.81	11.3	45.0	101	180	281	405	551	720
40	.194	0.774	3.10	12.4	49.6	112	198	310	446	607	793
45	.211	0.845	3.38	13.5	54.1	122	216	338	487	662	865
50	.229	0.916	3.66	14.7	58.6	132	235	366	528	718	938
60	.264	1.06	4.23	16.9	67.6	152	271	423	609	828	1,082
70	.300	1.20	4.79	19.2	76.7	173	307	479	690	939	1,227
80	.335	1.34	5.36	21.4	85.7	193	343	536	771	1,050	1,371
90	.370	1.48	5.92	23.7	94.8	213	379	592	853	1,161	1,516
100	.406	1.62	6.49	26.0	104	234	415	649	934	1,272	1,661
110	.441	1.76	7.05	28.2	113	254	452	705	1,016	1,383	1,806
120	.476	1.91	7.62	30.5	122	274	488	762	1,097	1,494	1,951
125	.494	1.98	7.90	31.6	126	284	506	790	1,138	1,549	2,023

Based on 100% coefficient of flow. For well-rounded entrance multiply values by 0.97. For sharp-edged orifices a multiplier of 0.65 may be used.

This table will give approximate results only. For accurate measurements see ASME Power Test Code, Velocity Volume Flow Measurement.

Values for pressures from 1 to 15 psig calculated by standard adiabatic formula.

Values for pressures above 15 psig calculated by approximate formula proposed by S. A. Moss: $w = 0.5303\, aCP_1 / \sqrt{T_1}$ where w = discharge in lb per sec, a = area of orifice in sq. in., C = coefficient of flow, P_1 = upstream total pressure in psia, and T_1 = upstream temperature in deg F abs.

Values used in calculating above table were C = 1.0, P_1 = gage pressure + 14.7 psi, T_1 = 530 F abs.

Weights (w) were converted to volumes using density factor of 0.07494 lb. per cu. ft. This is correct for dry air at 14.7 psia and 70°F.

Formula cannot be used where P_1 is less than two times the barometric pressure.

Air System Safety

Safety in the workplace is a primary design consideration. In a compressed air distribution system there are several factors involved. We will consider here only the distribution system and not the air compressors that have their own built-in safety systems.

The pressure rating of all piping must meet or exceed the maximum pressure to which the system may be subjected. The pressure rating should take into account the maximum temperature to which the piping will be exposed. Any exposed piping at an elevated temperature and which may be contacted by a person, should be shielded from contact or a suitable warning displayed. Piping must be adequately supported and allow for thermal expansion. Shut-off valves should be installed to allow maintenance of the various pieces of equipment but a pressure relief valve must be installed between the compressor and a shut-off valve to prevent over pressurization when the valve is closed.

Any air receiver should meet ASME Code for Unfired Pressure Vessels and be complete with a pressure relief valve and a pressure gauge. Applicable federal, state and local codes also must be met.

Pressure containing piping should be located away from passageways where fork lift trucks and other vehicles could come into contact with it, but should be accessible for maintenance.

Air Distribution System Layout

The foregoing gives recommendations for sizing the required air compressors, air receivers, headers and piping systems for minimum pressure drops and optimum efficiencies. Valves and fittings offering the least resistance to flow should be selected, including long radius elbows.

Where possible, a header loop system around the work place is recommended. This gives a two-way distribution to the point where demand is greatest. The header size should allow the desired minimum pressure drop regardless of the direction of flow around the loop. The header should have a slight slope to allow drainage of condensate and drop legs from the bottom side of the header should be provided to allow collection and drainage of the condensate. The direction of the slope should be away from the compressor.

Piping from the header to each point of use should be kept as short as possible. As previously stated, air piping from the header to the point of use should be taken from the top of the header to prevent the inclusion of condensate. These pipes normally run vertically downward from an overhead header to the point of use and pressure drop from the header to the point of use should not exceed 1 psi during the duty cycle.

For a lubricant free compressor, it is recommended that the air distribution header and piping be corrosion resistant.

A primary air receiver should be located close to the compressor(s) to shield the compressor(s) from fluctuations in air demand, maintaining a more stable compressor discharge pressure. Consideration should be given to secondary receiver(s) close to point(s) of use where relatively large surges in demand may occur. This also helps to stabilize the air distribution system.

Isolating valves to facilitate maintenance of system components, and tapped connections for system measurement and analysis, should be provided throughout the system.

Cost of Compressed Air

Compressor types, accessories, systems and operating pressures vary widely, which makes cost of operation difficult to establish. However, there are some good guidelines that can be used. The most common condition for comparison is a pressure at the discharge of the air compressor of 100 psig. This also is the standard inlet pressure used for rating of compressed air dryers and will be used in the following cost estimates:

Air Compressors

The most common air compressor in industrial use today is the lubricant injected type rotary screw compressor. These are available in single or two-stage versions, the two-stage version being more efficient than the single-stage. Generally a single-acting air cooled reciprocating compressor is the least efficient while a multi-stage, double acting, water cooled reciprocating compressor is the most efficient. Dynamic compressors efficiencies are closer to the multi-stage water cooled reciprocating compressors.

A common means of comparison is expressed in bhp/100 cfm with a compressor discharge pressure of 100 psig. The compressor capacity in cfm, or acfm, is the amount of air delivered from the compressor but measured at prevailing ambient inlet conditions. While the bhp/100 cfm can range from about 18 for an efficient two-stage water cooled reciprocating compressor to 30 for a small air cooled reciprocating compressor, a typical single-stage lubricant injected rotary screw compressor has a power consumption of approximately 22.2 bhp/100 cfm.

If we then allow a drive motor full load efficiency of 92% and 1 bhp = 0.746 kW, the resultant power consumption in electrical terms is 18 kW/100 cfm. This is what will be used in these discussions. CAGI Performance Data Sheets require total package input.

It should be noted that this is based upon 100 psig at the discharge of the air compressor and not at the point of use. Pressure drops will occur through piping and components downstream of the air compressor, resulting in a lower pressure at the point of use. If it is necessary to raise the discharge pressure of the compressor to achieve a required pressure at point of use, it is necessary to make sure that the air compressor is capable of the higher discharge pressure. It also should be noted that for each 2 psi increase in compressor discharge pressure, the power consumption will increase by 1%.

In addition, air compressors normally do not operate continuously at full capacity. While the average amount of air used may be less than the capacity of the air compressor, the reduction in power requirement is not directly proportional to the actual capacity. Rotary air compressors utilizing inlet valve throttling do not provide a significant reduction in kW as the capacity is reduced. Although the mass flow of air is reduced in direct proportion to the absolute inlet pressure at the inlet to the air end, the reduction in inlet absolute pressure results in a corresponding increase in the ratios of compression, adversely affecting kW/100 cfm. Other types of capacity control are available but all result in a higher kW/100 cfm at reduced capacity conditions.

Compressed Air Dryers

Like air compressors, dryer ratings are based upon a quantity of air measured at ambient conditions but with a dryer inlet condition of 100 psig and 100°F and with an ambient temperature of 100°F. Different dryer types have different requirements but each will have a pressure drop as the compressed air passes through it. This should be taken into account with the required increase in discharge pressure from the air compressor and its additional power requirement as indicated above.

Refrigerant Type Dryers
Typical rated Dew Point +35 to +40°F

Based upon Manufacturers' Data Sheets giving total power input in amps at a specified voltage, a reasonable rule of thumb is a power requirement of 0.5-0.8 kW/100 cfm of the rated capacity of the dryer. The dryer rating, rather than the compressor rating should be used in determining the power requirement of the refrigerant type dryer. Smaller capacity refrigerant type dryers may require more power/100 cfm and large dryers less but the above approximation is good from 500 through 1,000 cfm, not including energy to cover pressure drop through the dryer, which normally is approximately 3-5 psi, an additional 0.27- 0.45 kW/100 cfm.

Regenerative Desiccant Type Dryers
Typical rated Dew Point -40°F

The amount of purge air required to regenerate the desiccant bed may vary, based upon the type of controls used. Standard heatless dryers of this type require approximately 15% of the rated dryer capacity for purge air. Therefore, the electrical power requirement of this dryer will be 15% of the air compressor electrical power requirement as the amount of air delivered from the dryer to the system will be reduced by 15%. Based upon the above, this represents 2.7 kW/100 cfm. Special controls are available for most dryers of this type to economize the amount of purge air required that can be reduced to 10% or less, which represents 1.8 kW/100 cfm. Again, an energy allowance for pressure drop of 3-5 psi must be made, an additional 0.27- 0.45 kW/100 cfm.

Internal or external heating can be used to reduce the amount of purge air required. Purge air can be supplied by a blower, rather than compressed air. Vacuum purge also can be accomplished. These may provide overall energy savings.

Heat of Compression Type Dryers
Typical rated Dew Point is variable, depending on ambient and/or cooling water temperatures.

These may be either twin tower regenerative type or single rotating drum type. Technically, there is no reduction of air capacity with this type of dryer since hot unsaturated air from the compressor discharge is used for regeneration, then cooled and some of the moisture removed as condensate before passing through the drying section to be dried. However, for this to occur, an inefficient entrainment type nozzle has to be used and an electric motor also is used to rotate the dryer drum. Considering pressure drop and compressor operating cost, it is estimated that the total power requirement is approximately 0.5 - 0.8 kW/100 cfm.

Deliquescent Type Dryers
Typical Dew Point Suppression 15°F

Since the drying medium is consumed and not regenerated, there is no requirement for purge air, therefore, pressure drop through the dryer (and any associated filtration) and loss of air volume during the drain cycle are the operating costs which, excluding the replacement cost of the drying medium, are estimated at 0.2 kW/100 cfm.

Membrane Type Dryers
Typical Dew Point +40°F

This type of dryer requires an amount of purge air, or sweep air, to displace the moisture that passes through the permeable membranes. For comparable dew point depression with a refrigerant type dryer, this will be in the range of 15 to 20% of the air capacity of the dryer. Therefore, 15 to 20% should be added to the air compressor power requirement. Based upon the above, this amounts to 2.7 - 3.6 kW/100 cfm. Energy for a pressure drop of 3-5 psi will add 0.27 - 0.45 kW/100 cfm.

5

Air Tools and Air Tool Safety

AIR TOOLS AND AIR TOOL SAFETY

Air tools are designed to provide increased productivity, long service life, and safe operation. But, as with any type of tool or machine, these advantages can be realized only through proper application, proper maintenance, and user training.

Application

Air tools should be selected by a person who is familiar with air tools and their proper application. Air tool selection must take into account the particular factors in a given job that will affect air tool and operator performance. Some factors to be considered are the workstation design, the man-tool-task relationship, and the environment in which the user will perform the task.

The workstation should be carefully designed so that the workpiece is held securely, so that there is sufficient light and ventilation, and so that means are provided for safely holding or suspending the tool. The workstation should also be arranged for operator convenience and comfort.

Maintenance

Air tools must undergo periodic inspection and routine maintenance to ensure that they are operating properly. Tool maintenance and repairs should be performed by authorized, trained, and competent personnel.

Premature failure and poor performance of air tools can often be attributed to the following:

1. Water or foreign materials such as rust or pipe scale in the air.
2. Inadequate or improper lubrication.
3. Worn parts.

Training

Air tools should be operated only by qualified personnel who have been trained in their proper operation and safe use. Employers must ensure that their employees read and understand the tool manufacturer's operating and safety instructions.

Users must always wear protective equipment and clothing. Impact resistant eye protection must be worn while operating or working near air tools. Additional information on eye protection may be found in the Federal OSHA Regulations, 29 CFR, Section 1910.133, *Eye and Face Protection*, and ANSI Z87.1, *Occupational and Educational Eye and Face Protection*. Hearing protection should be provided in high noise areas (above 85 dBA). The close proximity of additional tools, reflective surfaces, process noises, and resonant structures can substantially contribute to the sound level experienced by users. Proper hearing conservation measures, including annual audiograms and training in the use and fit of hearing protection devices may be necessary. Additional information on hearing protection may be found in the Federal OSHA Regulations, 29 CFR, Section 1910.95, *Occupational Noise Exposure*, and the American National Standards Institute, ANSI S12.6, *Hearing Protectors.*

Gloves and other protective clothing should be worn as required, unless they create a greater hazard. Improperly fitted gloves may restrict blood flow to the fingers and can substantially reduce grip strength. Loose fitting clothing should not be worn around air tools that rotate during their use. Avoid inhaling dusts or mists resulting from the use of any air tool. Wear an approved respirator or mask if the ventilation is inadequate. Respirators should be selected, fitted, used and maintained in accordance with Occupational Safety and Health Administration and other applicable regulations.

Cumulative Trauma Disorders make up a large percentage of occupational injuries and affect a wide and diverse sector of society. Users of air tools should be made aware of the symptoms of the most frequently occurring cumulative trauma disorders such as carpal tunnel syndrome, tendonitis, tenosynovitas, DeQuervain's Disease, low back pain, and vibration - induced Raynaud's Syndrome.

In the industries that use portable air tools, the employer is responsible for the job design and content, operator training, work methods, work pace, work environment, proper tool application, and other workplace factors. These, along with personal factors such as pre-existing conditions such as arthritic inflammations, circulatory disorders or neuropathies, hereditary wrist size, inexperience or aggressive work methods, and activities outside the job workplace may contribute to painful repetitive motion disorders of the shoulders, arms, wrists, hands, and fingers. To avoid these disorders, users should be trained to:

- Use a minimum hand grip force consistent with proper control and safe operation.
- Keep the wrists as straight as possible.
- Avoid repetitive movements of the hands and wrists.
- Notify their employer if any symptoms of wrist pain, hand tingling or numbness, or other disorder of the shoulders, arms, wrists, or fingers occurs.

Some users of portable air tools may be susceptible to disorders of the hands and arms when exposed to extended vibration. Although the causes of these disorders are not well known, certain non-occupational factors such as age, physical condition, fat-clogged arteries, injury to the hands, smoking, medications, and circulatory problems are thought to increase the risk. It is uncertain what vibration exposures are required to cause the disorder, including the vibration intensity, the frequency spectrum associated with the intensity, the daily exposure level, and total exposure. Also, factors such as ambient temperature, grip force, and intermittence of exposure may play a part in experiencing the disorder. To avoid the disorder, users should be trained to:

- Use a minimum hand grip force consistent with proper control and safe operation.
- Keep the body and hands warm and dry.
- Avoid continuous vibration exposure by using rest intervals free from vibration.
- Avoid anything that inhibits blood circulation such as tobacco, cold temperatures, and certain drugs.

Additional specific guidelines on the safe operation of portable tools are available from the Compressed Air and Gas Institute publication B186.1, *Safety Code for Portable Air Tools*, and should be followed closely.

Additional specific guidelines on the safe use of grinding wheels, available from the same institute, can be found in ANSI's publication B7.1, *Safety Requirements for Use, Care, and Protection of Abrasive Wheels*. These guidelines should be carefully implemented wherever abrasive wheels are being used.

AIR TOOLS

To manufacture the many products used in today's world, industry must receive raw materials, process those materials into parts, assemble the parts into a finished product, and package that product for shipment. The nature of the product, its size, the quantity to be manufactured, and the tolerances specified are a few of the many considerations that affect the selection of tooling to perform these operations. This chapter is designed to help identify the advantages of using air tools in their many applications and to familiarize the reader with the diversified types of air tools available today.

AIR TOOLS FOR INDUSTRY

Air tools have gained widespread acceptance in industry because of their many inherent advantages. Air motors provide compact, lightweight, smooth-running power sources for air tools, as well as for other applications. They cannot be harmed by overloading, unlimited reversals, or continuous stalling. Also, they are explosion and shock resistant. Air motors start and stop almost instantly and provide infinitely variable control of torque and speed within their capacity range. Air and hydraulic motors share in common the advantages of high torque per pound of weight and safe operation in hazardous environments. Air motors also have certain advantages over hydraulic motors. First, air tools do not require return lines, but rather exhaust into the atmosphere. There is no heat buildup when air motors are stalled for a considerable length of time. In addition, low-pressure lines of 60 to 125 psig are less expensive than high-pressure hydraulic lines of 1500 to 3000 psig. Leaks in air systems do not present the safety and housekeeping problems associated with hydraulic leaks. Air motors are unique in their ability to operate at any speed within their range for indefinite periods of time. Even continuous stalling has no ill effects on the motor, since there is no heat buildup and no danger of overloading. All these features contribute to the long service life associated with air tools. In addition, simplicity of construction makes the repair of air tools comparatively easy.

PRODUCTIVITY THROUGH AIR TOOLS

The use of air tools contributes directly to increased productivity. Even though the cost of compressed air is greater than that of electricity, economics favor air tools because of their weight advantage and longer service life. Laboratory experiments have shown that for every additional pound of hand-held weight there is a productivity loss of up to 7%.

Compact Size

One advantage of air tools over comparable electric tools is that of size. Air tools may be as little as one-half the physical size of the equivalent electric counterpart, as seen in Fig. 5.1. This smaller physical size generally makes them easier to handle than the corresponding electric tools and sometimes allows them to be used in confined spaces where electric tools will not fit.

Light Weight

Air tools are lighter than comparable electric models, often weighing as little as half as much as the corresponding electric tool. This lighter weight allows the tool to be handled more easily, reduces operator fatigue, and ultimately increases worker productivity.

Operating Pressures

Air power tools typically are rated at 90 psig. They may operate at other pressures with varying consequences. Lower pressures result in decreased power to weight ratios negatively impacting production. Higher pressures may result in increased performance with a corresponding decrease in durability, reducing tool life and increasing service and repair costs. The quality of the work piece may also be affected by varying air pressure due to the proportional relationship between torque and supply pressure at the tool inlet. Similarly, paint spray may be too sparse or too dense if the supply pressure at the gun fluctuates significantly. Close control of the air supply pressure to air tools can improve quality control and, if done properly, can improve the energy requirement of the compressed air system.

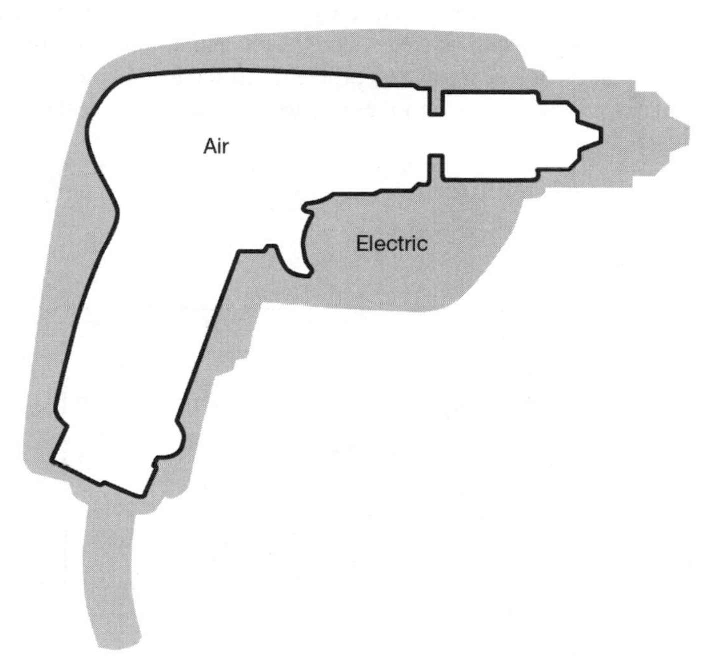

Figure 5.1 An air tool may be as little as half the physical size of the corresponding electric tool, as shown in the comparison above of two hand-held, quarter-inch drills.

Speed, Torque, and Power

Torques and speeds of air tools can be matched to the specific needs of an application. Standard air tools are available in a wide range of speeds, in excess of 100,000 rpm and torques of up to 50,000 ft.-lb.

The performance characteristics of an air motor can be expressed using torque and power curves like those in Fig. 5.2. These curves show the relationship among speed, torque, and horsepower for a given air motor operating at a given air pressure. The power curve shows that the horsepower of a motor with no governor increases to a maximum point as the speed increases and then decreases to zero as the speed continues to increase. This maximum horsepower of the nongoverned air motor is normally reached at roughly half the free speed, that is, the speed at which the tool runs when no load is applied.

As a load is applied to such a nongoverned air motor, the speed decreases and the torque and horsepower increase to a level where they match the load. As the load is increased, the horsepower produced by the motor continues to increase until the motor slows to roughly half of free speed. At this point, the motor has reached peak horsepower and will run at greatest efficiency. If the load is increased beyond this point, the torque will continue to increase to the stall point, but the horsepower will decrease.

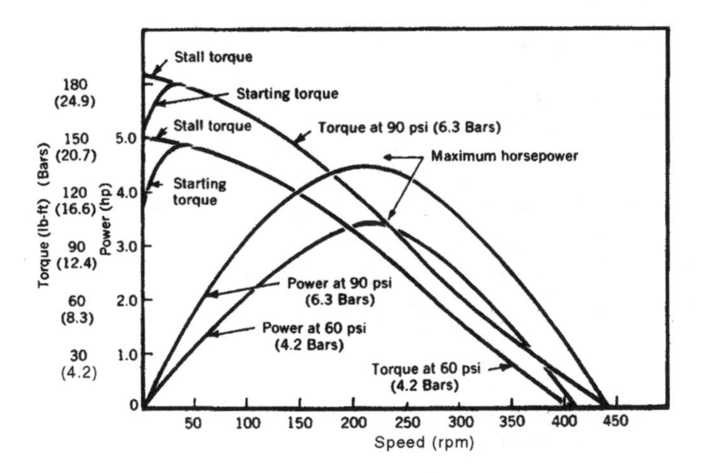

Figure 5.2 Torque and power curves for a typical air motor. Characteristics are similar to those of a series-wound DC motor; torque is maximum at zero speed, zero at free speed.

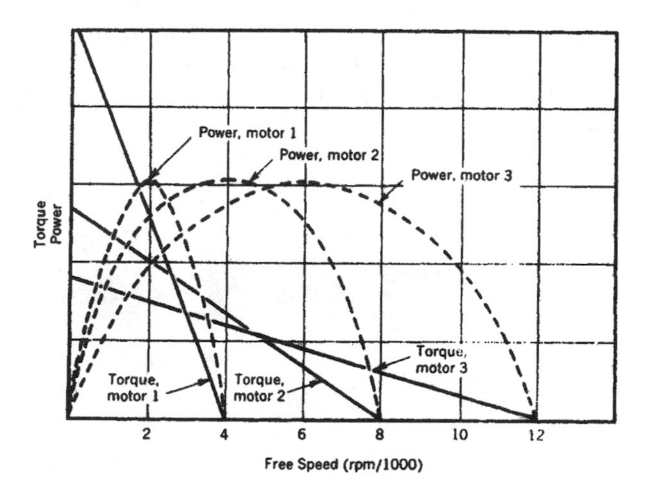

Figure 5.3 Torque and power curves for three motors having the same power rating. The motor with top speed of 12,000 rpm is much more susceptible to speed drop-off since its torque curve slope is shallow.

As the curve in Fig. 5.2 shows, an air tool will slow down to match the torque and horsepower requirements of the load, a factor that should be considered when selecting an air tool for a given application. This differs considerably from most AC electric motors, which tend to maintain a constant speed by consuming more energy. Nongoverned air tools use less air at peak horsepower than at free speed.

The major consideration in selecting an air motor for an application is the range of speed and torque required to satisfy the operating conditions. Different motors producing the same maximum power can have substantial variation in speed as a result of load change (Fig. 5.3). A 10% load variation will produce a speed variation of 435 rpm for a 3.1 hp motor with a 3200 rpm free speed. The speed of a motor of the same horsepower, but having a free speed of 2800 rpm, will vary only 360 rpm under the same load change. The steeper the torque curve, the less the speed variation will be with changes in load. Changes in basic free speed may be achieved through gearing.

Gearing

The nominal free speed of most air motors is higher than is suitable for many applications. To provide lower speeds and higher torque output, air tool manufacturers regularly incorporate reduction gears in a tool or air motor as an integral part of the unit (Fig. 5.4). By using different reduction ratios, it is possible to achieve a wide range of speeds. Whenever the speed of a motor is reduced by gearing, its torque is increased by the speed ratio and, at the same time, the torque curve slope increases. Increasing the torque curve slope decreases the effect of load change on speed. Geared units should be specified when minimum speed change is desired with a varying load. Many standard air motors can also be coupled to commercial gear boxes to provide low output speeds. The power and thrust required to drill certain metals are given in Table 5.1.

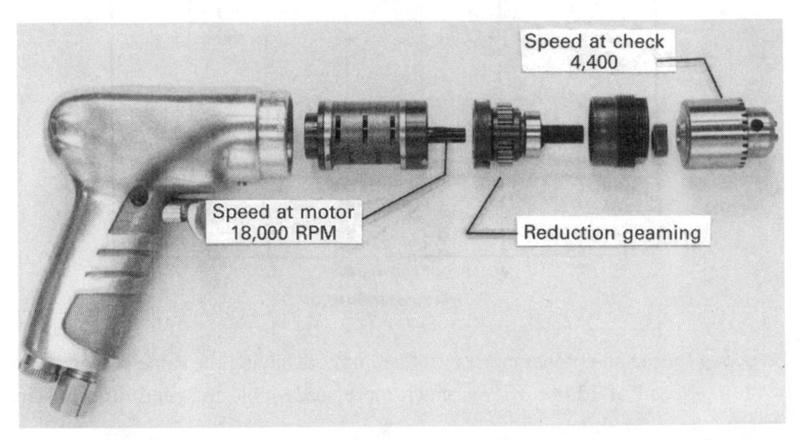

Figure 5.4 Reduction gears in an air tool provide for a lower speed at the output spindle than at the motor, with virtually no reduction in power.

Governors

Another way to modify speed is through the use of a governor. An air tool governor regulates the free speed to a predetermined limit and will keep the motor speed essentially constant through its normal working range as load is applied. This device counteracts an ungoverned air motor's tendency to lose speed as load is applied. Some manufacturers also furnish safety overspeed controls that can shut off the motor at a predetermined maximum speed. Another type of governor is a motor controller that regulates the air flow through the end plate into the motor to control the speed and power of the tool, and the motor controllers eliminate the need for additional overspeed shutoff devices. Speed-regulation devices to reduce available speed and power can be built into a motor or attached externally.

Durability

An air tool has some performance characteristics that an electric tool cannot match. An air tool can reach full speed in one-half revolution, and an air motor can stop very quickly after the valve is released. Also, in applications requiring greater control at start-up, an air tool offers variable control as standard so that no special valving is required. An electric tool, on the other hand, requires complex electrical circuits to achieve variable speed. Building variable speeds into an electric motor results in higher costs and creates a potential for heat problems, whereas very slow speeds with an air tool can be achieved by simple and inexpensive means without

Table 5.1 Drill Horsepower and Thrust

	Drill Size	1/16	3/32	1/8	3/16	1/4	5/16	3/8	7/16	1/2
	R.P.M.	4000	4000	4000	3000	2200	1800	1500	1300	1100
BRASS	FEED "/REV.	.0004	.0005	.0008	.0012	.0017	.0021	.0025	.003	.0035
	H.P.	.010	.012	.022	.047	0.10	0.15	0.20	0.25	0.30
	THRUST (LBS)	3	6	10	20	35	50	70	90	100
	R.P.M.	4000	4000	4000	4000	3000	2500	2000	1700	1500
ALUMINUM	FEED "/REV.	.0005	.0007	.001	.0015	.0020	.0025	.0030	.0035	.0040
	H.P.	.010	.020	.040	.080	.140	.220	.300	.375	.450
	THRUST (LBS)	2	3	6	12	20	40	60	80	100
	R.P.M.	4000	2850	2100	1400	1000	850	700	600	525
CAST IRON	FEED "/REV.	.0005	.0007	.0010	.0015	.002	.0022	.0027	.0030	.0035
	H.P.	.01	.02	.035	.060	.100	.125	.175	.210	.250
	THRUST (LBS)	6	12	30	45	90	100	140	180	225
B1112	R.P.M.	3600	2400	1800	1200	900	750	600	525	450
MILD	FEED "/REV.	.0005	.0007	.001	.0015	.0022	.0027	.0032	.0037	.0045
STEEL	H.P.	.015	.025	.040	.075	.125	.175	.250	.300	.365
	THRUST (LBS)	8	12	25	50	85	130	175	250	310
	R.P.M.	3600	2400	1800	1200	900	750	600	525	450
1045	FEED "/REV.	.0005	.0007	.0010	.0015	.0022	.0027	.0032	.0037	.0045
STEEL	H.P.	.03	.05	.09	.15	.22	.25	.40	.60	.65
	THRUST (LBS)	12	20	25	75	140	200	280	350	450

overheating. This quality in an air tool makes it ideally suited for driving screws where an operator sometimes needs to start a fastener slowly and then drive it full speed.

An air tool offers durability not found in an electric tool. An air tool will never burn out. If an air motor experiences an overload, it simply stops, and it can remain in a continuous stall indefinitely. In fact, when an air motor is running, the air flowing through it keeps it cool. If an electric tool is subjected to a sustained stall, burnout can result. Furthermore, if an electric tool is overloaded, it will heat up, and there will be a loss of speed and power.

An air tool can be switched from forward to reverse instantaneously without damage. Switching an electric motor in this way will cause an overload that may damage the unit.

Serviceability

Not only is an air tool more durable than an electric tool, but it is easier to service as well. An air tool usually has fewer parts than the corresponding electric tool, and its simple construction makes it easier to repair.

Remote Control

Like an electric tool, an air tool can be remotely controlled. This quality, coupled with an air motor's inherent nonsparking characteristic, makes it ideally suited for use in potentially hazardous or explosive environments.

Pilferage

Because air tools use compressed air, which is not readily available in most homes, as electricity is, theft is not as great a problem.

BASIC AIR MOTORS

Five basic categories of air motors will be discussed in this section:

1. Rotary Vane
2. Axial piston
3. Radial piston
4. Turbine
5. Percussion

Rotary-vane Air Motors

Rotary-vane air motors are the most common air motors used in industry today. They operate using blades that fit into radial slots in a rotor, as seen in Fig. 5.5. This rotor is mounted eccentrically inside a cylinder.

The rotary motion is a result of air pressure exerted against the exposed area of the blades. Thus, the force produced is transmitted through the rotor gearing to the output shaft or is transmitted directly, if no gears are used. The air is then discharged when it reaches the exhaust port (Fig. 5.6).

Rotary-vane motors are essentially high-speed units that deliver a high ratio of power to weight. They are produced as either reversible or single-direction models. Single-direction motors are available with clockwise or counterclockwise rotation.

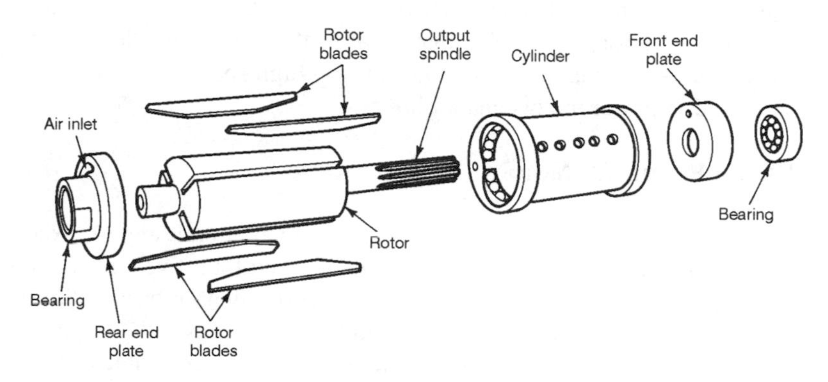

Figure 5.5 Parts which make up a typical, rotary-vane air motor.

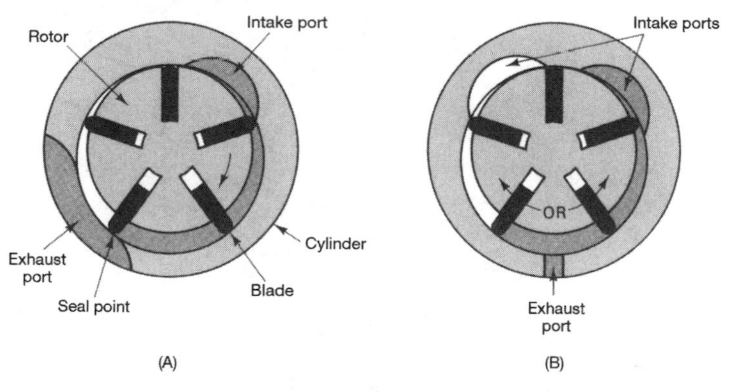

Figure 5.6A Single Direction Air Motor **Figure 5.6B** Reversible Air Motor

Figure 5.6 Direction of air movement through rotary-vane air motors.

Reversible motors are available in either balanced or unbalanced construction. Balanced motors produce equal power in either direction, whereas unbalanced motors produce more power in one direction than the other. Hoists are generally operated by unbalanced motors because more power is required to raise a load than lower it.

Some air tools that incorporate rotary-vane air motors are portable grinders, sanders, drills, tappers, screwdrivers, nutsetters, impact wrenches, electrode dressers, routers, rotary shears, saws, hoists, power motors, and fixtured, self-feed drills.

Axial Piston Air Motors

In the axial piston type motor, the pistons and piston rods move parallel to the center line of the spindle. Their axial movement converts a reciprocating movement into a rotary one through a wobble plate to turn the output spindle. Despite their relatively small size, axial piston motors develop high speed and power. They are typically used in power motors and air hoists.

Radial Piston Air Motors

This type of motor (Fig. 5.7) is similar to a radial aircraft engine, except that it is powered by air instead of fuel. It is available with four, five, or six cylinders. Output torque is developed by pressure on the piston within each cylinder. It is inherently a low-speed power source, operating at free speeds of 3500 rpm or slower. It can carry heavy loads at all speeds and is particularly adaptable to operations requiring slow speed, high starting torque, and smooth starting characteristics.

Figure 5.7 Sectional view of a typical radial piston air motor.

Turbine Air Motors

Turbine motors (Fig. 5.8) are the fastest of all air motors. They can be made to deliver speeds up to 300,000 rpm at comparatively low torques. Their high speed is achieved by expanding air to high velocities through carefully machined slots in the rotor or stator.

Turbine motors have long been used in portable die grinders, where their ultra-high speed enables them to perform fine finishing on production tools and dies. In recent years, they have gained increasing use in power motors. One manufacturer has incorporated a turbine motor in a self-feed drilling unit where its speed is utilized for drilling holes in hard synthetics.

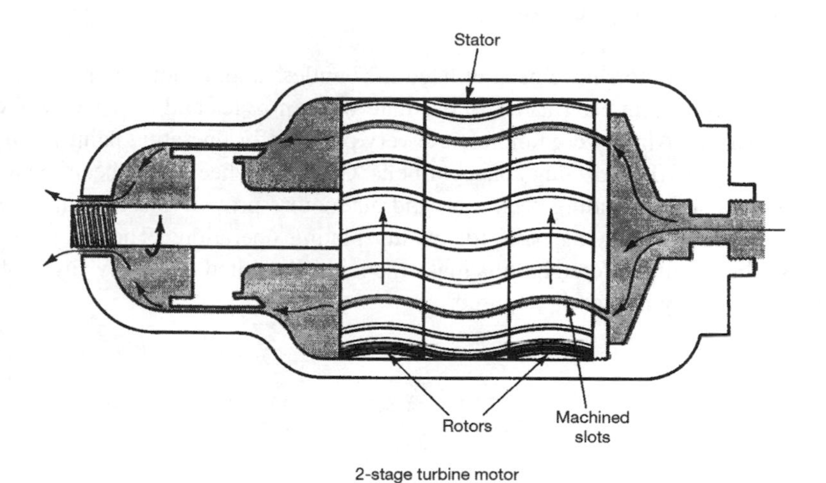

Figure 5.8 Movement of air through a two-stage turbine motor.

Percussion Air Motors

The percussion motor consists of valves, a cylinder, and a single piston (Fig. 5.9). By means of precise valve timing, the piston is given a reciprocating motion that can be used to deliver a powerful blow to an inserted or integral attachment on a tool.

Percussion motors are used where material must be broken up rapidly, chipped away, or impacted. Typical applications include breaking pavement, cleaning scale or sand from castings, removing rust, tamping sand in the foundry, and driving rivets or taper pins.

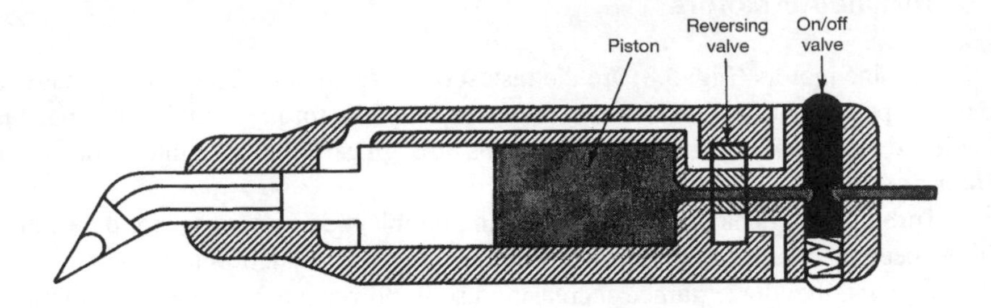

Figure 5.9 Sectional view of a percussion-type air motor.

Power Motor Application

Air-powered motors have been utilized in countless applications. For instance, they have been used in car washes to oscillate water nozzles and to power whitewall tire cleaners. Air-powered motors are very popular for operating paint-mixing propellers (Fig. 5.10) or driving conveyor belts. Grinding wheels have been mounted to air motors for grinding rock bits, and air motors provide the critical speed control necessary in moving parts for quality plating operations. The advantages and versatility of air-powered motors make them ideally suited for nearly any application where rotary motion is required.

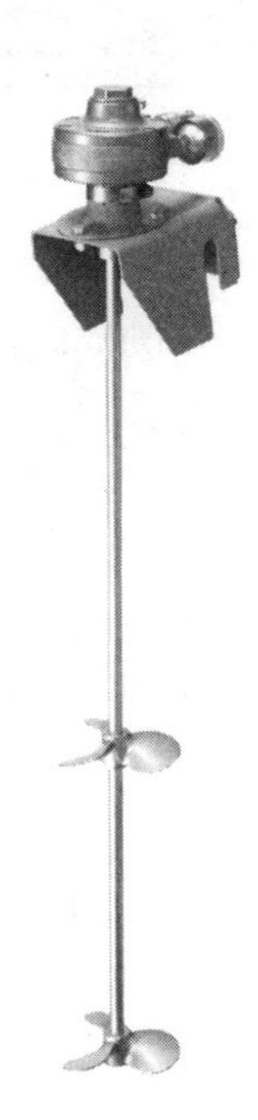

Figure 5.10 Paint mixer powered by a rotary-vane air motor.

Additional applications are as follows:

- Opening and closing large valves.
- Powering bridges on overhead cranes.
- Natural gas valve turning applications.
- Operation of core drills.
- Actuation of clamping devices.
- Operation of holding and positioning devices.
- Powering steel and plastic strapping machines.
- Mixing liquids and chemicals.
- Powering fuel hose take-up reels.

DRILLS

The air drill (Fig. 5.11) is one of the most commonly recognized portable power tools. Despite the continuing development of automatic production processes, there are still many applications in which only portable air drills can be used. This fact accounts for their widespread industrial use.

The operator can start the drill gradually, making it possible to be sure that the drill bit is properly located. Then the tool is brought to full speed to complete the operation.

In industry, air drills are generally preferred to electric drills because of their lighter weight, their ability to stall repeatedly without motor burnout, and their freedom from the hazard of electric shock.

Figure 5.11 A portable air drill used for putting holes in furniture.

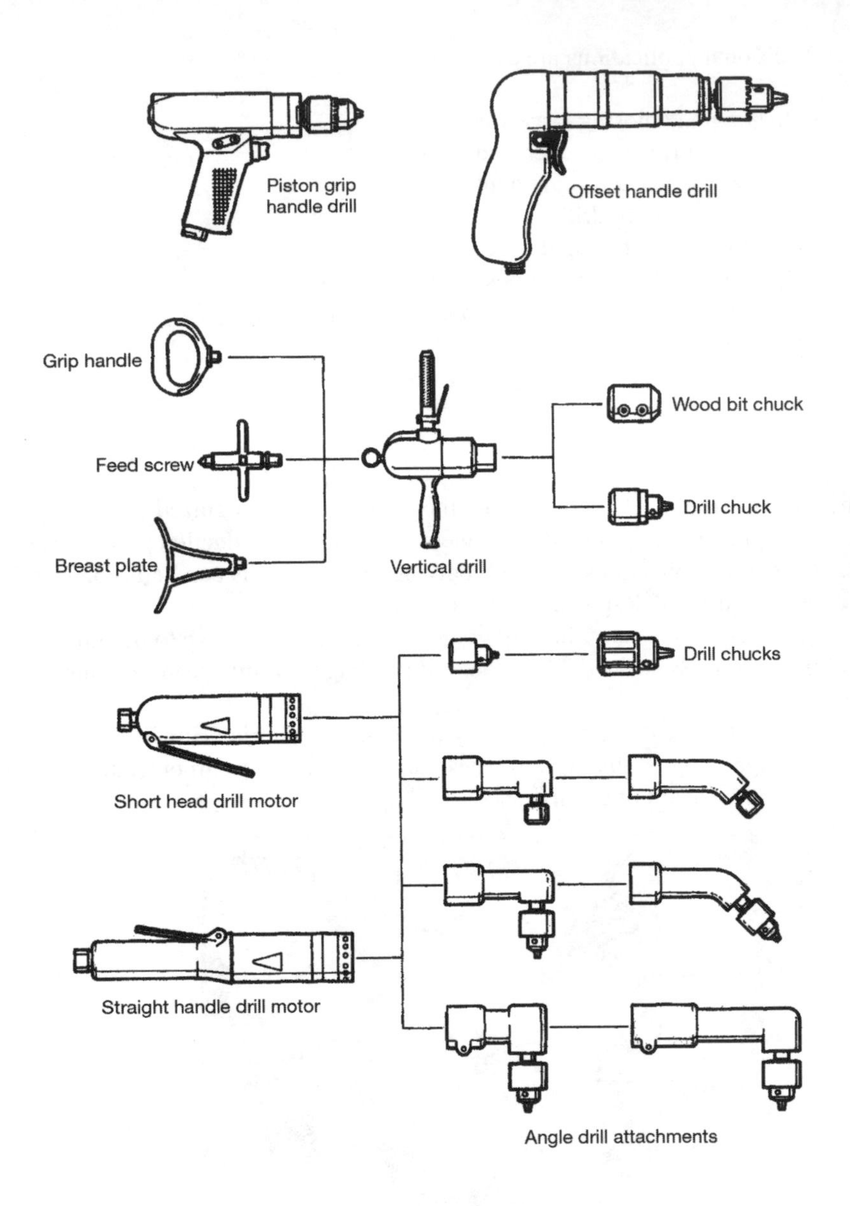

Figure 5.12 Various portable air drill combinations and attachments.

Drills are furnished in a variety of configurations, as shown in Fig. 5.12, to meet differing job requirements:

1. Pistol grip: the handle is set forward from the rear of the motor housing.
2. Offset handle: the handle is set near the extreme rear of the motor housing. The terms pistol grip and offset handle are often used interchangeably.

3. Spade handle: the handle is D-shaped with the trigger inside, and the drill is also equipped with a side handle.
4. Straight handle: the handle is in line with or part of the motor housing. This type of drill is used where vertical motion is required and has either a lever or button throttle.
5. Forty-five or 90 degree angle drills: these drills are used wherever work configurations or narrow space becomes a problem. Most drills in this category are light enough for manual use, although larger models can weigh from 60 to 80 lb.

Speed Range

Available speeds range from a few hundred rpm to approximately 19,000 rpm. The important factor is not so much the turning speed as the cutting speed in surface feet per minute. This cutting speed is determined by the diameter of the drill bit, the rotating speed, and the hardness of the material being drilled.

Weight

Weights range from less than 1 lb for small collet-type drills to 80 lb for heavy-duty, right-angle drills, with the most popular sizes being the 1/4- and 3/8-in. drills weighing from 2 to 2 1/2 lb. Larger sizes are only a small percentage of tile drills in current use.

Rotation

Most drills are right-hand, single-rotation models, but they can be furnished with reversible motors. Special chuck-retaining methods must be used with such models. Reversible models are used for tapping and for drilling in materials in which the drill bit tends to grab or become stuck when breaking through.

Capacity

The capacity of the drill is usually determined by the size of the chuck. The chuck size and the power are normally matched. Cutting-tool retainers range in size from 1/16- to 3/16-in. collets and 1/4- to 1/2-in. chucks. However, portable drills, both reversible and nonreversible, are available with the capacity for drilling, tapping, and reaming holes up to 2 in. in diameter.

Selection

For economical operation, a drill must produce a clean hole in the shortest possible time without undue effort by the operator. Selection of the drill is determined by the material to be penetrated and the size of the hole. To drill a 1/4-in. hole in

Table 5.2 Recommended Drills Speeds for Drilling in Various Materials

Material	S.F.M.	Recommended Cutting Speed Range							
		1/16 .062	1/8 .125	3/16 .187	1/4 .250	5/16 .312	3/8 .375	7/16 .437	1/2 .500
STEEL ALLOY 300-400 BRINELL	20-30	1250 1800	600 900	400 600	300 450	250 350	200 300	175 250	150 225
STAINLESS STEEL	30-40	1800 2500	900 1200	600 800	450 600	350 500	300 400	250 350	225 300
CAST IRON, HARD	30-40	1800 2500	900 1200	600 800	450 600	350 500	300 400	250 350	225 300
STEEL FORGINGS	40-50	2500 3100	1200 1500	800 1000	600 750	500 600	400 500	350 425	300 400
STEEL, TOOL ANNEALED .90-1.20 CARBON	50-60	3100 3700	1500 1800	1000 1200	750 900	600 700	500 600	425 525	400 450
STEEL, .40-.50 CARBON	70-80	4300 5000	2100 2500	1400 1600	1050 1200	850 1000	700 800	600 700	525 600
CAST IRON, MED. HARD	70-100	4300 6000	2100 3000	1400 2000	1000 1500	850 1200	700 1000	600 900	500 800
BRONZE, HIGH TENSILE STRENGTH	70-150	4300 9000	2100 4500	1400 3000	1000 2300	850 1800	700 1530	600 1300	500 1200
MALLEABLE IRON	80-90	5000 5500	2500 2800	1600 1800	1200 1400	950 1100	800 900	700 800	600 700
STEEL, MILD .20-.30 CARBON	80-110	5000 6700	2500 3400	1600 2300	1200 1700	950 1350	800 1150	700 1000	600 850
CAST IRON, SOFT	100-150	6000 9000	3000 4500	2000 3000	1500 2300	1200 1800	1000 1530	900 1300	800 1200
PLASTIC	100-150	6000 9000	3000 4500	2000 3000	1500 2300	1200 1800	1000 1550	900 1300	800 1200

Table 5.2 Continued Recommended Drills Speeds for Drilling in Various Materials

Material	S.F.M.	Recommended Cutting Speed Range							
		1/16 .062	1/8 .125	3/16 .187	1/4 .250	5/16 .312	3/8 .375	7/16 .437	1/2 .500
ALUMINUM	200-300	12000 18000	6000 9000	4000 6000	3000 4500	2400 3700	2000 3000	1700 2600	1500 2300
BRASS & BRONZE	200-300	12000 18000	6000 9000	4000 6000	3000 4500	2400 3700	2000 3000	1700 2600	1500 2300
MAGNESIUM	250-400	15500 25000	7500 12000	5000 8200	3800 6100	3000 4900	2500 4000	2200 3500	1900 3000
FIBERGLASS	300-400	18000 25000	9000 12000	6000 8200	4600 6100	3700 4900	3000 4000	2600 3500	2300 3000
WOOD	300-400	18000 25000	9000 12000	6000 8200	4600 6100	3700 4900	3000 4000	2600 3500	2300 3000

Table 5.3 Spindle Speeds (rpm) to Result in a Given Surface Speed

Drill Bit	Surface Feet per Minute (SFM)											
Diameter (in.)	30	40	50	60	70	80	90	100	110	200	300	400
1/8	917	1222	1528	1834	2139	2445	2750	3056	3662	6111	9168	12224
3/16	611	815	1019	1222	1426	1630	1833	2037	2241	4074	6111	8148
1/4	458	611	764	917	1070	1222	1375	1528	1681	3056	4584	6111
5/16	367	489	611	733	856	978	1100	1222	1345	2445	3666	4888
3/8	306	407	509	611	713	815	917	1019	1120	2037	3056	4074
7/16	262	349	437	524	611	698	786	873	960	1746	2619	3492
1/2	229	306	382	458	535	611	688	764	840	1528	2292	3056

soft materials such as wood or aluminum, a small, high-speed drill is ideal. But in hard metals, the small drill will stall repeatedly, and its high speed will burn up the bit. A tool with more power and slower speed is required for the harder metals.

As an aid in selecting a suitable speed, Table 5.2 lists the suggested surface speeds for high-speed steel drill bits in various materials. If there is a choice between tools of the same speed but different sizes, final selection is based on performance for a lighter weight tool or one with more power to maintain speed under load. Table 5.3 is included for convenience.

Drills with governor control of speeds are available so that maximum surface speed can be maintained under load.

TAPPING TOOLS

A portable tapping tool is sometimes the most practical means of producing threads in a predrilled hole. Portable tappers are low-speed tools, available in either reversible drills with a tapping chuck or single-direction drills with a special push-pull tapping attachment (Figs. 5.13 and 5.14).

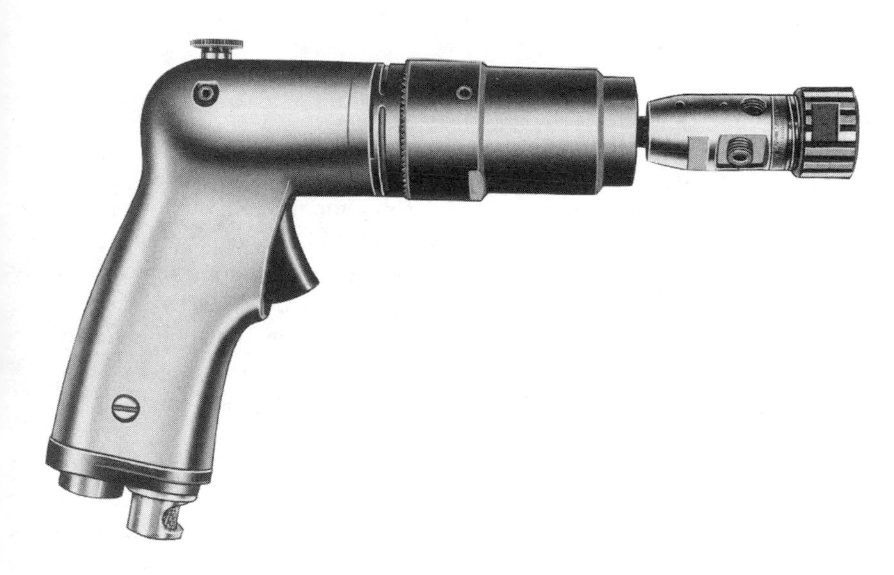

Figure 5.13 Portable Compressed-air Tapping Tool

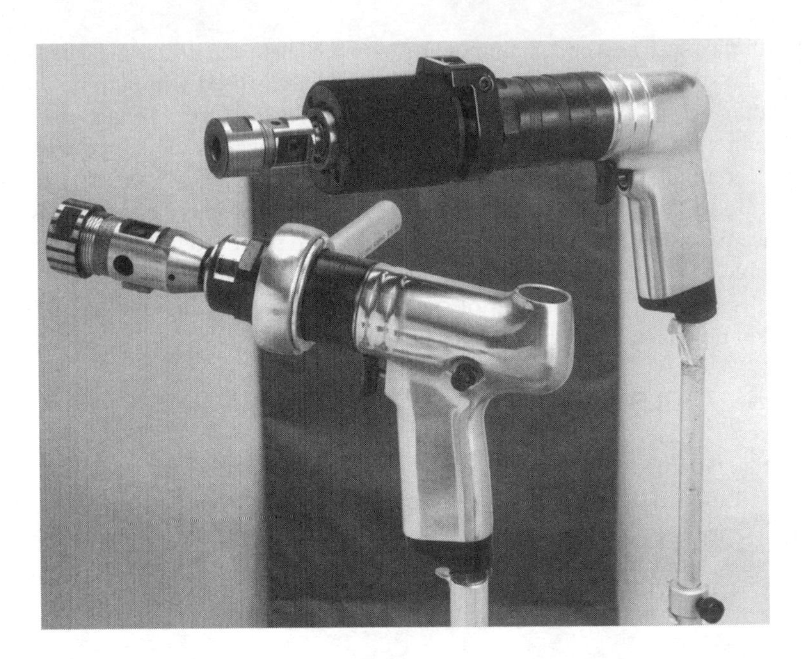

Figure 5.14 Two air-powered tapping tools; the upper model in the photograph is a push-pull tapper.

ASSEMBLY TOOLS

Air tools used for assembly applications are popular for high output and cost reduction. The controllable speed and torque of an air motor, its ability to tolerate repeated stalling without harm, its instant starting capability, and its high power per pound make it ideally suited to assembly work. Air tools perform many assembly operations that would be nearly impossible with tools powered by other means. Many of these applications will be discussed in later paragraphs in this section.

Screwdrivers and Nutsetters

The basic difference between a screwdriver and a nutsetter is in the output spindle. Screwdrivers generally have female spindles (Fig. 5.15) in 1/4- or 5/16-in. hex size to hold commercially available screwdriver bits. They may also have a nonrotating finder to help locate the bit on the screw head. Bits and finders are made in a wide variety of sizes and lengths to fit any standard screw head configuration. Nutsetters usually have male square-drive spindles (Fig. 5.15) to which sockets can be attached, although they may also be equipped with integral sockets or with larger hexagonal chucks to receive hex-shank nutsetting attachments. Screwdrivers and nutsetters will drive fasteners in a range of sizes from number 0 screws through 1/2-in. bolts. They weigh from 1/2 to 5 lb. Power output varies from

1/8 to 1 hp, with output speeds of 250 to 5000 rpm. The range of tools, attachments, and optional equipment makes it possible for off-the-shelf purchase of the correct tool for almost any application.

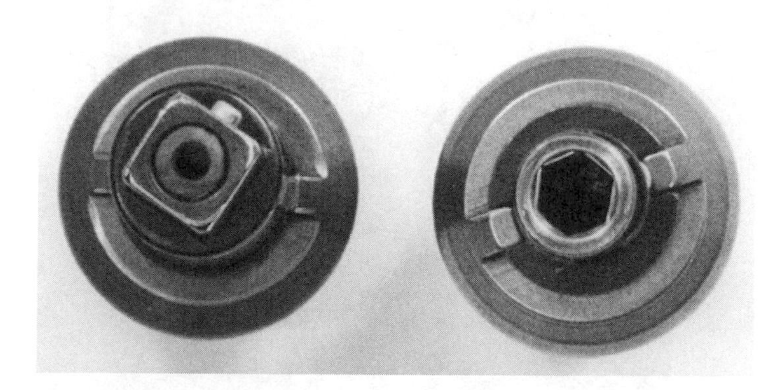

Figure 5.15 Output spindle of a nutsetter (left) and screwdriver (right).

Screwdrivers and nutsetters are classified by handle style, type of clutch, mode of throttle operation, type of drive, and torque capacity. Many models are furnished with a choice of single-direction or reversible motors, those with reversible motors being more popular.

Offset-handle or pistol-grip tools (Fig. 5.16) are normally used for horizontal work or where considerable torque reaction is transmitted to the operator. These tools can be actuated by either a trigger or by operator pressure against the fastener. Straight-handle tools are usually used vertically where reaction torque is low. They are normally balance suspended, as shown in Fig. 5.17. Lever, button, or push-start options are available on straight tools. A straight-style air wrench for applications in which space is limited is seen in Fig. 5.18.

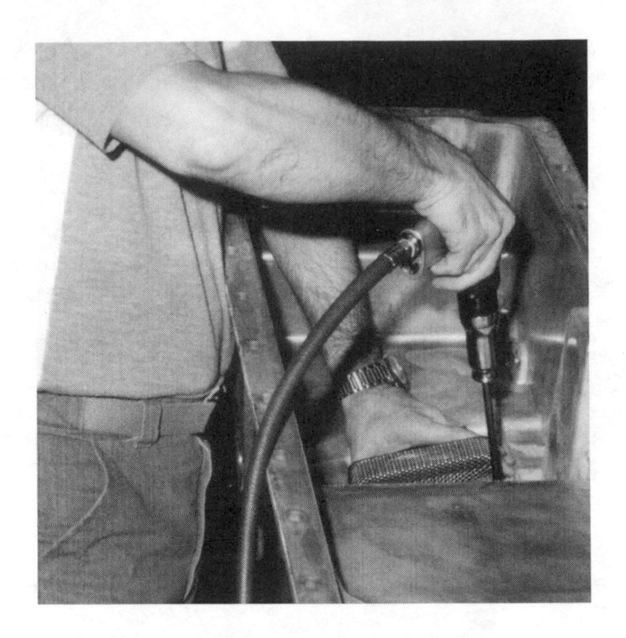

Figure 5.16 Pistol-grip screwdriver in use in assembly.

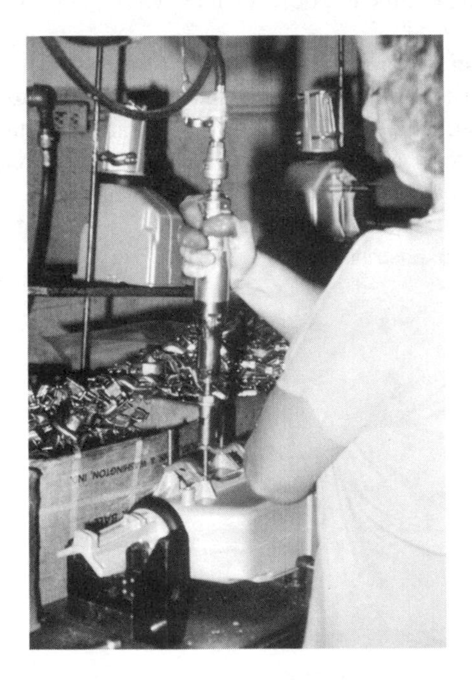

Figure 5.17 A balancer-suspended, straight style air screwdriver.

Screwdrivers may be driven directly or be fitted with one of several types of clutches. Torque repeatability is the factor that most often determines the type of clutch to be used. The more sophisticated the clutch, the less skill is required from the operator to achieve the desired torque.

Direct-drive tools are the simplest type, but their use is usually limited to the application of sheet-metal and wood screws where torque control is not critical and recessed head screws are used. They are typically utilized in stall-type applications.

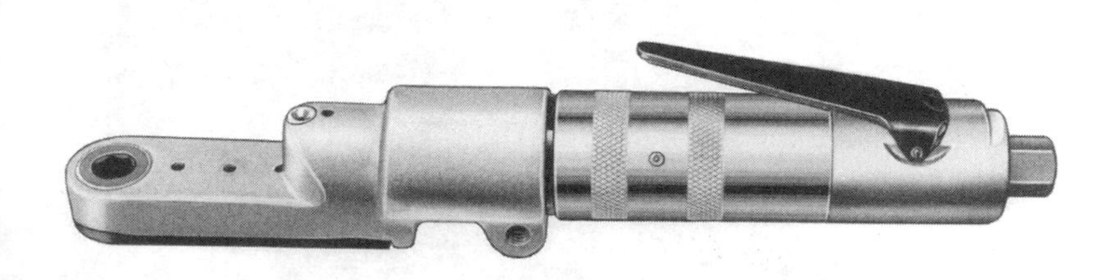

Figure 5.18 A flat angle, straight style air wrench used where limited access space precludes the use of other types.

Clutches for screwdrivers fall into three categories: positive, cushion, or shut-off. The positive clutch is the simplest and least expensive. It is a jaw-type, spring-loaded device that remains disengaged until the bit is pushed into the fastener head. When torque builds up in the fastener, the clutch acts as a ratchet, signaling that the fastener has reached the desired torque so that the operator will shut the tool off. This clutch is best suited to driving wood screws. The cushion clutch, also referred to as an adjustable clutch or adjustable ratcheting clutch, is usually composed of two stages. One is similar to the positive clutch described previously, except that the jaw faces are perpendicular and do not tend to disengage under torque. This allows the bit to be applied to and removed from the fastener with the motor running, since the bit remains stationary until axial pressure is applied to the fastener. The other part of the clutch consists of a spring-loaded pair of jaws with sloping faces or balls in detents. Torque setting is controlled by spring preload and is not affected by operator pressure. When fastener torque reaches the desired value, the second clutch ratchets. The selection of motor size and proper clutch adjustment for the particular fastener being driven present further torque buildup once the clutch begins ratcheting.

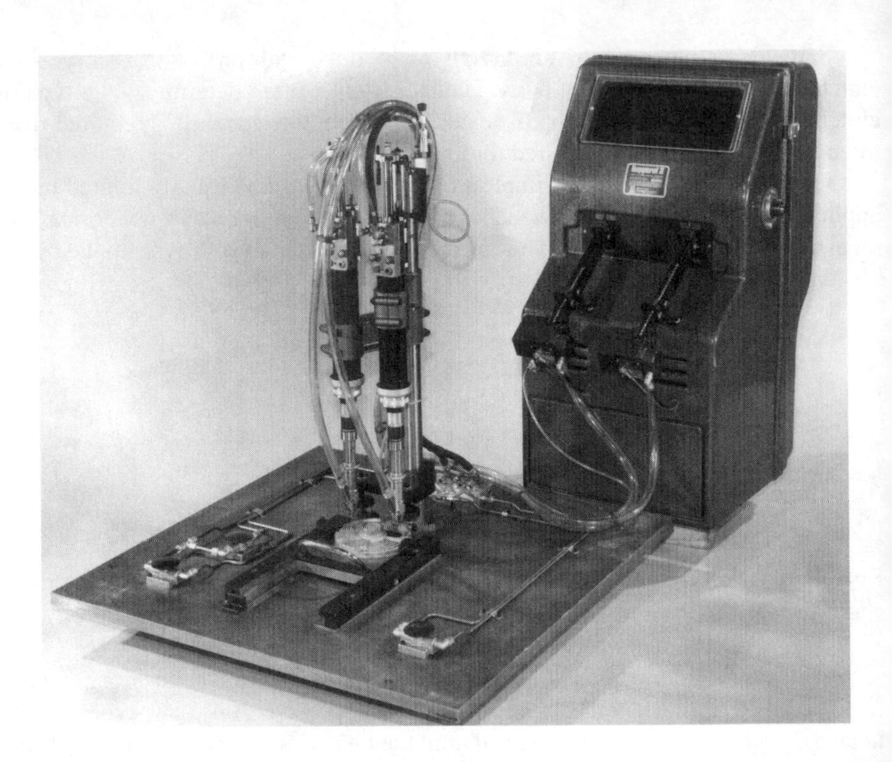

Figure 5.19 Air screwdrivers in an assembly station equipped with automatic screw feeders.

Shutoff and kickout-type clutches offer the best torque control and usually require less maintenance than other types. These clutches have a spring-loaded jaw or detented rolling element to control torque, but they either disengage, rotate freely, or shut off the tool when set torque is reached. Because they do not ratchet when the preset torque is reached, they generally offer the best torque control. For work in close clearances, a variety of attachments are made for screwdriving and nutsetting. These attachments may be positioned with the driving spindle at 45 or 90° to the tool, for many have a flat extended spur gear train with the driving spindle at 90, 75, or 60° to the tool, or parallel to the tool. Plain angle tools may be equipped for fastening with a square drive or with integral flush sockets.

Extended tools have either a square drive or an integral socket. Angle tools are available with clutches or as stall-type models. On work requiring higher torque angle tools are frequently used to reduce operator effort even when no clearance problems exist.

Screwdrivers can be adapted to certain applications by adding a hopper that feeds the screws automatically into position under the driver bit (Fig. 5.19).

Angle Nutsetters

An angle nutsetter, also referred to as an angle nut runner or angle wrench, is an air tool with straight, in-line motor housing and the output spindle arranged to drive at right angles to the axis of the housing (Fig. 5.20). While particularly suitable for applications in which space is limited, they are widely used for bolt and nut tightening. The right-angle configuration of the tool gives the operator a lever to make it easier to absorb the torque reaction.

Figure 5.20 An angle nutsetter being used in the assembly of an automobile hood latch.

Angle nutsetters are available with motors ranging from about 1/2 to 1 3/4 hp and are generally used for fasteners ranging in size from 1/4 to 3/4 in. Normally, a square drive output spindle is furnished for driving hex sockets, but integral flush sockets and other optional spindles are available. These tools are geared, and most models are offered in various speeds and torque output combinations. Many are available in reversible models.

Both stall-type and torque-control angle nutsetters are offered and provide accuracy and reliability. Stall-type tools are run until the fastener stops turning and the motor stalls. At a certain air pressure, a given-size air motor and set of gearing will always stall at the same torque. Adjustment of torque output is made by controlling the air pressure with an air-line pressure regulator. However, much of the effectiveness of this method depends on the technique of the operator, who must hold the tool firmly until the tool stalls, not ease off on the throttle or ratchet the tool.

Torque-control-type tools (Fig. 5.21) are probably the most common today because they are more accurate, and they eliminate most of the operator influence. A built-in control device shuts the motor off at a preset torque. This device can be an air shutoff valve that senses motor pressure, a mechanically disengaging clutch, a speed-sensing governor, or combinations of such devices. Adjustment of torque output is made either at the tool or with an air-line pressure regulator, depending on the design. Torque control is more precise with this type of tool, and sustained torque reaction to the operator is minimized. Both torque-control and stall-angle nutsetters can be provided with extensions to lengthen the lever arm or with reaction bars to further minimize or eliminate torque reaction (Fig. 5.22).

Nutsetters coupled with electronic torque-monitoring equipment are gaining wider acceptance. To be compatible with such systems, special angle nutsetters are available with built-in electronic options. These include torque transducers and angle encoders.

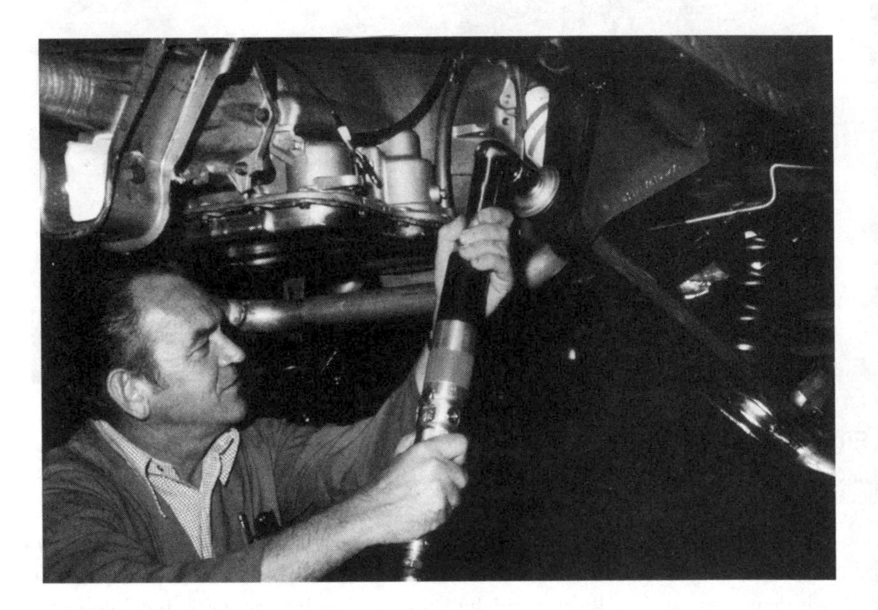

Figure 5.21 Torque-control angle nutsetters offer increased accuracy over hand-held stall tools. Here a torque-controlled nutsetter is used in automobile assembly.

Figure 5.22 A torque reaction bar on a nutsetter or other tool can reduce or eliminate torque reaction transmitted to the operator. Here such a reaction bar is in use with a tube expander.

Ratchet Wrenches

Ratchet wrenches are limited-access, production-assembly tools that are now used throughout the industry for applications where access to the fastener is so limited that conventional power tools cannot be used. By means of a unique and compact ratcheting mechanism, these tools transmit their power to a drive socket in the small end of the attachment. Such attachments are available with widths as small as 9/16 in. and thicknesses as small as 5/16 in.

Some ratchet wrenches may be used in reverse when they are turned over so that the socket runs in the opposite direction. Some even have a reversing mechanism in the valving section. Ratchet wrenches are obtainable as stall-type tools or with the shutoff valve options found in angle nutsetters.

Impact Wrenches

The rapid acceleration of an air motor and its insensitivity to stall make it an ideal power source for an impact wrench (Fig. 5.23). Air impact wrenches weighing only 2 1/2 lb can tighten bolts to a torque of 100 ft.-lb. Other models provide torque as high as 80,000 ft.-lb. Wrenches used for tightening high-tensile-strength bolts in steel structures to torques of 3000 ft.-lb weigh less than 34 lb and are easily handled by one operator.

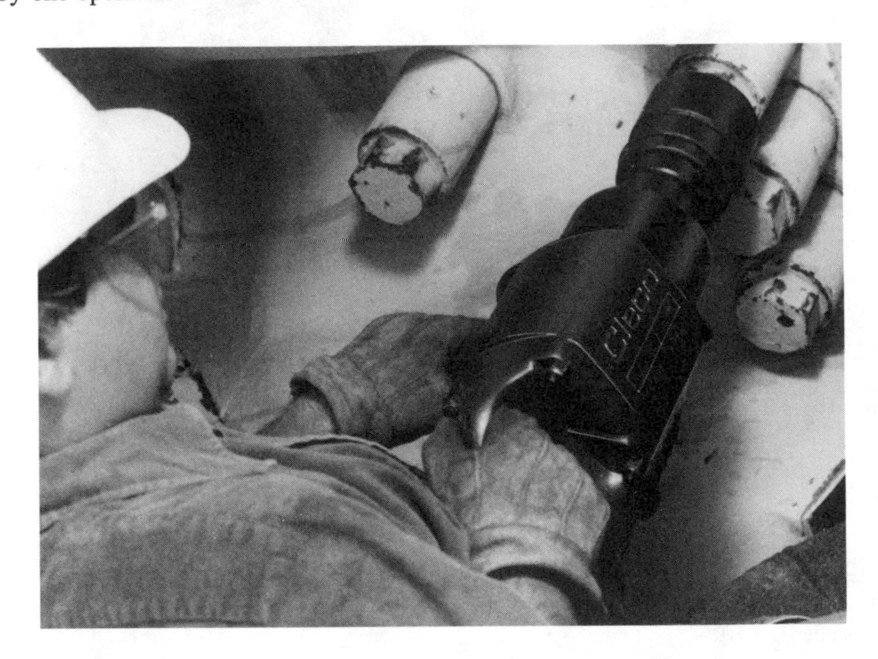

Figure 5.23 An impact wrench generates very high torque in a relatively small tool.

Such wrenches give fast, free-speed rundown before impacting. An impact wrench consists of a rotary-vane air motor and a special clutch mechanism through which energy from the motor is converted to a series of rapid, high-intensity rotary blows. These are transmitted from the wrench spindle to the fastener through a socket or other attachment.

Because of the nature of the impact mechanism, impact wrenches transmit little or no torque reaction to the operator. An operator can hold a heavy wrench on a fastener with one hand while tightening the fastener to as much as 3000 ft.-lb.

Wrenches are available in sizes to handle bolts as small as 1/4 in. in diameter or as large as 12 in. in diameter. Straight handles with lever or button throttles are furnished in sizes from 1/4- to 1/2-in. square drive. Pistol-grip or closed-type handles are furnished on sizes from 1/3- to 1 3/4-in. square drive. Larger tools with square drives as large as 3 1/2 in. are generally furnished in a straight configuration with a throttle handle on one side and a dead handle on the other side.

Wrenches are also available with 45 and 90° angle heads for applications with limited access. Considerable power is lost in the use of angle head attachments, or universal joints with impact wrenches. For safe operation, only attachments designed and manufactured for use with impact wrenches should be used with them. These attachments, such as sockets, universal joints, and adapters, are usually finished in black rather than the bright surface normally found on hand-tool sockets. Hand-tool sockets can shatter if used on impact tools. All modern impact wrenches are reversible and are used for disassembly as well as assembly. Most models provide a choice for square drive shanks and hex, quick-change chucks. Larger sizes have spline drives for heavy-duty application.

Sockets and attachments are retained on the driver or shank of the tool in several ways. Smaller impact wrenches with square drives up to 5/8 in. usually have pin or ball retainers mounted in the driver. Larger tools with square drives 3/4 in. and larger usually have a hole through the driver. Until recently, accepted practice was to retain the attachment on the driver with a through pin and keep the pin in place with an 0-ring that fit into a groove around the attachment. An alternate retaining method now available consists of a one-piece flexible plastic ring and a pin that fits into the holes on the attachment and driver and into the groove around the attachment.

Torques and bolt diameter capacities are approximate and depend on a variety of factors, such as thc following:

1. Condition and lubrication of the threads.
2. Air pressure.
3. Duration of impacting.
4. Type of material being fastened (i.e., metal to metal or metal to a resilient gasket material).
5. Condition of sockets or attachments and how well they fit the tool and the fastener head.

Variations in any of these factors can significantly affect the torque obtained in any particular case.

Some manufacturers rate impact wrenches by bolt diameter capacity, while others rate them according to torque ranges and working torques. In the square drive, the size on impact wrenches corresponds approximately to the bolt capacity. However, because of the many variables involved in impact wrench applications, it is advisable to try a tool under actual operating conditions to ensure success on a particular tool. Most tool manufacturers provide various types of built-in tightening controls on their wrenches. Some include actual torque-responsive devices that monitor the bolt resistance at each blow and automatically shut off the air supply when resistance reaches a predetermined level. Others have built-in timers that admit air to the motor for a prescribed period of time and then shut off the air. Some have an adjustable air restriction that limits the speed and impacting power of the wrench. Another has a torque-limiting bar that is an extension between the impact wrench and socket which, because of its length and diameter, will twist and keep the torque within a predesigned range. A combination torque and turn control wrench is also available. While bolt tightening cannot generally be controlled as closely with an impact wrench as with a torque-controlled nutsetter, good control can be achieved by the operator in many applications by careful attention and skilled use.

Multiple Nutsetters

Wherever, two or more fasteners must be driven in a high production assembly, the use of multiple nutsetters should be considered. These consist of a group of specially designed nutsetter motors, one for each fastener, mounted on a common frame and equipped with a common air supply manifold. An entire group of fasteners can be driven simultaneously, as shown in Fig. 5.24, with a multiple nutsetter drastically cutting assembly time and improving quality control.

Units with as many as 30 spindlcs havc bccn built and uscd successfully. Multiple nutsetters are custom designed for each specific assembly operation and can be furnished either as portable units, as components of stationary assembly machines, or as transfer-line components (Fig. 5.25). Portable units may be suspended from balancers or air cylinders, either of which can be trolley mounted.

Figure 5.24 Multiple nutsetters allow two or more nutsetters to be driven simultaneously to reduce assembly time and improve quality control.

Portable models are designed to incorporate a mounting plate, an air manifold, a throttle handle, support handle, an adjustable suspension ball, and a steel enclosure around die motor units lined with a sound-absorbing material. Torque reaction is transmitted through the fixture plate and is absorbed by other spindles rather than by the operator. To facilitate their use, multiple nutsetters can be furnished with horizontal swivel suspension, offset motor spindles, push-off devices, and extended spindles. When fastener patterns are compatible, it is sometimes possible to build a multiple unit in which the spindles can be readily adjusted to two or more positions.

Multiple nutsetters designed as components for larger assembly machines are usually furnished without handles, but with provision for slide mounting, and they may be equipped with special spindles to accommodate fastener feeding devices. Motors used in multiple nutsetters are most commonly of the stall type, but some incorporate torque-control clutches or retorquing devices. Clutches are used primarily on multiple nutsetters designed for smaller fasteners with relatively low torque limits. A retorquing device is common on multiple nutsetters for high-torque applications.

Figure 5.25 A multiple nutsetter, which is part of a stationary assembly line.

These devices apply successive follow-up torques to a fastener after the initial setting torque to take up the residual softening of torque in the fastener. Excellent quality control can be achieved in most multiple applications because of the elimination of many adverse factors that are present in a single, hand-held tool.

The motors are equipped with special square-drive or screwdriver-type spindles that are spring loaded to provide axial travel. This enables the spindles to accommodate fasteners that are not exactly the same height and ensures that all fasteners will be engaged before driving commences. The motor housing includes a mounting flange perpendicular to the axis of the motor and located toward the spindle end.

Optional mounting arrangements are frequently utilized. Motors are available in a large selection of horsepower ratings, speeds, and torque capacities. To satisfy the requirements for driving multiple fasteners in restricted places, standard-angle nutsetters and ratchet wrenches can be mounted together in multiple units.

Because multiple spindles are fixture mounted, it is relatively easy to add devices to indicate when each spindle has completed its driving function. This signal may be air, electric, or mechanical and can be used to operate lights, a bell, or other devices to indicate to the operator that all spindles have operated satisfactorily. To satisfy the increasingly important requirement to show proof of due care in fastening applications, the signal may be used to generate a computer readout or some other permanent record of the operation (Fig. 5.26).

Figure 5.26 Multiple nutsetters are sometimes connected to monitoring equipment, which can print out and/or display the torque applied to each fastener.

Riveters

Portable riveters fall into two general categories: rivet hammers and rivet squeezers. The latter are commonly referred to as compression riveters. Rivet hammers, as shown in Figs. 5.27A and B, are applied to one end of a rivet, and a heavy metal object known as a bucking bar is held against the other end. Thus, they can be used to drive rivets anywhere in an assembly, providing there is space to get the hammer and the bar on the rivet. Squeeze riveters are limited to driving rivets located within a certain distance from the edge of the assembly, since the yoke of the riveter must pass over the edge to engage the rivet between a stationary die on one face of the yoke and a moving die that advances from the other face (Fig. 5.28).

Practically all rivet hammers are air-operated since air cylinders and valves provide the necessary, fast-acting, reciprocating motion of the hammer in a compact, cool-running assembly. Several types of rivet hammers are available. The most common is the small, aircraft-style riveter, which delivers a series of rapid blows as long as the throttle is held down. One-shot rivet hammers, which deliver one hard blow each time the throttle is opened, are used where rivet material tends to work-harden under repeated blows.

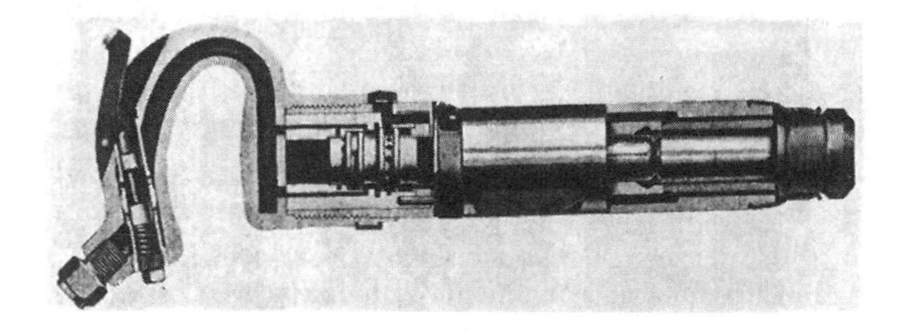

Figure 5.27A Cutaway view of a riveting hammer.

Figure 5.27B Portable riveter used in aerospace assembly.

Figure 5.28 A compression riveter with a C-yoke used to fasten a hinge to a metal frame.

A rivet squeezer, or compression riveter, consists of an air cylinder with one or more pistons that drive a cam or toggle mechanism. This, in turn, drives a ram in which a rivet die is positioned. The cam or toggle arrangement multiplies the force exerted by the piston, resulting in a force at the rivet die that is many times greater than the force that could be developed by the piston and cylinder alone.

Squeeze riveters can be equipped with C-type yokes in which one die is located on one leg of die C with the ram advancing from the other leg. They are also available with alligator-type yokes, as shown in Fig. 5.29, in which the moving jaw becomes the ram. Location and orientation of the rivet are the factors that determine the type and size of yoke that must be used.

Figure 5.29 A squeeze riveter with an alligator-type yoke.

Riveters are available in sizes to drive rivets from 3/32 to 1 1/4 in. in diameter. They can be hand held, balancer suspended, bench mounted, or pedestal mounted and can be equipped with a hand throttle on the tool or with foot valve actuation. They are used for steel fabrication, railroad car building, ship building, aircraft manufacturing, truck building, and the manufacture of many other products that require a strong, permanent assembly.

ABRASIVE TOOLS

Air tools are widely used for grinding, sanding, wire brushing, burring, rotary filing, sawing, and other similar applications. They can be fitted with a variety of attachments (Fig. 5.30), depending on the requirements of the job. Care must be taken to ensure that current safety standards are followed with regard to wheels and attachments, as well as mounting flanges, arbor lengths, diameters, throttles, and guards.

Figure 5.30 Forgings, castings and weldments often have burrs or other surface irregularities removed by grinding, sanding or wire brushing.

Safety Standards

Recommended safety standards are published by the American National Standards Institute, Inc. (ANSI), www.ansi.org. The two publications applicable to abrasive tools are:

1. B7.1: *The Use, Care, and Protection of Abrasive Wheels*
2. B186.1: *Safety Code for Portable Air Tools*

Both publications are evaluated and updated periodically. The most up-to-date copy should be used as a reference.

These codes enable manufacturers to design their products in accordance with clearly defined standards and enable users to establish proper procedures and rules for the use and care of these products. Users should refer to grinding wheel blotters for the maximum safe rpm of their wheels (Fig. 5.31), in addition to making absolutely sure that the other requirements for the safe operation of abrasive tools are strictly adhered to.

Figure 5.31 The maximum safe operating speed of a grinding wheel should never be exceeded.

Abrasive Tool Speeds

Maximum tool speeds must be controlled with close limits. Abrasive tool speeds are maintained by the following:

1. *Governor control:* Larger tools are usually designed with built-in governors or speed controllers that restrict the flow of air to the motor as the speed increases to the rated speed of the tool. Overspeed shutoff devices linked to the governor are available on some tools. Their use results in the tool shutting off if the rated speed of the tool is exceeded.

Figure 5.32 Die grinders with built-in speed regulator.

2. *Adjustable speed regulators:* Sometimes found on smaller grinders, the adjustable speed regulator is usually a needle valve or similar air flow restriction (Fig. 5.32). Care must be taken to ensure that wheels or attachments larger than those rated for the maximum safe speed of the tool are never used.

3. *Air-line pressure regulators:* An air-line pressure regulator can be used to change the inlet pressure of the air supply to the tool. This change in pressure will result in a corresponding change in speed. This method, however, will also affect the power produced by the tool (Fig. 5.33).

Figure 5.33 The speed of an air tool can be changed by adjusting the air-line regulator.

Abrasive Tool Throttles

A variety of types of throttles are available, the one selected depending on the design and purpose of the tool. In almost all cases, throttles must be self-closing. The only exceptions are throttles on small wheel tools under conditions prescribed in the *Safety Code for Portable Air Tools*, B186.1, referred to earlier. The use of lock-off-type throttle levers (Fig. 5.34) is becoming more prevalent. These throttles prevent inadvertent starting of the tool while it is attached to an air-line but is not in use. Other available throttles are self-closing button or thumb types and self-closing grip types.

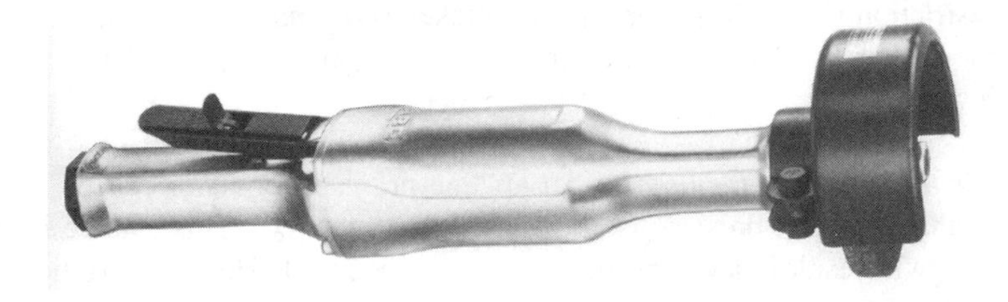

Figure 5.34 A lock-off throttle on a grinder helps prevent inadvertent startup of the tool.

Abrasive Tool Guards and Options

Current safety standards call for the use of guards on all large grinders (Fig. 5.35). The guard is intended to protect the operator from wheel fragments if the wheel breaks while running, deflects sparks and chips, limits wheel size, and keeps the operator from accidentally touching the wheel. Exhaust can be from the front, rear or side. Piped away exhaust attachments are available on many tools for jobs where low noise levels are required or where contaminants in the exhaust air are objectionable. Vacuum pickup of material removed by the abrasive wheel is also available on some tools.

Figure 5.35 A wheel guard protects the operator from sparks and chips, and from wheel fragments if a wheel breaks.

TYPES OF ABRASIVE TOOLS

Die Grinders

Die grinders (Fig. 5.36) are the smallest and fastest tools available. They are used for the finest kind of tool and die finishing, rapid metal and burr removal, and other comparable work. They are usually furnished with 1/8- to 1/4-in. female collet spindles, although male-threaded arbors and guards are available so that small-diameter wheels can be used. Care must be exercised to select small wheels and burrs that have speed ratings compatible with the high speeds of these grinders. There are two types of die grinders: turbine-powered and vane-powered.

Turbine-powered die grinders rely on their high speed, up to 100,000 rpm, to achieve metal removal. They are usually hand-held, as in Fig. 5.37, but are sometimes fastened by an appropriate fixture to a bench or to the turret of a lathe. Because of their high speed, turbine grinders are equipped with high-precision collets. Grinders of this type are generally fitted with a twist-type or screw-type throttle. This feature frees the operator from the need to manipulate a throttle. Some manufacturers furnish turbine grinders with a length of air hose and a cartridge filter to give extra protection to the motor and bearings by removing airborne contaminants.

Most turbine grinders have a single-stage rotor driven by compressed air. Some grinders (Fig. 5.38) have two rotors separated by a stator. This feature allows more efficient use of the air and provides increased torque.

Figure 5.36 Die grinders are used for tool and die finishing, rapid removal of metal and burrs, and some weld finishing.

Figure 5.37 Turbine grinders are used for light metal removal where precision is important.

Figure 5.38 Some turbine grinders use a two-stage turbine to gain additional power.

Vane-powered die grinders have speeds ranging from 10,000 to 60,000 rpm. They are usually hand held. They are more widely used than turbine grinders because of the torque available at optimum wheel speeds. As load is applied to a vane-motor grinder, its speed does not decrease as rapidly as that of a turbine-driven tool.

Small Wheel Grinders

Small wheel grinders normally use straight arbor-hole wheels or shank-mounted cone wheels (Fig. 5.39). Speeds may be as high as 40,000 rpm and are usually governor controlled. Weights range from 1 1/2 lb to approximately 6 lb. There is no firm delineation between these tools and the smaller die grinders or the larger horizontal grinders, but they do have many applications in industry and construction and warrant separate classifications as abrasive tools. These tools can also be used with rotary files or wire wheels, and models are available with extended spindles for use in places with limited access.

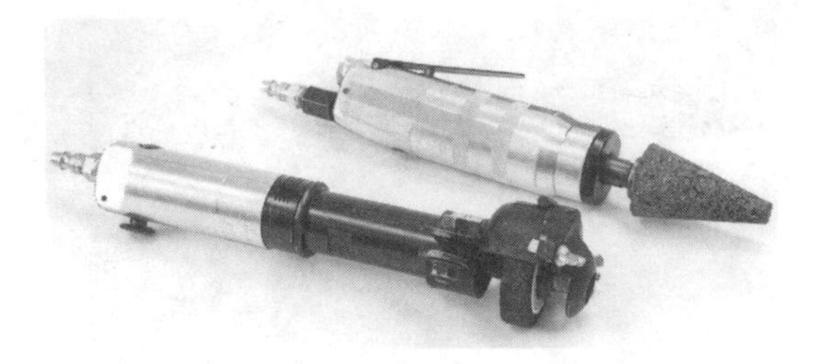

Figure 5.39 Small-wheel grinders, equipped with plug, cone, type-l or wire brush wheels are used in a variety of operations, primarily foundry work.

Angle Grinders

Angle grinders (Fig. 5.40) with speeds up to 20,000 rpm, have spindles mounted at right angles to the motor housing. Power is transmitted to the spindle through a set of gears in the angle head. As with die grinders, they are furnished with either female collets from 1/8 to 1/4 in. or threaded male arbors and guards. Weights range from 1 to 3 1/2 lb. Most angle grinders are in the small, ungoverned, higher-speed sizes, although larger sizes having a governed motor, and speeds as high as 7500 rpm are available.

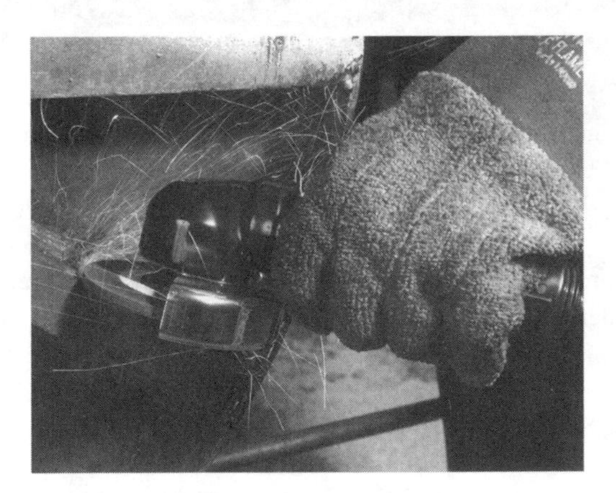

Figure 5.40 Right angle grinder being used to remove weld spatter from a fabricated part.

Large Wheel Grinders

These are heavier-duty tools that may be classified as either horizontal grinders or vertical grinders. The horizontal tools have the motor and handle aligned and are used for fast, heavy-duty metal removal (Fig. 5.41). Speeds range from 3100 to 10,800 rpm. Weights without wheels average 10 lb. Arbors are male threaded for use with wheels from 5 to 8 in. in diameter. The handle is usually either of the spade type or the straight type with an integral lever throttle. Figure 5.42 shows a horizontal grinder with a spade-type handle and self-closing throttle.

Figure 5.41 Large casting being cleaned up with a horizontal grinder.

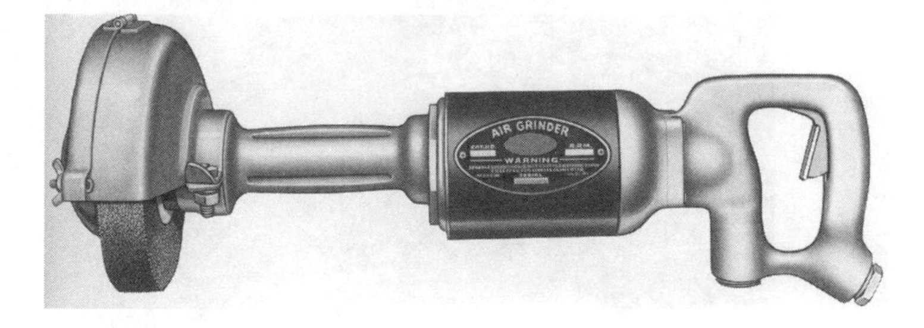

Figure 5.42 Horizontal grinders are usually equipped with a straight or spade-type handle. A spade handle is pictured here.

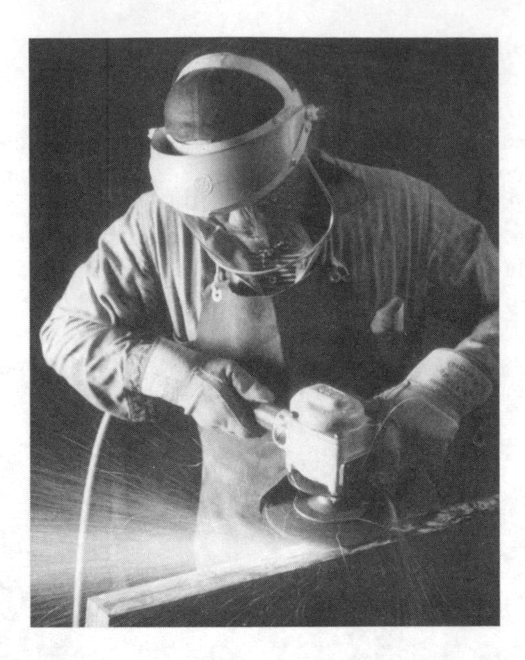

Figure 5.43 Vertical grinders are considered to be the most versatile and widely used of large-wheel grinders.

Figure 5.44 A vertical grinder being used to remove surface flaws and burrs from a large casting.

Vertical grinders (Figs. 5.43 and 5.44) are perhaps the most widely used for all large-wheel grinders. They are so named because the motor is mounted in a vertical position to grind surfaces perpendicular to the axis of the motor. They are most often used for rapid removal of metal from castings and weldments, but can be used with the proper wheels and guards for finishing concrete and other heavy-duty applications. Cup-type wheels up to 6 in. in diameter and depressed-center or flat-disc wheels up to 9 in. in diameter, with the proper guards and flanges, are most often used with these tools. Most vertical grinders are ungeared and rated at 4500 to 6000 rpm, although speeds of 12,000 rpm are available. Weights range from 4 to 12 lb; the average weight without wheel is approximately 8 lb. Most vertical grinders have a dead handle, and many such handles can be positioned in one of two angular locations for ease of operation.

Belt Grinders

A belt grinder uses a contact wheel and sometimes a platen or flat plate to support a continuous abrasive belt driven by a governed motor (Fig. 5.45). The belt is usually 2 in. wide. A front exhaust helps blow away the metal dust created by the grinding operation. These tools are useful for grinding welds, forgings, and castings and for removing surface imperfections.

Figure 5.45 A belt grinder is a versatile tool used on plate stock, welds, forgings and castings.

SANDERS AND POLISHERS

Sanders use coated abrasive materials in the form of discs, belts, or sheets. Support for such materials is provided by flexible backup pads, sponge-rubber pads, or platens. The proper selection of a portable air tool for sanding or polishing is determined by the nature of the work, such as metal removal, feather edging, wet sanding, or dry sanding. In selecting the speed, the user should be guided by the recommendations of the abrasive manufacturer.

A polisher is basically a sander equipped with a polishing bonnet (Fig. 5.46) instead of an abrasive disc. A maximum speed of 4800 rpm is recommended for polishing, with those running in the 1800- to 2500-rpm range being more generally used.

Figure 5.46 A polisher being used to put a final luster on a bathtub enclosure unit.

It is a generally accepted standard that the most efficient sanding speed is 9500 surface feet per minute on the perimeter of the disc. Such a speed is considered to provide the best combination of economical disc life and effective material removal.

There are four general styles of sanders available to industry: vertical, right angle, belt, and palm.

Vertical Sanders

Vertical sanders (Fig. 5.47) are so named because the motor and spindle are in a vertical position relative to the pad and coated abrasive disc. A vertical sander is normally equipped with two handles. One has an air inlet and throttle, and the other is called the support handle. Some vertical sanders are designed so that the support handle may be adjusted to several radial positions relative to the throttle handle. Live handles on some tools can be repositioned to place the throttle at the bottom or on either side.

Figure 5.47 Vertical sanders are capable of a high rate of material removal from composite materials. They are also used to finish metal products.

The speed of a vertical sander is established by gear reduction or controlled by a governor or air regulator. The nature of the work usually determines the choice between gearing or governing. Where rough material removal is required, a geared tool would probably be the best choice because of its high-torque output. But, if it is more desirable to maintain a uniform peripheral speed within narrow limits, a governor-controlled tool would probably be more suitable. Tools are available that are both geared and governed to achieve both of these advantages.

Belt Sanders

Belt sanders (Fig. 5.48) use wheels and sometimes a platen or flat plate to support a continuous abrasive belt. The width of the belts ranges from 1/2 to 3 in. These produce a fine-quality, in-line finish free of swirl marks. They are usually equipped with a forward exhaust stream that blows away the dust and grit, giving the operator a clear view of the work surface.

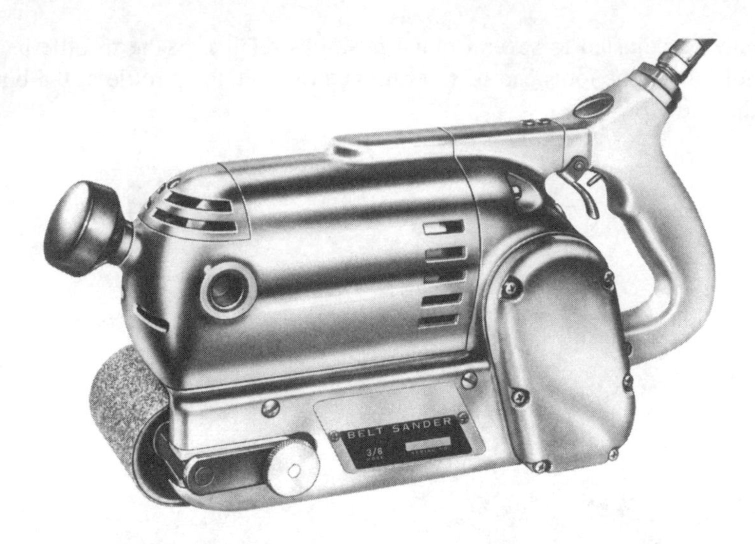

Figure 5.48 Belt sanders are used for fast removal of material, usually from wood parts.

Palm Sanders

Palm sanders are available in two types: orbital sanders and random orbital sanders. Orbital sanders (Fig. 5.49) are small, one-hand tools with a platen and clips to hold several pieces of abrasive paper. The tool moves the abrasive rapidly in a small orbit from 3/16 to 3/8 in. in diameter. Speeds are as high as 10,000 cycles per minute.

Figure 5.49 Palm sanders provide easy, single-hand control for finishing work on wood, metal and plastic composite parts.

Random orbital sanders (Fig. 5.50) provide a combination rotary and orbital movement. These tools are used for fine finishing primarily in the woodworking, automotive, and plastic industries and can provide a swirl-free finish. Wet sanding and vacuum pickup attachments are available.

Adhesive-backed abrasive discs are available for quick attachment and removal.

Figure 5.50 Random orbital sanders have a good operator control and provide a smooth, swirl-free finish.

SPECIAL SANDING ATTACHMENTS

Drum Sanders

An air-powered grinder, suitably adapted, can be used with a sanding drum and adapter. The diameter of the drum is usually no larger than 3 in., but the length of the drum may be as much as 10 in. The largest tool of this type, used in the woodworking industry, is commonly referred to as a rolling-pin sander (Fig. 5.51). The drum is equipped with a handle at one end and is powered at the other end by an air motor.

Figure 5.51 Drum sanders are widely used in the woodworking industry to sand irregular surfaces.

Belt Sanding Attachment

This is an extension arm fitted with an idler wheel attached to the front end of a straight grinder. The unsupported area of the belt between the two wheels is applied to convex surfaces. Sanding can also be done with the power wheel to sand concave surfaces.

Mandrel

A mandrel can be attached to a straight grinder and fitted with a cylindrical abrasive sleeve. It is used on tight-access work or a double-contoured surface. The mandrel is made of rubber and is slotted in such a way that the slots expand under centrifugal force. Such mandrels are available in a variety of sizes.

Radial Flap Wheel

The radial flap wheel (Fig. 5.52) carries a series of radially inserted abrasive papers. Under load, the papers fall back to present a powerful abrasive face. Such wheels are used primarily in the woodworking industry for sanding complex shapes that would otherwise be difficult to finish.

Figure 5.52 Sanding of double curved contours is made easy using a die grinder equipped with a flap wheel.

PERCUSSION TOOLS

Chipping Hammers

The chipping hammer is used for gouging and cutting away base material. It can be used for such operations as slag removal, peening, rock breaking, core busting, riser cutoff, fin removal, and weld-bead removal. Chisel inserts have hex, slotted, or round shanks.

Scaling Hammers

Scaling hammers (Fig. 5.53) are high-speed, lightweight percussion tools used for removing flux and spatter after welding. They deliver a somewhat lighter blow than chipping hammers. Other uses are to chip heavy paint, remove scale and rust, break sand molds, and process sheet metal. They are preferred where a light blow is required. Some manufacturers offer reduced sound and vibration models (Fig. 5.54).

These hammers are furnished as straight or pistol-grip models. They can also be furnished without a manual throttle, delivering their reciprocating action when pressure is applied to the chisel.

Figure 5.53 A scaling hammer is being used here to remove flux and spatter from a weldment.

Figure 5.54 Recent designs of scaling hammers have features which reduce sound levels and vibration.

Needle Scalers

The needle scaler is a hammer in which the blow is transmitted through a group of floating needles or metal rods (Fig. 5.55). It is especially well suited to cleaning away weld flux, rust, scale, paint, weld spatter, and other residues from irregular surfaces in shipyards, fabrication shops, refineries, foundries, and construction sites.

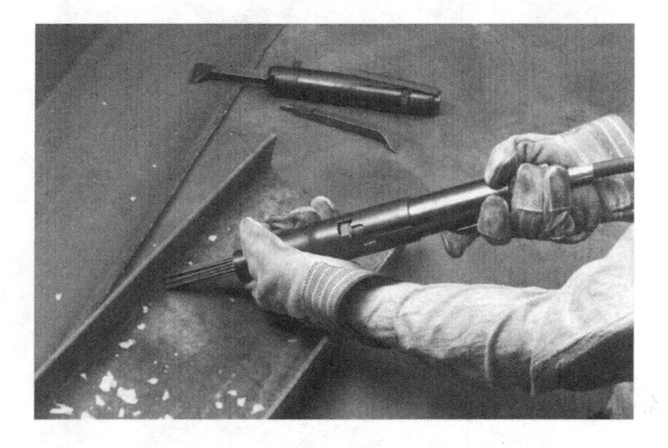

Figure 5.55 Needle scalers are well suited for cleaning weld flux, scale, paint and weld spatter.

Sand Rammers

In the preparation of molds for casting, it is necessary to compact the sand to a proper density. The sand rammer is used for this purpose and is available in a variety of sizes. It has an integral piston and rod to which the tamping device is attached. These tools can also be used for backfilling and for tamping refractory material in furnaces.

SPECIAL PURPOSE AIR TOOLS

Many industrial applications cannot be handled by standard air tools. Whenever these unusual requirements arise, modifications and specially designed tools often solve the problem. A wide variety of such tools is available in industry. Some of these will be described in detail.

Saws

Several air tool manufacturers offer standard motor housings with special sawing attachments. The most common tool in this category is the circular saw (Fig. 5.56) used in smaller sizes for cutting metal, plywood, and synthetics. Larger, specially designed saws with 8- and 12-in. blades are used for cutting lumber at construction and railroad sites. These saws, with the proper blades, are also used to cut a variety of metallic and nonmetallic material.

Figure 5.56 Circular saw being used to cut metal stock.

Routers

Routers are high-speed rotary tools (Fig. 5.57) fitted with standard cutting bits and used for the trimming and edging of wood, metal, synthetics, and other materials.

Figure 5.57 Routers are available for cutting wood, metal, and fiberglass.

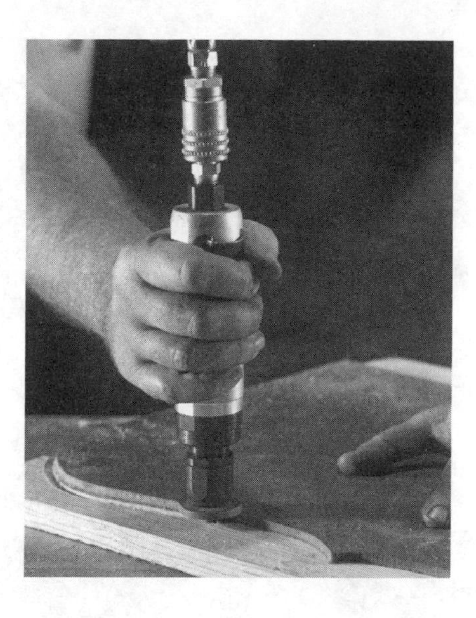

Figure 5.58 A stylus router is a type used for cutting a pattern, and for cut out work in aluminum.

They are available with a selection of handles and throttle controls and with speeds ranging from 750 to 25,000 rpm.

One type of router (Fig. 5.58) resembles a die grinder and is often called a stylus router. The nose housing accepts a template guide bushing through which a cutting bit protrudes. A metal template is laid against the work. The operator then follows the edge of the template, with the template guide bushing of the router producing a clean, precise cut. Special guide attachments are also available for use in trimming plastic laminates.

Shears

Air-powered shears are usually standard air tools fitted with various kinds of cutting attachments. One type of attachment (Fig. 5.59) has a scissors action for cutting and trimming sheet metal. Its cutting rate may vary from 8 ft./minute for 10-gauge metal to 16 ft./minute for light-gauge metal. Some air-powered shears, as shown in Fig. 5.60, have a rotating knife-edge blade that cuts textiles such as upholstery materials, synthetics, plastics, leather, fiberglass fabric, canvas, drapery goods, and similar materials.

Figure 5.59 Air powered reciprocating shears are used to cut and trim sheet metal.

Figure 5.60 Rotary shears speed the production cutting of textiles, leather, fiberglass fabric, sheet rubber goods, and similar materials.

Wire-wrapping Tool

A wire-wrapping tool is used to wrap a solid conductor wire around a terminal. The configuration of such a terminal is always square or rectangular. The wire is wrapped so tightly around the terminal that no soldering is required. This is one of the most reliable electrical connections known today. These tools are used in the computer, telephone, business machine, and communications industries. They are also used by makers of television sets, radios, automobiles, helicopters, airplanes, submarines, and space vehicles.

Single-acting Air Gun

The axial piston air tool (Fig. 5.61) is a single-shot gun. When its throttle is depressed, it produces a single, powerful blow without delivering any reaction to the operator. It is used for driving pins, center punching, wire cutoff, cutting and crimping, cutting and bending, and driving plastic rivets. Smaller models deliver a static force of 65 lb and larger models deliver a maximum static force of 300 lb.

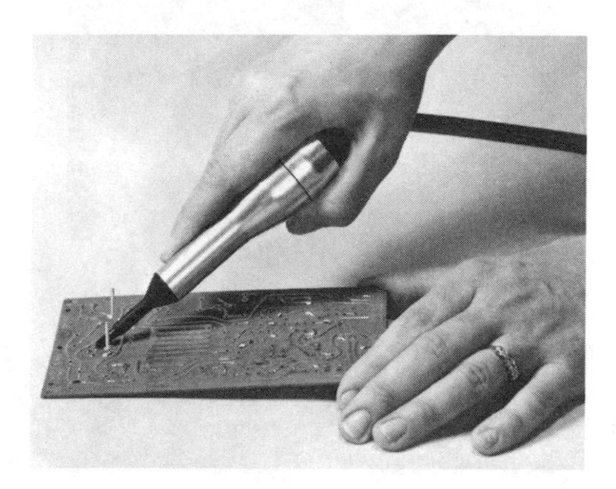

Figure 5.61 A single-shot gun designed to cut electrical wire for circuit boards.

Concrete Vibrators

Concrete vibrators (Fig. 5.62) are used in construction work. The air motor and eccentric weight are mounted in the vibrator head and attached to a long hose handle. When inserted in freshly poured concrete, the vibrator will cause the mixture to settle rapidly for greater density.

Figure 5.62 Concrete vibrators cause freshly poured concrete to settle and achieve greater density.

Core Knockout Tools

The core knockout tool (Fig. 5.63) is functionally the reverse of a vibrator. In a foundry, after a casting has cooled, the knockout tool clears away the sand that has been used in the core.

Figure 5.63 Core knockout tool being used to remove sand from a casting.

Electrode Dressers

Electrode dressers (Fig. 5.64) are designed to recondition electrode tips. They are available with a wide variety of interchangeable cutters to match the specific shape of the electrode being dressed. An electrode dresser eliminates the need for removing the electrode. It can be reconditioned on the spot by a welder.

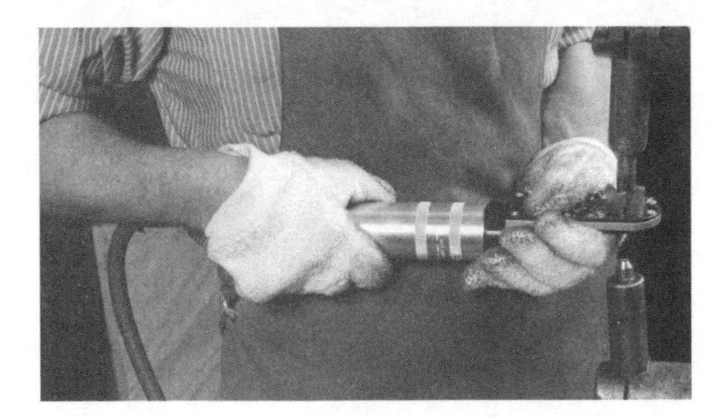

Figure 5.64 Electrode dressers quickly reshape electrodes to their original form for more effective welding.

Air Starters or Cranking Motors

An air starter is an air motor and reduction gear that drives a pinion engaged during startup with the engine's ring gear. It provides the high torque and speed required to start the engine.

Maintenance costs are low, and reliable hot-engine restarts as well as cold-weather starts are achieved. Gas-sealed starters for use with a natural gas energy source are available. Air starters are used for starting highway trucks, off-highway construction vehicles, buses, stationary power plants, pumps, locomotives, marine engines, and oil-field equipment.

AIR-POWERED AUTOMATIC PRODUCTION TOOLS

An air-powered production tool is one that automatically feeds a cutting tool into the workpiece and then retracts it. It uses an air or electric motor for rotary motion, an air piston or lead screw for forward and reverse feed of the cutting tool, and internal or external valving to control the forward and retract modes. Automatic production tools (Fig. 5.65) are among the most versatile production tools available to industry. They are used for drilling, boring, tapping, countersinking, counterboring, and spotfacing. They are used in place of manual operations to achieve greater uniformity, speed, and accuracy, to reduce operator fatigue, and to perform multiple operations simultaneously, and they can be part of tooling that is built around nearly any part. These air-powered units are more compact than their electric-powered counterparts, and they can be utilized in any required position. Many units can be employed in a limited space, and they can be permanently mounted or portable with fixtured mountings (Fig. 5.66).

The four basic categories of air-powered automatic production tools are fully air operated, air hydraulic, air mechanical, and air electric.

Figure 5.65 Automatic production tools replace manual operations to achieve greater consistency, speed and accuracy, to reduce operator fatigue, and to do multiple machining operations simultaneously.

Figure 5.66 Air tool being connected to a portable fixture for an automated drilling operation.

Fully Air Operated

This design incorporates a single- or double-acting air piston, an air motor, reduction gearing to achieve speeds from 450 to 19,000 rpm, and a drill chuck or tapping chuck up to 3/4-in. capacity. Units with built-in valves to control air supplied to the piston and motor are called valve-in-head tools (Fig. 5.67).

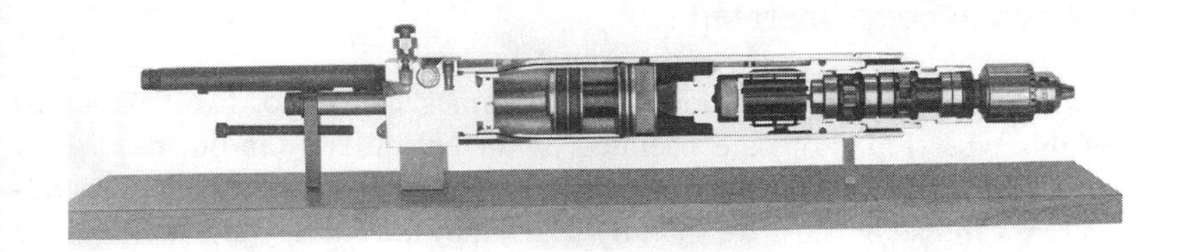

Figure 5.67 Cross-sectional view of a fully automatic production tool.

Those requiring external valves are referred to as through-head tools. Both types have needle valves that control the feed rate of the tool by regulating the exhaust air from the piston. Valve-in-head models also have needle valves to control the retract rate. External hydraulic-check assemblies are available for use with air-operated units to provide hydraulic feed control to produce a higher-quality hole with minimal breakout. Fully air operated units are considered general-purpose tools, and they find application in wood, composite plastics, and sheet, plate, and cast metals.

A specifically equipped automatic production tool that falls into the fully air operated category is the peck drill. These tools are used for deep-hole drilling where hole depth exceeds four times the drill-bit diameter. The peck drill feeds rapidly to the work, drills for a preset stroke, retracts to clear chips, feeds, and then continues to recycle until the total preset hold depth is drilled. Speeds range from 500 to 17,000 rpm and maximum drilling capacity is 5/16 in. in mild steel.

Air-Hydraulically Operated

Air-hydraulic tools are similar to the fully air operated units described previously, but instead of an external hydraulic check assembly, they have a built-in hydraulic oil chamber with an adjustable orifice to control the rate of feed (Fig. 5.68). This feature allows rapid traverse and skip-feed operation. These tools incorporate internal valves for feed and retract modes, while rotational speeds are controlled by gears. They find application primarily where feed control is critical, as in the control of breakthrough.

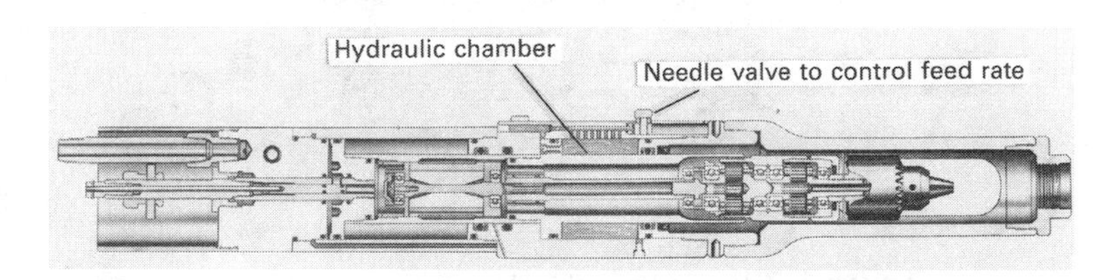

Figure 5.68 Cross-sectional view of an air-hydraulic operated automatic production tool.

Air-Mechanically Operated

Air-mechanically operated tools are sometimes called positive-feed tools. Drills of this type (Fig. 5.69) are operated by air but differ from other self-feed drills in that the drive spindle is fed mechanically. Mechanical-feed drills generally have a screw thread on the drive shaft by which they are fed forward and, in some cases, retracted. Speeds range from 50 to 1200 rpm, and feed rates range from 0.0002 to 0.16 in. per revolution. These drills generally find application where large holes are drilled in thick common metals or where holes of any size are drilled in exotic composite plastics and high-tensile-strength metals.

For tapping units, mechanical feed is provided by a lead screw and nut (Fig. 5.70). The thread of the lead screw is precision ground to the same pitch as the cutting tap. In operation, it is fed forward and retracted through a stationary nut by a reversible air motor. The air motor is controlled by valves within the tool. Lead screw and nut assemblies are interchangeable and are available from 11 through 80 threads per inch and in metric sizes.

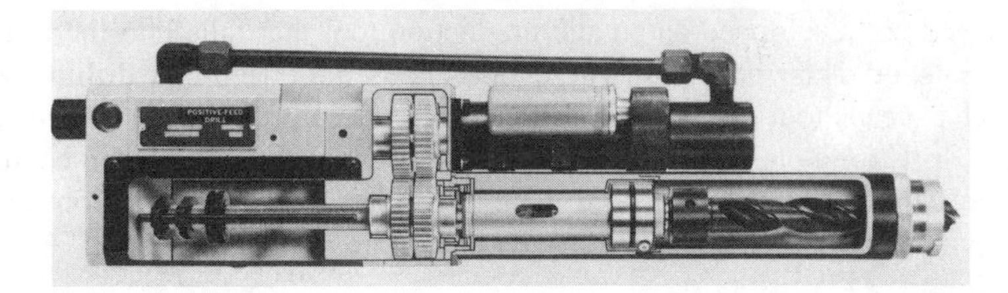

Figure 5.69 Cross-sectional view of an air-mechanically operated automatic production

Figure 5.70 Cross-sectional view of a tapping unit.

Air-Electrically Operated

The air-electric design incorporates an electric motor for rotational motion and an air piston for feed and retract modes. Internal valves control the air flow to the air piston, while electric switches turn the motor on and off.

Optional Head Assemblies

Automatic production drill units can be provided with head assemblies other than standard drill chucks. Most models can be equipped with collets, taper shank adapters, other jaw-type chucks, fluid chucks, and multiple-spindle drill heads. The multiple-spindle heads are capable of drilling two to six holes simultaneously (Fig. 5.71).

An offset drill head assembly (Fig. 5.72) places the standard drill chuck approximately 2 1/2 in. out from the axis of the tool. This permits grouping of tools to achieve holes with extremely close centerline distances.

Some drill units come equipped with self-reversing jaw-type or clutch-type tapping heads to accommodate from number 0 to 1/2-in. tap sizes.

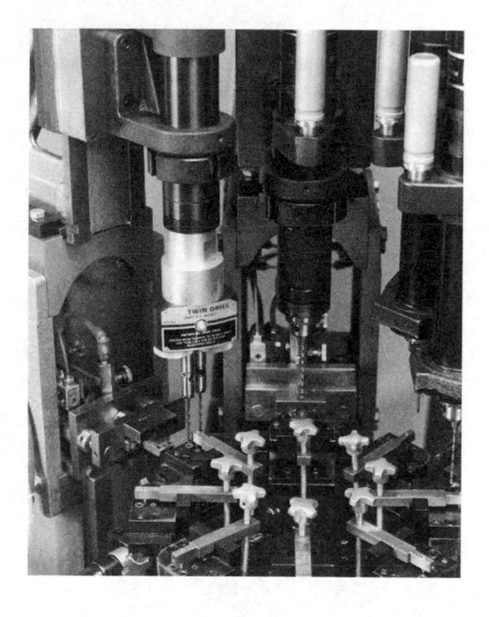

Figure 5.71 Multiple spindle drills allow two or more holes to be produced by one automatic production drill.

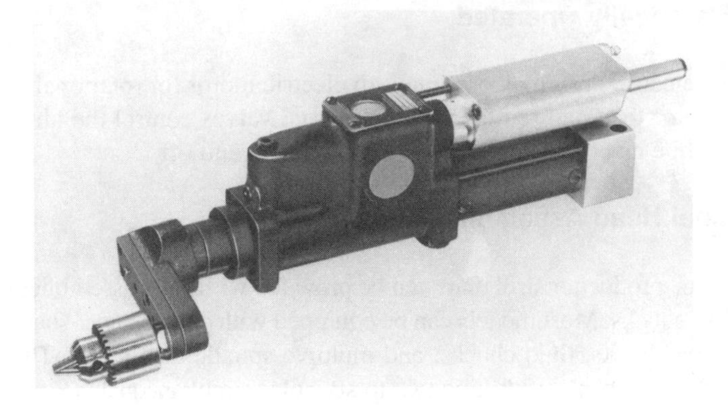

Figure 5.72 Offset drills allow closer tool groupings where holes are close together.

Mounts

Automatic production tools can be mounted in a variety of ways. A nose housing (Fig. 5.73) provides a means of mounting the automatic production tool to the fixture plate and, in some cases, to hold a guide bushing for the drill bit. The nose housing can be mounted to a fixture plate either by threading or by a portable bayonet-type mount.

Split block clamps, collet-type foot clamps, and standard and heavy-duty foot brackets can be used to mount the drills to machine plates. Tubular mounting systems (Fig. 5.74) provide the flexibility for mounting multiple tools at any orientation from one machine plate.

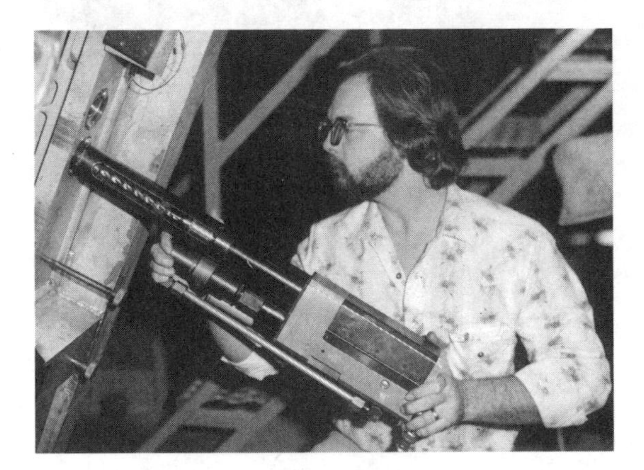

Figure 5.73 A nose housing provides a convenient means of mounting a tool where the part is too big to move to a machine.

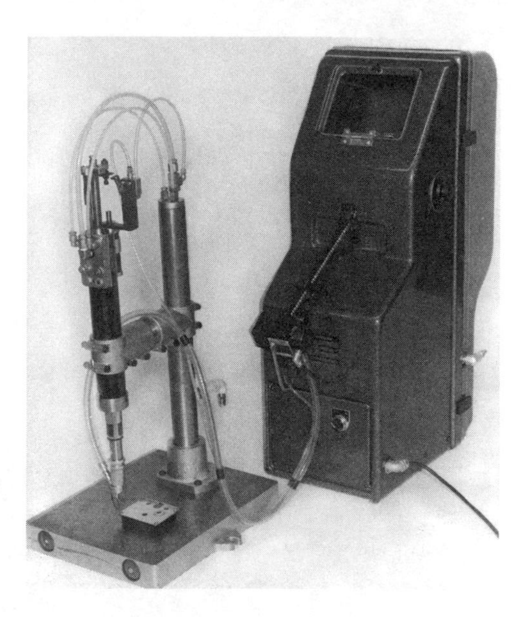

Figure 5.74 Tubular mounting systems provide for easy mounting, and flexibility in tool position.

OVERHEAD AIR HOISTS

Manufacturing depends heavily on transfer and handling of materials. There are many kinds of material-handling equipment available, but one of the most versatile and widely used types is the overhead hoist (Fig. 5.75). Although overhead hoists are available in hand-chain, electric, and air-operated models, air hoists offer many inherent advantages over hand-chain and electric hoists. When lifting loads, hand-chain hoists require substantial human effort, resulting in operator fatigue. This is not the case with air hoists. Furthermore, air hoists offer much faster lift and descent rates than those offered by hand-chain hoists.

Air hoists also offer two major advantages over electric hoists. First, air hoists are self-cooling and will not burn out even when overloaded to a stall. This advantage also makes them applicable in extremely hot environments. Second, standard air hoists offer continuously variable lift and descent speeds, while standard electric hoists offer no more than two different speeds in a given hoist.

Figure 5.75 Overhead hoists are widely used to move and position parts in nearly any operating environment.

Hoist Safety

Air-operated hoists are designed to provide increased productivity, long service life, and safe operation. But, as with any type of hoist or material-handling equipment, these advantages can only be fully realized through proper application and maintenance. To be sure that an air hoist always provides maximum performance and operator safety, the following practices should be followed:

1. Air hoists should be selected by a person who is knowledgeable about hoists and their proper application. Hoist selection must take into account the particular factors in a given application that will affect hoist performance.

2. Air hoist suspension is extremely important. Before a hoist is hung from a jib, crane, or structural member, the maximum load-carrying ability of the device should be evaluated carefully and never exceeded.

3. Air hoists should be operated only by qualified personnel who have been trained in their proper operation.

4. An air hoist should never be used to lift or transport humans.

5. No attempt should be made to lift a load greater than that for which the hoist was designed.

6. Air hoists, like any other hoist, must undergo periodic inspection and routine maintenance to ensure that they are operating properly.

These guidelines are to be considered only general in nature. Additional specific guidelines on safe hoist installation and use are available from the American National Standards Institute, publication B30.16, *Overhead Hoists (Underhung).*

Air Hoist Capabilities

Air hoists are available in rotary vane, axial piston, and rotary piston construction. These three styles offer broad application capabilities. They can lift loads up to 10 tons and, at the rated loads, can lift at speeds from 106 ft./minute for smaller hoists down to about 10 ft./minute for larger hoists.

Continuously variable control of lift and descent is possible on air hoists with either of two available control methods, a pendant control or a pull chain. The pendant uses pressurized air controlled by a hand-held pendant to shift the valve inside the hoist. The pull chain control is attached to an external lever that shifts the valve. If it is important to control either the lift or descent to a certain speed, adjustable exhaust valves on an air hoist will keep the speed within set limits. Air hoists can be hung from a stationary hook, from a manually traversed trolley, or from a powered trolley. Drum brakes, disc brakes, and worm-gear brakes are available to stop effectively and hold the rated load suspended, even when there is a loss of power (Fig. 5.76). Although all air hoists have an inherently spark resistant motor, some are available that offer added spark-resistant features, such as stainless steel load chain and copper-beryllium wheels. All air hoists offer a relativey long service life.

Figure 5.76 Air hoists are available with disc, drum or worm gear brakes.

Rotary-Vane Air Hoists

Rotary-vane air hoists (Fig. 5.77) are the most common type of air hoist and are the smallest of any powered hoists. They utilize a rotary-vane motor and are capable of lifting loads of up to 5 tons. At rated load and 90-psig (6.2-bar) air pressure, a rotary-vane hoist is capable of lifting at a rate of up to 106 ft./minute. Control is exercised by the operator using either the pull chain or pendant control. Speed control within specified limits can be set using adjustable exhaust control valves.

A load is lifted by a cable, roller chain, or link chain. The cable method uses a grooved drum powered by the motor, which raises and lowers the cable. The roller chain method uses a sprocket on the shaft of the motor to raise and lower the roller chain. The link-chain type uses a pocket wheel instead of the sprocket to move the chain. Each rotary-vane air hoist uses one of the three previously explained braking methods: disc, drum, or worm gear, and uses either a hook, manual trolley, or power trolley suspension.

Figure 5.77 Rotary vane air hoist being used to lift a machined part.

A rotary-vane air hoist is comparatively light, and because it has few parts it can be repaired quickly and easily.

Piston Air Hoists

Piston air hoists use compressed air to drive pistons that then rotate a drive shaft. Piston air hoists (Fig. 5.78) can be one of two types: axial piston or radial piston. The radial-piston air hoist has the pistons mounted perpendicular to the drive shaft. The axial-piston air hoist has its pistons mounted parallel to the drive shaft, with axial motion converted to rotary motion in the drive shaft by the use of a socket plate.

Like rotary-vane air hoists, piston air hoists can operate in contaminated environments unsuitable for electric hoists because air is brought in by hose and then exhausted from the hoist. The piston air hoist can operate in hot environments, it can be worked constantly, and when too heavy a load is attempted it will simply stall, causing no motor damage. The operator can control hoist speed at the control pendant and, with the piston hoist, the operator has superior jogging capabilities. In fact, it is the best hoist to use in very critical jogging applications.

Figure 5.78 Radial piston air hoist being used to lift a heavy machine assembly.

The air piston hoist is usually specified in applications of 3 tons or more where variable-speed control or jogging capabilities are desired, or where very heavy tonnage must be lifted and the danger of sparks from an electric hoist is unacceptable.

Mounting or Installing Hoists

Hoists can be mounted by any of several methods depending on the requirements of a specific application. For stationary applications, hoists are usually hook mounted. The hook allows the hoist to pivot so it is properly oriented relative to the load.

For applications where the load must be transferred, the hoist can be trolley mounted to an I-beam, jib, or crane bridge. The trolley allows the hoist to travel on the beam to move the load to the desired location. Power trolleys and tractors (Fig. 5.79) are available to move the load with minimal human effort.

The air supply can be connected to the hoist in a number of ways. On stationary hoists, the air hose need only be long enough to allow the hoist to swivel. On trolley-mounted hoists, the hose must be long enough to allow the hoist to travel along the trolley. The hose can be supported on separate hose trolleys or wound up on a take-up reel to keep it from hanging down in the way. Another common practice is to use self-storing hose coiled around a length of plastic-coated cable that runs along the beam (Fig. 5.80). Special air supply devices such as these are used for air tools as well as hoists.

Figure 5.79 Power tractor-mounted hoist frees the operator from having to pull or push the load.

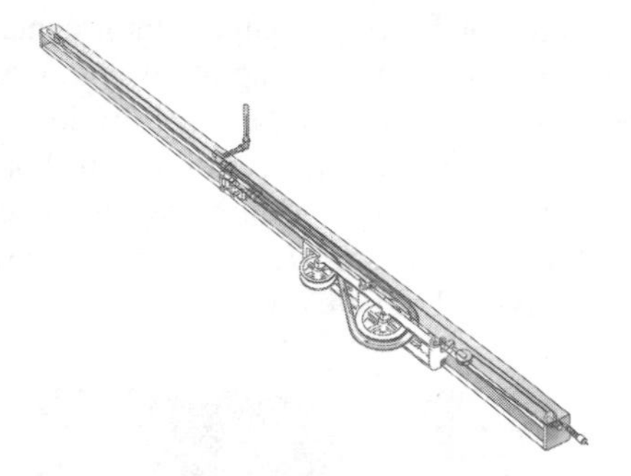

Figure 5.80 Freedom of movement of air-driven devices, including air hoists, is greatly increased by a carrier which takes up slack in the hose as the outlet point is moved.

Applications of Overhead Hoists

Applications for air-powered overhead hoists are limited only by the availability of compressed air and the maximum capacity of such hoists. All manufacturing plants need to raise and lower loads in many different departments. For instance, heavy bar stock or large castings must be placed in machines (Fig. 5.81) for secondary operations. Often, they are too heavy or awkward to be handled by an operator alone, so a hoist is used. Another area where hoists are popular is in maintenance departments for lifting heavy repair equipment and bulky components that must be removed, repaired, or replaced.

Figure 5.81 Overhead hoist being used to load a large casting on to a machine for secondary operations.

In plating, where it is important to prevent operator contact with caustic solutions directly or through splashing, the inching ability of an air hoist is invaluable (Fig. 5.82). An air hoist allows the load to be carefully lowered into the solution while the operator is a safe distance away. In addition, air hoists are more suitable than electric hoists in plating operations because they are inherently nonsparking.

Other applications for air hoists include raising and lowering parts such as castings, heavy ladles, molds, and large car parts.

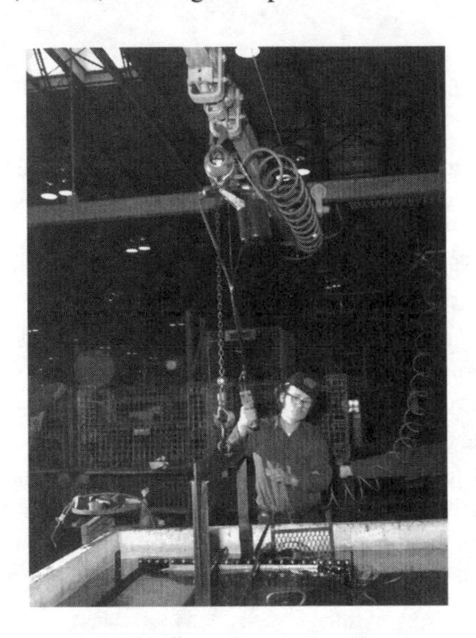

Figure 5.82 Air hoists are ideal for work in plating areas, where splashing must be avoided and where spark hazard is intolerable.

AIR WINCHES AND UTILITY HOISTS

Air winches and utility hoists are available in many sizes with capacities ranging from 500 to 10,000 lb of single-line rope pull. They incorporate spool-type rope drums with capacities for up to several hundred feet. They are mounted on substantial bases that allow them to be installed on floors, walls, or ceilings. This type of construction also makes them portable. The smaller varieties are available for mounting on poles or rock drill columns for construction, mining, and oil derrick operations. The hoist can be used for pulling in almost any direction approximately perpendicular to the drum axes for dragging, hauling, or hoisting (Fig. 5.83). The direction of the pull may be determined by guide sheaves or snatch blocks, and the pulling capacity may be increased by the use of multiple sheaves. Units are rated at 90 psig at the motor inlet, and speeds for rated rope pull range from 40 to approximately 200 ft./minute.

Figure 5.83 Utility hoist on the drill floor of an offshore drilling rig to handle drilling tools.

Essentially, a winch or a utility hoist consists of an air motor, a gear drive, and a spool hoist drum mounted on a substantial base. The air motor construction characteristics and control are similar to those of overhead piston hoists except that hand-lever-operated throttles are commonly used. Dynamic braking or recompression is a normal means of controlling the descent of a suspended or overhauling load. A hand-lever-operated locking brake operating on one flange of the rope drum is usually included on most makes. An automatic, air-release, spring-applied brake controlled by the main throttle, as on an overhead hoist, can be furnished as a substitute. Motor-disengaging clutches are available on most units to permit the rope to be pulled out from the drum for attachment to the load without losing the coils left on the drum. Some units incorporate a clutch to permit starting the motor before applying power to the drum or to permit lowering the suspended or overhauling load under control of the hand brake only.

ROCK DRILLS AND ASSOCIATED EQUIPMENT

The latter part of the nineteenth century saw mechanized rock drilling in the final stages of superseding the old hand methods. This important scientific advancement was accelerated by the need for tunnel driving in the rapidly developing railroad systems and by the depletion of bonanza metal deposits in the West with the consequent desirability of working ore deposits of lower grades. Engineers of that time estimated that hand drilling for blasting had constituted more than two thirds of the cost of earlier mining operations. The power rock drill was not entirely new, even in those days. It was plain that cutting the costs of mining lay in complete mechanization of the rock-drilling phase of the work.

As early as 1813, Trevithick and Brunton, business associates in Cornwall, England, had tried mechanical drilling; and Brunton had even suggested the use of compressed air, maintaining that the exhaust from the boring machine would aid in ventilating mine passages. In America, Couch and Fowle of Boston had taken out a patent on a percussion drill with the steel clamped to the piston, perhaps one of the earliest examples of the piston drill.

Contemporary developments in Europe were the patenting of a compressed air drill by Fountainemoreau in 1855 and the further work by Bartlett on a machine that was tried out in the Mount Cenis Tunnel. Bartlett's drill, improved by Sommeiller, was used in the final boring of that tunnel. It was necessary for the drillers to keep on hand 200 of the Bartlett-Sommeiller machines in order to maintain 16 in constant operation. The vast improvements made since then are indicated by the experience of any user of a present-day rock drill.

About the same time the Mount Cenis Tunnel was constructed, the St. Gotthard Tunnel was being bored with drills developed by Haupt and Taylor. The same design of machine was improved by Burleigh and was used in the Hoosac Tunnell in Massachusetts, completed in 1875. In the West, the importance of power drilling was demonstrated by the performance of Burleigh drills in the Sutro Tunnel of the Comstock lode, this project having been finished in 1874. By that year, other makes of drills were coming into use, one of which was instrumental in the boring of the Musconetcong Tunnel.

All the early piston drills were of the same general type, consisting essentially of a cylinder in which a piston reciprocated, with the drill steel clamped in a chuck that formed the forward end of the piston. Steam operation was restricted to drilling on the surface. Compressed air was the sole medium for powering underground drills with the exception of a few electric models beginning about the year 1950. The air was admitted into the cylinder by a mechanically actuated valve. In the early 1980's, the air-thrown valve came into use; this type had no metallic contact with the piston. Rotation of the piston and drill steel was accomplished through the now-familiar rifle bar and nut combined with the ratchet and pawl.

One of the most important improvements in the early history of rock drill design was the introduction of the self-rotating hammer drill (Fig. 5.84) by J. George Leyner in 1897. Refinements have been made by many designers and engineers, some of whom are still continuing this work.

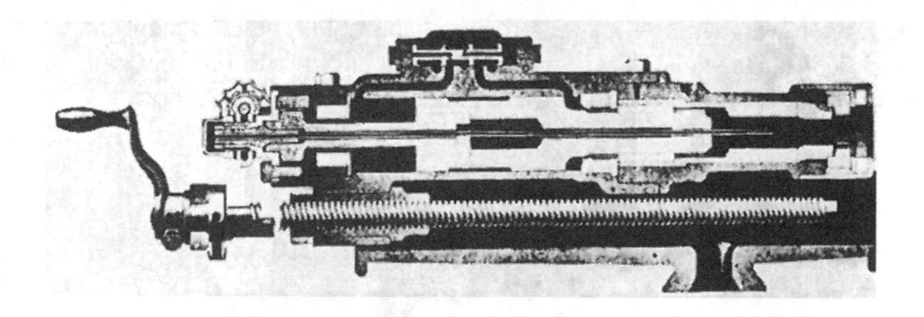

Figure 5.84 Section through a self-rotating hammer drill of the type introduced by George Leyner in 1897.

The most recent advance in rock drilling is the use of hydraulically powered rock drills (Fig. 5.85). The design of a hydraulic drill makes it possible to transmit the energy more effectively to the bit. The result is that these drills can normally drill 50% faster than the equivalent air drill. The other advantages of hydraulic drills are that they are less noisy, create no oil mist or fog, and, because of the more effective energy transfer, drill steel and bit life are, in many cases, improved.

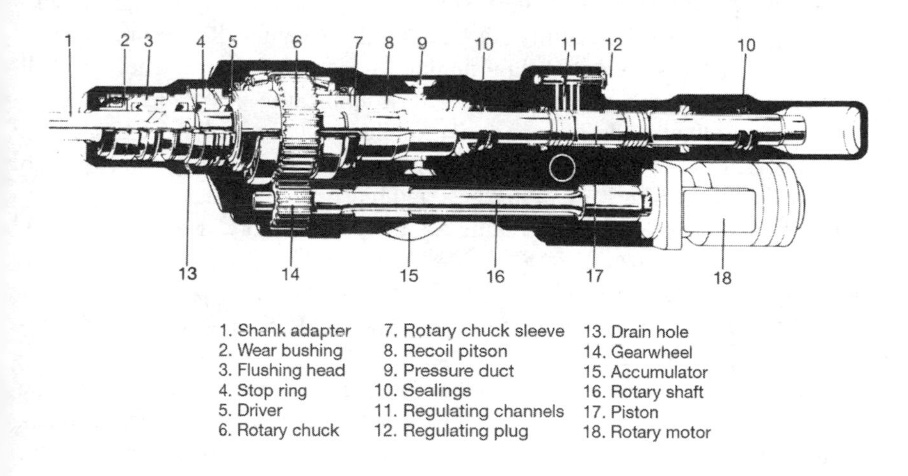

Figure 5.85 Typical parts of a hydraulic rock drill.

Hand-held Rock Drills

The hand-held, self-rotating hammer drill (Fig. 5.86) is commonly known in the United States as a sinker or rotator, depending on the locality and the usage. In some localities, the name sinker is applied only to the heavier machine used for shank sinking. In quarry language, the smaller machines are frequently referred to as block-hole drills.

Common weight classifications for the hand-held type of machines are the 30 to 40 lb class for the smaller types used for hitch cutting, drilling anchor holes, and coal mining; the 41 to 54 lb class for secondary more difficult jobs.

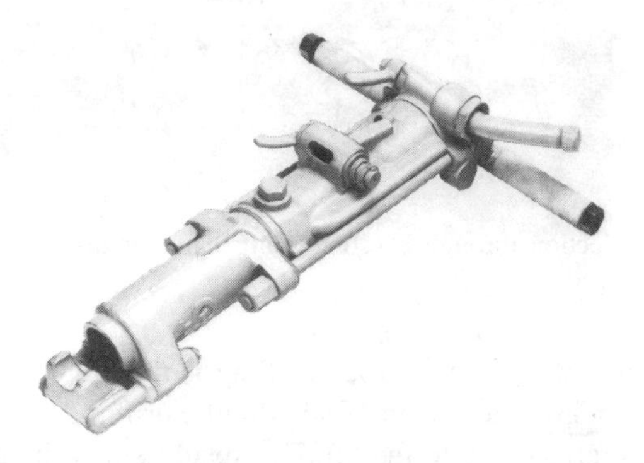

Figure 5.86 Hand-held sinker drill in the 30 to 50 lb weight range.

The utility drill, better known as the plug drill, weighing usually about 20 lb, is used for plug and feather work in dimension-stone quarries or for light block-hole drilling. Lighter-weight utility hammer drills weighing 8.5 lb are used to install fastener anchors and for general maintenance work. A large variety of drill bits and chisel accessories are available for these applications.

The 30 to 40 lb class drill is ordinarily applied to a depth of approximately 6 to 8 ft. The 41 to 54 lb class is good for depths up to 16 ft., and the 55 to 64 lb machine will drill to 24 ft. if suitable steels and bits are provided.

The heavier shank sinkers, weighing about 62 lb or more, are not ordinarily used for depths greater than 25 ft. Steels 7/8 in. or 1 in. hexagon, quarter-octagon, or round types are most frequently used in these drills. The more powerful hammer blows of the 62 lb sinker fit it for use with larger steels. The 1-1/4 in. round is commonly utilized with this larger drill.

By far, the most common shanks for hand-held rock drills are the collared type in sizes 7/8 x 3-1/4 in., 7/8 x 4 1/4 in., and 1 x 4-1/4 inch, either hexagon or quarter-octagon. Hand-held drills are generally classified into two groups, dry and wet, depending on the method used for cleaning for drill hole.

Dry Machines

Dry machines are further divided into three types:

1. The plain dry machine is the accepted type for general surface operations and shallow holes where good hole cleaning is not difficult to accomplish. The hole cleaning is done by supplying a pulsating flow of air through the steel during drilling. The air is supplied in puffs, once or twice during each cycle of the piston hammer.

2. The blower machine is used for the same application as the plain dry machine particularly where deeper holes or more difficult conditions require a stronger blowing action. The blower machine, while drilling, supplies a constant stream of air into the drill steel through a tube running down the central axis through the hollow rifle bar and piston. Some dry and blower machines are equipped with a choker valve that permits maximum blow air to be passed down the drill steel to get maximum hole-cleaning effect with the hammering action stopped.
3. Dry drilling with air feed leg drills is accomplished either by pulling the dust through a special bit and through the hollow drill steel into a dust-collecting box or bag, or pulling the dust through the annulus of the hole into a flexible cup at the collar of the hole, which is connected to the dust-collecting system.

Wet Machines

Wet machines are universally used in underground operations, particularly where dust conditions are objectionable, in extremely hard rock, or where water flushing is more efficient than dry blowing for cleaning the holes. The wet machine resembles the blower type, but water rather than air is supplied under pressure from an outside source through the central tube or through a rotary swivel at the front-head of the drill. In some situations, normal water flushing or dry drilling is unacceptable. In these cases, mist drilling is done using a small quantity of water injected into the flushing air.

Air Feed Leg Drills

Another type of mounting for hand-held drills is a feed leg (Fig. 5.87). This is a telescoping air feed having its own control throttle. It is attached to the drill with a friction-loaded knuckle joint at the point of attachment. For horizontal drilling, it may be adjusted and controlled so that it both supports and feeds the drill. The operator needs only to hold it steady while drilling and to adjust the thrust of the feed. The point, or sinker, has a spur designed to dig in and take the thrust of the feed when drilling lifters. Advantages claimed for the feed leg include its ability to bottom a drift or stope hole without a steel change when using carbide insert type bits. As the bit penetrates the rock, the feed is collapsed and advanced to take a new bit each time it runs out to its limit until the hole is bottom. The time required for setting up and tearing down is reduced to the minimum. In headings that are too small for a jumbo, in inaccessible workings such as sublevels, and in some types of stopes, the sinker and feed leg are practically universally applied.

Figure 5.87 Feed leg mounting for a hand-held drill.

Feed leg drills have become integrated units in which the leg and drill are compatible designs developed especially for feed leg drill applications.

Stopers

Stopers are used, as the name implies, for stoping, raising, roof bolting, or any work on which the holes look up at a steep angle. These rock drills are available in weights ranging from 75 lb to around 100 lb and are generally used with 7/8 or 1 in. hollow steel and bits from 1-3/8 to 1-5/8 in. gage. For use in narrow seams, such as coal mines, telescopic, offset-leg stopers are available. Here the rock drill is attached to the side of the air feed instead of being an extension of it to shorten its overall length. A stoper application is illustrated in Fig. 5.88.

In most stopers, the piston of the air feed remains stationary and the cylinder attached to the rear of the rock drill propels it forward and feeds the bit into the rock. In some reverse-feed designs, the cylinder is the stationary member. A sharp point, referred to as a stinger, keeps the feed from slipping against the supporting rock or timber.

To reduce spare parts inventory, many stopers can be converted to air-leg machines, drifters, or sinker-type drills.

Figure 5.88 Stoper drill in use in a roof bolting operation.

Also, muffled versions can be had, which considerably decrease the noise heard by the miner. These machines are equipped with automatic water valves to do wet drilling and provide for dust suppression. The water is automatically turned on with the throttle. Stopers with feed legs of various lengths, as well as telescopic feed legs, are available to suit varying needs. Some telescopic versions extend up to 64 in.

In selecting a set of steel for use with stopers, attention must be paid to the length of travel of the stoper feed. The permissible length of change will be less than travel by the length of steel in the stoper chuck. This is usually 6 to 7 in. For example, a 25 in. feed will use an 18 in. steel change. Stopers can be equipped with frontheads for use with shanked or shankless steel. With shanked steel, the fronthead has a steel holder to assist in removing the steel from a tight hole.

Noise Suppression

There has been a trend in recent years to develop noise suppression devices on both aboveground and underground drilling equipment. Underground mine regulations now require that all air drills be equipped with mufflers to reduce the air exhaust noise.

Drifter Drills

In simplest terms, the drifter drill (Fig. 5.89) is a rock drill especially adapted to putting horizontal holes in mining operations, drifting or cross cutting, and tunnel driving. It is also used on crawler-mounted drills for work in quarries or on contract jobs involving rock excavation.

Figure 5.89 Silenced air drill on feed.

While primarily designed for horizontal holes, it is equally effective for downhole or angular drilling and for primary ring or fan pattern blastholes, when suitably mounted.

Air drifters are commonly made in bore sizes from 3 to 6 in. They range in weight from 125 to at least 575 lb, depending on the type, size, and mounting. Hydraulic drills are comparable in size and weight. In mines, the once common method of supporting drifter drills by a screw or air column and cross arm or by a shift bar has been generally supplanted by automated, hydraulically operated booms mounted on a mobile or self-propelled chassis, either rail, crawler, or rubber tire mounted. The boom type of support, automatically adjusted by either an air or hydraulic mechanism, has become increasingly popular on mine jumbos (Figs. 5.90 and 5.91), owing to the fact that it saves both time and labor. The crawler mounting in Fig. 5.92 meets with favor in some underground operations, such as underground limestone quarries where drill patterns require high maneuverability of the carrier.

For large tunnels, the familiar multiple-drill jumbo (Fig. 5.93) is built to suit the size and section of the tunnel. The modern jumbo is now fitted with adjustable-type booms with which the holes may be spotted much more quickly and with less manual effort than with the conventional cross-arm mounting. In some cases fully automated computer controlled jumbos are utilized for high speed tunnelling and face drilling. A complete headings drilling pattern can be programmed into the computer. This will allow one operator to control several booms simultaneously.

Drill feeds are available to accommodate for single-pass hole depths of 2 to 24 ft. With the general use of tungsten carbide insert bits resulting in reduced bit gage wear, longer drill steel changes are common. Feeds are of either screw, chain, or cable design and can accommodate steel up to 20 ft. in length for fully utilizing the capabilities of present-day bits.

Figure 5.90 Three-boom air drill jumbo in mine application.

Figure 5.91 Mechanized drill rig drilling long production holes.

The 3 and 3-1/2 in. class drifter drills may be removed from their guide shells and fitted with spring handles for use as heavy-duty, hand-held drills in shaft sinking, or for general rock excavation where the nature of the ground requires the utmost in cutting ability. Air drifters are usually provided with supplementary airline oilers having a much larger capacity and requiring less frequent servicing. Special rock drill lubricant is used to assure proper lubrication. Hydraulic drifters are lubricated by grease in the mechanism, or by a pressurized air mixture.

Figure 5.92 Crawler mounted drill jumbo in limestone mine.

Figure 5.93 Two-boom electro-hydraulic jumbo in tunnel application. Unit has onboard air compressor for hole flushing.

Hydraulic Booms

The larger, heavier drifter drills have, for the last decade or more, been increasingly mounted for underground drilling and tunneling on an automated, hydraulically operated positioning device referred to as a boom. This consists of the main boom to which the drill and feed are attached, which is moved or actuated in the horizontal and vertical planes by hydraulic cylinders.

Most modern booms are also equipped with a device capable of rotating the boom, and hence the drill and feed, up to a full 360°. They can also be equipped with an automated feed extension device, which, in effect, increases the effective hole depth that can be drilled with the fixed-feed mount thereon. Some boom-feed systems incorporate automatic parallel holding, which assures that, once a setup is made, each hole will be parallel to the next.

Booms are available in fixed lengths ranging from 6 to 15 ft. They are also available with extendible features consisting of a telescoping extension that may increase the effective length, reach, and face coverage of the boom up to 5 ft.

Hydraulic booms are equipped with a pedestal for attachment to mobile mountings; rubber tired, crawler or rail mounted. Controls are usually grouped for remote operation of all features. Controls may be grouped for one-man operation of several booms or drills.

MOBILE MOUNTINGS FOR DRIFTER DRILLS

Jumbos and Mobile Mountings for Underground Applications

The equipment in many earlier mines was rail mounted; and when the drilling operations were finally mechanized, the booms were attached to rail cars or rail-mounted frames. This type of equipment is still widely used, particularly in smaller mines where rail-mounted haulage equipment is in use.

When mines began to utilize larger-sized openings and to increase productivity, a change was made to rubber-tired loading and haulage equipment, and this led to the need for rubber-tired drill jumbos. For units working in a captive area, an air-driven traction system is common. Where special maneuverability is required due to terrain or drilling patterns, crawler-mounted, air-powered carriers are sometimes used.

The use of hydraulic-powered drills has led to the introduction of electrohydraulic and diesel-hydraulic units, as well as a combination of both. Completely self-contained units are also used. These can be completely diesel-powered where conditions permit and can carry a supply of water if only mist drilling is being done.

On most jumbos, the controls for all booms, feeds, and drills are usually grouped in a console for one-man operation where this arrangement is feasible and allowed.

Crawler and Other Mountings for Surface Applications

The increasing range of sizes of drifters and the wider range of applications have brought extensive changes in the mobile mountings available for surface drilling. These changes have been made in order to attain increased capability, larger and deeper blastholes, more extensive mechanization of drilling operations, and thereby a reduction of physical effort.

Figure 5.94 Crawler drills working on highway excavation.

The self-propelled, crawler-mounted drill (Fig. 5.94) is the most commonly used machine for surface drilling in the 30 to 50 ft. hole depth, 2-1/2 to 6 in. hole size. Drifters that are available are of either the rifle bar or independent rotation type. The main thrust in the drifter development has been in the area of hydraulics. Variable stroke, sophisticated feedback systems that are able to sense ground conditions and automatically adjust the drifter operating parameters have made major inroads. Air or hydraulic motors drive the tracks, which can be operated independently for tractor-type steering. These machines have the ability to move from hole to hole under their own power over the roughest terrain and to tow a portable compressor as the job proceeds. Some all-hydraulic crawler drill rigs incorporate an onboard compressor for the hole flushing air.

Hydraulic-powered cylinders are used to raise and lower the boom on which the chain feed and drifter are mounted. Hydraulic swing, tilt adjustments, and feed lift are usually available so that all positioning adjustments for drilling are accomplished hydraulically. Medium-weight, self-propelled track drills are usually used for feeds up to 14 ft.; these units are mounted with 4-1/2 in. and larger bore air drifters or hydraulic drifters. Heavy-weight, self-propelled track drills are usually used for feeds up to 22 ft. and carry 5-1/2 in. and larger bore air drifters or hydraulic drifters (Fig. 5.95). Automatic steel-handling devices (Fig. 5.96) now make these larger drilling units truly one-man machines.

Both the medium and heavy units have controls for moving, positioning, and drilling, located for the operator's convenience and to save setup time. Due to environmental regulations in most areas, most units now come equipped with detergent tanks permitting a fine spray of detergent to be introduced with the air to reduce dust or with a dust collector (Fig. 5.97) which permits dry drilling.

Figure 5.95 Crawler drills with large air drifters in a quarry application.

Figure 5.96 Automatic steel-handling device on hydraulic track, drill working on airport.

Figure 5.97 Dust Collector mounted on side of track drill, permitting dry drilling.

Most drifters are generally offered with stop and reverse rotation to facilitate the use of sectional drill rod. Thus, deep holes can be drilled with 12 ft. drill steel without requiring the driller to handle long, heavy lengths of steel.

Chain feeds and drifters as used on track drills are available for mounting on trucks, tractors, and other types of mobile equipment, making almost any specialized equipment available, even for unusual operating conditions.

Also available for drilling holes 3 to 9 in. in diameter are crawler-mounted rigs with down-the-hole drills (Fig. 5.98).

Figure 5.98 Crawler mounted rig for down-the-hole drilling.

These are available in mobile mounting and feed units requiring separate air supply, or as completely self contained units incorporating engine-driven compressors.

Compressed air is also used to clean the hole and, on air rigs, to propel the unit to new drilling locations by means of air motors. This type of machine is used in the mining, quarrying, and construction industries for primary blasthole work. Normally, these drills are powered by compressors operating at pressures of 175 to 300 psig. If the same air supply is used to operate the rig, the pressure may be reduced through a pressure-reducing valve for use with the rig functions.

The down-the-hole drill actually goes in the hole with the bit attached directly to it. Several advantages result, the principal one being that the drill piston strikes directly on the bit. There is virtually no loss due to energy absorbed by a long length of drill steel. For long holes, over 75 ft., the down-the-hole drill provides less hole deviation and a higher penetration rate than drifter drills. The noise level is lower because the noise sources are underground. Down-the-hole drills utilize carbide insert bits up to a 9 in. gage. The piston strikes directly on the bit, which can be a four point (cross or X-type) or a button bit.

APPLICATIONS OF DRIFTER DRILLS

In the early 1950s, the drifter drill was adapted to construction blasthole drilling by mounting it on a feeding mechanism that was moved and positioned on an air-operated carrier. During the same period, development of reliable, thread coupled sectional steel made it possible to drill relatively deep holes with such machines. As hole sizes and depth increased, the demand for greater rotational power increased. There also was the need for reversing rotation to disconnect the threaded rod strings. Another aid to loosening the tightly connected drill string was to be able to hammer the string with no rotation. These needs led to the development of bigger and more powerful conventional rifle bar drifters with reverse rotation features and to the development of the independently rotated drifter wherein power for rotating the chuck, steel, and bit was supplied by an air or hydraulic motor through a gear train.

About the same time, for drilling smaller holes, primarily horizontal, in underground mining and tunneling, a machine was developed that consisted of an air-percussion hammer with hydraulically powered rotation and an air-powered feed mechanism. This drilling machine is known as the rotary percussion drill. Each function is an independently controlled-force of blow by air pressure to the hammer, rotational torque by hydraulic pressure, and volume to the hydraulic motor and preload by air pressure to the feed mechanism.

The dimensional stone industry has used a variety of drilling systems utilizing drifters. Slot cutting to separate a section of the quarry from the surrounding rock was done earlier by drilling holes close together, approximately 12 ft. deep, using drill bit guides. This procedure has been replaced in many quarries by jet burners and wire saws, but with the introduction of hydraulic drills, the old method may become more attractive. Hydraulic drills do not create the noise and dust of jet burners and can do the job faster. Once the slot is cut, drifters mounted on quarry bars or crawler rigs are used to drill vertical holes approximately 12 in. apart and to the depth of the slot. At the same time, drifters mounted on quarry bars drill horizontal holes at the base of the block about 12 in. apart and up to 50 ft deep.

A form of line drilling, also now referred to as presplit drilling, is done on construction work, mainly designed to leave a clean, smooth wall in a rock cut or cut-off wall. This reduces the danger of spalling or rock falls later as the blasted rock wall weathers. It is also used in foundation work involving rock excavation where blasting is not permitted by the engineers because of the effect of backbreak on adjacent structures or ground. Submarine drills are rock drilling machines used for deepening channels in harbors and rivers and for other miscellaneous subsurface marine drilling operations. Various types (Fig. 5.99) are used, depending on the depth of water and the rock formation encountered. The same drifter drill that is used on crawler rigs for land work can be used for this type of drilling in shallow depths. The drill is usually mounted on a derrick or slab-back guide that can be moved along the side of the boat. A percussive drill, especially designed for operation underwater, is used for the deepest channel work of this type. The drill is mounted on a ladder, which in turn is mounted on a tower that can be moved along the side of the boat. The drill works under water, thus eliminating the necessity of long steels.

Figure 5.99 Submarine drill system with four drills in operation.

Down-the-hole drills are extensively used for underwater drilling, using bits up to 9 in. in diameter.

Down-the-hole Drills

Down-the-hole drills (DHDs) can be used for drilling holes ranging in size from 3 inches to 60 inches. DHDs are commonly chosen when hole size exceeds 5 inches and straight non-deviating holes are required. DHDs use the air consumed in powering the percussive motor to clean and flush the hole of rock cuttings. Holes larger than 30 inches in diameter are commonly drilled using specialized cluster drills which are groups of smaller DHDs ganged together in a common can.

A DHD should be selected to be a close match to the hole diameter being drilled. Overextending a DHD to drill a hole much larger than the hammer not only causes a loss in drilling efficiency but can also lead to shanking. Shanking occurs when the bit cracks through the head to shank change in section. The larger the section change the more a bit will be prone to shanking.

DHDs can be powered by compressors ranging from 120 to 350 psi. Smaller and less production-intensive applications will commonly use low pressure (150 - 170 psi) single-stage screw compressors. Applications requiring higher production will commonly use two high pressure (300-350 psi), two-stage compressors. Deep holes such as those common in oil, gas and mineral exploration will commonly use low-pressure compressors coupled to a supplementary booster allowing pressures to exceed 500 psi.

Except for some low pressure applications, a DHD should be selected to consume the entire volume the compressor can deliver to prevent the compressor from unloading. The majority of DHDs are equipped with optional air bypass chokes to allow additional air to be bypassed to improve hole cleaning or reduce operating pressure. The use of bypass chokes should be avoided as they tend to backpressure the DHD and cause a loss in performance. Many DHDs are available as deep-hole or low-pressure designs. These drills are designed to perform well under conditions of elevated back-pressure, which is common in holes making water or those that are simply very deep.

The mountings used for carrying 3 to 6 inch drills (Fig. 5.100) can either be a track drill with a portable compressor or with compressor on board. Holes ranging in size from 5 to 24 inches are commonly drilled using rotary-blasthole drills (Fig. 5.101) which are track or tire mounted, depending on the level of mobility required. Larger holes, commonly for construction purposes, can be drilled using custom built top-head drive equipped cranes or cranes equipped with rotary tables and kelley bar drives.

Figure 5.100 CM695D and smaller DHD equipped track drill.

Figure 5.101 DMM and T4W

Because DHDs deliver percussion energy directly on the bit they typically have higher energy transfer efficiencies than what is common on drifter drills, which must deliver energy through long steel and coupling assemblies. The deeper the hole the greater the advantage of a DHD. Another advantage to DHDs is that they produce less noise than drifter drills. Due to bore size limitations on DHDs (the hammer must fit in the hole!) drifter drills commonly deliver more power, and penetrate faster for short holes less than 5 inches in diameter. DHDs gain a penetration advantage on all holes 5 inches and greater. However, due to the surplus of air consumed by DHDs beyond what is required to clean a hole, DHDs commonly have reduced accessory life on small holes due to erosion caused by very high up-hole velocities.

The majority of DHDs commercially available have valveless air cycles (Fig. 5.102) for directing air to and from the working chambers of the drill. Valveless drills use the position of the piston to port supply air to or exhaust air from the working chambers and offer very simple construction. Recent advancements in air cycle technology have led to the introduction of a hybrid DHD air cycle (Fig. 5.103), which optimizes air cycle efficiency through the use of a valved drive chamber and a valveless return chamber.

Figure 5.102 SF6 or other common DHD.

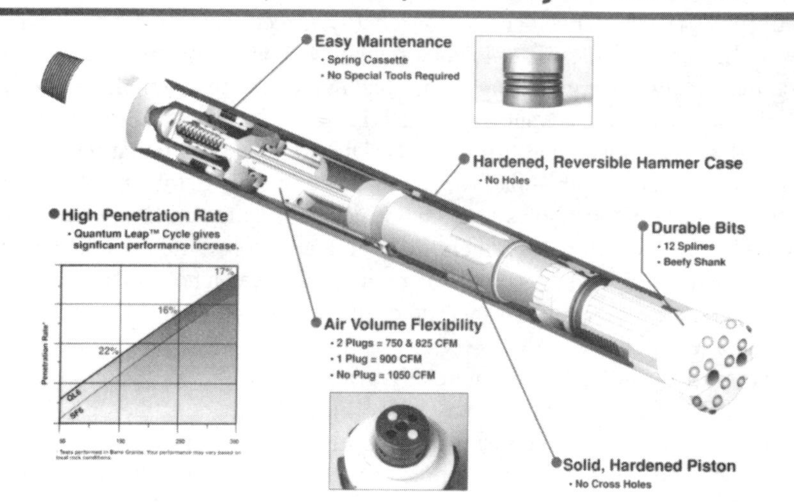

Figure 5.103 Hybrid DHD Air Cycle

DHD Bits

As with threaded bits, DHD bits are available in a wide variety of sizes and designs that suit various applications. Some of the variables and the applications that they would be suited for are listed below:

Head shape

Flat face

- General use bit particularly suited for hard rock applications.
- Heavy duty robust head is structurally very sound.
- Not good for hole deviation in broken or unconsolidated formations.
- Face slots improve hole cleaning.

Concave face

- Best design in terms of maintaining hole straightness as the concave portion centers the bit as its drills.
- Face slots provide improved hole cleaning for increased penetration rate.

Convex face

- Best design for very abrasive applications where the abrasive blast tends to wear away the body material supporting the gage structure of the bit.
- Double gage row of buttons is common where inner row helps protect outer row from excessive wear.

Carbide

Tough

- Compromises wear resistance for toughness.
- Carbides are less prone to breakage when not properly serviced/sharpened.
- Common in many larger bits and iron ore applications.

Standard grade (6% Co)

- Best general balance of toughness and wear resistance.

Buttons

Extra service gage buttons

- Longer buttons in gage row allow bit to operate to a reduced diameter before gage buttons are torn out.
- Longer inserts reduce fit related problems and socket bottom fatigue failures from button holes.

Ballistic inserts
- Only suited for soft rock applications.
- Provide extremely high drilling rates with low pressure, higher pressures tend to bury ballistic button bits.

PAVING BREAKERS

Hand-operated air paving breakers (Figs. 5.104 and 5.105) have a wide variety of uses in general construction and demolition work and in industrial plants. Breakers in the heavy class are used to break concrete pavement, to demolish concrete foundations and walls, to cut pavement and sub base, and to remove slag from furnaces and ladles. They are also used for trenching in hard ground and for breaking boulders in mill grizzlies or elsewhere when it is impractical to blast. Breakers in the medium class are applied to breaking light concrete pavement and floors, macadam, frozen ground or gravel, and, in general, to paving-breaker work of average nature.

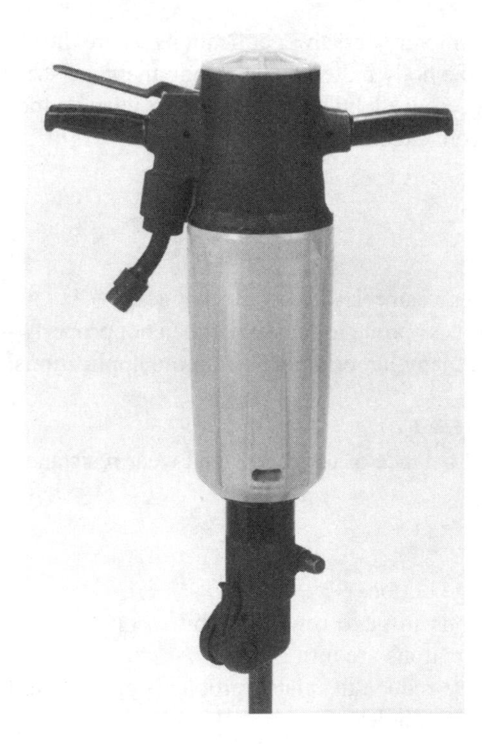

Figure 5.104 Silenced, vibration damped paving breaker.

Figure 5.105 Silenced paving breakers being used on highway repair job.

Breakers in the light class are useful for light demolition work on floors, paving, and masonry walls. They are particularly adaptable for work from a scaffold, for trimming, and for similar jobs requiring the operator to handle the tool continually. The very light class breakers are utility tools for use on light trimming, in steel mill maintenance and in close-quarter work.

Heavy-class breakers may be converted into sheeting drivers by removing the front head and substituting a sheeting-driver attachment. Thus modified, they are used to drive 2, 2-1/2, and 3 in. sheeting, light steel piles, and spiling in mine shafts and drifts. Round tools of small diameter may be driven by shaping the ends to fit the driving attachment.

Both heavy-weight and medium-weight breakers may be converted into railroad spike drivers by substituting a spike-driver head in place of the regular fronthead. Lightweight paving breakers may be similarly converted by substituting suitable fronthead parts.

Paving breakers use solid steel accessories and are not equipped with automatic rotation. Various designs of tool retainers are used, such as the latch and spring loop, located on the front heads. Chuck sizes vary according to the class of breakers from 3/4 x 2-3/4 in. on the very light machines to 1-1/4 x 6 in. on the heavy class. All types and classes include automatic built-in lubricators located in the handles. In addition to moil points, other paving breaker tools can be furnished, such as narrow and wide chisels, digging blades, frost wedges, asphalt cutters, clay spades and scoops, tamping pads and pile, and pipe or sheeting driver heads. The very light breakers usually use 1/2 in. hose, while the medium and heavy breakers require 3/4 in. except where lengths exceed 100 ft., in which case, the next larger hose size, 3/4 in. or 1 in., respectively, should be used.

As for all air tools, it is important for paving breakers to be properly lubricated. Although they are equipped with built-in lubricators, the use of a line oiler is recommended.

Mufflers to lower the air exhaust noise are available for most paving breakers, either as an integral part of the tool or as an easily attached device. The process noise, the tool striking the material to be broken, and other limitations or devices to further silence the breaker itself means that the noise levels are still somewhat above allowable levels for eight hours of exposure. Care should be taken to be sure that operators use approved ear-protection devices.

Advancements are being made in the reduction of hand/arm vibration transmitted to the operator. Methods used to reduce the vibration normally include isolation devices in the handles of the breaker. These isolation devices may utilize mechanical or air spring systems to reduce the vibration transmitted to the operator.

Boom-mounted, heavy-duty breakers for concrete breaking, demolition work, and boulder breaking have gained wide use where high productivity is required. These boom-mounted breakers can be powered pneumatically or hydraulically, and the booms can be mounted on a mobile carrier or a stationary pedestal. The energy per blow and per unit of time is considerably higher for these tools than for hand-held breakers. Breaking is faster and harder materials can be broken. In some applications where the transmission of the vibration to the surroundings is a problem, these breakers may be limited in use.

BACKFILL TAMPERS

The backfill tamper (Fig. 5.106), originally adapted from the air floor hammer used in foundries, is still finding wide use in construction and utility work. In the tamper, the piston rod connects the butt, available with diameters ranging from 3 to 6 in., rigidly to the piston, which lifts it off the ground on its return stroke to carry it down and full force on the power stroke.

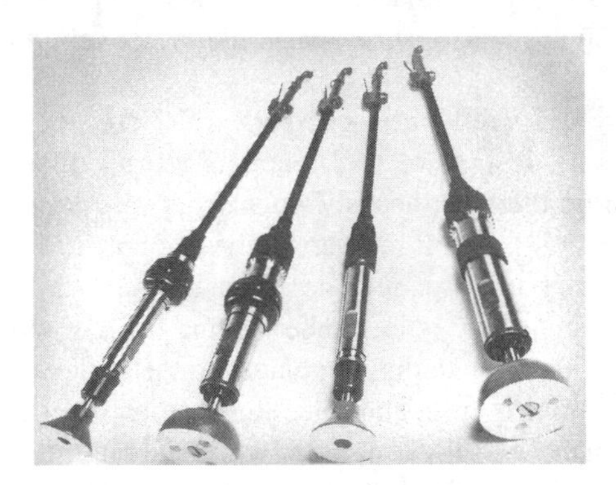

Figure 5.106 Backfill tampers with various tamping pads.

Tampers are classified by weight and are currently available in three main weight classifications: light, up to 25 lb; medium, 26 to 35 lb; and heavy, over 35 lb. Application depends to a great degree on the type of soil and the depth of fill to be compacted. The tamper is generally ideally suited for compacting backfill in lifts of 6 to 9 in. in trenches, service openings, and around foundations, poles, and footings. Compacting with tampers enables immediate repairs to be made to pavement, sidewalks, and the like, after excavation and installation of services, without fear of further settling of the fill and necessity of returning to do repairs.

Where wider areas of backfill require tamping, the triplex tamper will provide savings in time and manpower. The triplex tamper consists of three single tampers on a common frame with a handlebar and single throttle valve for one-man operations.

DETACHABLE BITS

Practically all rock drilling with feed-mounted drills, whether in underground or open-cut work, on tunnel or mine jumbos, on both hydraulic and air drills, is done with detachable, tungsten-carbide-insert bits. Detachable bits are available in both threaded and tapered-socket types. Threaded bits are supplied for all available drill steel sizes and types. Taper socket bits, which are mostly used on air feed-leg drills and hand-held drills, are mainly restricted to use on 7/8 and 1 inch steel.

Some hand-held rotary drilling is occasionally done in soft, non-abrasive formations with detachable, one-pass, steel bits. A considerable amount of feed-leg drilling is done with integral steel – a drill rod that has a carbide insert bit, either of chisel – or button type, forged integrally with the shanked drill rod (Fig. 5.107). However, even on hand-held and feed-leg drills, much drilling is done with detachable bits (Fig. 5.108).

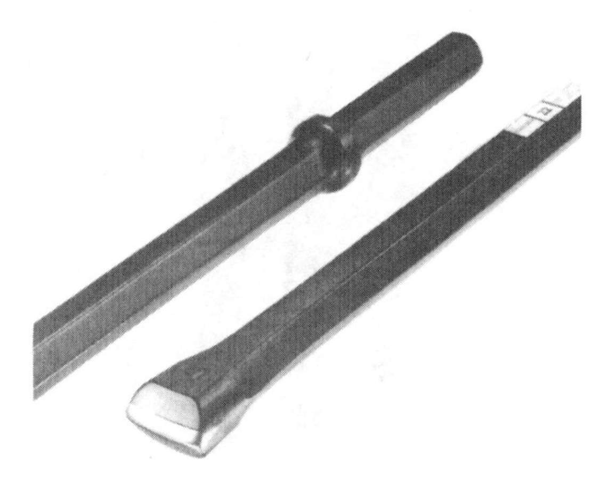

Figure 5.107 Integral Steel

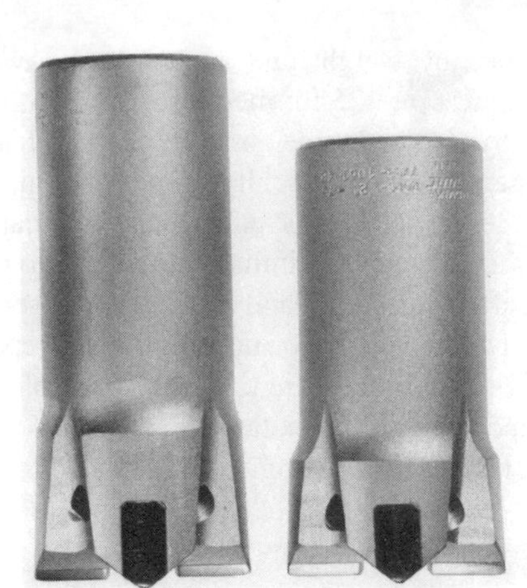

Figure 5.108 Detachable, Carbide Bits

Thread types vary depending on the application. For example, threads on bits and drill rods for hand-held rotary drilling are quite different from the threads used on larger bits for track drill work. Certain thread types can be uncoupled with less force, these are often employed on automated drill rod feeders. International standards are not necessarily available for the different threads in common use, but the threaded products supplied by most major manufactures can be expected to be compatible with each other. Several types of threaded drill rod for use with detachable bits, are shown in Fig. 5.109.

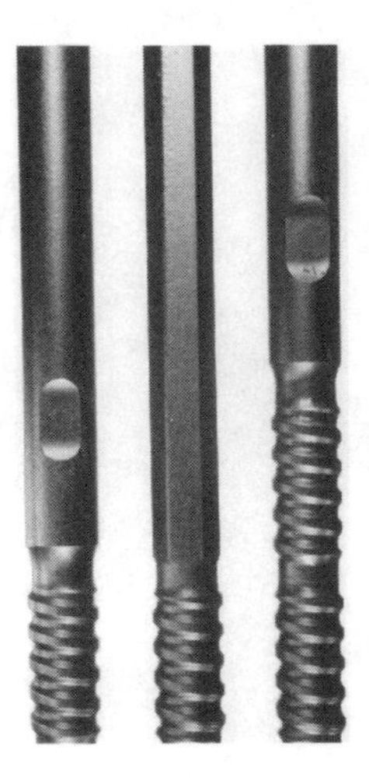

Figure 5.109 Several types of threaded drill steel.

Economical bit design configurations for various rock types and formations, and for specific requirements such as minimum hole deviation, are available from all major manufacturers. These designs are usually established by development testing under realistic field conditions. Design considerations include factors such as diameter and depth of hole to be drilled, the power of the rock drill, resistance to gage or face wear, the type of flushing (air or water) employed, and penetration rate required. In addition, tungsten carbide inserts of varying degrees of hardness and toughness are required for maximum drilling performance in different geological formations and rock types.

Carbide-tipped detachable bits are also supplied in bottom- and shoulder-drive types. In bottom-drive bits, the percussive force is transmitted through the drill rod to the base of the head of the bit. This design is the most efficient and is used on all sizes of sectional steel of the threaded-extension type. In shoulder-drive bits, the percussive force is transmitted through an upset shoulder on the drill rod, to the outer edge of the bit skirt. Shoulder-drive bits are in very limited use today and are found only in diameters under 2-1/2 inch, with smaller section rods 7/8 and 1 inch and smaller bore drills.

Button bits containing domed cylindrical carbide inserts that are mechanically fitted are in almost universal use. A range of button bits is illustrated in Fig. 5.110. The usual size range of threaded button bits is from about 1-1/4 to 4-1/2 and even 6-inch diameter.

Button bits generally offer several advantages over chisel-insert types:

- Longer service life with the use of more wear-resistant carbide.
- More consistent penetration rates.
- Less frequent sharpening and bit changing; experience has shown that occasional dressing (restoring the original button shape) extends bit life considerably.
- Smoother rotation, resulting in ease of hole collaring, and less wear on associated equipment.

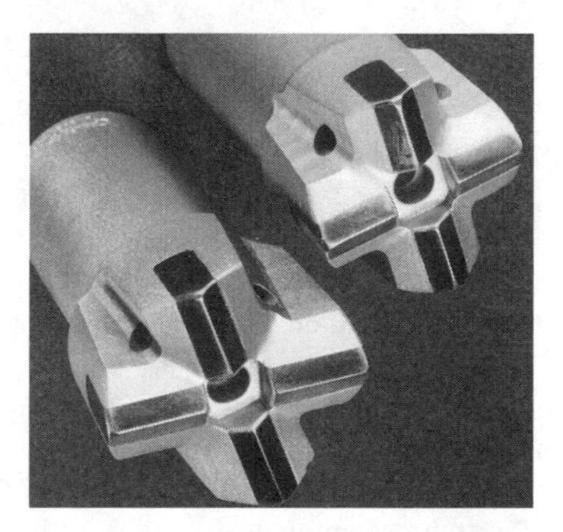

Figure 5.110 X Type and Cross Type Carbide Insert Bits

In chisel-type carbide insert bits, the face or cutting area design is of one of two types; the cross type used mainly horizontal or vertical, and the X-type, mainly used for holes over 2-1/2 in. and particularly for the deeper holes, drilled with track drills. These two types are shown in Fig. 5.108. The X-type bit is most often used in certain formations in which drilled holes are prone to rifling with standard bits.

Limited success has been achieved in designing a tungsten carbide insert bit that may be run to destruction without the necessity of resharpening. Such a bit would not only eliminate the cost of resharpening, but more significantly, would lower the drilling cost by increasing availability.

In certain rock formations some success has been experienced with small diameter one-pass bits that contain small carbide inserts. These bits are discarded when the inserts are blunt.

For down-the-hole drilling, special bits are required, with the shank integral to the bit. The bit shank fits directly into the drill. Shanks vary in design depending on the manufacturer of the drill. Several suppliers offer down-the-hole bits for use with drills of respective manufacturers. Down-the-hole bits almost invariably are of the button-insert type, but chisel-insert bits may occasionally be available (Fig. 5.111).

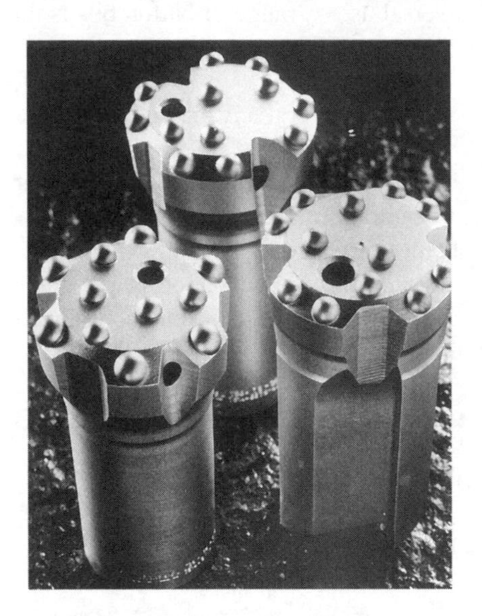

Figure 5.111 Button bits using carbide buttons.

With all percussion drills, whether of the external or down-the-hole type, productive drilling generally depends on the suitable selection of drill rod and bit. In the down-the-hole method the drill bit is an integral part of the rock drill, thereby eliminating the drill rod or tube as an agent for transmitting energy between drill and bit.

A down-the-hole bit consists of a cutting head containing carbide inserts and flushing holes, and a splined shank. The splined shank serves as an anvil for the drill piston and for rotating the bit. The flushing holes allow for the passage of air to remove rock chips from the hole. Down-the-hole bits are designed to produce a minimum clearance between drill body diameter and hole. Typically this clearance is of the order of 1/4 inch, but the figure can vary greatly depending on rock conditions. Down-the-hole bit face design and button placement patterns vary from one manufacturer to another. Most designs have been developed empirically and by incremental improvement during production drill testing. The major bit manufacturers have developed bit shapes and designs for specific requirements, such as for hard and for soft rock drilling respectively. As is the case for smaller diameter detachable bits, several other factors have to be considered in the selection of a down-the-hole bit for a particular application. Service representatives of reputable suppliers have considerable experience in evaluating customer requirements and are a good source of information. It is important to stress that the performance of a bit will vary depending on several interacting factors such as rock drill rotation rate, percussion energy and feed pressure levels, ground competence, and rock hardness. The final choice of bit design can therefore, often only be made after field testing of several designs.

BIT GRINDERS FOR SHARPENING DETACHABLE BITS

Most tungsten carbide-tipped bits require to be resharpened at regular intervals, depending on the hardness, toughness, and abrasiveness of the rock drilled, and also on operating conditions such as rock drill efficiency and adequacy of flushing. The use of effective resharpening techniques are most important in prolonging the life of the bit and in reducing bit cost per unit drilled.

A number of manufacturers offer air and electrically operated grinders for sharpening detachable bits. Bit grinders vary from simple hand-operated air grinders fixed to a workbench or to the drill carriage itself, to more sophisticated semi-automatic (Fig. 5.112) and fully automatic stationary grinders for high-volume bit sharpening. Grinder selection will depend on the location of the job, bit type and size, and the volume of bits to be sharpened. Where relatively infrequent sharpening is required, a hand-operated grinder is often used right on the job site. For higher volume sharpening, a larger, sophisticated semi-automatic or automatic grinder is generally indicated. Automatic grinders are normally installed in a central workshop, and drill bits are brought in at regular intervals for sharpening.

Figure 5.112 Semi-automatic bit grinder

For face dressing of button bits, hand-held air grinders with special cup-shaped, diamond-impregnated grinding pins are available to restore worn buttons to their original shape.

Bit resharpening practice of both button as well as chisel bits is well described in technical bulletins published by the various manufacturers and suppliers, who will also provide technical support and on-site training.

DRILL STEEL

In all rock drilling, the drill rod or tube not only positions the bit in the hole, but it transmits or dissipates the impact energy between rock drill and drill bit. The demand for higher production rates and smaller diameter blast holes has required the development of a competent drill rod. Modern rock drill rods are produced to quite rigid requirements. High quality, low-alloy steels rolled to specific dimensional tolerances are used in the production of hollow drill rod. During fabrication, the rod ends may be forged, the drill rod threads machined and the rods heat treated and finished to exacting specifications. The resultant products are specifically suited to the respective applications, whether in small diameter underground drilling, or large diameter blast-hole benching.

A range of drill rod lengths and sections are available, depending on the required application. Significant standardization of section sizes and lengths has given the customer an advantage in competitive purchasing of drill rods. Many well-run construction and mining firms which use drilling accessories maintain accurate performance records. This enables them to evaluate the cost per unit drilled, thus allowing them to base purchases on value and not solely on price. All large manufacturers and suppliers have available a full range of drill rods and can supply catalogues on request. Such catalogues often contain other information on the use and handling of drill rods as well as other drilling accessories.

Drill Steel Servicing

Today, much of the steel offered on the market is not serviced at the job but in strategically located shops operated by suppliers of drill steel, bits, and drilling accessories, or in local shops set up to service drill steel and accessories for a number of users in one area.

COUPLINGS

Standardized couplings are available for all common thread sizes. Couplings are made from high quality steel to the same exacting standards as drill rods. Couplings are necessary to join the respective components of a drill string, namely striking bar (or shank adapter) and drill rod. In long-hole and in bench drilling, for example, several drill rods may have to be coupled together to drill to the required depth. Coupling properties are finely tuned to that of the drill rod thread, to provide high fatigue resistance, long life and ease of uncoupling.

SHANK ADAPTERS OR STRIKING BARS

Shank adapters or striking bars vary in design depending on the drill model they are intended to fit; a large variety of shapes and sizes are therefore, available. Striking bars are required to transmit the drilling energy while supporting the entire drill string.

High strength, low-alloy steel is commonly used in the manufacture of striking bars. A significant amount of machining is required to shape the striking bar, after which the product is heat treated in modern equipment with close quality control. The product is often finished by grinding and/or shot peening before a corrosion protectant is applied.

Drill manufacturers will usually specify the correct shank adapter for the rock drill supplied. Manufacturers catalogues list a range of striking bars. Service and application brochures provide the recommended practice to follow to obtain the maximum performance from drill strings in general, and striking bars in particular. They are available from suppliers and manufacturers.

SUMP PUMPS

Air sump pumps are built as centrifugal (Fig. 5.113), diaphragm (Fig. 5.114), or venturi-ejector pumps. They are normally used to remove water from sumps, basements, manholes, pools, ship holes, and the like, or wherever water interfaces with the progress or work, as seen in Fig. 5.115.

The centrifugal pump consists of a housing enclosing an air motor, generally of the vane type, that drives an impeller. The flow capacity of this type of pump ranges up to 370 gpm and hydraulic head up to 200 ft., depending on the type of impeller. This type of pump can also be used for pumping other lower viscosity fluids such as oil and gasoline. Bronze parts may be required when pumping explosive fluids.

The diaphragm or venturi pump is suitable for pumping extremely polluted, viscous, highly abrasive and flammable fluids and is not hindered by large obstacles in the fluid. A flow capacity in excess of 100 gpm can be attained at under 30 ft. of hydraulic head. Reduced flow rates are possible up to about 200 ft. of head.

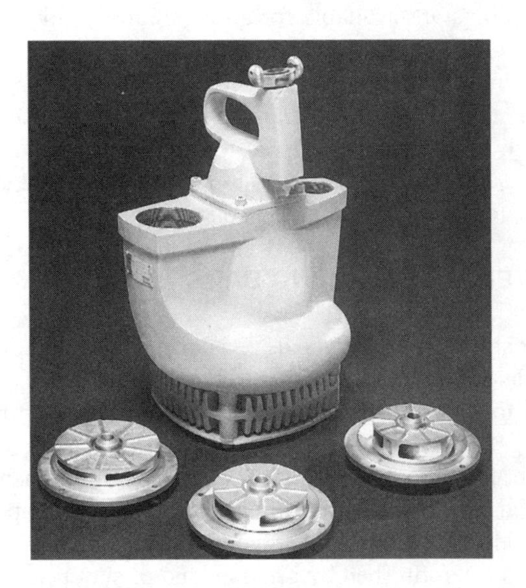

Figure 5.113 Centrifugal Sump Pump

Figure 5.114 Diaphragm Sump Pump

Figure 5.115 Centrifugal sump pump in use in construction application.

MISCELLANEOUS DATA

Line Oilers

The importance of an adequate continuous flow of oil into and through air operated rock drills, air tools, and associated equipment cannot be overemphasized. While almost all hand-held contractor's tools (rock drills, breakers, diggers, tampers, etc.) have built-in lubricators, the very small oil capacity and lack of adjustment for the rate of oil flow dictates the need for larger-capacity, adjustable-flow lubricators.

The line oil consists of an oil reservoir with suitable arrangement for filling, an oil flow adjustment (a central channel through which air passes enroute to the tool), and a means of picking up a measured or controlled amount of lubricant and delivering it into the air stream.

Line oilers are available in sizes from 1/3 pint for small air tools to 5 and even 10 gallon capacity for the large-bore drifters on crawler drills and for the drifters on multi-drill jumbos.

The most common lubricator for hand-held tools in the simple cast oiler seen in Fig. 5.116, which can operate in any position and will use practically all the oil in the reservoir.

Figure 5.116 Line Oiler

The automatic shut-off line oiler ensures that tools will not be operated without lubrication. When the oil reservoir is emptied, the flow is shut off.

A type of oiler now coming into rather common use is the separate oil reservoir type. It can be attached to a portable compressor, or built into the equipment, as in the case of a crawler-mounted drill (Fig. 5.117).

Figure 5.117 Oiler mounted in a portable compressor.

The quality of lubricating oil is as important as the quantity. For best drill performance and long life with low maintenance, the use of special rock drill lubricant is strongly recommended. Table 5.4 is the minimum lubricant specification for various groups of air tools. Recommended specifications for rock drill lubricants are shown in Table 5.5.

MINE AND CONSTRUCTION HOISTS

Air-powered Double- and Triple-drum Scraper Hoist or Slushers

Since slushers or scraper hoists are used principally in mines to handle the production by rock drills, they are included in this chapter. Slusher or scraper hoists are supplied in both double- and triple-drum configurations. They are constructed so as to allow power to be applied to each drum separately (Fig. 5.118).

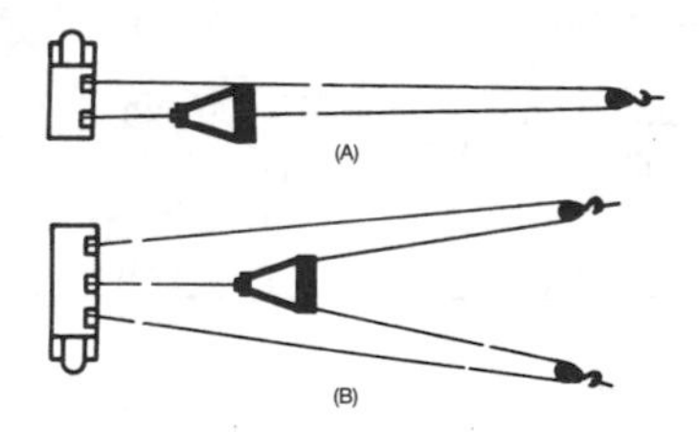

Figure 5.118 Diagrams of typical double and triple drum slusher hoist systems: (A) The most used underground layout using double-drum slushers and heavy duty sheave blocks. (B) A triple-drum slusher and two tail ropes put the scraper anywhere the operator chooses. This layout also can be adapted to around the-corner-mucking.

Table 5.4 Lubricant Specifications

1. Chipping hammers, paving breakers, clay diggers, trench diggers	Lubricant Rock Drill Oil Grade No. 10 Light		
Viscosity (SSU at 100 F)	175 – 225		
Flashpoint (C.O.C)	350°F mm		
Carbon residue (Conradson)	0.20% max		
Film strength (psi)	12,000 min		
Falex wear test	3,000 lb. min		
Steam emulsion no.	1200 min		
Pour point	10,000 SSU max at min operating temp.		
2. Sinkers	Rock Drill Oil Below 30°F Grade No. 10 Light	Rock Drill Oil Above 30°F Grade No. 30 Medium	
Viscosity (SSU at 100 F)	175-225	600-850	
Flashpoint	350°F min	380°F min	
Carbon residue (Conradson)	0.20% max	0.30% max	
Film strength (psi)	12,000 min	12,000 min	
Falex wear test	3,000 lb min	3,370 lb min	
Steam emulsion no.	1200 min	1200 min	
Pour Point	10,000 SSU max at min operating temp.		
3. Drifter	Rock Drill Oil Below 30°F Grade No. 10 Light	Rock Drill Oil 30-80°F Grade No. 30 Medium	Rock Drill Oil Above 80°F Grade No. 40-50 Heavy
Viscosity (SSU at 100 F)	175-225	500-850	1500-2000
Flashpoint (C.O.C.)	350°F min	380°F min	395°F min
Carbon residue (Conradson)	0.20% max	0.30% max	0.40% max
Film strength (psi)	12,000 min	12,000 min	12,000 min
Falex wear test	3,000 lb min	3,370 lb min	…………..
Steam emulsion no.	1200 min	1200 min	1200 min
Pour point	10,000 SSU max at min operating temp.		

Table 5.5 Recommended Specifications for Rock-drill Lubricants

Viscosity	
For ambient temperatures below 20°F	SAE 10
For ambient temperatures 20-40°F	SAE 20
For ambient temperatures 40 80°F	SAE 30
For ambient temperatures 80 110°F	SAE 40
For ambient temperatures above 110°F	SAE 50
Emulsibility	
ASTM steam emulsion number	1200 sec
Film Strength	
Pin Tester test	300,000 psi
The value indicated is the desired minimum. Rock-drill oils must display definite load-carrying ability. High film strength is mandatory for drilling conditions involving heavy rotational load.	

In double drum slushers, one drum is used for pulling the loaded scraper and is called the haul drum or pull-rope drum. The other drum is called the tail-rope drum. The rope from the haul drum is attached to the bail or drawbar of the scraper, as shown in Fig. 5.116. It is then passed over a sheave or pulley, which is anchored in the stope or drift somewhere above and behind the material to be moved and attached to the back of the scraper.

By engaging the clutch on the haul drum, the operator pulls the scraper forward, enabling it to dig into the material and gather its load. On reaching the unloading point, the clutch on the haul drum is disengaged and the tail rope drum clutch engaged, thereby returning the scraper to the muck pile or the material to be moved. Thus, the movement and the productive action of the scraper is controlled by the alternate action of the hoist at the will of the operator.

In mining (Fig. 5.119), the slusher is still used to scrape muck (broken rock or ore) into chutes or mill holes from which it is usually loaded by gravity into rail ears on a lower level. Muck also may be pulled up a ramp and loaded directly into cars. When the ramp is mobile and the slusher is mounted on it, the device is usually referred to as a scraper loading slide.

In triple-drum slushers (Fig. 5.120), two drums are tail rope drums. They are usually used with one scraper but tow sheaves, enabling the operator to scrape material from a much wider area (Fig. 5.116)

Air slushers are usually rated as to horsepower at 90 psig pressure. A variety of sizes is offered, ranging up to 35 hp. Drum capacities range from 190 ft. of 5/16 in. rope to over 375 ft. of 1-1/4 in. rope. Rope speeds vary widely with the gearing, the amount of rope wound on the drums, and other factors. Rated at half drum capacities, air slushers have line pull ratings of 1000 lb to over 4000 lb, with available line speeds generally between 150 and 250 ft. per minute. Since the scraper is empty when returning to the muck pile, tail rope speeds are usually 25 to 35% faster than rated haul rope speed.

Figure 5.119 Double-drum slusher in operation in a mine slope.

Figure 5.120 Triple drum slusher with air motor drive.

Air slushers are supplied in a variety of constructions: with the control levers in different positions in relation to the motor and with the rope winding off the top or bottom of the drum, whichever is more convenient for the specific application.

Different types of controls are also offered, including remote controls. The air motors used are normally of the multiple-cylinder-piston type, with some smaller models powered by vane air motors.

Air-driven scraper hoists or slushers offer safety features that eliminate shock and spark hazards, an advantage over electric-motor-driven slushers in dusty, wet, or gaseous mining conditions. They are much more readily set up or moved from working place in a mine since compressed air lines are usually available there for the drilling cycle. Air motors can be overloaded to the point of stalling without damage. They can accurately be controlled at almost any speed within the gear ratio, up to and including the point of stalling.

Single-drum Winches

Most manufacturers offer a variety of air-operated, single-drum utility winches, or tuggers (Fig. 5.121) for hoisting or pulling. They are used extensively in mines and by contractors for both underground and surface work. They also are widely used on oil and gas drillings (Fig. 5.122), in refineries and petrochemical plants, and by maintenance crews in general industry.

Air winches offer the same safety advantages described for air slushers. In addition, they are popular in marine applications due to the air motor drive resistance to the corrosive effects of a salt water environment.

Figure 5.121 Single-drum utility winch or tugger.

Figure 5.122 Single-drum winch on an off-shore oil drilling rig.

6

Blowers

ROTARY BLOWER HISTORY

The rotary blower evolved from a design of a water wheel invented by the Roots brothers for their woolen mill. The device consisted of two counter-rotating wooden paddles or impellers inside a casing. Water, guided through the machine, forced the impellers to rotate, turning the shaft and driving the machinery until the wood swelled and the impellers stuck. After it had been re-built and prepared for operation, the action of the rotors forced a quantity of air through the unit. The brothers decided that they had a better blower than a water wheel, which led to the rotary positive-displacement blower, now known throughout the world (sometimes referred to as the "Roots blower"). Today there are several major blower manufacturers worldwide.

The first major use was in foundry cupola furnaces and the first two experimental machines were tested in foundries. Many blowers were later manufactured for this service, and, in 1872, the *Engineering Journal* recorded a large blower for the West Cumberland Hematite Iron Works in England. Another major use was for mine ventilation, a relatively new field. Rotary positive displacement blowers constituted some of the largest installations of the 19th century. By 1870, several were already in use for ventilating mines of the Comstock Lode.

Two of the largest blowers ever built were installed in England in 1877. The impellers each had a diameter of 25 feet, 13 feet wide, and a blower capacity of 200,000 cubic feet per minute. A London engineering journal concluded that the rotary positive displacement blower was the most efficient type thus far. Still other early uses were in pneumatic conveying, aeration and agitation of liquid, supercharging and scavenging of large diesel engines, and for various vacuum processes.

Of the many services to which the blowers were applied, perhaps the most spectacular was in a subway constructed in 1867 beneath Broadway Avenue in New York City. The blower had an iron casing 22 feet high and impellers 16 feet long. The "Western Tornado," as the blower was called, provided the power for a 22-seat passenger car which ran on tracks between Murray and Warren Streets. The car was literally blown to one end, then sucked back on its return trip by the action of the unit. The system was the subject of an article in *The American Heritage of Invention & Technology.*

Technical Details

The rotary, positive displacement blower works on a very simple principle. See Figure 6.1. As the drive shaft is rotated, the impellers turn in opposite directions with very finite clearances between each other and between the rotors and the casing. As each impeller passes the inlet, a measured quantity of air is trapped between the impellers and the casing. As the shafts continue to rotate, this "pocket" of air is transported around the casing to the discharge side of the machine, where it is then expelled through the port, against the pressure prevailing in the discharge line. When this occurs, a back flow of air into the "pocket" from the higher pressure discharge line produces a constant volume pressure rise, causing a pressure pulse resulting in noise. As a "pocket" of air is expelled four times with each revolution of the drive shaft, or twice with each impeller, the fundamental frequency of the pressure pulse is four times the shaft speed.

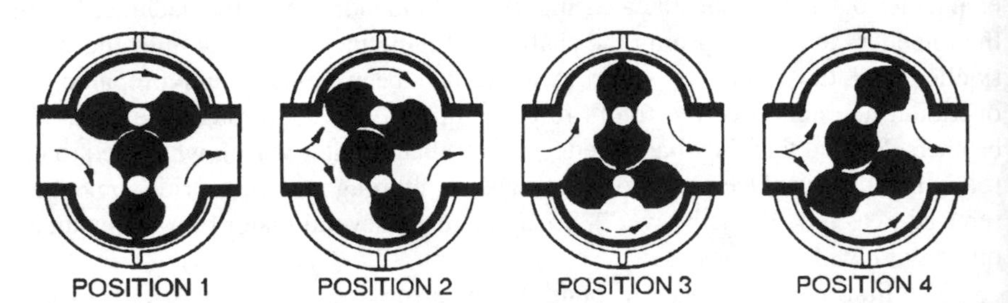

Figure 6.1 Blower Operating Principle

The displacement, or swept volume, of the blower *(not considering any leakage losses)* is the theoretical volume, in cubic feet per revolution, which the unit will transport from inlet to discharge in one revolution of the drive shaft. The term is usually expressed as "cfr". Figure 6.2 shows the area "**A**" which, when multiplied by the impeller length, represents the volume of each of the four "pockets" as the machine rotates. Therefore, this volume times four is the displacement or "cfr". It is calculated by the following equation:

$$Displacement\ (cfr) = K * GD^2 * CL/100$$

where:

cfr = cubic ft. per revolution
K = Constant, depending upon impeller geometry
GD = Timing gear pitch diameter, in inches
CL = Cylinder (impeller) length, in inches

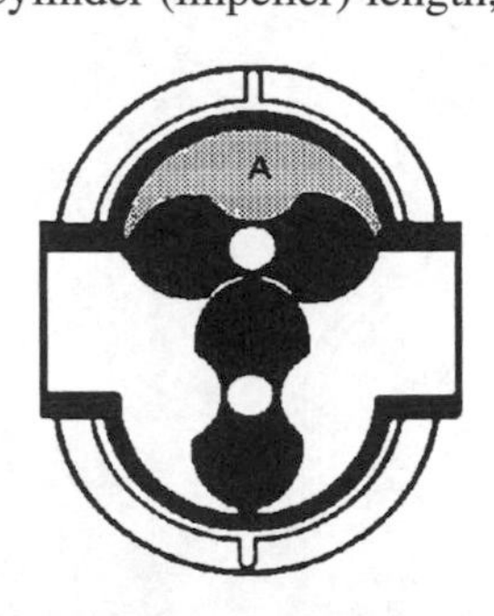

Figure 6.2 Theoretical Volume

It is important to note that the cfr multiplied by shaft speed (RPM) equals the ***gross*** **displacement** of the machine in cubic feet per minute. ***Net*** **displacement** will be discussed later.

While most applications for rotary positive displacement (PD) blowers involve handling *air*, the machine also is capable of handling any number of gases, from hydrogen to steam to natural gas to ethylene or, of course, nitrogen. Proper attention must be given to seals as well as performance calculations and limitations, all of which will be discussed later. In addition, they perform very well under vacuum conditions as well as pressure.

Further comments will be made later, relative to vacuum operation.

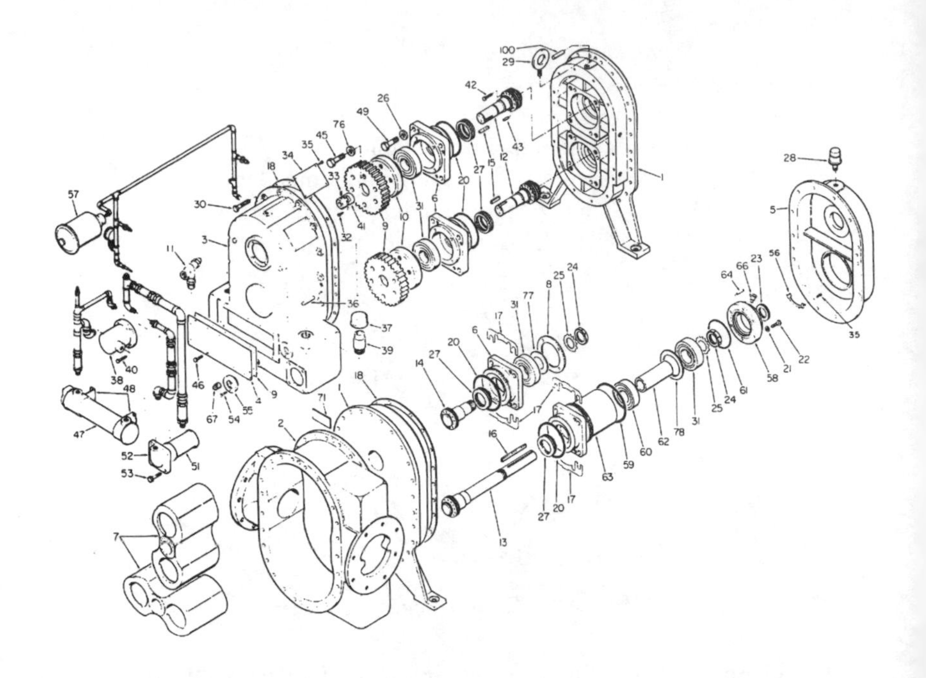

Figure 6.3 Typical Positive Displacement Blower Assembly

The typical PD blower is constructed almost entirely of cast iron, with the exceptions of shafts, bearings and timing gears. As in most products, there is some special nomenclature (See Figure 6.3). The terms impellers and rotors are used interchangeably. A rotor generally is considered as an assembly made up of impeller and shaft(s) (Items 7, 12, 13, 14). There are different impeller designs or profiles for special functions. The impellers are supported on anti-friction bearings (Item 31) by the end plates or, as they are more commonly called, headplates (Item 1). In turn, the headplates on most models support the entire blower with feet at their base. The cylinder (Item 2), which surrounds the impellers, is held in place by the headplates, completes the enclosure for the impellers and also contains the inlet and discharge ports.

There is some flexibility in the configuration of a rotary blower in terms of the locations of gas connections and drive shaft. Depending on the model, the gas connections can be either on the top and bottom of the unit, on the sides or a combination. When the connections are on the top and bottom, the unit is designated as a "Horizontal" machine and when the connections are on the sides, it is a "Vertical" machine. The designation has nothing to do with the direction of gas flow but comes from the orientation of the blower shafts, i.e., when a plane intersecting the shafts is horizontal, the unit is so designated. Conversely, when the plane is vertical, the machine is designated as vertical. Some units can have the inlet connection on the top and the discharge on the side. Since a plane through the shafts establishes the definition of the orientation, this would be a "Horizontal" machine. This issue is critical when designating the specifics of a blower as some manufacturers define the orientation differently.

Figure 6.4 Horizontal Blower

Figure 6.5 Vertical Blower

Timing gears (Item 9) on one end of each rotor outside of the blower chamber are critical parts, as they are responsible for maintaining the finite clearances between the rotating impellers. These gears are typically of AGMA 11 or 12 quality. They are attached to the shafts by various means, depending upon the design. Attachment must prevent any slippage, which would allow the impellers to go out of timing, allowing rotor to rotor contact; a serious condition, eventually causing failure.

Timing gears and gear-end bearings are oil lubricated and are enclosed in a gearhouse or gearbox (Item 3). At the opposite end of the unit the bearings may be either oil or grease lubricated, depending upon the design. Figure 6.3 shows a pressure-lubricated unit with oil lube at both ends. The drive shaft on smaller units normally is extended for either direct coupling or v-belt drive from the end opposite the gears.

In the areas where the shafts pass through the headplates, lip-type oil seals (Item 27) normally are used to restrict leakage of lubricant, although other types of seals sometimes are used. Air from these blowers is oil free. As a precaution against lubricant entering the air-stream, and in order to relieve the pressure across the lip seal, which is not designed for any significant pressure differential, an area vented to atmosphere is located between the lubricant seal and the air seal. In the case of some gas blowers the lip seals are replaced by mechanical seals, which are designed to seal against the gas pressure. In this case the vent areas are plugged to prevent gas leakage to atmosphere. More information about gas seals will be presented later.

Typically, rotary blowers are designated by their gear diameter and impeller, or cylinder, length. Thus a 6 x 12 (or 612) has a 6" diameter timing gear and a 12" impeller or cylinder length. The impeller diameter is somewhat greater than the gear diameter, depending on the specific design. Generally, the impeller OD is 1.5 to 1.625 times the gear diameter. Practically, the impeller length is no longer than 4 times the gear diameter. Any greater length increases displacement, to the point where there are problems in physically getting the volume of air in and out of the casing connections without serious aerodynamic losses and difficulty in maintaining stability in the casing.

Theoretically there is no limit to the size of a PD blower, however the practical aspects of the design impose certain limits. The largest size currently in production in the world is the 36 x 72 which handles about 58,000 inlet cubic feet of gas per minute, is rated at 12.5 psi pressure differential, runs at 513 rpm and requires a 3500 hp motor. Standard sizes stop at 20" - 26" gear diameter, depending on the manufacturer, and are rated at about 43,000 cfm and 10 - 15 psi differential. The smallest units are rated at less than 10 cfm, requiring fractional hp motors.

Practically speaking, the PD blower is limited to a compression ratio of about 2:1, due to temperature limits that we will discuss later. A 2:1 ratio means approximately 15 psig on air pressure service or about 15" Hg vacuum on vacuum service. As will be discussed later, there are ways to overcome this ratio limit with special designs. In terms of working pressure, these units normally are limited to 25 psig, as a standard. Under certain conditions, or with specific sizes or designs, however, this rule also can be circumvented.

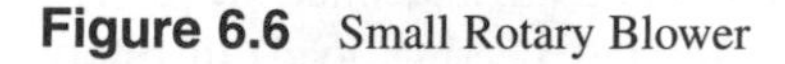

Figure 6.6 Small Rotary Blower

Figure 6.7 Large Rotary Blower

As discussed earlier, the two parameters used in blower sizing are gear diameter and cylinder, or impeller length. These must be considered in the design phase. Looking at a blower, it is obvious that the impellers are simple continuously loaded beams, supported on both ends by the bearings. The loading across the beam is the differential pressure between discharge and inlet. The bearings, of course, take all of the radial load and, theoretically, there are no axial thrust loads unless timing gears are helical type. Obviously, the larger the impeller/shaft diameter, the more loading the rotor can take. Conversely, the longer the impeller, or the wider the bearing span, the less loading the rotor can take without undue deflection. The pressure differential acting on the impeller creates a loading with a value of the pressure differential times the area of the impeller. A "square" blower is one in which the gear diameter is equal to the impeller, or cylinder length. These considerations are used in the design concept of a PD blower called "square loading." The design square loading of a blower line, simply stated, is the maximum pressure differential that a "square" blower can withstand without harmful deflections that would exceed the clearances within the unit and/or seriously detract from its life. As stated earlier, the load capability is directly proportional to the diameter and inversely proportional to the length of the impeller. Since the impeller diameter is proportional to the gear diameter, the maximum load is directly proportional to the gear diameter. Therefore, for any given size of unit, the following relationship is useful:

$$\text{Max. Press Dif} = (\text{DSL})\,(\text{GD})\,/\,(\text{IL})$$

where:

Max. Pressure Differential is in psi
Design Square Load is in psi
Gear Diameter is in inches
Impeller Length is in inches

Thus, within a line designed with a square loading of 18 psi, a size 10 x 10 would be capable of a maximum pressure differential of 18 psi, while a 10 x 30 in the same line would be rated at 6 psi. The design considerations that come into play when determining the square loading of a line of blowers are:

Shaft size, material and stiffness
Bearing size, rating and location
Impeller profile and stiffness
Gear design and location
Gear attachment design

As the bearings on a PD blower carry the radial load, and since the radial load is directly proportional to the pressure differential, bearing load, therefore, is dependent on the design square load.

Thermodynamic Cycle Theory

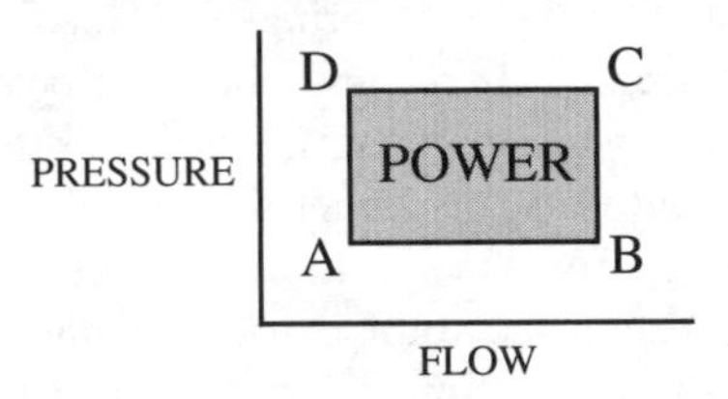

Figure 6.8 Thermodynamic Cycle Theory

A PD blower theoretically operates on a "square card" cycle. That is to say that the power required is dependent upon the flow and the pressure. As Figure 6.8 shows, moving from A to D around the diagram, air enters the blower (A to B), compresses it (B to C), expels it to the discharge (C to D). The cycle then is repeated with air entering at inlet pressure. The power is represented by the area inside the square, showing that the power is proportional to the flow and pressure, or:

$$P \propto 144 \times \text{Pressure (psi)} \times \text{Flow (cfm)}$$

Dimensionally:

$\text{ft-lb/min} = \text{lb/ft}^2 \times \text{ft}^3\text{/min}$

and:

$$\begin{aligned} \text{hp} &= \text{ft-lb/min} \times 1/33000 \\ &= 144 \div 33000 \times \text{Pressure (psi)} \times \text{Flow (cfm)} \\ &= .00436 \times \text{Pressure Differential} \times \text{Inlet Flow} \end{aligned}$$

or:

$$\text{hp} = .00436 \times \text{Pressure Differential} \times \text{cfr} \times \text{rpm (excluding frictional hp)}$$

Limitations

As with any machinery, there are limits to the manner in which a PD blower can be applied. Specifically, these limits are:

Temperature rise
Discharge temperature
Pressure differential
Working pressure
Shaft speed

Temperature Rise

Temperature rise is defined as the difference between inlet and discharge temperature. The increase in temperature is caused by the heat of compression, in accordance with the gas laws, including inefficiencies. Going back to the basic blower construction, it is evident that the impellers "see," alternately, inlet and then discharge temperature. Their average temperature should be the average between those two values. The casing, on the other hand, will "see" not only inlet and discharge temperature, but will also "see" ambient temperature on the external surfaces. The cylinder, therefore, will be at a lower average temperature than the impellers. Therefore, the impellers will tend to grow more, and faster, than the cylinder, creating a differential expansion, which causes the end clearance between the impeller and the headplates to reduce. Further temperature rise due to increased compression ratio will result in impeller-headplate contact, which will ultimately cause failure of the blower.

Discharge Temperature

Several problems occur when the discharge temperature increases to levels higher than recommended. Firstly, due to heat transfer, the lubricant begins to break down at temperatures above its limits. When this occurs, not only does the lubricant lose its qualities, but carbon begins to form, causing further problems to oil flow and rubbing surfaces. The higher temperatures also can result in reduced bearing life, as the bearing materials begin to break down due to the impaired lubricity. Gear life can be affected for the same reasons. There also could be damaging thermal distortion of parts, causing misalignment and, perhaps, even failure, due to the extreme temperatures. Finally, seal failure likely would occur as the materials in the seal parts are subjected to the higher thermal stresses.

Pressure Differential

There are specific problems related to exceeding the pressure differential design ratings. We have related bearing life to pressure differential, and that the bearings take the radial load caused by the pressure differential. Added load, therefore, will bring about reduced bearing life. Increased pressure differential also causes increased shaft stresses and there must be limits to avoid excessive deflections and possible fracture of the shafts.

The torque requirements of a rotary blower vary directly as do the pressure differential. Since the timing gears see this torque, increased pressure differential means increased loads on the gears. The gears have a finite design life, dependent on pitch line speed and loading. Therefore, gear life is dependent upon pressure differential.

Due to the deflections resulting from increased pressure differential, the clearances on the inlet side of the unit, between the impellers and the cylinder, begin to close up. This happens on the inlet side, because the force that causes the deflection is equal to the pressure differential from discharge to inlet times the projected area of the impeller. The pressure is higher on the discharge side, thereby forcing the impellers to deflect toward the inlet. If deflection causes the available clearances to be taken up, contact is made and the resulting rub can cause failure.

Increased compression ratio causes increased temperature rise. Assuming that the inlet pressure remains constant, an increase in pressure differential means an increase in compression ratio and, therefore, an increase in temperature rise. Excessive temperature rise causes the end clearances to close up, causing the eventual rub and, thence, failure.

As previously stated, there is a definite pressure pulse that occurs four or six times per revolution depending on number of lobes per rotor. This pulse is proportional to the pressure differential and causes vibration and noise. Therefore, as the pressure differential is increased, noise and vibration also increase. These vibrations and noise levels are reasonable when held within limits, but excessive values can be destructive.

Working Pressure

The normal maximum working pressure of a typical PD blower is 25 psig. This limit basically is a result of the casing design. Exceeding the design value can result in potentially damaging distortions, internal rubs and possibly even casing fracture. As with any vessel containing air or gas, one must be extremely cautious about controlling internal, or working, pressures.

Shaft Speed

Every blower is limited on shaft rotational speed. Displacement and torque are proportional to rotational speed, affecting both bearing life and gear life. In addition, noise and vibration will increase, perhaps to dangerous levels, if the shaft speed is allowed to rise beyond reasonable values. Also, as the rpm is increased, the stresses on the rotating parts go up, approximately as the square of the rpm. Excessive rpm, then, can result in break-up of the unit.

Slip

This term in blowers refers to the leakage flow from discharge to inlet. Finite clearances must be maintained within the PD blower for it to continue operating at design efficiency.

As the blower impellers rotate, carrying the "pockets" of air from the inlet to the discharge, one side of each impeller is exposed to discharge pressure while the other side sees inlet pressure. There is a clearance, or leakage path, between the two impellers and between the impellers and the casing. These leakage paths, along with the pressure differential, allow a certain amount of air from the pocket to "slip" back through from discharge to inlet. The ends of the impellers, in close proximity to the headplates, are further leak paths. The total of these leak paths must be accounted for in performance calculations. It has been shown for specific applications that about 2/3 of the slip occurs between the impellers and between the impellers and cylinder and 1/3 around the ends of the impellers. We will discuss the specifics of this relationship later.

A feature of some air-moving and compressing devices is a ratio called "Volumetric Efficiency", or "VE." We can apply this characteristic to the rotary blower and slip becomes a part of the definition. VE is defined as the ratio, expressed as a percentage, of the actual to theoretical (or ideal) displacement (flow), or:

VE(%) = 100 x Actual Flow/Theoretical Flow (Displacement) where Actual Flow is defined as actual inlet flow and Theoretical Flow is cfr x rpm.

Since slip detracts from the theoretical displacement and since slip can be expressed as flow, the above becomes:

VE(%) = 100 x (Theoretical Flow-Slip Flow)/Theoretical Flow.

Gross and *net* displacement were discussed previously. The Theoretical Flow from the above is *gross* displacement and the Actual Flow is *net*. Typical VE's for a rotary blower range from about 95% for a large machine at maximum rpm/minimum pressure differential to as low as 65% for a small machine at low rpm/high pressure differential.

A means of reducing slip is to inject a controlled amount of water or other fluid into the inlet of the blower along with the gas stream. This liquid disperses inside the casing and tends to fill up the clearances, effectively reducing the available slip paths. This concept is used most effectively on vacuum service where compression ratios, and therefore slips, are relatively high when operating dry. The addition of the liquid also reduces temperature rise and tends to flush out any contaminants, which might otherwise build up on the internals. Certain precautions must be considered, however, and these will be treated later.

DESIGN FEATURES

Some of the blower's design features have already been discussed , however there are many variations, not only from model to model, but also from manufacturer to manufacturer. These include:

Impellers

The most important part of the blower is the impeller. Impellers normally are cast grey iron with a typical variation being cast *ductile* iron for greater strength. In extreme cases, where corrosion or erosion resistance is required, they also can be cast or fabricated of stainless steel or coated with various materials. The other variation in impellers is in the profile. The typical impeller lobe profile is generated by an involute curve. This curve allows the impellers to rotate in conjunction with and in close proximity to one another without coming into contact. It is an efficient profile and relatively easy to form on various machine tools. When the involute impeller rotates, however, there is a point at which the clearance locus jumps from one point to another. As this occurs, there is a brief instant when a small volume is trapped between the impellers. While this causes no problems on normal air service, if there are *significant* or uncontrolled amounts of liquid entrained in the gas stream, the effect is an attempt to compress this liquid as the impellers continue to rotate. Liquid does not compress, so the impellers are forced apart by the liquid for an instant. This can be damaging to the unit as it may cause internal contact. Where there is a possibility of significant liquid entrainment, slightly different profiles can be selected.

These water-handling profiles are designed to allow a rolling or sweeping motion to the clearance loci rather than the jumping motion as with the involute. Therefore, there is no opportunity for the liquid to cause the deflection as with the involute profile. Various profiles for this purpose do not provide as much displacement as the involute. In addition, they are more difficult to form in the manufacturing process. These profiles are very useful, however, when applying the PD blower on vacuum service. In this service, a liquid, usually water, is purposely sprayed into the blower inlet to aid in sealing the clearances to make the machine more efficient and, as a bonus, the water keeps the unit cool, as it carries away the heat of compression, allowing the unit to achieve a higher compression ratio.

Staging these units permits the compound machine to meet requirements as low as 27" Hg. vacuum (3" Hg. absolute) or to a level near the vapor pressure of the sealing liquid. While these special impeller profiles have been developed for water-sealed applications, recent efforts to utilize the involute profile with a reduced, controlled water flow have been quite successful.

Three standard profiles are utilized by most manufacturers - Involute, High Displacement Liquid Passing, and Cycloidal, the latter two typically used for water-sealed vacuum service. The three profiles are shown in Figure 6.9. Recalling the previous discussion on displacement, which is calculated as:

$$\text{Displacement (cfr)} = K \times GD^2 \times CL/100$$

The "K" for each profile varies, being greatest for the involute profile and smallest for the cycloidal. The cycloidal, however, provides an impeller having a high degree of tolerance for water flow. In fact, the cycloidal profile also has been used as a *water pump.*

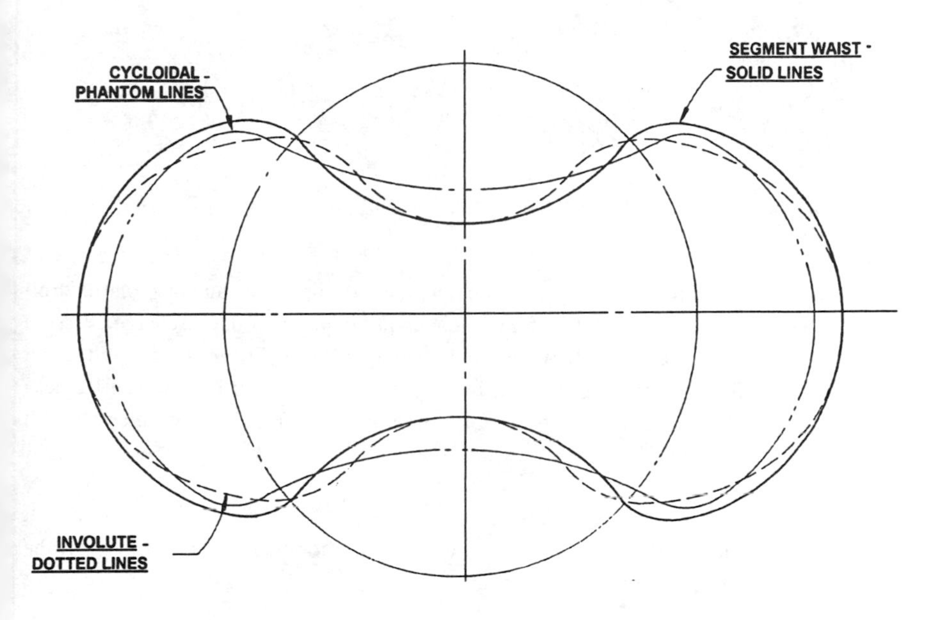

Figure 6.9 Comparison of Rotor Profiles

Three Lobe Blowers

More recent designs include three lobe blowers with some changes in operating characteristics.

The basic operation of a rotary three lobe type position displacement blower with precompression passage is illustrated (Figure 6.10) as follows:

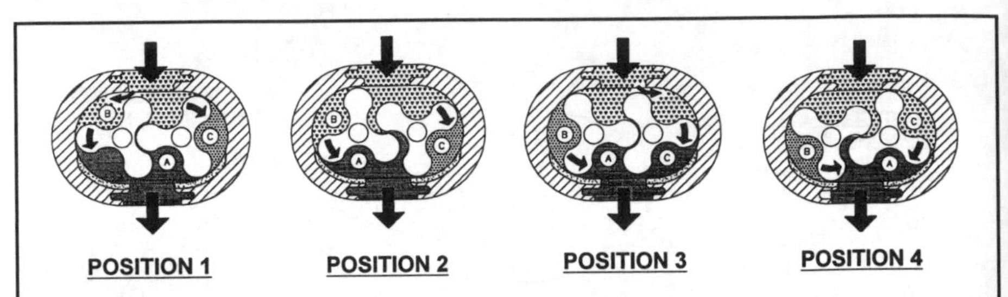

In position 1
Flow is top/bottom from blower inlet to discharge, the left impeller is mounted on the driving shaft then rotates counterclockwise and delivers a volume into the discharge pressure side. At the same time, space (B) between the impeller and the cylinder wall is filling with another and equal volume at inlet pressure. It is about to be sealed off by the counterclockwise rotation of this impeller. The space (C) is progressively filled and compressed with discharge volume through the precompression passage.

In position 2
The inlet area sealed in (B) and discharge pressure starts to enter space (A) through the precompression passage as rotation continues.

In position 3
Volume (C) is delivered to the discharge pressure side in the same manner as volume (A). Because of the almost complete pressure equalization through the precompression passage, no sudden shock will occur.

In position 4
The inlet area volume sealed in (C) and the discharge pressurized volume entering space (B) start to enter through the precompression passage in (C). Volume (A) is delivered to the discharge side and the volume in (B) will also start delivering with (A) as soon as the rotation reaches the outlet side.

Figure 6.10 Basic Operation of Three Lobe Blowers

Note: All blowers are available with either side/side or top/bottom connection.

Bearings

The bearings, which support the impellers, typically are of the anti-friction type, either ball or some form of roller. Depending upon such parameters as square loading, rpm, lubrication options and others, the manufacturer attempts to design the most appropriate bearing into the unit, keeping in mind life, size, bearing clearance, cost, availability and load carrying capability. Also important to the bearing system is control of thrust or side movement of the impellers. There is no inherent thrust generated by the impellers as they rotate but, due to the expansion process brought about by the temperature rise within the blower, other considerations must be taken into account to control axial movement of the rotors. Additionally, when operating at the higher speeds of today's PD blowers, a slight contact of impeller to headplate likely will continue in a degenerative mode, with contact causing temperature, which, in turn causes further expansion and more contact, leading eventually to failure. A positive control of the axial position of the impellers is mandatory and some thrust capability in the chosen bearing system is necessary.

The use of a V-belt drive system causes severe side loads on the drive shaft and bearings and such operation must be considered in the bearing design. Bearing life can be grossly affected when V-belt drives are used in lieu of direct-connection. Some manufacturers use an extra bearing on the drive shaft when such drives are to be applied. Another solution is to use a "jack-shaft" for the blower sheave and direct couple the shaft to the blower with a flexible coupling. This is a complicated arrangement but does accomplish the desired result of reducing the drive shaft bearing load.

Lubrication

Because the blower is equipped with rolling element bearings and gears, some form of lubricant must be provided. Generally, the gears are located on the end opposite the drive and are lubricated with oil, along with the bearings on that end. The bearings on the drive end may be either grease or oil lubricated, depending on the design and the temperatures at which the blower is expected to operate. Due to the action of the gears, typically there is adequate splashing action on the gear end so that no auxiliary slinger is needed to throw the oil onto the bearings. If the blower is oil lubricated on the drive end, a slinger usually is required to obtain the proper splashing action to direct the oil onto the bearings. On so-called splash-lubricated machines, there is no need for pumps, coolers, piping, relief valves, etc. in the lube system. Such a system, therefore, is quite simple and, other than periodic oil changes, is also fairly maintenance free. Operation at high temperatures for extended periods of time will cause deterioration of the lubricant, requiring more frequent changes.

Figure 6.11 Blower Equipped with Splash Lubrication

A pressure lubrication system, while more costly to purchase and maintain, is considered better in the long run as it overcomes the disadvantages of the splash system. It assures that the blower always will be supplied with a cool, clean flow of oil to the critical components. The oil will last longer and the bearings and gears will have a longer life expectancy. A pressure system can be furnished with an oil cooler so that oil temperatures are held to ideal levels, even when the blower is operating at high compression ratios with high ambient temperatures. Safety devices, such as low oil pressure and high oil temperature switches can be supplied. On critical service, where an unplanned shutdown cannot be tolerated, an automatic auxiliary oil pump, which comes on at a reduced oil pressure indicating a loss of the main pump, can be fitted into the system.

Figure 6.12 Blower Equipped with Pressure Lubrication

Casing

The casing, made up of the headplates and cylinder, performs the function of containing the gas and supporting the rotors, while also keeping the entire system in alignment. The materials of construction normally are cast grey iron and the normal design working pressure is 25 psig. With other materials, such as ductile iron, and other designs, the casing pressure rating can be enhanced. Another consideration is noise level and it has been demonstrated that the thicker the casing the lower the noise.

Timing Gears

Timing gears must be of high quality with good tolerance control to properly maintain the impeller to impeller clearances. A small inaccuracy in the gear profile can bring about contact between the impellers. Poor quality gears also will cause noise, although the gear pitch line velocities are relatively low at about 5000 fpm, maximum. The gear type normally is straight spur, however helical designs are used by some manufacturers. Helical gears produce a definite axial thrust on the rotors and must be considered in the design and life calculations of the bearing system. Most gears today are hardened and ground and of an AGMA quality of 11-12. Several methods are used to attach gears to shafts, but the most common is either some form of taper fit or a dual tapered ring system which, when pulled up axially, places an inward radial force on the gear hub, thus securing it to the shaft. Both methods allow the impellers to be easily "timed" or adjusted in the circumferential direction when assembled, or during repair procedures. Older types of attachments utilize radial or axial pins to locate the gears to the shafts with holes drilled and reamed on assembly. This method does not permit more than one or perhaps two repairs before the shafts and gears must be replaced, due to the successive numbers of holes through the parts.

Timing gears typically are mounted on the end of the blower opposite the drive end to allow easier maintenance and re-timing. However, with the gears mounted on the drive end of the blower, the drive torque is split at the gears prior to being transmitted through the drive rotor, thus improving the torsional characteristics of the rotor. Larger size blowers typically have the timing gears on the drive end.

Seals

Sealing systems for air blowers are relatively simple and the air seals usually involve single or multiple labyrinths where the shafts pass through the headplates with the adjacent space being vented to atmosphere. The lubrication seals generally are of the lip type to restrict leakage of the lube oil or grease. The vent area between the air seal (labyrinth) and the lip seal allows the air being handled to bleed to (or from, in the case of vacuum service) the atmosphere, thus keeping the lubricant away from the air being handled. PD blowers, therefore, are considered to be "oil-free." This term must be used carefully, as the blower does not *remove* any hydrocarbons which enter with the inlet air, nor can one guarantee that the unit, under all operating conditions, will not present some trace of oil in the discharged air-stream. It can be stated, for most practical applications, that the PD blower can be considered *essentially* "oil-free."

For gases which must be contained within the casing due to their cost or hazardous nature, special seals, usually of the mechanical face variety, are used. Some frames require the addition of a pressure lube system to provide cool, clean oil, under pressure, to make the seal function properly. A typical mechanical seal is pressurized by the unit's pressure lube system. This seal has been found to be very reliable, provided the oil pressure is maintained and the oil is kept cool and clean.

Shafts

Shafts, which support the impellers, are of several varieties. Some, especially those which have a lower design square loading, are of steel and are pressed through the impellers and then pinned radially through both parts. The disadvantage of this design is that the shaft diameter is restricted, in order to have sufficient material in the narrow, or waist section of the impeller to hold it to the shaft. One alternative is to cast the impeller and shaft in one piece of a stronger material, usually ductile iron, in which case the shaft can be made the full diameter of the waist section of the impeller. This design typically is used on blowers having a design square loading of 25 psi and greater. Another approach is to flange mount a stub shaft to each end of the impeller with high strength bolts and use taper pins for locating purposes. An advantage of this design is that it permits replacement of the shafts without replacing the entire impeller when only the shafts are damaged. The stub shaft design is used for square loading up to about 40 psi.

PERFORMANCE CALCULATIONS

The process of calculating the performance of a PD blower is relatively simple. Certain facts needed for a blower selection are:

Inlet Pressure (psia)
Inlet Temperature (degrees R)
Molecular Weight of the Gas Being Handled *
C_p/C_v (K value) of the Gas Being Handled **
Inlet Volume Requirement (icfm)
Discharge Pressure (psia)
* If air, calculate, using RH %. Also can use S.G. (MW/28.97; Dry Air = 1.0)
** If dry air, $C_p/C_v = 1.395$

Slip

The term "slip" is the amount of air that "slips" back from discharge to inlet through the clearances and is expressed in icfm. It is a loss that must be accounted for in the shaft speed required to achieve the specified amount of delivered flow. Slip depends upon the blower inlet conditions and discharge pressure. It also depends upon the clearances within the blower, but for a given size, this parameter is a constant. Each blower model, at any given set of conditions, can be expected to "slip" a certain amount of air back to inlet. In order to calculate for the specific set of conditions where the unit is expected to operate, we start with a known quantity and then correct that figure for the actual conditions. The known quantity, established through testing, is called 1 psi slip. This quantity is the rpm at which the blower must operate, with the discharge blocked off, to establish a pressure of 1 psi on a manometer connected to the discharge port. The significance of this is that, with the discharge blocked, all the air coming in at the inlet port is "slipping" back to the inlet. Therefore, since the displacement times the rpm equals the flow rate, and the displacement is constant for a given size, the slip can be expressed in rpm and referred to a standard set of test conditions. The 1 psi slip then can be calculated as follows:

Inlet Pressure	= 14.7 psia
Inlet Temperature	= 68°F (528°R)
Inlet S.G.	= 1.0
Pressure Differential	= 1.0 psi

To correct for actual or specified conditions, the following equation applies:

$$\text{SLIP} = 1\text{ psi SLIP} \times (\Delta P \times PS \times T1/S.G./P1/TS)^{1/2}$$

where:

1 psi SLIP	=	Blower rpm required to provide 1 psi on air at standard conditions with a blanked discharge.
ΔP	=	Operating pressure differential in psi.
S.G.	=	Specific Gravity of gas compared to dry air at 1.0.
T1	=	Operating inlet temp. in °R (°R = °F + 460)
TS	=	Standard Inlet Temp. (528 °R)
P1	=	Operating Inlet Press. (psia)
PS	=	Standard Inlet Press. (14.7 psia)

Slip can be reduced if the unit is water injected, as will be discussed later.

Inlet Volume/Shaft Speed

It has been pointed out that the PD blower's capacity is expressed as displacement times rpm. The rpm, as used here, is "effective" rpm, or the actual shaft speed less the slip rpm. The method of calculating either actual rpm for a given inlet flow or flow for a given shaft rpm is shown in this equation:

$$Q_1 = cfr \times (rpm - slip)$$

where:

Q_1 = Volume Flow of Gas at Blower Inlet in cfm
cfr = Blower Displacement in ft^3/revolution
rpm = Blower Shaft Speed
slip = Calculated Slip in rpm

Conversely, for a given flow:

$$rpm = \frac{Q_1}{CFR} + \frac{Slip}{R.PM}$$

Brake Horsepower

As discussed earlier, a PD blower operates on a "Square Card" cycle. The process for calculating the brake horsepower also includes a factor to take friction losses into account. Thus, brake horsepower is a combination of two components:

$$\boldsymbol{bhp = ghp + fhp}$$

where:

bhp = Brake Horsepower at the Blower Shaft
ghp = Gas Horsepower
fhp = Friction Horsepower

As explained previously, ghp is calculated as follows:

$$ghp = K \times rpm \times \Delta P \times cfr$$

where:

K = Constant, depending upon pressure or vacuum, gear speed & blower type (from empirically-established curves and approx. = .00436)
rpm = Shaft Speed
ΔP = Operating Differential Pressure in psi
cfr = Displacement in ft^3/revolution

Friction horsepower, fhp, is dependent upon rpm and model/size of blower and is generally a calculated number, using an empirical figure as a starting point, and corrected for rpm or gear speed by some exponent of the ratio of the standard to the actual rpm. The actual number is available from curves. If the unit is water injected, a further amount of HP must be added to pump the water.

Temperature Rise

Often it is necessary to estimate discharge temperature and sometimes an absolute requirement to know delta-T, to insure that the application does not exceed the maximum rating of the blower in question. Following is the equation for this calculation, ***for air service***:

$$\Delta T = bhp \times T_1 \times TRF/P_1/Q_1/.01542$$

where:

Δ	=	Temperature Rise in °R (or °F)
bhp	=	Calculated brake horsepower
T_1	=	Inlet temperature in °R
TRF	=	Temperature Rise Factor (Varies by Product Line, Approx. = .90)
P_1	=	Inlet Press. in psia
Q_1	=	Inlet Volume Flow in ft^3/min

Modifications in the above are required if the blower is handling a gas other than air or if the unit is water injected.

BLOWER PERFORMANCE CHARACTERISTICS

The basic characteristic of PD blowers is that they provide a relatively constant inlet volume under variable pressure conditions. These blowers provide high volumetric efficiency and incorporate many design features that contribute to job versatility and economical installation. There are no valves or vanes, reducing maintenance and replacement costs. Completely enclosed construction achieves compactness, provides protection against dust and makes the unit suitable for either indoor or outdoor installations.

Pressure is not developed inside the blower, but by the demand of the system. Discharge pressure, therefore, varies to meet the system's load conditions of pressure vs. flow. Relief valves normally are required to relieve pressure when flow is restricted to the point that the pressure can rise beyond design limits of the blower.

Volume is almost constant with pressure except for a slight increase in slip as the pressure rises. Horsepower is nearly proportional to pressure differential and rpm. At a given pressure differential, the hp is proportional to shaft speed (rpm). At a constant rpm, hp is proportional to pressure differential. Thus, torque varies directly with pressure differential. The unit is defined as a "constant torque" machine for purposes of determining the necessary characteristics for drivers, especially variable speed motors. This is opposite from centrifugal compressors, which are classified as "variable torque" devices.

The characteristics of the PD blower make it ideal for low pressure applications and vacuum service. It delivers *essentially* oil-free air and drivers can be sized accurately to the pressure or vacuum requirements.

In two-stage operation, where the discharge flow of one higher displacement machine is directed into the inlet of another, smaller displacement unit, higher pressures or lower vacuum levels can be achieved at significant power savings. The application of two-stage blowers normally is a job for the more advanced technician and these selections generally are done at the factory. The basic concept of two-staging is that the *actual* flow from the first stage must be compressed to a level that can be accepted by the smaller displacement second stage. Until the flow from the first stage is compressed to a flow that the second stage can handle, the pressure continues to rise across the first stage. It is critical that the first stage does not exceed its maximum pressure differential. Therefore, the relative sizes of the two stages are critical. As a vacuum pump, a two-stage blower typically utilizes an inter-stage bypass to allow the excess flow to circumvent the second stage until the compression ratio across the first stage is sufficient to permit the second stage to accept the flow.

SELECTION PROCEDURES

There are several methods used to select and determine the performance of a blower for a given set of conditions. These are:

Performance charts
Performance curves
Computer Programs

Charts or curves are established from the basic calculations discussed earlier, using the standard conditions in the process. As the calculation procedure allows for variations in inlet temperature, pressure, S.G., etc., it is the best method to use if accuracy is important. It can be a laborious task. However, computer programs today simplify the process considerably, allowing any number of different sets of conditions to be calculated with ease and rapidity.

This term scfm is used in many customer specifications but often is the source of confusion. scfm means "Standard cubic feet per minute" but must be defined properly since different people use different standards. (The definition of Standard Air (or a Standard atmosphere) adopted by the Compressed Air & Gas Institute, conforming with PNEUROP and ISO standards is:

Pressure = 14.5 psia (1 bar); Temperature = 68 °F (20 °C); Relative Humidity = 0%

ASME has used the definition of Standard air as:

Pressure = 14.7 psia; Temperature = 68 °F; Relative Humidity = 36%

This gives air having a density of 0.075 lb/ft^3. This is the definition for Standard air used in this Section for PD blowers.

To convert from scfm to Inlet, or Actual flow (acfm) the standard and the actual conditions prevailing at the point of measurement, in our case the inlet of the blower, must be known. The equation for this conversion is as follows:

$$acfm = scfm \times \{[P_s-(RH_s \times PV_s)] \times T_a \times P_b\} / \{[P_b-(RH_a \times PV_a)] * T_s \times P_a\}$$

where:

P_s = Standard Pressure (psia)
P_b = Atmospheric Pressure or Barometer (psia)
P_a = Actual Pressure (psia)
RH_s = Standard Relative Humidity (Expressed as a Decimal)
RH_a = Actual Relative Humidity (Expressed as a Decimal)
PV_s = Saturated Vapor Pressure of Water at the Standard Temperature (psi)*
PV_a = Saturated Vapor Pressure of Water at the Actual Temperature (psi)*
T_s = Standard Temperature (°R) (°F + 460)
T_a = Actual Temperature (°R)
*From Steam Tables

Relative Humidity, Temperature, Pressure and Barometer can have a profound effect upon the blower selection, but most importantly, on the horsepower calculated, since the bhp depends on the inlet volume used in the calculation. It is essential that performance data sheets spell out acfm and also provide the actual pressure, temperature and relative humidity conditions at the blower inlet.

Noise/Vibration Control

We have discussed the pulsation characteristic of the PD blower and the fact that the discharge pressure pulse generated causes both noise and vibration. A very effective way to reduce this inherent problem is the use of a wrap-around discharge plenum and a "jet" opening that allows the pressure equalization within the blower to take place in a controlled manner. By controlling the equalization, the internal pressure pulsations are reduced, resulting in less audible "popping" and less shock loading to the bearings and other blower components.

Compared with the conventional PD blower, the jet-assisted unit operates at noise levels up to 5 dB quieter at comparable speeds and pressures. The jet-assisted blower allows higher gear speeds at lower noise levels, permitting selection of smaller size units at comparable noise. In fact, a conventional blower must run at approximately two-thirds the jet-assisted unit speed for a similar noise level.

Obviously the vibration levels also are reduced, and life of bearings, gears and other components is enhanced. Treating piping, silencers and other components with sound absorbing materials helps further.

Gear Speed

"Gear Speed" has not yet been addressed in this guide. It is a significant criterion, not only in noise considerations but in other rotary blower issues. Gear speed is defined as the velocity of the timing gear at the pitch line diameter of the gear. It is calculated by the following formula:

$$\text{GEAR SPEED} = \frac{\pi}{12} \times \text{GEAR DIAMETER} \times \text{rpm}$$

where:

GEAR SPEED is in ft/min
GEAR DIAMETER is in inches
rpm is blower drive shaft speed

The selection of silencers depends upon the blower gear speed as well as other criteria. Gear speed also is a parameter used in the calculation of bhp, in that Friction Horsepower and the Gas Horsepower Factor, discussed earlier under Performance Calculations, require a calculation of gear speed in order to establish these elements.

Further reductions of noise and vibration can be effected by following these guidelines:

Operate at slower rpm (or Gear Speed)
Supply a motor rated at 5 dBA less than the final desired noise level
Wrap silencers and piping with acoustic material
Isolate the blower and baseplate

Complete enclosures, while effective for noise reduction, are sometimes inconvenient from a maintenance standpoint. However, in certain sensitive situations, this may be the only solution.

WATER-SEALED VACUUM PUMPS

We have noted previously the option of liquid injection to improve a rotary blower's performance. We now will address that subject for vacuum pumps, both single and two-stage.

GENERAL

Injection of a liquid into the inlet of the blower performs three functions: 1) Reduces the heat of compression, 2) Partially seals the clearances between the impellers and the casing to reduce slip, and 3) Tends to clean out the internals to help prevent build-up of solids. Another function of liquid injection can be to neutralize the effects of an acidic or basic element in the gas stream with the addition of a neutralizing agent into the injected liquid. This is a very special application, however, and requires a very thorough understanding of the chemistry of the situation. We consider this option to be beyond the scope of this chapter, however one should be aware of its possibilities.

The typical injection liquid is water so we will restrict the discussion to water. Even tap water, which is entirely suitable for drinking, will contain certain amounts of chemical compounds, which can precipitate out from the water under certain temperature conditions. Scale typically is calcium carbonate. When water containing significant amounts of calcium carbonate is injected into the blower, given the increase in temperature through the unit, the calcium carbonate can precipitate out and adhere to the internals surfaces, the clearances inside the machine can be eliminated and rubbing can occur. In operation, this soft, wet build-up is generally wiped off as the unit rotates and there is no problem. However, when the system is shut down and the build-up dries out, it expands, becomes hard and adheres more strongly to the blower parts. Given enough time, the unit can be totally locked up and any attempt to start the driver can cause serious damage to the driver, coupling or blower.

Specifications have been developed by blower manufacturers, which spell out the minimum requirements for seal water, including pH, hardness, maximum amounts of various elements, turbidity and other critical concerns.

Many vacuum pump applications are in processes where there can be varying amounts of water already in the gas stream, which will carry over to the vacuum pump. Examples of such applications are vacuum filters and paper machine vacuum. Although typically there is equipment in the systems, such as separators, to eliminate water carry-over, both for protection of the vacuum pump system and for conservation of the process water, generally there is some water left in the gas. This water also will contain contaminants, which can cause a build-up and must also be dealt with.

In addition to the *existence* of contaminants in the carry-over liquid, the amount of the carry-over must also be considered. Pumping the seal water requires a certain additional horsepower, which can be calculated. This power is relatively small when the injected amount is within the recommendations for the vacuum pump or blower. When the total amount of injected and carry-over liquid is considerable, however, the power to pump the liquid can exceed the capabilities of the driver or another element of the vacuum pump system, including the pump itself. Precautions must be taken to restrict the amount of carry-over liquid, typically with inlet separators. The inlet separator will require a means of draining, since the separator is under a vacuum. Draining the separator normally is accomplished with either a pump or a "barometric leg" when sufficient height is available to provide the water column necessary to overcome the vacuum level. The ingress of air at such drains must be avoided or the vacuum level will be affected.

The injection liquid must be directed into the vacuum pump so as to disperse within the pump and reach all the internals to do its job. A special nozzle (or sets of nozzles in the largest units) is mounted at the inlet opening of the vacuum pump. This nozzle atomizes the liquid and disperses it to all corners of the internals of the pump. In order to ensure the proper control and cleanliness of the water, a seal water control package will include the following items:

Solenoid valve
Strainer
Flow meter with bypass valves
Flow switch

The solenoid valve normally is inter-connected with the driver so that the water begins to flow when the driver is started and is turned off when the driver is de-energized. This assures a flow of sealing water during operation. The strainer is to ensure that there is no contaminant in the seal water supply that may clog the flow nozzle or meter. The flow meter is used initially to set the proper amount of water flow into the unit and is recorded in the instruction manual. After setting the appropriate flow, the valves adjacent to the meter are adjusted so that the flow is through the bypass and not through the meter. This is a precaution to protect the flow meter and also to reduce the amount of contaminant build-up inside the meter, which eventually may cause it to be unreadable. The flow switch is interconnected with the driver starter system so that if the flow ceases during operation, the driver will shut down.

It is good practice to operate the machine dry and unloaded for a period of time just prior to shutdown in order to allow the internals to dry out to assist in preventing lock-up during shutdown.

PERFORMANCE CALCULATIONS

Single-Stage

The process of determining speeds, horsepowers, etc. for the water-sealed blowers or vacuum pumps is very similar to dry blowers. The differences lie in the calculation of slip, hp and discharge temperature. Gas and friction horsepower (ghp & fhp respectively) are calculated in the same manner as dry units, however an additional amount of hp must be added to allow for pumping the water. Discharge temperature is calculated from compression ratio across the unit and knowledge of the steam tables for purposes of determining the relationship between temperatures and vapor pressures of water at those temperatures.

Slip

Actual inlet flow is calculated in accordance with the manufacturers' procedures, but the process roughly approximates the dry procedures, except for the reduced slip resulting from the water sealing.

Horsepower

The power calculations for a water-sealed unit are identical to the calculations for a dry machine except the power required to pump the water must be considered. Therefore,

$$\text{bhp} = \text{ghp} + \text{fhp} + \text{whp}$$

where:

bhp, ghp and fhp are as previously defined.
whp is the power to pump the amount of inlet water at the gear speed of the unit.
The value depends on the manufacturer.

Note that whp can be significant. The total flow, including any from the system, must be considered in this calculation and the user must be cautious about possible overloads due to water flows significantly higher than the normal seal water flow. Such overloads can shut down the driver on high power or, if extremely high, they can damage the pump or other drive train components.

Discharge Temperature

Normal calculation of estimated discharge temperature assumes that the inlet and outlet flows are saturated with water vapor. From thermodynamics, at saturation the partial pressure of the water in the air stream is equal to the vapor pressure of steam at the temperature of the mixture. Therefore, assuming saturation of the air stream at inlet and discharge, and knowing the saturation pressures of steam from the steam tables, one can calculate the discharge temperature. The partial pressure of the vapor in the inlet gas goes through the same compression as the air from inlet to discharge and some of the seal water evaporates, thus maintaining saturation at the discharge. Therefore, the partial pressure of the vapor on the discharge side will be the partial pressure at inlet temperature multiplied by the compression ratio of the vacuum pump. The corresponding temperature from the steam tables gives the estimated discharge temperature of the mixture.

$$VP_d = CR \times VP_i$$

where:

VP_d = partial pressure of steam at discharge temperature
CR = compression ratio (P_d/P_s in absolute pressures)
VP_i = partial pressure of steam at inlet temperature

From the steam tables, the discharge temperature of the mixture is the saturation temperature at VP_d.

The foregoing provides the basics for calculating performance for a water-sealed single-stage rotary vacuum pump. The data sheets, curves, etc. referred to are provided in the manufacturer's data books for the products in question. In addition, individual performance charts and curves, developed from these procedures are also available for estimating purposes.

Two-Stage Rotary Blowers and Vacuum Pumps

By arranging two PD blowers in series, the total differential pressure can be increased.

Similarly, vacuum capability can be increased and power requirement decreased with a two-stage arrangement.

HELICAL LOBE BLOWER

The helical lobe blower is a two-rotor, positive displacement blower, providing constant volume, variable pressure, oil free delivery of air. This is shown in Figure 6.13 and is commercially available in capacities ranging from 50 cfm to 6000 cfm and pressures to 20 psig and with special designs to 35 psig. They can be configured as two-stage blowers for pressures to 40 psig.

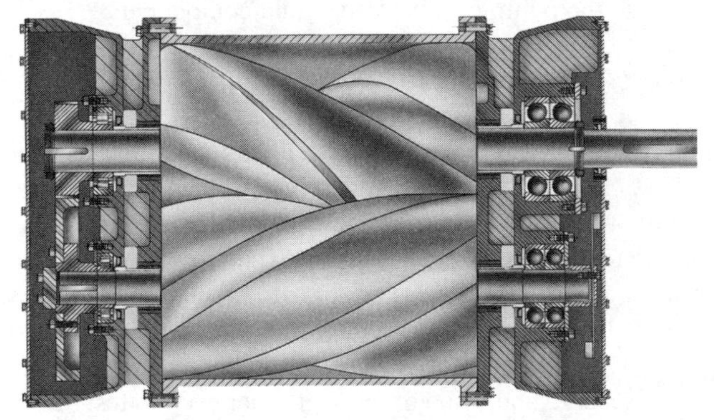

Figure 6.13 Helical Lobe Type Blower

The helical lobe blower employs two counter-rotating rotors, but unlike straight lobe blowers, the rotor profiles are different and the number of lobes are not necessarily equal. The meshing of the two helical lobe rotors provides controlled compression (with internal volume reduction). Timing gears prevent rotor to rotor contact.

The compression cycle, Figure 6.14, begins as rotation produces an increasing interlobe cavity on the inlet side due to unmeshing of the rotor lobes. Air or gas is drawn into the rotor chamber until the length of the interlobe cavity is filled. As further rotation continues, meshing of the rotors reduces the volume of the interlobe cavity. As the gas is moved axially along the rotor length, in the ever decreasing interlobe cavity, internal compression occurs until the rotors uncover the discharge port, releasing the gas to the discharge line.

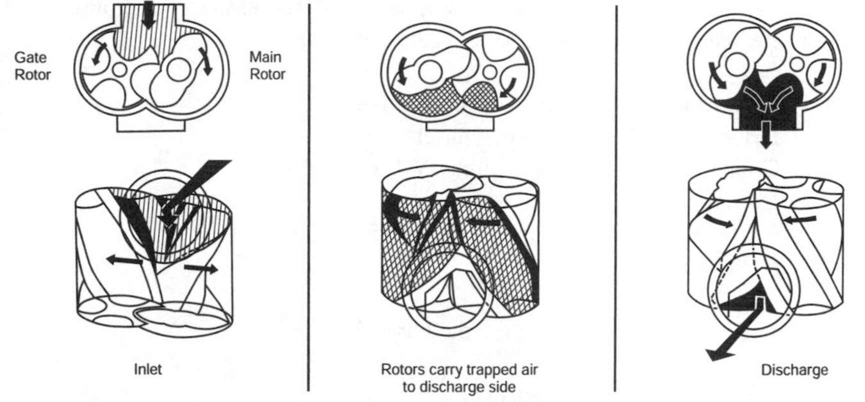

Figure 6.14 Compression Cycle

The timing gears maintain close rotor clearances. The rotors do not touch each other, the housing, or the bearing carriers. Although the clearances are small, lubrication in the compression chamber is not required, thus insuring oil-free air delivery.

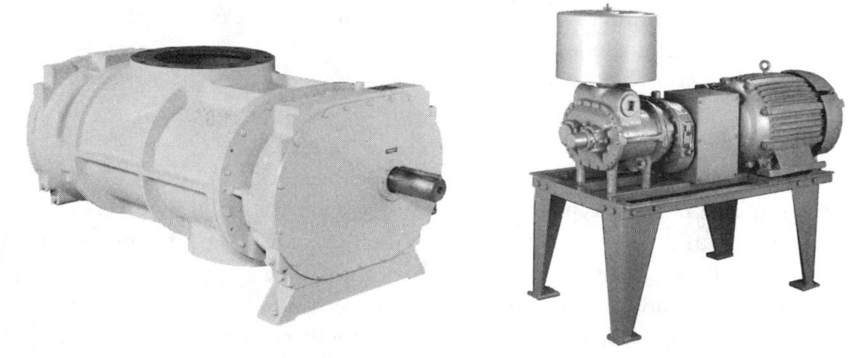

Figure 6.15 Typical Helical Blower

Figure 6.16 Typical Blower Package

DEVELOPMENT TRENDS

The trend today is for the design of this type of blower to become similar to the oil-free rotary screw compressor. The supporting shafts usually are made integral with the rotor body and of either ductile iron or steel. The rotor tip speeds can approach 100 m/s with either internal step-up gear drives, or high input shaft speeds.

Typical applications include: pneumatic conveying, chemical processing, dry bulk transport, waste water treatment, process gas and air separation.

Accessories

A blower requires a number of accessory items in the system to make a working installation. In addition, there are other accessories that sometimes are specified to make the system more automated, quieter, easier to maintain, more reliable, etc. The necessary accessories, generally, are:

Driver (Motor, Engine, Turbine, etc.)
Driver control (e.g., Motor Starter)
Connection to driver (Coupling, V-Belts, etc.)
Drive guard
Baseplate or other mounting device (Soleplates)
Relief valve

Other accessory items normally furnished are as follows:

Inlet and discharge silencers
Inlet filter (If on air service)
Expansion joints for air connections
Various pressure & temperature gauges

Additionally, any number of controls, safety switches, isolation valves, monitoring equipment and other items can be furnished at the discretion of the user. The list will be somewhat different if the blower is handling a gas, as opposed to air. Necessary accessories include:

Driver

The device most often used to drive the blower is an electric motor. Depending upon the required bhp and blower speed, the motor may be sized to direct-connect to the blower shaft or may be arranged for V-belt or gear drive. The motor manufacturer generally will want to know how the motor will be applied. In any case, the worst conditions of blower rpm and pressure differential must be allowed for in the selection of motor horsepower so that the motor will not be under-sized for any operating conditions of the system. The maximum rated blower rpm must not be exceeded. Too low a blower speed can result in a blower temperature rise, which exceeds the limits, due to the loss of blower efficiency at slow speeds. In addition, it is usually prudent to allow some tolerance in horsepower when sizing the motor. Blower manufacturers typically will quote bhp with a ± 4% tolerance. This tolerance percentage should be considered in motor sizing. Also, in an induction motor, there is a small amount of "slip" in the motor, e.g., an 1800 rpm induction motor will actually run at about 1775 rpm when placed in service under load. Synchronous motors, as the name infers, operate at the motor's synchronous rpm without slip. Synchronous motors seldom are used on PD blowers except in some high horsepower, low rpm applications, and where there is a need to adjust the plant's over-all power factor.

When a variable speed motor is to be used, one must consider the speed range of the system and not operate the unit too slowly due to potential temperature rise. When using a variable speed driver, a torsional analysis may be required to locate the torsional critical speeds so that considerations can be made in the application of the drive system.

The method of starting the blower also should be considered when selecting the motor. Although not recommended, should the blower be required to start up under load, the increased starting torque should be considered in the motor specifications. There also are recommended limits for maximum motor torque. Other factors affecting the motor selection, such as voltage, enclosure, insulation, etc. are usually a requirement of the user and are beyond the scope of this discussion.

Other driver choices, such as steam turbines, engines, gas turbines or others may also be considered. Generally, as far as the blower is concerned, the same considerations should be made, as with the motor drive. There are some special concerns with engine drivers, however, especially with respect to the system torsional responses. In such cases, a torsional analysis should be performed by the engine manufacturer.

Controls for the driver must be coordinated with, and preferably be supplied by, the vendor of the driver to ensure complete compatibility. The user will likely have some input into the controls selection and there also will be some blower requirements, especially if there are safety devices tied into the driver control.

Driver Connection

Devices to connect the blower to the driver are, for example:

Flexible couplings
V-belt Drives
Speed-increasing (or decreasing) Gearboxes
Universal Joints

When a blower is "direct-coupled" to its driver, the rpm of the driver must not exceed the maximum speed rating of the blower. A "flexible" coupling such as the grid type should always be used to connect the driver and blower shafts. It is extremely important to install the coupling in accordance with the blower and coupling manufacturers' instructions in order to have a successful installation. Some installations require Limited End Float couplings when the unit is driven by a sleeve bearing motor. Always consult the blower manufacturer in these instances. It is very important to carefully level and align the blower/driver when direct-driven, due to the limitations of the coupling regarding mis-alignment.

V-belt Drives

It is not always possible to select a blower that can be economically direct-connected, due to limitations of motor speeds and blower frames available. When this is the case, it is usually convenient to use a V-belt drive and, with the proper selection of sheaves and motor speed, one can come within a few rpm of the blower speed required for the specified blower flow. Use the appropriate service factors as spelled out in the drive manual in order to have a reliable long-lasting system. Installation of the driver is less critical than for direct coupling, but its shaft must be level and parallel with the blower shaft. The driver should be mounted on the inlet side of a vertical blower (horizontal piping) and on the side nearest to the shaft on a horizontal blower. The driver must also be mounted on an adjustable base to permit installing, adjusting and removing the belts. The blower sheave should be no more than 1/8 inch from the blower drive end cover so as to keep the overhang to a minimum. Proper tension is critical and should be accomplished in accordance with the drive manufacturer's instructions. Excessive tension can cause bearing failures while too little tension can cause slippage and premature belt failure.

Drive Guards

Any type of rotating drive connection should be protected by a stationary guard to prevent injury. OSHA has defined certain standards for drive guards and most blower suppliers furnish their guards in accordance with those standards. Depending upon the installation, one may select a standard or a weather-protected guard.

Blower Mounting Device

A fabricated baseplate normally is supplied for mounting of the blower and driver. An option is to mount the units independently on soleplates. In either case, it is recommended that the mounting device be grouted into a concrete foundation. Follow the blower manufacturer's instructions in that regard. Anchor bolts of the type shown in Figure 6.17 are recommended for tying the soleplates or baseplate into the foundation. Note also, the method of installing the grout, as shown in this figure. Level and alignment are critical and the manufacturer's instructions should also be followed here. If the unit was shipped from the blower manufacturer mounted on the baseplate along with the driver, it likely was aligned at the factory. However, due to the possibility of the assembly being twisted during shipment, thus disturbing the original alignment, the unit should be checked for alignment and level before starting. The blower instruction book provides the procedure to make these checks.

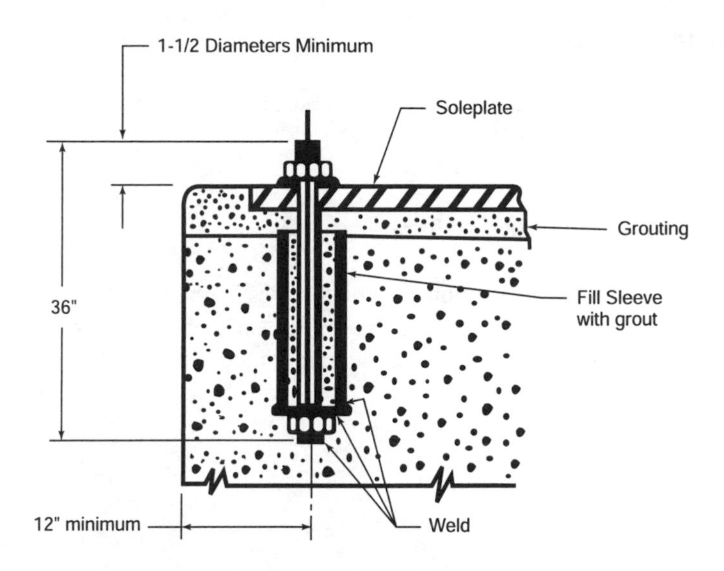

Figure 6.17 Typical Anchor Bolt

Relief Valve

Because the PD blower is a constant volume device, if the discharge line is restricted for any reason, the unit will continue to "pump" air into the system, thus raising the pressure in the discharge line and increasing the pressure differential across the blower. Eventually, if the driver has sufficient power, the pressure differential will reach a level that exceeds the rating of the blower. To preclude this occurrence, the discharge line must be equipped with a pressure relief valve that is set to open at a pre-set pressure well within the maximum allowable pressure differential of the blower and sized for the full capacity of the blower. When the set pressure is reached, the valve will begin to open, thus allowing the flow to escape to the atmosphere and reducing the pressure. If the restriction is a complete blockage, the valve will open fully and pass the entire flow of the blower. However, due to the "accumulation" characteristics of the valve, the final pressure could be as much as 10 or 25 % above the set pressure. The blower should be sized, therefore, to withstand the accumulation pressure. The valve should be as close to the blower discharge as possible. Note that, on vacuum service, the valve is on the inlet of the blower.

The relief valve is not designed to function as a pressure regulator and should not be used to set the pressure at which the system is to operate. There are other types of devices for this purpose, such as pilot-operated, diaphragm-actuated valves and these devices should be used for this function. The relief valve should be considered only as a safety device.

Other Accessories

In addition to the above accessories considered as necessary for the successful, safe operation of the blower system, the following may be considered desirable for convenience, safety and/or comfort.

Inlet and Discharge Silencers

We have discussed the noise characteristics of the PD blower and the ways to reduce the level. The most popular method is by use of silencers or mufflers. Nearly every blower installation has *at least* an inlet or a discharge silencer as a part of the package. There are several different styles and models of silencers and they are selected based upon the following criteria:

Blower Capacity
Blower Discharge Pressure *
Blower Speed and Type
* Or inlet vacuum in case of a vacuum pump

There are four basic silencer types:

- Absorption - unimpeded flow passage lined with acoustical material behind a perforated annulus.
- Chamber - internal baffles forming two or three unlined chambers inter-connected by perforated tubes, tuned to subdue flow pulsations.
- Combination - internal baffles forming unlined chambers inter-connected by perforated tubes with extended acoustically packed sound absorbing section.
- Silencer/Separator- chamber type silencer with drain allowing 95% water removal from the air stream.

It has been found that the blower gear speed has an influence on the silencer selection. Generally, at gear speeds below 3300 ft./min for the inlet silencer and 2700 ft./min for the discharge silencer selection, a chamber type is satisfactory. Above those speeds, it is necessary to add an acoustical packed section or use the combination type silencer, which has the pack built in. The packing absorbs the higher frequencies, which tend to cause ringing of the silencer shell and/or piping. In any case, the silencers should be mounted as close to the blower as possible, with the acoustical pack section closest to the connection of the blower.

The selection charts are relatively simple to use, knowing the gear speed, flow and pressure. Each manufacturer publishes its own list of the various types, so the first thing to do, after determining the gear speed, is to select the type. Next, enter the chart for inlet size selection with the inlet volume of the blower. Read directly the inlet silencer size. The discharge silencer size is also based on the inlet volume, but the discharge pressure is also a parameter. Find a silencer size, which will handle the blower inlet volume at the operating pressure.

Sizing a silencer for vacuum service is similar, except the inlet silencer usually is not required.

Silencers are available with different connection configurations to fit various piping arrangements. Select the straight-through or other configuration, which best fits the piping system design.

Typically, gas pumps are equipped with only discharge silencers as the inlet is usually piped to a system and does not cause noise problems.

Inlet Filter

There are several reasons to use an inlet filter on a pressure blower air application. Often, it is important to keep foreign material out of the system downstream of the blower, due to the system's inability to handle contaminants. An example is the fine bubble diffusers utilized in wastewater treatment plants. These diffusers, often made of porous ceramic material, are subject to clogging and must be protected from foreign material. There are other considerations, such as clogging of instrumentation, contamination of product, erosion of various parts of the blower system, build-up on valves and other parts, etc.

Industrial dry-type filters are recommended on PD blowers. The filter medium should be rated, as a minimum, 90% efficient on 10 micron particles and larger. However, 99% efficiency is very easy to achieve with dry-type filters.

Intake filters are sized according to the face velocity across the filter medium. The higher the face velocity, the faster the medium will load and require cleaning. Therefore, ratings of dry-type filters are on the basis of differential pressure across the filter, and vary according to manufacturers, but are approximately 1/2 - 1" H_2O when clean. As the filter medium becomes loaded with dust or dirt, pressure drop across the filter is increased, causing a lower inlet pressure to the blower, resulting in increased pressure differential across the blower and a drop in capacity. A manometer or differential pressure gauge should be placed across the filter to determine when the filter should be cleaned or replaced.

Weather hoods are available to keep rain, snow and other foreign particles away from the filter element.

Sizing of inlet filters is relatively simple and is similar to inlet silencers. The only information required for filters mounted on the blower, is the inlet volume of the blower. Should air be drawn from outside in an industrial atmosphere, where there is a risk of heavier contamination, a heavy duty inlet filter should be considered.

Inlet filter-silencers also are available but in general, they do not do as good a job of silencing.

Expansion Joints

Installation of spool-type expansion joints or flexible connectors at the blower inlet and discharge is recommended to allow for any pipe movement due to expansion from temperature change or incorrect piping alignment. The blower casing is not constructed to absorb pipe strain and should therefore be protected from such forces. In addition to the expansion joint on the discharge, it is also recommended that control elements be added to minimize piping vibrations and to protect the expansion joint.

Figure 6.18 Typical Blower Package Including Accessories

7

Gas & Process

Chemical and petroleum processes have resulted largely from the ingenuity of the people who originated them. Ingenuity is defined as skill or cleverness in devising or combining. In this chapter it is hoped that a real appreciation of such ingenuity will be developed. One hundred and eight basic elements are available in nature. There are a few basic rules for combining these elements into compounds, all well defined in organic and inorganic chemistry Pressure and temperature control and the use of catalysts have let us manufacture the many chemical products that are now on the market.

An interesting observation, as an example, is that a certain process initially devised to produce a certain compound required very high reactor pressures. Consequently, the designers of gas compressors were challenged to produce equipment that would deliver the new component gas to a reactor at 40,000 psi. Twenty or thirty years later the same product required only one-hundredth of the original pressure. Nevertheless, compressor manufacturers have met and will continue to meet new challenges of many kinds in the design of compressors to supply gas at the required pressure, twenty-four hours per day, with a minimum of interruption imposed on the process by the compressor equipment.

The proper selection of the type of compressor for the service should be carefully analyzed. Compare past experience, efficiency at all operating flows and pressures, cost of installation, and compatibility between the compressor design and the gas to be compressed. After all, it is the product of a plant produced at the required rate that justifies the plant's existence.

The purchaser must accurately define the composition of the gas to be compressed, including any entrained liquids. Vapors that will condense out during the compression cycle must be part of the gas analysis. In many instances the gas analysis will vary, and this variation must be defined. Any characteristic of the gas that would influence the reliability of the compressor must be studied, and any known designs for similar problems should be investigated. For instance, chlorine and hydrogen chloride are successfully compressed, but the gas must be void of water and free of solids. Coke-oven gases can be successfully compressed, but in most cases require scrubbers to remove tarry substances first, and then static electric precipitators to remove fly ash.

Compressors handling gases other than air are broadly divided into two groups: process gas compressors and oil and gas field compressors. Process gas compressors are employed extensively in petroleum refining and in the chemical industry. Oil and gas field compressors have somewhat different characteristics from process gas compressors and are used throughout the oil and natural gas industry for producing oil and natural gas and in various phases of the treatment and transport of natural gas.

A relatively new use for gas compressors is in tertiary recovery and enhanced oil recovery. High energy prices, declining reserves, escalating exploration costs, pricing, and tax legislation have created incentives for enhanced oil and gas recovery. In this application, compressors are used to transport carbon dioxide, nitrogen, or hydrocarbon gases through pipelines for injection into wells.

In this chapter, typical applications of compressors in process services and in the oil and gas industry are discussed. These applications illustrate the potential and flexibility of gas compressors and demonstrate that the modern compressor has the capability to provide virtually any type of gas-compression service.

OIL REFINERY PROCESSES

Refineries are a part of many large cities and communities, such as Los Angeles, San Francisco, and Chicago. The cleanliness of the environment around refineries is improving and every effort is being made to reduce pollutants. Hydrocarbon vented gases in the refinery are collected, compressed to the necessary pressure, treated, and used in the refinery fuel gas system.

The raw material of crude oil contains polluting substances such as sulfur and chlorides, including salt. The modern refinery has processes that extract sulfur commercially, thus producing a needed chemical and, at the same time, reducing emissions into the atmosphere. Figures 7.1 and 7.2 outline the general flow and functions of the refinery processes.

The initial process plant in all refineries is the crude oil distilling unit. Here the crude oil is heated to its boiling point and separated by means of fractionating columns into various fractions from heavy to light (Fig. 7.1).

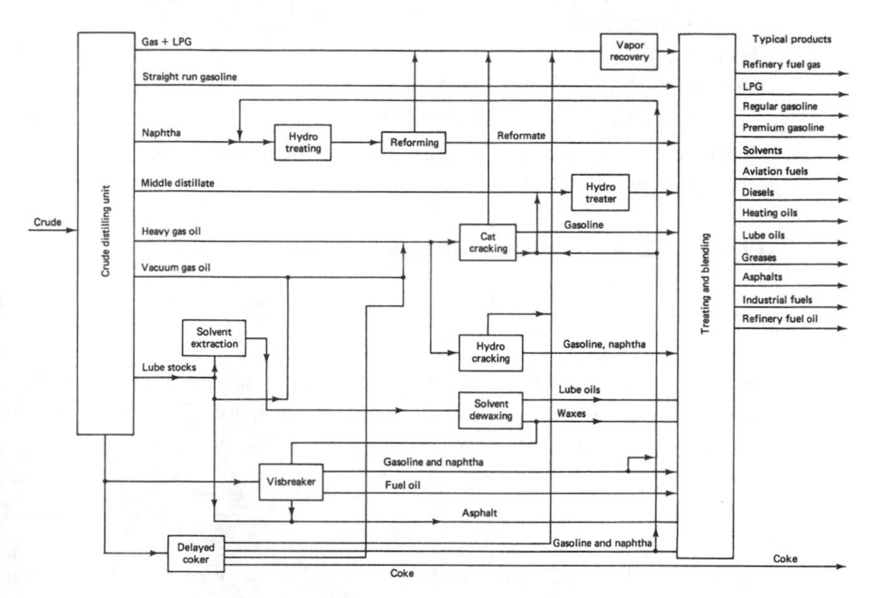

Figure 7.1 Flow diagram for a refinery.

Refinery Processes

1. Breaking Hydrocarbons Down
   Hydrogenation – Adding Hydrogen
   a. Thermocracking
      1. Coking (Delayed) – Flexicoking
      2. Visbreaking
   b. Catalytic Cracking
   c. Hydrocracking Fixed Bed Catalytic Process
   d. Hydrotreating
2. Building up Hydrocarbons
   a. Polymerization – Gaseous Olefins to Liquid Hydrocarbon
   b. Alkylation – Gases to Liquid
3. Modifying – Altering – Upgrading – Rearranging – Dehydrogenation
   a. Reforming
   b. Isomerization – Low Octane Liquid to High Octane Liquids
4. Separating Hydrocarbons – Light Ends Recovery
   a. Extraction – Physical Separation
   b. Distillation
      1. Atmospheric
      2. Vacuum

5. Cleaning Hydrocarbons
    a. Desalting
    b. Treating to Reduce H_2S, Acids, Etc.
        1. Hydrodesulfurization (HDS)
    c. Hydrotreating to Reduce Unsaturated Hydrocarbons
        1. Catalytic Hydrofining

Figure 7.2 All of these refinery processes utilize compressors.

Heavy Fractions

The heaviest fractions are boiled under a vacuum to make a product suitable for the manufacture of coke in the delayed coker. These heavy fractions are also used for heavy fuel oils, such as No.6 oil.

Gas Oil Fractions

The gas oils from the crude unit are feed stocks for gasoline and diesel fuel. These oils are sent to the catalytic cracking unit and the hydrocracking unit for making various forms of gasoline and light fuel oil and diesel oil. The fractionating columns in these plants separate these products after the cracking reaction takes place so that they can be segregated for final treatment.

Naphtha Fractions

The light naphtha components from the crude unit are sent to the reformer unit for making gasoline with a high-octane number. This product is mixed with gasoline from the catalytic cracking unit and hydrocracker to obtain the desired octane numbers of the various grades of gasoline: regular, premium, lead free, and so on. The alkylation unit also produces a high-octane component for blending into gasoline.

Liquefied Petroleum Gas (LPG) Fractions

The lightest fractions from the crude unit are propane and butane, which are used for LPG or bottled gas. They are mixed with the lighter components from the other process units to form the final products.

Lube-base Stocks

Only certain types of crude oil are suitable for manufacturing lubricating oil. The lube stock is sent to a solvent dewaxing process where it is mixed with methyl ethyl ketone, which dissolves the wax contained in the oil. The mixture of lube stock and solvent is refrigerated to from 0°F to -30°F, as it passes through tubes that have a rotating scraper. The outside of the tube is the evaporator of the refrigeration system. The cooled solution causes the wax to solidify, and the mechanical scrapers keep the solidified wax mixed with the oil. The lube stock then is pumped to a large rotating filter. The filter drum is constructed with radial segmented chambers. As the drum turns, the filter cloth is alternately pressurized or is exposed to a vacuum of 23 in. of mercury and then a positive pressure of 4 psig. As the tilter cloth passes the vacuum segment, the wax is extracted from the oil and is held to the cloth by its vacuum. As the drum turns the pressurized segment, the wax is blown off the cloth and removed from the system, and the oil is purified and wax free. A typical installation of this solvent dewaxing system uses horizontal, heavy-duty reciprocating ammonia refrigeration compressors. Practically all refining processes utilize centrifugal or reciprocating compressors.

It is important to note that each refining unit has a unique processing scheme determined by the equipment available, operating costs, and product demand. The optimum flow pattern for any refinery is dictated by economic considerations, and no two refineries are identical in their operations. The preceding description, however, will give some idea of the general nature of refinery processing.

Distillation of Crude Oil

Distillation of crude oil is the first major processing step in the refinery. It is used to separate the crude oil by distillation into fractions according to boiling point so that each of the downstream processing units will have the feedstocks that they require to meet the product requirements of the refinery. By changing the amount of heat and the temperature levels for the distillation process, variable boiling point materials can also be produced.

In many cases, the distillation is accomplished in two steps:

1. By fractionating the total crude supply at atmospheric pressure, a process called atmospheric distillation.
2. By feeding the heaviest fraction, called topped crude, from the atmospheric distillation to a second fractionator operating at a high vacuum, a process called vacuum distillation.

Atmospheric Distillation

Figure 7.3 shows the atmospheric distillation unit. The crude is normally pumped to the unit from storage at ambient temperature. After being heated in exchangers, it enters the desalting vessel at 250°F. Most crudes contain salt. The salt must be removed to minimize the fouling and corrosion caused by salt deposition on heat-transfer surfaces.

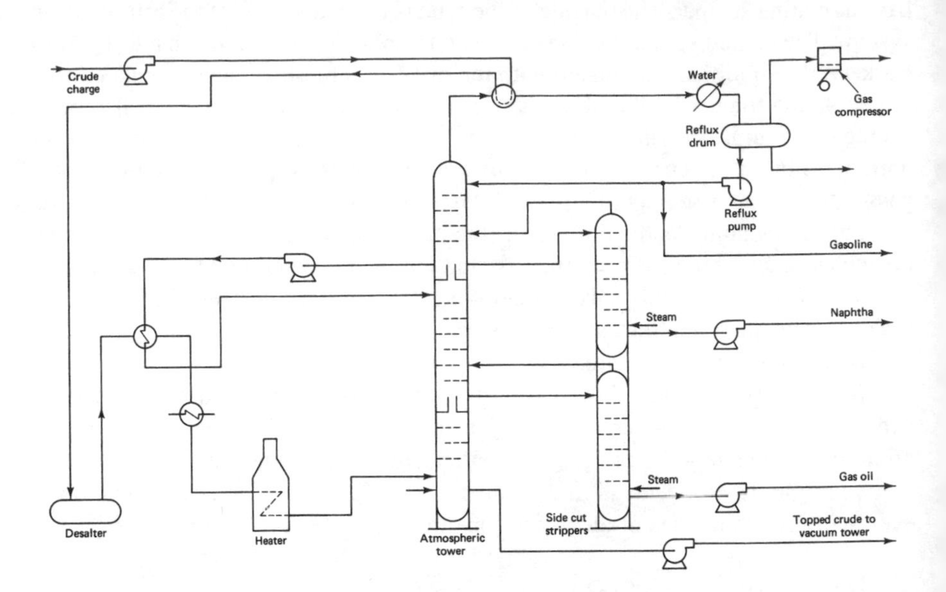

Figure 7.3 Atmospheric distillation unit for crude oil.

Desalting is carried out by emulsifying the crude oil with water. The salts are dissolved in the water, and in the water and oil phases, separated by using chemicals to break the emulsions or by developing a high potential electric field across the settling vessel to coalesce the water droplets.

The desalted crude is then heated by heat exchange from hot product and then further heated in a direct-fired furnace to about 750°F. At this temperature, it enters the flash zone of the fractionating column, where separation into fractions or "cuts" takes place. The fractionating column is a tall cylindrical vessel fitted with a large number of specially designed trays.

The mixture of vapor and liquid enters the fractionator near the bottom and rises gradually through the opening in the trays. As it rises, it cools, and a certain amount condenses on each tray until the tray is full of liquid up to the level of the overflow.

The level on each tray is kept just above the holes in the bubble caps so that all the vapor has to pass through the liquid. Each tray is a little cooler than the one below it, and lighter and lighter products will be present on each succeeding tray as the vapor passes up through the column. As the vapor bubbles through the liquid on the trays, that part of the vapor will condense that has the same boiling point as the liquid on the tray.

The temperature throughout such a column is controlled at the bottom by the furnace, which heats the incoming crude oil, and at the top by pumping back a certain amount of material, which leaves the top of the column after condensing. This material pumped back is called reflux.

In this way, by controlling the temperature at the top and bottom, the temperature variation throughout is kept under control so that the temperature of the trays varies gradually from the bottom to the top. The amount of liquid pumped back to the top of the column can be varied as required, in order to give the correct temperature at the top. This, in turn, controls the final boiling point of the gasoline leaving the top. Reflux can also be withdrawn and pumped back into the column at intermediate levels below the top for better control of temperature and distillation.

The atmospheric fractionation normally contains 30 to 50 fractionation trays. Side drawers are located to obtain the desired boiling range products from the fractionator; that is, gasoline off the top, topped crude off the bottom, and intermediate cuts, such as naphtha, gas, oil, and kerosene in between.

The liquid sidestream withdrawn from the tower will contain some material with a lower boiling point than the bulk of the material on the tray. These light ends are stripped from each sidestream in a separate, small stripping vessel containing four to five trays. As the liquid flows down over these trays, it meets an upward stream of steam, injected at the bottom of the stripper. The steam boils off the light ends, and this narrows the boiling range of the bulk of the liquid drawn off the side of the column.

From the side cut strippers the various liquids are pumped to storage or to the next downstream process-naphtha to the catalytic (cat) cracker or hydrocracker. The gasoline off the top of the column is treated and pumped to the gasoline pool for blending with other gasolines. Usually, at least four side strippers are provided in a plant to produce extra cuts, such as kerosene and diesel oil.

Some atmospheric crude units require a compressor to handle the gases from the reflux drum at the top of the column. As the gasoline vapors from the top of the column condense, the lightest vapors remain in the gaseous state and accumulate above the liquid level in the reflux drum. A compressor is required to pump this gas to the gas-treating unit for final disposition, either to fuel gas or other product. Because of the relatively small volumes, a reciprocating-type compressor is utilized and is called the vent-gas compressor. The molecular weight of this gas is 60 to 70 and it is normally pumped up to 100 to 200 psig pressure level.

Hydrocracking

Hydrocracking is a refinery process for making gasoline out of heavier feedstocks from the crude unit. The interest in hydrocracking was caused by two factors:

1. Economic demand for fewer, more efficient refineries to produce gasoline with an increase in the ratio of barrels of gasoline per barrel of crude oil.
2. Large quantities of by-product hydrogen were available from the many cat reformers that had been built, and hydrocracking requires a large amount of hydrogen.

In a number of refineries, cat cracking and cat re-forming work together. The cat cracker takes the more easily cracked oils as feed, while the hydrocracker can crack those oils that are not easily cracked in a cat cracker.

Although they work as a team, the processes are completely dissimilar. The cat cracker reaction takes place at a low pressure, 25 to 35 psig without hydrogen, and in the presence of a fluidized bed of catalyst. Hydrocracking occurs at a pressure of 1000 to 2000 psig, with a high concentration of hydrogen and in a fixed bed of catalyst.

The hydrocracking process is flexible. In addition to making gasoline from middle-distillate oils, the process is used to make distillates and light oils from residual oil. Of course, the catalyst and operating conditions are different; however, the same plant can be designed to operate in the alternate modes.

As in cat re-forming, the feedstock must be hydrotreated prior to entering the hydrocracker to remove the sulfur, nitrogen, and oxygen compounds, which are harmful to the catalyst. Thus, a hydrotreater may be required in conjunction with a hydrocracker, unless hydrotreating capacity exists within the refinery. In a two-stage hydrocracker, the first stage may perform the function of a hydrotreater.

A number of hydrocracking processes are available for licensing:

Isomax	Chevron and Universal Oil Products
Unicracking/SHC	Union and Exxon
H-G Hydrocracking	Gulf and Houdry
Ultracracking	Amoco
Hy-C, H-Oil	Hydrocarbon Research and
Shell	Cities Service

With the exception of H-Oil and Hy-C processes, all hydrocracking processes in use today are fixed-bed catalytic processes with liquid downflow through the reactors.

The hydrocracking process contains the most sophisticated rotating and reciprocating machinery that is found in refineries or petrochemical plants. Equipment for hydrocrackers can be classified as follows:

1. Barrel-type centrifugal compressors for hydrogen recycle service, 2000 psi pressure.
2. Large, balanced, opposed reciprocating compressors for hydrogen makeup, 2000 psi pressure.

Process Description

Figure 7.4 shows a typical two-stage hydrocracker. Reaction temperature is approximately 800°F and pressure in each reactor system is about 2000 psig. The fresh feed is pumped up to 2000 psi from the crude unit or cat cracker at elevated temperature of 600°F or higher. The feed mixes with makeup hydrogen, which is compressed into the plant from the hydrogen plant or the cat reformer. Since hydrogen is consumed in the hydrocracking reaction, it must be continuously added to the system. In addition, recycle hydrogen is mixed with the feed to supply the hydrogen atmosphere required by the reaction.

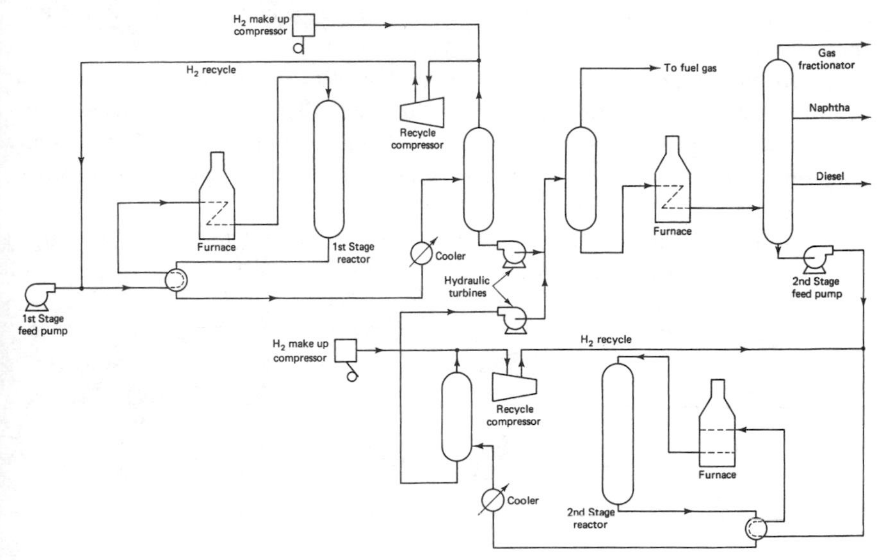

Figure 7.4 Two-stage Hydrocracker

The hydrogen recycle stream is circulated through the system by the recycle compressor, a harrel-type centrifugal machine. The mixture of feed, recycle hydrogen, and makeup hydrogen is heated in a furnace prior to entering the catalyst beds in the reactor.

From the first-stage reactor, the effluent travels through a bank of exchangers into a high-pressure separator where the hydrogen recycle gas is separated from the liquid product and fed to the recycle compressor for recirculation back through the first-stage reaction system. The liquid product from the high-pressure separator is let down through a hydraulic turbine into a low-pressure separator. The hydraulic turbine is utilized to drive the feed pumps.

From the low-pressure separator, the product is pumped to a fractionation column, where the gasoline comes off the top and is pumped to storage. The heavier fractions from the bottom of the column are pumped into the second-stage reaction system as feed.

The second stage is similar to the first, except that it operates at higher temperatures in order to crack the unconverted oil from the first stage. The second-stage product is combined with the first-stage product prior to fractionation. Thus, the second stage handles first-stage product plus some recycle of its own product.

Hydrogen Make-up Compressor

Make-up hydrogen is required in large quantities and in a relatively pure state (i.e., molecular weight of 2 to 3). It is usually supplied from two sources, hydrogen plant and a cat re-former, both at pressures in the vicinity of 200 psi. Thus, the hydrogen has to be boosted to a 2000 psi hydrocracker pressure level, and reciprocating compressors are normally required for the job.

Hydrogen consumption is about 200 scf (standard cube feet) per barrel of feed. Thus, on a 20,000 bpd (barrel per day) hydrocracker, 40 million standard cubic feet per day (scfd) of hydrogen must be compressed into the hydrocracker. For the same 20,000 bpd barrel per day unit, the horsepower required could be as high as 18,000, or 9000 for each 50% machine.

The following table indicates the horsepower required for typical hydrogen makeup to various-sized units, assuming two 50% machines:

Hydrocracker Capacity, bpd	Makeup Hydrogen, scfd	Approximate Compression Horsepower, bhp
5,000	10,000,000	5,000
10,000	20,000,000	9,000
20,000	40,000,000	18,000
30,000	60,000,000	27,000

The balanced, opposed compressor (Fig. 7.5) is the type required for this service. There have been different types of compressor arrangements for supplying the hydrocrackers with hydrogen.

Figure 7.5 Three 5400 hp reciprocating compressors on hydrocracking gas service. The compressors are driven by engine type sychronous motors. The discharge pressure is 2873 psig.

Several plants have centrifugal compressors located within the hydrogen plant upstream of the CO_2 removal equipment. At this point, the gas has a molecular weight of 18, and two barrel compressors in series can reach 2000 psi pressure level. The CO_2 is removed at the higher pressure level and the hydrogen leaves the hydrogen plant in pure form at 2000 psi.

Hydrogen Recycle Compressor

The hydrogen recycle compressors are barrel-type centrifugal machines similar to cat re-former machines except for the higher pressure ratings of 2000 psi.

Catalytic Re-forming

The demand of today's automobiles for high-octane gasoline, without the addition of lead, has stimulated interest in catalytic (cat) re-forming. The major function of the cat re-former is to produce a high-octane product that, when blended with other gasoline streams from the cat cracker and hydrocracker results in an overall gasoline octane number within market specifications. As lead is legislated out of gasoline, more cat re-forming capacity is required to make up for the octane rating improvement obtained with lead.

In catalytic re-forming, the hydrocarbon molecules are not cracked, but their structure is rearranged to form higher-octane products. The reaction resulting in the molecular rearrangement takes place in a series of three or four reactors at a temperature of 900°F to 1000°F in the presence of a metallic catalyst containing platinum and in a hydrogen atmosphere. The feed stock is a straight-run gasoline or naphtha from the crude distillation unit. The feedstock is hydrotreated prior to entering the cat re-former to remove the sulfur, which is harmful to the platinum catalyst. Thus, a hydrotreating plant is usually built along with a cat re-former, and this unit will be described in another section.

There are several proprietary catalytic re-forming processes, all with somewhat similar operating conditions and all using some type of platinum catalyst. These include the following:

1. Platforming (Universal Oil Products)
2. Powerforming (Exxon)
3. Ultraforming (Amoco)
4. Houdriforming (Houdry)
5. Catalytic re-forming (Englehard)
6. Rheniforming (Chevron)

These processes utilize large, barrel-type, centrifugal compressors. There are approximately 1000 catalytic re-formers such as this throughout the world and many more are proposed as the lead phase-out occurs.

Equipment for catalytic re-forming can be classified as follows:

1. Barrel-type centrifugal compressors for hydrogen recycle service.
2. Barrel-type centrifugal compressors for hydrogen makeup service.
3. Single-stage overhung or horizontally split centrifugal compressors for regeneration gas circulation during catalyst regeneration.

Process Description

For illustration, the platformer process of Universal Oil Products is used for the process description in Fig. 7.6. Although operating conditions are slightly different, the other cat re-forming processes are similar in principle.

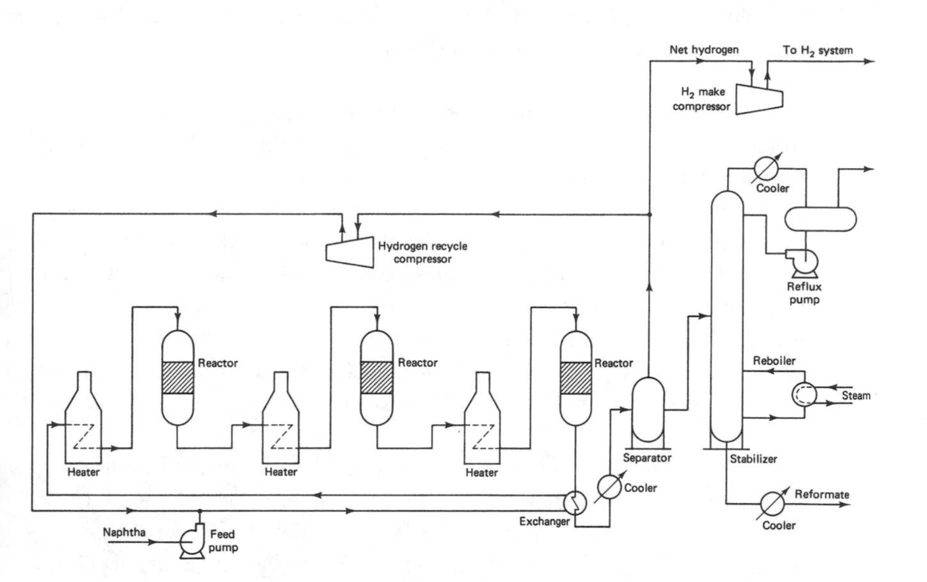

Figure 7.6 Catalytic Reforming Unit

The reaction takes place in three reactors filled with the platinum-bearing catalyst pellets. The mixture of hot oil and hot hydrogen is recycled through the beds to effect the molecular rearrangement. The reactors are steel vessels lined inside with refractory to insulate the metal from the 1000°F reaction temperature.

The process flow is as follows: the pretreated feed and recycle hydrogen are heated to a temperature of from 900 to 1000°F before entering the first reactor. In the first reactor, the major reaction is dehydrogeneration of naphthenes to aromatics and, as this is strongly endothermic, a large drop in temperature occurs. To maintain the reaction rate, the gases are reheated before being passed over the catalyst in the second reactor. Usually, three reactors are sufficient to provide the desired degree of reaction, and heaters are needed before each reactor to bring the temperature up to the desired reaction temperature.

As the mixture of product and hydrogen leaves the last reactor, it is cooled, and the hydrogen is separated from it in the hydrogen separator. The liquid products are condensed and sent on for further processing in the stabilizer column, where high-octane reformate (final product from the re-former) is accumulated and pumped to the gasoline pool.

The hydrogen leaving the hydrogen separator splits into two streams; one is recycled through the process to mix with feed going to the reactors, and the excess hydrogen is pumped away by compressors to be used in other processes such as hydrocracking. Since hydrogen is manufactured in the cat re-former process, there is always a sizable stream leaving the unit. It is one of the advantages of cat re-forming, since hydrogen is an expensive product.

Unfortunately, the catalyst deteriorates over a period of time and has to be regenerated. There are three types of regeneration.

1. *Semiregenerative:* Here, the plant is shut down after a 6-month to 2-year run, depending on the severity of operating conditions, and the catalyst is regenerated by high-temperature oxidation and chlorination. The hydrogen recycle compressor is used in this regeneration to pump air through the reactors to furnish oxygen for the regeneration reaction.
2. *Cyclic regenerative:* In the cyclic process, an extra reactor is installed in the train and it is used as a swing reactor. Every 24 to 48 hours, one reactor is valved out of service and the catalyst in it is regenerated. In this type of plant, a separate compressor is installed for circulation of air for regeneration. In many plants, a single-stage overhung centrifugal compressor is utilized.
3. *Continuous regenerative:* A development by Universal Oil Products features continuous regeneration. The reactors are stacked one on top of the other and the catalyst is continuously circulated; regeneration occurs almost continuously in a separate regenerator vessel. Again, a separate compressor is used for regeneration. It is a single-stage unit, but it handles gases at 900°F and requires special metallurgy.

Hydrogen Recycle Compressor

The hydrogen recycle compressor is the heart of the reaction system since it circulates the large quantities of hydrogen through the reactors and furnaces. The recycle compressor is a barrel-type compressor with approximately six or more wheels.

The first unit operated with pressure at approximately 1000 psi in the reactors, but, at present, the pressure level has been decreased to a level between 150 and 250 psig. The compressor differential pressure is between 100 and 150 psi, and the molecular weight of the recycle gas, which is about 90% hydrogen and 10% hydrocarbons, is from to 5 to 8. The polytropic head on a recycle compressor is usually less than 100,000 ft.-lbs/lb.

To ensure against hydrogen leakage, the barrel-type casing is used. Seals are of the oil-film type, with the small quantity of oil leakage discarded via the high-pressure oil trap in the seal oil system.

Drivers are electric motors or steam turbines. Since control of the recycle gas stream is not critical, a single suction throttling system on the motor unit and hand-set variable speed on the turbine are adequate. Small variations in recycle flow do not affect the reactions that occur. Of course, the less recycle, the lower the power consumption for the process.

There is some variation in molecular weight with time as the catalyst deteriorates between regeneration, especially on semiregenerative units. Care must be taken to make sure the minimum molecular weight is considered in establishing compressor head requirements and the maximum molecular weight in calculating the bhp for driver sizing. On these units, the variable-speed turbine is an energy saver. Suction throttling on the motor unit, unfortunately, is not efficient at lower system head requirements associated with high molecular weights.

On the cyclic regeneration systems, where the swing reactor is regenerated daily, the molecular weight stays fairly constant and there is little choice between motor and turbine from the standpoint of energy consumption.

The following is a tabulation of compressor sizes and types for various-sized cat re-formers. The units are based on a molecular weight of about 8, a suction pressure of 125 psia, and a discharge pressure of 250 psia. These conditions will vary for the different types of cat re-formers. The recycle rate on most cat re-formers varies between 4000 and 6000 standard cubic feet of gas per barrel of feed depending on process conditions.

Cat Re-former Capacity, bpd	Recycle Compressor, acfm	Approximate Compression Horsepower, bhp
5,000	2,500	2,500
10,000	5,000	3,500
20,000	10,000	6,000

Cat Re-former Capacity, bpd	Recycle Compressor, acfm	Approximate Compression Horsepower, bhp
30,000	15,000	8,500
40,000	20,000	12,000
50,000	25,000	14,500

On small units, reciprocating compressors are applied and electric-driven, horizontal-balanced opposed units are utilized. Two compressors are usually furnished.

Hydrogen Make-up Compressor

The hydrogen make-up compressor transfers the excess hydrogen from the system of the hydrotreaters and other processes that use hydrogen. The amount of hydrogen that is made varies between processes but, in general, is about 1200 standard cubic feet of hydrogen per barrel of feed. Thus, on a 35,000 bpd unit, 40 mm scfd of hydrogen must be compressed for transportation to other plants.

The design of the make-up compressor is similar to that of the recycle unit, but the unit is of a smaller size. The following table estimates the compressor size and types for various-sized cat re-formers. It is assumed that the compressor suction pressure is 230 psia. Horsepowers are not shown because they depend on discharge pressure, which varies for each plant. Usually, nine or ten wheels are required for the compressor.

Cat Re-former Capacity, bpd	Hydrogen Make-up Compressor, acfm	Compressor Designation
5,000	375	Reciprocating
10,000	750	Reciprocating
20,000	1500	Centrifugal
30,000	2200	Centrifugal
40,000	3000	Centrifugal
50,000	3750	Centrifugal

Regeneration Recycle Compressor

In cyclic regenerative plants, the regenerator recycle compressor is usually a single-stage, centrifugal type of unit. Following is a rough estimate of the sizes of units for various cat re-former capacities. The gas handled by the compressor is, for the most part, air with an equivalent molecular weight of 28. Suction pressure is about the same as the recycle compressor suction pressure, and the increase in pressure through the compressors is about 50 psi to satisfy the system pressure drop.

Cat Re-former Capacity, bpd	Regeneration Compressor, acfm	Recycle Compressor Designation
10,000	1250	Centrifugal
20,000	2500	Centrifugal
30,000	3750	Centrifugal
40,000	5000	Centrifugal
50,000	6250	Centrifugal

Hydrotreating

Hydrotreating (Fig. 7.7) is a process for removing objectionable elements from products or feedstocks by reacting them with hydrogen in the presence of a catalyst. The objectionable elements include sulfur, nitrogen, oxygen, halides, and trace metals. Hydrotreating is also referred to as hydrodesulfurization (HDS).

Hydrotreating serves two major purposes in a refinery.

1. It removes impurities, such as sulfur, nitrogen, oxygen, halides, and trace metals, from marketable products, such as fuel oils, distillates, and lube oils.
2. It removes these same impurities from feedstocks to cat crackers, cat re-formers, and hydrocrackers. Since these impurities are detrimental to the catalysts in these processes, hydrotreating plays a vital role in refinery production.

Many refineries have a number of hydrotreaters to perform the preceding functions. With environmental regulations dominating industrial processing, hydrotreaters will continue to be in demand.

Although a large number of hydrotreating processes are available for licensing, most of them have similar process flow characteristics and the flow diagram in Fig. 7.7 illustrates a typical process.

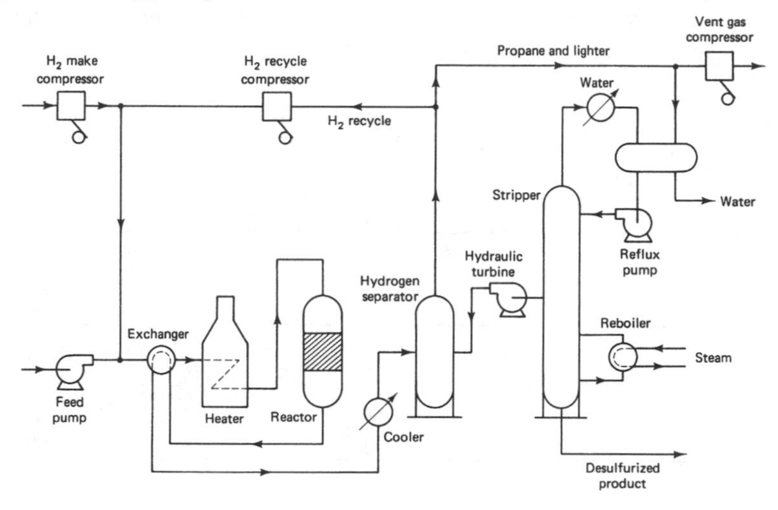

Figure 7.7 Hydrotreater

The charge pump pressurizes the feedstock up to about 1200 psi. The feedstock joins a stream of hydrogen recycle gas, and the mixture of hydrogen and oil is heated by exchangers in a direct-fired heater up to the reaction temperature of 700 to 800°F. The hot mixture enters the reactor, which is a pressure vessel with a fixed bed of catalyst. In the presence of the catalyst, the hydrogen reacts with the oil to produce hydrogen sulfide, ammonia, saturated hydrocarbons, and traces of metals. The trace metals remain on the surface of the catalyst, while the other impurities leave the reactor with the hydrogen-oil mixture. This mixture leaving the reactor is known as reactor effluent.

The reactor effluent is cooled before entering the high-pressure separator where the hydrogen-rich gas comes off the top and the oil off the bottom. The hydrogen-rich gas is treated to remove hydrogen sulfide and is used again in the process. It is recycled into the front end of the plant by the recycle compressor.

The oil from the bottom of the separator is throttled to a lower pressure and enters the stripper, where any remaining impurities are removed. In some plants a hydraulic turbine replaces the throttling valve.

The reaction consumes hydrogen at the rate of about 500 scf per barrel of feedstock. This figure varies with the process and with the amount of impurities that are to be removed. It can be as low as 200 or as high as 800 scf per barrel. The quantity of hydrogen-rich gas that is recycled is about 200 scf per barrel of feedstock.

There are three compressor applications on most hydrotreaters:

1. Recycle hydrogen-rich gas, molecular weight 6 to 8, with suction pressure of 1000 psig and discharge pressure of about 1150 psig.
2. Hydrogen makeup gas, molecular weight 2 to 6, suction pressure of 200 psig, and discharge pressure of 1150 psig.
3. Vent gas, small quantities at low suction and discharge pressure, with molecular weight varying from 20 to 40.

Quantities of recycle and makeup gas are estimated as follows for various-sized plants (since it is difficult to generalize on vent gas capacity as a function of plant capacity, it is not included).

Hydrotreater Capacity, bpd	Recycle Gas		Makeup Gas	
	scfm	acfm	scfm	acfm
10,000	15,000	225	3,500	250
20,000	30,000	450	7,000	500
30,000	45,000	675	10,500	750
40,000	60,000	900	14,000	1000
50,000	75,000	1125	17,500	1250

In most plants, all three compressor applications use reciprocating machines. The makeup hydrogen capacity may be large enough on larger plants, but the compression ratio is out of the range of the centrifugal compressor. The recycle compression ratio is low enough for a centrifugal application, but on plants smaller than 30,000 bpd, the capacity is too small. On larger plants, centrifugal recycle compressors should be evaluated against reciprocating units. Many centrifugal compressors are used in this operation.

Hydrogen Plants

Installation of hydrocrackers and hydrotreaters has resulted in a large demand for hydrogen in refineries. Cat re-formers, on the other hand, produce hydrogen as a by-product, but usually not in sufficient quantities to supply the bydrocracker and hydrotreaters in a refinery. Thus, supplemental hydrogen is often required. Two processes are available for hydrogen production:

1. Partial oxidation of heavy hydrocarbons, such as heavy fuel oil.
2. Steam re-forming of methane (from natural gas).

Since steam methane re-forming is currently much more widely used than partial oxidation, it will be described in this section.

Process Description

The flow diagram is shown in Fig. 7.8. The process takes place in three steps, as follows:

1. *Re-forming:* Natural gas (methane, CF_4) is pumped into the plant at approximately 200 psig pressure along with a supply of steam. With the addition of heat, the steam-gas mixture reacts in the presence of a catalyst to produce carbon monoxide and hydrogen:

$$CH_4 + H_2O \rightarrow CO + 3H_2$$

This is the re-forming reaction and it takes place at 1500°F. The reaction is carried out by passing the gas and steam mixture through a bank of catalyst-filled furnace tubes The furnace consists of one or two rows of numerous vertical tubes, fired on each side, to obtain even heat distribution around the tube and along the length of the tubes because of the extremely high tube wall temperature 1700 to 1800°F. Special alloys are used for the furnace tubes.

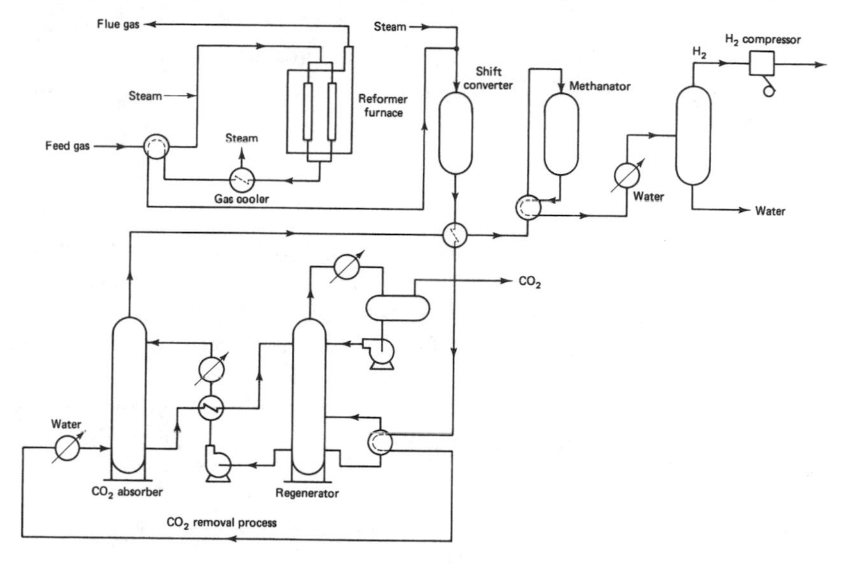

Figure 7.8 Hydrogen Plant, Process Flow Diagram

2. *Shift conversion:* The product from the re-forming reaction is then cooled to about 650°F and enters the shift converter along with additional steam. In the shift convertet, a fixed-bed catalyst reactor, the carbon monoxide in the gas reacts with the steam in the presence of shift catalyst to produce more hydrogen and to convert the CO into CO_2:

$$CO + H_2O \rightarrow CO_2 + H_2$$

The CO_2 is then removed from the gas by absorbing it in a special solution and boiling off from the CO_2 solution to the atmosphere or to a CO_2 collector for further use or sale.

3. *Methanation:* The remaining small quantities of carbon monoxide and carbon dioxide are converted to methane in another fixed-bed catalytic reactor for the final purification step in the process.

These process steps are required to obtain very pure hydrogen as a product, since impurities are detrimental to the expensive catalyst in the hydrocracker. Purities of 95% and higher are obtained in the steam methane reformer.

From the methanator, the gas, with a molecular weight of 2 plus, leaves the plant at about 150 psig and is ready to he compressed to the 2000 psig pressure level for supply to the hydrocracker and hydrotreater.

Compression Equipment

There are two major requirements for compressors in a hydrogen plant:

1. Compressors to move the methane into the process.
2. Compressors to move the nearly pure hydrogen from the process to the hydrocracker and hydrotreater.

Methane Compressor

Methane, or natural gas, is usually available at about 25 psig or higher and must be compressed to the 200 psig level in the steam methane re-former. Since hydrogen plants are sized by the volume of hydrogen produced (5 million, 10 million, or even 75 million sfcd), the size of the methane compressor for a given plant is not evident from the plant size. However, by calculating how many mols of methane are required for the reaction to produce 1 mol of hydrogen, the quantity of methane feed gas can be determined. The following table should be helpful in estimating compressor size.

H_2 Plant, scfd	Methane, scfm	Methane acfm, at 15 psig
10,000,000	3,300	1200
25,000,000	8,000	3000
50,000,000	16,000	6000
75,000,000	24,000	9000

From this table, based on capacity, centrifugal compressors are applicable for plants as small as 10 million scfd. With respect to head, a compression ratio of 5 1/4 is attainable in a single casing since the molecular weight of the methane is 20. A compression ratio of 5 1/4 will produce a discharge pressure of about 215 psia, which is adequate for most plants.

Conventional, horizontally split centrifugal compressors, motor or turbine driven, are suitable. Metallurgy is standard since the gas is noncorrosive, and sealing with carbon rings or oil seals is satisfactory.

The one problem with the centrifugal compressor application is turn-down ratio. The hydrogen plant capacity often fluctuates, especially if it is operating in parallel with one or more cat re-formers to furnish hydrogen to the refinery system. It is important to make sure that an adequately sized bypass line is available for partial-flow operation.

On plants smaller than 10 million scfd, reciprocating equipment should be considered. Heavy-duty, balanced opposed compressors would be applied to this service. With multiple compressors, the turn-down problem becomes less troublesome.

Hydrogen Compression

The compression of hydrogen from the hydrogen plant to the hydrocracker is discussed in the hydrocracker description. Normally, reciprocating compressors are required to pump the hydrogen, with a molecular weight of 2, from the 125 to 200 psig intake pressure to the 2000 psig pressure level required at the hydrocracker.

INDUSTRIAL REFRIGERATION

Chemical plants, refineries, dry-ice and meat-packing plants, along with breweries, all need refrigeration compression equipment. This refrigeration cycle is driven by screw or reciprocating compressors, and the power consumed adds considerably to the cost of the items produced.

Selecting the right compressor requires that thc following be considered:

1. Refrigerant to be used
2. Refrigeration cycle:
   a. Whcthcr single-stage or multi-stage
   b. Whether a cascade system will be selected (two refrigerants used in series)
3. Loading demand of the system
4. Monitoring of the refrigeration cycle loop to ensure that a malfunction within the cycle loop will he discovered and corrected

The refrigerant used in most industrial systems is ammonia because of certain thermodynamic properties. The low molecular weight of ammonia reduces the friction losses in the loop, and the saturation pressure is such that lower pressure is required to liquefy the ammonia in the condenser.

Typical ammonia [specific gravity 0.594 at 60°F] and 14.7 psia cycle conditions are as follows:

Evaporator: pressure, above atmospheric in the range of 25 psig; temperature, - 28°F and higher.

Condenser: pressure, 150 to 200 psig; temperature, 86°F.

The approximate value of bhp/ton of refrigerant is as follows for reciprocating compressors:

Ammonia	1.172
Refrigerant 11	1.111
Refrigerant 12	1.23
Propane	1.256
Propylene	1.262
N-butane	1.193

The Refrigerant 11 shows a lower bhp/ton; however, the price of R11 and R12 is much more than that of ammonia. Since most industrial systems are large, the cost to charge a system with other gases is higher than that of ammonia.

While R11 is used in some refrigeration cycles because of its specific gravity [4.78 at 60°F and 14.7 psia], the systems are more costly to install than those using ammonia.

The typical performance of an ammonia refrigerant system is as follows:

	Performance, Based on an Ammonia Refrigerant					
Evaporator temperature	-15°F	-1°F	+12°F	-26°C	-18°C	-11°C
Evaporator pressure	21 psia	30 psia	40 psia	1.5 bars	2.1 bars	2.8 bars
Condenser temperature	96°F	96°F	96°F	35.6°C	35.6°C	35.6°C
Condenser Pressure	200 psia	200 psia	200 psia	13.8 bars	13.8 bars	13.8 bars
Comp. Ratio	9.5	6.7	5.0	9.5	6.7	5.0
BHP/ton, stage 1	1.89	1.48	1.22	1.89	1.48	1.22
BHP/ton, stage 2	1.74	1.48	1.22	1.74	1.48	1.22

A typical ammonia compressor is shown in Fig. 7.9.

Figure 7.9 A heavy frame, 2500 hp electric-driven ammonia compressor.

CHEMICAL INDUSTRY PROCESSES

Synthetic Ammonia (NH_3) Production

Large quantities of ammonia are produced throughout the world, since it is an essential ingredient of chemical fertilizers. Nitrogen is necessary for living cells to produce protein. Very few living plants can produce amino acids that contain the elements carbon, hydrogen, nitrogen, and other elements in complex combinations that are essential foods for man and animals. The few living plants that can produce nitrogen are called legumes, including clover, alfalfa, and beans. These plants have a bacteria that fixes, or extracts, nitrogen directly from the air. Large quantities of nitrogen are formed during severe storms where lightning produces nitrogen from the air and this nitrogen is then absorbed by rain.

The first attempt to produce nitrogen utilized a high-voltage discharge and was made in 1780 by Lord Cavendish. A commercial plant using electric discharge was established at Nottoten, Norway, in 1900 and at Niagara Falls, New York, in 1902. Haber, a German scientist, was the first to synthesize ammonia by combining three parts hydrogen and one part nitrogen at 100 atm and cooled to 1112°F. The process used a catalyst material of asmium. The chemical equation $3H_2 + N_2 \rightarrow 2NH_3 + 26{,}000$ cal describes the reaction. The commercial development of Haber's discovery was carried out by Bosch. A chemical firm was established called Badische Anilin und Soda Fabrik (BASF). By 1913, it proceeded to construct a large synthetic ammonia plant. The main problem was to obtain the two constituents in a free state.

Nitrogen is obtainable from an air-separation plant. This same process produces oxygen as well. The oxygen by-product is then reacted with methane and steam to produce hydrogen. The chemical equation is $CH_4 + 2H_2O \rightarrow CO_2 + 4H_2$ for this process.

Other sources of hydrogen can be natural gas, water gas (steam passed over glowing coal or coke), chlorine cells, coke-oven gas, water electrolysis, liquid hydrocarbons, and refinery gas, such as catalytic re-forming as described previously. The most practical processes for obtaining hydrogen and nitrogen components are the partial oxidation system and the steam or natural gas re-forming system, shown in Figs. 7.10 and 7.11.

The partial oxidation process gave many problems in the design of furnace equipment to handle high temperatures and high-pressure gases or oils. In 1954, Texaco introduced its synthesis gas generation process. This process had the advantages of complete independence from hydrogen sources, whether gas or liquid, and operated at line pressure of 400 psig. Its biggest disadvantages were the requirement for an air-separation plant to provide oxygen for the furnace and its relatively short burner life. The furnace temperature was 2000°F, thus requiring a water quench as shown in Fig. 7.10.

The natural gas or steam re-forming process (Fig. 7.11) requires the feedstock he heated to 1200°F. At the secondary reformer the reaction is endothermic. The nitrogen component enters the process as a component of air. The water quench, plus nickel catalyst, converts the hydrocarbons to carbon dioxide and free hydrogen or, in a concurrent reaction, carbon monoxide and free hydrogen. The shift converter reacts carbon monoxide with additional steam to form carbon dioxide and free hydrogen. The methanator reacts residual CO and CO_2 with hydrogen and thus forms methane and water vapor. Methane is less objectionable as an impurity than CO or CO_2.

High design reactor pressures will increase the conversion rates and reduce the size of the converter required.

Early synthetic ammonia compressors installed in the United States in 1925 were designed for final discharge pressure of 4500 psig. In 1939, a new process named the Claude (French) process used a design final pressure of 15,000 psig. In 1980, a low-pressure system was introduced that required pressures of only 400 to 600 psig. This new system has a rating of 1000 tons. The new process design utilized dynamic compressors, while prior to 1970 nearly all compressors were reciprocating, positive-displacement units. The new low-pressure process has not, however, replaced many of the existing high-pressure plants.

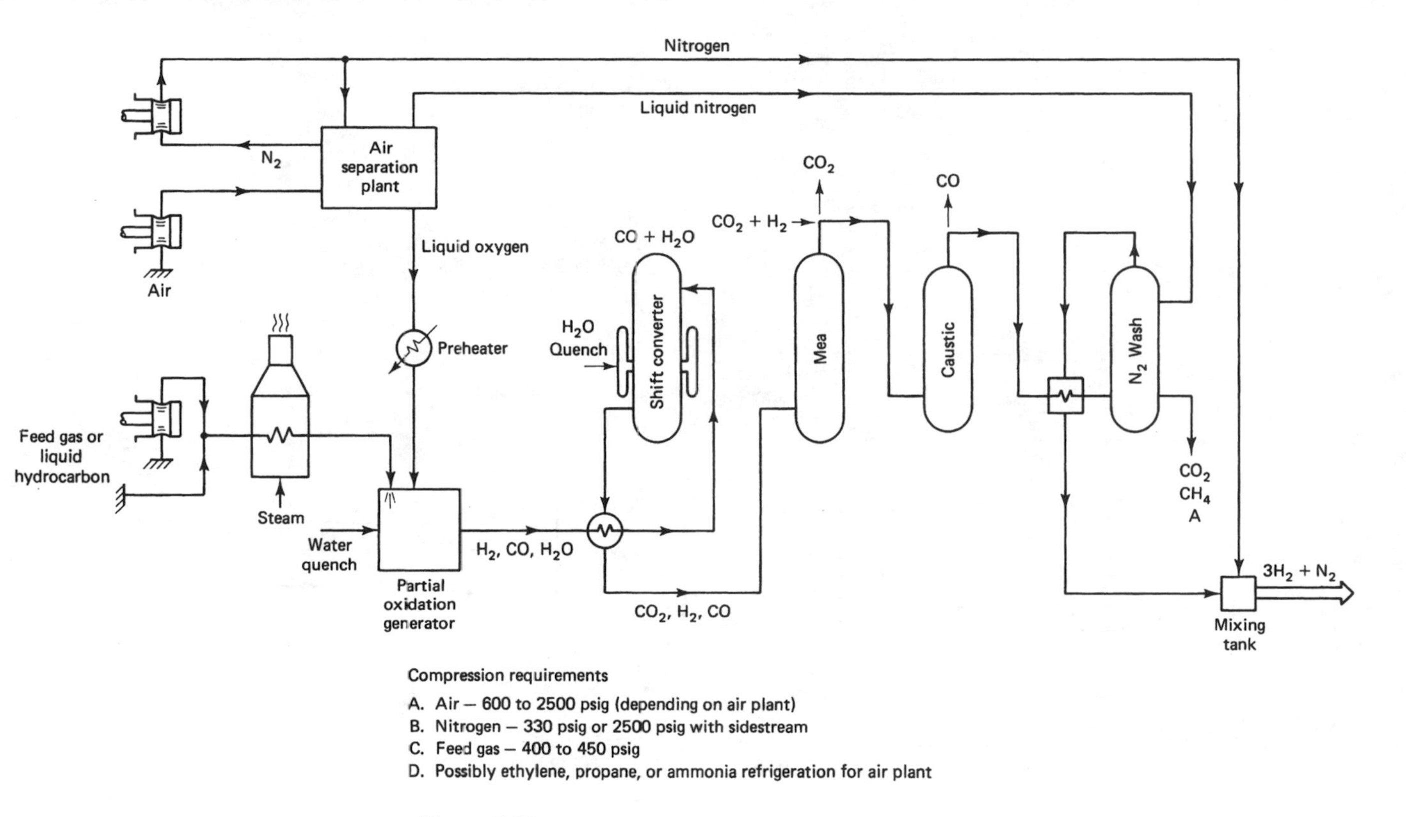

Figure 7.10 Flow Diagram for the Partial Oxydation Process

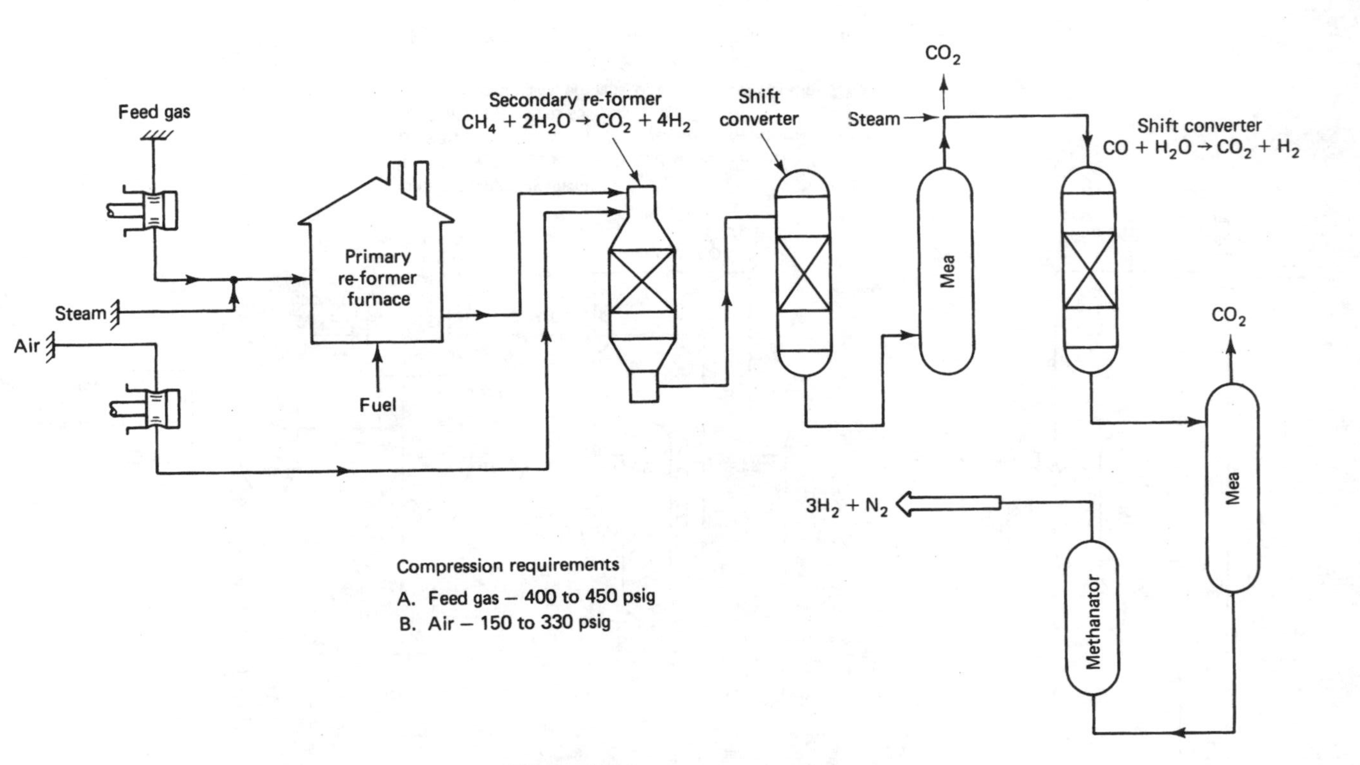

Figure 7.11 Natural Gas Reforming

Urea Production

The urea compound has a high nitrogen content and is an excellent fertilizing agent. Combining urea and formaldehyde yields a useful plastic. The formation of the compound is accomplished by the reaction of ammonia and carbon dioxide. The reaction pressure varies from 2000 to 6000 psig depending on the process.

The chemical reaction is described by the chemical formula $CO_2 + 2NH_3 \rightarrow CO(NH_2)_2 + H_2O$. The initial product is carbonate solution and the process is exothermic. By holding the solution in the reactor under elevated pressure and temperature, most of the carbonate is converted to urea. The conversion process is endothermic, but more heat is formed in the carbon dioxide and ammonia reaction than is absorbed in the conversion process. The higher the reactor pressure, the greater is the conversion. What carbonate is not converted is decomposed at a lower pressure. The ammonia gas is compressed to a pressure at which it will condense. This ranges between 200 and 270 psig. The ammonia feed and ammonia recovered from the decomposition is pressurized to the reactor pressure by using high-pressure reciprocating pumps.

Urea can be crystallized into pellets by a falling-film evaporation process. Urea for agricultural purposes also can be used as a liquid. Combining urea and formaldehyde yields a very useful plastic. The characteristic of this plastic is that it has brighter colors, is oil resistant, and is odorless and tasteless.

Heavy-duty, horizontal, electric-driven reciprocating compressors are used for urea production. These compressors often handle the carbon dioxide and ammonia service on the same compressor frame. A typical urea process is shown in Fig. 7.12.

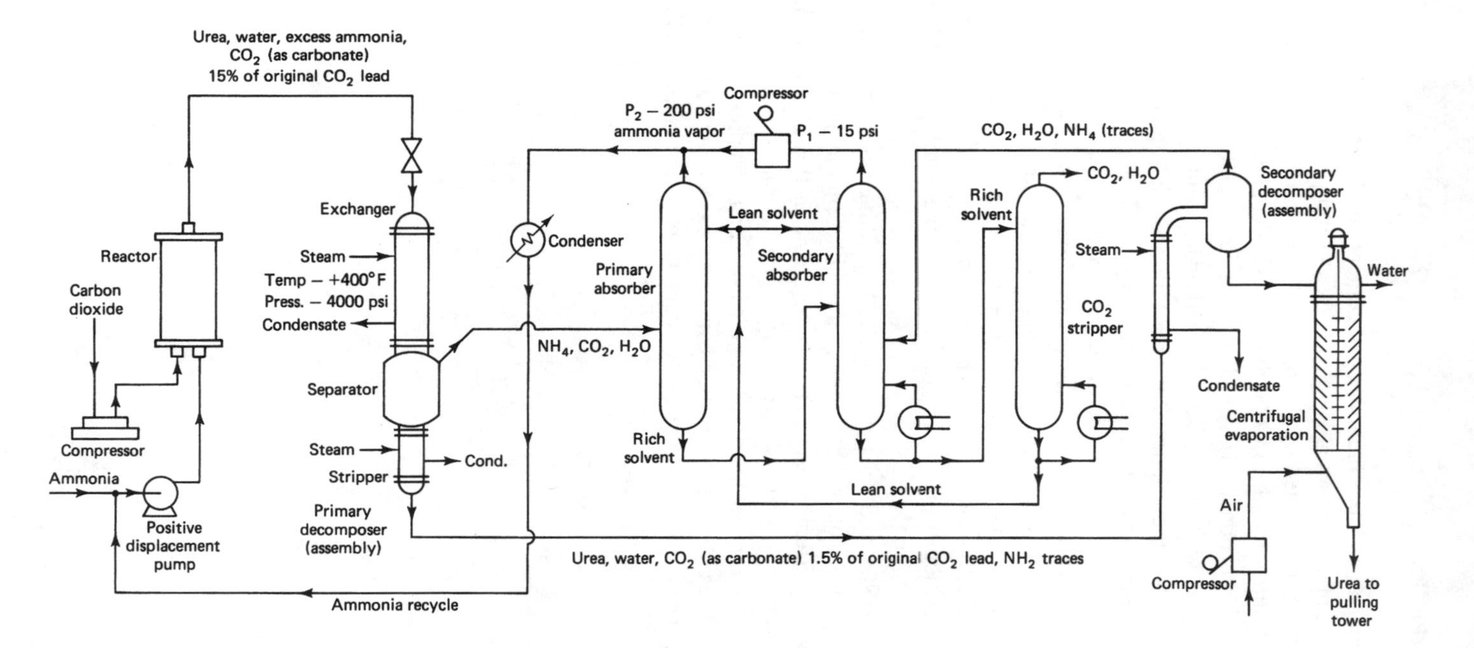

Figure 7.12 CPI-Allied Chemical Company, Urea Process

HALOGEN COMPRESSORS

The halogen family consists of elements that occur in salt beds or in sea water. The word halogen is a Greek word meaning salt forming. Chemically these elements are very active and are therefore not found in the free state, but in combination with other elements. The following table lists some of the important properties of the halogens.

Physical Properties of Halogens

Formula	Element	Color and State	Molecular Weight	Melting Point, °C	Boiling Point, °C
F_2	Fluorine	Pale yellow gas	19	-223	-188
Cl_2	Chlorine	Greenish-yellow gas	35.46	-103	34.7
Br_2	Bromine	Dark red liquid	79.92	-7.2	58.0
I_2	Iodine	Purplish-black solid	126.92	113	183.0

Fluorine

Fluorine occurs in nature combined with calcium as an insoluble salt (CaF_2), in igneous rocks such as cryolite (Na_3AlF_6), and as a double salt, apatite [$Ca_5(PO_4)3F$].

Chlorine

The main source of chlorine is common salt. It is not found in the free state because of its active chemical behavior. Chlorine is separated from sodium by electrolysis of brine.

$$2NaC1 + 2H_2O + \text{direct current} \rightarrow Cl_2 + H_2 + 2NaOH$$

Many small chlorine plants utilize single-cylinder, single-stage reciprocating compressors. A very common arrangement is to have a multiple number of first-stage units in one row and, on the opposite side, a row of second-stage units. This provides flexibility in the production rate since the number of cells on-stream can vary. Plants that are designed for output of 250 tons per day or more use centrifugal units. The chlorine leaves the cell from the anodes at 176°F, saturated with water vapor. The moisture is removed by a cooling and countercurrent bubble cap tower, with concentrated sulfuric acid scrubbing; the chlorine leaves the tower at about 68°F. The gas is pulled through the drying tower by the slight vacuum created by the first-stage compressor(s). The chlorine gas is then compressed in the required number of stages to condenser pressure at which the chlorine is liquefied, at 120 psig if water is used.

The oceans of the world contain an estimated 18 trillion tons of chlorine. Salt beds, salt domes, and volcanoes, all formed during past geological ages, represent the major commercial sources.

Bromides

Bromides are found in sea water as compounds. The dissolved bromides yield a pint of bromide per 7 1/2 tons of sea water. Salt wells in Michigan and other areas of the world are rich in sodium bromide.

Iodine

Iodine can be extracted from tissues of seaweed (kelp). The more commercially viable sources are found in brines, in oil wells, and as sodium idodate $NaIO_3$ as a 0.2 percent impurity in the sodium nitrate mines of Chile.

Chemically, any halogen will oxidize any halogen ion that comes later than itself in the periodic table. Commercially, bromide is produced chemically in this way: $C1_2 + 2NaBr \rightarrow 2NaCl + Br_2$.

Hydrogen Chloride

The HCl gas produced by burning pure hydrogen and chlorine contains contaminants that form deposits on hot surfaces such as valves used in reciprocating compressors. The operating life of these parts is extended if the gas is scrubbed and filtered. The design is similar to that of chlorine compressors. The HCl gas entering the compressor has more impurities and deposits than is usual for chlorine gas (HCI is highly hydroscopic). The major use of HCI is as basic feedstock to make polyvinyl chloride plastics, silicone rubber products, urethane foams, and the like.

Synthetic Polymers

A variety of synthetic polymers is produced throughout the world. The chemical industry has classified different polymers as follows:

Ordered polymer: A very large molecule in which the arrangement of the atoms is closely controlled.

Condensation polymerization products: The material produced by the chemical reaction condenses or solidifies at a given pressure and temperature. These are characteristics of nylon, Dacron, Mylar, and polyurethane.

Additional polymerization products: Materials similar to Saran, Orlon, polyvinyl chloride, Teflon, polystyrene, polyethylene, and polypropylene.

Gas compressors are necessary for feeding gases to the reactor (Fig. 7.13). Gases usually encountered in the production of synthetic polymers are ethylene, hydrogen, hydrogen chloride, methyl chloride, phosgene, and butane. In some cases, the gas is compressed to a pressure at which it can be liquefied and then pumped into the process.

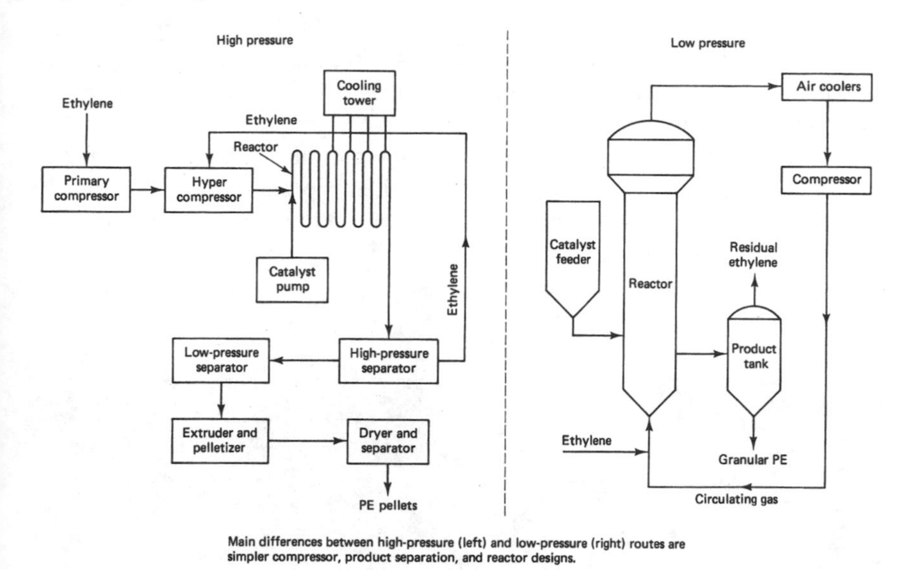

Figure 7.13 Synthetic Polymer Reactor

Polyethylene Production

Polyethylene polymer is, as stated, an ordered polymer. There are basically three varieties of this polymer presently manufactured. The low-density polyethylene product was originated by Imperial Chemical Industries in England in 1933. By 1938, 1 ton of polymer had been produced. In 1977, the world production, excluding the USSR, was 22 billion lb/year. The U.S. production alone was 8.3 billion lb/year. The polyethylene produced has a density of from 0.91 to 0.935 g/cm^3.

The low-density polyethylene process uses large reciprocating compressors compressing ethylene gas to 15,000 to 35,000 psig generally and as high as 50,000 psi in at least one case. The high-pressure ethylene in contact with the catalyst in a timed exposure period by flow rate is converted to a polymer. The rate of conversion is about 20 to 25%. The gas and molten polymer are separated. The unreacted ethylene is piped back to the inlet of the high-pressure compressor. In each plant there will be a primary or feed gas compressor, a secondary or hyper compressor (Fig. 7.14), and generally a small booster compressor. The normal operating pressures of a polyethylene plant are as follows:

Unit	Inlet (psig)	Discharge (psig)
Booster	0 – 70	150 – 500
Primary	150 – 500	1700 – 5000
Hyper	1700 – 5000	19,000 – 35,000

The end uses of low-density polyethylene are molded articles, approximately 14%; electrical insulation, approximately 11%; film, approximately 34%; pipe, approximately 15%; paper coatings, approximately 8%; bottles and tubes, approximately 11%; and miscellaneous, approximately 6%.

Various properties of polyethylene are important, depending on the application, and the product is manufactured and controlled to promote the best obtainable quality of one or more specific characteristics: crystallinity, relative rigidity, softening temperature, tensile strength, elongation, impact strength, and density. The high-pressure, low-density polyethylene process produces a material having the characteristics needed for most markets.

Figure 7.14 A 5000 hp heavy duty horizontal, electric driven, high pressure (hyper) compressor for the production of polyethylene.

Resin properties are controlled by varying throughput rates, the method of initiating conversion or reactor configuration. Polyethylene is formed by stringing the ethylene molecules together in a long chain and then constructing cross branches to tie these chains together. Normally, an increase in reactor pressure increases the number of cross branches. The chain branching with especially short branches is probably the single most important factor that affects the physical properties of polyethylene.

An entirely different process produces high-density polyethylene. This compound is only a linear arrangement of the chains. The fewer the cross branches, the closer together the strings will lie and, therefore, the higher the density will be. High-density polymers are produced at a process operating pressure of 15 to 3000 psi. The product has a density of 0.945 to 0.970 g/cm³. The polymer produced is ideally suited as a synthetic fiber material since the crystallinity is 75% or higher. Copolymerization of ethylene with relatively small amounts of other olefins such as *I*-butane and propylene produces polymers with slightly lower densities and degrees of crystallinity. As ordered polymers, it is very possible for such plastics to possess a range of properties at higher pressures. On the other hand, the feed gases of propylene and f-butane are more expensive than ethylene.

Thirty-five percent of ethylene is produced by cracking ethane and 31% by cracking propane. The remainder comes from cracking heavier hydrocarbon material, such as naphtha or gas field condensates, off-gas from catalytic cracking units in petroleum refineries containing ethylene, and other light hydrocarbons. Ethylene production has increased to the point that there now exist ethylene pipelines extending over long distances in Louisiana and Texas. Pipeline compressors are nonlubricated, reciprocating, positive-displacement machines.

Production of Low-density Polyethylene Using a Low-pressure Process

In this process (Fig. 7.13), gaseous ethylene and a solid, supported, chromium-containing catalyst are fed to the reactor. Gas is recirculated through the reactor to keep the growing polymer particles fluidized. The recirculating gas passes continuously through coolers to remove reaction heat. The polymer is withdrawn as a dry granular product, which passes directly to the pelletizing extruder. Reaction pressure is 100 to 300 psi, and reaction temperature is controlled at less than 100°C (212°F). Only one compressor is required to feed the fluidized reactor. This process has energy savings and capital-cost advantages.

OIL AND GAS INDUSTRY

Gas Gathering

Gas gathering is defined as the collection of natural gas from the well head and moving it to a gas plant or to a major transmission line (Figs. 7.15 and 7.16). This application can involve either individual, small well-head compressors as small as 5 horsepower or larger-power central units of several thousand horsepower handling several wells or an entire field. As the field pressure drops due to depletion, the suction pressure at the compressor will drop correspondingly, increasing the ratio of compression. The discharge pressure can also vary greatly depending on the flowing pressure of transmission pipelines in the area. Typical ranges of pressure for a reciprocating compressor in gas-gathering services are suction from about 10 to 150 psig and discharge from 150 to 1200 psig. Machines are expected to handle a wide range of these conditions and keep the driver reasonably well loaded.

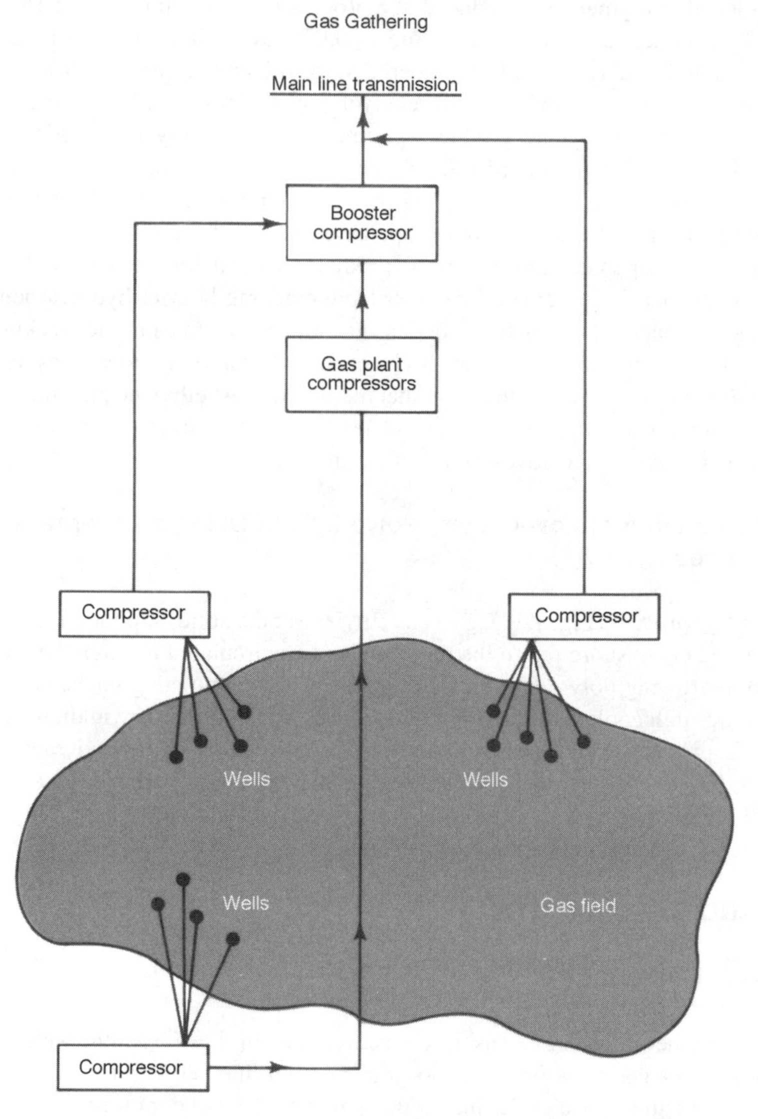

Figure 7.15 Gas gathering is the moving of natural gas from the well head to a gas plant or a major transmission line.

Gas-gathering applications probably account for the majority of installed reciprocating compressor horsepower in the oil and gas industry. Compressor units are generally sold to the user; however, the rental of compressor units is growing rapidly and is itself becoming a viable service industry.

Figure 7.16 Field installation of a balanced opposed compressor in gas gathering service.

The nearer the compressor is to the well head, the more likely the gas will be saturated. Some tolerance of this liquid is desirable in the design of a field unit. Additionally, even the standard machine should be capable of compressing gas with as much as 2 percent H_2S without serious corrosion problems. Reciprocating compressors are generally utilized in gas-gathering applications; however, some rotary positive units are used in low-pressure applications.

Gas Lift

Gas lift (Fig. 7.17) utilizes natural gas to produce oil. Bottom hole pressures are maintained by forcing gas down the well casing. Oil is then lifted upward through the production tubing. The gas-oil mixture passes through a separator where the gas is separated from the oil, and the gas is then piped to the compressor suction, where it is recompressed. Typical conditions of service are suction pressures ranging from 25 to 65 psig and discharge pressures from 800 to 1200 psig.

Gas lift is utilized where electricity to run pumps is not practical or economical and gas is readily available. Reciprocating compressors are generally utilized, such as those in Fig. 7.18.

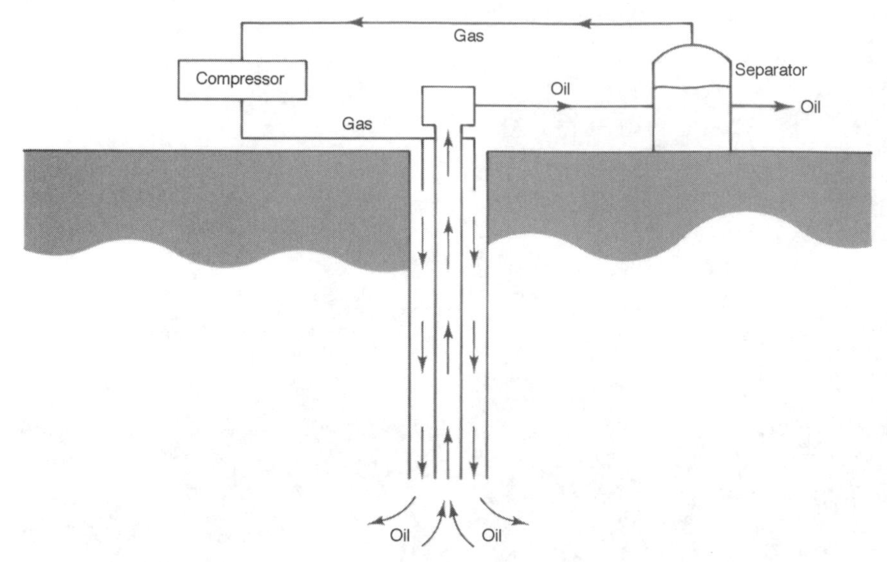

Figure 7.17 Gas lift application utilizes gas to produce oil.

Figure 7.18 Field installation of a balanced opposed, two-stage compressor in gas lift service.

Reinjection

Reinjection (Fig. 7.19) is the compressing of natural gas for injection into the gas cap of a formation to maintain field pressures and prevent salt water from moving into the oil field. This helps in the production of the oil and provides an alternative to flaring when there is no market for the gas that is being produced. When the oil reaches the surface, the gas is separated and sent to the compressor, where it is recompressed. These applications often require high pressure since field pressures may be high when the field begins producing. Typical pressure conditions are suction from 80 to 400 psig and discharge from 3000 to 4000 psig.

Distribution

This service, although not actually a gas field application, is the moving of gas after it leaves the main transmission line. Often, some compression is required to deliver the gas to industries or residences, or both. This service is relatively easy, since the gas is clean, at moderate pressures such as 200 to 400 psig, and does not require a great deal of compression. Generally, reciprocating compressors are utilized.

Transmission

Transmission is the movement of gas across country from the originating field to population centers where it is used. Generally, this involves large-diameter pipe with typical pressures from 900 to 1200 psig.

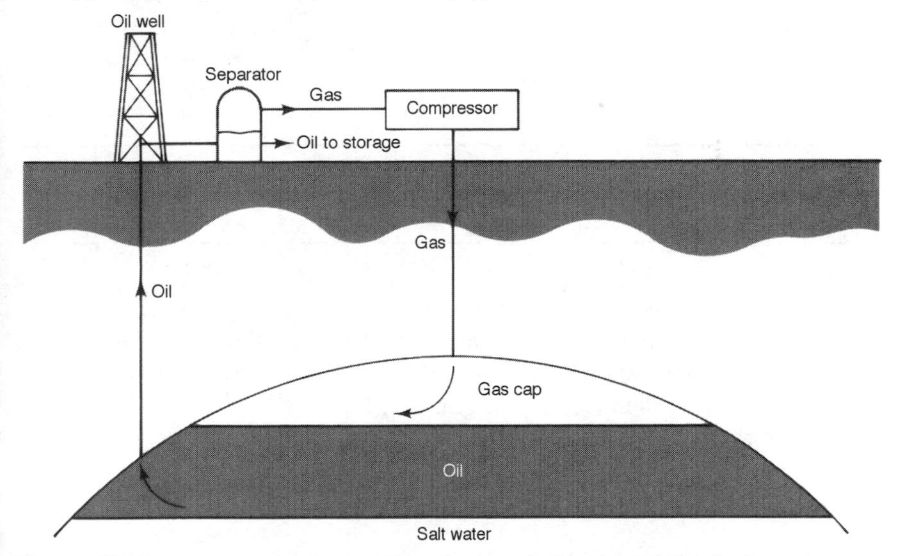

Figure 7.19 Reinjection aids in the production of oil and provides all alternative to flaring.

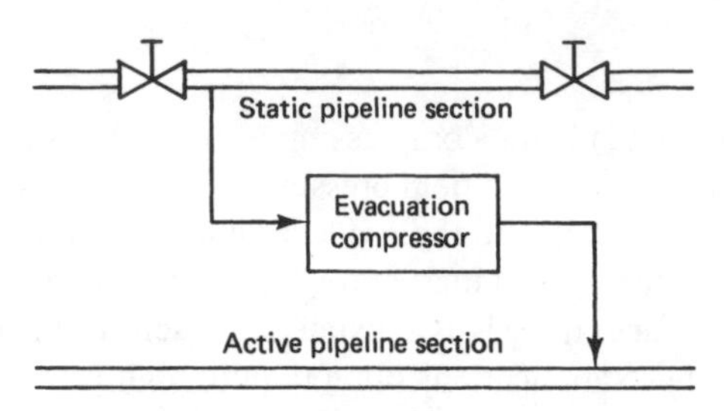

Figure 7.20 Pipeline evacuation.

Compressor stations are required at intervals to maintain pressure and flow rates. These stations must handle high volumes of gas at low ratios of compression. The horsepower requirements are large and employ either gas-turbine-driven centrifugal or integral engine-driven equipment. Efficiency becomes more important as the gas moves farther from the field toward the end user.

Pipeline evacuation evolved with the rapid rise in natural gas prices during the early 1980s. Prior to that time, it was common practice to vent the gas in a static section of pipeline to atmosphere while repairs were made to the pipeline. The static pipeline section could be anywhere from about a quarter of a mile to many miles in length. As gas prices increased, venting became economically unattractive.

Pipeline evacuation involves the transfer of gas from a static section of pipeline to an active section of pipeline (Fig. 7.20). This is accomplished by a reciprocating compressor that can handle wide variation in suction pressures while compressing against a constant discharge pressure. As gas is evacuated from the static section, the suction pressure declines, the compression ratio increases, and the compressor capacity decreases.

Applications can vary widely, but the following range of conditions can be considered typical:

	Initial Pressure	Intermediate Pressure	Final Pressure
Intake (psig)	850	450	50
Discharge (psig)	850	850	850

Packaged compressor systems specifically designed for this application feature multi-stage compressors that can maintain high driver loading throughout a wide range of compression ratios. Most units of this type are driven by natural gas engines. The entire system (compressor, engine, cooler, etc.) is generally mounted on a trailer for portability to various field locations.

Storage

Since the utility demand for natural gas is much greater during the winter months, it is necessary to store the gas produced during warm weather in reservoirs and then retrieve it to satisfy peak cold-weather requirements. The compressor must not only be able to handle filling the reservoir but also the return of the gas.

This dual service requires operating pressure flexibility and is provided best by the reciprocating compressor. Typical pressure conditions are suction from 35 to 600 psig during injection, 300 to 800 psig during withdrawal, and discharge from 600 to 4000 psig during the injection phase and 700 to 1000 psig as the gas is withdrawn from the reservoir and fed to the transmission line.

Enhanced Recovery

Since up to 60% of the discovered oil remains in place after primary and secondary recovery, further (tertiary) methods of recovery may be economically attractive. Three major categories of enhanced oil recovery are now in use.

1. ***Thermally enhanced recovery.*** These methods are generally applicable to heavy oils and require steam injection or *in situ* combustion. Steam-injection processes are either steam soaking or steam drive. In the soaking process, the steam is allowed to remain in the strata for a few days or weeks, after which the well is returned to production, with the oil having been heated to a viscosity enabling it to flow under the action of lift equipment. The process is then repeated.

   Steam drive requires a continuous injection of steam to form a steam flood front. Fluids move from the injection well to the producing well. This process does not require compression, as the steam is injected into the formation at boiler pressures.

   In situ combustion involves an underground fire flood. Compressed air to support combustion is injected down the well into the producing strata to maintain the fire. The heat makes the oil less viscous and more easily recovered. Generally, reciprocating compressors are utilized. Typical discharge pressures are from 1500 to 5000 psig.
2. ***Chemically enhanced recovery.*** Polymer flooding is a chemically augmented water flood in which small concentrations of chemicals are added to injected water to increase flood front effectiveness in displacing oil.

   Alkaline flood involves the addition of sodium chemicals. These solutions will react with constituents present in oil to form detergentlike solutions. This material reduces the ability of the formation to retain oil.

   Micellar-polymer is new and expensive but has the ability for efficient oil recovery. This system requires no compression equipment.
3. ***Gas-enhanced recovery.*** Carbon Dioxide or Nitrogen. This method involves miscible and immiscible applications of compressed CO_2 or N_2. In miscible processes, the injected gas mixes with the oil and forms a single oil-like liquid that can flow through the reservior. CO_2 is often selected for certain formations due to its ability to mix with oil and its sweep characteristics through the formation. CO_2 also reduces viscosity, density, and

surface tension, as well as distilling the light ends. It has also proved to be less expensive than natural gas in many cases.

Complete miscibility or mixing of CO_2 with crude oil depends on reservoir temperature, pressure, chemical nature, and oil density. Economics of CO_2 injection dictate that large quantities must be available within 300 or 400 miles when naturally produced or available as a result of flue gas separation from plants where CO_2 is a by-product.

The selection of N_2 or CO_2 as a medium for injection depends on the formation structure. In some reservoirs, nitrogen is utilized for pressure maintenance or flood application where CO_2 is not available. Nitrogen is often the product of an air-separation plant.

Both CO_2 and nitrogen miscible flooding generally involve significant amounts of compression by either reciprocating or centrifugal compressors. CO_2 is taken at pressures below 200 psig, compressed above 1500 psig, and put in a pipeline. At the field, it is injected into the ground at field pressure, which frequently is above 2000 psig. When the oil is recovered, the CO_2 is recycled into the field.

Several processes allow the producer to reclaim the CO_2 from the gas and reinject it into the producing formation. These processes are refrigeration, cryogenics, membrane separation, and absorption. Each of these processes has a different removal efficiency. Approximately 60% of the CO_2 can be removed from a natural gas stream by the use of the membrane system. Membrane units consist of a vessel filled with thousands of hollow fibers. The natural gas containing the CO_2 is passed over the fibers, and the natural gas, being more absorbent than the CO_2, diffuses through the membrane material to the hollow center.

Furthermore, CO_2 absorption can be accomplished through the absorption process using methyldiethanolamine (MDEA) and diethanolamine (DEA). These two processes in series with a membrane system will remove the balance of the CO_2. The MDEA process will remove approximately 38% and the DEA process another 2 percent.

The refrigeration/cryogenic process uses a cooling medium to condense the CO_2 out of the gas stream. These processes require some form of dehydration in the system to preclude freezing in the process. After the CO_2 is removed, it can be dehydrated, compressed, and reinjected into the formation.

When air-separation plants are installed and N_2 is used, compression is required to inject the N_2 into the oil field. Like CO_2, the N_2 produced can be reclaimed in a nitrogen rejection plant. A cryogenic process is used to separate the nitrogen from the associated gas stream. The cold box, which separates the N_2 from the residue gas stream, can be a separate addition to an existing gas plant, or it is possible to design an integrated NRU (nitrogen rejection unit) and NGL (natural gas liquids) processing plant. If the combined plant is selected, it is possible to design an integrated plant that optimizes the N_2 and gas liquids recovery, leaving the methane as salable gas. The biggest advantage of an integrated plant is the overall thermal efficiency, which is high enough to provide savings in the total energy required to operate the plant. Once the N_2 is separated, it can be recompressed and injected or vented as required by the overall production requirements.

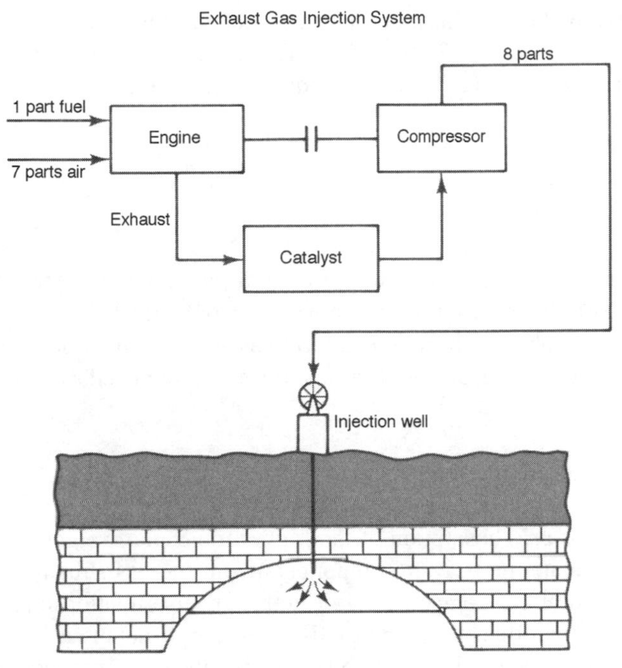

Figure 7.21 Engine exhaust can yield a noncorrosive inert gas.

Exhaust Gas. Engine exhaust gas, when properly treated by a catalyst, can yield a noncorrosive inert gas (Fig. 7.21). This treated gas acts as an economical substitute for natural gas in oil-field applications in pressure maintenance and miscible flooding. Injection gas is typically 88% N_2 and 12% CO_2 Injection pressure is dependent on the individual field. Generally, the volume of fuel gas required to injection gas is 1 to 8. The engine is used to drive a reciprocating compressor, which in turn compresses the treated exhaust gas to injection pressure.

Fuel Gas Boosting for Prime Movers

Reciprocating compressors are often used to increase the pressure of the hydrocarbon fuel gas used for operating gas engines or gas turbine drivers. Suction pressures range from 10 to 50 psig (low end in landfill gathering systems; high end in refinery or utility distribution headers) and discharge pressures range from 40 to 400 psig (low end, engines; high end, gas turbines). Horsepower requirements are generally less than 800 hp; however, systems requiring up to 1500 hp have been proposed. Control systems employing clearance pockets, cylinder end unloading, and automated bypass are required to control the gas flow required by gas engine or gas turbine generator sets.

With more emphasis placed on overall energy costs, cogeneration is quickly becoming a more important field. Cogeneration is the production of electricity while harnessing the waste heat of the prime mover. This process can provide overall thermal efficiencies of 75 to 80%. Fuel gas boosters are common accessory requirements to provide fuel gas at pressure for the prime mover.

GAS PLANTS

General Information

The term gas plant, as used in the natural gas industry, generally refers to a general class of plants designed to remove the more valuable, heavier components from the less valuable methane in a natural gas stream. These heavier components (ethane, propane, butane, and heavier gasolines) are removed by four basic means, as described next.

Absorption Process

This process uses absorption of the higher molecular weight gases into an oil medium to separate the heavier components from the gas stream (Fig. 7.22). Oil will absorb natural gas components in roughly inverse relationship to those same components in the oil. Thus, if the partial pressure of butane is reduced in an oil stream (which is then called lean oil), it will absorb a proportionately higher amount of butane from the gas stream. This oil can then be heated to drive off the gas components and thereby accomplish the separation.

Conventional Refrigeration Process

This process of separating the gas components from the gas stream uses cooling of the gas stream. The heavier components will condense first, thus allowing separation. Cooling can be done by conventional refrigeration systems in the range from +20 to -30°F, using Freon, butane, or propane refrigerant. A simplified schematic diagram of this system is shown in Fig. 7.23.

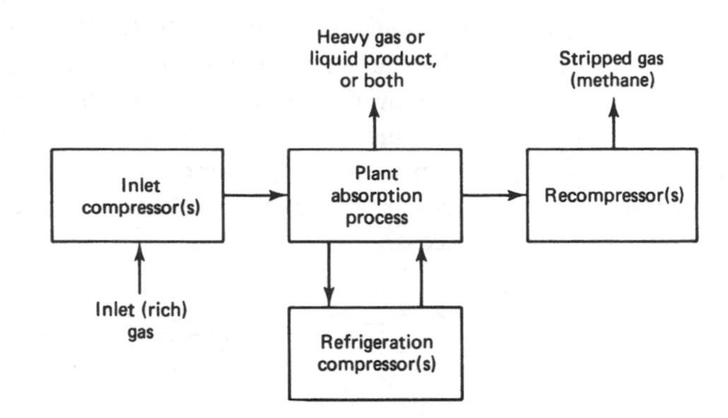

Figure 7.22 Absorbtion Process

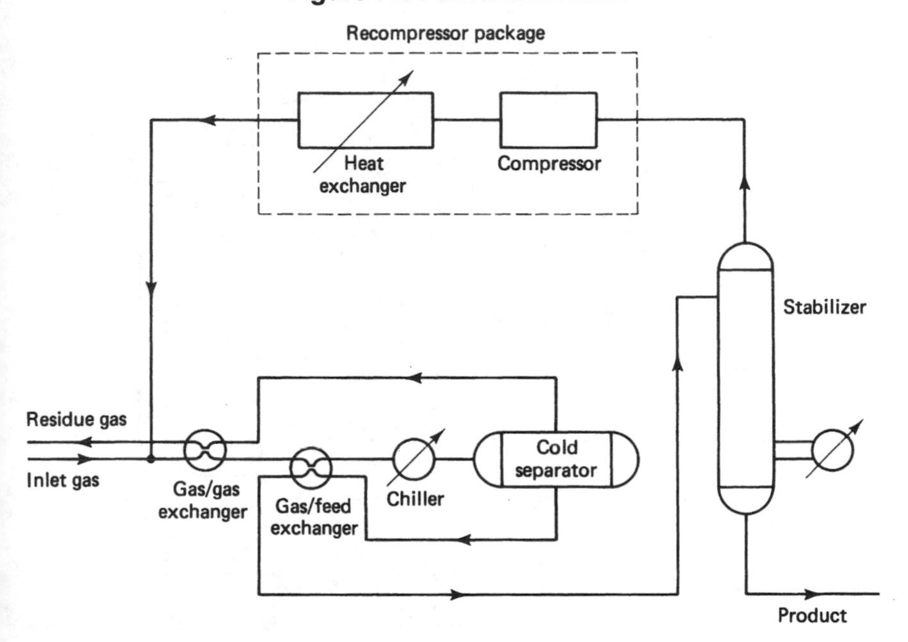

Figure 7.23 Refrigeration System

Expander Refrigeration Process

This process of separation is also a refrigeration process, and it utilizes expansion of the gas stream, as usual, to obtain the refrigeration. Utilizing a turboexpander or an expansion valve, the system achieves temperatures of -100 to -180°F. As ethane became more valuable as feed stock for plastics and chemical plants, it became economically justifiable to achieve the much colder temperatures necessary to separate this component.

The condensing temperature of ethane (C_2H_6) is -130 to 150°F. Expanding the gas stream through a valve or mechanical energy extracting device (reciprocating engine or centrifugal turbine) for recovering energy provides temperatures low enough to recover ethane. Figure 7.24 shows a typical expander plant. The majority of plants built today over 20 million scfd are expander plants. The turboexpander can be used to drive a centrifugal compressor and recover 10 to 12% of the pressure lost through the expansion process.

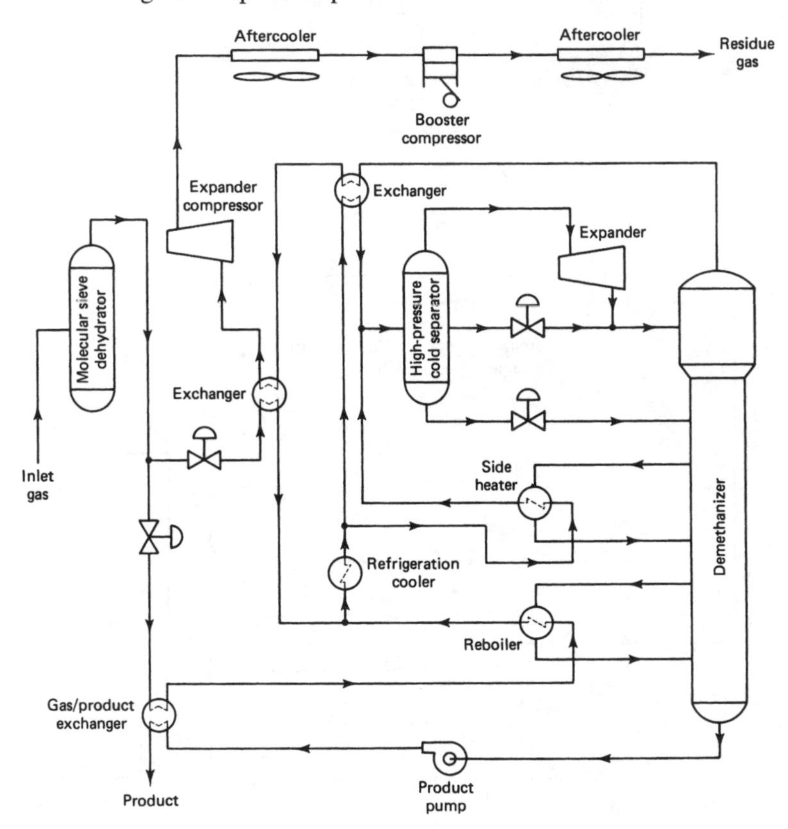

Figure 7.24 Typical Expander Plant

Membrane Separation Process

The membrane separation process is relatively new. Gas separation by membranes is based on the principle that some gases permeate much more rapidly than others due to their solubilities in a given membrane material. Use of membranes, to separate gases should be viewed as a bulk separation process, whereby a gas mixture is split into two enriched streams, each containing a portion of all the original components. Membrane separation is generally considered within the following limitations:

1. Availability of low- to medium-sized feed streams (150 million scfd to 200 million scfd).
2. Moderate concentrations of the more permeable gas in the feed stream (10 to 85 mol percent).
3. Moderate- to high-feed pressure (250 to 2000 psig).
4. Moderate-feed temperature (30 to 150°F).
5. Acceptability of moderate recoveries (85 to 95%) and product purities less than 95%.

Membrane separation can be tailored to meet specific requirements of purity and recovery and optimization of costs between separators and compressors is possible. In general, high purity means modest recovery, and low differential pressure across the membrane requires more membrane surface, and a high differential pressure requires more compression equipment. Either reciprocating or centrifugal compressors may be used depending on the pressure and the flow requirements. A simplified schematic diagram of this system is shown in Fig. 7.25.

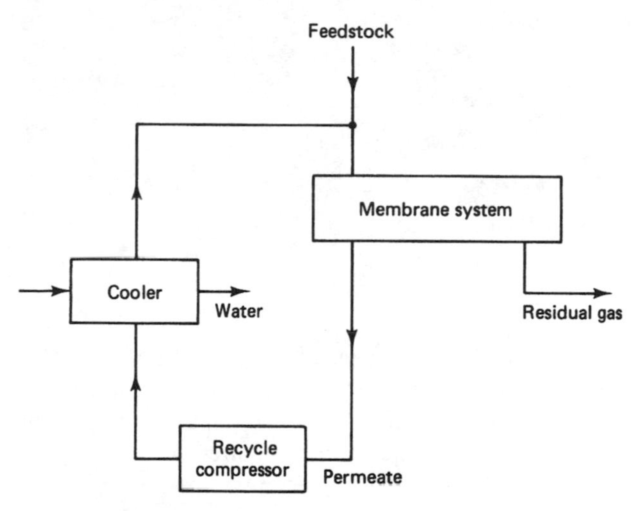

Figure 7.25 Membrane Separation System

With increasing interest in enhanced oil and gas recovery using carbon dioxide and nitrogen flooding of reservoirs, there is escalating interest and need for separation processes to separate methane from carbon dioxide, carbon dioxide from ethane and heavier liquids, and to separate carbon dioxide and hydrogen sulfide. Various companies and individuals have developed several specialized processes to achieve these separations. Several have been patented in recent years. All these processes require compressors with the specific size and types depending on the particular plant size and the process being used.

Mini-plants

Plants have become smaller over the years. Both refrigeration and absorption plants are now compact enough to be delivered on a single truck (Fig. 7.26). Modularizing these mini-plants has allowed manufacturers to make them on a mass-production basis in a factory as opposed to a job site fabrication in the field.

Figure 7.26 Expander Plant

Elements of the plant can be packaged separately and assembled in the field (Fig. 7.27). Reciprocating, or centrifugal compressors, or both, and turboexpanders are commonly used in these mini-plants, with the capacity and pressure conditions varying widely depending on the application requirements.

Figure 7.27 Gas Plant

Compressors in Gas Plants

Compressor requirements for a gas plant, like that in Fig. 7.28, vary widely depending on the type and size of the plant (100 to 1000 million scfd) and the makeup of the gas stream. The two primary elements that determine the type of compression used are as follows:

1. The performance flexibility required.
2. The energy balance of the plant.

These two items usually overshadow first cost, making it a distant third. Larger plants tend to use centrifugal compressors with turbines, either gas or steam, as drivers. Large-capacity and relatively stable gas conditions make the choice of centrifugal compressors practical on the basis of efficiency and installed cost.

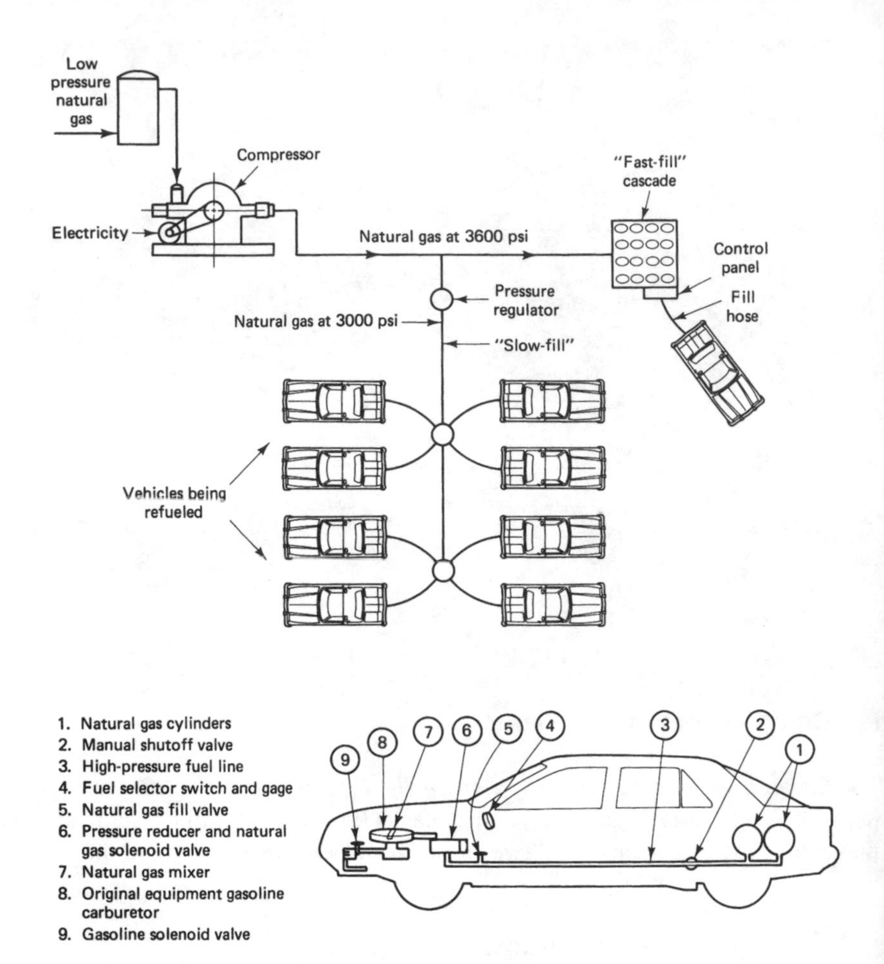

Figure 7.28 Components of a CNG combination refueling system.

As plants get smaller reciprocating compressors become more practical. Internal combustion engines powered with natural gas are the obvious choice for drivers. Recently, environmental pressures are causing the choice of electrical motors to become more prevalent. New developments in electronic controls are giving speed flexibility to electric motors, making them more acceptable for compressor drivers. Likewise, new developments in natural gas combustion technology have resulted in the availability of clean burning gas engines that meet most current environmental requirements, have relatively good fuel economy, and produce acceptable operating records.

Inlet and Recompression Compressors

Treating gas by passing it through a process plant requires a pressure drop. This pressure drop sometimes must he recovered before the gas enters or reenters a pipeline. This pressure loss, typically 50 to 100 psig across a gas plant, is made up by an inlet compressor or a recompressor, or both, depending on the initial inlet pressure of the gas and what is done with the gas after processing.

An expander plant takes a much greater pressure drop, typically 400 to 600 psi. A good rule of thumb for the most economical design is the pressure drop through the expander will be one-half of the final discharge pressure to the pipeline. If the pressure is low going into the plant, an inlet compressor will be required to compress the gas up to about 800 to 1000 psig, which is required to economically operate the expander. A recompressor may be required after the expander to raise the gas pressure up to the pipeline level or for reinjection.

Normally, in an expander plant that discharges to a sales pipeline, a total of 5000 to 6000 horsepower will be required per 100 million scfd of inlet gas. Plants requiring less than 10,000 horsepower will normally use gas engine drivers in lieu of gas turbines unless the waste heat from the turbine can be efficiently used in the process. This is due to the better fuel efficiency of the reciprocating gas engine compared to that of the smaller gas turbines.

As regenerative cycle gas turbines become more proved and accepted, this condition may change. Due to the high horsepower often required, environmental considerations may necessitate gas turbines, electric motors, or clean-burning gas engines with systems to reduce NOX (nitrogen products from combustion), CO, and hydrocarbon pollutants. Regulations vary from state to state, and the laws and regulations covering the installations should be thoroughly researched.

Refrigeration Compressors

Refrigeration compressors are required in absorption plants to drop the lean oil temperature. Smaller mini-plants normally use commercially available Freon systems for this refrigeration. For larger gas plants, propane refrigeration is most frequently used. Small to medium (5 to 100 million scfd) plants typically use a natural-gas, engine-driven reciprocating compressor for this application. Turbines may be economically advisable if the waste heat can be used. Propane refrigeration units require no gas cooling. Scrubbers are usually part of the refrigeration plant and not located on the compressor skid. Many times in the smaller plants the refrigeration is combined on one unit.

De-ethanizer Overhead Compressors

Once the desired components are separated from the methane, either by absorption, refrigeration, or expansion, they may he further divided by condensing them in towers specifically designed for the purpose. Each is designated by the component it is to remove – de-ethanizer for ethane, depropanizer for propane – and so on. In some large plants, the residue gas from these towers may require a small reciprocating compressor to raise the pressure of the gas to move it on through the process or to sales.

The de-ethanizer overhead gas on any type of plant may require compression. This gas will typically be at 40 to 60°F and 300 to 500 psig. Many times this is chilled in the process and pumped as a liquid product.

Still Overhead Compressors

This phase of refining is most commonly done with the still overhead in an absorption plant. When the inlet gas stream has contacted the lean oil, it becomes rich with the heavy components it has absorbed. This rich oil is then heated, and the valuable heavy ends are driven off in the still. This residue is then cooled, causing further liquification. The liquid product is handled by pumps, and is sold. However, some gas will remain and must be compressed to be further processed or sold. This service typically requires 300 to 500 horsepower per 100 million scfd of inlet gas transferred to the plant and is generally a reciprocating compressor application. Pressure ratios are small, normally 10 to 50 psi pressure rise, at process pressures of 400 to 800 psig.

Vapor Recovery

Hydrocarbon liquids of all types are stored in specially designed storage tanks and transported in tank cars. During storage or transit, the liquids are constantly generating hydrocarbon vapors. It is necessary to have a vapor-recovery compression system to collect the hydrocarbon vapors that have boiled off the hydrocarbon liquids, compress these vapors to a design pressure, and then cool the vapors. During the cooling process, liquids condense, and are collected and pumped back into a storage tank or pipeline.

Storage Tanks

During the storage or handling of hydrocarbon liquids with a low vapor pressure, there is a constant need to maintain the pressure in the tank at a safe value. A vapor-recovery system is designed to prevent leakage of vapors into the atmosphere and to prevent pressure buildup that could cause structural damage to a storage tank. Care must also be taken to maintain a slight positive pressure (normally in inches of water) in the tank to prevent inward collapse of the tank or entry of air into the tank.

Tanks used for storing volatile hydrocarbon liquids are connected together through a manifold, and a vapor-recovery system is used to maintain the tank battery vapor pressure at the design pressure. These applications require compressor systems that can compress vapors with entrained liquids. Typical compressors used are the sliding-vane, liquid-ring, and reciprocating types.

These compressor systems, which average less than 25 hp, usually have very low suction pressures and can operate at a slight vacuum. The service is normally not continuous duty. Condensation, which occurs in the compressor during frequent shutdowns, dictates the need for liquid-handling characteristics to be designed into the compressor.

Tank Car Unloading

The requirement for volatile liquid unloading is similar to the storage tank service described previously. If feasible, the vapor-recovery system is used to maintain the design pressure in the tank being charged or discharged. A balance line is connected between the storage tank and the tank car during loading or unloading. The receiving tank is always loaded from the bottom to minimize vapor formation, and the vapor-recovery system takes care of any vapors that do form.

COMPRESSED NATURAL GAS FOR VEHICLE FUEL

History

The use of methane as a vehicular fuel was first introduced in Italy in the 1930s. Other countries subsequently using this fuel were Canada, New Zealand, the United Kingdom, Holland, Iran, and Australia. At this time, however, the ready availability and low price of gasoline have delayed significant developments in the compressed natural gas (CNO) market.

Interest continued to lag, until air pollution concerns prompted the search for cleaner-burning fuels. Some public utilities in the United States converted their vehicles to liquified natural gas (LNG) in order to demonstrate methane's potential as a clean-burning, nonpolluting fuel. During the same period, developmental work on an experimental CNG system was being carried out. The CNG system proved itself to be satisfactory and made possible the conversion of several hundreds of vehicles to methane fuel during the late 1960s and early 1970s. The latter half of the 1970s saw sharp curtailment of vehicle conversion due to publicity about natural gas shortages, whether real or contrived. In the 1980s, because of many factors, natural gas has again become more abundant, and a resurgence in its use as a vehicle fuel has occurred.

Systems

Generally, CNG systems are designed in three basic types: quick fill, slow fill, and combination. Components of the combination system are seen in Fig. 7.28.

1. *Quick fill:* This system utilizes a fairly large reciprocating compressor and a storage volume that can be transferred to the vehicle's storage bottles in approximately the same amount of time required to fuel a gasoline-powered vehicle.
2. *Slow fill:* Using this system, vehicles are charged directly from the compressor and may take several hours to fill. This is an unattended operation and is shut down when the required tank pressure is achieved. Slow-fill systems find their greatest potential in home use where an overnight fill is not objectionable.
3. *Combination:* This system provides a limited quick-fill capability in conjunction with the normal slow-fill operation. Small fleets or families having several vehicles may find the combination system most advantageous.

Compression

Suction pressures may vary from 1 to 50 psig, with discharge pressures generally around 3600 psig. These conditions will probably dictate the use of four stages of compression with intercooling. Power requirements can range from 3 to 150 bhp, depending on the size of the fueling system and the suction pressure available. The compressor must be a positive-displacement type, adequately sized, and properly instrumented to give trouble-free unattended service, as shown in Fig. 7.29.

Figure 7.29 Compressed natural gas for vehicle fuel.

Operation

In operation, gas is taken from either a well or a supply line and compressed to a sufficiently high pressure to fill the storage tanks in the vehicle. The natural gas remains there in a vapor state until needed for fuel. After having been reduced in pressure and mixed properly with air, the gas is admitted to the engine's combustion chamber, where it is ignited. Engine performance drops slightly when the vehicle is tuned to CNG, but due to power reserves of most models, this had not been particularly noticeable.

Potential

As fueling centers become more widespread, the use of compressed natural gas, as shown in Fig. 7.29, is likely to increase. This will be due to its availability well into the twenty-first century, while petroleum liquid fuels will be subject to potential shortages until stability in the Middle East is achieved. Additionally, there are the advantages of reduced emissions, low cost, flexibility, longer engine life, and reduced dependence on imported products.

LANDFILL GAS APPLICATIONS

The generation of methane gas, a by-product of waste decomposition, including the anaerobic decay of organic material in municipal landfills, has been viewed in the past as somewhat of a problem. In some of the larger landfills in the United States, it was necessary to collect the gas to prevent its migration into adjacent commercial or residential properties where it could become an explosion hazard. It then had to be disposed of by incineration which was wasteful. The advent of natural gas shortages and escalating energy prices solved this problem, and today many landfill sites around the country are producing methane gas quite economically for commercial or industrial use.

A typical landfill must digest itself, or decompose, for 5 to 10 years before a commercially attractive quantity of gas is produced. To produce the gas, numerous wells, approximately 6 in. in diameter, are drilled to between 100 and 150 ft. in depth, and a gathering system is installed to collect the gas to a central location for scrubbing, dehydration, compressing, and processing. The expected life of most landfills is between 15 and 20 years.

The gas produced in a landfill can vary, but generally it is composed of about 52% methane and 46% carbon dioxide, with the remainder being nitrogen, oxygen, and miscellaneous other gases. The gas comes from the wells saturated with water at about 120°F and the heat content is around 500 Btu/ft^3.

To make the gas salable or usable, it first must be treated to remove offensive odors, water; and particulates. It can be used as low-energy fuel gas for gas engines, gas turbines, or gas-fired steam boilers and sold to a commercial on-site power generation or cogeneration project, or be further processed by molecular sieves to remove the carbon dioxide and enrich the Btu content to make the gas salable to a pipeline transmission company or utility.

Generally, reciprocating compressors are utilized in the compression of landfill gas; however, rotary positive displacement units can be used in lower discharge pressure applications.

Gas	Typical Pressures
Low Btu	
Suction (psig)	-1 to 0
Discharge (psig)	60 to 100
Enriched Btu	
Suction (psig)	-1 to 0
Discharge (psig)	300 to 600

AIR DRILLING

Rotary drilling with air results in the fastest penetration rate while cleaning the hole and removing cuttings for drilling 2000 to 10,000 ft. oil and gas wells. Unfortunately, hole instability and formation fluids allow air drilling of only about 10% of the footage in the United States, and 10,000 ft. is the normal maximum depth because of potential hole problems.

The earliest significant use of air for rotary drilling of gas wells occurred around 1956 in Pennsylvania, when several drilling companies started experimenting with various techniques and equipment. Simultaneously, other people in the Rockies, Arkansas, and West Texas found conditions adaptable to air drilling. Mining equipment and compressor manufacturers responded rapidly to supply the need for equipment adaptable to the rigors of oil field use.

Until the advent of portable, rotary-table-type rigs in the mid- 1960s, capable of drilling 3000 to 10,000 ft., most drilling was done with jackknife-type skid rigs with sufficient power at the tail shaft to run two small reciprocating air compressors. Normally, 1600 to 2400 cfm is required and pressures of up to 1000 psi were frequently required. A separate engine-driven reciprocating booster compressor was needed for almost all operations.

During the conversion of the drilling industry to diesel power, portable rotary rigs for under 10,000 ft. replaced skid rigs, thus creating a demand for new diesel-driven primary (low-pressure) reciprocating or rotary positive-displacement compressors. Drilling in the western United States and overseas evolved into requiring 2400 to 4000 scfm of air with capability of 1000 to 1500 psig discharge. This equipment is frequently rented by the operator of the well for the drilling contractor and usually is furnished by a rental or service company, complete with personnel. Air is frequently used in only the first 6000 to 10,000 ft. of a deep well, thus requiring more air and higher pressures, since larger hole sizes are required.

In the Appalachians, shallower drilling is prevalent and the wells are, for the most part, completed with air by contractors who own and operate their own air compressors. Hole diameter is generally smaller, and operating techniques are different, resulting in the normal complement of air being 1600 to 2400 scfm, capable of 500 to 600 psig pressure.

Simultaneously, high-pressure, rotary-screw compressors, with 200 to 400 psi discharge pressure have started to fill a niche in two types of applications. For shallower holes, throughout the air drilling areas of the oil fields, hydraulic, top-drive, pulldown-type rigs are becoming competitive. These rigs have on-board, high-pressure, rotary-screw compressors and can be used with an auxiliary compressor for 3000 to 4000 ft. wells. Reciprocating boosters are occasionally required. Where pressures of 1000 to 1500 psig are required, two-stage boosters must be supplied with air at 200 to 300 psig for satisfactory staging.

With the frequent moves required, averaging 5 to 10 days for portable rigs and 30 to 90 days for deep rigs, the package size will continue to dictate the type of equipment used. Road limits and location size almost mandate a unit no larger than 8 ft. wide, 8 ft. 6 in. tall, 40 ft. overall length, and 35,000 lb maximum weight. Operators are seeing increasing use of air, with improved techniques for using soap, foam, and stiff foam, all of which provide increased penetration rates over mud or fluid drilling. Portable and skid mounted rigs are discussed and illustrated in Chapter 1.

Air compressors are being used increasingly in other oil field applications, such as air notching, well servicing, and clean-out.

Positive-Displacement Gas Compressors

TYPES, ARRANGEMENTS, AND DESIGN DETAILS

Positive-displacement gas compressors are mechanical devices designed to meet a wide range of flow requirements, with inlet and discharge pressures ranging from vacuum to thousands of pounds per square inch. The various types of positive-displacement machines are defined in Chapter 2.

Process Compressors

The reciprocating type of compressor used in compressing gases has a distinctively different design from a 100 psig air compressor. To customize the compressor design for variations in pressure levels and capacity, the gas compressor cylinders can be arranged in any of the following configurations:

Type L (Fig. 7.30)
Type Y (Fig. 7.31)
Type W (Fig. 7.32)
Multiple crank throw, horizontal opposed (Fig. 7.33)

For process applications, the multiple-throw, horizontal-opposed construction provides maximum flexibility. A multiple number of stages for one or more than one service can be combined on one machine. Services requiring a large intake volume can be arranged in a multiple number of compressor cylinders. Figure 7.33 shows two first-stage cylinders for a three-stage, 8000 hp carbon dioxide compressor on enhanced recovery service.

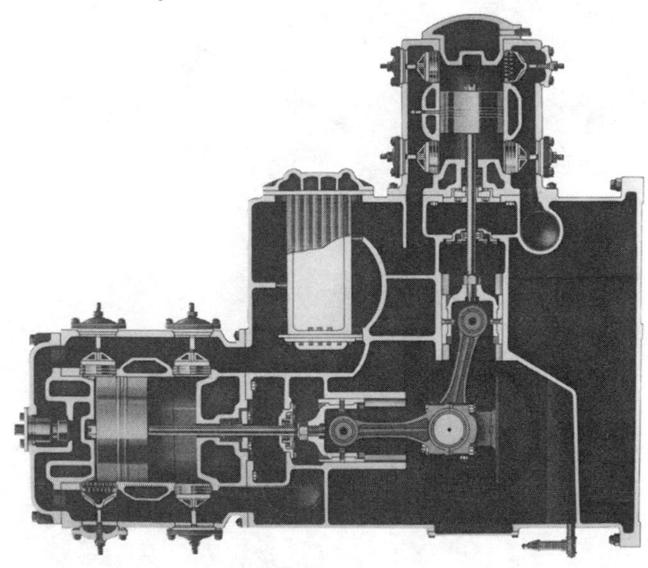

Figure 7.30 Type "L"

Figure 7.31 Type "Y"

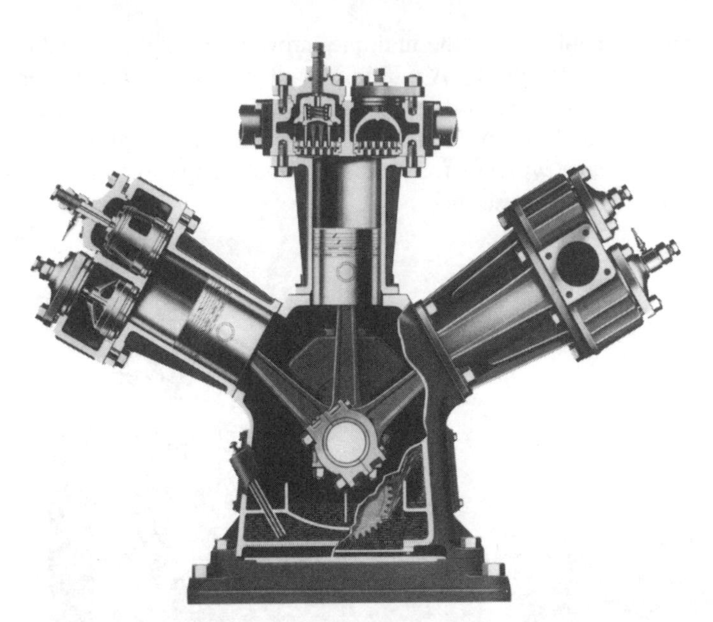

Figure 7.32 Type "W"

Figure 7.33 Three-stage, 6,000 hp balanced opposed carbon dioxide compressor.

The driving systems of large process reciprocating compressors are of various types. The most common is the synchronous motor. Frame arrangements of synchronous motors are as follows:

1. Two-bearing synchronous motor: In this design, the motor manufacturer supplies the motor shaft and bearings within the motor frame.
2. Engine-type, low-speed motor (Fig. 7.34): The engine-type synchronous motor is the most frequent choice for slow-speed, high-horsepower process reciprocating compressors. The motor is furnished without shaft or bearings. The compressor supplier furnishes the motor extension shaft and outboard bearing.

The large diameter and narrow width of the synchronous motor permits mounting of the rotor on the shaft extension. This provides considerable rotating inertia, eliminating the need for, or reducing the size of, a compressor flywheel. The motor stator and outboard bearing are supported on the compressor foundation. The compressor manufacturer should furnish the motor because it is an integral part of the design and engineering of the compressor system.

Induction motor drives can be direct connected or coupled, flange mounted, V-belt, or gear reducer drive. The responsibility to design the drive system to keep current pulsation within the required limits generally is assigned to the compressor manufacturer. If a gear reducer is applied, the drive system must be designed and analyzed for torsion to prevent torque reversals and excessive peak torques at the gear teeth. Negative torque would cause gear teeth separation, resulting in noisy gears and premature wear of the gear teeth.

Figure 7.34 Engine type synchronous slow speed compressor driving a five-stage compressor.

Electric motors are available in many enclosures. The proper enclosure is a function of the environment where the motor will be installed. Motors can be designed with many different enclosures, such as open, drip proof, weather protected I or II, inert gas filled, and forced ventilated.

The compressor manufacturer determines the necessary additional amount of flywheel or rotor weight required. The normal maximum permitted current pulsations as specified by American Institute of Electrical Engineers (AIEE) standards is 66%, but the specific requirements should be specified by the user to suit the particular system requirements.

Torsional vibration analysis is an important design consideration in the selection of couplings and gearing for drive trains of large reciprocating compressors. The torsional reciprocating compressor analysis consists of taking the various masses of the complete shafting system of the installation and evaluating how they turn against each other. Engine and compressor manufacturers have been doing these types of analyses for over 50 years.

The objective of a reciprocating compressor analysis is to determine:

1. The various torsional natural frequencies that exist within the complete shafting system.
2. The frequencies and magnitude of fluctuating torques within the system.
3. The critical speeds of the system, where the frequency of a torque fluctuation coincides with a susceptible natural frequency (e.g., resonance) where torsional failures could occur.
4. That the critical speeds are moved outside the design operating speed range. (The system is tuned by varying the size of the masses [e.g., fly wheels, counterweights, etc.] or the rings [selection of couplings, V-belt sizes, etc.])

The following paragraphs are general guidelines for the torsional vibration analysis of complex or new combinations of reciprocating compressors or for equipment operating over a relatively broad speed range.

The natural frequencies of all modes of vibrations up to 20% above the highest known exciting frequency should be determined. Guides for exciting frequencies are as follows:

Engines and reciprocating compressors: twelve times basic speed.
Motors and generators: two times basic speed.
Turbines and other rotating equipment: two times basic speed.

There are other potential exciting frequencies (e.g., the number of gear teeth on each gear times the gear speed, or the number of turbine blades times the turbine speed); however, these frequencies are generally so high that it should be safe to ignore them.

There are exciting frequencies at which the magnitude of the exciting torques is not known or not readily determined; therefore, resonance with these frequencies should be avoided. At resonance with a responsive mode of vibration, the effect of a small force can be magnified to dangerous limits. Therefore, lacking a more complete analysis, natural frequencies should be avoided that fall within the range of 10% above the maximum operating speed to 10% below the minimum operating speed of the following equipment:

Reciprocating equipment.
Lobed or vaned rotors (up to eight lobes or vanes) – the number of lobes or vanes times the rotative speed.

A deeper investigation may reveal that the natural frequency in question may not be responsive.

Natural frequencies for the lower modes of vibrations should not fall within 10% of each other. This is to ensure that two modes of vibration do not couple and mutually excite each other.

Except in the case of synchronous-motor-driven equipment, there is no such thing as a constant-speed compressor. Whenever the speed of the driver can be varied (i.e., the speed of the engine or turbine), then the permissible speed limits for the installation must be clearly defined and the torsional analysis must ensure that the installation is torsionally safe within these limits. For compressors considered as constant-speed units for which no fixed speed limit can or will be defined, then the analysis should ensure that the installation is torsionally safe within the 10% range of the rotative speed.

The degree of analysis required from the torsional vibration aspect will vary with any project. The following factors will increase the complexity of the analysis:

1. Broadening the permissible operating speed range.
2. Broadening the differences in the rotative speed between driven and driving equipment.
3. Increasing the sources of excitation (i.e., reciprocating equipment driving reciprocating equipment).

Oil and Gas Field Compressors

Oil and gas field applications require compressor systems that are compact and can be easily moved from one location to another. These machines are similar in configuration to the process machines discussed previously. The normal drives for these compressors are coupled gas engines or electric motors. These reciprocating units are called "separables" in the oil and gas industry. For compactness, such units feature a short stroke while operating at higher rotative speeds. The piston speed is maintained typically from 750 to 1200 fpm (Fig. 7.35). They are usually assembled into a readily portable package consisting of mounting skid, driver, cooling equipment, piping and controls (Fig. 7.36).

Figure 7.35 Two stage, two throw horizontal opposed, 200 hp (reciprocating gas booster compressor.

Figure 7.36 Two stage, 13 hp portable package compressing 60,000 cfd of flash gas from oil well separator into 400 psig gas gathering system.

Integral Gas Engine Compressor

Another type of reciprocating compressor is the integral gas engine compressor. The typical configuration consists of a crankcase with a horizontal crankshaft. The power cylinders deliver power through combustion of natural gas. These cylinders may be mounted vertically in-line, in a V-type vertical configuration, or horizontally, delivering the power through connecting rods to the crankshaft. Compressor cylinders are horizontally mounted to the same crankcase and crankshaft as are the power cylinders.

The integral gas engine compressor is built in sizes from 25 to 12000 horsepower. Primary oil and gas field applications are natural gas storage plants, natural gasoline, liquefied petroleum gas plants, and natural gas transmission pipe-lines.

There are two basic types of gas engines, the two-stroke cycle and four-stroke cycle, and either type can be turbocharged. The two-cycle engines require less displacement for the same rating. The differences in performance between the two-cycle and four-cycle engines are small, especially in the turbocharged models.

Rotary Positive-Displacement Gas Compressors

Sliding-Vane

The rotary sliding-vane compressor is primarily a 100 psig air compressor. Chapter 2 has further information on this device and its applications. It has been utilized quite successfully, however, on natural gas applications at lower pressures. The gas must not contain components that would deposit on the rotor, especially any sticky material such as tar. Such impurities would prevent free movement of the vanes in and out of the rotor.

Helical-screw-type Gas Compressors

The largest use of this compressor in handling gas is for industrial refrigeration. This chapter has discussions of both the reciprocating compressor and helical-screw compressor, including applications to refrigeration. Efficiency must be a consideration and must be evaluated. The helical-screw machine is different from other positive-displacement compressors since its efficiency curve is the island type. The maximum efficiency is at the designed fixed compression ratio. If, for example, the built-in compression ratio were 3:1, the adiabatic efficiency would have to be 82%.

At compression ratios of 3.6:1, 4.0:1, and 4.8:1, the approximate compression efficiencies would be 79.5%, 78.5%, and 76%, respectively. The adiabatic efficiencies of the reciprocating compressors at the same ratios are approximately 87%, 88% and 89.5%, respectively. This is not a complete evaluation between the two compressor designs, nor would the comparison be of much value for a service that operated continuously at the same compression ratio.

To compress gases other than air, certain adaptations must be made in the design. Light gases such as ammonia, helium, and hydrogen require significantly higher rotor tip speeds or oil-flooded designs. By injecting oil between the rotors, leakage of gas back to the inlet side of the compressor is reduced. A secondary advantage is that the oil removes a great deal of heat of compression. The removal of heat extends the allowable compression ratio per stage. The heat absorbed by the oil is removed in an external heat exchanger. Liquids other than oil may be used for injection in some applications.

Some designs of helical-screw compressors provide a means of varying the capacity. The device consists of a sliding control valve that changes the point at which the rotor length begins compression. This provides a stepless variation in capacity downward to about 20%.

Reciprocating Compressors: Type and Construction

Reciprocating compressors are built in a multitude of horsepower and speed ranges. The following covers the various types and construction details.

A. Frames

1. Single-throw

The single-throw is available in both the horizontal and vertical arrangements and is generally applied up to about 300 horsepower, with an upper limit of about 500 horsepower. Single cylinders or multiple tandem cylinders can be used with the single crank, which is normallybelt driven, but may be direct through reducing gears driven by slow-speed motors or by high-speed turbines or motors. The single-crank design yields unbalanced inertia forces that must be absorbed by the skid and foundation (Fig. 7.37).

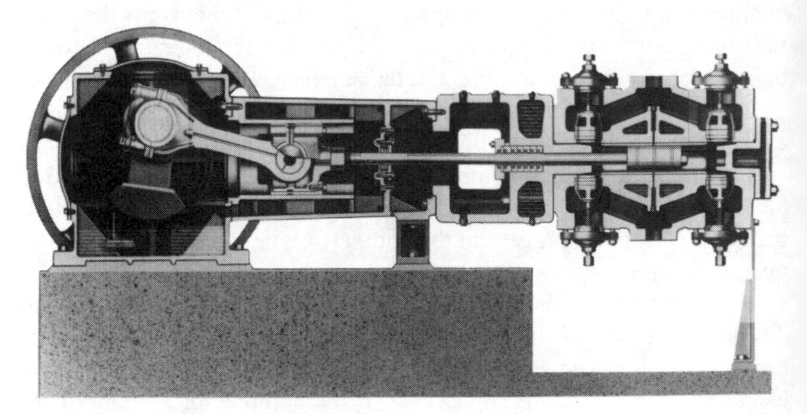

Figure 7.37 Horizontal, Single-throw Reciprocating Gas Compressor

2. Multi-throw, Horizontal, Balanced-Opposed Frame

As discussed earlier, this design is extremely flexible, lending itself to multiple cylinders arranged opposite each other on either side of the frame. For process services, several stages or even different services can be placed on the same frame. When two cylinders with equal reciprocating weights are located on opposite sides of a frame and are powered by a double-throw crankshaft with cranks set at 180°, all primary and secondary inertia forces developed mutually cancel each other, as may be seen in Fig. 7.38. Only couples are transmitted to the foundation. As many as five pairs of crank throws can be arranged on one compressor frame.

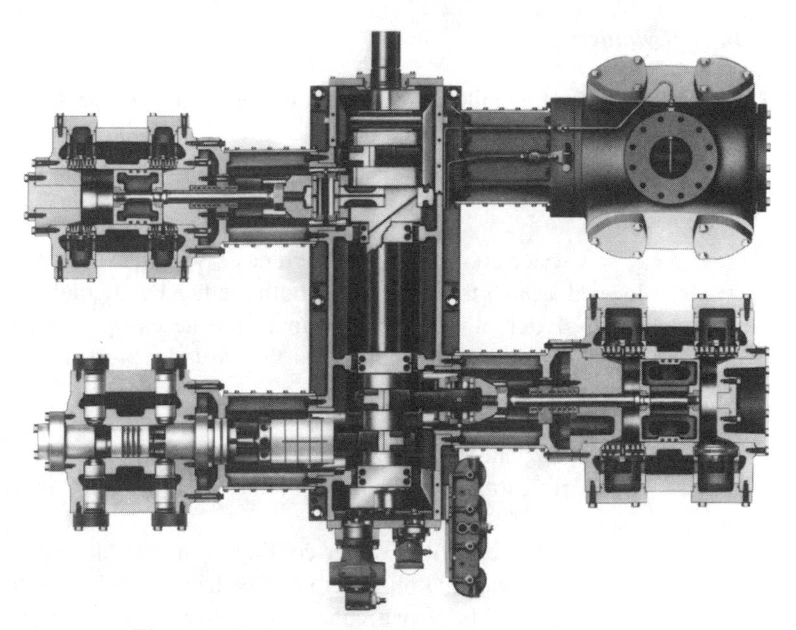

Figure 7.38 Balanced Opposed Compressor Design

Unbalanced forces cause mechanical vibrations that can result in alignment, piping, and vibration problems. These unbalanced forces areproduced by two masses, rotating and reciprocating. Rotating masses consist of the crankpin, crankpin web, and approximately two-thirds of the mass of the connecting rod. The centrifugal force produced by these masses is an exciting force and has the same intensity throughout the revolution.

Reciprocating masses consist of the piston, piston rod, cross-bead,crosshead pin, and the remaining one-third of the connecting rod mass. Acceleration and deceleration of the reciprocating masses cause reciprocating forces, which act along the axis of the cylinder and exert a variable force on the crankpin. The forces resulting from the rotating and reciprocating masses are generally resolved into force systems consisting of two parts, primary and secondary forces. These are expressed in both the horizontal and vertical direction. In the case of multicrank compressors, there are also moments or couples.

The more compact oil and gas field compressors utilize the principles just described, the principal difference being that they have shorter strokes and higher rotative speeds than is the case with process compressors. To reduce inertia forces, the stroke of the piston is in the range of 3 to 8 in., compared to process electric-motor-driven compressors, which have strokes of 9 to 18 in. The advantages of the higher-speed unit are its compactness, portability, and driver flexibility.

B. *Cylinders*

Two basic types of cylinders are used on reciprocating compressors in industry, water cooled and noncooled.

1. Water-cooled Cylinders

A water-cooled cylinder includes a waterjacket that normally is located around the bore and in both the head and crank-end cylinder heads. External passages eliminate the necessity of water passing through gasketed internal joints. Water, under pressure, is circulated through these jacketed areas:

a. To reduce thermal stresses.
b. To maintain a relatively cool cylinder bore.
c. To carry away a small amount of the heat of compression.

Another benefit of water cooling is the maintenance of a high-viscosity lubricating oil film to reduce friction and wear in the cylinder bore. This is of particular importance in corrosive gas services with large-diameter, long-stroke cylinders. The jacketcooling water should always be 10 to 20°F above the suction gas temperature to avoid condensation within the cylinder bore.

A thermosyphon cooling system for a water-cooled cylinder (Fig. 7.39) may be supplied where the gas discharge temperatures are between 190 and 210°F or the adiabatic gas temperature rise in compression is less than 150°F. Static cooling of water-cooled cylinders may be supplied when the gas discharge temperatures are less than 190°F or the adiabatic gas temperature rise in compression is less than 150°F. A mixture of equal parts glycol and water is usually employed except in extreme cold-climate outdoor applications.

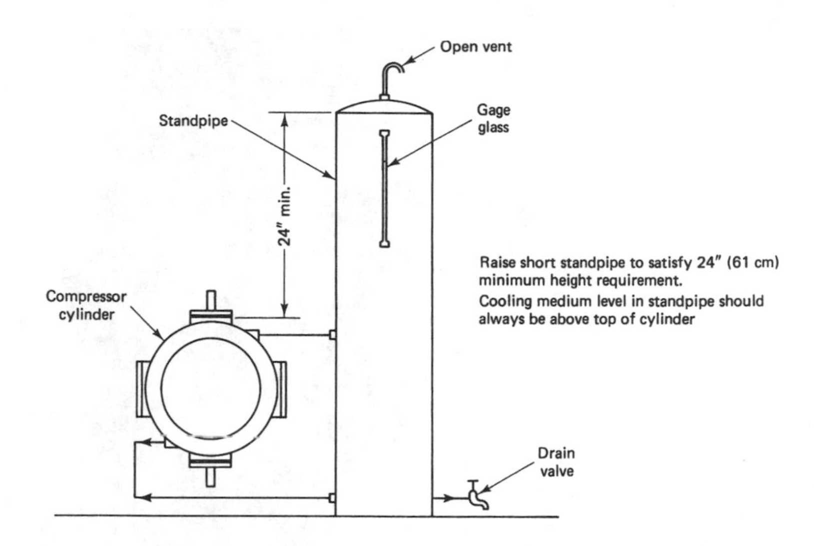

Figure 7.39 Thermo-syphon system with standpipe. Note: for use with gases having a *k* value (C_p/C_v) of 1.26 and below; with cylinder discharge temperature (adiabatic) over 210°F, up to and including 230°F.

2. Non-cooled Cylinders

Non-cooled compressor cylinders are used more frequently in a variety of oil and gas field applications. Perhaps the best known application is pipeline transmission, for which the compressor cylinders operate with low compression ratios and, consequently, experience only a small temperature rise, with discharge temperatures in the range of 140 to 180°F.

A noncooled cylinder (Fig. 7.40) is identical in basic design to a water-cooled cylinder except that there are no jacket-water cooling passages.

Most compressor units today are offered as packaged systems with gas engine prime movers (Fig. 7.86). Most applications for non-cooled cylinders occur where the gas inlet temperatures are 120°F and below, and the only means of cooling is to use gas engine jacket water at a temperature of 160 to 180°F. There is reasonable doubt as to the quantity of the heat rejected.

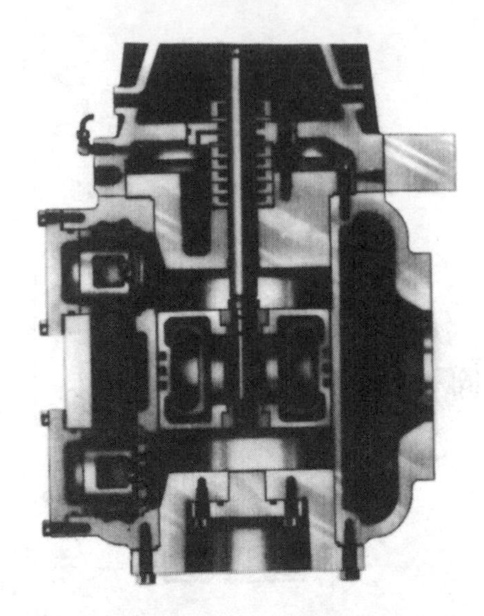

Figure 7.40 Non-cooled Cylinder

3. Compressor Cylinder Designs

Three cylinder designs are used in the process and gas industry. Horizontal cylinders with top intake and bottom discharge are preferred for saturated gases. Most cylinders can be fitted with liners or alternately must be suitable for up to 0.125-in. diameter of reboring without affecting the maximum allowable working pressure.The maximum allowable working pressure shall exceed the rated discharge pressure by at least 10% or 25 psi, whichever is greater.

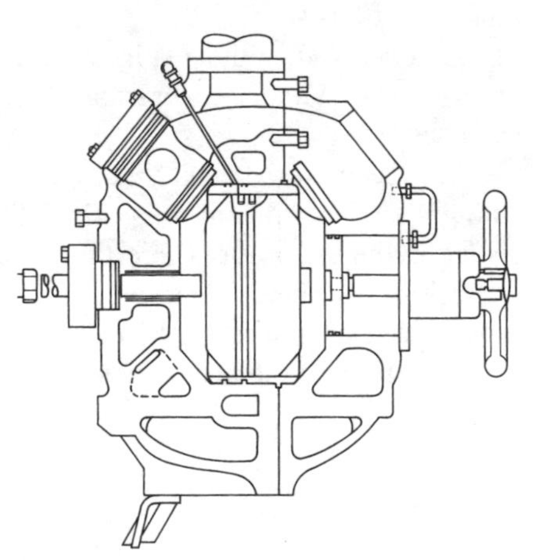

Figure 7.41 Two-piece clam-shell (gas field) low pressure cylinder.

4. Compressor Cylinders

a. Low Pressure (Gas Field)

Two-piece, clam-shell-type cylinders, with liner and with valves in both heads, and three-piece cylinders with valves in the barrel are used for many applications (Figs. 7.41 and 7.42). The piston is of two-piece construction to assure accurate wall thickness and to eliminate core plugs, which might work loose during compressor operation. The piston is anodized aluminum impregnated with TFE particles to provide resistance to ring groove wear or bore scuffing.

TFE-filled pistons and packing rings should be used where the gas contains entrained liquids and traces of H_2S. The packing is vented, fill-floating, segmental-ring type with each cup containing one radial and one tangential ring, except for the last cup, which has double tangential rings to seal in both directions (Fig. 7.49).

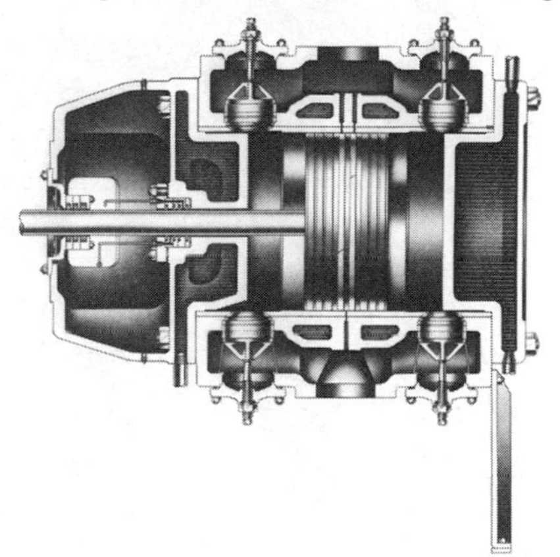

Figure 7.42 Three-piece low to medium pressure cylinder (gas field and process).

b. Low to Medium Pressure (Gas Field and Process)

The cylinder materials for a three-piece cylinder designed for medium pressure can be either gray iron, nodular iron, cast steel, or fabricated steel, depending on bore size and pressure. A general guide is as follows; it should not he used in place of regular design procedures:

	Maximum Allowable Working Pressure psig
Gray iron	1600
Nodular iron	2600
Cast steel	2600
Forged Steel	Over 2600

Pressure-containing castings of ASTM A278 for gray iron, ASTM A395 for nodular iron, and ASTM A216 for cast steel are suggested. Ring materials are similar to low-pressure designs. Pistons may be aluminum, gray or nodular iron, or steel, depending on weight and differential pressure.

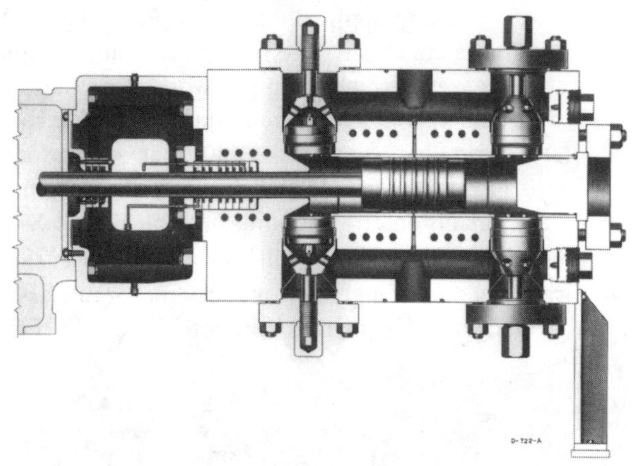

Figure 7.43 High Pressure, Forged Steel Cylinder

c. High Pressure

Forged-steel cylinders are required for pressures above approximately 2600 psig (Fig. 7.43). In Fig. 7.43, the cylinder and packing box are water cooled.

5. Compressor Cylinder Arrangements

a. Single Acting

In a single-acting compressor, gas compression takes place on only one end of the cylinder. The other end is vented to a lower pressure. Care should be taken to ensure that a crosshead pin reversal is achieved to provide proper crosshead pin bearing lubrication and long life. The load on the crosshead pin must reverse direction relative to the pin during each full crankshaft rotation.

b. Double Acting

In a double-acting compressor, gas compression takes place on both the crank end and head end of the cylinder.

c. Divided Cylinder

The outer end (head end) is the low-pressure stage of compression, and the frame end (crank end) of the cylinder provides the high-pressure stage of compression. A forged-steel cylinder may be used with separate suction and discharge passages machined into each end of the billet.

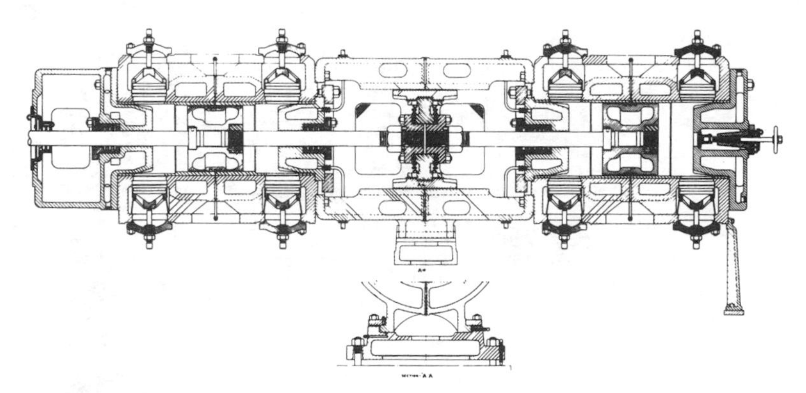

Figure 7.44 Double-acting, tandem cylinders, two-stage with high pressure cylinder adjacent to frame.

d. Tandem Cylinders

Two or more cylinders are arranged on a common piston rod for a single compressor throw. Cylinders can be double-acting (Fig. 7.44) or single-acting, depending on frame load and crosshead pin reversal design criteria (Fig. 7.45).

e. Steeple

The steeple type is defined as two or more stages utilizing a common step-type or trunk-type piston. Normally, the low-pressure stage is double-acting, while the subsequent stages are single-acting. This arrangement will permit up to five stages on a single compressor throw.

Figure 7.45 Non-cooled tandem cylinders, two-stage, single-acting, low-pressure stage, adjacent to frame.

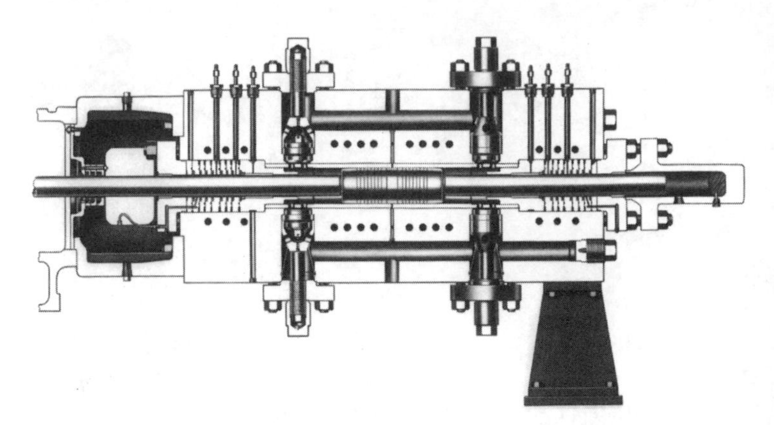

Figure 7.46 High-pressure double-acting tailrod cylinder. The forged steel cylinder above uses a tailrod with forged steel guard for higher pressures.

f. *Double-acting Tail Rod*

Both cylinder ends compress gas and the piston rod is extended into the outer end of the cylinder, thus providing equal flow to both ends. This arrangement is utilized where a crosshead pin reversal problem might exist or as a means to reduce frame load on a conventional double-acting cylinder arrangement (Fig. 7.46).

g. *High-pressure Plunger*

High-pressure plunger cylinders are designed for pressures up to 60,000 psig and are primarily used in the petrochemical industry. The forged-steel barrel is made of aircraft-quality forged steel and the plungers are usually nitrided steel or tungsten carbide materials (Fig. 7.47). The plunger reciprocates through a cooled packing; piston rings are not used.

h. *Non-lubricated Cylinders*

The elimination of hydrocarbon lubricants in compressor cylinders has long been a requirement in many process applications where contamination or possible combustion is not tolerable. The strongest example would be in compressing oxygen. The oil and oxygen would ignite. In the case of oxygen, no hydrocarbon compound can be allowed to come in contact with the gas. In other cases, it may be that lubricating oil will discolor the final product or contaminate the process. With the advent of self-lubricating materials such as TFE for piston rings, piston wear bands, packings, and special valves, the applications of non-lubricated compressors has increased and become widespread.

Compressor cylinder designs may be defined as:

(1) Fully lubricated
(2) Minimum lubricated, or mini-lube
(3) Non-lubricated

(1) A fully lubricated cylinder is supplied with one or more points of lubrication in the bore, where oil is introduced to provide a protective film. This oil is injected under pressure from an auxiliary pump.

(2) A mini-lube cylinder has no injection of oil to the cylinder bore. The only lubrication is that which is carried from the lubricated pressure packing.

(3) Non-lubricated cylinders operate truly without the benefit of any lubricant. They depend solely on non-metallic piston rings, riders, and special valves. Extra distance pieces or compartments may be provided, as required, along with oil slingers to preclude absolutely any possibility of oil traveling along the piston rod into the cylinder. Slightly higher power requirements may be expected with non-lubricated compressors.

i. Labyrinth-type Cylinders

A labyrinth-type piston design can be used for non-lubricated, dry gas compressor service. The compressor cylinder or cylinders are located vertically above the crankshaft. The piston uses TFE-filled guide rings and utilizes a captive-ring design principle. The captive ring principle restrains the piston ring within the piston so that the resulting radial force from the pressure difference over the piston ring is no longer able to press the ring against the cylinder wall, thus eliminating the main cause of wear. The clearance between the cylinder wall and the piston ring is approximately 1 to 2 mils, thus reducing gas leakage between the high- and low-pressure side during compression.

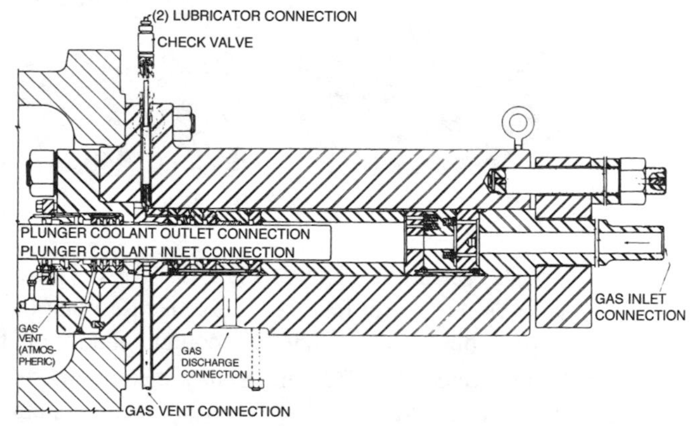

Figure 7.47 High Pressure Single-acting Plunger Cylinder

The piston tail rod runs in a TFE-filled bushing above the piston, which, assisted by the captive rings, ensures an extremely stable piston. (Fig. 7.51).

6. Cylinder Components

a. Valves

Valve designs utilizing moving, flexing, metallic strips or channels are employed on low-pressure, dry, noncorrosive applications. Circular-plate-type valves are most often used for high-pressure and corrosive services.

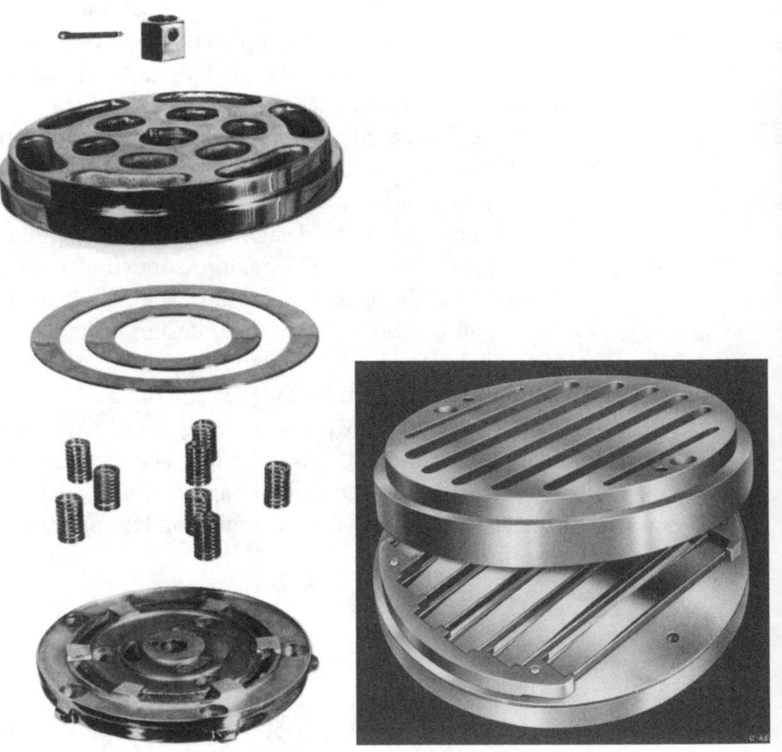

Figure 7.48 Valves

The selection of valve materials for operation in various gases depends on several factors. The series designation for stainless-steel strips or plates depends on H_2S content and lubrication requirements. Nonmetallic valve plate materials are sometimes used for these difficult services. Valve spring materials include stainless steels, cadmium-plated steel alloys, Inconel, and Hastelloy.

Valve seats and guards (Fig. 7.48) are gray or nodular iron for low-pressure applications and steel for high-differential-pressure service.

b. Piston Rod Packing

Most reciprocating compressors are double acting. To seal the frame end of the cylinder where the piston rod travels through the cylinder head, piston rod packing is used.

Piston rod packing is furnished in a great many materials, depending on the service and application. Typical packing ring materials are non-metallic, such as filled TFE and bronze; packing cases are made of iron, steel or bronze. Typical packing ring configurations include tangential and radial sealing rings; the number of sets of packing rings (Fig. 7.49), is a function of cylinder pressure.

For critical high-pressure service such as hydrogen or non-lubricated applications, the packing rings may he cooled. Water or oil is pumped through the packing case jackets, thus cooling the packing rings to ensure longer packing life.

c. Piston Design

The piston design for reciprocating compressors is critical to the operation of the compressor. There are many designs used, depending on application and service. Pistons are made of gray or nodular iron, aluminum, steel, and fabricated steel. One-, two-, and three-piece piston designs (Fig. 7.50) are utilized depending on application criteria.

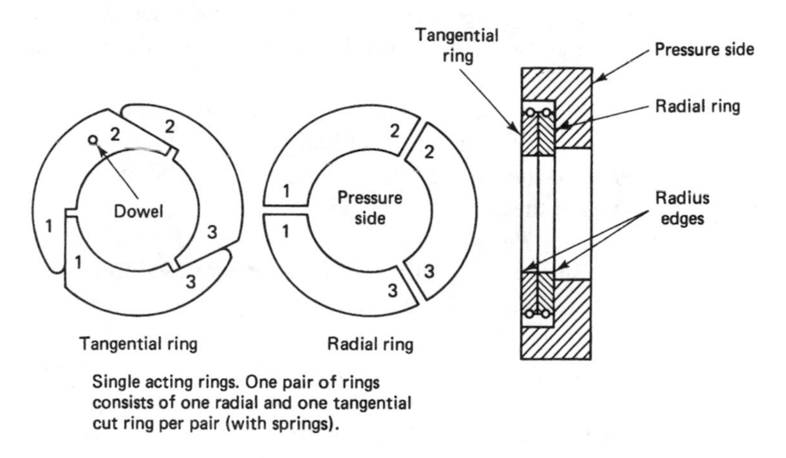

Figure 7.49 Piston Rod Packing

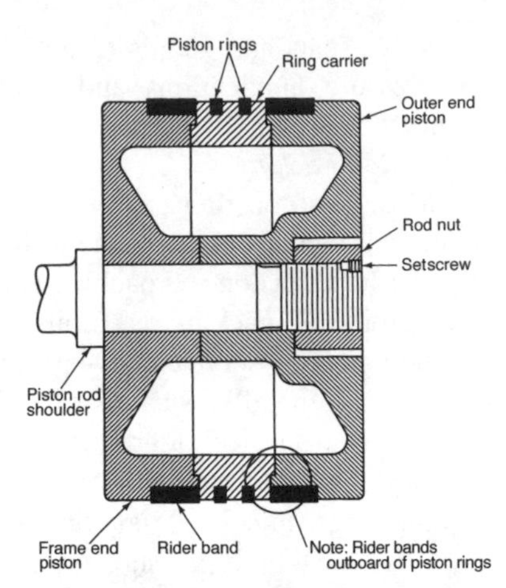

Figure 7.50 Typical Three-piece Piston Design

For applications using large-diameter pistons or critical services where lubrication may be difficult, or for non-lubricated services, a rider or bearing surface is frequently provided to assist in break-in or to assure long life. The rider band is usually a filled Teflon-type material.

d. Piston Rods

A surface hardness of RC 50 minimum is preferred on piston rods in the area that passes through the packing. Special alloys and rod coatings are often applied for sour gas service. Threads may be either cut or rolled, depending on root stress design load criteria.

e. Piston Rings

Many types of piston rings are used, depending upon cylinder pressure, temperature, and bore size. Piston and rider rings are generally made of filled TFE materials, metallic materials, or combinations of these.

f. Distance Pieces

Since reciprocating compressors are used to handle many gases that are flammable or toxic, many types of distance pieces are used to ensure that these gases do not get into the compressor frame and that lubricating oil from the compressor frame is not mixed with cylinder lubrication. In addition, distance pieces are usually provided between the frame and the cylinder to allow maximum accessibility to the piston rod packing and the frame-end wiper packing.

Distance pieces with open-end compartments allow non-hazardous gas to be vented to the atmosphere. If the gas being compressed is hazardous, the distance piece has gas tight covers and is frequently vented or purged with inert gas to prevent contamination of the atmosphere and the crankcase oil by the compressed gas.

Water-cooled or oil-cooled packing is sometimes required in high-pressure service. A variety of distance pieces and configurations is available so that the user may choose exactly the right cylinder-distance system for this application. The following configurations are available (Fig. 7.51):

Type A': Short, close-coupled, single compartment for lubricated service when oil carryover is not objectionable and where it is desired to keep overall width to a minimum for ease of transportation.

Type A: Standard, short, single-compartment, for lubricated service where oil carryover is not objectionable.

Type B: Long, single compartment, for non-lubricated and lubricated service. To prevent crankcase oil from entering the cylinder. No part of the piston rod shall alternately enter the crankcase and the cylinder pressure packing. Normally, an oil slinger is attached to the rod to ensure that oil does not travel along the rod.

Type C: Long, two-compartment distance piece for flammable, hazardous, or toxic gases. No part of the piston rod shall alternately enter the wiper packing, intermediate seal packing, and the cylinder pressure packing. The compartment next to the cylinder can be purged with an inert gas to prevent the hazardous gas from entering the crankcase. Normally, an oil slinger is attached to the rod to ensure oil does not travel along the rod. Slingers may be used in either or both compartments.

Type D: Short, two-compartment distance piece for flammable, hazardous, or toxic gases. No part of the piston rod shall alternately enter the wiper packing and the intermediate seal packing. The compartment next to the cylinder can be purged.

7. Capacity Control

Several different control devices are available to vary compressor flow rate to match system demands.

a. Clearance Pockets

Adding clearance volume to a cylinder end changes the suction valve opening as it is related to the position of the crankshaft. Clearance pockets can be used to reduce compressor flow and horse power or to permit operation at alternate pressure and flow conditions. They are generally added to the head end of the cylinder, as seen in Fig. 7.52.

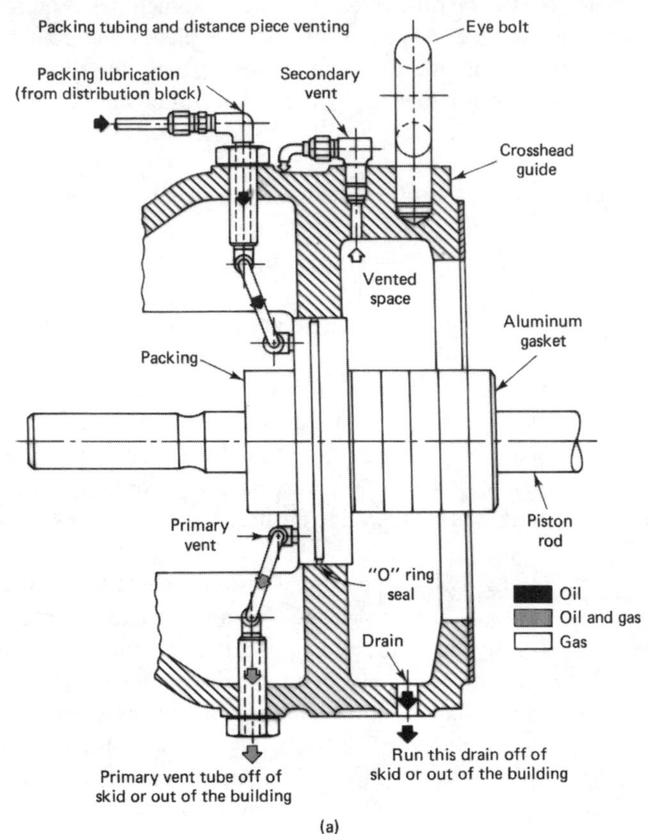

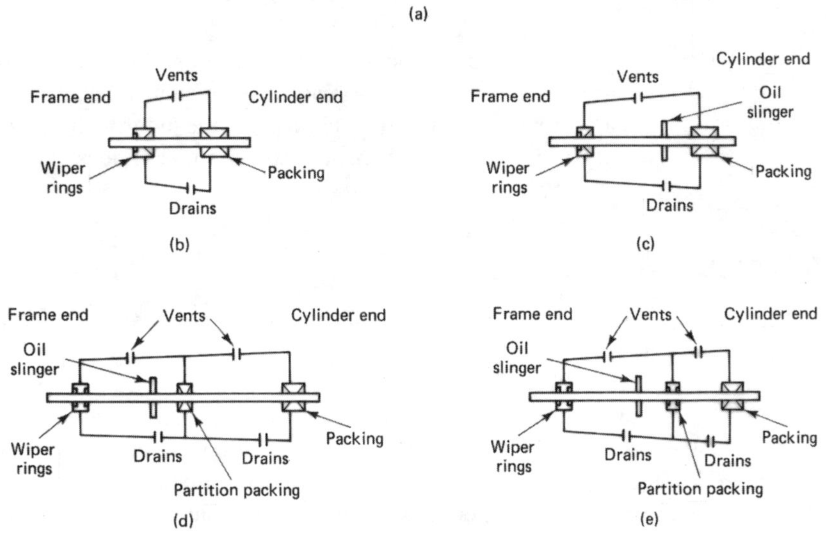

Figure 7.51 Several distance piece configurations.

(1) Fixed-volume clearance pocket: reduces compressor capacity a definite amount or a single step. Manual or pneumatic actuation is available.

(2) Variable-volume clearance pocket: provides capacity reduction in an infinite number of steps over a given reduction range. It is normally for manual actuation only; however, automatic operation may be provided but is generally complicated and expensive.

b. *Split-valve Yokes*

Clearance can be added to a cylinder by elevating the compressor valve from its cylinder seat. This requires a two-piece (split) valve yoke that permits the valve either to sit on the cylinder seat (minimum clearance) or to be sandwiched between the parts of the two-piece yoke (added clearance). The compressor must be shut down to make these physical changes (Fig. 7.53).

c. *Clearance Rings*

A clearance ring, equivalent to a thick gasket, can be inserted between the cylinder barrel and head to add clearance volume. These modifications are generally limited to the outboard cylinder head, as inboard modifications affect the piping. The compressor must be shut down to make this physical change.

d. *End Unloading*

For start-up and capacity control, unloading devices or valve removal can be utilized to vent a cylinder end continually to suction pressure. In general, end unloading is not used for inlet pressure over 2500 psi. If one cylinder end is unloaded, the flow and horsepower are reduced by approximately 50%.

(1) *Depressors* (Fig. 7.54): End unloading can be accomplished by depressing (holding open) suction valve sealing elements that vent the cylinder end to suction pressure. Unloading devices are required over all suction valves and are available for manual or pneumatic actuation.

Figure 7.51 Several distance piece configurations

a. Close coupled single-compartment distance piece.
b. Standard, short single-compartment distance piece. (May be integral with crosshead guide or distance piece).
c. Long, single-compartment distance piece.
d. Long, two-compartment distance piece of sufficient length for oil slinger if required. Partition pecking may be lubricated in each compartment.
e. Short, two-compartment distance piece with partition packing and oil slinger if required. Partition packing may be lubricated in frame-end compartment.

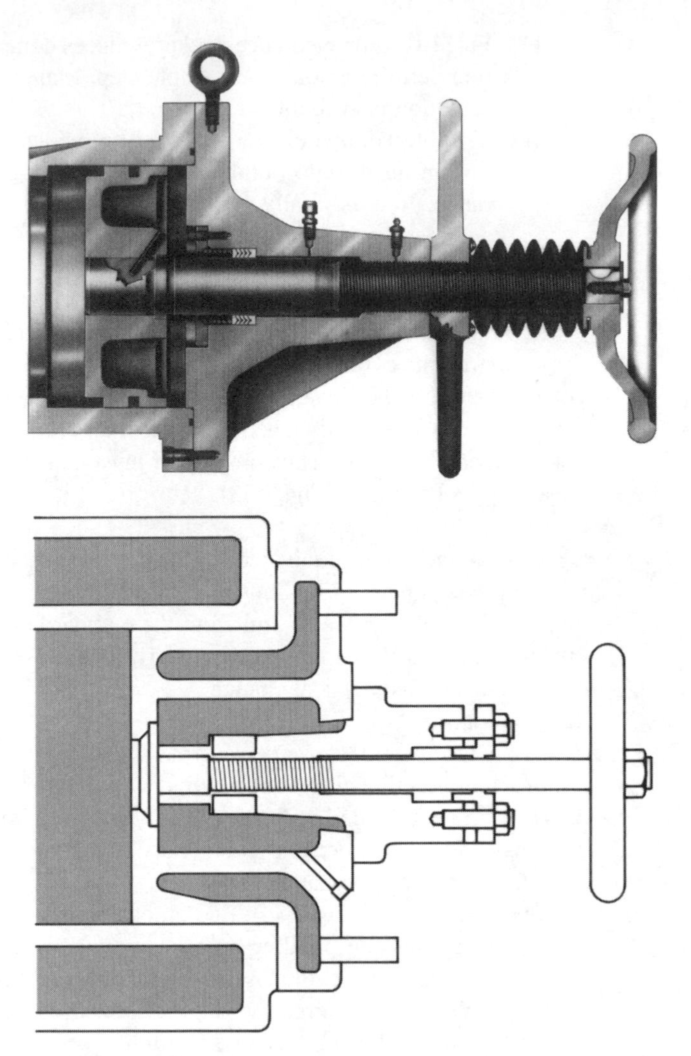

Figure 7.52 Manual fixed volume clearance pocket and manual variable clearance pocket.

(2) *Plug* (Fig. 7.54): A cylinder end may be unloaded by venting directly to the suction gas port through a separate opening either in the cylinder barrel or in the center of a circular plate type suction valve. This device does not interrupt normal intake valve operation and is available for manual or pneumatic actuation.

e. Gas Bypass

Changes in process flow demands can be compensated for by recirculating gas around the compressor. A portion of the discharge gas is redirected (bypassed) through a heat exchanger and then expanded back to suction pressure, where it is mixed with the normal suction gas stream. The other capacity control devices listed previously achieve flow reduction with an almost proportionate reduction in horsepower. With gas bypass systems, the horsepower requirement remains constant because the compressor throughput is not reduced.

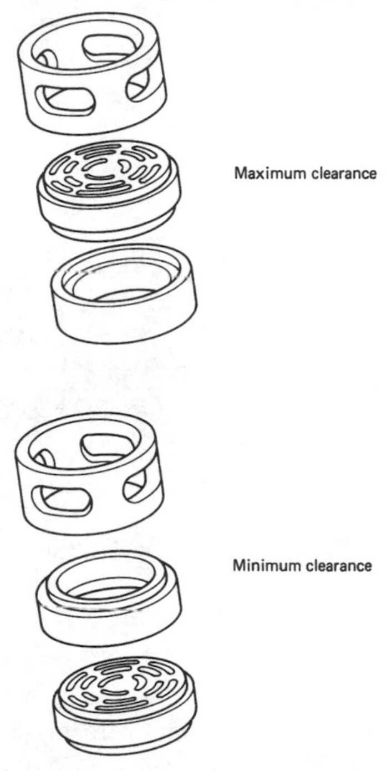

Figure 7.53 Split valve yoke, maximum and minimum clearance.

f. Speed Control

Capacity is almost proportional to the compressor's rotating speed. Special attention to torsional natural frequencies is mandatory when designing variable-speed Systems. Special flywheels, quill shafts, and couplings are often required. Oversized or auxiliary frame oil pumps may be necessary when speed reduction is used.

(1) *Engine drivers:* Speed reductions of 50% are possible.

(2) *Turbine and gear:* Speed reductions of 25% are commonly used.

(3) *Electric motors:* Speed reductions of up to 50% are possible with ac variable-speed controllers and DC. Initial capital costs, stiff electrical system grids, and motor efficiency versus power correction factor limit the practicality of AC systems.

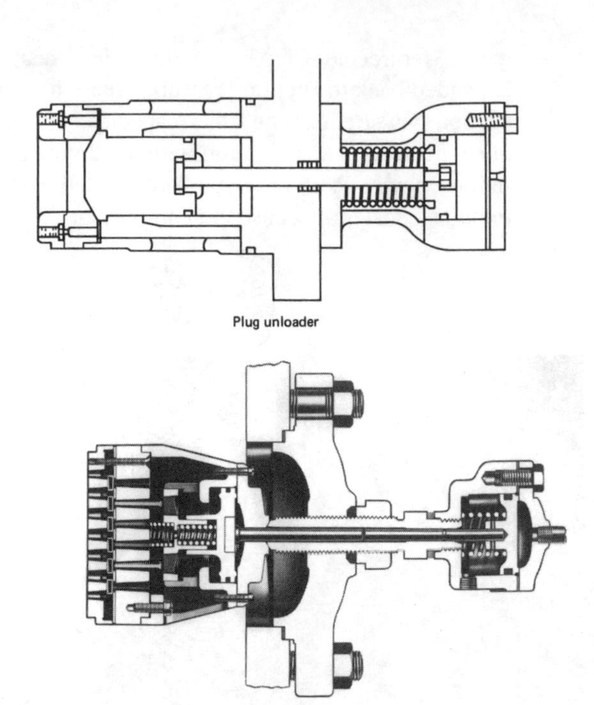

Figure 7.54 Suction valve depressor and plug unloader.

8. Sour Gas

Sour gas can be defined as gas having substantial amounts of hydrogen sulfide (H_2S) (above 1/2 mol percent). Sulfides result in a highly corrosive condition requiring special selection of cylinder materials. The most critical material selection is any part that is made of steel. Any steel in an H_2S atmosphere is susceptible to stress corrosion cracking. Such factors as yield strength, hardness, design stress, and temperature all influence the performance of these parts in H_2S atmosphere. Various experiences and testing have developed these guidelines.

The selection of component material is best done by the compressor manufacturer so that the material will suit the gas being handled and will be compatible with other wearing parts within the cylinder.

- Materials with yield strengths of 60,000 to 90,000 psi are acceptable.
- All test results indicate that hardness levels should not exceed RC22. The softer the material, the less susceptible it is to corrosion cracking and embrittlement.
- Design temperature must be kept low to prevent acceleration of corrosion fatigue.

a. *Compressor Cylinder Material Selection for Sour Gas Cylinder Body:*

Gray iron and nodular iron cylinders have been successfully used on many sour gas applications with both low and high H_2S concentrations. When steel cylinders are required, castings or forged-billet-type cylinders are applied. Special attention must be given to yield strength, carbon percentage and hardness.

Liner Material: Centrifugally cast gray iron liners have been successfully used with a hardness level of 217 BHN, minimum. A higher hardness level, which exhibits good antiwear properties, is preferred. All these applications are lubricated machines, and the lubricant film on the liner prevents corrosion. In some cases, Ni-resist has been used for anticorrosion reasons.

Piston: Cast iron and aluminum have been successfully used.

Rider Rings: Filled TFE has proved to be the best in this environment. Split or band types have been used successfully.

Piston Rings: Filled TFE gives the most consistent results.

Piston Rods: A number of materials that have been used with various degrees of success include the following:

(1) Stainless steel, No.410, with a core hardness of RC22 to RC28, induction hardened in the packing travel to RC40.
(2) SAE 8620 or AISI 4140 with a maximum core hardness less than RC22, with a chrome-plated or tungsten carbide overlay in the packing travel area.
(3) Carpenter Custom 450 with an RC40 hardness and 17-4 PH with a maximum hardness of RC3 have been used in sour gas applications.

Distance Piece: Must be closed, with proper venting.

Packing Rings: Filled TFE packing rings with metallic backup rings have proved to be most successful.

Packing Cases: Gray and nodular iron have been used up to 1500-psig discharge pressure. In excess of 1500 psig, AISI 4140 material has been used successfully. Packing cases must be cooled at pressures in excess of 1500 psig when filled TFE rings are used.

Valves: The use of steel valve components must comply with the design criteria outlined previously. Gray or nodular steel seats and guards are used for medium differential pressure applications. The use of inert, nonmetallic valve plates (thermoplastic) is well suited for sour gas service.

9. Material Requirements for Special Applications

Table 7.1 gives typical materials for a few special gas compressor applications.

Table 7.1 Typical Compressor Cylinder Material Usage

Compressor Application	Hydrogen	Carbon Dioxide	Hydrogen Sulfide (sour gas)	Oxygen	Low-Temperature Applications below -60°F	Chlorides
Cylinder and heads	Gray iron Nodular iron Steel	Gray iron Nodular iron Steel	Gray iron Nodular iron Steel	Gray iron Nodular iron	Stainless steel Iron alloy	Gray iron Nodular iron
Liners	Gray iron Ni-resist	Iron Ni-resist	Iron Ni-resist	No liner	Gray iron Ni-resist	Gray iron
Piston	Gray iron Aluminum Steel	Gray iron Aluminum Steel	Gray iron Aluminum	Gray iron Stainless steel Monel	Stainless steel Iron alloy	Gray iron Stainless steel
Piston rods	Steel Stainless steel	Steel Stainless steel	Stainless steel	K-monel	Stainless steel	Stainless steel
Piston rod nuts	Steel Stainless steel	Steel Stainless steel	Stainless steel	Everdur	Stainless steel	Stainless steel
Valve seats and guards	Gray iron Nodular iron Steel Stainless steel	Gray iron Nodular iron Stainless steel	Gray iron Nodular iron Stainless steel	Gray iron Nodular iron Stainless steel	Stainless steel Iron alloy	Gray iron Nodular iron
Valve plates	Stainless steel Nonmetallic	Stainless steel Nonmetallic	Stainless steel Nonmetallic	Stainless steel Nonmetallic	Stainless steel	K-monel Nonmetallic
Valve springs	Steel Stainless steel Chrome-Vanadium	Steel Stainless steel Chrome-Vanadium	Inconel Stainless steel Chrome-Vanadium	Stainless steel Chrome-Vanadium	Inconel Chrome-Vanadium	K-monel Chrome-Vanadium
Packing cases	Gray iron Steel Stainless steel	Gray iron Steel Stainless steel	Gray iron Stainless steel	Bronze Stainless steel	Stainless steel	Gray iron Stainless steel
Packing rings	Nonmetallic[a] Metallic (bronze)	Nonmetallic[a] Metallic (bronze)	Nonmetallic[a]	Nonmetallic[a]	Nonmetallic[a]	Nonmetallic[a]
Piston rings	Nonmetallic[a] Metallic (bronze)	Nonmetallic[a] Metallic (bronze)	Nonmetallic[a]	Nonmetallic[a]	Nonmetallic[a]	Nonmetallic[a]
Rider rings	Nonmetallic[a] Metallic (bronze)	Nonmetallic[a] Metallic (bronze)	Nonmetallic[a]	Nonmetallic[a]	Nonmetallic[a]	Nonmetallic[a]
Valve seats	Soft steel	Copper Soft steel	Soft steel	Copper	Steel Steel	Monel
Valve cover gasket	O-ring	O-ring	O-ring	O-ring	O-ring	O-ring
Studs	Steel	Steel	Steel (4140/B714 stainless steel)	Steel	Stainless steel	
Type distance piece	A	A, B	C, D	D	C, D	C, D
or	A	AB	CD	D	C	CD

[a]Typical filler material includes fiberglass, carbon, and bronze; appropriate selection depends on the service.

COMPRESSOR PERFORMANCE

The following section on compressor performance has been reprinted with permission from the *Gas Processors Suppliers Association Engineering Data Book, Eleventh Edition,* with minor editorial changes including the elimination of references to sections that are not included. A few footnotes have been added, as well as a few cross references to other sections of the *Compressed Air and Gas Handbook.* Figure numbers have been changed to the system used in this handbook.

Depending on application, compressors are manufactured as positive-displacement, dynamic, or thermal type (Fig. 7.56). Positive displacement types fall into two basic categories: reciprocating and rotary.

The reciprocating compressor consists of one or more cylinders each with a piston or plunger that moves back and forth, displacing a positive volume with each stroke.

The diaphragm compressor uses a hydraulically pulsed flexible diaphragm to displace the gas.

Rotary compressors cover lobe-type, screw-type, vane-type, and liquid ring type, each having a casing with one or more rotating elements that either mesh with each other, such as lobes or screws, or that displace a fixed volume with each rotation.

The dynamic types include radial-flow (centrifugal), axial-flow and mixed-flow machines. They are rotary continuous-flow compressors in which the rotating element (impeller or bladed rotor) accelerates the gas as it passes through the element, converting the velocity head into static pressure, partially in the rotating element and partially in stationary diffusers or blades.

Ejectors are "thermal" compressors that use a high-velocity gas or stream jet to entrain the inflowing gas, then convert the velocity of the mixture to pressure in a diffuser.

Figure 7.57 covers normal range of operation for compressors of the commercially available types. Figure 7.58 summarizes the difference between reciprocating and centrifugal compressors.

ACFM = actual cubic feet per minute (i.e. at process conditions)
A_p = cross sectional area of piston, sq in.
A_r = cross sectional area of piston rod, sq in.
Bhp = brake or shaft horsepower
C = cylinder clearance as a per cent of cylinder volume
C_p = specific heat at constant pressure, Btu/(lb · °F)
C_v = specific heat at constant volume, Btu/(lb · °F)
D = cylinder inside diameter, in.
d = piston rod diameter, in.
F = an allowance for interstage pressure drop, eq. 7.4
Ghp = gas horsepower, actual compression horsepower, excluding mechanical losses, bhp
H = head, ft · lb/lb
h = enthalpy, Btu/lb
ICFM = inlet cubic feet per minute, usually at suction conditions
k = isentropic exponent, C_p/C_v
MC_p = molal specific heat at constant pressure, Btu/(lb mol · °F)
MC_v = molar specific heat at constant volume, Btu/(lb mol · °F)
MW = molecular weight
mm cfd = million cubic ft/day
N = speed, rpm
N_m = molal flow, mols/min
n = polytropic exponent or number of mols
P = pressure, psia
P_c = critical pressure, psia
PD = piston displacement, ft³/min
P_L = pressure base used in the contract or regulation, psia
pP_c = pseudo critical pressure, psia
P_R = reduced pressure, P/P_c
pT_c = pseudo critical temperature, °R
P_V = partial pressure of contained moisture, psia
p = pressure, lb/ft²
Q = inlet capacity (ICFM)
R = universal gas constant

$$= 10.73 \frac{\text{psia} \cdot \text{ft}^3}{\text{lb mol} \cdot {}^\circ\text{R}}$$

$$= 1545 \frac{(\text{lb/ft}^2) \cdot \text{ft}^3}{\text{lb mol} \cdot {}^\circ\text{R}} \text{ or } \frac{\text{ft} \cdot \text{lb}}{\text{lb mol} \cdot {}^\circ\text{R}}$$

$$= 1.986 \frac{\text{Btu}}{\text{lb mol} \cdot {}^\circ\text{R}}$$

r = compression ratio, P_2/P_1
s = entropy, Btu/(lb · °F) or number of wheels
SCFM = standard cubic feet per minute measured at 14.7 psia at 60 °R
stroke = length of piston movement, in.
T = absolute temperature, °R
T_c = critical temperature, °R
T_R = reduced temperature, T/T_c
t = temperature, °F
V = specific volume, ft³/lb
VE = volumetric efficiency, per cent
W = work, ft · lb
w = weight flow, lb/min
X = temperature rise factor
y = mol fraction
Z = compressibility factor*
Z_{avg} = average compressibility factor = $\frac{Z_s + Z_d}{2}$ *
η = efficiency, expressed as a decimal

Subscripts
d = discharge
is = isentropic process
p = polytropic process
S = standard conditions, usually 14.7 psia, 60°F
s = suction
t = total or overall
1 = inlet conditions
2 = outlet conditions
L = standard conditions used for calculation or contract

*To avoid possible confusion, many authorities refer to Z as supercompressibility, reserving compressibility to mean departures from liquid, or incompressible behavior. For further discussion of this topic see Chapter 8 under the heading *Superconductivity.*

Figure 7.55 Nomenclature

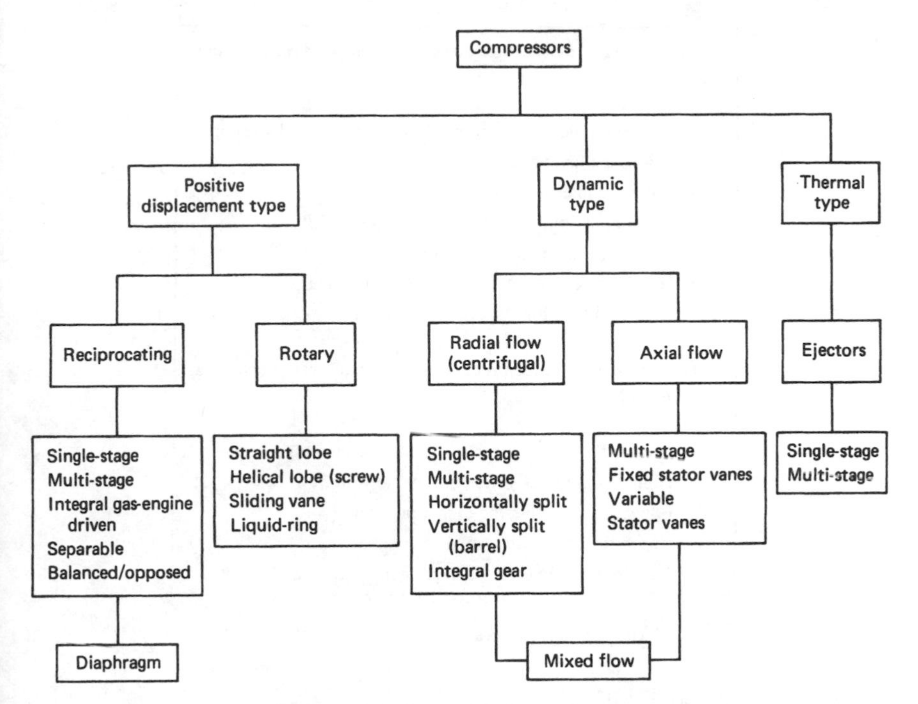

Figure 7.56 Types of Compressors

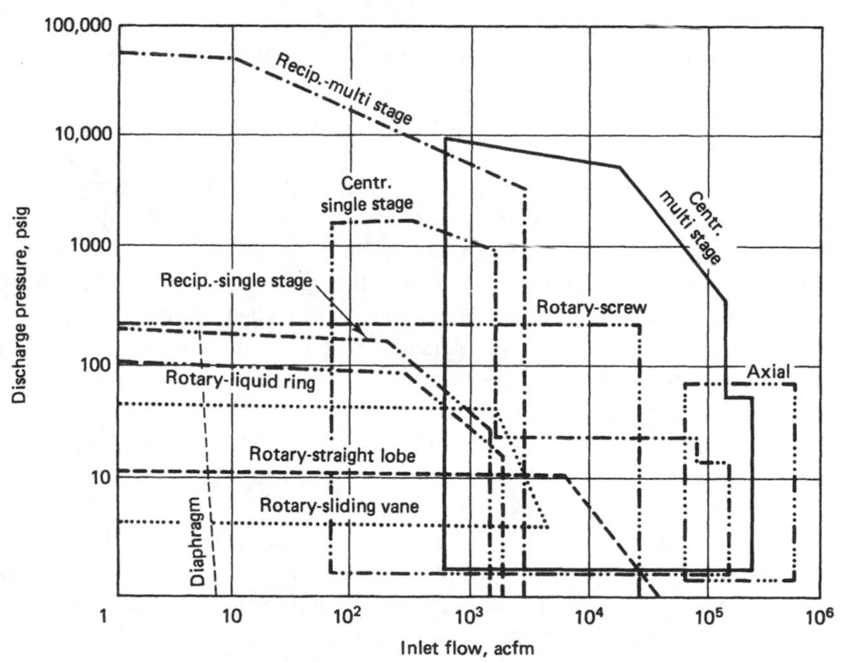

Figure 7.57 Compressor Coverage Chart

The advantages of a centrifugal compressor over a reciprocating machine are:

a. Lower installed first cost where pressure and volume conditions are favorable.
b. Lower maintenance expense.
c. Greater continuity of service and dependability.
d. Less operating attention required.
e. Greater volume capacity per unit of plot area.
f. Adaptability to high-speed low-maintenance-cost drivers.

The advantages of a reciprocating compressor over a centrifugal machine are:

a. Greater flexibility in capacity and pressure range.
b. Higher compressor efficiency and lower power cost.
c. Capability of delivering higher pressures.
d. Capability of handling smaller volumes.
e. Less sensitive to changes in gas composition and density.

Figure 7.58 Comparison of reciprocating and centrifugal compressors.

RECIPROCATING COMPRESSORS

Reciprocating compressor ratings vary from fractional to more than 20,000 hp per unit. Pressures range from low vacuum at suction to 30,000 psi and higher at discharge for special process compressors.

Reciprocating compressors are furnished either single-stage or multi-stage. The number of stages is determined by the overall compression ratio. The compression ratio per stage (and valve life) is generally limited by the discharge temperature and usually does not exceed 4, although small-sized units (intermittent duty) are furnished with a compression ratio as high as 8.

Gas cylinders are generally lubricated, although a non-lubricated design is available when warranted, for example, for nitrogen, oxygen, or instrument air.

On multi-stage machines, intercoolers may be provided between stages. These are heat exchangers which remove the heat of compression from the gas and reduce its temperature to approximately the temperature existing at the compressor intake. Such cooling reduces the actual volume of gas going to the high-pressure cylinders, reduces the horsepower required for compression, and keeps the temperature within safe operating limits.

Reciprocating compressors should be supplied with clean gas as they cannot satisfactorily handle liquids and solid particles that may be entrained in the gas. Liquids and solid particles tend to destroy cylinder lubrication and cause excessive wear. Liquids are noncompressible and their presence could rupture the compressor cylinder or cause other major damage.

Performance Calculations

The engineer in the field is frequently required to:

1. Determine the approximate horsepower required to compress a certain volume of gas at some intake conditions to a given discharge pressure.
2. Estimate the capacity of an existing compressor under specified suction and discharge conditions.

The following text outlines procedures for making these calculations from the standpoint of quick estimates and also presents more detailed calculations. For specific information on a given compressor consult the manufacturer of that unit.

For a compression process the enthalpy change is the best way of evaluating the work of compression. If a P-h diagram is available (as for propane refrigeration systems), the work of compression would always be evaluated by the enthalpy change of the gas in going from suction to discharge conditions. Years ago the capability of easily generating P-h diagrams for natural gases did not exist. The result was that many ways of estimating the enthalpy change were developed. They were used as a crutch and not because they were the best way to evaluate compression horsepower requirements.

Today the engineer does have available, in many cases, the capability to generate that part of the P-h diagram he requires for compression purposes. This is done using equations of state on a computer. This still would be the best way to evaluate the compression horsepower. The other equations are used only if access to a good equation of state is not available.

Some practical references continue to treat reciprocating and centrifugal machines as being different so far as estimation of horsepower requirements is concerned. This treatment reflects industry practice. The only difference in the horsepower evaluation is the efficiency of the machine. Otherwise, the basic thermodynamic equations are the same for all compression.

The reciprocating compressor horsepower calculations presented are based on charts. However, they may equally well be calculated using the equations in the centrifugal compressor section, particularly Eqs. (7.25) through (7.43). This also includes the mechanical losses in Eqs. (7.37) and (7.67).

There are two ways in which the thermodynamic calculations for compression can be carried out by assuming:

1. Isentropic reversible path: a process during which there is no heat added to or removed from the system and the entropy remains constant, pv^k = constant.
2. Polytropic path: a process in which heat transfer or changes in gas characteristics during compression are considered pv^n = constant.

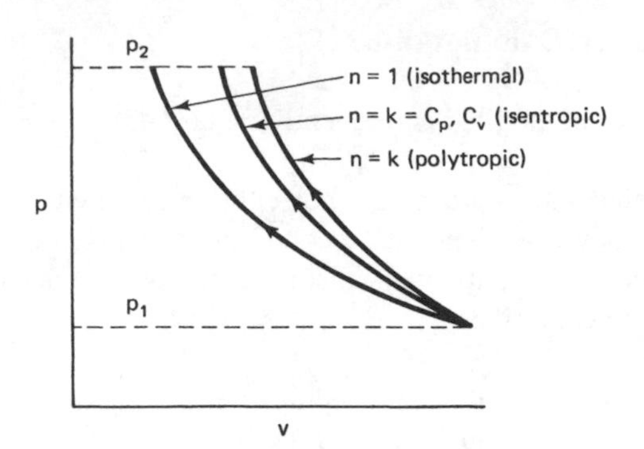

Figure 7.59 Compression curves.

Figure 7.59 shows a plot of pressure versus volume for each value of exponent n. The work, W performed in a flow process from p_1 to p_2 along any polytropic curve (Fig. 7.62) is:

$$W = \int_1^2 V \cdot dp = \int_{p1}^{p2} V \cdot dp \tag{7.1}$$

The amount of work required is dependent upon the polytropic curve involved and increases with increasing values of n. The path requiring the least amount of input work is $n = 1$, which is equivalent to isothermal compression, a process during which there is no change in temperature. For isentropic compression $n = \mathrm{k} =$ ratio of specific heat at constant pressure to that at constant volume.

It is usually impractical to build sufficient heat-transfer equipment into the design of most compressors to carry away the bulk of the heat of compression. Most machines tend to operate along a polytropic path which approaches the isentropic. Most compressor calculations are therefore based on an efficiency applied to account for true behavior.

A compression process following the middle curve in Fig. 7.59 has been widely referred to in industry as "adiabatic." However, many compression processes of practical importance are approximately adiabatic. The term adiabatic does not adequately describe this process, since it only implies no heat transfer. The ideal process also follows a path of constant entropy and should be called "isentropic," as will be done subsequently in this chapter.

Equation (7.3), which applies to all ideal gases, can be used to calculate k.

$$MC_p - MC_v = R = 1.986 \text{ Btu/(lb mol} \cdot {}^\circ\text{F)} \tag{7.2}$$

By rearrangement and substitution, we obtain

$$k = \frac{C_p}{C_v} = \frac{MC_p}{MC_v} = \frac{MC_p}{MC_p - 1.986} \tag{7.3}$$

To calculate k for a gas we need only know the constant pressure molal heat capacity (MC_p) for the gas. Figure 7.60 gives values of molecular weight and ideal-gas heat capacity (i.e., at 1 atm) for various gases. The heat capacity varies considerably with temperature.

Gas	Chemical formula	Mol wt	Temperature							
			0°F	50°F	60°F	100°F	150°F	200°F	250°F	300°F
Methane	CH_4	16.043	8.23	8.42	8.46	8.65	8.95	9.28	9.64	10.01
Ethyne (Acetylene)	C_2H_2	26.038	9.68	10.22	10.33	10.71	11.15	11.55	11.90	12.22
Ethene (Ethylene)	C_2H_4	28.054	9.33	10.02	10.16	10.72	11.41	12.09	12.76	13.41
Ethane	C_2H_6	30.070	11.44	12.17	12.32	12.95	13.78	14.63	15.49	16.34
Propane (Propylene)	C_3H_6	42.081	13.63	14.69	14.90	15.75	16.80	17.85	18.88	19.89
Propane	C_3H_8	44.097	15.65	16.88	17.13	18.17	19.52	20.89	22.25	23.56
1-Butene (Butylene)	C_4H_8	56.108	17.96	19.59	19.91	21.18	22.74	24.26	25.73	27.16
cis-2-Butene	C_4H_8	56.108	16.54	18.04	18.34	19.54	21.04	22.53	24.01	25.47
trans-2-Butene	C_4H_8	56.108	18.84	20.23	20.50	21.61	23.00	24.37	25.73	27.07
iso-Butane	C_4H_{10}	58.123	20.40	22.15	22.51	23.95	25.77	27.59	29.39	31.11
n-Butane	C_4H_{10}	58.123	20.80	22.38	22.72	24.08	25.81	27.55	29.23	30.90
iso-Pentane	C_5H_{12}	72.150	24.94	27.17	27.61	29.42	31.66	33.87	36.03	38.14
n-Pentane	C_5H_{12}	72.150	25.64	27.61	28.02	29.71	31.86	33.99	36.08	38.13
Benzene	C_6H_6	78.114	16.41	18.41	18.78	20.46	22.45	24.46	26.34	28.15
n-Hexane	C_6H_{14}	86.177	30.17	32.78	33.30	35.37	37.93	40.45	42.94	45.36
n-Heptane	C_7H_{16}	100.204	34.96	38.00	38.61	41.01	44.00	46.94	49.81	52.61
Ammonia	NH_3	17.0305	8.52	8.52	8.52	8.52	8.52	8.53	8.53	8.53
Air		28.9625	6.94	6.95	6.95	6.96	6.97	6.99	7.01	7.03
Water	H_2O	18.0153	7.98	8.00	8.01	8.03	8.07	8.12	8.17	8.23
Oxygen	O_2	31.9988	6.97	6.99	7.00	7.03	7.07	7.12	7.17	7.23
Nitrogen	N_2	28.0134	6.95	6.95	6.95	6.96	6.96	6.97	6.98	7.00
Hydrogen	H_2	2.0159	6.78	6.86	6.87	6.91	6.94	6.95	6.97	6.98
Hydrogen sulfide	H_2S	34.08	8.00	8.09	8.11	8.18	8.27	8.36	8.46	8.55
Carbon monoxide	CO	28.010	6.95	6.96	6.96	6.96	6.97	6.99	7.01	7.03
Carbon dioxide	CO_2	44.010	8.38	8.70	8.76	9.00	9.29	9.56	9.81	10.05

Exceptions: Air, Keenan and Keyes, Thermodynamic Properties of Air, Wiley, 3rd Printing 1947. Ammonia, Edw. R. Grabl, Thermodynamic Properties of Ammonia at High Temperatures and Pressures, Petr. Processing, April 1953. Hydrogen Sulfide, J. R. West, Chem. Eng. Progress, 44, 287, 1948.

Figure 7.60 Molal Heat Capacity MC_p (ideal-gas state), Btu (lb mol · °R) (1) Data source selected values of properties of Hydrocarbons, API Research Project 44.

Since the temperature of the gas increases as it passes from suction to discharge in the compressor, k is normally determined at the average of suction and discharge temperatures.

For a multicomponent gas the mol weighted average value of molal heat capacity must be determined at average cylinder temperature. A sample calculation is shown in Fig. 7.61.

The calculation of pP_p and pT_c in Fig. 7.61 permits calculation of the reduced pressure $P_R = P/pP_c$ mix and reduced temperature $T_R = T/pT_c$ mix. The compressibility* Z at T and P can then be determined using the charts in Chapter 8.

If only the molecular weight of the gas is known and not its composition, an approximate value for k can be determined from the curves in Fig. 7.62.

Estimating Compressor Horsepower

Equation (7.4) is useful for obtaining a quick and reasonable estimate for compressor horsepower. It was developed for large slow-speed (300 to 450 rpm) compressors handling gases with a specific gravity of 0.65 and having stage compression ratios above 2.5.

Caution: Compressor manufacturers generally rate their machines based on a standard condition of 14.4 psia rather than the more common gas industry value of 14.7 psia.

Due to higher valve losses the horsepower requirement for high-speed compressors (1000 rpm range, and some up to 1800 rpm) can be as much as 20% higher, although this is a very arbitrary value. Some compressor designs do not merit a higher horsepower allowance and the manufacturers should be consulted for specific applications.

$$\text{Brake horsepower} = (22)\,(\text{ratio/stage})(\text{no. of stages})(\text{mm cfd})(F) \qquad (7.4)$$

where mm cfd compressor capacity referred to 14.4 psia and intake temperature

F = 1.0 for single-stage compression
1.08 for two-stage compression
1.10 for three-stage compression

Equation (7.4) will also provide a rough estimate of horsepower for lower compression ratios and/or gases with a higher specific gravity, but it will tend to be on the high side. To allow for the tendency use a multiplication factor of 20 instead of 22 for gases with a specific gravity in the 0.8 to 1.0 range; likewise, use a factor in the range of 16 to 18 for compression ratios between 1 and 2.0.

*[To avoid possible confusion, many authorities refer to Z as *supercompressibility*, reserving *compressibility* to mean departures from liquid, or incompressible behavior. For a further discussion of this topic, see Chapter 8 under the heading *Supercompressibility*.]

Example gas mixture		Determination of mixture mol weight		Determination of MC_p, Molal heat capacity		Determination of pseudo critical pressure, pP_c, and temperature, pT_c			
Component name	Mol fraction y	Individual Component Mol weight MW	$y \cdot MW$	Individual Component MC_p @ 150°F*	$y \cdot MC_p$ @ 150°F	Component critical pressure P_c psia	$y \cdot P_c$	Component critical temperature T_c°R	$y \cdot T_c$
methane	0.9216	16.04	14.782	8.95	8.248	666	615.6	343	316.1
ethane	0.0488	30.07	1.467	13.78	0.672	707	34.6	550	26.8
propane	0.0185	44.10	0.816	19.52	0.361	616	11.4	666	12.3
i-butane	0.0039	58.12	0.227	25.77	0.101	528	2.1	734	2.9
n-butane	0.0055	58.12	0.320	25.81	0.142	551	3.0	765	4.2
i-pentane	0.0017	72.15	0.123	31.66	0.054	490	0.8	829	1.4
Total =	1.0000	MW =	17.735	MC_p =	9.578	pP_c =	667.5	pT_c =	363.7

$MC_v = MC_p - 1.986 = 7.592$ $k = MC_p/MC_v = 9.578/7.592 = 1.26$

* For values of MC_p other than @ 150°F, refer to Fig. 7.60

Figure 7.61 Calculation of k.

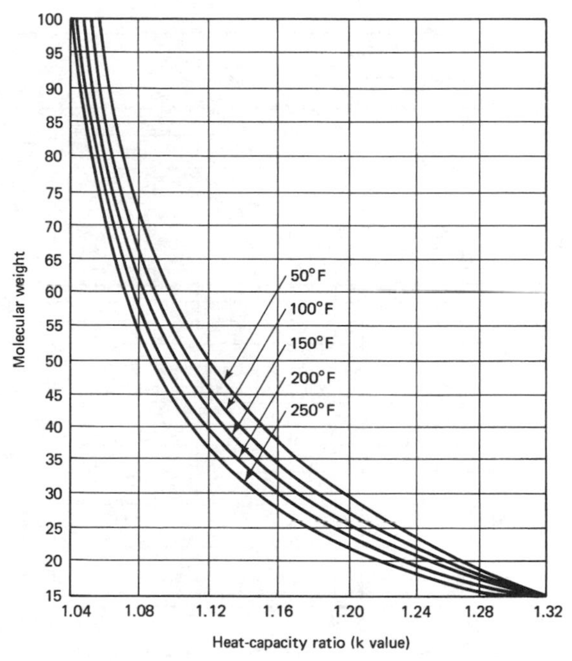

Figure 7.62 Approximate heat-capacity ratios of hydrocarbon gases.

Curves are available that permit easy estimation of approximate compression-horsepower requirements. Figure 7.63 is typical of these curves.

Example 7-1

Compress 2 mm cfd of gas at 14.4 psia and intake temperature through a compression ratio of 9 in a two-stage compressor. What will be the horsepower?

Solution steps

$$\text{Ratio per stage} = \sqrt{9} = 3$$

From Eq (7.4) we find the brake horsepower to be

$$(22)\ (3)\ (2)\ (2)\ (1.08) = 285 \text{ Bhp}$$

From Fig. 7.63, using a k of 1.15, we find the horsepower requirement to be 136 Bhp/mm cfd or 272 Bhp. For a k of 1.4, the horsepower requirement would be 147 Bhp/mm cfd or 294 total horsepower.

The two procedures give reasonable agreement, particularly considering the simplifying assumptions necessary in reducing compressor horsepower calculations to such a simple procedure.

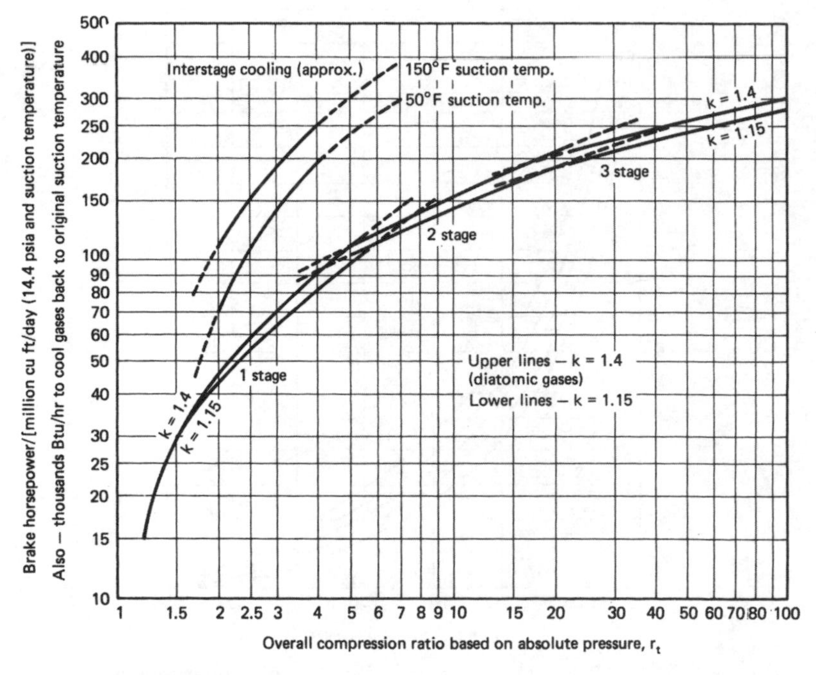

Figure 7.63 Approximate horsepower required to compress gases.

Detailed Calculations

There are many variables which enter into the precise calculation of compressor performance. Generalized data as given in this section are based upon the averaging of many criteria. The results obtained from these calculations, therefore, must be considered as close approximations to true compressor performance.

Capacity

Most gases encountered in industrial compression do not exactly follow the ideal gas equation of state but differ in varying degrees. The degree in which any gas varies from the ideal is expressed by a supercompressibility factor, Z, which modifies the ideal gas equation:

$$PV = nRT$$

to

$$PV = nZRT$$

Supercompressibility factors can be determined from charts in Chapter 8 using the pP_R and pT_R of the gas mixture. For pure components such as propane, compressibility factors can be determined from the P-H diagrams, although the user would be better advised to determine the compression horsepower directly using the P-H diagram.

For the purpose of performance calculations, compressor capacity is expressed as the actual volumetric quantity of gas at the inlet to each stage of compression on a per minute basis (ICFM).

From scfm,

$$Q = \text{scfm}\left(\frac{14.7}{520}\right)\left(\frac{T_1 Z_1}{P_1 Z_L}\right) \tag{7.7}$$

From weight flow (w, lb/min)

$$Q = \frac{10.73}{\text{MW}}\left(\frac{wT_1 Z_1}{P_1 Z_L}\right) \tag{7.8}$$

From weight flow (w, lb/min)

$$Q = \left(\frac{379.5 \cdot 14.7}{520}\right)\left(\frac{N_m T_1 Z_1}{P_1 Z_L}\right) \tag{7.9}$$

From these equations inlet volume to any stage may be calculated by using the inlet pressure P_1 and temperature T_1. Moisture, when in the form of a gas, should be handled just as any other component in the gas.

In a reciprocating compressor, effective capacity may be calculated as the piston displacement (generally in cfm) multiplied by the volumetric efficiency.

The piston displacement is equal to the net piston area multiplied by the length of piston sweep in a given period of time. This displacement may be expressed as follows:

For a single-acting piston compressing on the outer end only,

$$\text{PD} = \frac{(\text{stroke})(N)(D^2)\pi}{(4)(1{,}728)} \tag{7.10}$$
$$= 4.55\,(10^{-4})(\text{stroke})(N)(D^2)$$

For a single-acting piston compressing on the crank end only,

$$\text{PD} = \frac{(\text{stroke})(N)(D^2 - d^2)\pi}{(4)(1{,}728)} \tag{7.11}$$
$$= 4.55\,(10^{-4})(\text{stroke})(N)(D^2 - d^2)$$

For a double-acting piston (other than tail rod type),

$$\text{PD} = \frac{(\text{stroke})(N)(2D^2 - d^2)\pi}{(4)(1{,}728)} \tag{7.12}$$
$$= 4.55\,(10^{-4})(\text{stroke})(N)(2D^2 - d^2)$$

Volumetric Efficiency

In a reciprocating compressor the piston does not travel completely to the end of the cylinder at the end of the discharge stroke. Some clearance volume is necessary and it includes the space between the end of the piston and the cylinder head when the piston is at the end of its stroke. It also includes the volume in the valve ports, the volume in the suction valve guards and the volume around the discharge valve seats.

Clearance volume is usually expressed as a percentage of piston displacement and referred to as percent clearance, or cylinder clearance, C.

$$C = \frac{\text{clearance volume, in}^3}{\text{piston displacement, in}^3}\,(100) \tag{7.13}$$

For double-acting cylinders, the percent clearance is based on the total clearance volume for both the head end and the crank end of a cylinder. These two clearance volumes are not the same due to the presence of the piston rod in the crank end of the cylinder. Sometimes additional clearance volume (external) is intentionally added to reduce cylinder capacity.

The term volumetric efficiency refers to the actual volume capacity of a cylinder compared to the piston displacement. Without a clearance volume for the gas to expand and delay the opening of the suction valve(s), the cylinder could deliver its entire piston displacement as gas capacity. The effect of the gas contained in the clearance volume on the pumping capacity or a cylinder can be represented by

$$VE = 100 - r - C\left[\frac{Z_s}{Z_d}(r^{1/k}) - 1\right] \tag{7.13}$$

where:

r = P_d/P_s (pure ratio, not expressed as a percentage)
P_d = cylinder discharge pressure, psia
P_s = cylinder suction pressure
Z_d = supercompressibility factor at discharge conditions
Z_s = supercompressibility factor at suction conditions

Volumetric efficiencies as determined by Eq. (7.14) are theoretical in that they do not account for suction and discharge valve losses. The suction and discharge valves are actually spring-loaded check valves that permit flow in one direction only. The springs require a small differential pressure to open. For this reason, the pressure within the cylinder at the end of the suction stroke is lower than the line suction pressure and, likewise, the pressure at the end of the discharge stroke is higher than line discharge pressure.

One method for accounting for suction and discharge valve losses is to reduce the volumetric efficiency by an arbitrary amount, typically 4 percent, thus modifying Eq. (7.14) as follows:

$$VE = 96 - r - C\left[\frac{Z_s}{Z_d}(r^{1/k}) - 1\right] \tag{7.15}$$

When a non-lubricated compressor is used, the volumetric efficiency should be corrected by subtracting an additional 5 percent for slippage of gas. This is a capacity correction only and, as a first approximation, would not be considered when calculating compressor horsepower. The energy of compression is used by the gas even though the gas Slips by the rings and is not discharged from the cylinder.

If the compressor is in propane, or similar heavy gas service, an additional 4 percent should be subtracted from the volumetric efficiency. These deductions for non-lubricated and propane performance are both approximate and, if both apply, cumulative.

Figure 7.64 provides the solution to the function $r^{1/k}$. Values for compression ratios not shown may be obtained by interpolation. The closest *k* value column may be safely used without a second interpolation.

Volumetric efficiencies for high-speed separable compressors in the past have tended to be slightly lower than estimated from Eq. (7.14). Recent information suggests that this modification is not necessary for all models of high-speed compressors.

In evaluating efficiency, horsepower, volumetric efficiency, and so on, the user should consider past experience with different speeds and models. Larger valve area for a given swept volume will generally lead to higher compression efficiencies.

Equivalent Capacity

The net capacity for a compressor, in cubic feet per day at 14.4 psia and suction temperature, may be calculated by Eq. (7.16a), which is shown in dimensioned form:

$$\text{mm cfd} = \frac{PD\dfrac{\text{ft}^3}{\text{min}} \cdot 1440\dfrac{\text{min}}{d} \cdot \dfrac{VE\%}{100} \cdot P_s\dfrac{\text{lb}}{\text{in.}^2} \cdot 10^{-6}\dfrac{\text{mmft}^3}{\text{ft}^3} \cdot Z_{14.4}}{14.4\dfrac{\text{lb}}{\text{in.}^2} \cdot Z_s} \qquad (7.16\text{a})$$

which can be simplified to Eq. (7.16b) when $Z_{14.4}$ is assumed to equal 1.0.

$$\text{mm cfd} = \frac{PD \cdot VE \cdot P_s \cdot 10^{-6}}{Z_s} \qquad (7.16\text{b})$$

For example, a compressor with 200 cfm piston displacement, a volumetric efficiency of 80%, a suction pressure of 75 psia, and suction compressibility of 0.9 would have a capacity of 1.33 mm cfd at 14.4 psia and suction temperature. If compressibility is not used as a divisor in calculating cfm, then the statement "not corrected for compressibility" should be added.

Compression Ratio	1.10	1.14	1.18	1.22	1.26	1.30	1.34	1.38	1.42
	k, isentropic exponent C_p/C_v								
1.2	1.180	1.173	1.167	1.161	1.156	1.151	1.146	1.141	1.137
1.4	1.358	1.343	1.330	1.318	1.306	1.295	1.285	1.276	1.267
1.6	1.533	1.510	1.489	1.470	1.452	1.436	1.420	1.406	1.392
1.8	1.706	1.675	1.646	1.619	1.594	1.572	1.551	1.531	1.513
2.0	1.878	1.837	1.799	1.765	1.733	1.704	1.677	1.652	1.629
2.2	2.048	1.997	1.951	1.908	1.870	1.834	1.801	1.771	1.742
2.4	2.216	2.155	2.100	2.050	2.003	1.961	1.922	1.886	1.852
2.6	2.384	2.312	2.247	2.188	2.135	2.086	2.040	1.999	1.960
2.8	2.550	2.467	2.393	2.326	2.264	2.208	2.156	2.109	2.065
3.0	2.715	2.621	2.537	2.461	2.391	2.328	2.270	2.217	2.168
3.2	2.879	2.774	2.680	2.595	2.517	2.447	2.382	2.323	2.269
3.4	3.042	2.926	2.821	2.727	2.641	2.563	2.492	2.427	2.367
3.6	3.204	3.076	2.961	2.857	2.764	2.679	2.601	2.530	2.465
3.8	3.366	3.225	3.100	2.987	2.885	2.792	2.708	2.631	2.560
4.0	3.526	3.374	3.238	3.115	3.005	2.905	2.814	2.731	2.655
4.2	3.686	3.521	3.374	3.242	3.124	3.016	2.918	2.829	2.747
4.4	3.846	3.668	3.510	3.368	3.241	3.126	3.021	2.926	2.839
4.6	4.004	3.814	3.645	3.493	3.357	3.235	3.123	3.022	2.929
4.8	4.162	3.959	3.779	3.617	3.473	3.342	3.224	3.116	3.018
5.0	4.319	4.103	3.912	3.740	3.587	3.449	3.324	3.210	3.106
5.2	4.476	4.427	4.044	3.863	3.700	3.554	3.422	3.303	3.193
5.4	4.632	4.390	4.175	3.984	3.813	3.659	3.520	3.394	3.279
5.6	4.788	4.532	4.306	4.105	3.925	3.763	3.617	3.485	3.364
5.8	4.943	4.674	4.436	4.224	4.035	3.866	3.713	3.574	3.448
6.0	5.098	4.815	4.565	4.343	4.146	3.968	3.808	3.663	3.532
6.2	5.252	4.955	4.694	4.462	4.255	4.069	3.902	3.751	3.614
6.4	5.406	5.095	4.822	4.579	4.363	4.170	3.996	3.839	3.696
6.6	5.560	5.235	4.949	4.696	4.471	4.270	4.089	3.925	3.777
6.8	5.713	5.374	5.076	4.813	4.578	4.369	4.181	4.011	3.857
7.0	5.865	5.512	5.202	4.928	4.685	4.468	4.272	4.096	3.937

Figure 7.64 Values of $r^{1/k}$.

In many instances the gas sales contract or regulation will specify some other measurement standard for gas volume. To convert volumes calculated using Eq. (7.16) (i.e., at 14.4 psia and suction temperature) to a P_L and T_L basis, Eq. (7.17) would be used:

$$\text{mm scfd at } P_L, T_L = (\text{mm cfd from Eq. 7.16})\left(\frac{14.4}{P_L}\right)\left(\frac{T_L}{T_s}\right)\left(\frac{Z_L}{Z_s}\right) \tag{7.17}$$

Discharge Temperature

The temperature of the gas discharged from the cylinder can be estimated from Eq. (7.18), which is commonly used but not recommended. (*Note:* The temperatures are in absolute units, °R or K.) Equations (7.31) and (7.32) give better results.

$$T_d = T_s\left(r^{(k-1)/k}\right) \tag{7.18}$$

Figure 7.65 is a nomograph that can be used to solve Eq. (7.18). The discharge temperature determined from either Eq. (7.18) or Fig. 7.65 is the theoretical value. While it neglects heat from friction, irreversibility effects, etc., and may be somewhat low, the values obtained from this equation will be reasonable field estimates.

Rod Loading

Each compressor frame has definite limitations as to maximum speed and load-carrying capacity. The load-carrying capacity of a compressor frame involves two primary considerations: horsepower and rod loading.

The horsepower rating of a compressor frame is the measure of the ability of the supporting structure and crankshaft to withstand torque (turning moment) and the ability of the bearings to dissipate frictional heat. Rod loads are established to limit the static and inertial loads on the crankshaft, connecting rod, frame, piston rod, bolting, and projected bearing surfaces.

Good design dictates a reversal of rod loading during each stroke. Nonreversal of the loading results in failure to allow bearing surfaces to part and permit entrance of sufficient lubricant. The result will be premature hearing wear or failure.

Rod loadings may be calculated by the use of Eqs. (7.19) and (7.20).

$$\text{Load in compression} = P_d A_p - P_s (A_p - A_r) = (P_d - P_s) A_p + P_s A_r \quad (7.19)$$

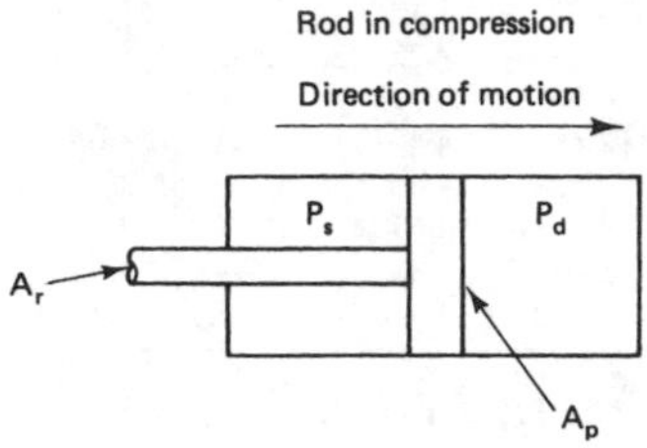

Rod in compression

Direction of motion

P_s P_d

A_r

A_p

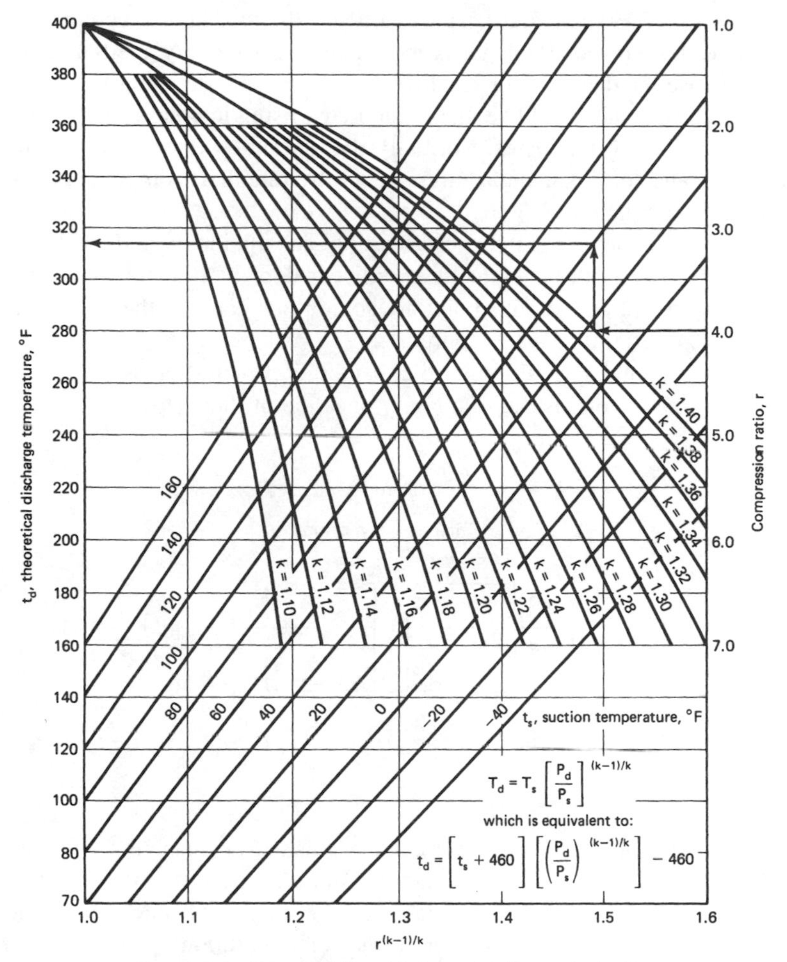

Figure 7.65 Theoretical discharge temperatures, single-state compression. Read r to k to t_s to t_d.

$$\text{Load in tension} = P_d(A_p - A_r) - P_sA_p = (P_d - P_s)\,A_p - P_dA_r \tag{7.20}$$

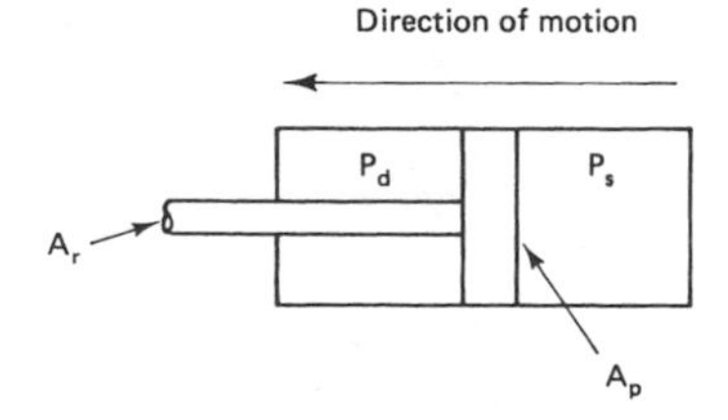

Using Eqs. (7.19) and (7.20), a plus value for the load in both compression and tension indicates a reversal of loads based on gas pressure only. Inertial effects will tend to increase the degree of reversal.

The true rod loads would be those calculated using internal cylinder pressures after allowance for valve losses. Normally the operator will know only line pressures and because of this manufacturers generally rate their compressors based on line-pressure calculations.

A further refinement in the rod-loading calculation would be to include inertial forces. While the manufacturer will consider inertial forces when rating compressors, useful data on this point are seldom available in the field. Except in special cases, inertial forces are ignored.

A tail-rod cylinder would require consideration of rod cross-section area on both sides of the piston instead of on only one side of the piston, as in Eqs. (7.19) and (7.20).

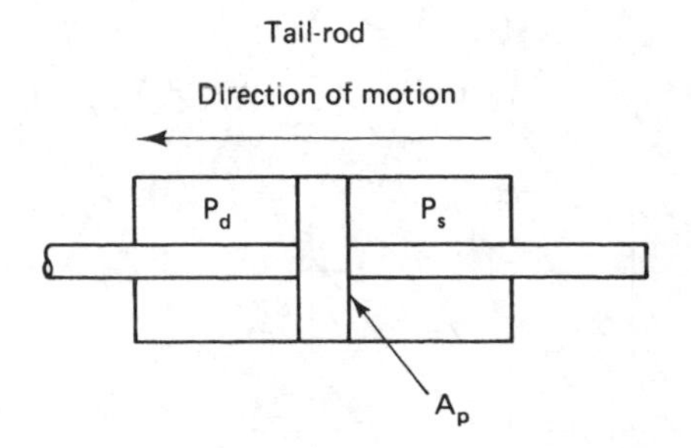

Horsepower

Detailed compressor horsepower calculations can be made through the use of Figs. 7.66 and 7.67. For ease of calculations, these figures provide net horsepower, including mechanical efficiency and gas losses. Figures 7.68 and 7.69 are included for modifying the horsepower numbers for special conditions.

Proper use of these charts should provide the user with reasonably correct horsepower requirements that are comparable to those calculated by the compressor manufacturer. For more detailed design, the engineer should consult a compressor manufacturer.

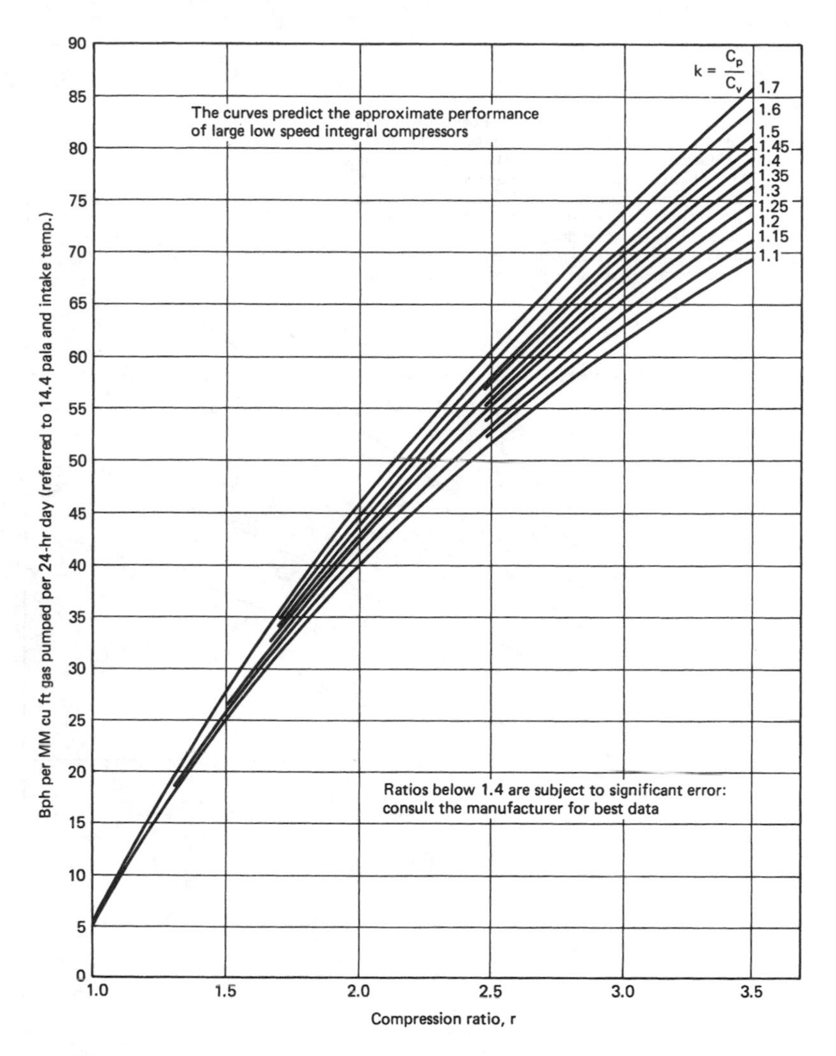

Figure 7.66 Bhp per million curve, mechanical efficiency-95%. Gas velocity through valve-3000 ft./min (API equation).

Volumes to be handled in each stage must be corrected to the actual temperature at the inlet to that stage. Note that moisture content corrections can also be important at low pressure and/or high temperature.

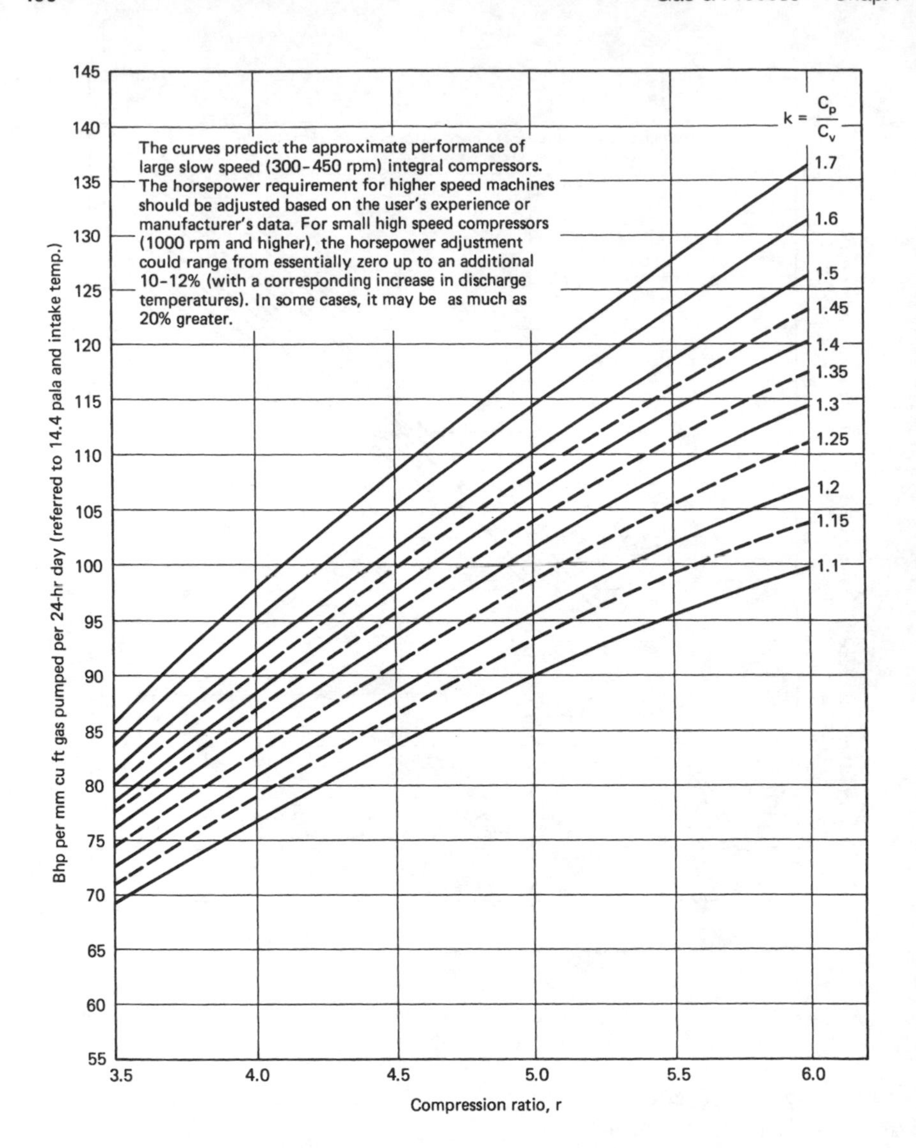

Figure 7.67 Bhp per million curve, mechanical efficiency-95%. Gas velocity through valve – 3000 ft./min (API equation).

When intercoolers are used, allowance must be made for interstage pressure drop. Interstage pressures may be estimated by:

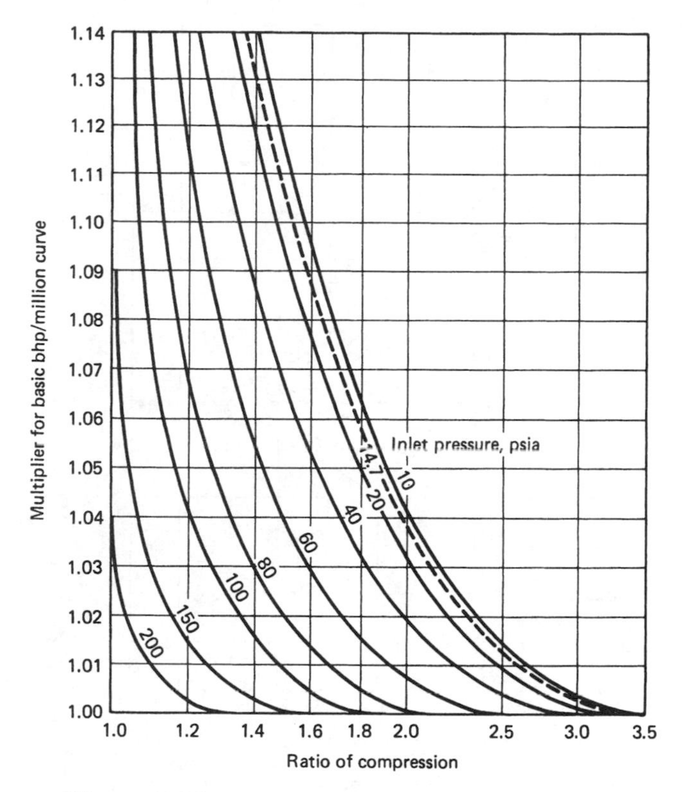

Figure 7.68 Correction factor for low intake pressure.

1. Obtaining the overall compression ratio, r_t.
2. Obtaining the calculated ratio per stage, r, by taking the s root of r_t, where s is the number of compression stages.
3. Multiplying r by the absolute intake pressure of the stage being considered.

This procedure gives the absolute discharge pressure of this stage and the theoretical absolute intake pressure to the next stage. The next stage intake pressure can be corrected for intercooler pressure drop by reducing the pressure by 3 to 5 psi. This can be significant in low pressure stages.

Horsepower for compression is calculated by using Figs. 7.66 and 7.67 and Eq. (7.21).

$$\text{Bhp} - (\text{Bhp/mmcfd})\left(\frac{P_L}{14.4}\right)\left(\frac{T_s}{T_L}\right)(Z_{avg})\,(\text{mmcfd}) \qquad (7.21)$$

Bhp/mmcfd is read from Figs. 7.66 and 7.67, which use a pressure base of 14.4 psia.

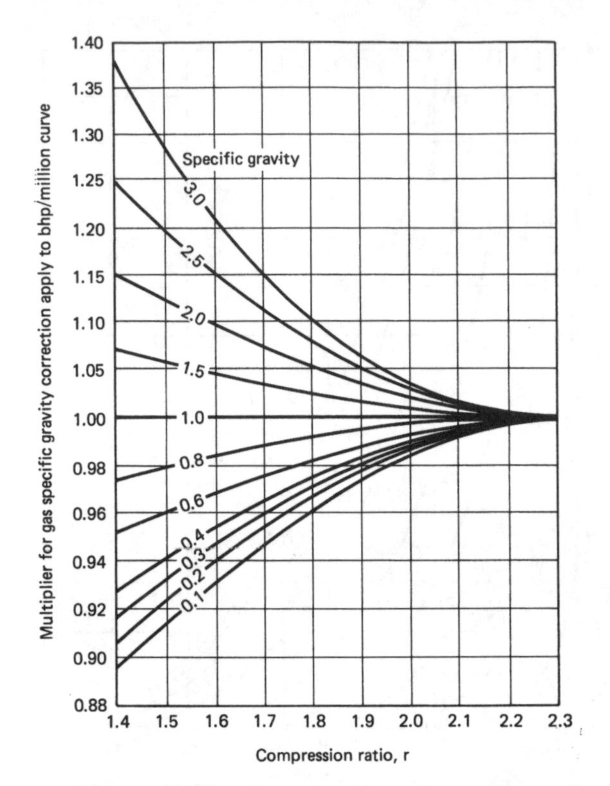

Figure 7.69 Correction factor for specific gravity.

Figures 7.66 and 7.67 are for standard valved cylinders. Caution should be used in applying conventional cylinders to low compression-ratio pipeline compressors. For low ratio pipeline compressors a high clearance type cylinder permits valve designs with higher efficiency. The compressor manufacturer should be consulted for bhp curves on this type cylinder.

Figure 7.68 provides a correction for intake pressure. The correction factor, as read from the curve, is used as a multiplier in the right hand side of Eq. (7.21) to obtain the corrected brake horsepower.

Figure 7.69 provides a correction factor for gas specific gravity. The correction factor is used as a multiplier in the right-hand side of Eq. (7.21) to obtain the corrected horsepower.

Data presented in Figs. 7.66 and 7.67 are for slow-speed integral compressors rather than the high-speed separable compressors. To adjust the horsepower for the high-speed unit, the values obtained from Figs. 7.66 and 7.67 may be increased by the following percentages:

Gas Specific Gravity	Per Cent Horsepower Increase for High-speed Units
0.5 – 0.8	4
0.9	5
1.0	6
1.1	8
1.5 and propane refrigeration units	10

Because of variations by different manufacturers in specifying valve velocities for high speed as opposed to slow speed compressors, a given unit may differ from the horsepower corrections shown. Experience with compressors from a specific manufacturer will serve to guide the user and give confidence in utilization of the correction factors shown. For applications which are outside typical ranges discussed here, compressor manufacturers should be consulted.

Example 7.2

Compress 2 mm scfd of gas measured at 14.65 psia and 60°F. Intake pressure is 100 psia, and intake temperature is 100°F. Discharge pressure is 900 psia. The gas has a specific gravity of 0.80. What is the required horsepower?

1. Compression ratio is

$$\frac{900 \text{ psia}}{100 \text{ psia}} = 9$$

This would be a two-stage compressor; therefore, the ratio per stage is $\sqrt{9}$ or 3.

2. 100 psia x 3 =300 psia (first-stage discharge pressure)
   300 psia – 5 = 295 psia (suction to second stage)

where the 5 psi represents the pressure drop between first stage discharge and second stage suction.

$$\frac{900 \text{ psia}}{295 \text{ psia}} = 3.05 \text{ (compression ratio for second stage)}$$

It may be desirable to recalculate the interstage pressure to balance the ratios. For this sample problem, however, the first ratios determined will be used.

3. From Fig. 7.62, a gas with specific gravity of 0.8 at 150°F would have an approximate k of 1.21. For most compression applications, the 150°F curve will be adequate. This should be checked after determining the average cylinder temperature.
4. Discharge temperature for the first stage may be obtained by using Fig. 7.65 or solving Eq. (7.18). For a compression ratio of 3, discharge temperature = approximately 220°F. Average cylinder temperature = 160°F.

5. In the same manner, discharge temperature for the second stage (with r = 3.05 and assuming interstage cooling to 120°F) equals approximately 244°F. Average cylinder temperature = 182°F.
6. From the physical properties section in Chapter 8, estimate the compressibility* factors at suction and discharge pressure and temperature of each stage.

* To avoid possible confusion, many authorities refer to Z as supercompressibility, reserving compressibility to mean departures from liquid, or incompressible behavior. For further discussion of this topic see Chapter 8 under the heading *Superconductivity*.

1st stage:	Z_s	= 0.98
	Z_d	= 0.97
	Z_{avg}	= 0.975
2nd stage:	Z_s	= 0.94
	Z_d	= 0.92
	Z_{avg}	= 0.93

7. From Fig. 7.66 Bhp/mmcfd at 3 compression ratio and a k of 1.21 is 63.5 (first stage). From Fig. 7.66, Bhp/mmcfd at 3.05 compression ratio and a k of 1.21 is 64.5 (second stage).
8. There are no corrections to be applied from Fig. 7.68 or 7.69, as all factors read unity.
9. Substituting in Eq. (7.21),

$$\text{1st stage: Bhp/mmscfd} = 63.5\left(\frac{14.65}{14.4}\right)\left(\frac{560}{520}\right)0.975 = 67.8$$

$$\text{Bhp for 1}^{\text{st}}\text{ stage} = \text{2mm scfd} \times 67.8 = 135.6$$

$$\text{2nd stage: Bhp/mmscfd} = 64.5\left(\frac{14.65}{14.4}\right)\left(\frac{580}{520}\right)0.93 = 68.1$$

$$\text{Bhp for 2nd stage} = \text{(2 mm scfd) (68.1)} = 136.2$$

$$\text{Total Bhp for this application} = 135.6 + 136.2 = 271.8$$

Note that in Example 7.1 the same conditions result in compression power of 285 Bhp, which is in close agreement.

Limits to Compression Ratio Per Stage

The ratio of compression permissible in one stage is usually limited by the discharge temperature or by rod loading, particularly in the first stage.

When handling gases containing oxygen, which could support combustion, there is a possibility of fire and explosion because of the oil vapors present.

To reduce carbonization of the oil and the danger of fires, the safe operating limit may be considered to be approximately 300°F. Where no oxygen is present in the gas stream, temperatures of 350°F may be considered as the maximum, even though mechanical or process requirements usually dictate a lower figure.

Packing life may be significantly shortened by the dual requirement to seal both high-pressure and high-temperature gases. For this reason, at higher discharge pressures, a temperature closer to 250° or 275°F may be the practical limit.

In summary, and for most field applications, the use of 300°F maximum would be a good average. Recognition of the above variables is, however, still useful.

Economic considerations are also involved because a high ratio of compression will mean a low volumetric efficiency and require a larger cylinder to produce the same capacity. For this reason a high rod loading may result and require a heavier and more expensive frame.

Where multi-stage operation is involved, equal ratios of compression per stage are used (plus an allowance for piping and cooler losses if necessary) unless otherwise required by process design. For two stages of compression the ratio per stage would approximately equal the square root of the total compression ratio; for three stages, the cube root; etc. In practice, especially in high-pressure work, decreasing the compression ratio in the higher stages to reduce excessive rod loading may prove to be advantageous.

Cylinder Design

Depending on the size of the machine and the number of stages, reciprocating compressors are furnished with cylinders fitted with either single- or double-acting pistons; see examples in Figs. 7.70 through 7.72.

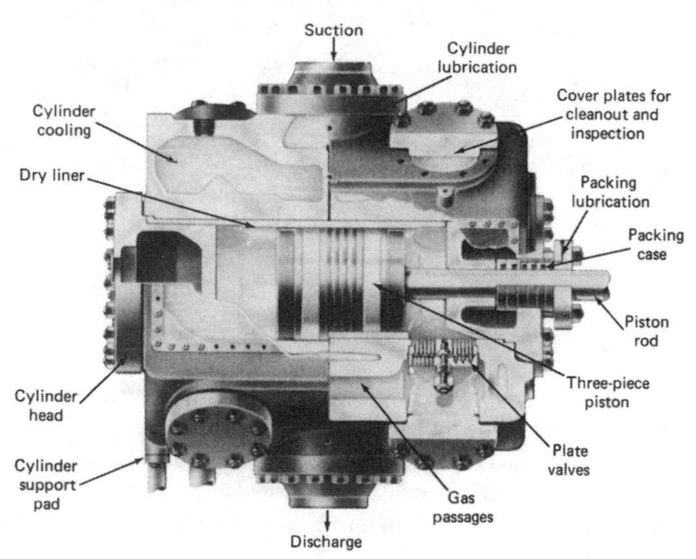

Figure 7.70 Low pressure cylinder with double-acting piston.

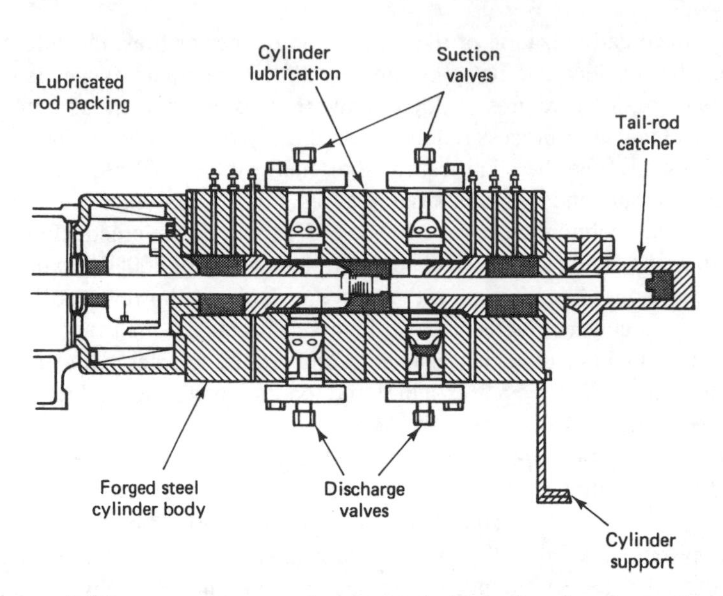

Figure 7.71 High pressure cylinder with double-acting piston and tail-rod.

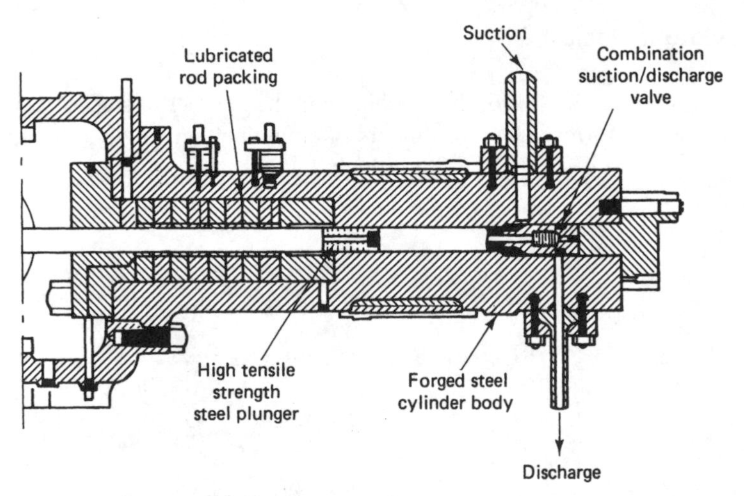

Figure 7.72 Single-acting plunger cylinder designed for 15,000 psig discharge.

In the same units, double-acting pistons are commonly used in the first stages and often single-acting in the higher stages of compression.

Cylinder materials are normally selected for strength; however, thermal shock, mechanical shock, or corrosion resistance may also be a determining factor. The following table shows discharge pressure limits generally used in the gas industry for cylinder material selection.

Cylinder Material	Discharge Pressure (psig)
Cast iron	Up to 1200
Nodular iron	About 1500
Cast steel	1200 to 2500
Forged steel	Above 2500

API standard 618 recommends 1000 psig as the maximum pressure for both cast iron and nodular iron.

Cylinders are designed both as a solid body (no liner) and with liners. Cylinder liners are inserted into the cylinder body to either form or line the pressure wall. There are two types. The wet liner forms the pressure wall as well as the inside wall of the water jacket. The dry type lines the cylinder wall and is not required to add strength.

Standard cylinder liners are cast iron. If cylinders are required to have special corrosion or wear resistance, other materials or special alloys may be needed.

Most compressors use oils to lubricate the cylinder with a mechanical, force-feed lubricator having one or more feeds to each cylinder.

The non-lubricated compressor has found wide application where it is desirable or essential to compress air or gas without contaminating it with lubricating oil.

For such cases a number of manufacturers furnish a "non-lubricated" cylinder (Fig. 7.73). The piston these cylinders is equipped with piston rings of graphitic carbon or plastic as well as pads or rings of the same material to maintain the proper clearance between the piston and cylinder. Plastic packing of a type that requires no lubricant is used in the stuffing box. Although oil-wiper rings are used on the piston rod where it leaves the compressor frame, minute quantities of oil might conceivably enter the Cylinder on the rod. Where even such small amounts of oil are objectionable, an extended cylinder connecting piece can be furnished. This simply lengthens the piston rod so that no lubricated portion of the rod enters the cylinder.

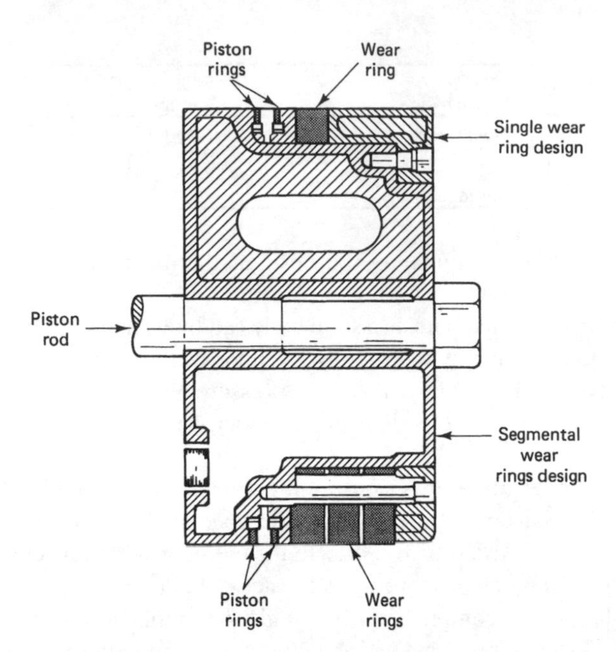

Figure 7.73 Piston equipped with Teflon® piston and wear rings for a single-acting non-lubricated cylinder.

A small amount of gas leaking through the packing can be objectionable. Special distance pieces are furnished between the cylinder and frame, which may be either single-compartment or double-compartment. These may be furnished gastight and vented back to the suction, or may be filled with a sealing gas or fluid and held under a slight pressure, or simply vented.

Compressor valves for non-lubricated service operate in an environment that has no lubricant in the gas or in the cylinder. Therefore, the selection of valve materials is important to prevent excessive wear

Piston rod packing universally used in non-lubricated compressors is of the full-floating mechanical type, consisting of a case containing pairs of either carbon or plastic (TFE) rings of conventional design.

When handling oxygen and other gases such as nitrogen and helium, it is absolutely necessary that all traces of hydrocarbons in cylinders be removed. With oxygen, this is required for safety; with other gases, to prevent system contamination.

High-pressure compressors with discharge pressures from 5,000 to 30,000 psi usually require special design and a complete knowledge of the characteristics of the gas.

As a rule, inlet and discharge gas pipe connections on the cylinder are fitted with flanges of the same rating for the following reasons:

1. Practicality and uniformity of casting and machining.
2. Hydrostatic test, usually at 150% design pressure.
3. Suction pulsation bottles are usually designed for the same pressure as the discharge bottle (often by federal, state, or local government regulation).

Reciprocating Compressor Control Devices

Output of compressors must be controlled (regulated) to match system demand. In many installations some means of controlling the output of the compressor is necessary. Often constant flow is required despite variations in discharge pressure, and the control device must operate to maintain a constant compressor capacity. Compressor capacity, speed, or pressure may be varied in accordance with the requirements. The nature of the control device will depend on the regulating variable, whether pressure, flow, temperature, or some other variable, and on type of compressor driver.

Unloading at startup. Practically all reciprocating compressors must be unloaded to some degree before starting so that the driver torque available during acceleration is not exceeded. Both manual and automatic compressor startup unloading is used. Common methods of unloading include discharge venting, discharge to suction bypass, and holding open the inlet valves using valve lifters.

Capacity control. The most common requirement is regulation of capacity. Many capacity controls, or unloading devices, as they are usually termed, are actuated by the pressure on the discharge side of the compressor. A falling pressure indicates that gas is being used faster than it is being compressed and that more gas is required. A rising pressure indicates that more gas is being compressed than is being used and that less gas is required.

A common method of controlling the capacity of a compressor is to vary the speed. This method is applicable to steam-driven compressors and to units driven by internal-combustion engines. In these cases the regulator actuates the steam-admission or fuel-admission valve on the compressor driver to control the speed.

Electric-motor-driven compressors usually operate at constant speed, and other methods of controlling the capacity are necessary. On reciprocating compressors up to about 100 hp, two types of control are usually available. These are automatic start-and-stop control and constant-speed control.

Automatic start-and-stop control, as its name implies, stops or starts the compressor by means of a pressure-actuated switch as the gas demand varies. It should be used only when the demand for gas will be intermittent.

Constant-speed control permits the compressor to operate at full speed continuously, loaded part of the time and fully or partially unloaded at other times. Two methods of unloading the compressor with this type of control are in common use: inlet-valve unloaders and clearance unloaders. Inlet-valve unloaders (Fig. 7.74) operate to hold the compressor inlet valves open and thereby prevent compression. Clearance unloaders (Fig. 7.75) consist of pockets or small reservoirs which are opened when unloading is desired. The gas is compressed into them on the compression stroke and expands into the cylinder on the return stroke, reducing the intake of additional gas.

Motor-driven reciprocating compressors above 100 hp in size are usually equipped with a step control. This is in reality a variation of constant-speed control in which unloading is accomplished in a series of steps, varying from full load down to no load.

Five-step control (full load, three-quarter load, one-half load, one-quarter load, and no load) is accomplished by means of clearance pockets. On some makes of machines inlet-valve and clearance control unloading are used in combination.

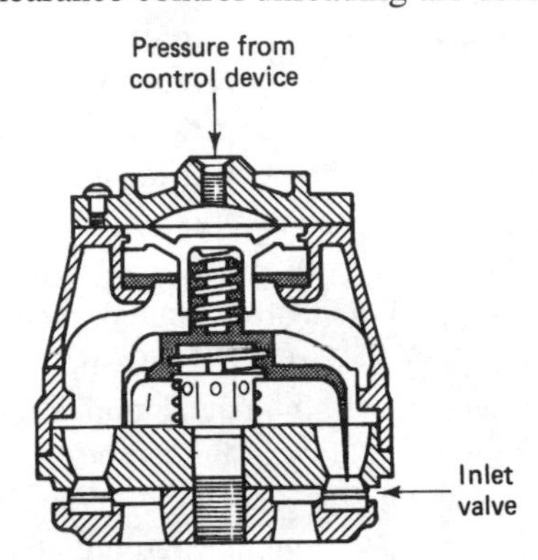

Figure 7.74 Inlet Valve Unloader

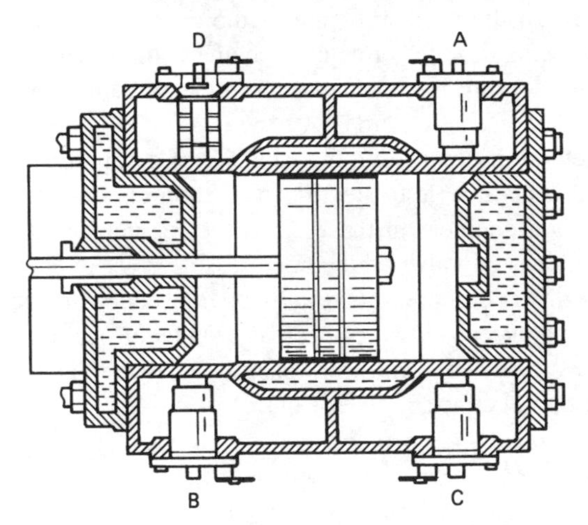

A, B, C, D are pockets referred to in Fig. 7.76

Figure 7.75 Pneumatic valves controlling four fixed pockets in compressor for five-step control.

A common practice in the natural gas industry is to prepare a single set of curves for a given machine unless there are side loads or it is a multiservice machine.

Figure 7.76 shows indicator cards that demonstrate the unloading operation for a double acting cylinder at three capacity points. The letters adjacent to the low-pressure diagrams represent the unloading influence of the respective and accumulative effect of the various pockets as identified in Fig. 7.75. Full-load, one-half and no-load capacity is obtained by holding corresponding suction valves open or adding sufficient clearance to produce a zero volumetric efficiency. No-capacity operation includes holding all suction valves open.

Figure 7.77 shows an alternative representation of compressor unloading operation with a step-control using fixed volume clearance pockets. The curve illustrates the relationship between compressor capacity and driver capacity for a varying compressor suction pressure at a constant discharge pressure and constant speed. The driver can be a gas engine or electric motor.

The purpose of this curve is to determine what steps of unloading are required to prevent the driver and piston rods from serious overloading. All lines are plotted for a single stage compressor.

The driver capacity line indicates the maximum allowable capacity for a given horsepower. The cylinder capacity lines represent the range of pressures calculated with all possible combinations of pockets open and cylinder unloading, as necessary, to cover the capacity of the driver.

Starting at the end (line 0-0) with full cylinder capacity, the line is traced until it crosses the driver capacity line at which point it is dropped to the next largest cylinder capacity and follows it until it crosses the driver line, etc. This will produce a "sawtooth" effect, hence the name "sawtooth" curve. The number of teeth depends on the number of combinations of pockets (opened or closed) required for unloading.

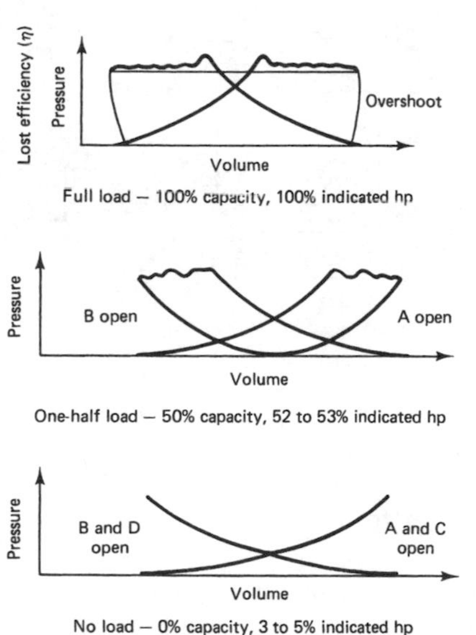

Figure 7.76 Indicator diagram for three load points of operation.

The same method is followed for multi-stage units. For each additional stage another "sawtooth" curve must be constructed (i.e., for a two-stage application); two curves are required to attain the final results.

Although control devices are often automatically operated, manual operation is satisfactory for many services. Where manual operation is provided, it often consists of a valve, or valves, to open and close clearance pockets. In some cases, a movable cylinder head is provided for variable clearance in the cylinder (Fig. 7.78).

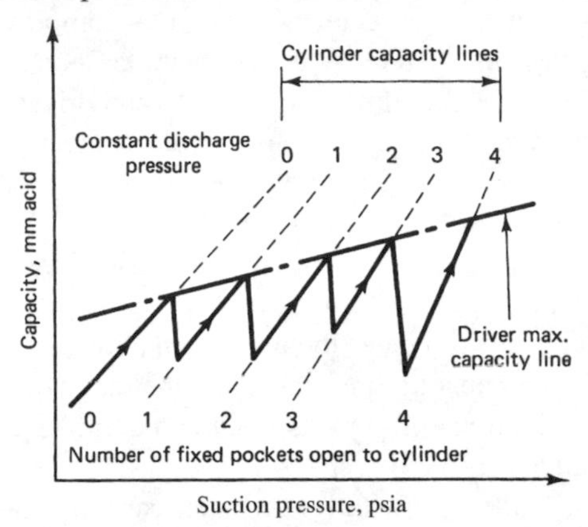

Figure 7.77 "Sawtooth" curve for unloading operation.

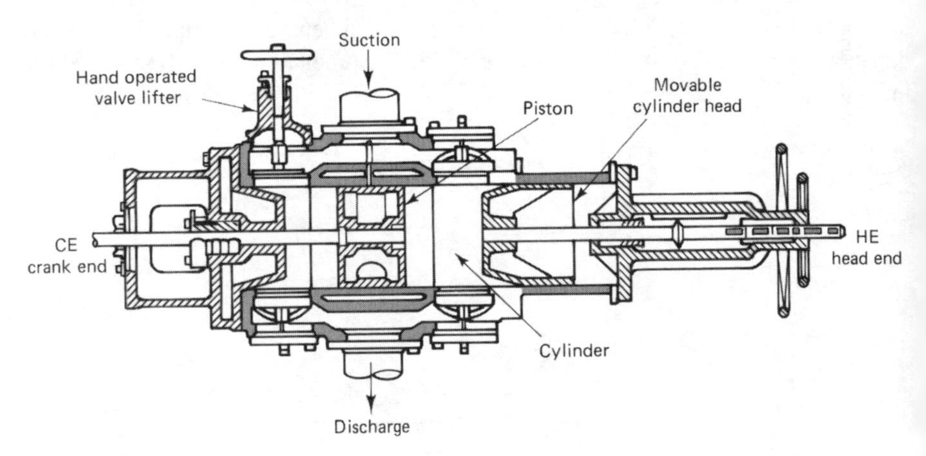

Figure 7.78 Sectional view of a cylinder equipped with a hand-operated valve lifter and variable-volume clearance.

Gas Pulsation Control

Pulsation is inherent in reciprocating compressors because suction and discharge valves are open during only part of the stroke. Pulsation must be damped (controlled) in order to:

1. Provide smooth flow of gas to and from the compressor.
2. Prevent overloading or underloading of the compressors.
3. Reduce overall vibration.

There are several types of pulsation chambers. The simplest one is a volume bottle, or a surge drum, which is a pressure vessel, unbaffled internally and mounted on or very near a cylinder inlet or outlet. A manifold joining the inlet and discharge connections of cylinders operating in parallel can also serve as a volume bottle. Performance of volume bottles is not normally guaranteed without an analysis of the piping system from the compressor to the first process vessel.

Volume bottles are sized empirically to provide an adequate volume to absorb most of the pulsation. Several industry methods were tried in an effort to produce a reasonable rule-of-thumb for their sizing. Figure 7.79 may be used for approximate bottle sizing.

Example 7.3

Indicated suction pressure = 600 psia
Indicated discharge pressure = 1400 psia
Cylinder bore = 6 in.

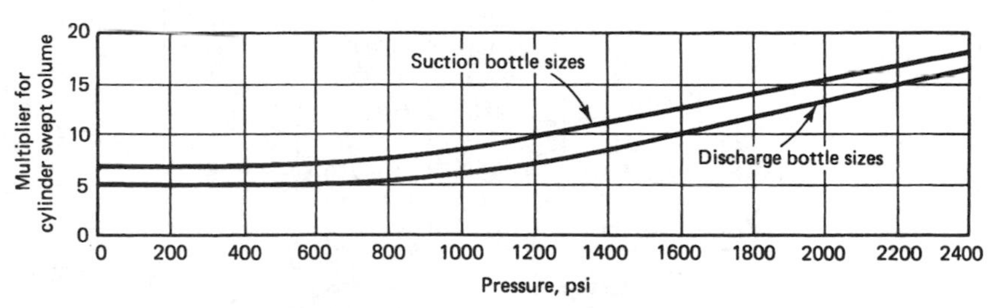

Figure 7.79 Approximate bottle sizing chart.

Cylinder stroke = 15 in.
Swept volume = $\pi\ (6^2/4)\ (15) = 424\ \text{in.}^3$

From the chart, at 600 psia inlet pressure, the suction bottle multiplier Is approximately 7.5. Suction-bottle volume = (7.5) (424) = 3.180 in.^{3}.

At 1,400 psi discharge pressure, the discharge bottle multiplier is approximately 8.5. Discharge-bottle volume = (8.5) (424) = 3600 in.^{3}.

Note: When more than one cylinder is connected to a bottle, the sum of the individual swept volumes is the size required for the common bottle.

For more accurate sizing, compressor manufacturers can be consulted. Organizations which provide designs and/or equipment for gas-pulsation control are also available.

Having determined the necessary volume of the bottle, the proportioning of diameter and length to provide this volume requires some ingenuity and judgement. It is desirable that manifolds be as short and of as large diameter as is consistent with pressure conditions, space limitations, and appearance.

A good general rule is to make the manifold diameter one and a half times the inside diameter of the largest cylinder connected to it, but this is not always practicable, particularly where large cylinders are involved.

Inside diameter of pipe must be used in figuring manifolds. This is particularly important in high-pressure work and in small sizes where wall cross section may be a considerable part of the inside cross section. Minimum manifold length is determined from cylinder center distances and connecting pipe diameters. Some additions must be made to the minimum thus determined to allow for saddle reinforcements and for welding of caps.

It is customary to close the ends of manifolds with welding caps, which add both volume and length. Figure 7.80 gives approximate volume and length of standard caps.

	Standard weight, Schedule 40		Extra strong, Schedule 80		Double Extra strong	
Pipe size	Volume, cu in.	Length, in.	Volume, cu in.	Length, in.	Volume, cu in.	Length, in.
4"	24.2	2 1/2	20.0	2 1/2	15	3
6"	77.3	3 1/2	65.7	3 1/2	48	4
8"	148.5	4 11/16	122.3	4 11/16	120	5
10"	295.6	5 3/4	264.4	8 3/4		
12"	517.0	6 7/8	475.0	6 7/8		
14"	684.6	7 13/16	640.0	7 13/16		
16"	967.6	9	911.0	9		
18"	1432.6	10 1/16	1363.0	10 1/16		
20"	2026.4	11 1/4	1938.0	11 1/4		
24"	3451.0	13 7/16	3313.0	13 7/16		

Figure 7.80 Welding Caps

Pulsation Dampers (Snubbers)

A pulsation damper is an internally baffled device. The design of the pulsation damping equipment is based on acoustical analog evaluation which takes into account the specified operating speed range, conditions of unloading, and variations in gas composition.

Analog evaluation is accomplished with an active analog that simulates the entire compressor, pulsation dampers, piping and equipment system and considers dynamic interactions among these elements.

Pulsation dampers also should be mounted as close as possible to the cylinder; and in large volume units, nozzles should be located near the center of the chamber to reduce unbalanced forces.

Pulsation dampers are typically guaranteed for a maximum residual peak-to-peak pulsation pressure of 2 percent of average absolute pressure at the point of connection to the piping system, and pressure drop through the equipment of not more than 1 percent of the absolute pressure. This applies at design condition and not necessarily for other operating pressures and flows. A detailed discussion of recommended design approaches for pulsation suppression devices is presented in API Standard 618, Reciprocating Compressors for General Refinery Services.

As pressure vessels, all pulsation chambers (volume bottles and dampers) are generally built to Section VIII of ASME Code and suitable for applicable cylinder relief valve set pressure.

Suction pulsation chambers are often designed for the same pressure as the discharge units, or for a minimum of two-thirds of the design discharge pressure.

Troubleshooting

Minor troubles can normally be expected at various times during routine operation of the compressor. These troubles are most often traced to dirt, liquid, and maladjustment, or to operating personnel being unfamiliar with functions of the various machine parts and systems. Difficulties of this type can usually be corrected by cleaning, proper adjustment, elimination of an adverse condition, or quick replacement of a relatively minor part.

TROUBLE	PROBABLE CAUSE(S)
COMPRESSOR WILL NOT START	1. Power supply failure. 2. Switchgear or starting panel. 3. Low oil pressure shut down switch. 4. Control panel.
MOTOR WILL NOT SYNCHRONIZE	1. Low voltage. 2. Excessive starting torque. 3. Incorrect power factor. 4. Excitation voltage failure.
LOW OIL PRESSURE	1. Oil pump failure. 2. Oil foaming from counterweights striking oil surface. 3. Cold oil. 4. Dirty oil filter. 5. Interior frame oil leaks. 6. Excessive leakage at bearing shim tabs and/ or bearings. 7. Improper low oil pressure switch setting. 8. Low gear oil pump by-pass/relief valve setting. 9. Defective pressure gage. 10. Plugged oil sump strainer. 11. Defective oil relief valve.
NOISE IN CYLINDER	1. Loose piston. 2. Piston hitting outer head or frame end of cylinder. 3. Loose crosshead lock nut. 4. Broken or leaking valve(s). 5. Worn or broken piston rings or expanders. 6. Valve improperly seated damaged seat gasket. 7. Free air unloader plunger chattering.
EXCESSIVE PACKING LEAKAGE	1. Worn packing rings. 2. Improper lube oil and or insufficient lube rate (blue rings). 3. Dirt in packing. 4. Excessive rate of pressure increase. 5. Packing rings assembled incorrectly. 6. Improper ring side or end gap clearance. 7. Plugged packing vent system. 8. Scored piston rod. 9. Excessive piston rod run-out.
PACKING OVER-HEATING	1. Lubrication failure. 2. Improper lube oil and/ or insufficient lube rate. 3. Insufficient cooling.
EXCESSIVE CARBON ON VALVES	1. Excessive lube oil. 2. Improper lube oil (too light, high carbon residue). 3. Oil carryover from inlet system or previous stage. 4. Broken or leaking valves causing high temperature. 5. Excessive temperature due to high pressure ratio across cylinders.
RELIEF VALVE POPPING	1. Faulty relief valve. 2. Leaking suction valves or rings on next higher stage. 3. Obstruction (foreign material, rags), blind or valve closed in discharge line.
HIGH DISCHARGE TEMPERATURE	1. Excessive ratio on cylinder due to leaking inlet valves or rings on next higher stage. 2. Fouled intercooler piping. 3. Leaking discharge valves or piston rings. 4. High inlet temperature. 5. Fouled water jackets on cylinder. 6. Improper lube oil and or lube rate.
FRAME KNOCKS	1. Loose crosshead pin, pin caps or crosshead shoes. 2. Loose worn main, crankpin or crosshead bearings. 3. Low oil pressure. 4. Cold oil. 5. Incorrect oil. 6. Knock is actually from cylinder end.
CRANKSHAFT OIL SEAL LEAKS	1. Faulty seal installation. 2. Clogged drain hole.
PISTON ROD OIL SCRAPER LEAKS	1. Worn scraper rings. 2. Scrapers incorrectly assembled. 3. Worn scored rod. 4. Improper fit of rings to rod/side clearance.

Figure 7.81 Probable causes of reciprocating compressor trouble.

Major trouble can usually be traced to long periods of operation with unsuitable coolant or lubrication, careless operation and routine maintenance, or the use of the machine on a service for which it was not intended.

A defective inlet valve can generally be found by feeling the valve cover. It will be much warmer than normal. Discharge valve leakage is not as easy to detect since the discharge is always hot. Experienced operators of water-cooled units can usually tell by feel if a particular valve is leaking. The best indication of discharge valve trouble is the discharge temperature. This will rise, sometimes rapidly, when a valve is in poor condition or breaks. This is one very good reason for keeping a record of the discharge temperature from each cylinder.

Recording of the interstage pressure on multi-stage units is valuable because any variation, when operating at a given load point, indicates trouble in one or the other of the two stages. If the pressure drops, the trouble is in the low-pressure cylinder. If it rises, the problem is in the high-pressure cylinder.

Troubleshooting is largely a matter of elimination based on a thorough knowledge of the interrelated functions of the various parts and the effects of adverse conditions. A complete list of possible troubles with their causes and corrections is impractical, but a list of the more frequently encountered troubles and their causes is offered as a guide in Figure 7.81.

SYSTEM COMPONENTS

The process and oil and gas industries have basic API (American Petroleum Institute) specifications that apply to the applications and design of the compressors and system components. In the process industry, API-618 provides reciprocating compressor design criteria. API-11P is the general purchase specification relating to packaged reciprocating compressors for oil and gas production services. Each of these specifications addresses system components like the compressor, driver, instrumentation, pulsation control, coolers, and so on. It is common for the API specifications to refer to other detailed specifications, such as TEMA for coolers, ANSI for piping, and NEMA for motors. ASME B19.3 covers safety standards for process compressors.

Instrumentation

The typical oil field compressor has a basic protection and monitoring control system. Sensors monitor oil and water temperatures, process pressures and temperatures, vibration levels of major components, plus oil and water pressures. Control systems have a first-out function to indicate to the operator the cause of equipment shutdown. On engine-driven compressors, the controls are powered by the ignition system or are pneumatic, using field-gas or instrument air.

Figure 7.82 shows a pneumatic control panel for an oil field compressor.

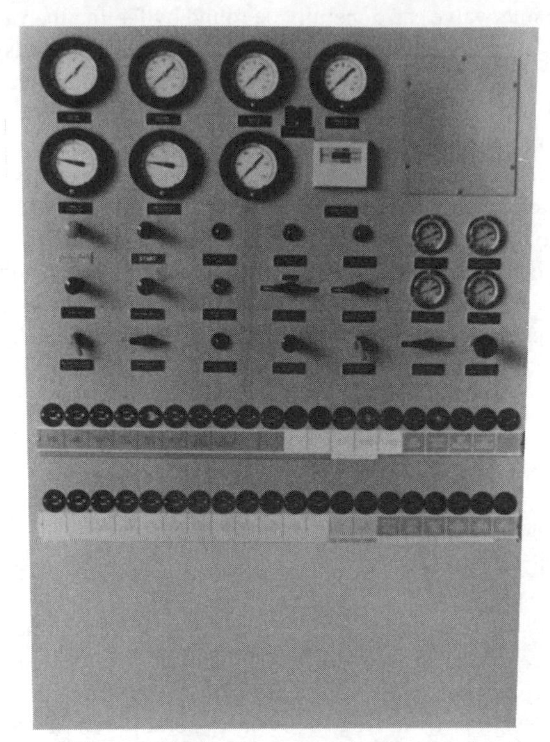

Figure 7.82 Pneumatic Control Panel

Electric-motor-driven compressors usually use electric power for the controls; however, pneumatic controls are sometimes used. In nonhazardous locations, electric controls are generally used; however, pneumatic controls are sometimes used. In hazardous locations, pneumatic controls are generally used.

Control Systems on process machinery are often much more complex. It is common to utilize all logic control systems that will handle full start-up and loading procedures of the machine, plus automatic capacity adjustment. The increased use of microprocessors provides another dimension to instrumentation. Microprocessors can perform calculations with the recorded data, and more precise process control is possible.

Gas Piping

Process gas piping for compressor applications is generally designed and fabricated in accordance with ANSI B-31.3. Compressor piping should be carefully arranged to avoid strains that may cause machinery misalignment. Misalignment is a frequent cause of vibration, and most misalignment is directly traceable to piping strains. On packaged gas field compressors, piping runs are normally short enough that pipe strain due to thermal expansion generally does not create a problem.

Piping must be sized to minimize pressure drops. On oil field applications with 0.65-specific gravity natural gas, compressor packages typically use a design velocity limitation of 2500 to 3000 ft./min. Package compressors normally utilize pipe clamps to support and hold the piping in place.

On process machines, which normally involve longer pipe runs, care must be taken to allow for reduction in strain on the machine due to (1) the dead weight of the piping itself, or (2) expansion or contraction of the pipe as it undergoes temperature change. Prevention of excessive dead loads on the equipment flanges is normally handled with flexible-type supports. Strain due to expansion or contraction is controlled by means of a bellows-type expansion joint on low-pressure applications. High-pressure installations utilize an expansion loop or pipe bend adjacent to the flange of the compressor to control pipe strain.

Pulsation Control

Since the flow into and out of a reciprocating compressor is cyclic, pressure pulses are formed in the suction and discharge gas piping. Pulsations are also directly related to rotative speed, compressor stroke, and gas passage volume. In some cases, high rpm, short stroke volume bottles can be eliminated entirely. However, if pulsations are not minimized and controlled, severe vibration or compressor valve breakage, or both, can occur, resulting in a hazardous operating condition. Additionally, the efficiency of the compressor may be affected.

One method of pulsation control uses volume bottles (with no internal parts) mounted on both the inlet and discharge of each cylinder. Bottles should be mounted as close to the cylinder as possible. Figure 7.83 shows typical pulsation bottles mounted on a compressor cylinder. Historically, a rule of thumb formerly used, hoping to achieve a design peak-to-peak pulsation level of about 5 percent for a 0.65-specific gravity natural gas at less than 1500 psi, was to make the suction bottle equal to 12 times the swept volume of the cylinder and the discharge bottle 10 times the swept volume. It should be pointed out that this is not a very sure method and does not assure avoidance of resonant conditions and high peak-to-peak pulsations. For critical services, or where it is preferred to supply smaller bottles to achieve a given degree of pulsation control, vessels with internal baffles and choke tubes are furnished. Analog and digital modeling techniques are available to simulate pulsation levels throughout a system. These simulations are done regularly in the process industry to provide vessel and piping designs for reliable, long-term compressor and piping system operation. Typical pulsation levels with an acoustic analysis are 2 percent peak-to-peak or less. Generally, it is recommended to do a mechanical and acoustical study for each installation.

Figure 7.83 Pulsation bottles mounted on a cylinder.

Pulsation bottles are considered unfired pressure vessels and are generally designed and built to ASME Code Section VIII. It is a good design practice to use reinforcing pads on all nozzle connections for pulsation bottles.

Analog analysis. The entire piping and compressor system can be studied by an electronic analog device. The analog simulates the flow of gas through piping as if it were alternating current. The piping system is simulated by an equivalent electrical system. The voltage drop through the system due to electrical resistance is correlated to the pressure drop through piping and heat exchangers. Electrical resistors are used to simulate flow friction through an orifice, the orifice being represented as a linear device with the correct pressure drop at the average flow rate. The compressor system, including pulsation bottles, coolers, and gas piping on the machine and the customer's piping from the machine to the first major vessel of the user's system, is scanned by the analog. If large pulsations are detected, the piping system is modified on the analog hoard, simulating physical changes to the system that would correct the gas flow pattern of the actual system.

The cathode tube gives an image of the pressure-volume card inside the compressor cylinder. The waveform through any section of the piping can he viewed. If acoustical resonance in the piping system is at the same frequency as a major harmonic of compressor operating speed, or if the vessels are too small, pulsation amplitude may be excessive. Pulsation levels can be reduced by changing the size and configuration of the vessel, adding orifices to the piping, or modifying the length and diameter of pipe in the system.

The pressure-volume diagram of each compressor cylinder can be viewed. The card will be distorted if the piping to and from the compressor cylinder causes high pulsation levels. It is very possible that the waveform is such that it adversely affects the opening or closing of the valve(s). Whether this distortion is detrimental can only he determined by calculating the effect of the pulsations on the valve operation. If distortion occurs in the actual compressor, the valves may be slammed closed, resulting in failure of valve plates or valve springs, or both. Pulsations can develop dynamic wave action, which could cause supercharging or starvation of gas to the cylinder. Capacity may he increased or decreased accordingly. Part load operation of the compressor should be scrutinized by the analog method.

Digital analysis. A digital analysis can be used to provide piping system analysis similar to that of the analog method. Two methods are commonly used; each has advantages and disadvantages relative to the other and to the analog simulation.

The analog and impedance digital methods treat the system as a series of pipes connected together. The flow in each pipe is assumed to be one dimensional. In a digital analysis using the impedance method, the response of each pipe to the frequency of interest is calculated. By combining these, the response of the complete system is obtained. When this system is stimulated by pulsations of the frequency and amplitude produced by the cylinder, the pulsations in the piping system are obtained. This is a linear analysis, and the various harmonics of interest are calculated separately. The assumptions made are the same as those made in an analog study. The advantages of this method over the analog are that (1) specialized equipment is not needed and the size or number of jobs that can he analyzed is not limited by the special equipment available, and (2) the time taken to set up a digital calculation is less than that required to build an analog of the piping system. The advantage of the analog is that once the analog is set up on the board, runs representing variable-speed or unloading conditions can he made instantaneously, whereas separate runs are required with a digital analysis.

The assumption that all effects are linear, which is inherent to the analog and impedance methods, precludes the accurate simulation of losses through orifices, compressor valves, and other nonlinear components.

In the time-domain method of digital analysis, a small time step (representing a few degrees of crankshaft rotation) is chosen and the solution built up by successively calculating pressures and temperatures in the cylinder and piping at each time step through the cycle. This method is nonlinear and can accurately simulate conditions in the cylinder, the pressure drop, and the dynamics of the compressor valves, orifices, pipe friction, and other nonlinear losses. It does, however, require significant amounts of computer time and is usually best used for a check of the final design, including investigation of the effect of the pulsations on the valve dynamics after the design has been developed using the impedance or analog methods.

It is also possible to combine an impedance analysis of piping with a time-domain solution for the cylinders and valves. It is then necessary to iterate to obtain the interaction between the two.

Mechanical analysis. The acoustic analysis of the piping system determines the predominant gas pulsation frequencies present in the system. The mechanical analysis is completed to assure that these frequencies do not excite the system's mechanical natural frequencies to cause excessive piping vibration. A review of the type and location of supports is included with the mechanical analysis.

Separation

Reciprocating compressor valves are not very tolerant of liquids in the gas stream, so proper separation of liquids from the gas stream is essential to ensure successful operation. The two most common types of scrubbers used in the oil and gas industry for compressors are a stainless-steel mesh pad or vanes. Figure 7.84 shows a typical oil field compressor scrubber. The principle of both designs is the same; they use a velocity reduction and a sharp change in direction to drop out large liquid droplets and particulate contamination. The gas then makes repeated changes in flow direction as it passes through a 4 to 6 in. mesh pad or through vanes, depending on the design. This change of direction causes entrained liquid to coalesce and drop out of the gas stream. The stainless mesh pad has the advantage of being cheaper than the vanes and has a wider capacity operating range. However, the efficiency of the vane-type scrubber is higher and a smaller diameter scrubber can be designed. The vanes are especially attractive for larger separators.

These same scrubbers are used in the process industry along with many specialized separators for individual processes. When there are particulate contaminants, as well as liquid to be separated from the gas stream, multi-stage coalescing filter-separators are used. These have replaceable or cleanable filter elements, or both, and provide a high degree of protection.

Centrifugal-action separator elements are often used for process applications. Locating the separator element as part of the suction pulsation vessel provides additional protection close to the cylinder. This ensures that most entrained liquids are removed before the gas enters the cylinder. Scrubbers are considered unfired pressure vessels and are usually designed to ASME Code Section VIII. It is also recommended that reinforcing pads be used in all major nozzle connections for scrubbers.

Figure 7.84 Oil Field Gas Scrubber

Cooling

During the compression of gas, heat is generated. For compression ratios greater than 3, intercooling of the gas stream may be required. Cooling of the gas stream has two objectives: (1) to reduce the gas temperature and volume, and (2) to condense and remove water vapor or other condensable constituents. Both of these objectives tend to reduce power requirements. Cooling after thc last stage of compression (aftercooling) does not reduce power, but is done to promote safety and to liquefy condensables that might otherwise deposit under undesirable conditions.

Water cooling is required to cool auxiliary support systems such as engine jacket water or lube oil and cylinder jacket water, when required. Water also can be used as a primary cooling medium for compression.

The two main types of coolers used are air cooler and water coolers.

Air-cooled designs. Air coolers are of fin-tube construction with the compressed gas passing through the tube. A cooler may consist of a single cooler section or several sectionalized assemblies. Air coolers are most common in the oil field where cooling water is scarce or of poor quality. Air coolers transfer the heat from the compressed gas or water to the atmosphere. On engine-driven packages, the cooler fan would typically be driven from the engine, and the engine cooling would be in a closed system using a radiator section in the cooler to discharge the engine's heat to the atmosphere. Figure 7.85 shows an air-cooled cooler used for oil field applications.

Figure 7.85 Air-cooled heat exchanger during transport to job site.

The two major types of air-cooled heat exchangers are induced draft and forced draft. Generally, the induced draft design is less expensive to fabricate; however, it does require more fan horsepower to provide the same amount of cooling. Project economics will dictate which design is applicable.

Water-cooled designs. When cooling water of good quality is available, shell and tube coolers are more economical and also much quieter. There are again two basic designs of water-cooled heat exchangers: (1) gas in the tubes, water in the shell, and (2) gas in the shell, water in the tubes. There are advantages and disadvantages to both designs. Proper selection for a particular application depends on cooler size, pressure limitations, and relative cleanliness of the gas and water.

Process compressors usually utilize shell and tube coolers as they are normally installed in process plants or refineries. API-618 is a reference for the major specification that would apply to cooler designs for process compressors.

Utility Piping

Utility piping consists of piping for oil, water, vents, drains, fuel, or gas engine supply. In general, care must be taken in design and fabrication of utility piping to ensure the following:

1. Cleanliness.

2. Ease of operation and maintenance (i.e., drain and fill valves on all systems should be placed for easy access by the operator). High points on water lines should have vent valves, and low points should have drain valves. Break flanges, unions, or isolation valves should be placed around common maintenance items like auxiliary water pumps.
3. Line sizes should be adequate to minimize pressure drops in water and oil systems.

Package

Skid. Process compressors are normally not skid mounted. In oil field applications, however, they are normally skid mounted packages. The driver, compressor, cooler, and all the necessary items are mounted and aligned on a skid. Thus, a properly designed and fabricated skid is of utmost importance. Figure 7.86 shows a typical oil field compressor skid. The skid must be designed to withstand the loading and handling of the equipment during the move from the fabricating facility to the job site. It must also withstand moves from one job site to another, as packaged compressors are frequently relocated when operating conditions change. It must be large enough to support all the auxiliary equipment and still provide adequate working space around the machine for maintenance and operation. At the same time, it must be compact enough to allow shipment to a job site without disassembly. On smaller units, less than 800 hp, skids are often concrete filled. This minimizes the foundation requirements, and in some cases the units can be installed on a firm, level underbase such as packed gravel.

Figure 7.86 Separable Oil Field Compressor Skid

Transportation costs will increase due to additional weight. Proper alignment of the engine and compressor on the skid is such a key fabricating consideration that most compressor packagers optically align these components on the assembly floor prior to final shimming or grouting of the engine or compressor to the skid. Extra time spent during assembly is paid back several fold by easier, quicker installation in the field.

Assembly. During the design and assembly of a compressor package, care must be taken to ensure that safety, ease of operation, and maintenance are given high priority. Inspection openings must he clear of obstructions so they can be properly utilized. Instrumentation should be placed and oriented so it is clearly visible to the operator standing at ground level. There must be sufficient space on the package so that the operator has normal access for maintenance on items such as lubricator pumps, separator dump valves, compressors, engines, and other routine maintenance items.

The package must be designed to provide maximum operator safety. All belts, couplings, and flywheels should be totally enclosed by guards that meet the necessary standards. Provisions should be made to pipe relief valves to a safe location (often to the top of the cooler). All shutdown instrumentation should be calibrated and tested. If the unit is to be unused or stored for a period of time, provision must be made for protection of exposed surfaces, instruments, and anything else that could be damaged by the elements.

Valving

All compressors require process valving to operate them safely. This valving can be furnished either as part of the package or as part of the installation. The following is the minimum valving recommended.

1. *Suction block valve:* A suction block valve is used to isolate the machine from the supply when maintenance is necessary.
2. *Relief valves:* Relief valves should be furnished in the suction line, in each interstage before the cooler, or in the discharge before the cooler. Relief valves are required by ASME, Section VIII, which is a minimum standard to which most pressure vessels are designed and built. Rupture discs are sometimes furnished in lieu of relief valves. The relief valve must be set to protect the component with the minimum design pressure, and it must be large enough to relieve the maximum predicted capacity.
3. *Discharge check valve:* A discharge check valve may be installed immediately downstream from the compressor unit. This protects the unit from being exposed to full line pressure when the machine is shut down.
4. *Discharge block valve:* A discharge block valve is installed downstream from the discharge check valve and provides positive isolation of the machine for maintenance or repair.

5. *Blowdown valve:* A blowdown valve should be installed preferably between the discharge check and discharge block valves. It is used to depressurize the machine for maintenance purposes.
6. *Bypass valve:* The bypass valve takes gas from compressor discharge upstream of the check valve and allows it to be recycled back to the suction of the machine downstream of the suction block valve. This valve can be manual or automatic. It also is used to unload the machine during start-up.

After the machine is up to idle or operating speed, the bypass valve can be slowly closed to load up the unit. This method of loading will increase the life of the engine driver.

The bypass valve also can be used for capacity-control purposes to maintain a minimum suction pressure or maximum discharge pressure. When used in this manner, it is necessary to use gas downstream of the aftercooler to prevent overheating due to the recycling of hot gas.

In a full logic control system, suction, bypass, blowdown, and discharge valves can all be automated to open and close as required. Figure 7.87 shows a schematic diagram of a standard process valve system for compressor installation.

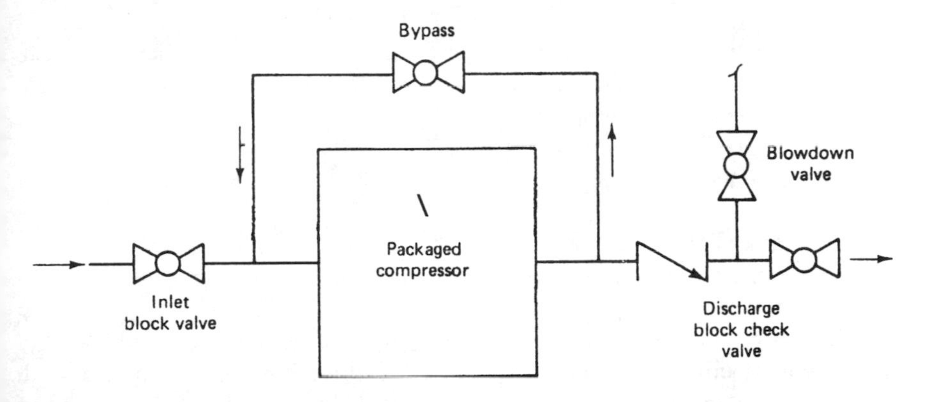

Figure 7.87 Schematic of bypass valve system. Only cooled gas generally is bypassed back to suction. Low percentage flow bypass or short duration bypassing may not require coding.

PRIME MOVERS

The prime mover or driver of a compressor is the main power source that provides the energy to drive the compressor. The driver, through the coupling or other connection, must provide the power to start the compressor, accelerate it to full speed, and keep the unit operating under any design condition of capacity and power.

In the selection of a driver for a positive-displacement compressor, two important factors should he taken into consideration. A complete technical analysis should be done, considering the application, in addition to a total economic analysis.

The technical conditions should include the following items:

1. Application and service requirements of the compressor.
2. Available power sources.
3. Compatibility of driver and compressor.

Commercial consideration should include the following cost items:

1. Driver first cost.
2. Fuel consumption or power cost.
3. Maintenance cost.
4. Installation cost (especially when comparing natural gas versus an electric motor).
5. Environmental considerations.

It is normally to the user's advantage for the compressor supplier to furnish both the driver and accessories so that total responsibility for a properly designed system is assumed by one source.

Natural Gas Engines

Gas engines are normally divided into two general categories related to speed. These categories are slow-speed engines (0 to 600 rpm) and medium-speed engines (600 to 2100 rpm).

Slow speed engines are commonly used in integral gas engine compressors. An integral engine utilizes a common crankshaft to drive both the compressor and power cylinders. Integral machines are further divided into two horsepower groups: small horsepower, 25 to 800 hp, and large horsepower, 800 to 7000 hp.

Small horsepower integral engines are generally used for oil field services such as gas gathering, gas injection, and small gas processing plants, and in some sizes, 25 to 100 hp, are available at up to 1400 rpm. Large horsepower units are used in process plants, main line gas transmission, gas injection, and large gas plants.

Figure 7.88 shows a typical main line gas transmission integral engine installation. Figure 7.89 is an integral gas engine packaged for offshore gas lift.

Figure 7.88 Gas transmission integral engine compressor.

Medium-speed gas engines (600 to 2100 rpm) are generally used to drive separable oil field compressors. Horsepower sizes range from 5 to 3600 hp. The smaller horsepower driver, 5 to 400 hp, is generally medium speed, 1400 to 1800 rpm, and can be direct connected to a compressor or used as a V-belt driver. Occasionally, a gear speed reducer is used.

Figure 7.89 Packaged integral gas engine offshore gas lift.

Large horsepower drivers, 300 to 3600 hp, are generally all direct connected and operate in speed ranges from 600 to 1200 rpm. There is a general industry trend to further increase driver speeds, consistent with increased compressor operating speeds. Figure 7.90 shows a typical, packaged, medium-speed, compressor installation.

Figure 7.90 Gas engine-driven two-stage compressor in a mini-gas plant.

The natural gas engine driver is currently the most common driver used in the oil and gas industry Care should be used when selecting a gas engine driver to be certain that the horsepower rating meets the required service. Engine manufacturers can rate their engines for maximum, intermittent, or continuous use, all of which are applicable, depending on the application. Some gas engines are rated on a Diesel Engine Manufacturers Association (DEMA) basis, which is a continuous-use industrial rating.

Recent exhaust emission legislation, both federal and state, makes it necessary to review the emission characteristics of the engine prior to purchase. Low-emission engines or catalytic converters are two potential solutions to exhaust emission requirements.

Electric Motors

Electric motors are the most widely used compressor drivers when considering air compressor and process compressor applications. As previously stated, natural gas engines are the primary oil and gas field drivers; however, electric motors are used more frequently than formerly due to environmental considerations. These motors do not have emission problems.

All major motor manufacturers in the United States build their motors to NEMA (National Electrical Manufacturer Association) standards. There are three basic types of motors available that apply to process and oil field compressors. These are induction motors, synchronous motors, and DC motors. Proper motor selection is a very critical decision and involves several key factors. Each motor selected as a driver should be analyzed from a technical and economic point of view.

Induction and synchronous motors are generally selected according to the chart in Fig. 7.91. This selection chart does not, however, take into account all the economic and technical factors that should be reviewed prior to final selection.

Economic considerations in selecting a motor should include the motor cost, maintenance cost, operating cost, and, most important, an analysis of the accessory equipment required to operate the motor. Some of the key accessories that should he reviewed are the motor starter, motor controls, any transformer requirements, and cost of power lines, if applicable.

The technical considerations include the following items:

- A proper motor-to-compressor speed match
- Torsional analysis, including the flywheel effect (WR^2 of the system)
- Proper motor enclosure for the area classification and weather conditions in which the motor will operate
- Voltage and frequency required
- Speed: torque requirements for starting and operation
- Current restrictions, including KVA inrush required for starting
- Analysis of power-factor corrections in the system
- Altitude of installation
- Ambient temperatures at job site
- Desired motor efficiency
- Number of starts per hour required for normal service
- Service factor of the motor
- Insulation requirements

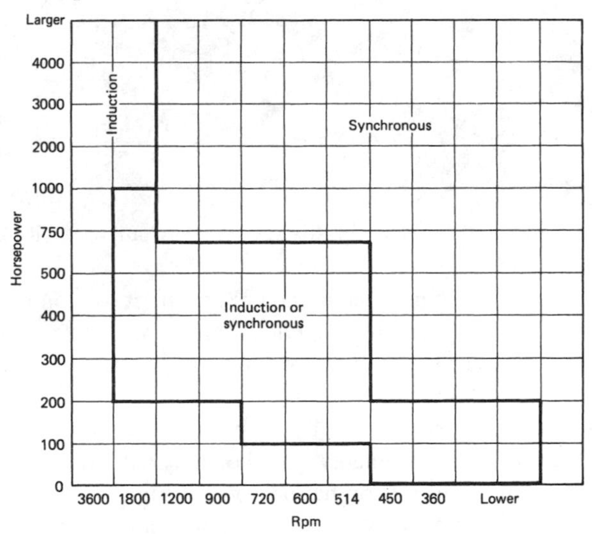

Figure 7.91 Motor selection chart for general areas of application of induction and synchronous motors.

It is very important to apply an electric motor properly. Motors must be designed to deliver a minimum full-load continuous horsepower. Good design practice dictates motor power at least 10% over the worst-case compressor horsepower requirements. This safety factor can be obtained in the service factor of the motor.

Induction motors. The most common compressor motor driver is the induction motor. Induction motors generally have good efficiency and excellent starting torque, but rather high inrush current requirements. Inrush current is the amount of current required to start the motor and driven equipment. Induction motors operate at speeds below synchronous speed by a value known as slip, which varies with the load. Full-load slip varies from 1 percent for very large motors to 5 percent for smaller motors. Induction motor efficiencies are in the high 80 or low 90 percentile, depending on the horsepower available. Smaller-horsepower induction motors are generally less efficient. Figure 7.92 shows an induction-motor-driven oil field compressor package.

Figure 7.92 Induction Motor-driven Oil Field Compressor Package

Synchronous motors. Synchronous motors are the most common com-pressor drivers used for higher-horsepower applications as may be seen in the chart in Fig. 7.87. These motors are typically more efficient than induction motors, with efficiencies in the range of 93 to 97%.

Synchronous motors must be carefully analyzed because of their lower torque characteristics. The torque requirements must be analyzed along with the starting current inrush requirements to assure that proper starting power will be available. Synchronous motors operate at synchronous speed and do not have the slip characteristics of induction motors. Figure 7.93 shows a typical synchronous-motor-driven process compressor.

Figure 7.93 Synchronous Motor-driven Process Compressor

DC motors. In the past few years, the use of DC motors as oil field compressor drivers has increased in popularity. The reason for these increases are threefold:

1. Availability of DC traction motors.
2. Variable-speed capabilities of DC motors to control compressor capacity.
3. Economic considerations of motor drive versus engine drive.

Offshore oil field compressors are using more DC motor drivers because of the added speed flexibility, lower initial cost, and projected lower maintenance costs. When utilizing DC motors in a hazardous atmosphere, it is necessary to provide a continuous positive air pressure in the motor enclosure to assure that no gas can get into the motor and be ignited by the motor. Figure 7.94 shows a DC-motor-driven offshore separable package.

There are numerous motor enclosures available to meet various operating requirements. Enclosures can be designed to prevent the entry of water or dirt and permit operation in a hazardous atmosphere.

There are several hazardous-area classification groups. These hazardous groups have been defined by the National Electric Code. Class 1, group D, is the general section that normally applies to compressor motors. This section is broken down into two further divisions:

1. Division 1 locations are those in which hazardous concentrations of flammable gas or vapors exist either continuously or periodically during normal conditions.

2. Division 2 locations are those in which flammable gases are handled, processed, or used. The gases will normally be confined within closed systems from which they can only escape in the event of accidental breakdown or abnormal pressure requirements.

Figure 7.94 Direct Current Motor-driven Separable Gas Compressor

A class 1, group D design is not explosion proof unless specifically certified by Underwriters Laboratories, Inc.

The improvement in the electronics control industry has greatly increased the potential for motors to be utilized as compressor drivers, especially in oil field applications. This has happened because of technological advances in motor controls. It is now economical to buy induction motors or synchronous motors with variable-speed controls to adjust the compressor operating speed. DC motors, having inherent variable-speed capability, already provide the needed variable speed with little further equipment needed. Variable speed to control compressor performance is a very desirable characteristic of a compressor prime mover.

Gasoline Engines

Gasoline engines are seldom used in the process and oil and gas industry because of high fuel cost. They are primarily used as drivers for standby compressors. The operating and application characteristics are similar to those of natural gas or diesel engine drivers.

Diesel Engines

Diesel engines are used rather infrequently as process and oil and gas industry compressor drivers. However, there are some applications, such as air drilling compressors, kick-off compressors (used to start an oil field gas lift), fire floods, or standby compressors, where diesel engines are the most economical drivers.

Diesel engines that can burn either diesel oil or natural gas also have limited applications. The dual-fuel configuration allows the operator to select the most economical fuel. For example, natural gas may be readily available in the summer but not economically available in the cold winter. Thus, the operator can use diesel fuel in the winter and gas in the summer. Some dual-fuel engines can be converted to spark ignition gas engines rather easily. Figure 7.95 shows a diesel-driven air compressor package used for air drilling.

Figure 7.95 Diesel Engine-driven Oil Field Air Compressor

Steam Turbines

Steam-turbine drivers are ordinarily used to drive positive-displacement compressors in process applications where steam is readily available as a power source. Generally, it is not economical to use steam as a driver fuel unless it is available as a result of a process in a refinery or process plant. Figure 7.96 shows a steam-turbine-driven process compressor.

Figure 7.96 Steam turbine-driven process compressor. Note gear box in drive train.

Careful analysis of the mechanical drive train is necessary. A complete torsional analysis must be done by the compressor supplier with particular emphasis on the coupling and quill shaft selection.

Steam turbines for mechanical drive service are available in horsepower ranges from 10 hp up to several thousand horsepower. Speed ranges vary with horsepower size. It is generally possible to find a turbine and compressor speed match so that a single or double reduction gear can be selected for the drive train.

Several types of steam turbines are available. The proper steam turbine should be selected from the various types available to suit the overall economics. Single-stage noncondensing, multi-stage condensing, or multi-stage noncondensing turbines are the types most generally used in process applications.

The steam conditions available will generally determine the type of steam turbine to be used. If the turbine exhaust steam can be utilized in the plant process, a back-pressure turbine should be used. If steam consumption is the prime economic factor, then a condensing turbine should be selected. Many specialty turbines are available, such as mixed-pressure turbines in which the steam not only drives the turbine, but is bled off at various pressures for process use.

A complete steam thermal balance analysis should be done to select the proper design of steam turbine. Steam costs versus alternate fuel costs should be reviewed prior to final selection.

Gas Turbines

Gas turbines used as prime movers for positive-displacement compressors have limited application in the process and oil and gas industry. The gas turbine is relatively new as a driver compared to the gas engine, motor, or steam turbine. However, there are some applications in which gas-turbine-driven reciprocating compressors may have an advantage. One application is offshore compression, where weight is an important consideration. Figure 7.97 shows a gas-turbine-driven, offshore packaged compressor. Another application could be in a refinery or process plant where the gas turbine exhaust heat could be utilized to improve the overall plant thermal efficiency.

Mechanical analysis of the drive train, including a proper torsional analysis and coupling/quill shaft selection, is very important. Other important considerations are the turbine controls and how they interface with the process or compressor control requirements. Proper care must be taken to be sure that the turbine horsepower available exceeds the horsepower required by the driven compressor, and that the torque is adequate throughout the compressor operating range. The loading and unloading steps must also be reviewed to be certain that no operating problems are encountered during normal start-up or shutdown or during emergency shutdown.

Figure 7.97 Gas-turbine-driven oil field compressor package for offshore application.

Hydraulic Turbines

An hydraulic turbine is basically a centrifugal pump operating in reverse. The application of an hydraulic turbine would be a specialty situation where sufficient high-pressure liquid exists in a refinery or process plant. By decreasing the liquid pressure across the turbine, the pressure of the liquid is reduced to a desirable level and power is recovered. When high-pressure liquid is available, this type of driver offers essentially free energy.

The drive train normally would involve a reduction gear or, in the case of small horsepower, a V-belt drive. A torsional analysis of the drive train is required, along with the proper coupling selection.

Environmental Considerations

The Federal Clean Air Act of 1963 and subsequent amendments provided funds to develop, establish, and maintain air-pollution control programs for state and local agencies. Each state has subsequently taken the basic federal guidelines and developed more in-depth, detailed requirements that suit its local population and environmental considerations. Most states have adopted laws that limit the maximum NOX and CO emissions to 250 tons per year per plant site. As previously mentioned, each state has its own regulations, and each individual application must be reviewed for conformity to current regulations.

In the oil field, two methods have been developed to deal with the emission control requirements for engine exhaust systems. One method is to use catalytic converters, which use proprietary precious metal combinations dispersed over a metal or ceramic subbase. When the engine exhaust gas is passed over the converter, it assists in converting these gases to acceptable alternatives and reduces the NOX in particular.

Some manufacturers use a pelleted catalyst instead of the metal or ceramic subbase. These catalytic converters generally tend to achieve 90% reduction of NOX, CO, and HC, with the most emphasis put on NOX. Several companies manufacture catalytic converters. Some of these companies can furnish specialized catalysts designed specifically to meet specified emission requirements and codes.

The second, and preferred method, is to utilize an engine with a modern-day, clean-burn combustion system that burns the fuel cleanly and emits very low levels of NOX, CO, and HC.

INSTALLATION AND CARE OF STATIONARY RECIPROCATING COMPRESSORS

Many installation requirements of positive-displacement gas compressors are the same as those for other positive-displacement compressors. To avoid repetition, the reader is referred to the section on installation and care of stationary, reciprocating compressors in Chapter 2. Included there are discussions of compressor location, foundations, subsoil characteristics, concrete, grouting, foundation bolts, alignment, leveling, and other topics.

The following further discussions include material that applies specifically to gas compressors, as well as some more general material discussed here for convenience.

Skid-mounted Units

The selection and preparation of a site for the installation of a skid-mounted unit should be made prior to the arrival of the unit.

1. The site should be selected to:
    a. Isolate the package from hazardous areas.
    b. Minimize the piping system.
    c. Allow space for transport truck, service equipment, and personnel.
    d. Keep sufficient distance from areas that are potential problems regarding noise and exhaust emissions.
    e. Provide adequate drainage.
    f. Optimize cooling.
2. The pad area should be solid, reasonably smooth, and level. Uniform contact between the support material and the underside of the skid is the most important consideration to prevent high centering (teetering) and resultant vibration problems. The manufacturer should be contacted for recommendations for the specific location being considered.

Platforms

Offshore platform design constraints must take into consideration the following: water depth, deck height above mean water level, wave conditions (height, speed, direction, and frequency), wind speed and direction, platform load, unbalanced forces, couples and vibrations transmitted by equipment to the platform, natural frequency of the structure, and sea bottom conditions (mud, gravel, rock, etc.). Advice of a consultant versed in platform design is recommended.

Site Considerations

Offshore. The location of a platform must take into consideration the following: water depth, sea bottom conditions, proximity to undersea facilities (wells, pipelines, control modules, etc.), as well as sea and wind conditions. The employment of a consultant familiar with code requirements, platform design, and local conditions is recommended.

On shore. Site considerations include proximity to the facility served (pipeline, refinery, industrial plant, etc.), soil conditions, availability of utilities, elevation, ambient temperature range, neighbors, noise levels, exhaust emissions, and ambient conditions (desert, swamp, woodlands, industrial park, etc.).

Packaged compressor installation guides. Portable packaged reciprocating compressors are highly popular in oil and gas field operations. The following suggestions may be helpful to the engineer planning or supervising the installation of a packaged unit (Figs. 7.98 and 7.99). These installation comments are not intended to replace specific requirements of the package or compressor manufacturer nor to supersede piping codes, safety regulations, and prevailing laws and ordinances that may be applicable to a particular installation.

Figure 7.98 Field installation, 32 Bhp, 1400 rpm integral engine-compressor in casinghead gas service.

Figure 7.99 400 Bhp, 1400 rpm, balanced-opposed three-stage compressor being trucked to dock for offshore platform casinghead gas service.

Selection and site preparation for packaged compressors. Prior to arrival of the compressor at the job site, certain preparations should be made.

Site. The compressor site should be selected in much the same way as for skid-mounted units. However, since the requirements are not completely identical, a separate list is given for convenience. The site should:

1. Isolate the package from hazardous areas, such as storage tanks, open flames, and vents.
2. Minimize length of piping runs to be made.
3. Allow space to maneuver the truck that will transport the unit to the site and set it on location. Other space considerations would be work space for mechanics and access for trucks that will deliver bulk oil.
4. Keep sufficient distance from area homes or facilities to minimize potential noise problems or safety hazards. (*Note:* Special silencers are available for restricted areas.)
5. Provide highest possible elevation to allow adequate drainage in bad weather.
6. Allow for maximum exposure to prevailing winds in warm climates, but shelter the unit in cold areas. In warm climates, face the cooler into the prevailing winds.

Responsibility. The compressor package can be handled in a conventional manner by oil field trucks. In most cases, transportation is the responsibility of the end user.

Arrangements. It is normally possible for the delivery trucker to set packaged compressors without additional equipment if the work area is of sufficient size. When ordering a truck, accurate compressor weight must be given and self-loading and unloading capability specified. If this is not done, additional costs for a crane or other trucks for the setting of the unit could be incurred. *Not all truckers have oil field winches; you must specify.*

Large Unit Requirements. All but the largest packages are shipped as a single haul. Very large units require two trucks and also require field installation and handling of coolers.

Setting. Once on location, the trucker will normally tailboard the unit on his trailer, depending on size and location, and set the unit where indicated.

If a site is dirt, sand, or other material that is not too hard, the trucker can back up on the pad area, lower one end of the compressor skid to the ground, and drag a flat surface on the soil. The unit will be set on this surface. In some cases, the setting and positioning of the unit and the weight of the truck cause ruts or depressions in the pad area. Once set down, the unit may be lifted slightly in order to replace compacted soil to provide an even surface. Again, an even contact with the underside of the skid is all-important. The package manufacturer can provide information on proper handling if the installation requires a crane.

The following points must be checked before the delivery truck leaves the site:

- Is the unit set in the proper direction according to climatic conditions?
- Is the surface underneath the skid smooth and free of voids?
- Has everything been unloaded from the truck? This includes parts, parts books, manuals, mufflers, and so on.
- Has any damage to the unit been noted on the bill of lading and acknowledged with the trucker's signature?

Connection. When planning suction, discharge, and fuel gas lines, routes should be considered that will minimize the obstruction of the work area around the compressor. Buried lines reduce personal injury hazards and allow better access to the unit.

Suction and Discharge Lines. Most companies run welded suction and discharge lines. In any case, line pipe must be suited to the pressure requirements.

Most units are generally furnished without suction and discharge companion flanges, but with studs, nuts, and gaskets. Flange sizes are noted on the specification sheet for each unit. It is important that piping is properly braced to avoid hanging excessive weight from these connections.

Each suction line requires a full opening block valve. It should be installed as close to the inlet flange as possible. The discharge line requires two valves. The first valve downstream from, and as close as possible to, the discharge flange is a full-opening check valve. A full-opening block valve should be installed downstream of the check valve. Positioning in this sequence permits easier removal of the check valve for repair, if required.

General Hints

1. Piping support: Be sure off-skid piping is adequately supported to minimize vibration and fatigue.
2. Spacers: Full-flow paddle spacers are generally installed between cylinder and scrubber flanges to permit installation of orifice plates, if required.
3. Pipe: The use of schedule 80 pipe and forged fittings in all screwed pipe assures that lightweight material will not find its way into hazardous service. This is a reasonably inexpensive safety precaution. The thickness of pipe must be specified by calculation, of course, but must be checked by actual measurement.
4. Glycol units: Glycol units installed directly downstream from the compressor are subject to some oil contamination and a 20°F temperature range above ambient.
5. Meters: Meters located close to compressors are susceptible to vibration or pulsation. Reasonable distances and proper bracing are recommended.

6. Scrubbers: Compressor skid-mounted scrubbers are not adequate for proper cleanup of extremely wet or contaminated field gas. Good separation prior to compression is very desirable and usually results in less expense. On-line time is drastically affected. A clean, dry stream will provide the most efficient operation.
7. Pressure controllers: Surges in wells can cause operation problems with compressors. Suction pressure controllers are available in various degrees of sensitivity. If a pressure controller is needed, it is important to get a good one with more sensitivity than actually needed. Slow response time is often the same as none at all if the compressor overloads and shuts down.
8. Bypass valves: Automatic bypass valves are available to maintain compressor operation when gas flow from the well is restricted below the capacity of the compressor. Without an automatic bypass valve, a sharp pressure decrease would shut the unit down due to low suction pressure. An automatic bypass valve can also be used on the discharge line to switch the unit to bypass operation if discharge pressure becomes too high. If volume changes are abrupt, the automatic bypass valve must be very sensitive. A less sensitive valve can be employed for gradual changes. Again, a better controller can justify the cost difference in increased on-line time.
9. Start-up screens: Start-up screens are sometimes placed at the suction flange of a compressor to catch welding slag, rock, and the like, which might be left in the flange line. Their purpose is to protect the compressor from damage. However, experience has shown that these screens are often forgotten and left in the line. In time, the gas flow can erode the screen, break it up, and send it into the unit. An alternative is to adequately purge all lines prior to final connection to the compressor and omit the screen.
10. Makeup oil supply: Because the unit can run only a very short time without makeup oil to the lubricators, the oil supply tank should be filled prior to startup. It is critical that the oil type and viscosity and other specifications be approved by the manufacturer. Additionally, it is very important to remove all contaminants from the tank, lines, and skid before allowing oil to flow into the compressor. These areas are often purged with air or gas. Some oil should be allowed to drain from the tank through the connection line. This initial oil should be collected in a receptacle and then discarded prior to final connection to the skid.
11. Coolant: If your compressor arrives without coolant, the initial fill should be made with water. This allows any leaks to be repaired without waste of more expensive coolants. As soon as possible after the start-up and the correction of any attending problems, the water should be drained and replaced with an antifreeze solution.

The use of one of the commonly available, premixed coolants available in 55-gal drums is strongly recommended. More specific recommendations can be furnished by the compressor manufacturer. These products are used full strength from the drum. The initial expense is offset by effectiveness of the product. Excess is then stored near the unit and later additions are pure, proper strength solutions, thereby eliminating later addition of "ditch water" or even salt water and other contaminants and minerals. Longer component life and less corrosion are the benefits, resulting in lower costs and more on-line time.

Starting a New Reciprocating Compressor

The following rules should always be observed in starting up a new reciprocating compressor in addition to those specified by the manufacturer in the operator's manual.

Check over the compressor to ensure that all parts are assembled properly and securely tightened. Remove the crankcase covers and clean out the interior thoroughly, being sure that the oil strainer is clean. Check the main and connecting rod bearing cap nuts to see if they are tight and the locking devices are in place. Bar the compressor over at least two complete revolutions to make certain that the moving parts are free from interferences. Watch the crankshaft to see that all the connecting rod bearings run true and free. Remove the lubricating oil filter cover, if an oil filter is provided, and clean out the inside and the filter elements and then reassemble. Fill the crankcase with the proper grade of lubricating oil to the required level. Liberally oil the running gear in the frame. On units equipped with pressure oiling systems for the frame running gear, the oil pump should be primed.

Fill the cylinder lubricator with the proper grade of compressor cylinder oil and open all feeds on the lubricator to permit maximum feed. Disconnect the lubricator oil lines at the check valves on the cylinders and operate the lubricator, filling the oil lines until the oil drips from them at the point where the lines are disconnected. Reconnect these lines and operate the lubricator a minute or so longer to inject oil into the cylinders to ensure lubricating oil will be present immediately upon starting the compressor.

Metallic packing of the segmental type should be installed in the packing gland, as follows: The packing is made up of numerous sections or cups, each cup containing two segmental rings. Each ring is divided into segments held to the rod by a spiral spring around the periphery of the packing rings. The parts of the rings and sections of the packing are numbered and lettered. The rings are installed on the rod by placing the spiral springs around the rod, then the segments under the springs.

Each pair of rings must be inserted in its proper packing section or cup. The lettered side of the rings is centered between any of the spaces in the straight cut rings or in a hole drilled for it in its companion ring; otherwise, rings will not enter their case or section. The segments of each ring are numbered 1-2, 2-3, 3-1, and rings must be assembled so that the numbers mate accordingly (i.e., 1 to 1, 2 to 2, 3 to 3). The rings are lettered A, B, C, D, and so on. The A rings should be placed nearest the cylinder bore, followed in sequence by rings bearing consecutive letters of the alphabet.

Check to ensure that the suction line leading to the cylinders is thoroughly cleaned, free from rust, pipe scale, welding shots, sand, water, or condensate and other foreign material that will damage the compressor cylinder if allowed to enter. Also, see that the suction and discharge lines are unobstructed and that all valves therein are properly set. On multi-stage units, it may be advisable, if the unit has been in storage or subjected to questionable handling during the shipment, to remove the tube bundles from the intercoolers to make certain that the shell and tubes are thoroughly cleaned and free from foreign material. The suction ports of the cylinder should also be thoroughly cleaned, removing any material that may have accumulated therein during the erecting of the unit.

Check the valves in the cylinders to be sure that the valves are placed properly in the cylinder ports, suction, and discharge, and securely tightened in place. Considerable loss of capacity and possible serious damage may result if the valves are placed in the incorrect ports and are not seated and held solidly in the valve ports.

Before starting the compressor, turn on the cooling water to the cylinders and to the coolers used with the compressor. Check to ensure that the cooling system is filled and that flow is indicated at the outlets. Adjustment of the amount of cooling should be made after the compressor has warmed up. In areas of hard water, the lining (or scaling) temperature of the water should be determined and, if feasible, the discharge temperature of the cooling water should be kept below this temperature to prevent scale buildup.

On a motor-driven unit, push the starting button and immediately push the stop button to observe the direction of rotation of the unit, thus making sure that the unit will run in the correct direction of rotation as indicated by the manufacturer.

After the correct rotation is established (motor driven or otherwise), the unit should be started and run for a period of several minutes. As soon as the unit is started, note immediately that the lubricator is feeding properly and that all parts are being lubricated. Also note that the unit operates without any noise or knocks. After this short run, stop the unit and feel all bearings to make sure that there is no tendency of parts to heat too rapidly and that the lubrication to all parts is adequate.

Start the unit again and allow it to run for a longer period. Repeat these run-in periods, increasing their duration until a continuous run-in of at least 2 hours is obtained without any overheating or knocking of the running gear, and it is observed that all parts are operating and wearing in properly. The pressure drop across the filter or strainer, or both, should be monitored, with corrective action being taken when the drop is excessive.

At this point, consideration should again be given to the cleanliness of the suction line. This is an extremely vital consideration, particularly in process compressors where the suction lines are usually long and the gas to be compressed is drawn from a closed system, from gas-generating or process equipment, towers, heat exchangers, formers, and the like. Suction equipment to clean and dry the gas to be compressed should be used before the gas is drawn into the compressor cylinders. Although every precaution has been taken in the thorough cleaning of the suction piping and vessels therein during the installation and connecting up of such equipment, foreign material may have entered these lines during installation. It cannot be stressed too strongly that if such foreign material is drawn into the compressor cylinders, excessive wear and failure of elements of the cylinder such as valves, pistons, and piston rings, and severe scoring of the cylinder bore may result.

Several methods are used in starting up a unit to prevent foreign material from entering the compressor cylinders. One method, although considered not fully effective, is to blow out the suction pipe line and equipment therein. If such a procedure is followed and a considerable amount of foreign material is blown from the line, it may be an indication that all may not be removed by this method. To ensure complete removal and cleanliness, sections of the suction lines and equipment therein should be dismantled and again cleaned thoroughly.

Another method, usually very effective, is to install a reinforced conical fine screen with sufficient flow area in the suction line as close as possible to the compressor cylinder suction flange. The screen must be readily accessible for periodic removal, cleaning, and reinstallation. The use of this screen in the suction line is suggested also even when the blow-back method is used. It is recommended that a pair of pressure gauges or a manometer be connected on either side of the screen for the purpose of checking the pressure drop through the screen. When the drop becomes excessive, the screen should be removed, cleaned, and replaced.

On multi-stage units, particularly where the gas compressed flows between stages through long fabricated lines, coolers, separators, and the like, similar screens should be installed at the suction flange of all stage cylinders. Generally, on a two-stage air compressor, the suggested methods apply to the first-stage suction only.

The valves, which have been left out during the idle run-in period, should next be installed. The unit should then be run with the suction being taken from the atmosphere with a free discharge. It may be necessary on a gas compressor to disconnect the suction and discharge lines temporarily from the system during the run-in period. Such an open suction line should be screened temporarily to prevent foreign material from being drawn in. The compressor should be run in this manner for about 1/2 hour, after which the unit should be brought up to full load by gradually increasing pressure in increments over a period of 4 to 8 hours. It is recommended during the break-in run (except on non-lubricated compressors) that oil be applied to the piston rod to facilitate the running in of the packing. This is especially true when metallic packing is used.

The additional time taken to run-in the unit to ensure that it is operating properly will be compensated for by the increased satisfaction resulting in the subsequent performance of the unit. In the running-in of an engine or turbine-driven compressor, the same procedure as outlined for a motor-driven compressor should be followed. However, in this case, the advantage of the variable-speed feature should be utilized by running the unit at the start at slow speeds and gradually building up to full-rated speed as the pressure and load are built up.

Regular Inspection

The manufacturer's manual should be used and followed for maintenance. Regular inspection should be made at definite intervals, at which time any necessary corrections may be made, such as replacement of worn parts or packing, adjustment for wear of working parts, and cleaning of valves and crankcases. Particular attention on air compressors should be paid to the suction filter to make certain that it is clean and unclogged at all times. The frequency of regular inspections will depend on the conditions prevailing at the installation. It is recommended that the unit be checked very frequently during the first few weeks of operation, and extension of the periods between inspection be dependent on the experiences observed during the earlier periods of operation.

The presence of any deposits on the valves indicates either that the intake is dirty or that too much oil or unsuitable oil is being used or that there are leaking cylinder valves or valve gaskets. All ports and passages should be examined and any obstructions, such as carbon and sticky oil, removed.

Oil used in the crankcase of a reciprocating compressor must be changed at intervals. Here, again, the frequency of changes will depend on the actual conditions prevailing, but the oil should be changed at least once a year. Before refilling, the crankcase should be thoroughly cleaned and care taken not to leave shreds of lint that would obstruct oil passages. It is inadvisable, in any case, to use waste for wiping out the crankcase, and gasoline or other flammable liquids should never be used for washing it.

Water jackets should be inspected and washed out as frequently as water conditions may require. On multi-stage units, particular attention should be given to the intercooler water surfaces, which must be kept as clean as possible in order to keep the compressor operating at peak economy. It must be remembered that inefficient intercooling will result in an increase in the horsepower required to drive the com-pressor and, consequently, an uneconomical unit.

It is recommended that the inlet and discharge line from each cylinder be equipped with a thermometer. The normal operating temperature should be observed during the early operating period of the unit. Any increase in the normal discharge temperature from a cylinder can indicate the presence of worn valves, incorrect speed, defective capacity controls, inadequate cooling-water quantity, excessive cooling-water temperature, excessive discharge pressure, inadequate cylinder lubrication, worn piston rings, or scored cylinders.

It is recommended that an operator's log be used to record the operation of the compressor. The log should show the operating conditions such as temperature, interstage pressures, and so on, at regular intervals and should include all normal maintenance performed on the machine, such as oil changes and oil added. The log should also contain a report of any unusual conditions or events such as power or water failure. A good oil analysis program is helpful in indentifying problem areas. Oils with viscosity and flash point on the high side are preferable for compressor cylinders working under high temperature conditions (i.e., single-stage machines with compression ratio of 7 or higher) and for cylinders handling refinery gas (i.e., methane and the like).

The preceding general comments will be applicable to most air compressor installations. For installations that involve gases, very high ambient temperatures, or high cooling-water temperatures or for those units operating at high discharge pressures and temperatures, it is advisable to consult the compressor manufacturer for recommended oil specifications.

Rate of Oil Feed for Compressor Cylinders

The amount of oil required will vary somewhat with the type of machine, the local conditions of operation, and the compressor service as well as the gas handled. In general, the best way to determine the maximum amount of lubrication to be fed is to remove valves from the cylinders periodically and examine the bores to determine the amount of oil present. A film of oil should be felt upon all parts, but there should be no excess oil present. If the elements feel dry, the feed should be increased. If oil lies in the bore with excessive quantities in the discharge ports, the feed should be reduced. This periodic examination should determine the final amount to be fed. Although it is difficult to predict the exact amount of lubrication necessary for all conditions, it has been observed that the rates will fall generally within the limits shown in Tables 2.8 and 2.9. The feeds given, the total to the cylinder bore and packing, are based on empirical formulas and may be varied to suit the particular conditions of service of the compressor cylinders and the gas compressed. The figures given are the suggested feeds when clean and dry conditions prevail in the compressor cylinder. Wet and dirty conditions of the air or gas compressed may require increased feeds as conditions may indicate.

In starting a new compressor, the oil to the various cylinders should be fed in liberal quantities, approximately twice the amount given in the tables. After a glazed surface is formed on the cylinder bore, the amount can be gradually cut down to that required, as determined by periodic examination.

Oil fed to the piston-rod metallic packing will also depend on the condition of the air or gas compressed and may vary from service to service. Under normal, clean, and dry conditions, two to three drops of oil per feed per minute should be satisfactory. Periodic examination of the piston rod to ensure that an oil film is present will indicate whether the oil feed is correct.

Frame and Bearing Lubrication

Where the same oil is used to lubricate both the compressor cylinder and the running gear, it must be selected primarily to suit the requirements for the compressor cylinder, except that consideration should be given to the ambient temperature in which the compressor is required to operate. If freezing ambient temperatures are experienced, a pour test of the oil is of great importance.

Selection of oil for the running gear depends on:

1. Size and speed of compressor.
2. Type of lubrication system used:
   a. Splash system, where some parts of running gear dip into the oil.
   b. Flood type, where an oil pump circulates the lubricating medium, flooding the bearings with oil.
   c. Pressure type, where an oil pump forces the lubricating medium under pressure into the bearings.

Oil used for frame and bearings only should be selected in accordance with specifications in Fig. 7.73. Because of wide variations in construction and in lubrication systems used with compressors, the manufacturer's recommendations are of great importance and should be closely followed.

Synthetic Lubricants

Several synthetic or fire-resistant lubricants are available. These lubricants are primarily intended for cylinder lubrication and are not normally needed for the crank-case because of the relatively slight hazard in the crankcase. Before these lubricants are used, it is recommended that their use be discussed with both the manufacturer of the lubricant and the manufacturer of the compressor.

The manufacturer of the lubricant should be familiar with the conditions of service, including such items as the gas being handled, the ratio of compression, moisture content of the gas handled, and the operating temperatures, as well as the intended use of the compressed air or gas. The compressor manufacturer should be consulted to make certain that the materials of construction as well as the lubricating system are suitable for using synthetic fluids. Because of the high specific gravity, the usual sight-glass fluids (diluted glycerine, salt solutions, etc.) cannot be used. The lubricants will sometimes soften and lift many common paints; therefore, any painting on surfaces that may come into contact with these lubricants should be done with a special paint having adequate resistance to the synthetic fluid. In general, it is recommended that no internal painting be done.

Synthetic lubricants tend to swell many types of rubber, including neoprenes and buna N rubber. It is necessary, therefore, to make certain that packings and gaskets are made of a material not affected by synthetic fluids.

The rate of feed of synthetic lubricant must be watched, just as with petroleum lubricants, to make certain that the cylinders and valves are properly lubricated. Experience has shown that the feed rate for synthetic lubricants should be increased to at least 1.5 times the petroleum lube rate.

Synthetic lubricants are not as compatible with water as are compounded petroleum oils and, therefore, efficient separation after any cooler must be exercised. In addition, cylinder jacket temperatures should be maintained relatively high at all times.

The synthetic fluids are classified as essentially nontoxic. It is recommended, however, that repeated or prolonged contact with the skin be avoided. Similarly, precautions must be taken to avoid oral ingestion of the fluids or accidental contact with the eyes.

Preventive Maintenance

Routine preventive maintenance is essential to the safe and efficient operation of the unit. The purpose of this guide is to provide suggestions concerning the most basic facets of a routine maintenance program and, consequently, to aid you in obtaining maximum on-line time. The most important point in any maintenance program is that it be carried out regularly and completely.

Accurate records of operating conditions are an important part of a good preventive-maintenance program. All readings should be recorded daily and compared frequently. This enables the operator to note changes that might indicate internal problems and can also be useful in developing a maintenance schedule.

Experience indicates that good operating techniques and an efficient inspection schedule will result in longer life of the engine or compressor, or both, and fewer periods of shutdown for repairs. The inspection and operating suggestions in this section have been found to be applicable at practically every installation. Careful observation of the unit during its initial period of operation will probably indicate other inspections to be made periodically. It is always recommended that one refer to the manufacturer's manual for detailed instructions.

Operating Routine

The manufacturer's manual should be followed at all times. The inspection and routine maintenance schedule listed next gives the minimum recommended time intervals between inspections. Many of the time intervals can be extended beyond those listed for a particular unit, depending on operating conditions and other related factors.

EVERY DAY

1. Inspect the engine's fuel gas scrubber to see that it is functioning properly.

2. Check the crankcase oil level; it should be at the full mark on the oil level gage. Add oil if necessary.
3. Fill each lubricator compartment with the proper grade of lubricating oil. This should be done each shift or every 8 hours.
4. If appropriate, examine the fuel-injection valves. If solid-type push rods are used, check the tappet clearances periodically. Check the clearance against those recorded after the cylinders were balanced.
5. Check both engine and turbine speed periodically.
6. Check lube oil pressure periodically. See manufacturer's operation data for correct value.
7. Turn the handle on the oil filter two or three complete revolutions every 8 hours if a scraper type oil filter is supplied.
8. Compressor suction and discharge pressures and temperatures should be checked periodically. The operator should occasionally measure the compressor cylinder valve cover temperature; an increase in this temperature usually indicates a leaky valve.
9. If compressor is multi-stage, check the suction and discharge pressures and temperatures of all stages as well as all the compressor valve cover temperatures occasionally. An increase in these temperatures usually indicates a valve leak.
10. Check the temperature of the lubricating oil entering the cooler. See manufacturer's operation data for correct value.
11. Check exhaust temperatures periodically. Compare them with temperatures that were previously recorded. Higher exhaust temperature in a cylinder usually indicates carbon deposits in the cylinder parts.
12. Check cooling-water flows and temperatures. See manufacturer's operation data for correct value.
13. Check traps or separators, or both, to be sure that they are functioning.

EVERY MONTH

1. Remove and clean spark plugs if necessary. Check the gap clearance and reset as required.
2. Check out safety shutdown switches and controls to be sure they are functioning correctly.
3. Check compression pressures. Low compression can be caused by sticking piston rings.
4. Inspect and service your particular ignition device as described in the manufacturer's instructions.
5. Check lubricator reduction gears; oil if necessary.
6. On motor-driven units, inspect the brush holders. They should move freely; clean if necessary.
7. Change oil and filter in small engines (see manufacturer's recommendations).

EVERY THREE MONTHS

1. On large engines, examine air starting valves and air check valves for wear. Check to see that they are operating correctly. In low altitudes where humidity is high and the starting air is likely to be moist, more frequent checking is advisable.
2. Inspect the fuel-injection valves and seats. If the valves are leaking due to worn valves or seats, it will be necessary to grind the valves and replace the seats. Correct seating of the valves and properly adjusted tappet clearances are necessary to keep the power cylinders balanced.
3. Clean the turbocharger blower wheel on engines so equipped. The time between cleanings will be different for each installation since the amount of dirt, soot, and oil vapors entering the engine air filter will vary.

EVERY SIX MONTHS

1. On large engines, examine the internal drive chains for correct tension. If there is excessive slack in the chains due to wear, adjust the tension as required.
2. Check the ignition device for signs of wear.
3. On large motor-driven compressor units, remove the oil filter cover or housing and clean the filter elements and case.

EVERY YEAR

1. Inspect heat exchangers for fouling or leakage, or both.
2. Examine the compressor cylinders as follows:
   a. Check piston rings and cylinder bore for excessive wear.
   b. Inspect the pressure and oil wiper piston rod packings; clean or replace as required.
   c. Check piston rings for wear and damage.
   d. Examine the valve seats, discs, springs, and stop plates; replace worn or damaged parts.
   e. Examine water jackets for scale and corrosion; clean if necessary.
   f. Replace any damaged gaskets.
3. Check unloaders and unloader controls, if used, for correct operation.
4. Engine and compressor analysis and maintenance equipment is available that can assist the operator to determine what part of the unitneeds maintenance, minimizing costly down time.

EVERY THREE YEARS

1. Check all the main bearings and the crankshaft for correct alignment. If bearing clearances are excessive, replace the bearing liners.

OTHER PERIODIC CHECK-UPS

1. Examine the compressor valves. The frequency of inspection depends on the type of compressor service. Keeping the valve seats in good condition is important. Reface the seats when necessary, making sure that the proper lift is maintained. Replace cracked, broken, or warped discs. Check the valve springs against a new one; they must have the same tension.
2. Check the foundation bolts for tightness.
3. Check the gas regulators in the engine's fuel gas line to see that they are maintaining correct gas pressure. The required gas pressure will vary with the Btu value of the gas.
4. Clean the lubricator reservoir with a suitable solvent as required.
5. Check the accuracy of the compressor gages. If service is severe, a check should be made each week.
6. Examine the air intake filter for excessive dirt and other foreign materials; clean when necessary Excessive pressure drop across the filter is evidence that cleaning is necessary.
7. Drain any condensation from the starting air system. The frequencyof draining will depend on the location of the installation.
8. Check the cooling water if it has been treated for use in a closed cooling system. Balance the concentration of treating chemicals as required.

OPERATING HINTS

1. Stop oil leaks as quickly as possible. The leaking oil may run down the foundation, seep under the grout, and destroy the grouting job.
2. Do not put a heavy load on a cold engine. Allow it to warm up first.
3. Never admit cool water to a hot engine. If you have forgotten to turn on the cooling water when starting up, shut down the engine and let it cool off.
4. Never attempt to start the engine without first unloading the compressor cylinders.
5. Check temperatures and pressures after the engine has been started. Do not permit an engine to continue to run if no reading is shown on the oil-pressure gage or if the pointer on the gage surges back and forth. Shut the engine down and investigate the problem.
6. Be sure the load is balanced on all the cylinders at all times.
7. Should the lubricator reservoir be allowed to run dry, it will probably be necessary to prime each pump again.

8. Break in new packing gradually, especially high-pressure packing.
9. Keep the scavenging air filter clean at all times. Low scavenging air pressure and low turbocharger speed indicate a dirty air filter or a dirty blower wheel. High scavenging air pressure indicates defective impeller blades or retarded spark, overspeeding the turbine.
10. When operating at maximum load, set the fuel supply pressure at a reading as low as possible.
11. Never stop an engine without unloading it and giving it a chance to cool down for at least 5 minutes, except in case of an emergency.

To close the section on reciprocating compressors, although it may be used for other types as well, the following compressor inquiry form, Fig. 7.100, is included. It may be used as a guide to the information that should be submitted with an inquiry about a new compressor.

( ) Any Customer Written Specifications Yes________No________(If "yes," attach).
( ) Description of Application:

( ) Duty Cycle________hrs/day. Average continuous operating time________minutes
( ) Gas Handled________Clean________Contaminated with________
Dry_____ Wet_____
( ) Gas Analysis:—(If available give "N" value, mol. wt., compressibility factor

( ) Any comments on previous experience or preference as to materials in piston rings, piston rod, cylinder liners, type stuffing box, type packing, valve material?

( ) Barometer________psia or Altitude________ft. above sea level
( ) Intake pressure________psig Intake Temp________°F Rel. Humidity________%
Possible variation of intake pressure from________psig to________psig
( ) Disch. Press.________psig. Possible variation from________psig to________psig
( ) Capacity required________cfm, cfh, cfd measured at________psig at________°F dry or not.
Acceptable variation of capacity from________cfm to________cfm
( ) Regulation required to control from________intake pressure________discharge pressure
( ) Automatic start and stop Cut-in________psig Cut-out________psig
( ) Constant speed control Cut-in________psig Cut-out________psig
( ) Type of Drive________
( ) Electrical Conditions________Steam Conditions________
( ) Type of Mounting________
( ) Location of Unit: outdoors or indoors; hot or cold ambient; ventilated or nonventilated space.
( ) Cooling Water: Temp________°F. Clean________Dirty________Salt Water________
Fresh Water________Corrosive?________
( ) Special accessories required:
Filter_____Aftercooler_____Receiver_____Type Starter_____Belt Guards_____
( ) Other pertinent information not covered above:

( ) Number of units required________Shipment needed________
( ) Only estimating information needed by________
( ) Firm quotation needed by________

Figure 7.100 Compressor Inquiry Sheet

Dynamic Process Compressors

Dynamic compressors are machines in which air or gas is compressed by the mechanical action of rotating vanes or impellers imparting velocity and pressure to the air or gas. In an axial compressor, as the name implies, flow is in an axial direction. In a centrifugal compressor, flow is in a radial direction.

DEFINITIONS

The following definitions will be helpful in understanding the construction and application of dynamic-type compressors:

Base plate is a metal structure on which the compressor is mounted and often the driver as well.

Blower is a term applied in the past to a compressor in a specific low-pressure application, for example, cupola blower, blast furnace blower.

Capacity is the rated maximum flow through a compressor at its rated inlet and outlet temperature, pressure and humidity. Capacity is often taken to mean the volume flow at standard inlet temperature, pressure and humidity.

Casing is the pressure containing stationary element that encloses the rotor and associated internal components, and it includes integral inlet and discharge connections (nozzles).

Diaphragm is a stationary element between the stages of a multi-stage centrifugal compressor. It may include guide vanes for directing the air or gas to the impeller of the succeeding stage. In conjunction with an adjacent diaphragm, it forms the diffuser surrounding the impeller.

Diaphragm cooling is a method of removing heat from the air or gas by circulation of a coolant in passages built into the diaphragm.

Diffuser is a stationary passage surrounding an impeller in which velocity pressure imparted to the flowing medium by the impeller is converted into static pressure.

Efficiency: Any reference to the efficiency of a dynamic-type compressor must be accompanied by a qualifying statement that identifies the efficiency under consideration. (See Chapter 8 for definitions of adiabatic, polytropic, etc.)

Exhauster is a term sometimes applied to a compressor in which the inlet pressure is less than atmospheric pressure.

Flange connection (inlet or discharge) is a means of connecting the casing to the inlet or discharge piping by means of bolted rims (flanges).

Guide vane is a stationary element that may be adjustable and that directs the flowing medium to the inlet of an impeller.

Impeller is the part of the rotating element that imparts energy to the flowing medium by means of centrifugal force. It consists of a number of blades mounted so as to rotate with the shaft. Impellers may be classified as follows:

a. *Open face* (Fig. 7.101A), without enclosing cover, may be cast in one piece, milled from a solid forging, or built up from castings, forgings, or plates.
b. *Closed type* (Fig. 7.101B), with enclosing cover and hub disk, which may be cast in one piece or built up from castings, forgings, or plates. Blades may be attached to the enclosing cover and hub disk with separate rivets, with rivets machined integral with the blades or by welding.

Impellers are further classified with respect to blade form, as follows:

a. *Radial bladed*, having straight blades extending radially.
b. *Backward bladed*, having straight or curved blades installed at an angle to the radius and away from the direction of rotation.

Inducer is a curved inlet section on an impeller (Fig. 7.101A).

Multi-casing compressor. When two or more compressors are driven by a single motor or turbine, the combined unit is called a multi-casing compressoror compressor train.

Multi-stage axial compressor is a machine having two or more rows of rotating vanes operating in series on a single rotor in a single casing. The casing includes the stationary vanes and the stators for directing the air or gas to each succeeding row of rotating vanes. These stationary vanes or stators can be fixed or variable angle, or a combination of both. This type of machine usually has two bearings, with the driver coupled to the compressor shaft (Fig. 7.102).

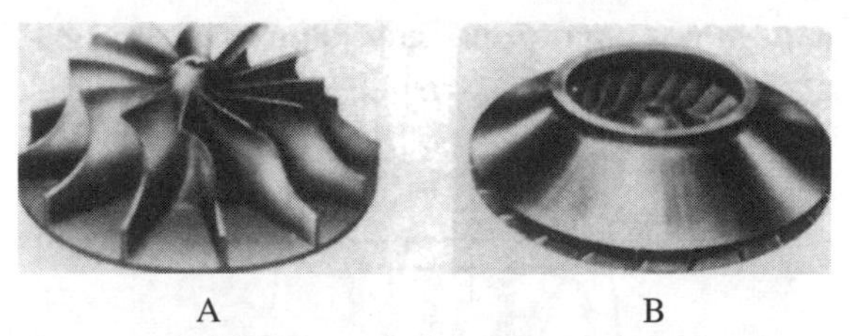

Figure 7.101 (A) An open, radial impeller. (B) A closed backward blade impeller.

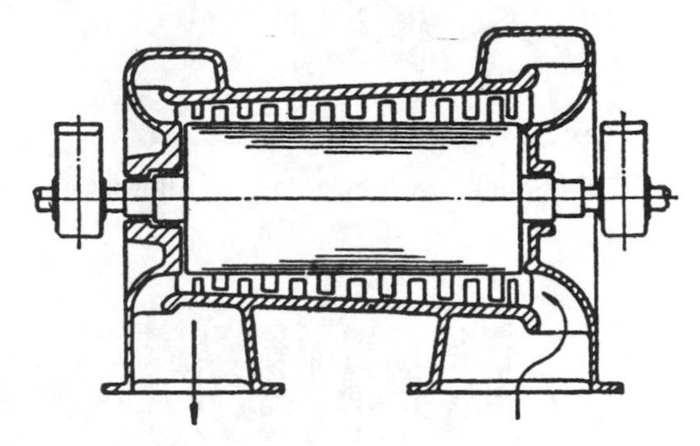

Figure 7.102 Multi-stage Single-flow Axial Compressor

Multi-stage centrifugal compressor is a machine having two or more stages. Such compressors may be described as:

a. *In-line:* All impellers are on a single shaft and in a single casing. A single-flow compressor of this type is seen in Fig. 7.103. A double-flow unit is seen in Fig. 7.104. Units may be further identified as internally or externally cooled, the latter shown in Figure 7.105.
b. *Integrally geared:* These units have bull gear drive with one or more pinions. Impellers are mounted singly at one or both ends of each pinion, and each impeller has its own separate casing. Normally used only on air and nitrogen service, these machines usually have provision for external cooling between stages. On small plant-air units, coolers and controls are often packaged on the same base as the compressor. These are described in Chapter 2.

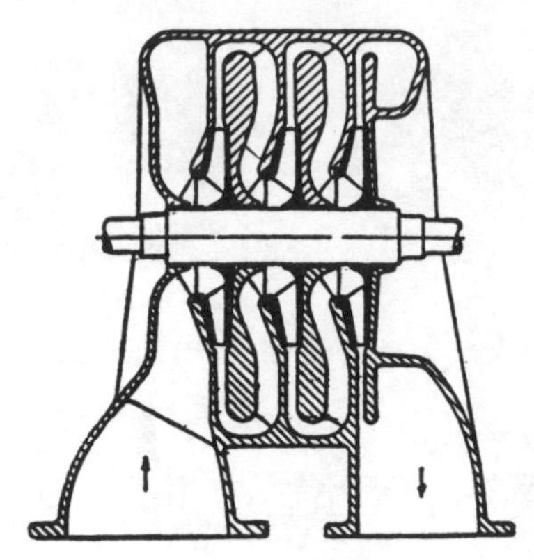

Figure 7.103 Multi-stage Single-flow Centrifugal Compressor

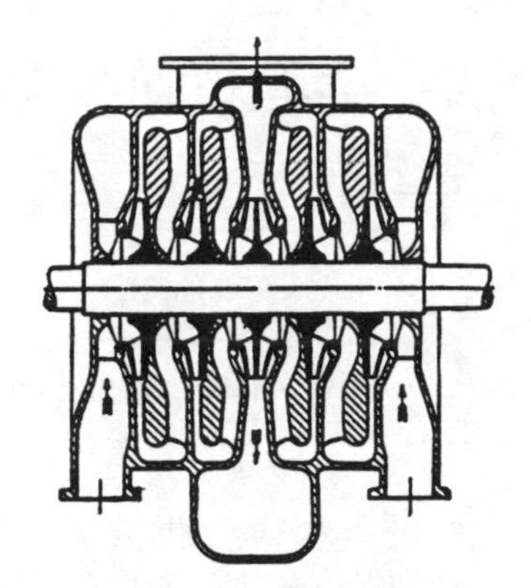

Figure 7.104 Multi-stage Double-flow Centrifugal Compressor

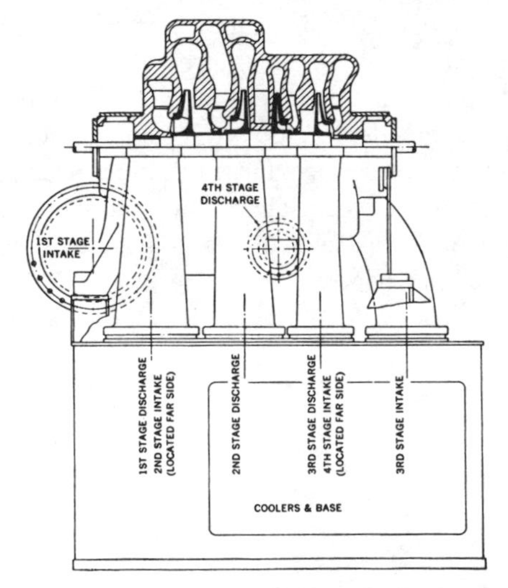

Figure 7.105 Multi-stage Single-flow Externally Cooled Centrifugal Compressor

Overhung-type centrifugal compressor denotes a single-inlet compressor with the impeller mounted on an extended shaft of the driver (i.e., one in which the compressor has no shaft of its own); for example, an extended pinion gear shaft.

Pedestal-type centrifugal compressor denotes a single-inlet compressor with the impeller mounted on a shaft supported by two bearings in a pedestal, with the driver coupled to the compressor shaft.

Performance curve is a plot of expected operating characteristics (e.g., discharge pressure versus inlet volume flow, or shaft horsepower versus inlet volume flow; see Fig. 7.106).

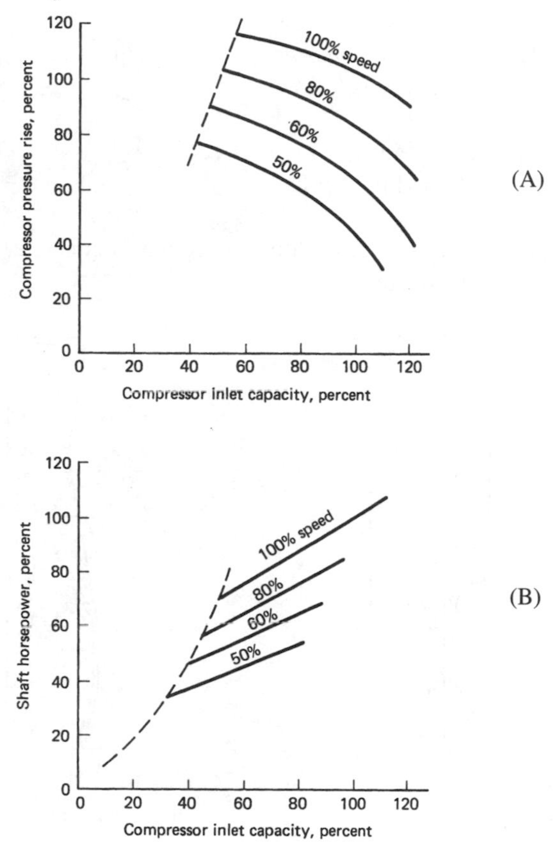

Figure 7.106 (A) Typical performance curves for a centrifugal compressor, either single-stage or multi-stage; (B) comparable performance curves for an axial compressor.

Rotor is the rotating element and is composed of the impeller or impellers and shaft and may include shaft sleeves, thrust bearing collar; and a thrust balancing device.

Seals are devices used between rotating and stationary parts to separate and minimize leakage between areas of unequal pressures. Basic types include clearance-type metallic labyrinths: single and multiple, injection-type labyrinths, eductor-type labyrinths, or a combination. Dry contact seals include carbon ring and synthetic materials such as Teflon. Liquid-injection types use water or oil seals.

Shaft is that part of the rotating element on which the rotating parts are mounted and by means of which energy is transmitted from the prime mover.

Shaft sleeves are devices that may be used to position the impeller or to protect the shaft.

Single-stage centrifugal compressors are machines having only one impeller. They may be classified as follows:

a. Single-flow (Fig. 7.107)
b. Double-flow (Fig. 7.108)

Sole plate is a metal pad, usually embedded in concrete, on which the compressor feet are mounted.

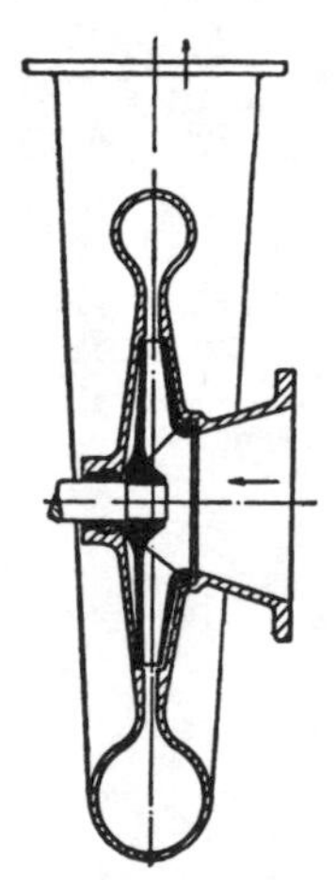

Figure 7.107 Single-stage, Single-flow Centrifugal Compressor

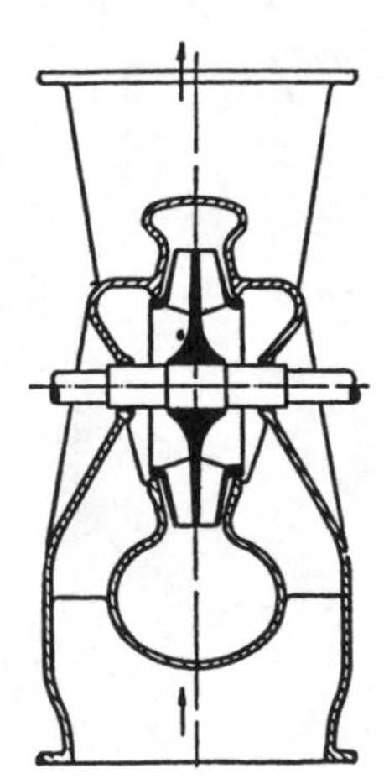

Figure 7.108 Single-stage, Double-flow Centrifugal Compressor

Stability or percentage stability is 100 minus the surge limit at rated discharge pressure, where the surge limit is expressed in percentage of rated capacity.

Surge limit (pulsation point) is the volume flow below which partial or complete cyclic flow reversal occurs, resulting in unstable aerodynamic operation.

Thrust balancing device (balance piston or drum) is the part of the rotating element that serves to counteract any inherent axial thrust developed by the impeller.

Volute is a stationary, spirally shaped discharge passage that converts velocity head to pressure.

CENTRIFUGAL COMPRESSOR CHARACTERISTICS

Compression of gas by means of a reciprocating compressor is easily pictured and is generally well understood by engineers and operators. The capacity of such a unit at constant speed is essentially constant, and the discharge pressure is that required to meet the load conditions.

A centrifugal compressor, on the other hand, develops its pressure within itself, independent of the load, but the load determines the flow to be handled. This is a generalized statement, of course, limited by the physical size of the unit and the size of the driver.

Both of the preceding statements are made on the assumption of constant speed and no controlling devices. On both centrifugal and reciprocating compressors, it is possible either to vary the speed or to provide integral regulatory means so that any desired pressure or flow requirement may be met, providing it is within the limits of the compressor and its driver (Fig. 7.108).

In its simplest form, a centrifugal compressor is a single-stage, single-flow unit with the impeller overhung on a motor. Such a unit is shown in Fig. 7.109 with a section cut away so that the flow of gas through the unit may be traced. This single-flow unit consists of the inlet nozzle, the impeller, the diffuser, the volute, and the driver. The passage of gas through the unit follows the order above. The gas enters the unit through the inlet nozzle, which is so proportioned that it permits the gas to enter the impeller with a minimum of shock or turbulence. The impeller receives the gas from the inlet nozzle and dynamically compresses it. The impeller also sets the gas in motion and gives it a velocity somewhat less than the tip speed of the impeller.

The diffuser surrounds the impeller and serves to gradually reduce the velocity of the gas leaving the impeller and to convert the velocity energy to a higher pressure level. A volute casing surrounds the diffuser and serves to collect the gas to further reduce the velocity of the gas and to recover additional velocity energy.

The maximum discharge pressure that may be obtained from a single-stage unit is limited by the stresses permissible in the impeller. Where the requirement for pressure exceeds that obtainable from a single-stage compressor, it is possible to build a centrifugal compressor with two or more impellers. This requires a return passage to take the gas leaving each diffuser and deliver it to the inlet of each succeeding stage.

Figure 7.109 Cutaway view of a single-stage, single inlet centrifugal compressor with closed-type impeller.

A typical multi-stage centrifugal compressor is shown in Fig. 7.110. It can be seen that the compression is accomplished by three impellers that act in series and are mounted on the same shaft. Flow of gas between stages is guided by the interstage diaphragms from the discharge of one impeller into the inlet of the next impeller. Sealing between stages is done by the labyrinth rings, which impose restriction on the flow between impellers at the shaft, at the impeller eye, and at the balancing drum.

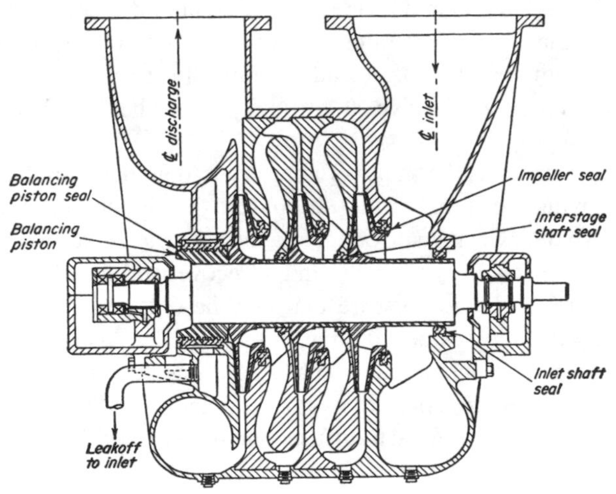

Figure 7.110 Vertical section drawing showing typical multi-stage centrifugal compressor.

Operating Characteristics

A typical set of performance curves for a centrifugal compressor is shown in Fig. 7.111A. Corresponding curves for an axial compressor are shown in Fig. 7.111B.

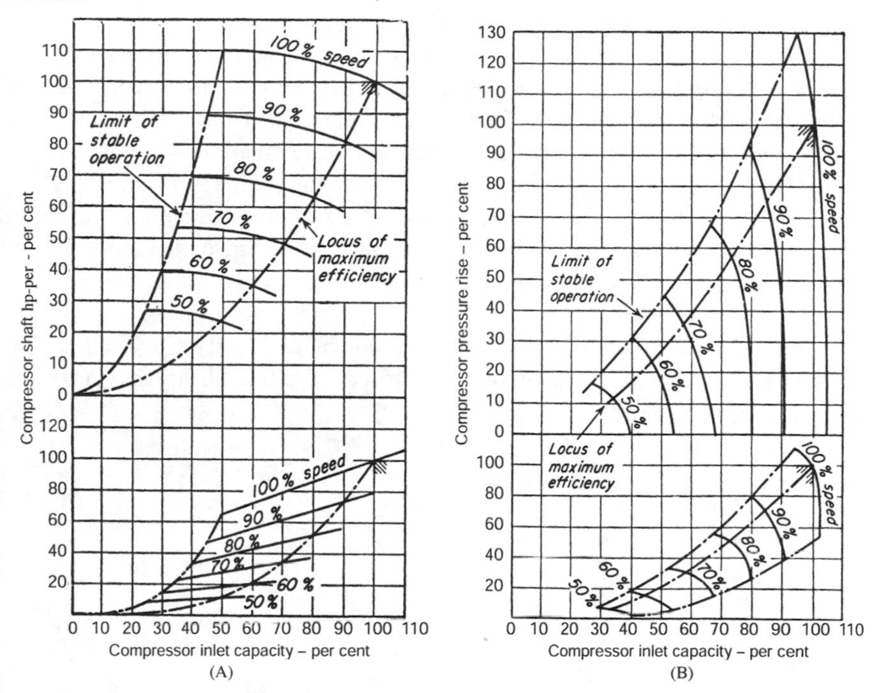

Figure 7.111 (A) Typical performance curves for a centrifugal compressor, either single-stage or multi-stage; (B) comparable performance curves for an axial compressor.

Demand Load

Regardless of the actual service to which a centrifugal unit may be applied, the general nature of the demand load may be divided into three classifications as follows:

1. Frictional resistance.
2. Constant head or pressure.
3. A combination of constant head or pressure plus frictional resistance.

The frictional resistance load is that which would be typically encountered in natural gas transmission. It is the pressure necessary to overcome the frictional resistance of flow through piping or associated equipment.

The fixed-head or pressure load is that which is required to overcome a liquid head or a controlled back pressure. In yeast or sewage agitation, for instance, air is blown into the bottom of a vat or tank, and the level of the liquid is held at a given value. This liquid head presents a fixed pressure regardless of the flow. Also, in certain processes, it is desirable to operate under a fixed pressure that is maintained by some external pressure controller at the exhaust from the process.

A third type of load, which is a combination of the above, is by far the most common. Virtually all loads are, to a certain degree, a combination. An example of this type of load is that of a blast furnace for which the majority of the pressure requirement is to overcome frictional resistance, but, in addition, some is required to maintain a pressure inside the furnace in conjunction with controlling the exhaust from the top of the furnace. Likewise, the above-mentioned yeast or sewage agitation generates to a degree a combination load, since there is some frictional resistance in the pipe between the compressor and the vats, as well as some pressure loss through the nozzles where the air enters the vats.

In Fig. 7.112, the three types of loads are shown graphically. Curve *CD* represents the purely frictional load, curve *AB* represents the fixed head, and curve *AE* represents the combination.

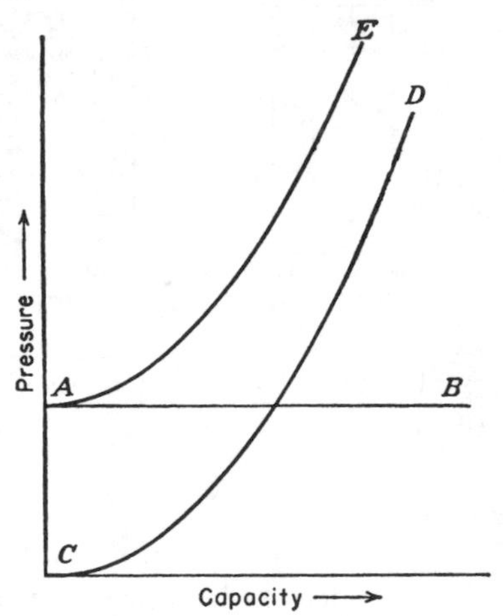

Figure 7.112 Typical curves illustrating three types of compressor loading.

Application to Load

As previously mentioned, a reciprocating compressor is essentially a constant-flow, variable-pressure unit. This is shown in Fig. 7.113 as line *JK*. Actually, because of the decrease in volumetric efficiency at increasing pressures, the reciprocating compressor will have a sloping characteristic, shown by line *JL*. A centrifugal compressor is essentially a variable-flow, constant-pressure unit as indicated by the line *FM*. Because of internal losses, the compressor characteristic is not a straight line but is similar to line *FG*.

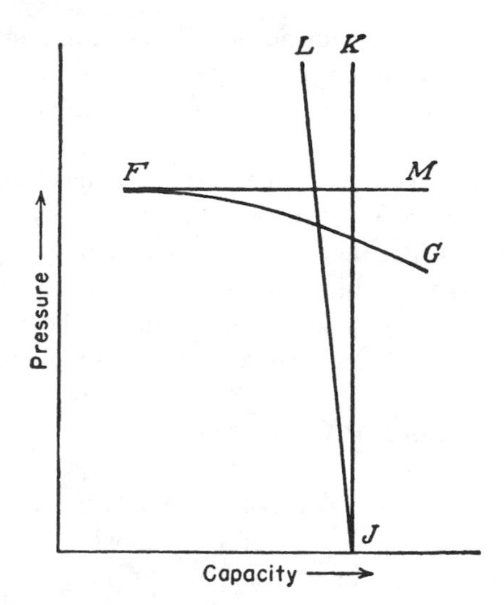

Figure 7.113 Performance characteristics of centrifugal versus reciprocating compressors.

It would be possible to select a centrifugal or a reciprocating compressor for the same flow and pressure as indicated by point *H*, and the characteristic curves are shown in Fig. 7.114 as *FG* for the centrifugal and as *LJ* for the reciprocating units.

To explain the application of a centrifugal unit to any given service, the performance curve, shown by line *FG*, is superimposed on a demand or load curve, *AE*. With this combination of compressor and load, the capacity handled will fall at the intersection of the two curves at point *H*. This is the only point at which the compressor will operate at that given demand requirement.

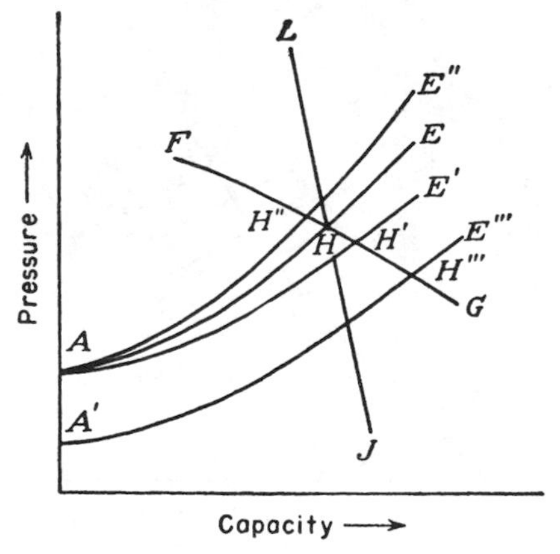

Figure 7.114 Characteristic curves of a centrifugal compressor and a reciprocating compressor superimposed upon demand-load curves.

For many applications, the demand curve will change, and this may result from varying the frictional resistance. This is shown by lines *AE'* and *AE"*, in which case the flow handled by the compressor is at the intersection of these two curves, or point *H"*.

In many processes, it is desirable to maintain a constant flow through the unit or to maintain a constant pressure delivered to the process. This must be done in spite of the fact that the actual demand requirements may vary from *AE* to *AE'*, and it would be desirable to maintain the constant flow corresponding to *H*. This could be done by means of partially closing a valve at the intake or discharge of the compressor so as to give a new demand curve passing through point *H*.

Controlling Pressure or Capacity

The use of a valve as a means of controlling pressure or flow at any given value is the simplest form of control. It is not efficient, however, since this artificially created resistance represents an irrecoverable loss of power. A more efficient way to control the unit for any given pressure or flow is to vary the speed. This creates a family of curves like those shown in Figure 7.115 by curves *FG, F'G', F"G"*, and so on. By varying the speed, it is possible to set the intersection of the compressor characteristic and the demand curve to any given required pressure or flow within the operating limits of the compressor and driver.

The method of controlling the compressor characteristics, therefore, depends on the type of driver. For steam-turbine drive, the normal method of control would be to vary the speed, permitting efficient operation and a wide range of control. In the case of motor drive, the picture becomes more involved, since the most commonly used motors are essentially constant-speed drivers. For motor drive, the following control possibilities may be considered:

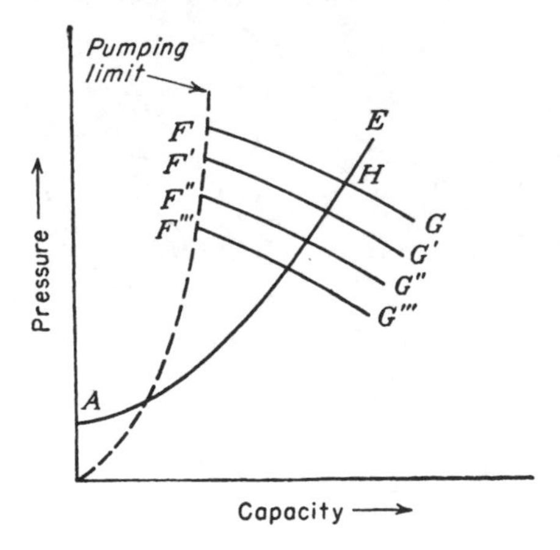

Figure 7.115 Characteristic curves of a centrifugal compressor at variable speed superimposed upon a demand-load curve.

1. Synchronous motor or squirrel-cage induction motor:
   a. Speed variation may be obtained by using hydraulic coupling between the motor and the compressor.
   b. Speed variation may be obtained by using an electric coupling between the motor and the compressor.
   c. Pressure or flow variation may be obtained by means of a butterfly valve or equivalent installed near the compressor inlet or near the compressor discharge, preferably the former.
   d. Pressure or flow variation may be obtained by adjusting the characteristic curve of a centrifugal compressor by use of adjustable inlet guide vanes or adjustable diffuser vanes.
   e. Static-type, variable-frequency control for variable-speed start for limited kVA inrush requirements with synchronous motor drive two-pole motors.
2. Wound-rotor induction motor: Speed variation may be obtained by varying the resistance in the rotor or secondary circuit using either a liquid rheostat or a step resistance (rarely used).

Briefly, a centrifugal compressor can be operated to meet a given pressure and flow requirement by varying either the demand curve or the compressor characteristic curve so that the intersection of these curves will be at the required point.

Effect of Varying Inlet Conditions

The pressure delivered by any given centrifugal unit depends on the density of the gas being compressed. This is demonstrated by Fig. 7.116 in which centrifugal characteristic curves for constant speed have been drawn for varying inlet conditions. Curve *FG* represents the characteristic for a centrifugal compressor designed to handle air at an inlet pressure of 14.4 psia, inlet temperature of 60°F, molecular weight of 28.95 (dry air), and *k* value of 1.398. This unit develops 15 psig at an inlet capacity of 20,000 cfm.

If the inlet temperature increases to 100°F, all other conditions remaining the same, the discharge pressure developed at 20,000 cfm is 13.7 psig. Likewise, if the inlet pressure drops to 12.4 psia, but the inlet temperature and other conditions remain as first specified, the discharge pressure at 20,000 cfm is 12.9 psig. Furthermore, if the molecular weight is the only variable, then with a 17.35 molecular weight gas and 20,000 cfm, the discharge pressure developed will be 8.2 psig. A decrease in the *k* value to 1.15, all other conditions unchanged, with result in a discharge pressure of 16.4 psig.

To illustrate this more precisely, Table 7.2 gives three sets of inlet conditions and the resulting discharge pressures. In each case given in Table 7.2, the compressor would develop substantially the same head, which means that the number of stages would be the same. Because of differences in pressure ratios and gas characteristics, the units for the preceding three conditions would not necessarily be physical duplicates.

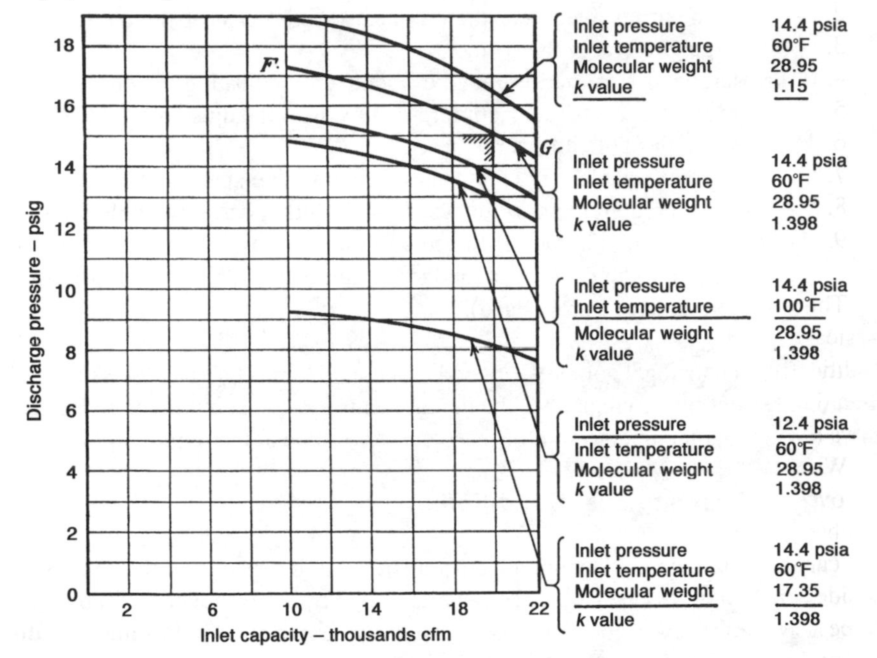

Figure 7.116 Characteristic curves for a given centrifugal compressor operating at constant speed under varying inlet conditions.

Table 7.2*

Barometer, psia	14.4	14.4	14.4
Inlet pressure, psig	- 0	- 0.65	+2.0
Inlet temperature, °F	60	110	60
Molecular weight	28.95	10.1	63.0
k value	1.398	1.36	1.11
Inlet capacity, cfm	20,000	20,000	20,000
Adiabatic head, ft-lb per lb	22,000	22,000	22,000
Discharge pressure, psig	15	3.22	64.8

* For metric equivalents, refer to Chapter 8.

Selection of Unit

The preceding study of the effect of inlet conditions on the characteristic of a compressor emphasizes two points in connection with its proper application. They are (1) the importance of investigating the operating conditions to be sure that the compressor is large enough to meet the job requirements, and (2) the necessity of a means of controlling the centrifugal compressor when variations in the operating conditions are such that a pressure would be developed in excess of the actual requirements.

1. Minimum inlet pressure
2. Maximum inlet temperature
3. Maximum molecular weight
4. Maximum k value
5. Maximum discharge pressure
6. Maximum inlet capacity
7. Maximum moisture content
8. Supercompressibility factor at inlet and discharge
9. Gas characteristics (e.g., analysis, corrosiveness, dirt content)

This information is necessary to ensure obtaining a centrifugal unit that will be physically capable of meeting the requirements under the most adverse operating conditions. Also, a knowledge of the range of these conditions is needed to ensure that a driver of sufficient size is furnished. Attention is called to Fig. 7.117, which shows a typical inquiry form.

Whereas a centrifugal compressor delivers practically constant discharge pressure over a considerable range of inlet flows, an axial compressor is characterized by substantially constant inlet flow over a considerable range of discharge pressure. This can be expressed in terms of stability; thus, a centrifugal compressor has a considerably greater range of stability than an axial compressor. This characteristic can be viewed from a different standpoint, however, because it also means that the flow of a centrifugal compressor must be greatly reduced to obtain any increase in pressure ratio. An axial compressor, on the other hand, can develop a very substantial increase in pressure ratio with a reduction of only 5 percent in flow rate.

In selecting a machine for a given application, the preceding characteristics must be carefully considered. The practical lower limit of capacity for which either compressor may be designed is determined by the specific requirements. In connection with low-capacity units, consideration must also be given to an economic comparison with reciprocating and rotary compressors capable of doing the same job, including cost and installation. In some instances, it may be economical to use an axial or centrifugal compressor for a portion of the required pressure range, followed by a reciprocating or rotary compressor for the remainder of the pressure range, followed by a reciprocating or rotary compressor for the remainder of the pressure range.

Compressor Data	Normal	Maximum	Minimum
1. Inlet flow*			
2. Mass flow			
3. Molecular weight			
4. Average compressibility (Z_{ave})			
5. Average K (C_p / C_v)			
6. Relative humidity %			
7. Barometer			
8. Inlet temperature			
9. Inlet pressure			
10. Discharge temperature			
11. Discharge pressure			
12. Speed required			
13. Power required (all losses included)			
14. Side load: Flow			
Pressure			
Temperature			
15. Indicate guarantee point			
16. Delivered flow			
17. Cooling water temperature			
18. % Turn down			
19. % Rise to surge			
20. Control method			
21. Inlet throttle device			
22. Type of intercoolers and aftercooler			

Gas analysis % volume

Constituent	Guar.	Alt.
Total	Total	100.0

*If inlet flow is listed in SCFM (standard conditions) indicate conditions under which flow was measured i.e. Temperature ________ ; Pressure ________ ; ________ % R.H.

Turbine Data

	Power	Speed	Steam conditions: Inlet pressure	Inlet temperature	Discharge pressure
Rated					
Actual (Guar.)					
Part load					
Part load					

Maximum induct / ext. flow __________ Induct / ext. pressure __________

Electric Motor Data

Site data:

Area: ☐ Class ______ Group ______ Div ______

☐ Non-hazardous

Unusual conditions: ☐ Dust

☐ Fumes

Altitude ________ Ambient Temperature ________

Enclosure type:

☐ TEFC ☐ TEWAC

☐ Weather protected ☐ TEIEF (using _____ gas)

☐ Force ventilated

☐ Open-drip proof

☐ Open

☐ Exp. proof; Class _______ Group _______

Motor type:

☐ Squirrel Cage

☐ Synchronous (☐ Brushless ☐ Slip ring)

☐ Wound rotor

☐ Direct current

☐ __________

Basic Data:

__________ Volts __________ Phase __________ Hertz

Nameplate power __________ Service factor __________

Synchronous rpm __________

Insulation class __________ Type __________

Temperature rise ________ °C above ________ °C by _____

Starting:

☐ Full voltage ☐ Reduced voltage ☐ Loaded ☐ Unloaded ☐ Wye-Delta ☐ Voltage dip _____%

Figure 7.117 Centrifugal compressor inquiry form.

Since the size of the required compressor and the horsepower of the driver are direct functions of inlet capacity, extreme care must be exercised in establishing actual inlet capacity. Standard capacity (scfm) is frequently used in specifications and, since there are several standards in common use, it is necessary to establish the particular standard involved. Air has been frequently specified at 14.7 psia, 60°F and dry. For all applications, the standard capacity must be corrected to the actual inlet pressure, inlet temperature, and moisture content to arrive at actual inlet capacity. Although examples are based upon the above common basis of measurement, CAGI now has adopted the ISO standard of 14.5 psia, 68°F and dry.

Approximate Selection Limitations

Although the following rule-of-thumb approach will vary among compressor manufacturers, a brief survey of the following points will serve to guide the selection of the type of compressor and, in some cases, will eliminate centrifugal units from consideration where conditions make it inherently not suited.

There is no definite minimum inlet capacity. However, single-stage and multistage compressors have been built for inlet capacities as low as 250 cfm.

Numerous commercial centrifugal compressors are in service at discharge pressures up to 2500 psig. For other applications, centrifugal compressors have been built for discharge pressures up to about 10,000 psig.

There are frequent applications involving an inlet capacity within the practical limits of a centrifugal compressor, but because of the high pressure ratio, the inlet capacity to the last stage may be so low as to preclude the use of a centrifugal compressor.

Because of thermal stresses and problems of alignment, a limit of 450°F is normally set as the maximum discharge temperature at design pressure and capacity for centrifugal compressors of standard design using standard materials. Where discharge temperatures must exceed this value, special designs and materials are used. Centrifugal compressors have been built for discharge temperatures up to about 1000°F. In cases where the pressure ratio would result in a discharge temperature in excess of 450°F and where this temperature is neither required nor desired, several alternatives may be used:

1. Multisection compressors with external coolers between sections of the same compressor.
2. Multicasing compressors with external coolers between compressors. If necessary; each compressor in the multicasing arrangement may be a multisection compressor.
3. Internal cooling of the, compressor as with injection of a coolant such as water or a condensed vapor into the gas stream in the diffuser sections of the compressor, or indirectly as with coolers mounted within the compressor casing or external to the compressor.

STAGE THEORY

This discussion will be concerned largely with the conventional compressor stage, that is, one with a radial inlet, closed impeller running at 800 to 900 fps tip speed, feeding a vaneless diffuser. However, sufficient attention will be given to such variations as inducer impellers and vaned diffusers that a general understanding of most combinations of commonly used components should result.

Discussion will center on the impeller and diffuser (Fig. 7.118), because these are the two key elements that determine the characteristic shape of the performance curve. Poorly designed inlets, collectors, or return channels can naturally affect performance, but their influence on characteristic shape is usually small and will henceforth be ignored.

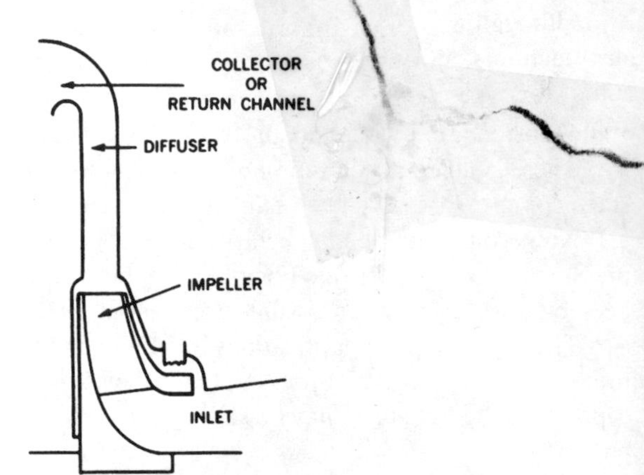

Figure 7.118 Impeller-diffuser arrangement in a centrifugal compressor.

This discussion is directly applicable to a single-stage machine and to each stage of a multi-stage machine, and the approach taken is largely qualitative.

Elements of the Characteristic Shape

Any discussion of characteristic shape must be divided into three parts. We have a basic slope of head versus flow, upon which a choke or stonewall effect must be superimposed in the overload region and a minimum flow or surge point in the underload region. The resulting overall characteristic will then be the basic slope, as altered and limited by choke at high flow, and as limited by surge at low flow in Fig. 7.119.

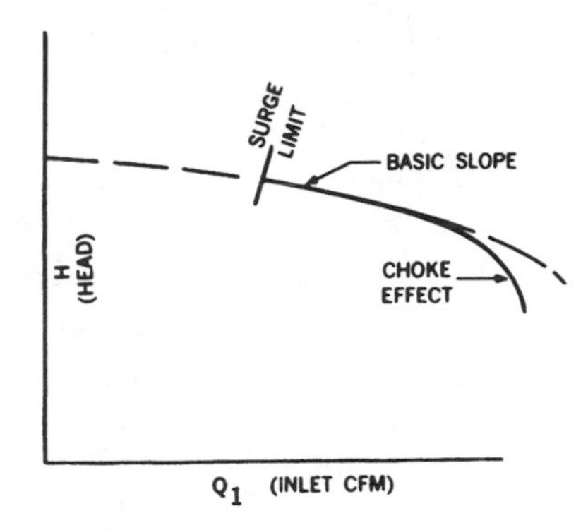

Figure 7.119 Surge and choke (or "stonewall") limitations on compressor flow.

Basic Slope

To understand basic slope, it is necessary to look at what is occurring at the impeller tip in terms of velocity vectors. In Fig. 7.120, V_{rel} represents the gas velocity relative to the blade. U_2 represents the absolute tip speed of the blade. The resultant of these two vectors is represented by V, which is the actual absolute velocity of the gas. (By vector addition, $U_2 + \rightarrow V_{rel} = V$.) It can be seen that the length of the vectors and the magnitude of the exit angle are determined by the amount of backward lean in the blade, by the tip speed of the blade, and by gas velocity relative to the blade, which is in turn dictated by tip volume flow rate for a given impeller.

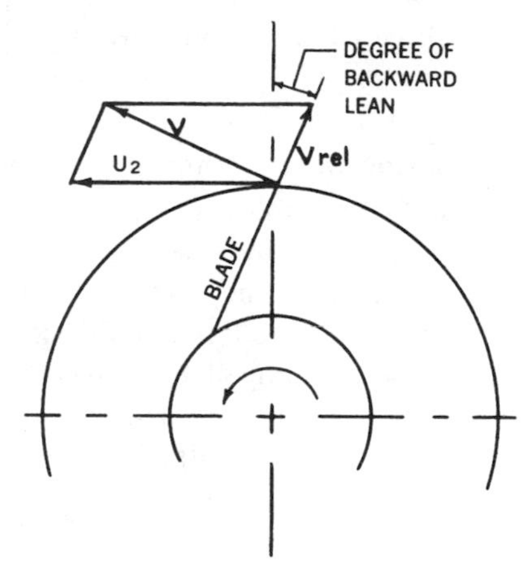

Figure 7.120 Velocity diagrams for gas and blade-tip velocities.

Having the magnitude and direction of the absolute velocity V, this vector may now be broken into its radial and tangential components, V_r and V_t, as in Fig. 7.121. The vector V_t is reduced somewhat by slip factor in a real impeller, an effect that can be ignored in a qualitative discussion such as this. The head output is proportional to the product of U_2 and V_t. For a given rpm, U_2 is constant; therefore, head is proportional to V.

The first question is what happens to the magnitude of the tangential component V_t as we vary the amount of flow passing through the impeller at constant rpm? As the flow is decreased, V_{rel} decreases. As V_{rel} decreases, angle decreases markedly. This makes V_t increase, which increases head output. This head increase with decreasing flow is the basic slope of the stage characteristic.

Blade Angle Is a Compromise

How does the degree of backward lean affect the steepness of the basic slope? One may picture a radial blade (zero backward lean). V_{rel} is now the same as V_r in

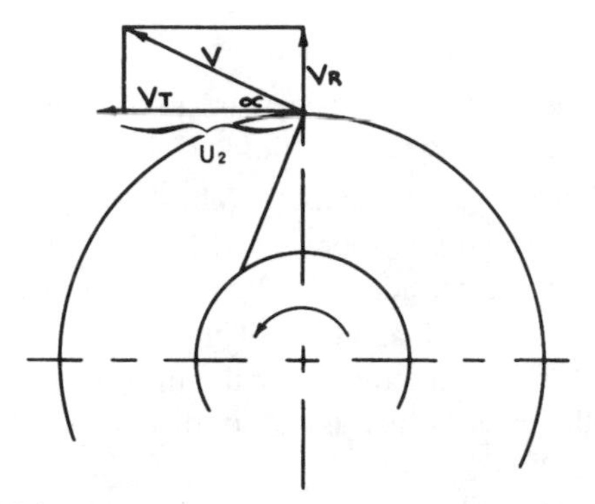

Figure 7.121 Blade velocity diagrams when flow is decreased.

Fig. 7.121, and V_t is now equal to U_2. As the flow is reduced in this impeller, V_r and α decrease as before. However, V_t remains constant. Therefore, head output remains theoretically constant regardless of flow. In a real impeller, of course, the head is reduced on increasing flow by a decrease in efficiency attributable to higher frictional losses. The resulting basic slope normally shows 2 to 3 percent head rise when going from design flow to minimum flow.

Now, for the opposite extreme, an impeller having a very high degree of backward lean (say 45 degrees off radial at the tip), it is seen that a change in flow, and therefore a change in the V_{rel} vector length, will cause very large changes in V_t, and therefore in head. Thus, such an impeller will typically produce a head rise of 20% or more when moving from design flow to minimum flow.

It is evident from the foregoing that the effect of backward lean on head output is minimized at low flow, and a high backward lean impeller will produce almost as much head at minimum flow as a low backward lean impeller running at the same tip speed. As one moves out toward design flow, however, the head difference becomes quite dramatic, as seen in Fig. 7.122. The normal industry standard for conventional closed impellers is represented by the middle line, which is 25 to 35 degrees of backward lean. This configuration is really a compromise between the high head obtainable at design flow with low backward lean blades and the steep basic slope obtainable with high backward lean blades.

One further point should be made concerning basic slope before closing the subject. In the foregoing discussion, the term flow was used without elaboration, the implication being that impeller-tip volume rate is dictated by inlet volume rate regardless of rotative speed and type of gas. This, of course, is not quite true, since gases, unlike liquids, are compressible. It is well known that a heavy gas will be compressed to a greater extent in a given stage than a light gas (i.e., the heavy gas has a higher volume ratio).

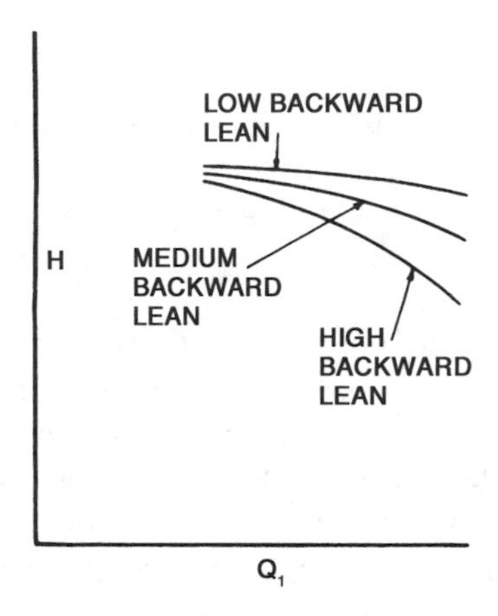

Figure 7.122 Effect of blade angle, or "lean," upon the head-flow characteristic curve.

Therefore, for a given inlet volume flow rate entering a given impeller at a given speed, the magnitude of V_{rel} is less for a heavy gas than for a light gas. If the impeller has backward lean, the magnitude of V_t will be greater for the heavy gas. Since head output is proportional to V_t, a given impeller running at a given speed will produce more head when compressing a heavy gas than when compressing a like volume of a light gas (both volumes expressed in inlet cfm). What is more, the magnitude of the difference increases as inlet flow increases, so the basic slope of a given backward lean impeller is actually less steep for a heavy gas than for a light gas. The higher the backward lean, the more pronounced this effect is (Fig. 7.123).

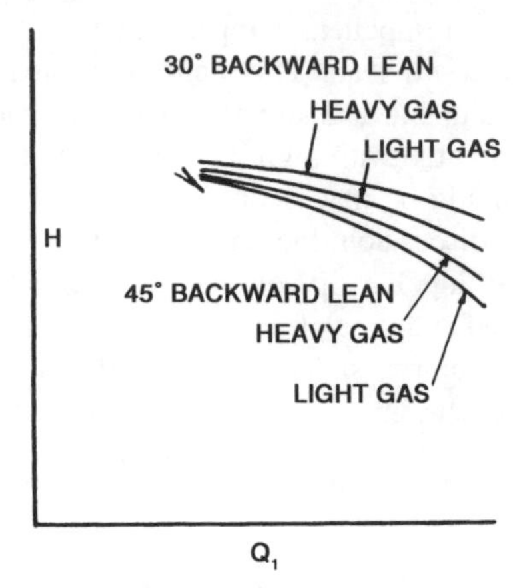

Figure 7.123 Effect of gas density upon blades of 30 degree and 45 degree blade tip angle.

Fan-law Effect

The effect of volume ratio upon what is known as fan-law is worthy of mention. Fan-law states that the cfm potential of a stage is proportional to rotative speed and that the head produced is proportional to speed squared. Reexamination of Figs. 7.120 and 7.121 will demonstrate the logic of this law.

If V_{rel} were truly proportional to inlet cfm, and both inlet cfm and speed were increased by 10%, then the head output would be 21% greater because the tip-vector geometry would maintain exact similarity. Because higher head produces higher volume ratio in a given gas, however, V_{rel} does not increase quite in proportion to speed and inlet cfm. By reasoning similar to that used in discussing heavy gas versus light gas, the head output of a backward leaning stage handling 10% more inlet cfm at 10% higher speed will increase somewhat more than 21%. By similar reasoning, if we reduce speed and inlet flow from 100 to 90%, the head produced will be slightly less than the 81% predicted by fan-law.

Fan-law received the designation of law front the fact that a fan is a low head compressor normally handling air, a light gas. Since volume ratio effects are extremely small when imparting a small head to a light gas, excellent accuracy can be obtained by the dimensional approach upon which fan-law is based. As a general rule, the higher the head, the heavier the gas, and the greater the backward lean, the poorer the accuracy will be. As a practical matter, speed changes up to 30 or 40% can be handled with sufficient accuracy for most purposes when the unit is a typical single-stage air compressor. A little more discretion must be used on multistage compressors handling heavy gases, however, because fan-law deviation can become quite significant for speed changes as small as 10%.

Choke Effect

The basic slope of the head flow curve has been discussed at some length, but the choke or stonewall effect that occurs at flows higher than design flow and which must be superimposed upon the basic slope (Fig. 7.119) has not yet been discussed.

Just as basic slope is controlled by impeller-tip vector geometry; the stonewall effect is normally controlled by impeller-inlet vector geometry; In Fig. 7.124, vector U_1 may be drawn to represent the tangential velocity of the leading edge of the blade similar to U_2 at the tip. Vector V may also be drawn representing absolute velocity of the inlet gas, which, having made a 90 degree turn, is now moving essentially radially (hence, the term *radial inlet*). By vector analysis, V_{rel}, which is gas velocity relative to the blade, has the magnitude and direction shown, where $U_1 + \rightarrow V_{rel} = V$. At design flow, the direction of V_{rel} essentially lines up with the blade angle as shown.

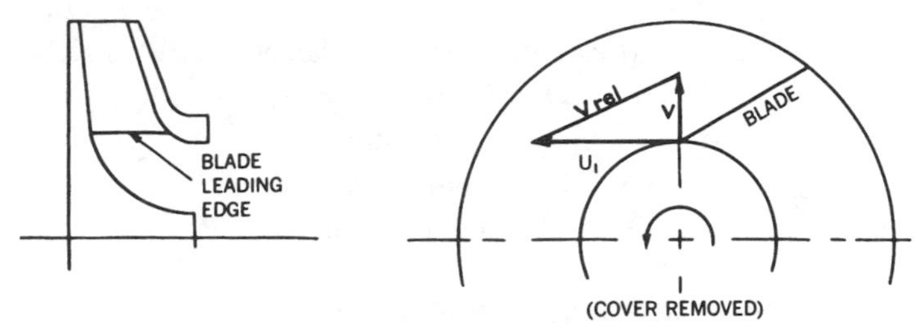

Figure 7.124 Impeller inlet geometry and velocity diagram.

Mach Number Considerations

The magnitude of V_{rel} compared to the speed of sound at the inlet pressure and temperature is called the relative inlet Mach number. It is the magnitude of this ratio that indicates stonewall effect in a conventional stage. While true stonewall effect should theoretically not be reached until the relative inlet Mach number is unity, it is conventional practice to limit the Mach number to 0.85 or 0.90 at design flow.

It is evident from Fig. 7.124 that, for a given rpm, the magnitude of $V\sim$ will diminish with decreasing flow, since V is proportional to flow. If V_{rel} decreases, then relative inlet Mach number decreases, so the stonewall effect is normally not a factor at flows below design flow It is also evident that at low flows the direction of V_{rel} is such that the gas impinges on the leading side of the blade, resulting in positive incidence, a factor that is not very detrimental to performance until very high values of positive incidence are reached.

Let us now increase flow beyond the design point. As V increases, so also does V_{rel} and relative inlet Mach number. In addition, V_{rel} now impinges on the trailing side of the blade, a condition known as negative incidence. It has been observed that high degrees of negative incidence tend to contribute to the stonewall problem as Mach number 1.00 is approached, presumably because of boundary layer separation and reduction of effective flow area in the blade pack.

Significance of Gas Weight

Since values of U_l are typically in the 500 fps range and values of V in the 250 fps range, it is obvious that, since the speed of sound for air at 80°F is 1140 fps, lighter gases suffer no true impeller stonewall problems as described, even at high overloads. Some head loss below the basic slope will be observed, however even in the lightest gases, due in part to increased frictional losses throughout the entire stage and in part to the extreme negative incidence at high overloads.

The lightest common gas handled by conventional centrifugal compressors for which stonewall effect can be a definite factor is propylene with a sonic speed of 740 fps at - 40°. In order of increasing severity are propane at 718 fps at -40°F, butane at 630 fps at - 20°F, chlorine, and the various Freons. The traditional method of handling such gases is to use an impeller of larger than normal flow area to reduce V, and run it at lower than normal rpm to reduce U_l, thus keeping the value of V_{rel}, abnormally low. This procedure requires the use of more than the usual number of stages for a given head requirement and sometimes even requires the use of an abnormally large frame for the flow handled.

Inducer Impeller Increases Head Output

Much development work has been done in recent years toward the goal of running impellers at normal speeds on heavy gases in order to reduce hardware costs to those incurred in the compression of light gases. One approach has been to use inducer impellers (Fig. 7.125). The blades on this impeller extend down around the hub radius so that the gas first encounters the blade pack while flowing axially. Figure 7.125 shows the vector analysis at the inducer outer radius. Assuming that the inducer radius is the same as the leading edge radius of the conventional radial inlet impeller, the vector geometries of the two are identical.

The advantage of the inducer lies in the fact that, as we move radially inward along the blade leading edge, the value of U_l and therefore of V_{rel} and Mach number, decreases. As we move along the leading edge of the conventional impeller, the vector geometry remains essentially constant. It can be seen, therefore, that while maximum Mach number for the two styles is the same, the average Mach number for the inducer is less for a given flow and speed. The inducer impeller can therefore be run somewhat faster, resulting in greater head output. The big disadvantage of the closed inducer impeller lies in the difficulty of fabrication. It is obviously more difficult to weld the longer and more curved blade path of an inducer impeller than that of a conventional impeller. Other disadvantages are the greater weight and the greater axial space required by the inducer impeller over that of the conventional impeller.

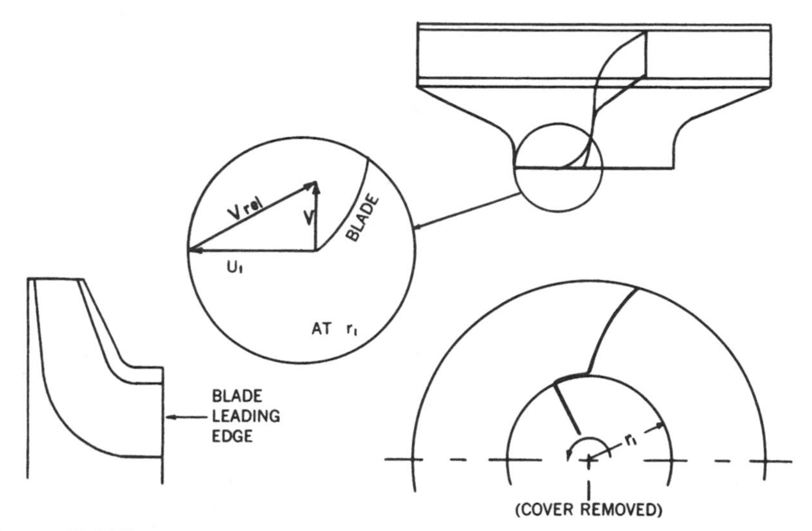

Figure 7.125 Velocity diagram for gas and blade at the outer radius of the inducer.

Another method of obtaining increased head output for a given Mach number is to reduce the backward lean. This expedient, however, has some disadvantages, not the least of which is the flatter characteristic curve that results.

Surge

Having discussed basic slope and choke, we are left with one major task, a discussion of minimum flow or surge. Surge flow has been defined by some as the flow at which the head-flow curve is perfectly flat and below which head actually decreases. This definition has a certain appeal, because straddling a surge flow, so defined, are many pairs of flow values producing identical heads, leading one to conjecture that the flow value is actually jumping back and forth between some such pair. However, since numerous centrifugal stages have been observed to run smoothly at flows below such a rate and others to surge at flows above such a rate, this definition must be considered imperfect at best. Unlike choke flow, which hurts nothing but aerodynamic performance, surge can be quite damaging physically to a compressor and should be avoided. The higher the pressure level involved, the more important this statement becomes.

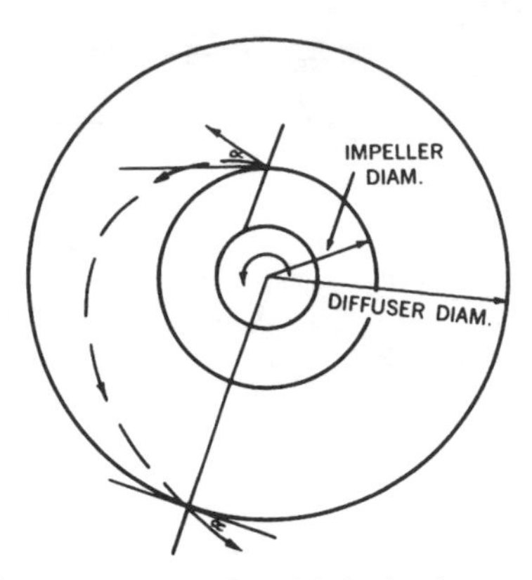

Figure 7.126 Velocity vectors at blade tips in a conventional stage.

To understand what causes surge in a conventional stage, one must refer back to the tip-vector geometry of Fig. 7.121. As flow is reduced while speed is held constant, the magnitude of V_r decreases in proportion and that of V_r remains constant for radial blades or increases for backward lean blades. As flow decreases, therefore, the value of the flow angle decreases. In the normal parallel-wall vaneless diffuser, this angle remains almost constant throughout the diffuser, so the path taken by a particle of gas is a log spiral (Fig. 7.126). The reason that angle remains constant in a parallel-wall diffuser is that both V_r and V_t vary inversely with radius V_r because radial flow area is proportional to radius and V_t, because of the law of conservation of momentum.

It is evident from Fig. 7.126 that the smaller the angle α, the longer the flow path of a given gas particle between the impeller tip and the diffuser outer diameter. When angle α becomes small enough and the diffuser flow path long enough, the flow momentum at the walls is dissipated by friction to the point at which pressure gained by diffusion causes a reversal of flow, and surge results. The angle α at which this occurs in a vaneless diffuser has been found to be quite predictable for various diffuser impeller diameter ratios. The flow and angle α at which surge occurs can be lowered somewhat by reducing diffuser diameter, but at the cost of some velocity pressure recovery.

Vaned Diffusers

Before we discuss the foregoing in more detail, let us briefly discuss vaned diffusion, a device sometimes used in high-performance air machines. Figure 7.127 shows the configuration of diffuser vanes. The vanes force the air outward in a shorter path than unguided air would take, but not so short a path as to cause too rapid deceleration with consequent stream separation and inefficiency.

The leading edge of the diffuser vane is set for shockless entry of the air at approximately design flow. It is evident that at flows lower than design the air impinges on the diffuser vanes with positive incidence. Conversely, at flows higher than design, negative incidence prevails. In a typical high-speed, high-performance air stage, positive incidence at the leading edge of the diffuser vane triggers surge on decreasing flow.

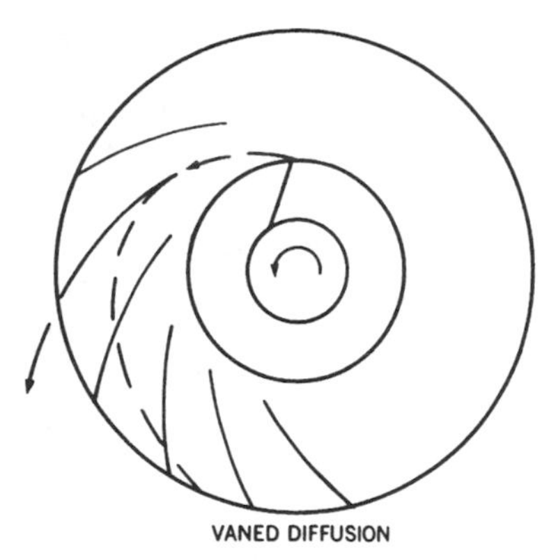

Figure 7.127 Arrangement of diffuser vanes in the centrifugal compressor casing.

On increasing flow, negative incidence at the inducer vanes can cause choking before impeller inlet stonewall is reached. In spite of this disadvantage, vaned diffusion is sometimes used for air because stage efficiency is improved 2 to 3 percent. The short flow-range problem can be alleviated by making the diffuser vanes adjustable.

Vaneless Diffusers

Having complete discussion of vaned diffusion, let us return to the more common vaneless diffuser. We have seen that when V_r and α become too small there will be surge. What can be done if the parameters are such that there is a low value for α at design flow? One may artificially increase V_r and α by pulling the diffuser walls together until it reaches the proper value at design flow. This brings to light an important distinction: head output, as discussed earlier, is controlled by vector geometry in the impeller tip largely irrespective of what happens in the diffuser. Surge point is controlled by vector geometry in the diffuser, largely irrespective of what occurred in the impeller. In the common case where impeller tip width and diffuser width are the same (Fig. 7.118), the two sets of vector geometry are the same, ignoring impeller blade solidity. If such a stage has poor stability, it is frequently possible to lower the surge point by narrowing the diffuser without markedly changing the basic slope of choke flow. This procedure can be carried only so far, however, because extreme positive incidence at the impeller inlet will eventually trigger surge, regardless of diffuser geometry

Just as we did when discussing choke flow, let us look at the effect of heavy gas compression on surge point. Since a heavy gas is compressed more at a given speed than is a light gas, it is evident that the critical value for a will be reached on decreasing flow at a higher inlet flow of heavy gas than that of a light gas. A given stage, therefore, has a higher surge flow or a lower stable range when compressing heavy gas than when compressing light gas at the same speed (Figs. 7.128 and 7.129).

By similar reasoning, a given stage compressing a given gas at varying speed will surge at somewhat different inlet flows than those predicted by fan-law When speed is 10% above design speed, for instance, surge flow will be more than 10% higher than surge flow at design speed. When speed is 90% of design, the stage will surge at less than 90% of design speed surge flow.

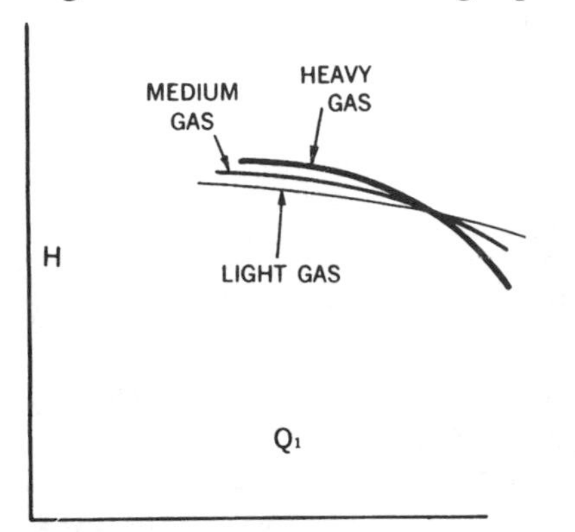

Figure 7.128 Head-flow characteristic, showing effect of gas density.

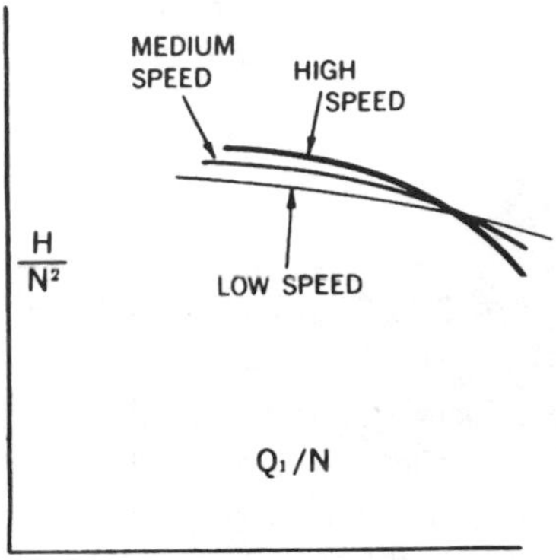

Figure 7.129 Effect of speed upon the operating characteristic of the centrifugal compressor.

GUARANTEES

Because of varying conditions of installation and operation, performance guarantees are subject to a tolerance. The limits of this tolerance are normally stated by the manufacturer. Certain chemical and petroleum standards require the capacity and head to be guaranteed with no negative tolerance.

In the case of variable-speed compressors, the specified capacity and pressure can be obtained by adjustment of speed that is not guaranteed, and the horsepower will be guaranteed within the specified tolerance.

In the case of constant-speed compressors, the specified capacity; head, and horsepower are subject to the specified tolerance, but the tolerances should be non-cumulative; that is, for a given head and capacity, the Bhp per 100 cfm shall be within the specified tolerance.

Because of difficulty in accurately predicting the characteristic curve of an axial or centrifugal compressor, only one capacity and one discharge pressure rating together with corresponding power input are normally guaranteed. The shape of the characteristic curve is seldom guaranteed.

SPECIFIC SPEED

The fundamental geometry of a centrifugal or axial compressor stage is dictated by the three variables: flow, head, and rotative speed. Specific speed is a useful tool for quickly evaluating the type of stage that will be required to relate these three variables properly for a given application. Mathematically, it is expressed by the following equation:

$$N_s = \frac{N\sqrt{Q}}{H^{3/4}}$$

where:

N_s = specific speed
N = design rotative speed, rpm
Q = design gas flow, icfm
H = design isentropic head per stage, ft-lb/lb

It is not necessary to grasp the physical significance of the definition of specific speed to use it intelligently. It should be considered to be a type characteristic of the impeller that specifies its general proportions and characteristics, rather than as an rpm for special conditions (Fig. 7.130).

In general, the radial impeller will be used for specific speeds between 400 and 950. For specific speeds of 800 to 1400, the mixed-flow impeller is used, and axial compressors operate in the 1300 and higher ranges.

Specific speed as previously defined, and as it has been traditionally used in American industry, is not truly dimensionless. A simple conversion into 51 units will, therefore, not yield the same parameters as those stated here. Until this matter has been restudied, it is suggested that the compressor variables involved be converted from SI to English Engineering units and use made of Fig. 7.131 and the parameter ranges given previously. Obviously, these ranges are overlapping, since specific speed is only a general guide to design. Its usefulness lies in providing an estimated evaluation of compatibility of compressor operating requirements and driver speed. The specific speed for a given set of conditions may be readily obtained from Fig. 7.131.

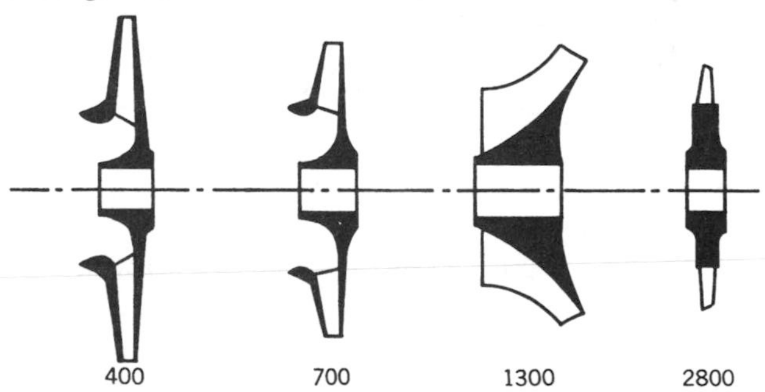

Figure 7.130 Typical impeller proportions for various specific speeds.

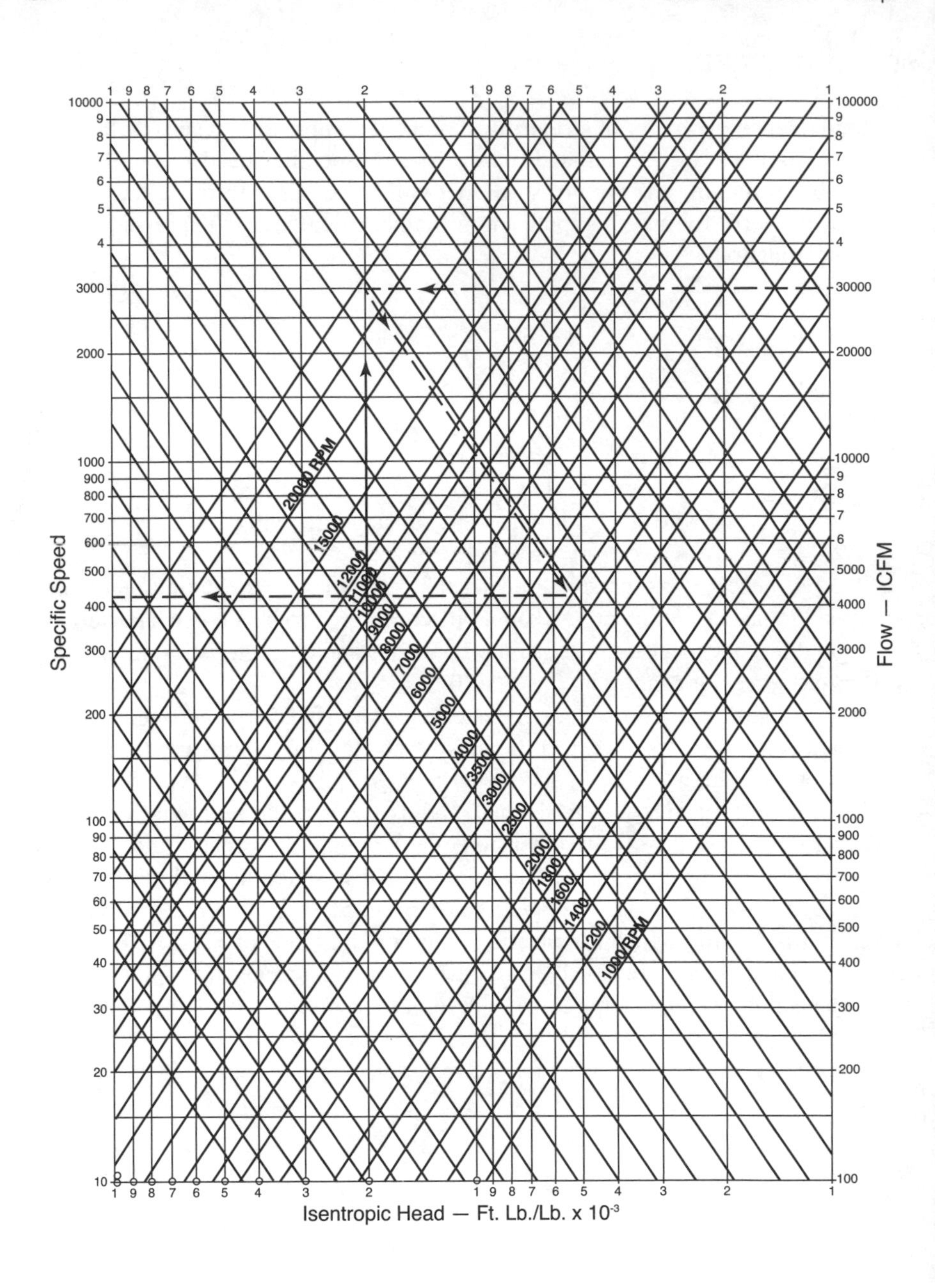

Figure 7.131 Specific speed versus flow, speed and head.

DRIVERS

The compressor driver, whether turbine, motor, or engine, is an important part of any compressor installation. It requires proper selection and matching to the compressor to ensure a satisfactory installation. It is preferable to have the compressor manufacturer select the driver and furnish a complete unit, including compressor, driver, lubrication system, seal system, coupling, and control, so that all elements may be properly coordinated. Thus one manufacturer assumes the responsibility for the complete package.

The selection of the type of driver is determined by the service requirement of the compressor, the conditions at the proposed site, and the availability and economy of electricity and various types of fuels or waste process gases.

The drivers commonly used with centrifugal compressors include:

1. Steam turbines
2. Electric motors
3. Gas expansion turbines
4. Combustion gas turbines
5. Internal combustion engines

Steam Turbines

Since centrifugal compressors are inherently high speed units with rpm sometimes exceeding 15,000 and with horsepower up to 100,000 hp, they are well suited for use with direct-connected steam turbines. Steam turbine drivers may be designed for practically any set of steam conditions and are, therefore, readily adaptable, within limits, to whatever steam may be available. Where low-pressure exhaust steam from the turbine can be utilized in the plant system, a short, sometimes single-stage, turbine may be used. Where steam consumption of the turbine is of prime importance, a multi-stage turbine may be selected. When low-pressure steam in small quantities is available continuously or intermittently, a mixed-pressure turbine may be selected. An automatic-extraction turbine can be designed to supply low-pressure steam when such steam is required.

For ratings up to 4000 or 5000 hp, turbines are generally equipped with a single, governor-controlled steam inlet valve supplemented by part load hand valves. For larger ratings and higher steam flows, 100,000 lb/hour or higher, turbines equipped with multiple, governor-controlled steam inlet valves result in better speed control and improved economy with varying loads and are generally preferred to the single-governor valve design.

Construction and materials of the turbine should be suitable for the maximum steam conditions, horsepower, speed, and other requirements as specified, in accordance with the latest edition of NEMA standards for mechanical drive applications. The turbine should be rated so that it will carry the maximum load requirement of the compressor under the most adverse operating conditions. When specifying the turbine, the following minimum information should be given:

Rated Conditions*

1. Horsepower.
2. Speed, rpm.
3. Initial steam pressure at throttle, psig.
4. Initial steam temperature at throttle, °F.
5. Steam pressure at exhaust, psig or in. Hg absolute.
6. Adjustable speed range, percent.
7. Speed variation, percent.
8. Service, continuous or intermittent.
9. Type of automatic speed control.
10. Altitude at place of installation.
11. Construction, indoors or weather-protected.

Electric Motor Drivers

Standard electric motors are limited to a maximum speed of 3600 rpm and, therefore, usually require a speed-increasing gear between the motor and the compressor. A number of centrifugal air compressors are manufactured with an integral or built-in speed-increasing gear, which allows direct connection of the compressor to a standard-speed motor. For a constant-speed compressor, an alternating-current, squirrel-cage induction or synchronous motor is normally used. For a variable-speed compressor, a wound-rotor induction or direct-current motor may be used. Where variable speed is required but where conditions indicate a constant-speed driver to be most suitable, a variable-speed transmission such as an electric or hydraulic coupling may be used to vary the compressor speed.

The selection of the type of motor and form of enclosure to ensure successful operation and reliable service should be made only after giving careful consideration to the following conditions:

1. Voltage, number of phases and frequency.
2. Speed-torque curves for both starting and operating conditions.
3. Flywheel effect, WK^2, of compressor, gear, and coupling, all referred to the motor shaft speed.
4. kVA inrush limitation of power source.
5. Need for system power factor correction.
6. Permissible motor overload (service factor).
7. Altitude at place of installation.
8. Ambient temperature.

*Conditions other than rated should be clearly specified.

9. Required speed range.
10. Continuous or intermittent service.
11. Desired motor efficiency.
12. Frequency of starting in normal service.
13. Presence of gritty or conducting dust, lint, oil vapor, salt air, corrosive fumes, flammable or explosive gases, or existence of any other ambient condition that may affect the successful operation or life of the motor.
14. Any other conditions that may affect the performance or life of the motor.

When other than a direct-connected motor is used, the choice of the speed for the motor is determined by selecting the most economical combination of motor and gear.

For geared rotary or dynamic compressors, the equivalent flywheel effect referred to the motor shaft is given by

$$WK^2_{eq} WK^2_{cr} = \frac{(\text{compressor speed in rpm})^2}{(\text{driver speed in rpm})^2} + WK^2_8$$

where subscripts *cr* stands for compressor rotor and *g* for gear.

Where a synchronous motor must be used and uninterrupted service is required, automatic resynchronization is normally furnished. This is accomplished by specifying 100% pull-in torque. Direct-current motors are sometimes used for compressor drives, but the speeds usually available make geared units necessary. Variable speed, controlled either manually or automatically, is obtained by a suitable field rheostat.

Motor Control

Motor control may be purchased separately if desired. The control must be properly coordinated if the installation is to work satisfactorily. There are frequently some restrictions in connection with the power supply system that will determine the type of motor control required. Both the motor manufacturer and the control manufacturer should be completely informed of what is contemplated.

The types of control available and most frequently used for the different types of motors are:

Squirrel-cage Induction Motors

1. Manual, full voltage.
2. Manual, reduced voltage.
3. Magnetic, full voltage.
4. Magnetic, reduced voltage.

Wound-rotor Induction Motors

1. Manual, secondary resistance.
2. Motor driven, secondary resistance.
3. Magnetic, secondary and primary control.

Synchronous Motors

1. Magnetic, full voltage.
2. Magnetic, reduced voltage.
3. Semi-magnetic, reduced voltage.

Gears and Variable-speed Couplings

Gears and variable-speed couplings, when used with centrifugal compressors, should be selected so that they are of ample continuous rating so as to meet the maximum power requirements of the compressor. When excess power is specified in the driver, such as a 15% service factor stipulated for the motor, they should have a continuous horsepower rating equal to the maximum continuous horsepower rating of the driver, if this is greater than the maximum specified power requirement of the compressor.

Gas Expansion Turbines

Gas expansion turbines, sometimes called expanders, are selected according to the same general specifications as outlined for steam turbines. Gas expansion turbines are adaptable only where an adequate supply of gas at elevated temperatures, up to 1400°F, is available and when this gas is at such a pressure that the necessary power can be developed by its expansion. An electric motor or steam turbine is often used with an expansion turbine to provide starting and supplementary power.

Combustion Gas Turbines

A combustion gas turbine includes an expansion gas turbine as well as a fuel combustion chamber. Due to the inherently high speed characteristic of gas turbines, they are well suited for driving dynamic compressors. Waste-heat recovery systems can be used with a combustion gas turbine to increase the operating efficiency of the gas turbine.

Gas turbines are generally of two basic designs; general industrial and jet-engine derivative. The latter utilizes a jet engine as a gasifier in conjunction with a separate power turbine that drives the compressor.

Internal Combustion Engines

Reciprocating engine drivers, because of their relatively low rotative speeds, require speed-increasing gears for centrifugal compressor drives. Such engines use natural gas from the pipeline as fuel. Torsional vibration studies, normally made on engine applications, usually indicate the need for a damping provision built into the gear or coupling. This is not a very common method of driving centrifugal compressors.

CONTROL OF DYNAMIC COMPRESSORS

Control systems for dynamic compressors, axial and centrifugal, may become quite complex; however, they have fundamentally only two functions to accomplish:

1. Antisurge control to prevent the compressor from operating below its stable operating range and thereby exposing either itself or the process equipment to damage.
2. Performance control to adjust the output of the compressor to the demands of the user's process.

All dynamic compressors have what is commonly called a surge limit or minimum flow point below which the performance of the compressor is unstable. This instability manifests itself in pulsations in pressure and flow, which may become severe enough to cause damage.

The surge limit and type of antisurge control are affected by:

1. Type of compressor
2. Design pressure ratio
3. Characteristics of the gas handled
    a. Inlet temperature
    b. Gas constant R
    c. k value
4. Speed

Generally, the gas handled is of constant composition, and the constants R and k can be neglected in the design of the anti-surge equipment. Where variation in suction temperature may be wide, temperature must be compensated for.

To keep a compressor from surging, all that is required is to maintain a flow greater than the safe minimum. If the consumer's requirements are not greater than the minimum stable flow, the difference must be either blown off or recycled to the suction of the compressor.

The control engineer's responsibility, then, is to match the compressor surge line with a control system characteristic so that compressor surge is never reached. To accomplish this, the engineer must know the surge characteristics of the compressor being designed.

Figure 7.132 shows the typical pressure-volume characteristics of an axial compressor, and Fig. 7.133 shows those of a centrifugal compressor. From these curves, it can be seen that an axial compressor at a given speed approximates a constant volume variable pressure characteristic. Thus, the axial compressor antisurge protection is most frequently oriented to pressure and that of the centrifugal unit to flow.

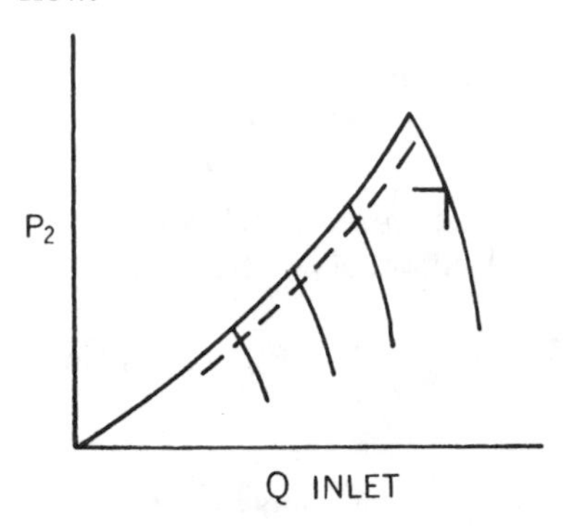

Figure 7.132 Typical performance characteristic curve for an axial-flow compressor with variable speed.

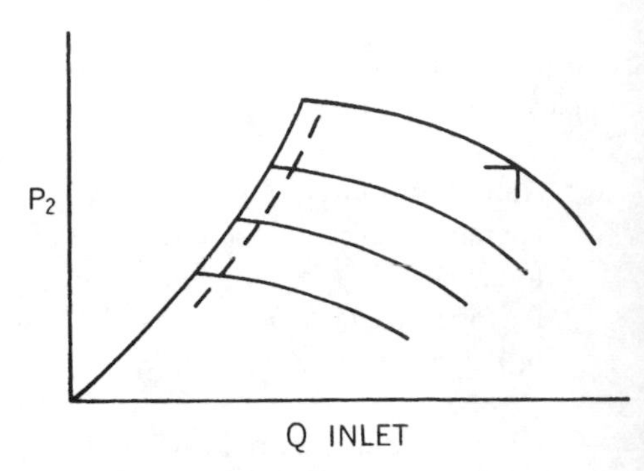

Figure 7.133 Typical performance characteristic curve for centrifugal compressor with variable speed.

Assume that the compressor is operating at the rated point at full speed (Fig. 7.134), and the process demand decreases with a pressure-oriented anti-surge control system. The discharge pressure will increase. The pressure transmitter, *FX*, will monitor the pressure, sending a signal to *SC*, which will open the blow-off valve, *BOV*, when the pressure reaches the set point or dashed line. If the compressor were driven by a variable-speed drive, a speed transmitter, *SX*, would supply a signal to *SC*, which would modify the set point. The dashed line is the locus of the speed-modified set point. In the same manner, a temperature transmitter, *TX*, could be used to modify the set point for temperature variations. If the performance control is by suction throttling and not variable speed, a pressure transmitter at the inlet will supply a signal to the *SC* to modify the set point. In this case, the *SC* is actually a pressure ratio controller.

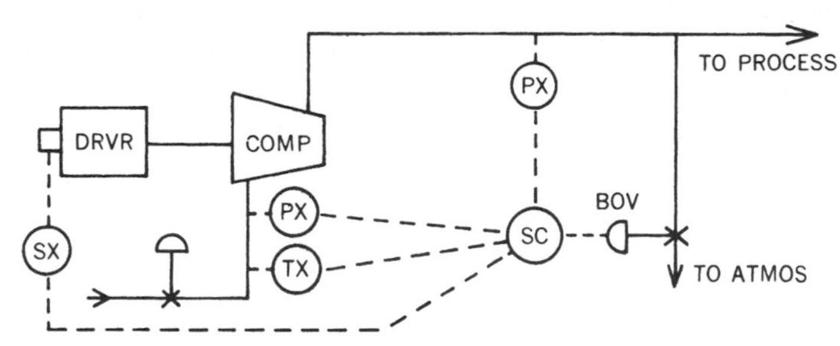

Figure 7.134 A typical pressure-oriented anti-surge control system.

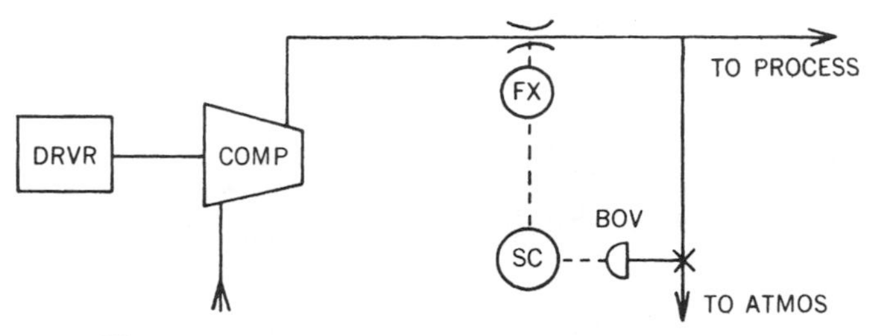

Figure 7.135 A typical flow-oriented anti-surge control system.

Assume again that the compressor is operating at the rated point at full speed (Fig. 7.135), and the process demand decreases. The flow will decrease. The flow transmitter, *FX*, will monitor the flow sending a signal to *SC*, which will open the blow-off valve, *BOV*, when the flow reaches the minimum value or set point. With flow monitoring at the discharge of the compressor, no modification of the set point is required with speed, since the flow element ΔP remains essentially constant for normal speed variations in a machine taking in atmospheric air.

This insensitivity of the set point to speed variation is illustrated as follows:

The ΔP across a flow element is proportional to ρV^2, where

ρ = density
V = velocity through the flow element

At the discharge, ρ is proportional to the pressure, which is proportional to the square of the speed, neglecting the minor effect of temperature. Therefore,

$$\rho d \sim N^2$$

The actual suction volume at which the compressor surges varies approximately directly with the speed. If the outlet condition to the compressor is essentially constant, the weight flow at which surge occurs is proportional to the volume and, therefore, varies directly with the speed. This would not be true if suction throttling performance control were used; however, suction throttling and variable-speed control are not normally both employed in the same control system. Therefore,

$$W \sim N$$

and

$$Qd \sim \frac{W}{\rho d}$$

But the velocity through the flow element is directly proportional to *Qp*. Therefore,

$$V \sim \frac{W}{\rho d}$$

Substituting these proportionalities into V^2 in terms of *N*,

$$\Delta P \sim \rho d$$

$$\Delta P \sim N^2\left(\frac{N}{N^2}\right)^2$$

$$\Delta P \sim \frac{N^2N^2}{N^4} = 1$$

or, rather, the ΔP is independent of the speed.

ANTI-SURGE CONTROL WITH SUCTION THROTTLING PERFORMANCE CONTROL, CONSTANT SPEED

Figure 7.136 shows the typical performance characteristics of a compressor controlled by suction throttling of the inlet at constant speed. The upper curve, 1, represents the performance with no suction throttling. Curves 2 and 3 are the performance at increasing degrees of suction throttling. The actual inlet volume to the compressor at surge is the same at points *a, b,* and *c*, and so is the pressure ratio. Flow, *Q*, as plotted in Fig. 7.136, therefore is the volume measured before the suction throttle valve.

There are three methods of protecting the compressor from surging when this mode of performance control is used.

1. Figure 7.137A illustrates a pressure ratio control. When pressure ratio control is used the ratio of the discharge pressure to the inlet pressure is calculated by the controller and maintained at an arbitrary value somewhat lower than the surge pressure ratio by opening the blow-off valve. The disadvantage of this system is that, if the compressor characteristic is flat, much of the range must be sacrificed (Fig. 7.137B).

The shaded area indicates the range that is not available. However, it is a very easy and inexpensive system to install.

2. A second method, which increases the operating range of the compressor, is illustrated in Fig. 7.138A. This system maintains a minimum ΔP as the set point. The operating range loss is illustrated in Fig. 7.138B. Again the shaded area indicates the range that is not available through this mode of control.

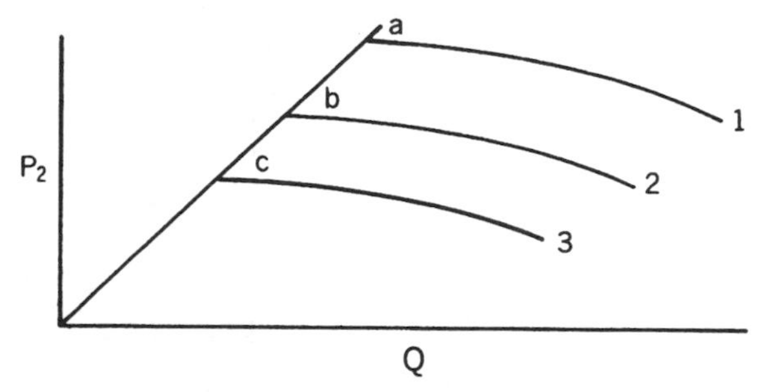

Figure 7.136 Performance characteristics of a compressor with inlet throttling control.

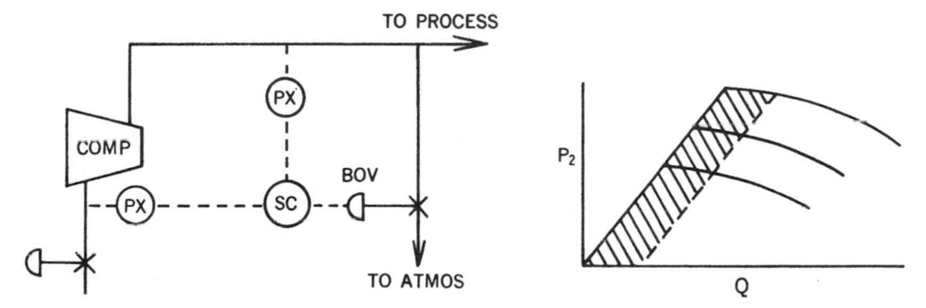

Figure 7.137 (A) Diagram of pressure-ratio control, and (B) the resulting performance characteristic curve.

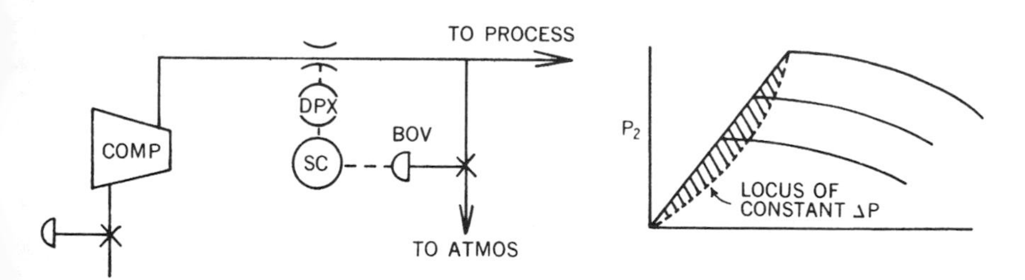

Figure 7.138 (A) and (B) Alternate method to increase the operating range of the compressor.

3. The system illustrated in Fig. 7.139A and B will match the suction throttle surge line exactly and provides the maximum possible operating range. In this mode of control, a minimum $\Delta P/P_2$ is maintained. Since a constant value of $\Delta P/P_2$ can be established that exactly matches the surge line, the maximum range of operation is available. In actual practice, a value of $\Delta P/P_2$ slightly to the right of the surge line would be used.

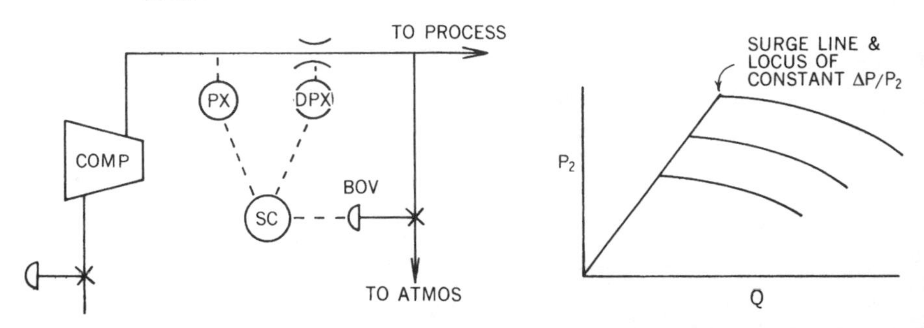

Figure 7.139 (A) and (B) Alternate control system to provide maximum operating range.

ANTI-SURGE CONTROL WITH ADJUSTABLE INLET GUIDE VANES

Compressors equipped with adjustable inlet guide vanes present an entirely different anti-surge control problem. The surge line of the compressor is a function of both the flow and the guide vane setting. Several schemes are used, and their complexity is dependent on how closely the surge line is to be matched. The simplest system is shown in Fig. 7.140A. The pressure drop, ΔP, across the flow element is continuously monitored. If the consumer's requirements decrease to a point where the flow element ΔP would be less than the set point, the surge controller opens the *BOV* to maintain the set point. Figure 7.140B shows the compressor characteristic and the control characteristic. Note, however, the extensive range that is lost with this mode of anti-surge control.

A wider useful range of the compressor characteristic can be retained with the anti-surge control system seen in Fig. 7.141. Both the flow element ΔP and the discharge pressure P_2 are monitored, and a minimum ratio of $\Delta P/P_2$ is maintained by the *BOV* if the consumer requirements are not sufficient. As can be seen, this mode increases the usable range.

The maximum usable range is obtained with the antisurge control system seen in Fig. 7.142. The differential signal from the flow element is modified by a signal that is a function of the inlet guide vane position. By proper shaping of a cam in the guide vane position transmitter *SX*, the set point can be modified to exactly parallel the compressor surge line and the loss in usable range held to a minimum.

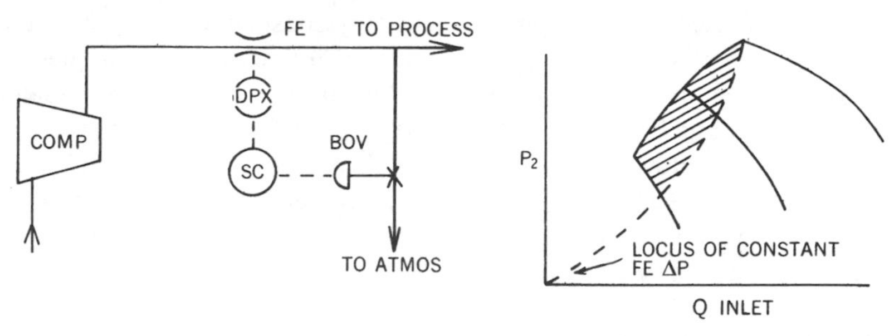

Figure 7.140 (A) and (B) Simplest anti-surge control with adjustable inlet guide vanes.

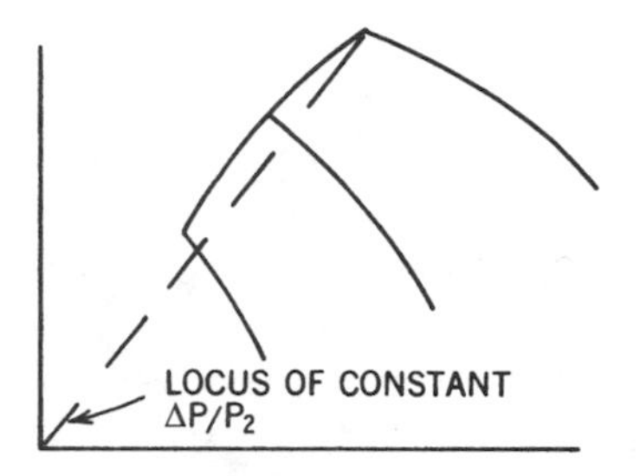

Figure 7.141 Alternate system of adjustable inlet guide vane control.

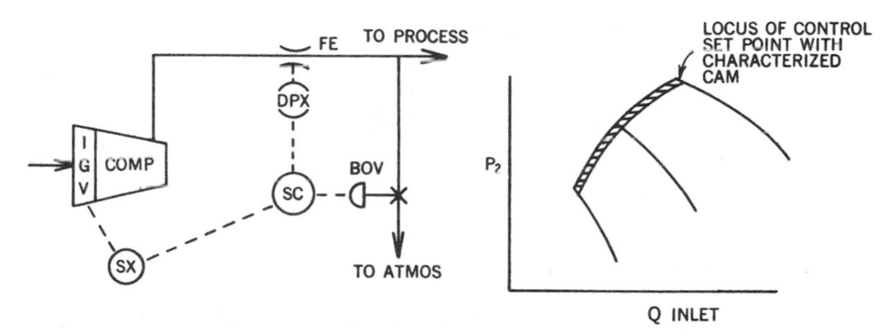

Figure 7.142 (A) and (B) System which results in maximum usable range.

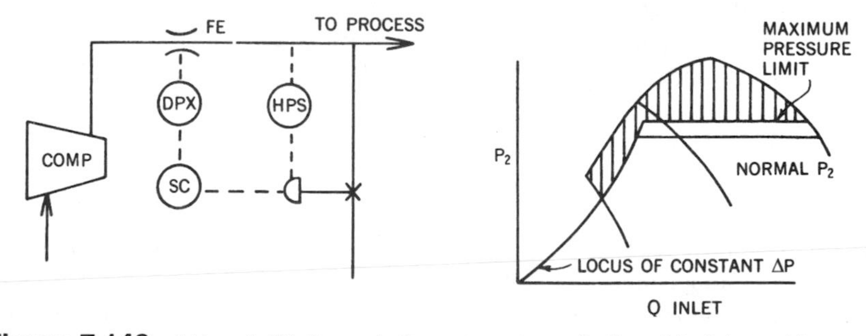

Figure 7.143 (A) and (B) Second alternate system of adjustable inlet guide vane control utilizing a pressure switch.

If the performance control is one of reasonably constant discharge pressure, a modification of the system (Fig. 7.140) is usually sufficient, is the simplest, and gives adequate range. It is frequently used in plant air anti-surge control systems.

In the mode in Fig. 7.143, the minimum set point ΔP is set lower, thus increasing the usable range. However, by so doing, the vane positions near full open are not protected from surge. By the addition of a pressure switch, this area can be protected. The pressure switch, set 5 to 10 psig higher than control pressure, directly opens the *BOV* on an increase in pressure above set point.

PERFORMANCE CONTROL

In the previous paragraphs, the basic methods of anti-surge control have been discussed. Anti-surge control is generally a passive control until the pre-established conditions have been reached, at which time it then controls the compressor to protect the system from surging. In addition to this protective control, further controls are necessary to adapt the compressor performance to the varying load requirements of the process it supplies.

A compressor control system can be designed to maintain a desired pressure to a process or a desired flow to a process. It cannot be designed to maintain both.

PRESSURE CONTROL

Figure 7.144 shows a typical system requiring constant pressure control. The process shown here might be a petrochemical process where the pressure P_2 at the process must be maintained at a fixed value, regardless of the flow through the process. It could also represent a plant air system where the plant air pressure must be maintained at 100 psig regardless of usage.

Constant pressure control can be accomplished by:

Variable speed
Adjustable inlet guide vanes or adjustable diffuser vanes
Intake throttling
Discharge throttling
Blow-off (recycle)

The methods are listed in decreasing order of efficiency.

In Fig. 7.144, the pressure is monitored, and a signal from the pressure transmitter (*PX*) is sent to a pressure controller (*PC*), which adjusts one of the above devices to maintain a constant pressure.

Each of the five methods of pressure control will be discussed. In the discussion, a centrifugal compressor will be assumed; however, the principles are the same for an axial compressor.

Variable-speed Constant Pressure Control

Figure 7.145 shows a typical set of variable-speed compressor characteristic curves. Each curve shows the pressure at which the compressor supplies a certain volume rate of flow, Q. If the compressor outlet is choked below a certain rate at any given speed, a pulsating flow, called compressor surge, will occur, as shown by the limiting curve.

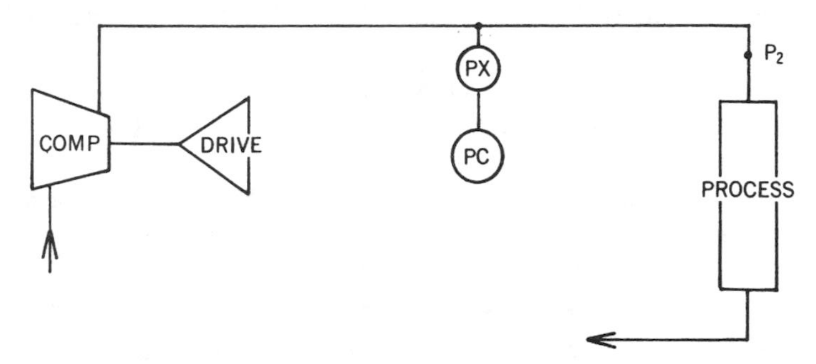

Figure 7.144 A typical system requiring constant pressure control.

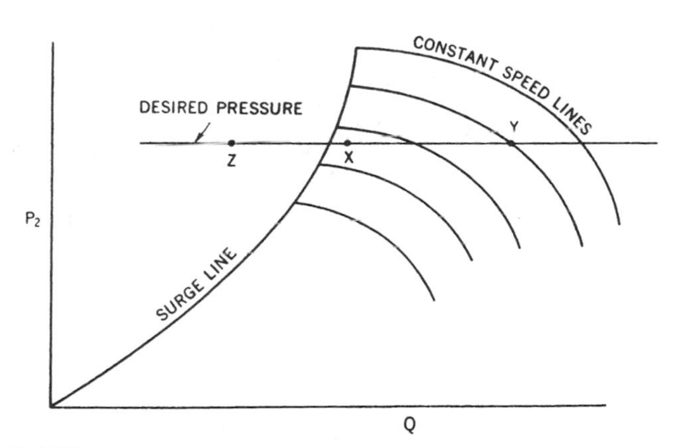

Figure 7.145 Characteristic curves for variable speed, constant pressure control.

We may assume that the compressor is operating at point Q_Y and the process requires a higher flow. The pressure immediately tends to fall as the operating point moves out to the right and downward along the characteristic curve for the given speed. However, the control system, sensing this drop in pressure, will increase the speed of the driver to return the pressure to the desired value. Conversely, had the flow decreased, the pressure would have tended to increase and the control system would decrease the speed at the driver until the desired pressure was reached. The flow could be reduced by speed reduction until point X was reached. The compressor must operate at this reduced speed in order to maintain the desired pressure. The anti-surge controls discussed earlier will prevent the operating point moving to left of a point on each speed curve similar to point *X*. If the process required a flow rate of only Q_z the volume $Q_x - Q_z$ would have to be blown off or recycled. This is accomplished by transferring from variable speed to blow-off control, which is the only suitable control when the process requires flows below the stable operating range. (For a more complete discussion see the paragraph on Blow-off, Constant Pressure Control).

ADJUSTABLE INLET GUIDE VANE CONSTANT PRESSURE CONTROL

Figure 7.146 shows a typical set of adjustable inlet guide vane characteristic curves. Let us assume that the compressor is operating at flow rate Q_Y and the process requires a higher flow. As the pressure tends to fall, the control system immediately opens the guide vanes to return the pressure to the set point. Conversely, the control system will close the vanes as the flow requirements decrease until point *X* is reached, at which time the anti-surge control comes into play. Further reduction in process requirements, as with variable-speed control, can only be accommodated by blow-off.

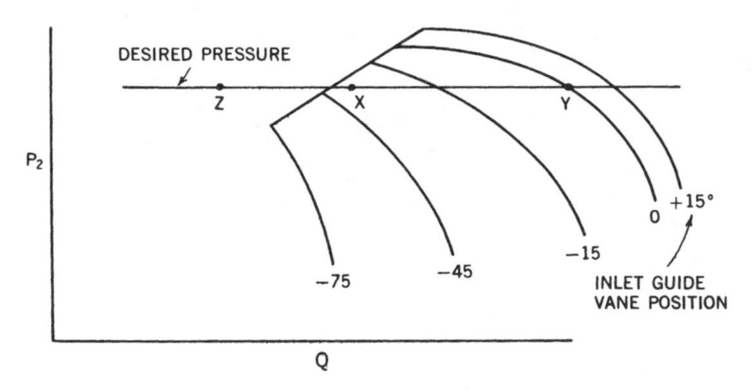

Figure 7.146 Characteristic curve typical of adjustable inlet guide vanes.

SUCTION THROTTLING, CONSTANT PRESSURE CONTROL

The two previous methods of control utilized the families of characteristic curves available with variable-speed or inlet guide vanes for constant pressure control. Intake, or suction, throttling control is usually used where the compressor is not equipped with inlet guide vanes and is driven by a constant-speed drive. It therefore has a single pressure-volume characteristic curve. However, variable performance can be achieved through suction throttling through valve TV-l.

Assume that the compressor is operating at point W on its unthrottled characteristic curve and there is a reduction in process requirements. If unthrottled, the pressure would increase to Y^1. However, by throttling across TV-1, the inlet pressure can be reduced and, although the compressor is operating at the pressure ratio and volume of Y^I the discharge pressure and volume flow to the process will be equivalent to point Y. An example will help to clarify this: Assume TV-1 is open:

$$Q_w = 100\% \qquad \frac{P_3}{P_{1w}} = 2.0$$

$$Q_{y1} = 80\% \qquad \frac{P_3}{P_1} Y^1 = 2.10$$

Inlet pressure P_1 14.7 psia = P_2 (no throttling)
Desired P_3 = 29.4 psia

At $(P_3/P_1)y'$, the pressure ratio is 2.10. To maintain the discharge pressure of 29.4, the inlet pressure P_2 must be reduced to 14.0 (29.4/2.1). The volume to the compressor at y1 is 80%, but the equivalent volume at 1 is less than the ratio of 80% x (14.0/14.7) or 76.3%. This is the actual volume of P_1 conditions delivered to the process.

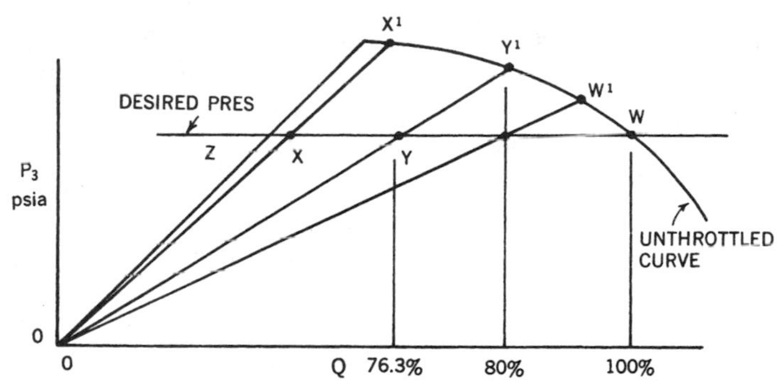

Figure 7.147 Typical throttled inlet characteristic curves.

The preceding calculation can be done by a graphical solution, if the pressure-volume characteristic is plotted with ordinates and abscissa oriented from zero, as seen in Fig. 7.147. Suction throttling control can reduce the flow to point X. Any further reduction in flow again requires blow-off.

DISCHARGE THROTTLING, CONSTANT PRESSURE CONTROL

As with intake throttling, there is only a single pressure-volume characteristic curve with discharge throttling. Pressure control is maintained simply by throttling actual compressor discharge pressure to the value desired. Discharge throttling requires more power than intake throttling for the same flow, as illustrated by the following example.

Referring to Fig. 7.147, we assume that the process requires 80% flow with discharge throttling, TV-1 would be in the discharge line, the compressor must operate at Y^1, and the gas must be throttled to the desired pressure. With suction throttling, the compressor operates at W^1 at a lower pressure ratio. Although the actual inlet volume to the compressor is higher with suction throttling, the weight flow to the process is the same; and since the pressure ratio is lower, the horsepower required is lower. For this reason, discharge throttling control is seldom used. A diagram of the controls involved in inlet throttling is shown in Fig. 7.148.

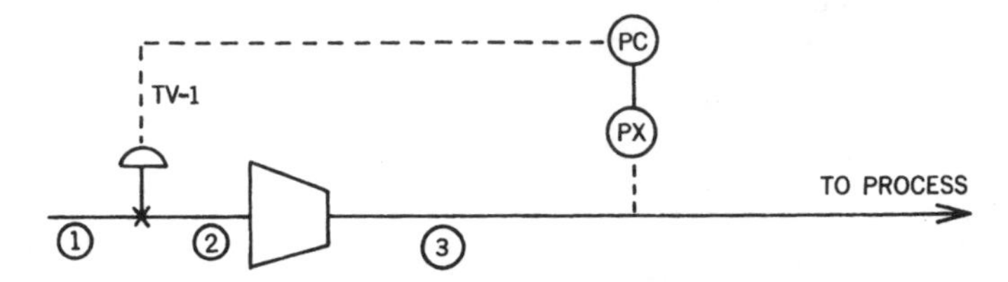

Figure 7.148 Schematic diagram of the controls involved in inlet throttling.

BLOW-OFF (RECYCLE), CONSTANT PRESSURE CONTROL

This is the least efficient method of control and is used only in conjunction with the more efficient control methods to extend their control range. If only blow-off control were used (Fig. 7.147), the compressor would always operate at point W regardless of the process requirements. The difference in flow between the process requirements and Q_w would have to be blown off and all the work expended on this extra flow wasted. For flows less than the surge limit, no other recourse is available, and blow-off must be used. The anti-surge control system utilizes blow-off control, as described earlier.

FLOW CONTROL

Figure 7.149 shows a typical system requiring constant volume control. The process shown might be a blast furnace where the flow delivered to the furnace must be held constant regardless of the varying resistance of the furnace and its charge.

As with pressure control, volume control can be accomplished by:

Variable speed
Adjustable inlet guide vanes (IGV) or adjustable diffuser units
Suction throttling (STV)
Discharge throttling (DTV)
Blow-off

In Fig. 7.149, the flow is monitored by the flow element *FE*, and a signal proportional to the flow is sent to the flow controller to adjust the necessary device to utilize one of the control methods above. Not all the equipment shown in Fig. 7.139B would be required for any given single method of control.

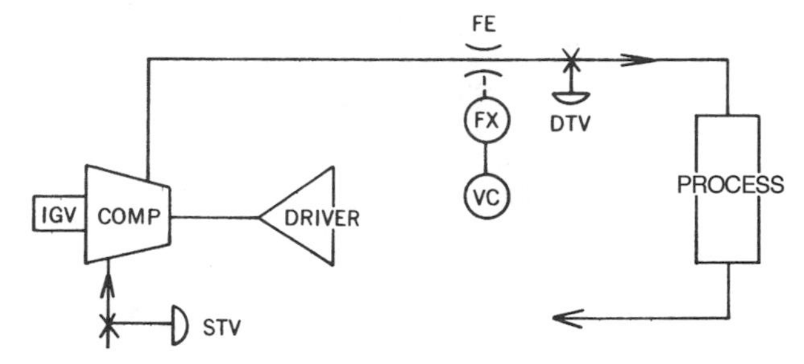

Figure 7.149 Typical system requiring constant volume control.

VARIABLE-SPEED, CONSTANT-FLOW CONTROL

Figure 7.150 shows a typical variable-speed compressor characteristic with constant-flow requirements superimposed on it. We will assume that the compressor is operating at point *Y* and the process resistance increases. The compressor will immediately tend to decrease in flow as the operating point thus moves up to the left along its characteristic curve. However, the control system sensing the decrease in flow will increase the speed until the desired flow at the higher resistance is again maintained at point Y^1. Conversely, if the resistance decreases and the flow increases, the speed will be reduced. Any desired flow may be chosen and controlled within the shaded area. If the compressor has flow-oriented, anti-surge control, the flow element and flow transmitter utilized for process volume control are the same as those used in the anti-surge system; and once the anti-surge system comes into play, flow control of the process is lost. If flow control were required in the area to the left of the surge line, separate flow elements and transmitters would be required, one serving the process control and the other the anti-surge system.

ADJUSTABLE INLET GUIDE VANE, CONSTANT-FLOW CONTROL

Figure 7.151 shows a typical adjustable inlet guide vane characteristic with constant volume requirements superimposed on it. We will assume that the compressor is operating at *Y* and that the process resistance decreases. The flow will begin to increase as the compressor operates lower on its characteristic curve. However, the control system will immediately sense the increased flow and close the *IGV* until the desired flow is reestablished at the lower pressure, point Y^1. The reverse occurs with an increase in resistance. The desired volume set point may be chosen anywhere to the right and below the surge line. However, the usable portion of this area is determined by the type of anti-surge control utilized. The reader is referred to the discussion of anti-surge control a few paragraphs earlier.

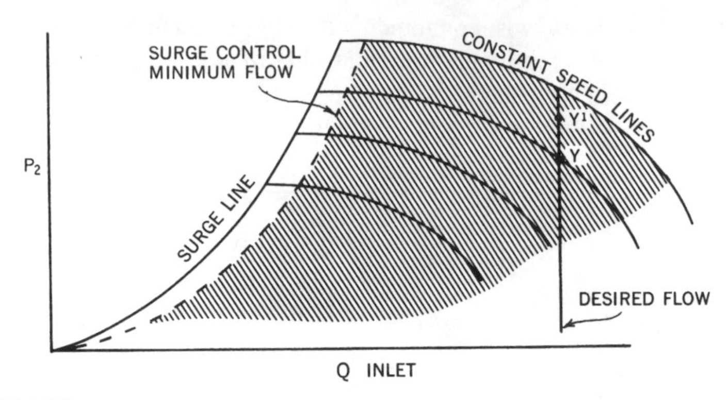

Figure 7.150 Variable speed characteristic curve with superimposed constant flow requirement.

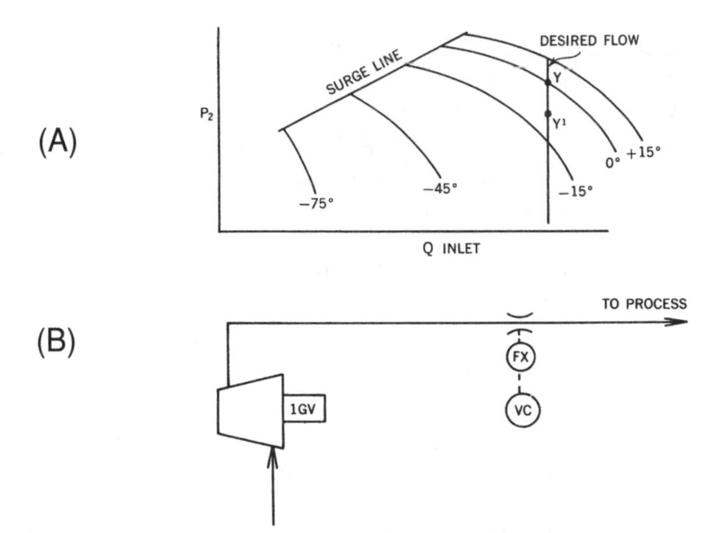

Figure 7.151 (A) and (B) Adjustable inlet guide vanes with constant volume requirements superimposed.

SUCTION THROTTLING, CONSTANT-FLOW CONTROL

Figure 7.152 shows a typical suction-throttling characteristic with constant-flow process requirements. Suction-throttling for flow control operates in exactly the same manner as it does for pressure control except that the throttle valve is operated in response to a change in flow, rather than a change in pressure. We may assume that the compressor is operating at point *W* on its throttled characteristic curve and that there is a reduction in process resistance. If unthrottled, the flow would increase toward Y^1 until the process resistance were again matched. However, the control system, sensing the increased flow, partially closes the suction throttle valve to compensate for the reduced process resistance. The control system then throttles the flow until it is equal to the desired flow, *Y*, as measured at (1) the actual inlet flow to the compressor at the reduced pressure and (2) is equal to the flow at Y^1.

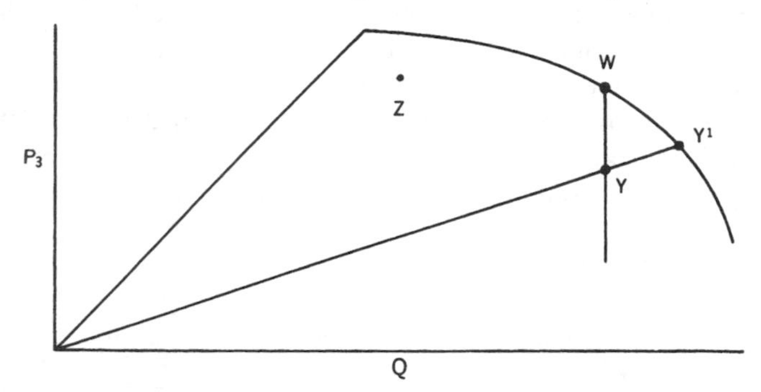

Figure 7.152 Suction throttling characteristic curve with constant-flow process requirement superimposed.

Discharge Throttling, Constant-flow Control

Constant-flow control can also be accomplished with discharge throttling; however, as with discharge throttling and constant-pressure control, it is less efficient and requires more power for the same flow than suction throttling. We will assume that the compressor (Fig. 7.152) is operating at *W* and that a reduction in process resistance occurs. If unthrottled, the flow will increase toward Y^1 until the process resistance is again matched. However, the control system, sensing the increased flow, will partially cover the discharge throttle valve, forcing the compressor operating point back up along the characteristic curve to *W* and to the original flow. With discharge throttling of the compressor for a given flow, the compressor will operate at a maximum power level regardless of the process resistance. Suction-throttling control allows the power level to reduce as the process resistance decreases. Discharge throttling, therefore, is seldom used as a method of flow control.

Blow-off, Constant-flow Control

As with constant-pressure control, blow-off control is used only to extend the operating range and as antisurge protection for more efficient control methods. As an example of the inefficiency of this method, let us assume that the process requires operation at point *Z*. The compressor will operate at point *W* on its characteristic curve, and the flow $Q_w - Q_z$ will be blown off, all the work done upon this extra flow being wasted.

CENTRIFUGAL COMPRESSOR SHAFT SEAL

Centrifugal compressors are inherently high speed machines. They compress gases as mild as air and those as toxic, flammable, and corrosive as hydrogen sulfide. They operate at all pressure levels from high vacuum to over 10,000 psig. As one might expect, the compressor shaft seal can take many forms, from a mere restriction to minimize gas losses to a complex buffered design that must be 100% leak-tight to minimize the possibility of fire or of harm to personnel. There can be five basic types of shaft seals applied in centrifugal compressors:

1. Labyrinth seal
2. Restrictive-ring seal
3. Mechanical (contact) seal
4. Liquid-film seal
5. Pumping liquid-film seal

These seals may be used independently or in combination with one another to satisfy various requirements.

Labyrinth-type Shaft Seal

The labyrinth-type seal, when applied in its simplest form, is not a true seal. It is merely a device to limit the loss of a gas without contact between the shaft and the compressor casing. However, the labyrinth can be sectionalized to provide one or more annuli, which are buffered or educted, or both, to eliminate the process gas loss or to channel its flow in a controlled manner.

The labyrinth consists of a series of straight or staggered, close-clearance, short-length restrictions that take on the appearance of knife edges. These thin restrictions can be machined from either aluminum, bronze, filled TFE, or other suitable material and mounted in the compressor end housing as shown in Fig. 7.153. The labyrinths may also be made from an 18-8 stainless-steel strip, 0.10 to 0.015 in. thick, formed into the cross-section of a J and caulked into the shaft with a soft, stainless-type wire (Fig. 7.154).

Labyrinths can be applied at any rotating speed because the physical contact area is nil, and a slight rub during acceleration or deceleration should not precipitate a serious failure. The knife edges merely wipe or bend. However, when the labyrinth is buffered or educted, proper care should be given to sizing the annuli and the associated system to ensure the effectiveness of the desired gas seal. In the case of a machine touch point seal, a babbitt or other similar mating material surface may be provided.

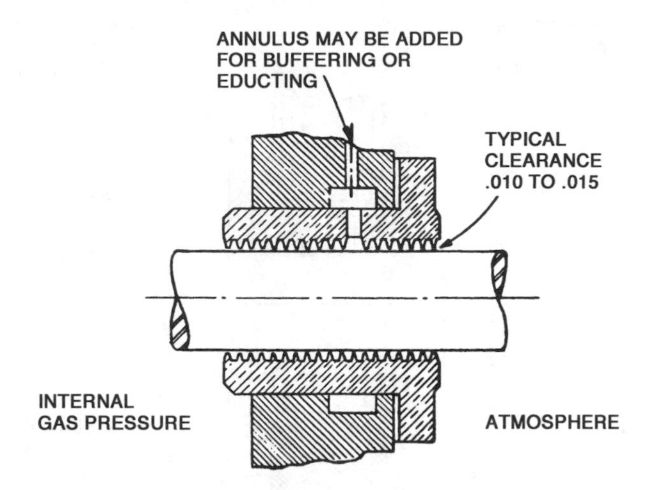

Figure 7.153 Labyrinth type shaft seal.

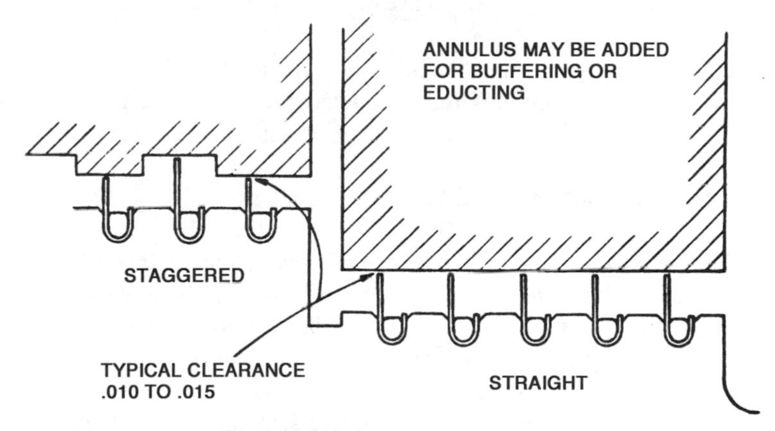

Figure 7.154 Labyrinth type shaft seal.

Restrictive-ring Shaft Seal

The restrictive-ring seal, like the labyrinth seal, is not a true seal when applied in its simplest form. It is a device to limit the loss of gas to a greater extent than the labyrinth without prolonged contact between rings and shaft. The rings may be arranged with one or more annuli for buffering or educting, or both (Fig. 7.155). The ring-type seal is considerably more effective per unit of length than the labyrinth seal. The clearances can be made smaller because the rings are usually free to float with the shaft to avoid binding or severe rubbing.

The rings are usually made of carbon, using either one piece or segmented construction. A garter-type spring is used with the segmented construction to hold the segments together and impart an axial force to the carbon through an angled joint.

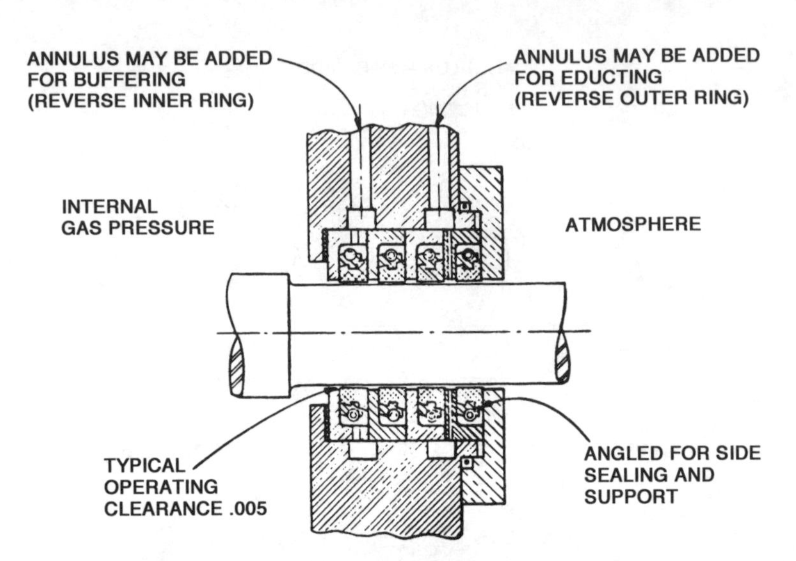

Figure 7.155 Restrictive ring type shaft seal.

The ring is thus held tight against the housing to maintain position and to avoid side leakage. When the rings are buffered or educted, the same precaution must be taken in sizing the annuli and the associated system as with the labyrinth-type seals.

Mechanical (Contact) Shaft Seal

The mechanical seal is the next step toward positive sealing in that it limits the loss of gas to almost a negligible amount because there is an extremely small clearance between the carbon and the mating shoulder (Fig. 7.156). The life of the seal is not very predictable when run dry. Therefore, most mechanical seals are buffered with a lubricating oil, which is supplied in sufficient quantity and at high enough differential pressure above the gas to ensure positive sealing and lubrication of the rubbing faces. The oil-buffered mechanical seal entails a second restrictive seal that limits the liquid sealant flow to the atmosphere to that amount required for heat carry-off. This clean-side leakage is returned to the reservoir.

There is a small amount of inner sealant leakage toward the gas. This leakage is gathered in a trap and is manually or automatically drained. This sealant is either discarded or reclaimed. The possibility of imperfect contact always exists with a mechanical seal. However, most designs inherently provide a positive seal at shutdown without a buffering fluid as long as the carbon face and mating shoulder are in intimate contact, a feature that makes such seals very attractive in refrigeration applications.

Efforts have been made to minimize the rubbing speed between the carbon face and the mating shoulder. One method that reduces the rubbing speed employs a carbon ring that rubs both the rotating seat affixed to the shaft and a stationary spring-loaded seat, like that in Fig. 7.157. The carbon ring rotates at a speed less than the shaft rotational speed.

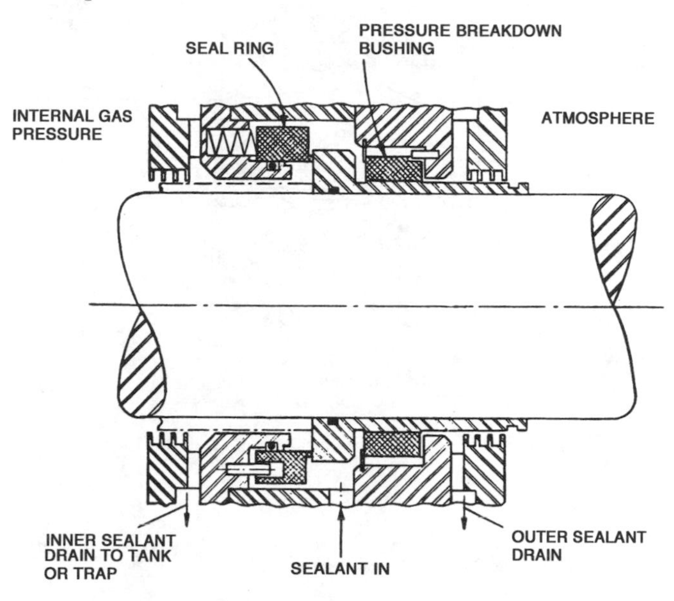

Figure 7.156 Mechanical contact type shaft seal.

Liquid-film Shaft Seal

A liquid-film seal employs double floating bushings or rings that have small clearances with respect to the shaft. They are buffered in between with a sealant, usually oil, to create a positive flow of sealant toward the gas within the compressor and toward the atmosphere. This type of seal is seen in Fig. 7.158. This seal is one of the simplest to manufacture and operate, if the inner sealant leakage is reusable, because the clearances can be made comparable to those in journal bearings. However, if the inner sealant leakage must be discarded or put through a clean-up process, the running clearances have to be made very small, and such small clearances can lead to mechanical problems.

Pumping Liquid-film Shaft Seal

The pumping liquid-film seal is essentially a buffered liquid-film seal with large clearances, backed up with a viscous-type pump positioned adjacent to the gas side of the inner bushing. The pump creates a back pressure that impedes the sealant flow toward the gas side of the seal during operation (Fig. 7.159).

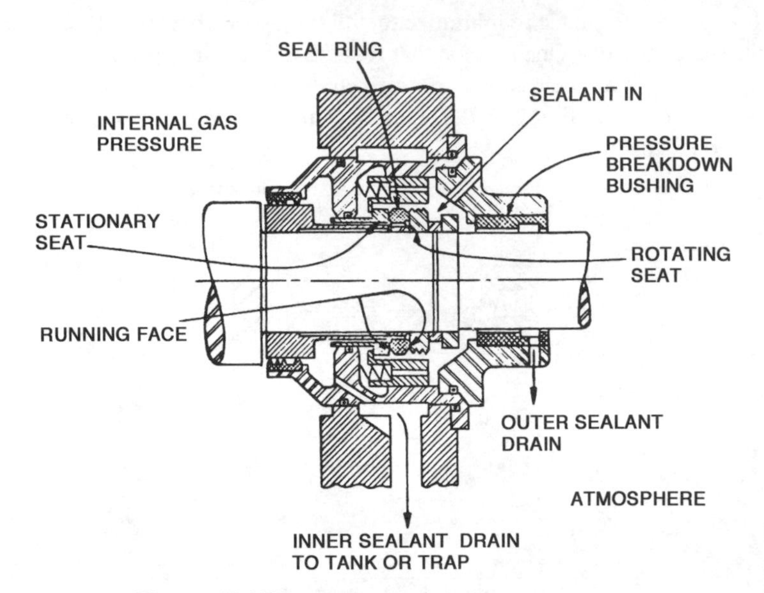

Figure 7.157 Mechanical contact type shaft seal.

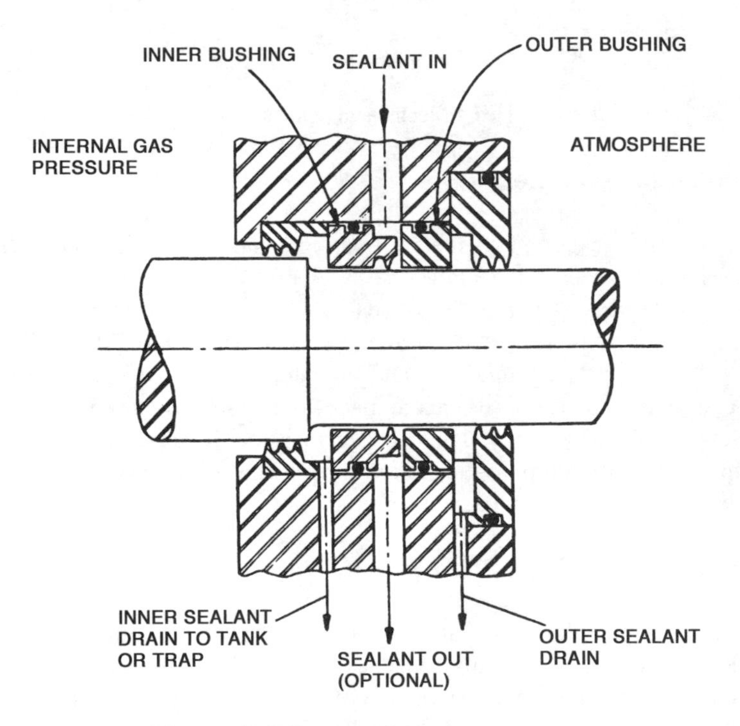

Figure 7.158 Liquid film type shaft seal.

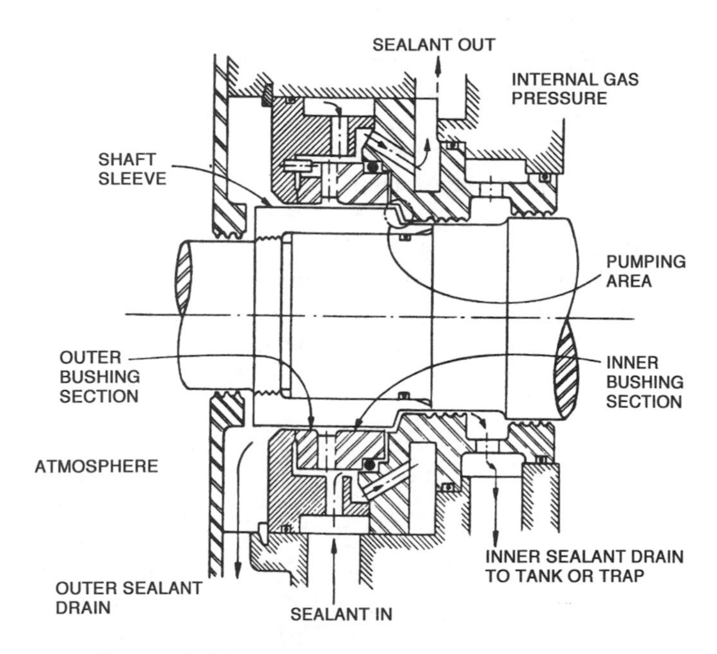

Figure 7.159 Pumping liquid film type shaft seal.

This back pressure is a function of speed. It counterbalances the oil-to-gas pressure differential during operation in the normal speed range of the compressor. It performs like a mechanical seal with respect to the amount of sealant reaching the gas side of the seal. This can be as little as only a few gallons a day. At standstill and at low-speed operation, it performs like the high clearance seal described previously. The running clearances within this seal between the bushing and the shaft are comparable to those in journal bearings, all others being several times this magnitude. The inherent advantage of this type of seal is its ability to operate with large clearances and yet experience low sealant leakage at operating speeds.

Selection of Compressor Shaft Seal and Associated Systems

Selection of the compressor shaft seal and the associated system depends primarily on gas composition, gas pressure levels, process requirements, and the need for reliability. However, where buffer gas or liquid-type seals are involved, the seals can perform no better than the associated sealant system. A seal selection should not be made until all aspects of the process and compressor are considered.

Among the process requirements and limitations to be considered are the following: The gas to be compressed, pressures at inlet and outlet, and whether a buffer gas or liquid sealant will be used. Disposal of the leakage gas must also be considered. Compressor details and operating cycles also influence seal selection or design.

Cost of the gas being handled is a factor, as well as its temperature, flash point, toxicity, and chemical activity or inertness. Pressures during operation and off design, during recycle, and at stagnation or holding are important, the last especially so in refrigeration compressors. Pressures during process upsets and safety relief valve settings may dominate the design considerations. The operating cycle, whether continuous or intermittent, and the plant turnaround cycle will also influence the design.

If there is to be a buffer gas, it must be compatible with the process gas, and its composition, whether it is inert or sweet, or sweet but flammable, must also be taken into account. Availability, cost, pressure level, and reliability must also be considered. In the case of a liquid sealant or buffer, the same factors enter the picture, and one may also need to look into the cost of recovering leakage to the gas side of the seal.

The leakage gas may be allowed to escape to the atmosphere, or it may be transferred to another part of the process. Its pressure, tolerance for air and moisture, and any educator and operating medium requirements must be taken into account.

In the compressor, the shaft diameter and peripheral speed must be known, plus the axial length of the seal, how isolated it will be from the compressor bearing, and how it will be influenced by critical speeds, either during normal running or while accelerating through the critical speeds.

In the case of buffered seals, it is imperative to have a quality sealant supply system; that is, the sealant system must contain auxiliaries, spares, or means for isolating and bypassing components so that they may be maintained or replaced without upsetting the pressure level or flow of the sealant. Power supplies must be from separate sources or must be different: steam, AC power, DC power, or gas. System monitoring has to be employed to warn of impending failure, actuate auxiliaries, and trip the entire unit to avoid extensive damage. Many applications require a large rundown tank referenced to the process gas pressure level, which provides the sealant for normal compressor operation for a specified time after a total sealant pump failure. After the unit is tripped, the run-down tank also provides sealant during the coastdown of the compressor, isolation of the compressor, venting of the compressor, and during any purge of the compressor in the case of a flammable or toxic gas.

PERFORMANCE CALCULATIONS

Within the ranges of pressure and temperature usually encountered, air follows the perfect gas laws. Some other diatomic gases also follow perfect gas laws, but hydrocarbon and other gases and many mixed gases deviate to a considerable extent. Since many applications involve performance calculations for air only, this section has been divided into parts, as follows:

1. Information required for compressor calculations, compression formulas, and step-by-step explanation of performance calculations.
2. Performance calculations for air with corrections for humidity.
3. Performance calculations for gases, including determination, from the gas analysis or applicable gas constants.

Introduction to Curves and Performance Calculations

Table 7.3 contains formulas for calculating centrifugal compressor head, discharge temperature, and horsepower. Derivations are based on commonly accepted thermodynamic relations for gases. Isothermal compression calculations are generally used only when extensive cooling is accomplished during the compression cycle. This cooling can take the form of shell and tube intercooling between stages of compression or some form of liquid injection. Expected overall isothermal efficiency must be determined by the designer of the compressor. Discharge temperatures depend on the type of cooling and location of coolers and cannot be determined directly from isothermal efficiency. Hence, isothermal compression calculations are of little use to the estimator.

Either adiabatic or polytropic relations can be used as the basis for comparing centrifugal compressor performance. In recent years, however, polytropic relations have generally replaced adiabatic relations for this purpose. When the performance of a given compressor or stage is known, application of this performance information to gases with different specific heat ratios is more straightforward with polytropic relations. Calculations and examples used in this chapter, therefore, will be based on polytropic compression.

Explanation of Centrifugal Compression Calculations

A. Information Required for Compressor Calculations

1. Physical properties of gas being compressed
   Molecular weight, MW
   Adiabatic exponent, k
   Supercompressibility factor, Z
   Relative humidity, RH
   (These can be determined from complete gas analysis)
2. Inlet conditions at compressor flange
   Capacity, cfm, lb/minute
   mm scfd (at defined conditions) Inlet temperature, °F
   Inlet pressure, p, (psia)
3. Discharge pressure at
   compressor flange, p_2 (psia)
4. Water temperature, if intercooling is used, t_w, °F
5. Process temperature limitations, if any

TABLE 7.3 Thermodynamics equations for compressor calculations.

	Adiabatic	Polytropic	Isothermal
Compression process	$Pv^k = C$	$Pv^n = C$	$Pv = C$
Determination of exponent	$k = \frac{c_p}{c_v}$	$\frac{n-1}{n} = \frac{k-1}{k} \times \frac{1}{\eta_p}$	
Theoretical discharge temperature, °F abs	$T_2 = T_1 r^{(k-1)/k}$	$T_2 = T_1 r^{(n-1)/n}$	$T_2 = T_1$
Discharge temperature, °F abs	$T_2 = T_1 + \frac{T_1[r^{(k-1)/k} - 1]}{\eta_{ad}}$	$T_2 = T_1 r^{(n-1)/n}$	$T_2 = T_1$
Head (H) (ft-lb/lb)	$H_{ad} = Z_{av} R T_1 \frac{r^{(k-1)/k} - 1}{(k-1)/k}$	$H_p = Z_{av} R T_1 \frac{r^{(n-1)/n} - 1}{(n-1)/n}$	$H_i = Z_{av} R T_1 \ln_e r$
Gas horsepower, ghp (using capacity)	$\text{ghp} = \frac{Q_1 P_1 \frac{Z_1 + Z_2}{2Z_1} [r^{(k-1)/k} - 1]}{229\eta_{ad}\left(\frac{k-1}{k}\right)}$	$\text{ghp} = \frac{Q_1 P_1 \frac{Z_1 + Z_2}{2Z_1} [r^{(n-1)/n} - 1]}{229\eta_p\left(\frac{n-1}{n}\right)}$	$\text{ghp} = \frac{Q_1 P_1 \frac{Z_1 + Z_2}{2Z_1} \ln_e r}{229\eta_i}$
Gas horsepower, ghp (using weight)	$\text{ghp} - \frac{WH_{ad}}{33{,}000\ \eta_{ad}}$	$\text{ghp} - \frac{WH_p}{33{,}000\ \eta_p}$	$\text{ghp} = \frac{WH_1}{33{,}000\ \eta_i}$
Brake horsepower, Bhp	Bhp = ghp + mech. losses	Bhp = ghp + mech. losses	Bhp = ghp + mech. losses

B. Preliminary Calculations

1. From inlet capacity in cfm, the approximate polytropic efficiency may be determined from Fig. 7.160. The actual value will vary due to speed, specific wheel design, compression ratio, and other factors. The compressor manufacturer must be contacted when accurate data is desired.
2. Polytropic ratio $(n - 1)/n$ may be found from the ratio of specific heats (k) and the polytropic efficiency (η_p) using Fig. 7.161.

$$\frac{n-1}{n} = \frac{k-1}{k} \times \frac{1}{\eta_p}$$

3. Compression ratio, r, may be found from inlet and discharge pressures.

$$r = \frac{p_2}{p_1}$$

C. Discharge Temperature

$$T_2 = T_1 \times r^{\frac{n-1}{n}}$$

The discharge temperature must be considered in the selection of materials and design of a compressor.

D. Polytropic Head

1. If supercompressibility factor is involved, discharge supercompressibility Z_2 may be found from the gas analysis. The discharge temperature and pressure may then be calculated.

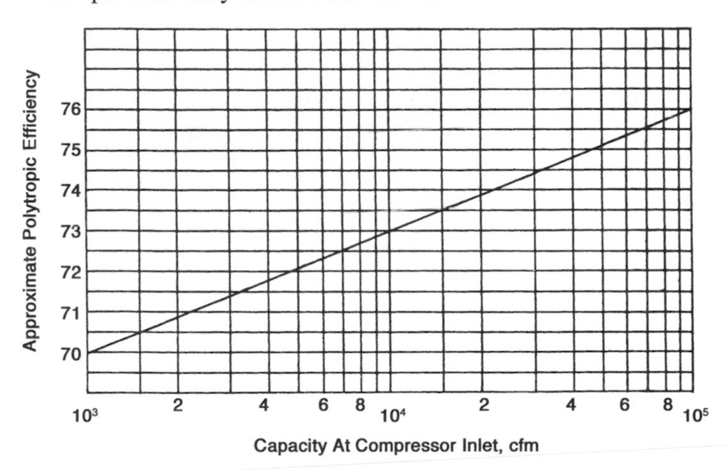

Figure 7.160 Approximate polytropic efficiency versus compressor inlet capacity.

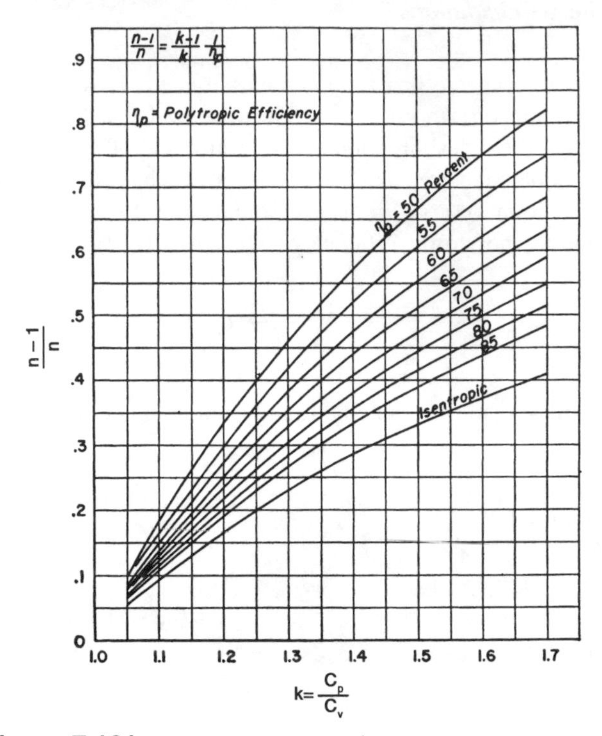

Figure 7.161 Polytropic ratio $\frac{n-1}{n}$ versus adiabatic exponent k.

2. Polytropic head:

$$H_p = \frac{Z_1 + Z_2}{2} RT_1 \frac{r^{(n-1)/n} - 1}{(n-1)/n}$$

The value of $r^{(n-1)/n}$ may be obtained from Fig. 7.162. Polytropic head is an indication of the number of impellers required for the conditions specified.

E. Gas Horsepower

1. From weight flow,

$$\text{ghp} = \frac{W \times H_p}{33{,}000 \times \eta_p}$$

2. Or, alternatively, from inlet capacity and inlet pressure,

$$\text{ghp} = \frac{Q_1 \times P_1 \times \frac{Z_1 + Z_2}{2Z_1} \times \frac{r^{(n-1)/n} - 1}{(n-1)/n}}{229 \times \eta_p}$$

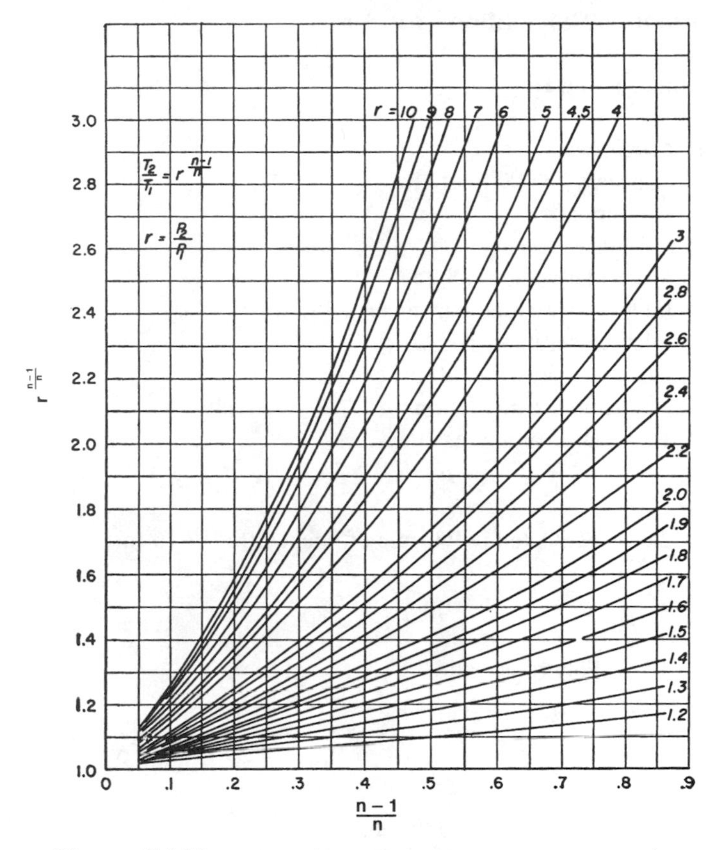

Figure 7.162 Polytropic temperature ratio versus ratio $\frac{n-1}{n}$.

F. Brake Horsepower

The gas horsepower (ghp) calculated is not the true input horsepower to the compressor. Mechanical and hydraulic losses to be considered are:

a. Bearing losses.
b. Seal losses.
c. Other losses that can be ignored in most estimating work include radiation losses, labyrinth-seal losses, and recirculation due to balancing devices.

In estimating bearing and seal losses, figures of 30 and 20 horsepower may be used, respectively. These figures are quite approximate since a given-size bearing will have variable losses depending on bearing load, rotative speed, oil temperature, and so on.

Performance Calculations for Air

1. Example Using Dry Air

A. Conditions

Capacity, 75,000 inlet cfm
Gas, dry air
k, 1.395
Molecular weight, 28.95
Inlet pressure (and barometer), 14.7 psia
Inlet temperature, 100 °F
Discharge pressure, 30 psig

B. Preliminary Calculations

1. Discharge pressure

$$p_2 = 30 + 14.7 = 44.7 \text{ psia}$$

2. Estimated polytropic efficiency (Fig. 7.160):

$$\eta p = 75.6\%$$

3. Polytropic ratio:

$$\frac{n-1}{n} = \frac{k-1}{k} \times \frac{1}{\eta_p} = \frac{0.395}{1.395} \times \frac{1}{0.756} = 0.378$$

4. $$r = \frac{p_2}{p_1} = \frac{44.7}{14.7} = 3.04$$

C. Discharge Temperature

$$T_2 = T_1 \left[r^{(n-1)/n} \right] = 560 \left(3.40^{0.378} \right)$$
$$= 560 \text{ x } 1.522 = 853 \qquad (853 - 460 = 393°\text{F})$$

D. Polytropic Head

$$H_p = RT_1 \frac{r^{(n-1)/n} - 1}{(n-1)/n} = \frac{1544}{28.95} \times 560 \frac{\left(3.04^{0.378} - 1 \right)}{0.378}$$
$$= 53.3 \text{ x } 560 \text{ x } \frac{0.522}{0.378}$$
$$= 41{,}200 \text{ ft-lb/lb}$$

Compressor will normally have three to five stages.

E. Horsepower

$$1.\quad \text{ghp} = \frac{Q_1 \times p_1 \times \dfrac{r^{(n-1)/n} - 1}{(n-1)/n}}{229 \times 0.756}$$

$$= \frac{75{,}000 \times 14.7 \times (0.522 / 0.378)}{229 \times 0.756}$$

$$= 8800 \text{ hp}$$

2. Bhp = ghp + mechanical losses (bearings)
   = 8800 + 30 = 8830hp

Humidity Corrections

Air or gas can hold varying amounts of water vapor, the amount depending on temperature and pressure conditions and the degree of saturation. All compressor calculations must take into account the presence of water vapor and include the resulting capacity corrections. In addition, centrifugal compressor calculations must also take into consideration any changes in the density and k value, both of which are affected by the water vapor.

A. To correct for water vapor, it is necessary to determine its partial pressure.
   1. If the mixture is saturated with water vapor, the dry bulb temperature of the mixture determines the vapor pressure. The vapor pressure is then the saturation pressure at mixture temperature. Dry bulb temperature, wet bulb temperature, and mixture temperature are then identical.
   2. If the mixture is not saturated, (i.e., the vapor is superheated, the dew point temperature or equivalent information must be known, as it is the temperature at which the mixture will be saturated with water vapor if cooled at constant total pressure.
   3. The amount of vapor in unsaturated mixtures may also be expressed in terms of relative humidity. Relative humidity is the ratio of the actual water vapor pressure to the partial pressure of the water vapor if the air were saturated at the dry bulb temperature of the mixture.

B. Use is made of the partial pressures in order to correct weights and capacities for water vapor present in the mixture.
   1. Whenever the weight of one constituent of a mixture of two gases is known, the weight of the other can be determined from the following form of the general gas equation.

$$W_2 = W_1 \times \frac{MW_2}{MW_1} \times \frac{p_2}{p_1}$$

Total pressure of mixture: $p = p_1 + p_2$.

For mixtures of air and water vapor, the preceding formula becomes

$$W_v = 0.622 \times W_a \times \frac{p_v}{p_a}$$

Total pressure = p_a, + p_v, where subscript a represents dry air and subscript v represents water vapor.

2. The volume of the mixture calculated by using the weight and partial pressure of any one of its constituents.

$$\frac{W_a \times R_a \times T_1}{p_a \times 144} = \frac{W_a \times 1544 \times T_a}{p_a \times 144 \times MW_a}$$

C. 1. The k value and molecular weight can he calculated from the mol percentage of the individual constituents making up the mixture (see Chapter 8).

2. Specific humidity (SH = lb of water vapor/lb dry air) is useful in determining k and MW for air-water vapor mixtures from Fig. 7.163.

$$\text{SH} = \frac{W_v}{W_a} = 0.622 \frac{p_v}{p_a}$$

Figure 7.164, which gives specific humidity of air-vapor mixtures at saturation, is also useful in calculating drop out or condensation, when mixtures are cooled between two or more stages of compression.

Example of Compressor Calculation Involving Humid Air

In many cases involving air, inlet capacity is referred to standard conditions, dry, which is another way of indicating the weight of dry air to be compressed. The actual capacity handled by the compressor must be corrected for temperature and pressure differing from the standard conditions as well as the relative humidity. Changes in the previous example illustrate this.

A. Conditions

Inlet capacity, dry air at 60 °F, 14.7 psia, 75,000 cfm
Inlet temperature, 100 °F
Inlet pressure (and barometer), 14.7 psia
Relative humidity, 60%
Discharge pressure, 30 psig

B. Preliminary calculations

1. Weight flow dry air $W = \dfrac{QP_1 \times 144}{RT}$

$$= \frac{75{,}000 \times 14.7 \times 144}{53.3 \times 520}$$

$$= 5730 \text{ lb/minute}$$

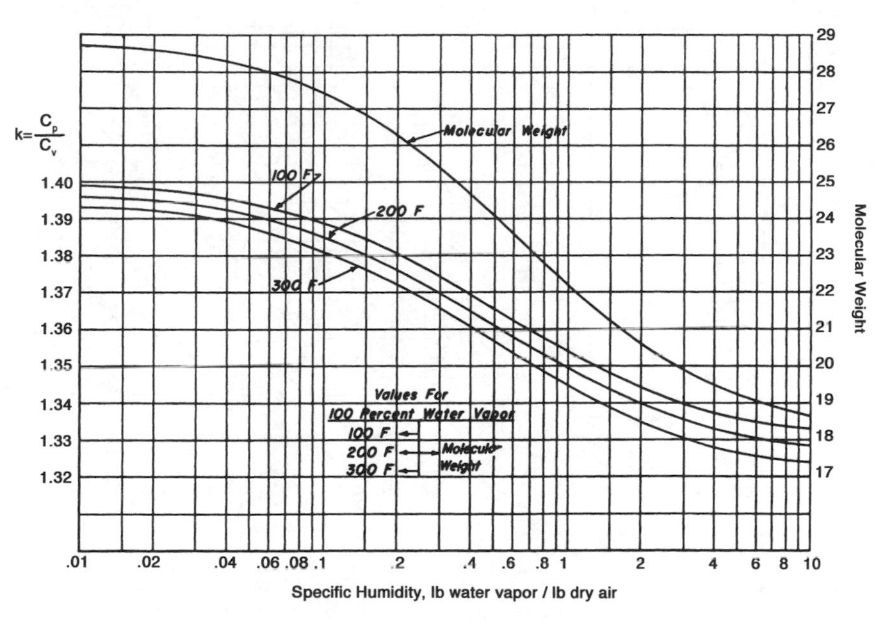

Figure 7.163 Molecular weight and adiabatic exponent for air vapor mixtures.

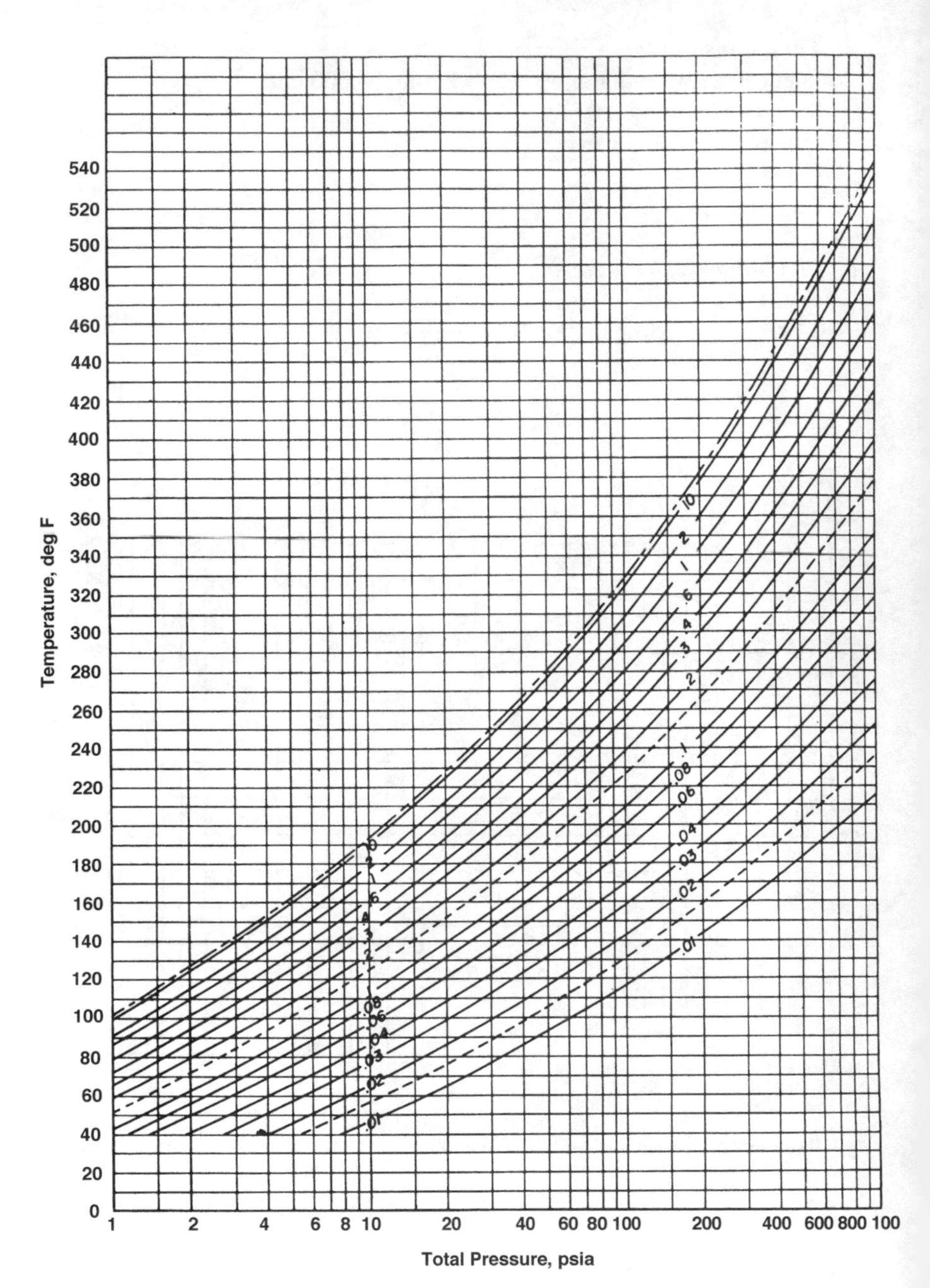

Figure 7.164 Pounds of vapor per pound of dry air at saturation.

2. Partial pressure of water vapor

   a. Saturation pressure at 100 °F from steam tables:

   $p_{vs} = 0.949$ psia

   b. p_v at 60% RH = 0.60 x 0.949 = 0.569 psia
   (p_a = 14.7 – 0.569 = 14.131 psia)

3. Volume of air and water vapor mixture

   a. $Q_1 = \frac{5730 \times 53.3 \times 560}{14.131 \times 144} = 84{,}000$ inlet cfm

   b. This may be checked by computing the weight of water vapor in the mixture, and then the volume occupied by that water vapor may he found from its partial pressure.

   $$W_2 = 0.622 \times 5730 \times \frac{0.569}{14.131} = 143 \text{ lb/minute}$$

   $$Q_1 = \frac{143 \times (1544/18.016) \times 560}{0.569 \times 144} = 84{,}000 \text{ inlet cfm}$$

4. Specific humidity = 143/5730 = 0.025 lb of water vapor/lb of dry air
5. Using Fig. 7.163 and specific humidity, one may determine k and MW of the mixture.

   $k = 1.393$ (at estimated $T_{av} = 250$°F)

   $MW = 28.5$ $\qquad R = \frac{1544}{MW} = 54.2$

6. Estimated polytropic efficiency from Figure 7.160 at 84,000 cfm is $\eta_p = 75.8\%$.

7. $\frac{n-1}{n} = \frac{0.393}{1.393} \times \frac{1}{0.758} = 0.372$

8. $r = \frac{44.7}{14.7} = 3.04$

C. Discharge Temperature

$$T_2 = 560 \times 3.04^{0.372} = 560 \text{ X } 1.512$$
$$= 848R \ (= 388°\text{F})$$

D. Polytropic Head

$$H_p = \frac{RT_1\left[r^{(n-1)/n} - 1\right]}{(n-1)/n} = 560 \times 54.2 \times \frac{1.512-1}{0.372}$$
$$= 41{,}800 \text{ ft-lb/lb}$$

E. Horsepower

1. $\text{ghp} = \dfrac{84{,}000 \times 14.7 \times (1.512-1)/0.372}{229 \times 0.758}$

   $= 9790$

2. Bhp = 9790 + 30 = 9820

Supercompressibility

Most gases show some departure from the simple perfect gas law when considered over a wide range of pressures and temperatures, and many gases show considerable differences even in a comparatively narrow range. If a satisfactory Mollier diagram is available, or, in other words, if all of the physical properties are shown for the particular gas in question, the Mollier diagram should be used to calculate real horsepower and other related items of performance as well as equivalent capacities at various temperatures and pressures. For gases the physical properties of which are only partially known, or not readily available, it is the usual practice to depend on specific information concerning the supercompressibility. A full discussion of this subject is found in Chapter 8 as it applies to all types of compressors.

Effect of Supercompressibility on Dynamic Compressor Calculations

Many gases normally handled in dynamic compressors cannot be treated with sufficient accuracy by perfect gas equations. Supercompressibility factors are shown in Figs. 8.1 to 8.6, and sample calculations involving Z factors appear on pages 649 to 656. For centrifugal compressor calculations of head and horsepower, it is reasonably accurate to assume that the calculations are affected directly by the average of inlet and discharge supercompressibility factors of the gas mixture,

$$\frac{Z_1 - Z_2}{2}$$

Polytropic formulas for real gases take the form

$$H_p = \frac{Z_1 - Z_2}{2} \times R \times T_1 \times \frac{r^{(n-1)/n} - 1}{(n-1)/n}$$

$$\text{ghp} = \frac{Q_1 \times p_1 \times \left(\frac{Z_1 + Z_2}{2Z_1} \right) \times \frac{r^{(n-1)/n} - 1}{(n-1)n}}{229 \times \eta_p}$$

The correction for supercompressibility in the horsepower formula differs from the head correction since inlet volume in the power formula has already been corrected for inlet supercompressibility. Since discharge supercompressibility is a function of discharge temperature, it will be necessary to estimate discharge temperature before calculating head and horsepower. It is to be understood that the final temperature of the gas after compression through a pressure range may be affected by the supercompressibility of the gas. For the purpose of estimating final temperatures necessary for the design of heat-exchanger apparatus, it is satisfactory to use figures based on the assumption that perfect gas laws may be used in most applications. This will result in figures that are slightly conservative. Sample compressor calculations for gas follow.

A. ***Gas:*** typical natural gas (Table 8.1)
Inlct pressure, 250 psia
Inlet temperature, 100 °F
Inlet capacity, measured at 14.7 psia and 60 °F, 102,000 scfm (cmh)
Discharge pressure, 593 psia

B. Preliminary Calculations

1. MW and k may be calculated from the gas analysis table, Table 8.1.

$$MW = 19.75 \qquad \left(R = \frac{1544}{MW} = 78.3 \right)$$

2. $r = \frac{p_2}{p_1} = \frac{593}{250} + 2.37$

3. Estimated discharge temperature:

$$\eta_p = 75\%$$

$$\frac{n-1}{n} = \frac{1.24-1}{1.24} \times \frac{1}{0.75} = 0.258$$

$$T_2 = 560 \times (2.37)^{0.258} = 560 \times 125$$

$$= 700°\text{F absolute or } 240°\text{F}$$

4. Inlet and discharge supercompressibility factors may be found using gas analysis, given conditions, and estimated discharge temperature (see Chapter 8 for calculations).

$$Z_1 \text{ at 250 psia, 100°F} = 0.97$$
$$Z_2 \text{ at 593 psia, 240 °F} = 0.97$$

5. Weight flow:

$$W = \frac{102{,}000 \times 14.7 \times 144}{520 \times 78.3} = 5320 \text{ lb/minute}$$

6. Inlet capacity: $Q = \dfrac{WZ_1RT_1}{p_1 \times 144}$

$$Q_1 = \frac{5320 \times 0.97 \times 78.3 \times 560}{250 \times 144}$$

$$= 6270 \text{ inlet cfm}$$

7. Estimated polytropic efficiency (Fig. 7.160):

$$\eta p = 72.4\%$$

C. *Discharge Temperature*

$$T_2 = T_1\left[r^{(n-1)/n}\right] = 560\left(2.37^{0.268}\right) = 560 \times 1.26$$
$$= 706R = 246°\text{F}$$

D. *Polytropic Head*

$$H_p = Z_{av}RT_1 \frac{r^{(n-1)/n} - 1}{(n-1)/n} = 0.97 \times 78.3 \times 560 \times \frac{0.26}{0.268}$$
$$= 41.200 \text{ ft-lb/lb}$$

E. *Horsepower*

1. $$\text{ghp} = \frac{Q_1 \times p_1 \times \dfrac{Z_1 + Z_2}{2Z_1} \times \dfrac{r^{(n-1)/n} - 1}{(n-1)/n}}{229 \times 0.724}$$

$$= \frac{6270 \times 250 \times 1.0 \times (0.26/0.268)}{229 \times 0.724}$$

$$= 9170 \text{ hp}$$

2. Bhp = ghp + mechanical losses (bearings and seals)
   9170 + 30 + 20 = 9220hp

The conditions have been selected for the preceding example so that the weight flow in pounds per minute and head in foot-pounds per pound are almost the same as for the sample dry air calculation. This results in approximately the same horsepower for the two examples, despite the great differences in other conditions and gas characteristics.

INSTALLATION OF COMPRESSORS AND THEIR DRIVERS

General

Proper installation is the most important requisite for satisfactory operation of high-speed rotary machinery. Although such machinery will operate under somewhat adverse conditions, it is always advisable to provide proper operating conditions if the machinery is to give maximum reliability at minimum operating cost.

Plant Layout

Machinery foundations, piping, electric wiring, and all necessary auxiliary equipment must be carefully arranged in the plant layout. Even though space is limited, skillful placing of the equipment will result in better installation, more reliable operation of the equipment, and a minimum of maintenance. Machinery that is easy to install, operate, and maintain will generally get better attention from the operating and maintenance personnel.

In planning a plant layout, the following are some of the points which should be kept in mind.

Crane facilities. Heavy machinery can best be handled with overhead crane facilities. This is true not only during installation but also during periodic maintenance. If crane facilities cannot be provided, some other arrangement must be made for handling heavy units.

Space. Ample space should be provided to permit easy handling during erecting. Floor space should be provided in the vicinity of heavy machinery where the top half may be placed during periodic inspection of the rotating element and internal parts. The designer must note carefully and make provision for clearance limitations specified on the outline drawing.

Accessibility. Machinery should be installed where it is easily accessible for observation and maintenance. In elevated locations or in pits, there should be stairways, catwalks, and the like. An internal combustion engine, of course, must never be operated in a pit in which carbon monoxide could accumulate.

Operating convenience. The equipment should be arranged so as to provide maximum accessibility to parts that require observation or attention during operation. Auxiliary equipment, such as oil coolers, gage boards, and pumps, should be located where they do not interfere with the routine inspection of the equipment.

Cleanliness. Rotating equipment and auxiliaries should be installed in clean locations. Equipment must not be subjected to unnecessary hazards from dirt or moisture. Outdoor installations require particular attention, and such requirements should be carefully noted in the specifications so that proper provisions may be made in designing the machinery.

Piping and wiring. The equipment should be so located as to permit a minimum of piping and wiring, and compromise, if necessary, should be made only in favor of ease and convenience in operating the equipment.

Instrument location. Instruments, mounted on gage boards or elsewhere, should be located within easy view of the operator when starting the machine.

Storage. If machinery is to be stored for even a short period, it should be properly flushed to protect it against corrosion. Machinery should always be stored in a clean, dry place. This is particularly important with respect to electric motors. In case of doubt, the machinery manufacturer may be consulted regarding proper handling.

Foundations

While the foundation for high-speed rotating machines need not be as massive as that for reciprocating machines, it must be sufficient to provide a permanently rigid, nonwarping support for the machinery. Foundations should be properly designed to avoid possible resonance with running frequencies of the compressor and drive train. To meet these requirements, all conditions surrounding the foundation should be uniform; that is, the foundation should rest entirely on natural rock or entirely on solid earth, but never on a combination of both. Foundations supported on piling must have a heavy, continuous mat over the piling. A noncontinuous mat may settle unevenly and result in misalignment of the machinery.

The temperature surrounding a foundation should likewise be uniform. If one section of a foundation is subjected to a substantially lower temperature, (e.g., open windows or doors in cold weather) than another section, there can be possible distortion and resulting misalignment of the machinery. An outline drawing of a rotating machine may include a suggested foundation arrangement, but this should not be construed to be an actual design. To design a foundation effectively and arrange the piping properly requires an intimate knowledge of local conditions with which the machinery builder cannot be familiar. Moreover, these local conditions may present problems beyond the scope of the machinery builder's experience. The user must necessarily take full responsibility for an adequate foundation. In general, it is desirable to have the foundation designed by a competent structural designer who has had experience with foundations for heavy machinery.

The following are some of the points that should be kept in mind:

1. The foundation must provide a permanently rigid, nonwarping support for the machinery.
2. The foundation substructure should rest on a uniform footing, entirely on bedrock or entirely on solid earth.
3. The temperature surrounding the foundation should be substantially uniform. If it is variable, the variations should be essentially the same for the entire foundation.
4. If more than one machine is to be installed in a given location, each should have a completely independent foundation supported from bedrock or solid earth. The foundations should be entirely free from building walls or other parts of the building that might transmit resonant vibration. Operating platforms must also be isolated from the machinery foundations.
5. If the foundation substructure rests on bedrock and if it is imperative that no resonant vibration be transmitted to adjacent structures, a vibration-damping material should be interposed between the substructure and the bedrock.
6. Where foundations must be supported from floor beams, whether there be one or more machines so supported, a vibration-damping material should be interposed between the beams and each foundation.
7. Where the foundation substructure rests on piling, the piling should be covered with a heavy, continuous mat.

Adequate and satisfactory foundations may be made of reinforced concrete, structural steel, or combinations of the two. Masonry, brickwork, or concrete foundations are permissible if the equipment is set at substantially ground level on a firm support and where the foundation structure is primarily necessary to provide mass. (This is not always the case. Structural integrity or rigidity may be equally important, or even more important, including some cases where there is a potential vibration problem as when dowel pins are overlooked and a pedestal breaks free from a mat.)

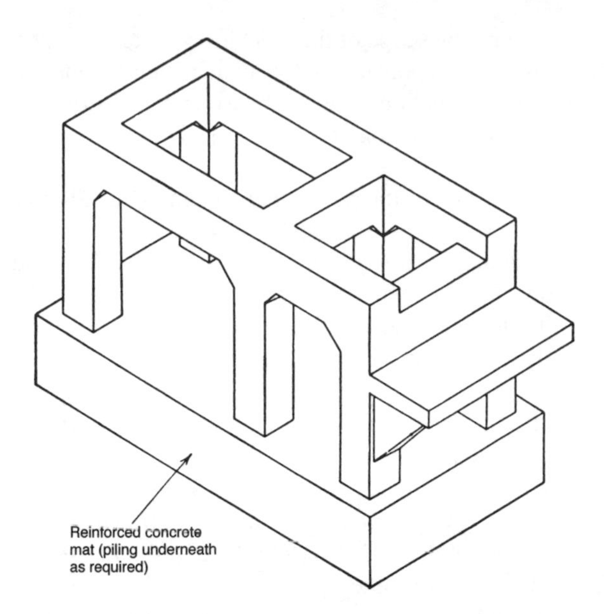

Figure 7.165 Typical reinforced concrete foundation suitable for any centrifugal compressor with either a base plate or soleplates.

Figure 7.165 shows a typical reinforced-concrete foundation structure suitable for rotating machinery of any size. This type of foundation structure permits the use of a base plate or soleplates, as preferred or dictated by the type of installation.

Figure 7.166 shows a structural-steel foundation with a continuous base plate spanning the six columns. This arrangement requires a substantial base plate. Each column must have a milled end pad, with all columns machined to equal length and with each end square with the axis of the column. The base plate is bolted and doweled to each column. Shims are normally provided between the top of the column and the bottom of the base plate to take care of slight inconsistencies. As an alternative, jack-screws may be provided at the bottom of each column to adjust the elevation of the base plate before grouting.

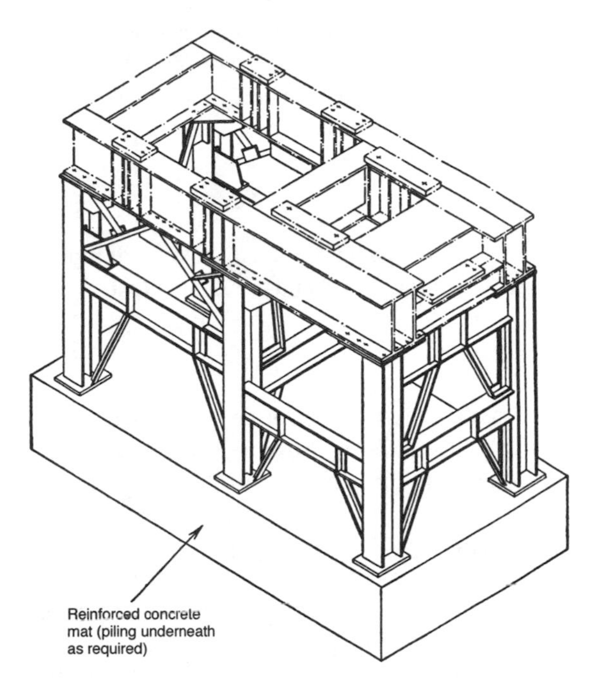

Figure 7.166 Typical structural steel foundation suitable for any centrifugal compressor having a substantial continuous base plate for spanning the six columns.

In addition to the cautions described, the following empirical design suggestions may be of interest:

1. The vertical loading may be calculated by adding not less than 50% to the total dead weight of the machinery.
2. The lateral loading may be taken as 25% of the total dead weight of the machinery, unless local earthquake requirements increase that figure.
3. The longitudinal loading may be taken as 10% of the total dead weight of the machinery.
4. All columns must be designed for the same unit stress to avoid distortion.
5. Center-line loading on vertical columns should be provided wherever possible. It is recognized that eccentric loading is unavoidable at times.
6. Total maximum vertical deflection of horizontal members should not exceed 0.015 in. for machinery operating at 3600 rpm. Deflection should be somewhat less for higher speeds and may be slightly more for lower speeds.
7. Concrete grout with a consistency barely permitting flow should be used for grouting soleplates or for getting base plates to reinforced-concrete structures.

8. If a structural-steel foundation includes a structural-steel deck on which the base plate is to be mounted, there may be need for metal grout to adjust for machining inconsistencies. In such cases, type metal with about 5 percent antimony may be used.

Piping

The reader is referred to Chapter 4 for a discussion of piping requirements, and particularly in Tables 4.5 through 4.9 and the related text.

TESTS

A number of different shop tests are performed on compressors in accordance with applicable standards. These cover hydrostatic, mechanical, and performance tests. The reader is also referred to Chapter 8 for further discussion of these tests.

Hydrostatic Tests

Compressor parts subject to pressure in operation are tested hydrostatically to 150% of the maximum pressure that can exist in the compressor or the casing section, or both, under the most severe operating conditions. Compressors having a maximum operating pressure not exceeding 50 psig are tested in accordance with manufacturer's standards. The compressor data sheet lists the actual hydrostatic test pressure used. The compressor parts subjected to hydrostatic test are maintained at test pressure for a minimum of 30 minutes to permit complete examination. Pressure vessels, filters, coolers, and so on, that are part of the lubricating and seal oil systems are hydrostatically tested in accordance with applicable codes or, where no codes apply, in accordance with manufacturer's standards.

Mechanical Tests

Each compressor is given a mechanical running test in accordance with manufacturer's standards to check general operation, vibration, and performance. Where practical, a turbine-driven compressor is short-time tested to 110% of the maximum continuous speed and at 100% for motor-driven units. Vibration measurements are taken throughout the operating speed range. Auxiliary equipment is tested in accordance with manufacturer's standards and, where practical, is used in the mechanical running test of the compressor. Noise-level tests, when required, should be made on the machine at the site and recorded in accordance with the *ISO 2151, Acoustics-Noise Test Code for Compressors and Vacuum Pumps – Engineering Method (Grade 2)*. Where practical, critical speeds of the compressor are determined during the running test. The compressor is designed so that its critical speeds shall not be detrimental to its satisfactory operation. The critical speeds of the driver shall be compatible with the critical speeds of the compressor, and the combination shall be suitable for the operating speed range.

Performance Tests

Each compressor is tested over its operating range to determine its performance. The test procedures used are in accordance with the latest edition of the ASME Power Test Code or any other applicable code that may be specified. Adequate readings are taken during the test, including discharge pressure, inlet pressure at compressor inlet flange, inlet temperature, relative humidity or ambient, discharge pressure at compressor discharge flange, compressor speeds, and brake horsepower. Readings are taken over the operating range and particularly at the specified operating point. The compressor manufacturer provides a test report, if specified, for each compressor showing the relative test data.

Witness Tests

Witness tests may be performed on any or all of the specific tests outlined previously.

INSPECTION AND MAINTENANCE OF DYNAMIC-TYPE COMPRESSORS

The purpose of an inspection and maintenance program for a piece of rotating machinery is to assure a higher unit availability. The investment in this program must begin with management decisions during the planning stages of a production facility and be continued throughout the life of the machine. Properly operating the compressor is an essential step in properly maintaining the unit, which is the responsibility of the user. Some of today's processes are very complex and impose undefined demands on the compressor in use. The maintenance program should be preplanned, but it should remain flexible enough to reflect additional needs as they develop throughout the operating life of the unit.

The design of a new compressor plant should go beyond the basic requirements of installation and operation. A well-engineered system must allow adequate space for dismantling, with provision for lifting and removal of assemblies and components to a laydown area. Employees must be available and be properly trained well in advance to ensure proper operation, as well as proper inspection and maintenance of the new unit. Proper tooling must be secured, along with spare parts and expendable maintenance materials in order to reduce the inspection time to a minimum.

Personnel

Certain skilled personnel are required to maintain an installation adequately. These personnel must be capable of:

1. Obtaining and analyzing maintenance data that will reflect the condition of the compressor before and after an inspection.
2. Scheduling the maintenance and inspection program based on operating data analyzing, including spare parts and materials.
3. Dismantling and reassembling the unit properly following the manufacturer's recommendations and adhering to sound mechanical practices.
4. Performing required nondestructive material tests, such as magnetic particle inspection and dye penetrant check to assure the soundness of critical components.

Maintenance Data

Specific data for maintenance evaluation should be included in the routine process data-monitoring system. Then these data should have guidelines established to call attention to a deteriorating performance level. Given the proper guidance, the plant operator can become an important link in the cycle of maintenance data evaluation. His or her alertness to changes in the data values will assist system maintenance supervision personnel in scheduling the proper corrective action. Some general maintenance data are common to all compressor installations, and these items can be used as a beginning for data collection. Additional data may be required if there are any unique process or application demands. The user of the equipment is generally in the best position to evaluate any process problems that may be experienced; therefore, the user must make the final choice on data procurement.

The following are examples of maintenance data that can be obtained by operating personnel and applied in the maintenance program. The listing is not a complete guide as objectives are different in each installation. However, once maintenance objectives are defined, the necessary data can be selected:

1. Load-bearing babbitt temperature from an embedded thermocouple or resistance temperature detector (RTD) can signal bearing distress by a gradual or sudden increase in temperature from previously established steady-state values.
2. Thrust bearing shoe temperature from a thermocouple or RTD embedded in the shoe near the babbitt surface can signal significant changes in thrust loading. Changes in thrust loading may come from internal or external causes. Damage to the compressor balance piston seals that reduces their effectiveness could reflect further damage within the compressor rotating element. Increased coupling friction caused by deteriorating coupling

teeth or a lack of lubricant can impose thrust loads high enough to exceed a thrust bearing capability;

3. Gas intercooler temperatures will reflect deteriorating cooler performance due to fouling from cooling water or process gas. The cooler approach temperature should not change over a period of time for a given set of cooling water and gas flow conditions.
4. The pressure differential across the lubricating system oil filter is directly related to the foreign material collected on the filter element. Foreign material can be forced through some elements if the differential pressure exceeds a safe limit.

The operator's contribution to the success of the maintenance program is not limited to written data. The operator can be of invaluable service by developing an alert sense toward noises and a keen eye for abnormal conditions. For instance, sound level or pitch changes are often forecasters of impending damage. Loose bolts on a stationary part can be retightened, if caught in time, to keep the part or bolt from falling into a rotating member.

Certain periodic data must be obtained by specially trained personnel for complete maintenance data evaluation. Here, requirements for special equipment and skills may direct the use of outside consultant for economic reasons. Regardless, the importance of this type of data to the successful program cannot be overestimated. The following are examples of this type of data.

Vibration Analysis

A meaningful vibration analysis must be accomplished with a good-quality analyzer. The instrument capability should be broad enough to include all vibration components from the lowest to the highest possible excitation frequency. Where speed increasers are employed, gear tooth mesh frequencies should also be considered in the spectrum. The analyzer should have provisions to determine the phase relationship of any unbalance for correction purposes. Vibration analysis is an effective tool in identifying the cause of unbalance in rotating elements, for example, erosion or unequal buildup of foreign material from the process gas. Many impending mechanical problems can be identified in their early stages by the keen vibration analyst. Loose rotating element components, misalignment, and coupling lockup are but a few that can be mentioned. Certain dynamic problems are not necessarily brought to light by vibration analysis, so this technique should not be used solely, to the exclusion of other techniques. Flexure at the hub of a large rotor due to the gyroscopic effect of the rotor is one example. Certain other vibrations have sudden onset, with little or no warning.

Lubricating and Seal System Oil Analysis

Oil contamination due to water, foreign particles, or reactions with the product gas have been blamed for many system failures. Early detection of this condition can avoid later failures. Samples of the lubricant should be taken frequently from the reservoir and thoroughly analyzed. The oil supplier usually furnishes a complete analysis service for any collected samples. He must be considered the authority on the proper lubricant additives for the particular application.

Mechanical Alignment of Components

Mechanical alignment of the entire compressor train must be periodically verified at the normal operating conditions of the compressor. Repeatable, accurate measurements must be taken to determine the actual thermal growth of each component in the train. The measurements must be corrected to reflect the actual movement of each coupling. Ambient temperature and process data should be recorded for future reference. Alignment corrections must be made where initial cold alignment offset does not fully compensate for actual thermal growth. The periodic alignment check should include a thorough inspection of all piping connected to the compressor for possible strains on the nozzles that could cause distortion or loss of alignment.

Cooling-water Chemical Analysis

Proper chemical treatment of cooling water can control the rate of fouling of heat-exchanger tubes. The appropriate treatment must be determined by a chemical analysis.

Compressor Performance Evaluation

Deteriorating compressor performance can be observed by periodic recording of key process data. A reduction in capacity may indicate material buildup within the rotating element or some internal damage. To be significant, comparative data should be obtained at the same process conditions. Some calculated corrections can be made for minor deviations from the original test conditions in the final comparison of performance.

The importance of proper maintenance data collection and analysis immediately after a unit inspection cannot be overemphasized. The post-inspection data will reveal the effectiveness of the work accomplished and set new criteria for future maintenance data evaluation.

Scheduling

The maintenance inspection timetable must be tailored to each individual installation to account for any unique process demands. The object is to inspect and correct abnormalities in the compressor before serious equipment damage occurs. The original schedule to meet this objective may be altered by experience and plant economics. For instance, the cost of downtime along with the reconditioning costs versus longer operating periods with possible increased reconditioning costs must be considered. It may be more economical to schedule inspections farther apart and replace major assemblies rather than smaller components more frequently. For example, the actual outage time span can be greatly reduced if a complete spare rotating element is on hand. Replacing it rather than reconditioning the existing element during the inspection may be preferable. The used element can probably be economically reconditioned to design specifications and returned to storage for the next scheduled outage. A further advantage to the major component replacement system is that it reduces the strain on repair shop facilities during the outage time.

The inspection schedule will also determine the type of spare parts, tooling, and special personnel required. The manufacturer's spare-parts list should always be considered as a guide for stock levels, tempered by the unique process demands and the objective of the inspection program. The alert planner will make use of his or her operating experience with a certain unit and begin developing packaged kits of parts including such small items as lock rings, O-rings, and gaskets. Frequently, it is the small item that extends the outage time. In a time of high labor costs, it may prove far more economical to expand the spare-part inventory on certain items when costs of repair and extended outage time are subtracted from the cost of the parts.

Dismantling, Inspection and Reassembling

The compressor manufacturer's recommended dismantling and reassembly procedures should always be followed closely. It is a wise investment to use a qualified serviceperson from the manufacturer's service organization for technical supervision during the inspection. It is the responsibility of the user, however, to properly supervise employees and assure the quality of the work required for a successful maintenance program.

It may not be economically feasible for all users of dynamic-type compressors to support a maintenance organization capable of accomplishing their maintenance objectives. Here, the manager must evaluate outside services that are available. Some manufacturers have well-established, factory-certified shops in local areas with complete capability and experience in servicing their compressors. Some have field service groups that can provide in-plant maintenance service. Factory design engineering services including field engineering services are also available from some manufacturers. Factory repair service also provides certain other advantages, since stock parts and special materials are readily available along with special engineering services and test facilities.

Accurate records and photographs of equipment in the as-found condition are invaluable tools in future inspection scheduling. The location of all damaged parts, internal rubs, erosion, or corrosion should be accurately recorded. Accurate records of all measurements on each component of the assembly should be tabulated each time the compressor is inspected. The first entry should always be the design-specified value for comparison with later measurements so that the wear trend can be easily evaluated. Every spare part received from the factory should be correctly identified and dimensionally checked, or it will not have been properly prepared for storage. The agonizing disappointment of removing a critical part from storage and finding it damaged from improper storage, or dimensionally incorrect, will eventually be experienced by the too casual storekeeper.

Accurate set-back measurements should always be recorded upon disassembly for use as checkpoints during reassembly. Field measurements should always be compared with the manufacturer's specified values. All internal clearances should be restored within the specified tolerance to assure the design performance level after inspection.

Proper cleaning and protection of parts is extremely important to prevent damage to critical areas. For example, the mechanical balance of a rotating element can be seriously impaired by improper cleaning methods, thereby jeopardizing the success of the entire inspection and maintenance program.

High operating speeds required in dynamic compressors demand extremely close tolerance and high-quality finishes in certain critical areas. Bearing journals and seal surfaces are ground to a superfinish to ensure proper seal and bearing life. Shaft fillets are highly polished to reduce stress concentration in critical areas. These surfaces must be protected during the dismantling and reassembly of a unit to prevent serious consequences later in the life of the compressor.

Nondestructive Material Tests

Highly stressed parts must undergo proper nondestructive material tests after cleaning to ensure their mechanical soundness. The use of one of the proven inspection methods such as dye penetrant, wet magnaglo, or magnaflux is recommended. Items such as impellers, blading, gear teeth in speed increasers, coupling hubs and sleeves, shaft keyways and fillets, and shrunk-on sleeves are examples of components that may require material tests.

Corrective Engineering

During the early life of the compressor, certain components may require attention more frequently than the rest of the assembly. Frequent part replacement may be due to a process demand that was undefined during the compressor design stage. The troublesome part can usually be redesigned to better withstand the demands imposed on it, thereby increasing its life expectancy in line with the remainder of the assembly. This maintenance step may involve the manufacturer's design engineer, who becomes a part of a corrective engineering team along with the maintenance engineer and operating engineer. The user's investment in corrective engineering will be small in comparison to the increased reliability and production throughout the life of the compressor.

Maintenance Through Operation

The intricacy of a compressor installation is usually dictated by the process demands. The equipment operator's responsibility also increases with the system complexity. For instance, an installation handling a dangerous gas would require a more elaborate seal system than a unit handling air. The operator must have a thorough knowledge of the system's limitations. It is the operator's responsibility to be thoroughly familiar with the design intent for each compressor unit and adhere to sound operating practices. The manufacturer's instruction book is available for this purpose as well as for instructing the maintenance organization in the upkeep of the unit. The operator should be familiar with the unit operating requirements well in advance of start-up. Casual or unplanned operating practice can never be justified and will eventually lead to trouble. Compressor reliability is linked as closely to good operating practice as to good maintenance procedures.

Instrumentation and Protective Devices

Instrumentation and protective devices frequently represent a significant share of the initial investment in a new compressor system. The importance of this equipment cannot be overemphasized. However, it is only as dependable as its calibration. Maintenance of these devices is usually the responsibility of a separate group from mechanical maintenance and, therefore, they are frequently neglected. The simple pressure gage is a valuable source of data when reliable, but becomes worthless if its calibration cannot be trusted. Instruments must be recalibrated at regular intervals and checked whenever calibration is doubtful. The recalibration and testing of instruments should be scheduled along with other compressor maintenance.

Protective devices used for alarm or trip functions must be kept in calibration and properly tested to ensure their functional reliability. Each of these devices should have a test arrangement to prove its integrity with the compressor on stream and without risking an accidental shutdown.

Many costly plant outages could have been avoided in the past if proper attention had been given to the protective devices on a piece of rotating machinery.

Preventive Maintenance

There is no single, established approach to an effective maintenance program for every compressor installation. The importance of tailoring the maintenance program objectives to each unique compressor installation has already been emphasized. The following preventive maintenance schedule is not intended as a complete guide, but only to suggest how to devise a plan for the particular installation in question. The items selected and the schedule itself must be tailored to the installation and carried out with common sense and good judgement.

A. *Daily Operator Maintenance Instructions*

1. Check the lubricating oil and seal oil reservoirs for possible water accumulation by draining a small sample from the reservoir low-point drains.
2. Check the lubricating and seal system oil filters for excessive pressure differential.
3. Verify that the oil levels in seals and lubricating oil reservoirs are within a safe operating range.
4. Check the operation of all process cooler and separator traps and seal system sour oil traps by observing the liquid levels in sight glasses or blowing off trap bypass valves to drain funnels.
5. Review all supervisory and process instruments such as those indicating oil pressures and temperatures, vibration, process pressures, and temperatures to ascertain that no unexplained deviations have occurred.
6. Listen for noise level and pitch changes around compressors, gears, and drivers.
7. Inspect visually for oil, gas, or water leaks and loose parts.
8. Check the differential pressures across intake gas filters, intercoolers, aftercoolers, and interstage separators for excessive differential that could signal plugging or other deterioration.
9. Observe the level in the seal oil drain sight flows and the bearing oil drain sight flows for abnormal level changes and check regularly to establish that the connections have not become clogged.

B. *Weekly Operator Maintenance*

1. Verify the calibration and operation of all protective alarm and trip devices through actual test. Test lockout arrangements should be provided on each device to allow safe test with the compressor on stream.

C. *Monthly Maintenance*

1. Make a vibration survey of each bearing housing, including shaft readings where possible. The data at each location should include unfiltered or total wave and rotational or filtered wave at rotating frequency components. If these two values do not agree, a thorough investigation, including a frequency search, should be made to

determine the cause. Any significant increase in vibration should be noted and corrected at the earliest possible moment to prevent permanent bearing or compressor damage.

2. Test the performance of all intercoolers, aftercoolers, and oil coolers to evaluate their efficiency. The rate of deterioration will determine the cleaning schedule.
3. Lubricate the linkage, pins, and slide bars of all control valves and valve positioners or guide vane positioners.
4. Obtain oil samples from lubricating and seal oil reservoirs for analysis by the lubricant supplier.

D. Major Maintenance Shutdown

1. Coupling Inspections

   Gear Type

   a. Dismantle, and remove all grease, taking note of the condition of the grease. If significant separation has occurred, the lubricant supplier should be consulted for further grease recommendations. The coupling grease should always be checked within the first month of initial operation of a new unit to verify that the lubricant has not separated or otherwise deteriorated. This practice should be followed until a lubricant is selected. The amount of sludge buildup in continuous oil-lubricated couplings may indicate the need for better or additional oil filtration at the coupling spray nozzles. Check the spray nozzle pattern on reassembly.
   b. Clean all hubs and sleeves thoroughly and inspect gear teeth for abnormal wear and broken or cracked teeth. The hub and sleeve teeth and hub keyway should be given a thorough magnetic particle or dye check inspection for evidence of cracks.
   c. Repack and replace the proper type and amount of grease using new gaskets or O-rings, where applicable. Follow the coupling supplier's recommended bolt torque values and bolt tightening sequence.

   Non-lubricated Type

   a. Inspect the flexible hinge member for cracked or damaged disks in the disk pack or diaphragm.
   b. Inspect coupling hubs and spacers at all high stress points for cracks using the magnetic particle or dye check method.
2. Verify the Alignment of All Couplings
   a. The cold offset alignment should be verified by actual thermal growth measurements at operating conditions. This procedure should be repeated whenever the process conditions are changed

significantly. The cold offset may require changes due to foundation settling or shifting.

b. The ambient temperature and the location of the dial indicator, driver, or driven shaft must be recorded on all alignment records for future reference.

c. A 12-inch diameter face plate temporarily bolted to each coupling hub will assist greatly in improving angular alignment accuracy on small-diameter, high-speed couplings.

3. Clean and Inspect All Journal Bearings

   a Remove and inspect each bearing for signs of babbitt damage.

   b. Measure and record the bearing-to-shaft clearance following the manufacturer's recommended procedure. Replace any bearings found to have clearances exceeding the manufacturer's specifications.

   c. Inspect the shoe-to-retainer contact point on all pivoted pad bearings for signs of fretting or excessive wear that could hamper the shoe pivot freedom.

   d. Thermocouple or RTD detector lead wires embedded in shoes must be installed to allow complete freedom of shoe movement.

   e. Inspect bearing seal rings, if used. Garter springs or other retainer springs should be replaced if weakened by worn spots.

4. Clean and Inspect Each Thrust Bearing Assembly

   Note: Record an axial set-back measurement to locate the rotating element in the casing before disturbing the thrust bearing. Check the manufacturer's drawing for reference dimensions. Always keep forward and rear thrust assemblies and related shims separated. These assemblies must not be interchanged.

   a. Inspect shoes for signs of loose or damaged babbitt.

   b. Inspect hardened contact buttons on self-aligning bearing shoe backs and leveling plates for flat spots or fretting. Slight wear spots should be removed by light stoning to restore the button crown or leveling plate curvatures.

   c. Check the thrust collar radial and face runout if a removable collar is used. Carefully follow the manufacturer's instructions for reassembly of the thrust collar if it is removed.

   d. Readjust the thrust bearing clearance to the proper value.

   Note: Forward and rear thrust shims must be adjusted together to keep the rotating element axial position in its correct relationship with the casing.

5. Inspect all oil baffles for signs of rub, plugged drain-back holes, or chipped touch points. Reinstall with specified radial and axial clearances.

6. Remove the casing cover for an internal inspection if performance level or maintenance schedule dictates this step.

   a. Properly clean and inspect all impellers or blading for erosion or corrosion. All highly stressed parts should be given a magnetic particle inspection or dye checked for cracks. Casing diffusers or stator blading should be cleaned and inspected to meet the same requirements as the rotating speed.

b. Interstage seals should be replaced if damaged or eroded to such an extent as to allow clearances to exceed specified values.

c. Rotating elements should be checked for balance if increase in vibration has been noted. The cause of any such lack of balance should be in-vestigated, such as loss of a nut or accumulation of sludge.

7. Lubrication and Seal Systems
   a. Drain and clean each system reservoir. Use squeegees or synthetic sponges for internal cleaning. No rags of any type should be used, since lint may get into the lubricating and seal systems.
   b. Centrifuge the oil before returning it to the reservoir or replace it with new oil. Replacement oil should be carefully strained to keep foreign material that may be present in the oil drums out of the system.
   c. Clean or replace lubricating and seal oil supply filters.
   d. Remove tube bundles from oil coolers and thoroughly clean the water and oil sides. Test the tube bundle of the cooler hydrostatically to its rated pressure before reassembling the cooler.
   e. Inspect all system controls and regulating valves for foreign material, sticking pistons or valve stems, and so on.
   f. Inspect the lubricating and seal oil pumps for abnormal wear at pump element bearings, shaft seals, and couplings.
8. Process Check Valves
   a. Remove and inspect each valve for wear at hinge pins, disk guide pins, disk return springs, and seals. All worn parts should be replaced.
9. Process Expansion Joints
   a. Inspect the internal bellows surface for pitting and erosion that could lead to failure. If lined, inspect the liner.
   b. Adjust pipe support hangers, if required, to position the expansion joint properly in its cold position.
10. Speed Increasers
    a. Inspect pinion and gear bearings for damage or wear and proper clearance. Carefully check the bottom of the gear base for metallic particles.
    b. Check the gear tooth contact pattern. Check gear teeth for signs of abnormal tooth wear and unequal tooth loading.
    c. Gear and pinion teeth should be subjected to thorough inspection by magnetic particle process or dye check.
    d. Clean all mesh spray nozzles and all internal oil supply passages. Make sure that the nozzles are properly secured on reassembly.
    e. Inspect all splash pans for possible fatigue cracks or loose fasteners.

11. Instrumentation and Protective Devices
    a. Check the calibration of all instruments used for monitoring operations or for obtaining maintenance data.
    b. Check the calibration and functioning of all alarm and trip devices.
12. Main Driver
    a. Follow the driver manufacturer's recommended inspection and CAGI's *Compressed Air and Gas Drying.*

INTRODUCTION TO GAS DRYING

Reference to Similarities to Air Drying

Compressed air is the most utilitarian compressed gas and, as such, has received the most attention of all the gases. Purification techniques have been perfected and specifications have been clearly defined for compressed air. Although other gases are seldom employed as a utility as compressed air is, there is a definite need for them to be dried. These gases are often raw materials for a process and must be reduced in water content to meet process requirements. An example is shown in Fig. 7.167. In addition, water, especially in liquid form, can present the same problems during the transmission of gas to its point of use in a process system, as would be expected in a compressed-air system.

The drying of compressed gases other than air may require specialized attention due to the particular properties of the gas. The technology, however, and, in fact, the equipment are often quite similar to that used for drying compressed air.

Most gases, unless the system temperature and pressure conditions are unusually high or low, behave like ideal gases, having the same relationship for volume, pressure, temperature, and molecular weight. For this reason, air and many other gases can be treated with the same type of equipment to reduce the water content.

Although the equipment used for drying air and other gases may be similar and utilize the same operating mechanisms, the size, choice of materials, and other specifications may be decidedly different because of the specific properties of the gases. These properties include specific gravity, specific heat, viscosity, thermal conductivity, explosive characteristics, toxicity, corrosion, and others. Also, in some cases a particular gas may have an effect on certain materials of construction and even on the properties of the water contained in the system because of their solubility in liquid water or their chemical reaction with water. These and other properties of a particular gas may have to be considered in the design of gas-drying equipment.

Figure 7.167 A fuel gas dryer for a petrochemical plant.

It is the purpose of this chapter to relate the drying of compressed air to the drying of other compressed gases and to define those areas of difference that require special consideration.

DESIGN CONSIDERATIONS

Contaminant Filtration, Separation, and Disposal

In any properly designed drying application, care should be taken to provide for contaminant filtration, if possible, to help achieve the dew point performance required. A contaminant can be considered as any component of the gas stream that, if not removed, will affect the dryer's performance on either a short- or long-term basis.

Water, in the form of liquid drops, droplets, aerosols or fog, is perhaps the most common contaminant. Dryers are designed to take out or remove water only as true water vapor as it can exist at the specific operating pressures, temperatures, and flow. Proper removal by means of knockout drums, centrifugal separators and coalescers is essential to good drying practices.

When the compressed gas contains oil vapors, aerosols, or condensed liquids, whether from a compressor or, in the case of some of the hydrocarbon gases, as a natural constituent, these should also be removed to minimize contamination of the dryer's desiccant. This is usually accomplished by means of scrubbers, coalescing filters filtering to 0.3 microns or smaller. As in water removal, any liquid droplets or aerosols are best removed by a separator, scrubber, and coalescing filter.

Other contaminants can affect drying performance. These can be constituents of the gas stream that compete with water vapor for its place in the desiccant. Other contaminants can plate out or coat the desiccant so that water vapor cannot be adsorbed into the desiccant's pores or capillaries. Proper selection of the desiccant, as well as the method of regeneration, can usually correct or minimize this condition.

Disposal of contaminants should be made in accordance with local Department of Environmental Resources (DER) requirements, and it is the customer's responsibility to conform to other local regulations as well. Condensate in the form of water can be sent to a drain. Oil condensates should be disposed of by methods prescribed locally, and other contaminants can possibly be disposed of by sending them to a flare or to a hazardous-waste-disposal system.

Design Pressures

Design pressures for gas-drying equipment should follow the ASME Pressure Vessel Code, Section VIII, or Section m if Nuclear Code is required. If the gas being handled is a lethal gas, the ultimate customer must advise the dryer manufacturer if the vessel is to be manufactured to the special ASME requirements for lethal materials. This is the customer's choice, and it depends on the application and insurance company requirements.

Special Materials

Construction materials for units drying gases other than air should be selected as applicable to the particular situation or as requested by the customer. These requirements may include anything from standard screwed brass valves and malleable iron fittings to steel valves, socket weld fittings, or stainless-steel construction through-out. Familiarity with the application will help in the initial quoting stages so that the proper type of construction will be used.

Control Requirements

As with air dryers, the control of gas dryers is largely a question of customer preference as to whether the dryer is to be manual or automatic and whether it is to be controlled by moisture load, time interval, or both. The location of the equipment will dictate whether it must be suitable for indoors, outdoors, hazardous, or explosion-proof locations or a combination of these.

Leakage Considerations

Particular care should be given to leakage requirements, which may be quite stringent in the case of gases such as hydrogen, helium, natural gas, and toxic or radioactive gases. Selecting the proper safety valve for a pressure vessel is also extremely important, taking into account whether the valve must be gas tight or whether its discharge must be piped to a safe location.

Purge and Purge Exhaust Disposal

Disposal of effluent or purge gas from dehydration systems requires that water be disposed of by effluent or gas purging. Since dehydrated gas may contain noxious or toxic contaminants, these contaminants may be in the effluent from a refrigeration or deliquescent system or in the purge gas of a regenerative system. The type of contaminant will determine what is required by the user to comply with plant, DER, and OSHA requirements. In many cases, the effluent or purge gas contains no contaminants. In such cases, the effluent can be disposed of through a regular sewage system. The purge gas, if valuable, can be recovered by introducing it into the intake side of the compressor once the liberated water is condensed and removed.

Specifications for Gas Quality

Gas quality is dependent on the customer's needs and the ultimate use of the product. Quality should be specified by the customer as to dryness required, what undesirable components must be removed, and what degree of filtration is necessary for desiccant dust removal.

TYPES OF DRYERS FOR COMPRESSED-GAS SERVICE

All types of dryers currently manufactured can be used in some phase of compressed-gas service. The following outlines the use of deliquescent dryers, refrigerant dryers, and adsorption desiccant dryers in compressed-gas industries.

Deliquescent Dryers

A deliquescent dryer can be used in situations where contaminants in the gas stream may be quite high and in hazardous areas where remote operation without electricity may be necessary. Normally, these dryers must be operated at inlet temperatures of 100°F or less. Deliquescent dryers consume no power while operating, do not use a portion of gas for purging, are capable of handling light and heavy hydrocarbons in the gas stream, and do not react with most contaminants. Outlet pressure dew point compensates for varying inlet temperature and pressure conditions. Their only moving parts are those required to discharge effluent. Since deliquescent dryers use a consumable material, periodic addition or replacement of the desiccant is required. Desiccant consumption is high, and dew point depression is minimal at higher inlet temperatures. Environmental conditions or restrictions must be considered, as with other types, in the removal of the solutions drained from the dryer.

Refrigerant Dryers

Refrigerant dryers can be used in compressed-gas service for applications that do not require pressure dew points below 35 to 40°F. Also, refrigerant dryers are often used for bulk water removal from many industrial gases, such as argon, nitrogen, and dissociated ammonia. Air-separation plants utilize refrigeration of the compressed gases to lower the energy consumption of the dual-tower drying system by removing nearly 70 or 80% of the compressed-gas water content.

Refrigerant dryers are also used to condense out heavy hydrocarbon gases such as butane, pentane, and hexane from natural gas without gas compression. To accomplish this, normally a deliquescent, regenerative, or glycol unit must be installed prior to the refrigeration unit to remove water vapor that otherwise could condense out also.

Adsorption Desiccant Dryers

Heat-reactivated dryers. Heat-reactivated, adsorption desiccant dryers are often used for drying almost any of the common gases, such as carbon monoxide, carbon dioxide, argon, oxygen, and nitrogen, as well as hydrocarbon gases such as ethylene, ethane, methane, and natural gases. Most of the above drying applications require an extremely low water content of 1 to 25 ppm by weight. The most common heat-reactivated compressed-gas dryers are discussed next.

Internal Heat-reactivated Dryers. Regeneration of this type of heat-reactivated dryer is accomplished by energizing the heaters and purging, either with part of the gas being dried or by dry air or dry inert gas introduced from an external source. With some gases, purge exhaust disposal must be considered because of safety and environmental restrictions. These internally steam- or electric-heated dual-tower Adsorption desiccant dryers are commonly used for flow rates ranging up to 200 scfm.

External Heat-reactivated Dryers. This type of dryer reactivates the wet tower by using a blower for recirculating a captive volume of the gas in a closed-loop circuit, thus eliminating purge losses. The reactivation heat is supplied by an external steam or electric heater. The heated gas, 350 to 500°F, enters the dryer tower, driving off the adsorbed moisture. The liberated moisture is then carried out of the bed into the water-cooled condenser where the water vapor is condensed and the liquid water is separated. At the end of the reactivation heating period, the desiccant is cooled down to approximately 125°F prior to tower switchover.

This type of dryer provides pressure dew points of as low as - 40 to - 100°F and reduces the dew point and temperature elevation spikes commonly experienced with internal heat reactivated dryers.

Split Stream Steam or Electric Heat-reactivated Dryers with an External Steam or Electric Heater. In the split-stream type, a throttling valve diverts some gas from the inlet side of the dryer. This gas is heated and then it strips moisture from the adsorbent during the regeneration. The gas is then cooled, water condenses, and is drained from the unit. The gas is then returned to the inlet without purge losses. Split-stream reactivation is commonly used to dry natural gas and many other hydrocarbon gases or other precious gases that cannot be purged to atmosphere. The distinct advantage of the split-stream reactivation versus a closed-loop reactivation is obviously the absence of the reactivation blower; however; pressure loss is incurred to operate the throttling valve and the regeneration system.

Heatless dryers. Heatless dryers are rarely used to dry compressed gases since a large amount of the gas being processed must be purged during the reactivation. It should be pointed out that heatless dryers can be utilized for dry, explosive, flammable, and precious gases, as long as a means of recovering the gas and returning it to the intake side of a compressor is provided. Perhaps better and more efficient compressed-gas drying can be accomplished by utilizing a closed-loop or a split-stream reactivation drying system.

Glycol dehydration. A common method of drying natural gas is glycol dehydration. Glycol dryers can obtain pressure dew points as low as - 40°F. They are generally used where the pressure of the gas is abovc 100 psig and inlet temperatures are below 120°F. The glycol dryer can he used where no electrical power is available, although a combustible gas must be used to heat the boiler.

APPLICATIONS FOR COMPRESSED-GAS SERVICE

The main reason for removing water from a gas is to limit the adverse effects it has on a system or to aid in its transportation. These adverse effects can be physical, chemical, or electrical, or a combination of these. The drying of a gas, whether at atmospheric or higher pressures, is done for the following reasons:

1. To control the relative humidity in industrial processing and in drying material at limited temperatures.
2. To control dew point and prevent the precipitation of water.
3. To minimize the chemical and corrosive effect of water.
4. To reduce the weight of gas for compression and transportation purposes.

Many industrial gases can be dried by adsorption; a partial list follows:

Acetylene	Freon	Natural gas
Ammonia	Furnace Gas	Nitrogen
Argon	Helium	Oxygen
Carbon dioxide	Hexafluoride	Propane
Chlorine	Hydrogen	Propylene
Cracked gas	Hydrogen chloride	Sulfur dioxide
Ethane	Hydrogen sulfide	Sulfur hexafluoride
Ethylene	Methane	

The degree of dryness and the type of desiccant dryer to be used depend on the application.

Natural Gas Drying

Some natural gas wells furnish gas of very high purity, that is, gas that is almost pure methane. However, most hydrocarbon streams are complex mixtures of hundreds of different compounds. A typical well stream is a high-velocity, turbulent, constantly expanding mixture of gases and hydrocarbons intimately mixed with water vapor, free water, solids, and other contaminants.

Contaminant-removal processes can be divided into two groups: dehydration and purification. The principal reasons for the importance of natural gas dehydration include the following:

1. Liquid water and natural gas can form solid, icelike hydrates that can plug valves, piping, and other parts of the system.
2. Natural gas containing liquid water is corrosive, particularly if it contains CO_2 or H_2S.
3. Water vapor in natural gas pipelines may condense, causing slugging flow conditions.
4. Water vapor increases the volume and decreases the heating value of natural gas, thus leading to reduced line capacity.
5. Dehydration of natural gas prior to cryogenic processing is an absolute requirement to prevent ice formation on low-temperature heat exchangers.
6. To prevent freezing after regulation.

Of these, the most common reason for dehydration is the prevention of hydrate formation in gas pipelines.

Natural gas hydrates are solid crystalline compounds formed by the chemical combination of natural gas and water under pressure at temperatures considerably above the freezing point of water. In the presence of free water, hydrates will form when the temperature is below a certain point, called the hydrate temperature.

Hydrate formation is often confused with condensation, and the distinction between the two must be clearly understood. Condensation of water from natural gas under pressure occurs when the temperature is at or below the dew point at that pressure. Free water obtained under such conditions is essential to the formation of hydrates, which will occur at or below the hydrate temperature at the same pressure.

During the flow of natural gas, it is necessary to avoid conditions that promote the formation of hydrates. This is essential since hydrates may choke the flow stream, surface lines, and other equipment. Hydrate formation in the flow stream results in a lower value for measured wellhead pressures. In a flow-rate measuring device, hydrate formation results in lower apparent flow rates. Excessive hydrate formation may also completely block flow lines and surface equipment. Thus, the need to prevent hydrate formation is obvious. The easiest way to eliminate hydrate formation is to substantially remove the water from the natural gas stream.

Another important application for desiccant drying is the liquefaction of natural gas. Methane is converted into a liquid in a cryogenic process at - 285°F and atmospheric pressure. There is a 600 to 1 reduction in volume. As a liquid, large volumes of methane can he easily transported and stored. Natural gas companies liquefy and store gas up to 20 million scfd when demand is low and use the stored liquid during periods of high demand. Natural gas found in remote areas can be liquefied and transported to places of demand. Because of the low dew points needed for the cryogenic production of LNG, molecular sieve dryers are used.

Refrigerated gas dryers are also used for the dehumidification of fuel gases, including landfill gas, digester gas, methane and natural gas. Drying to pressure dew points of plus 35 to 40°F will improve the heat value efficiency of the boilers and complete burning of the fuel gases. Dry gas also helps to meet the air quality requirements from the combustion exhaust gases. Dry fuel gas in an internal combustion engine improves the operating efficiency as well as reducing maintenance costs. Landfill and digester gases dried in refrigerated dryers are low cost alternative fuels for internal combustion engines and boilers.

Figure 7.168 Landfill/Digester gas outdoor installation, air-cooled refrigeration system for cooling. Mechanical separate for moisture separation. No-gas loss automatic drains for condensate removal.

Mechanical refrigeration cools and dehumidifies the fuel gas through the dryer. Automatic no-gas loss drainage system helps the operation of the dryer non-hazardous. A package custom designed for outdoor installation offers economical installation.

Chemical Process Industry Drying

Nitrogen usage in the United States is more than 550 billion ft^3/year. Twenty-five percent of this gas is used as blanketing atmospheres for the chemical processing industry. Nitrogen, in most cases, is dried by adsorption. This keeps moisture from causing adverse chemical effects or a possible chemical reaction from a system. These potential chemical effects with water present could include corrosion of pipelines and vessels, poisoning and consumption of valuable catalysts, and inhibition of other reactions, as well as other effects. This is also true of other inert gases.

An example of this is an inert cover gas used for polymer production. A dry inert gas blankets the polymer material being produced. If moisture is present in the gas, the polymer will undergo oxidative degradation and the desired product will not be obtained.

Refrigerants are also dried by adsorption. To operate properly, the moisture content of the refrigerant system must not be allowed to be above a certain maximum. Therefore, it is very important during the manufacturing and assembly that no moisture enter the system. Also, if any moisture enters during the operation of the system, it must be removed quickly. Sources of moisture in a system can be leakage at the water-cooled condenser, oxidation of certain hydrocarbons, wet oil, wet refrigerant, decomposition of motor insulation in hermetically sealed units, and low-side leaks, which introduce wet air. The effects of this moisture are valve failure, corrosion, freezing of the expansion valve, copper plating, formation of ice in the evaporator, and chemical damage to insulation or other system materials. Desiccant dryers are used to remove this moisture from the installed system. Desiccant drying is exceptionally well suited for this application, not only because of its water-removal capacity, but also because there is no reaction with the refrigerant, oil, or machine parts. It is used satisfactorily with all refrigerants.

Carbon dioxide is another gas that is dried by solid adsorption. Over 4 million tons per year are produced. It is recovered from synthesis gas in ammonia production, from refinery production of hydrogen, from fermentation processes, and from natural wells. The main uses of carbon dioxide are refrigeration, beverage carbonation, urea manufacture, and enhanced oil recovery. Carbon dioxide is dried for production purposes. A carbon dioxide gas dryer at a chemical plant is shown in Fig. 7.168. The presence of moisture in carbon dioxide can also cause process line freezing at high pressures, along with corrosion problems.

Other Chemical Process Uses

The unloading of chlorine tank cars and other transfers of chemicals is accomplished by padding the tanks with dried compressed gas. Moisture in contact with many chemicals will cause rapid corrosion, deterioration, polymerization, and oxidation.

Figure 7.169 A carbon dioxide gas dryer operating at 720 psig for a chemical plant.

Oxygen manufacturers require the dew point of the air be as low as possible. If water is present, freeze-ups will occur at the low temperatures necessary for the liquefaction of air.

PETROCHEMICAL PROCESS INDUSTRY

Alkylation

Alkylation is the union of an olefinic with a paraffinic hydrocarbon to obtain high-octane gasoline. The reaction is catalyzed by hydrofluoric or sulfuric acid. The olefin is injected into the paraffinic feed, and the combined streams are contacted with acid.

The paraffin concentration is kept in excess to prevent copolymerization. Acid alkylation is limited to isobutane with propylene, butylene, and pentylene. Ethylene must be combined using $AlCl_3$. Phosphoric acid is used as a catalyst to unite propylene and benzene to form isopropyl benzene. Aluminum chloride and HCI catalyzes ethylene and benzene to ethylbenzene. Alkylations using aluminum chloride and other halogen-activated catalysts must have dry feed streams. Reduction, dilution, or loss of the catalyst will take place if the feed is not dried.

Gas Plant

The C_4 naphthenes and lighter gases from various refinery operations are sent to this section of the refinery. All these gases require dehydration.

Catalytic Re-forming

Catalytic re-forming refers to the octane improvement of straight-run gasoline and cracked refinery naphthas. C_5 and C_6 naphthenes are isomerized and dehydrogenated to aromatics; paraffins are hydrocracked or cyclicized and dehydrogenated to aromatics. Catalytic re-forming is also a source of benzene, toluene, and xylene. Recycled hydrogen, inert gas, and regeneration gases used in this process must be dried by means of solid adsorption to ensure proper process reactions. Hydrogen is also produced in large quantities by catalytic re-forming and must be dried before other refinery uses or sale.

Isomerization

Isomerization is the conversion of normal butane, pentane, and hexane into their respective isomers. It is a fixed-bed, vapor-phase process that is carried out under dry hydrogen atmosphere. This prevents coke deposition and saturates any cracking products. In this process, both the paraffin and hydrogen feeds need to be dried.

Other petrochemical processes, such as hydrocracking and catalytic cracking, utilize adsorption drying. Recycle and makeup hydrogen regeneration gas must be dried for proper reactions.

Metals and the Metallurgical Industry

In the metals industry, a blanketing or protective gas is used in many heat-treatment processes. The drying of this gas is imperative to produce a more uniform metal. A required grade is made with precision, and the furnace in which the heat treating is done works with greater regularity.

Steel is sometimes annealed in a controlled atmosphere prepared by the combustion of natural gas. A gas formed this way is called an exothermic base gas. An exothermic base gas is an inert gas generated when natural gas is burned with a controlled amount of air, a process that produces mostly nitrogen and carbon dioxide with 0.1 to 0.5 percent combustibles and 0 to 0.1 percent maximum oxygen. During the combustion, a considerable amount of water vapor is formed. The gas is cooled and then dried by a desiccant dryer. The gas then blankets the heat treating of the steel to prevent oxidation.

In heat treating or annealing aluminum, an exothennic gas is also used. Too much moisture in the furnace atmosphere can cause oxidation of the alloying constituents. The amount of moisture is extremely critical whenever a metal is exposed to processing. This is true not only for heat treating but also for polishing, carburizing, and welding of titanium, stainless steel, and other alloys.

A hydrogen atmosphere is used in copper-brazing furnaces for annealing highly oxidizable metals. Where even a slight amount of moisture is very detrimental, nickel, nickel steel, and Monel wires must be annealed in these furnaces to avoid discoloration.

Another heat-treating atmosphere used to a large extent is cracked ammonia. Anhydrous ammonia is dissociated into gases, resulting in three parts of hydrogen to one part of nitrogen. Most cracking units are highly efficient, so that the degree of dissociation is usually 99.75 to 99.95%. Since one volume of ammonia yields two volumes of the mixed gas, the ammonia content is then 0.125 to 0.025 percent by volume, respectively, or 2,500 to 5,000 ppm by weight. A molecular sieve dryer is used to remove water and the undissociated ammonia.

A gas of increasing importance in the heat treating of metals is HN_x. This gas has excellent properties in the bright annealing process and is nearly neutral with regard to carburization when treating steels with different carbon contents. Thus, the gas can be used universally. This makes the gas distribution within the workshop easier and enables the purchase of bigger and more economical gas-production plants.

The following is a typical HN_x gas composition:

N_2	= 93% to 96% by volume
H_2	= 7% to 4% by volume
O_2	= 0.0005% by volume
$CH_4 + NO + NO_2 + CO_2 + CO$	= 0.05% by volume
Dew point	= -58°F

HN_x gas is produced by the controlled combustion of fuel gas. This gas must be dried before being used in a furnace.

Electrical Industry

Sulfur hexafluoride (SF_6) is a nontoxic, inert gas that is about six times as heavy as air. Usually, this gas is dried to a pressure dew point of - 60°F or lower. Its main use is as an insulating gas for arc suppression in electrical switch gear, circuit breakers, and so on.

Food and Beverage

Carbon dioxide (CO_2) is obtained as a by-product from air-separation plants and is recovered from the fermentation process in the making of beer, wines, and various alcohols. The carbon dioxide is recovered from these processes, cleaned up, and then dried so that it can be stored as a high-pressure gas or as a solid. Carbon dioxide is reintroduced into various food products for the carbonization of soft drinks, beer, and some wines. As a solid, carbon dioxide is known as dry ice and is used for short-term storage of frozen foods, such as in portable ice cream carriers where mechanical refrigeration is not possible.

Nitrogen is used as a blanketing gas to eliminate oxygen and prolong the storage life of fruits such as apples, grapes, and bananas. It is also used for the blanketing of wines and aseptic packaging of foods such as coffee and potato chips, as well as other packaged foods. Elimination of moisture and oxygen from all these products is essential to retard spoilage.

Semiconductors

Nitrogen is a colorless, odorless, tasteless diatomic gas that is completely inert. One of the more recent applications has been as a purging and blanketing gas used in the production of semiconductor and microprocessor chips in the electronic industry.

Helium, another inert gas, is also used in the semiconductor industry as a blanketing inert atmosphere in the production of electronic parts and the growing of germanium and silicon crystals.

Medical Industry

Oxygen, which is essential to life, is used for resuscitation and inhalation therapy for people with lung disorders such as emphysema, to ease recovery after a major operation, and as a heart stimulant for patients with heart disorders.

Nitrous oxide is a relatively mild anesthetic used mostly in dentistry, dental surgery, and medicine. It is a narcotic in high concentrations and has the side effect of a type of carefree hysteria, from which it has derived its common name, laughing gas.

Medical compressed air is compressed air that has been produced free of any contaminants such as oil, carbon monoxide, dirt, dust, and water vapor. It is used in the preparation of pharmaceuticals, as a breathing aid, and as an atmosphere in hyperbaric chambers where operations are performed at elevated pressures.

Carbon dioxide is used as a purging or blanketing atmosphere in the production of medicines and pharmaceuticals where elimination of oxygen is important. It is also used as an aerosol propellant in medication dispensers.

Helium provides an inert atmosphere necessary in the production of certain types of pharmaceuticals. It is also used, mixed with oxygen as a breathing gas to replace nitrogen, which is normally adsorbed into the bloodstream. In a hyperbaric chamber, this gas is used for the patient's breathing so that decompression will have none of the damaging side effects caused by nitrogen.

Nitrogen, as with most inert gases, is used in the production of pharmaceuticals. As a liquid, it is used in surgery and for the quick freezing of tissues for medical study.

8

General Reference Data

What is the capacity of an air compressor? How is the type and service limitation of a compressor defined? Will a drill bit purchased in New York fit a pneumatic tool purchased in San Francisco? Are compressors available to operate at pressures and capacities suitable for rock drills purchased from various sources of supply? If you order two compressors from two builders, will they work together in unison as a single source of air power? The answer is the adoption of *standards* – common language for performance and acceptability. These standards do not result in identical compressors, tools, or fittings, but they do make it possible for buyer and seller to be on a common ground of understanding. They are the "know-how" factors that reduce economic complexity to simple formula – solution to a recurring difficulty.

The Compressed Air and Gas Institute is heavily involved in the development of standards related to compressed air systems. The institute works with organizations, such as ASME, ANSI, ASTM, ISO, PNEUROP, and others providing input and expertise to establish equitable standards for the manufacturers and uses of pneumatic equipment.

Standardization on a voluntary basis, such as is advocated by the Compressed Air and Gas Institute, is more than an engineering function. Through a well-rounded program of sound standards, economy of operation and increased production inevitably result. By diminishing inventory and investment, by speeding up maintenance and shipments, by cutting down accidents, standards increase output, decrease cost, and are a benefit to buyer and seller alike.

Standard definitions, nomenclature, and terminology, which constitute a large portion of this section, represent either scientific fact or common usage. The purpose in presenting them is to make available a common language for the description of compressed air machinery and tools and to permit an accurate definition of performance. The acceptance of such definitions, nomenclature, and terminology by the builder and user of compressed air machinery and tools will avoid confusion, eliminate argument, and prevent misunderstanding, all in the public interest.

The Compressed Air and Gas Institute does not recommend standardization as to general design, appearance, performance, or overall interchangeability. Where interchangeability of parts used in connection with such machinery is desirable and necessary, they have for the most part followed the published recommendations of the ASA. For example, screw threads, pipe threads, companion flanges, pneumatic and rock-drill tool shanks are completely interchangeable for apparatus built by any manufacturer. Some of these standards are included in this section. These latter refer particularly to compressors and compressor practice.

A clear understanding of any subject depends primarily upon complete agreement on the definitions or all the important terms and values. This is true particularly in any consideration of such things as performance guarantees, test methods and procedure, and related subjects. The Compressed Air and Gas Institute has therefore adopted as standard the following definitions. They are in all cases rational, and do not violate scientific fact in any respect. In those cases where it has been necessary to choose between two or more possible definitions, each of which is valid, that definition which corresponds more nearly to established practice and usage has been adopted.

DEFINITIONS

See glossary.

Symbols

bhp brake horsepower
c_p specific heat at constant pressure
c_v specific heat at constant volume
C a constant; coefficient of discharge
D deviation of horsepower from the isentropic value for a real gas
f ratio of supercompressibility factors = Z_2/Z_1
ghp gas horsepower
H head of fluid in ft.-lb/lb
H_{ad} adiabatic head for ideal gas, ft.-lb/lb
H_{ac} adiabatic head for real gas, ft.-lb/lb
J Joule's constant, the mechanical equivalent of heat. J = 778 ft.-lb/Btu
k adiabatic exponent = ratio of specific heat at constant pressure to specific heat at constant volume = c_p/c_v
K radius of gyration of a rotor

M number of mols of a substance, that is, weight in pounds divided by the molecular weight, also in pounds
MCp molal specific heat at constant pressure
MW molecular weight
N rotative speed, rpm; number of stages
n polytropic exponent, in equation $Pt'' = C$ and related equations
p pressure, lb/in.^{2} absolute (psia)
P pressure, lb/ft^2
P_a adiabatic horsepower for perfect gas
P_{ae} adiabatic horsepower for real gas
P_c critical pressure; the pressure required to liquefy a gas at its critical temperature
P_R reduced pressure = p/p_c
d_v partial pressure of any component (usually of the vapor component) in a gas mixture, psia
q capacity, or flow; cubic feet per second, cfs
Q capacity, or flow; cubic feet per minute, cfm
r ratio of pressure = p_2/p_1; usually discharge pressure over intake pressure
R gas constant = 1544/MW for perfect gases
RH relative humidity
Subscripts 0, 1, 2, c refer to conditions of a gas in states 0, 1, 2, c
t temperature in degrees Fahrenheit, °F, or Centigrade, °C
T absolute temperature, equal to °F + 459.6, or °C + 273
T_c critical temperature
T_R reduced temperature, equal to T/T_c
v specific volume, ft^3/lb
V volume, ft^3, or m^3
w weight flow, lb/minute
W weight
x mol fraction of a constituent of a mixture
X adiabatic factor, equal to $\left(\frac{p_1}{p_2}\right)^{(k-1)/k} - 1$
y weight fraction of a constituent in a mixture
z supercompressibility factor; $pv = ZRT$
η efficiency (identified with proper subscript)

Subscripts

ad	adiabatic process
av	average
c	real gas
i	isothermal process
m	mixture
p	polytropic process
sv	saturated vapor
v	vapor
1	intake conditions
2	discharge conditions

SUPERCOMPRESSIBILITY

While a great many gases are well represented by perfect gas laws over fairly wide ranges of pressure and temperature, there are many others which show considerable variation from those laws. Even gases which are ordinarily treated as perfect gases require special representation in the neighborhood of the critical point.

Departure from perfect gas laws is referred to as supercompressibility, and is accounted for by means of a factor, Z, called the supercompressibility factor, introduced into the gas equations. Expressed mathematically,

$$Pv = ZRT \tag{8.1}$$

$$\frac{P_1 v_1}{T_1 Z_1} = \frac{P_2 v_2}{T_2 Z_2} \tag{8.2}$$

Stated simply, Z is a correcting factor which permits the application of ideal gas laws, with accuracy, to any known gas or gas mixture.

The numerical value of Z in Eqs. (8.1) and (8.2) is, of course, 1.0 for an ideal gas. In the case of actual gases, this value may be less than, equal to, or greater than 1.0, depending on the gas involved and the pressure and temperature conditions being considered.

In the measurement of air or gas passing through a nozzle it is often convenient to use equations of hydraulics, and to allow for the compressibility of ideal gases by a factor called *compressibility*. For real gases near the critical pressure, a further correction, called *supercompressibility*, allows for departure of real gases from ideal gases. Thus compressibility relates ideal gases to liquids, while supercompressibility relates real gases to ideal gases. The distinction is important, since compressor engineers may encounter both in the same problem.

The reader interested in pursuing this topic further is referred to the text *Thermodynamics* by Joseph Keenan for a discussion of supercompressibility. Compressibility factors, and a derivation of the compressibility equation, are given in the ASME *Fluid Meters Handbook.*

Determination of Supercompressiblilty Factor

The magnitude of Z depends upon the particular pressure and temperature at which the gas is being considered. Furthermore, the magnitude of Z at any stated pressure and temperature combination is different for practically every known gas; thus exact values for a given gas or gas mixture can be determined only by extensive laboratory tests of each gas or mixture. How can the compressor engineer determine the value of Z?

The most accurate source of graphical information is an authentic Mollier diagram applying to the particular gas or gas mixture being compressed. Unfortunately, this data is available for only a few of the many gases and mixtures which are commonly compressed.

An alternative approach is to approximate quite closely the value of Z for any gas or gas mixture by utilizing the law of corresponding states. This law or principle states that the magnitude of Z for a given gas at a specified pressure and temperature is definitely related to the critical pressure and temperature of that gas. The reduced pressure and reduced temperature of the gas may be stated mathematically:

$$P_R = \frac{p}{p_c} \tag{8.3}$$

$$T_R = \frac{T}{T_c} \tag{8.4}$$

The subscript R denotes the reduced function and p and T indicate the actual pressure (psia) and temperature (Rankine or Kelvin) of the gas for which the compressibility must be found.

T_c = Critical temperature, which is the maximum temperature (Rankine or Kelvin) at which a gas can be liquefied.

P_c = The critical pressure of the gas; that is, the absolute pressure, psia, required to liquefy a gas at its critical temperature.

Further, the law of corresponding states tells us that if various gases have their reduced pressures and reduced temperatures equal, then the supercompressibility factors of these gases are of about the same magnitude.

Consider three different gases, w, x and y, all existing at different pressures and temperatures. If this condition is true,

$$P_R = \frac{p_w}{p_{cw}} = \frac{p_x}{p_{cx}} = \frac{p_y}{p_{cy}}$$

and

$$T_R = \frac{T_w}{T_{cw}} = \frac{T_x}{T_{cx}} = \frac{T_y}{T_{cy}}$$

then the magnitude of the Z factor of each of the three gases is practically the same, even though the gases exist at widely different pressures and temperatures.

This fact permits the development and use of generalized supercompressibility charts using P_R and T_R as coordinates and such charts may be applied to any known gas or gas mixture with a high degree of accuracy to determine the compressibility factor of the gas or mixture at any condition.

The general practice of the compressor industry is to use one of the established equations of state for the computation of Z internally in computer programs. One such equation is the Redlic-Kwong equation, which can be used for pure gases and mixtures, except near the critical point. Figures 8.2 to 8.6 were derived from the Redlic-Kwong equation. Figure 8.1 is a sketch showing how these charts overlap and is included only to make the later, accurate charts more usable.

The value of Z at the intake and discharge condition of each stage of compression is readily found from these charts since the intake pressure and temperature and the discharge pressure are known and the discharge temperature is easily calculated from the isentropic formula previously given.

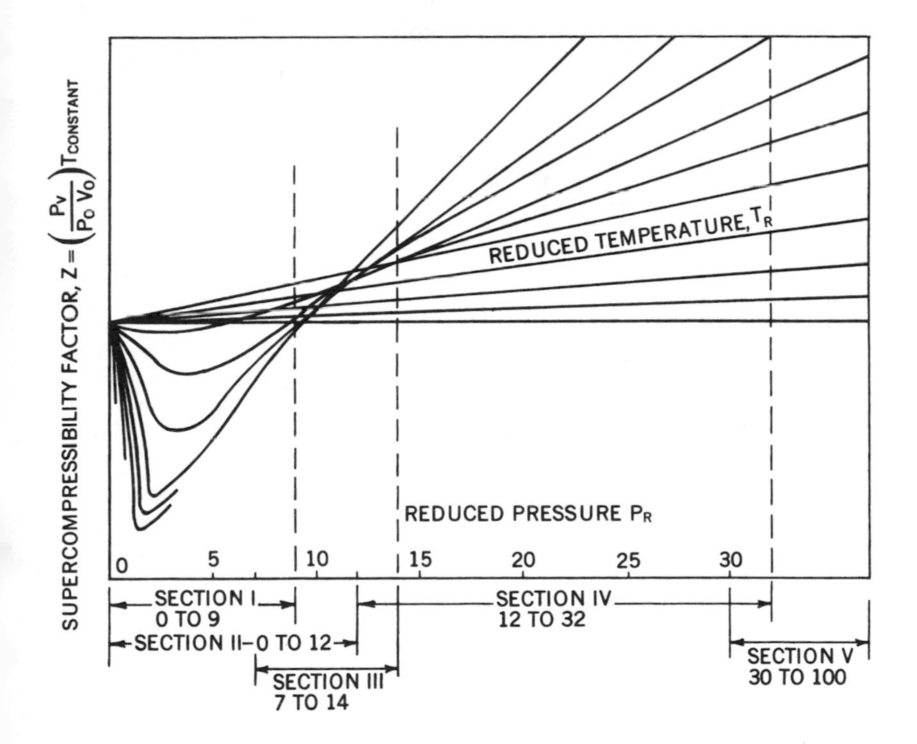

Figure 8.1 Supercompressibility factor vs. reduced pressure at varying reduced temperatures. The above curve is a composite sketch of all five sections of the compressibility plot. As shown above they overlap one another. The scales used on individual sections are arranged so as to maintain a consistent accuracy in reading.

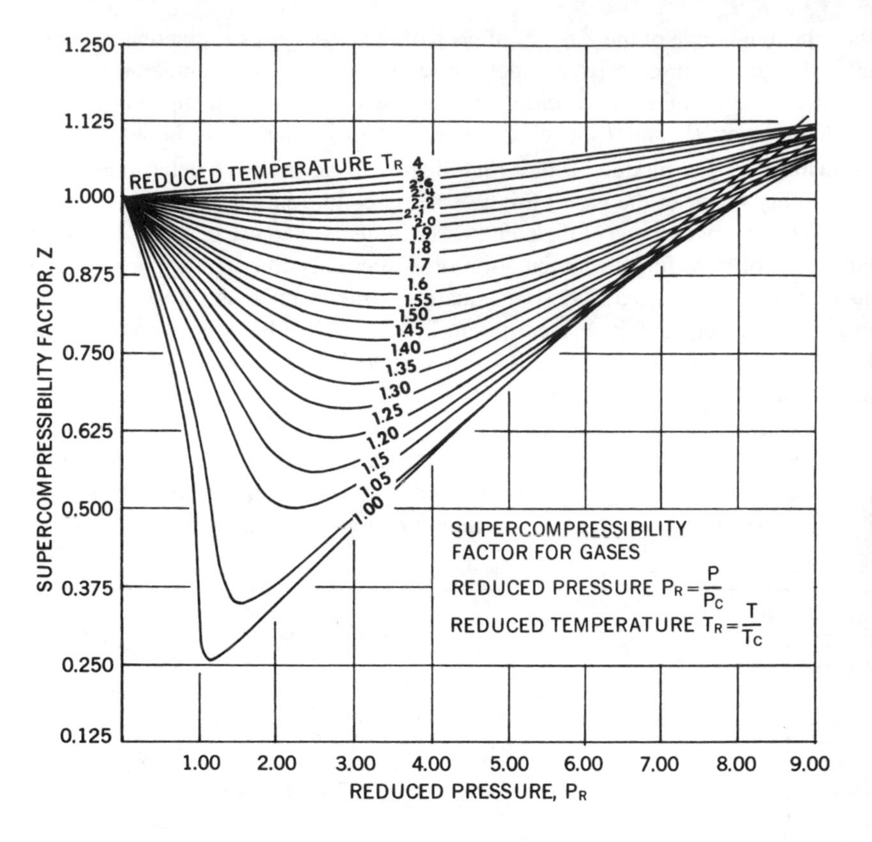

Figure 8.2 Supercompressibility factor for reduced pressure range, 0 to 9, inclusive.

Example 8.1

Find the volume of 1 lb of chlorine at 458 psig and 516~ Barometric pressure = 14.0 psia. From Table 8.39, p_c = 118 psia, T_c = 291°F, MW = 70.914.

$$T_R = \frac{516 + 459.6}{291 + 459.6} = 1.30, \qquad P_R = \frac{458 + 14.0}{118} = 4.00, \qquad \text{See Figure 8.3}$$

$$v = \frac{ZRT}{P} = \frac{(0.67)(1544)(516 + 459.6)}{(144)(70.914)(458 + 14.0)} = 0.21 \text{ ft}^3$$

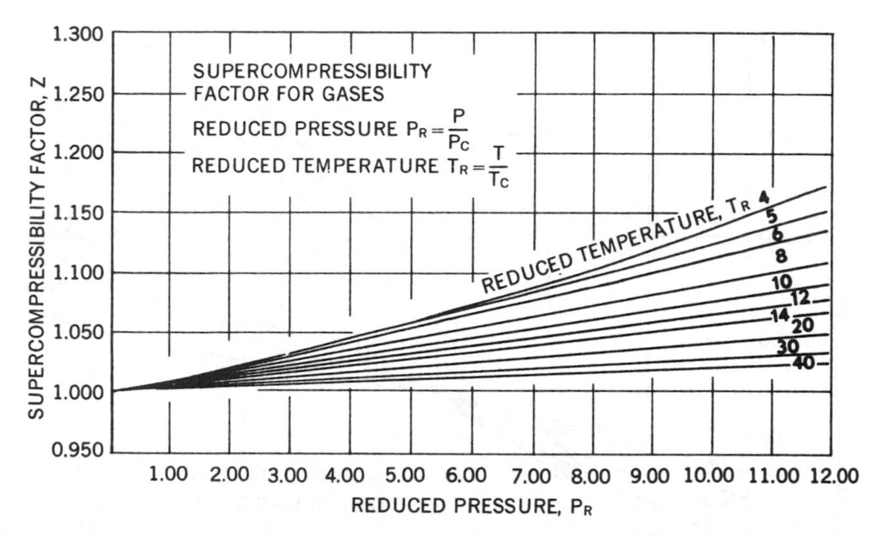

Figure 8.3 Supercompressibility factor for reduced pressure range, 0 to 12, inclusive. Note: In this range of reduced pressure, overlapping that of Figure. 8.2, the supercompressibility factor reaches a maximum at reduced temperature approximately equal 4. It then decreases with increasing values of reduced temperature. To avoid confusion in reading, which would result from the overlapping curves, values of Z for T_5 greater than 4 are shown in this separate graph.

PHYSICAL PROPERTIES OF GAS MIXTURES

If the chemical composition of a gas mixture is known, it becomes possible to determine the gas characteristics necessary to make compressor calculations through the application of the following relations:

$$W_m = W_1 + W_2 + W_3 + \cdots$$
$$M_m = M_1 + M_2 + M_3 + \cdots$$
$$x_1 = \frac{M_1}{M_m} \qquad x_2 = \frac{M_2}{M_m} \qquad x_3 = \frac{M_3}{M_m}$$
$$y_1 = \frac{W_1}{W_m} \qquad y_2 = \frac{W_2}{W_m} \qquad y_3 = \frac{W_3}{W_m}$$

And therefore,

$$x_1 + x_2 + x_3 + \cdots = 1.0 \text{ and } y_1 + y_2 + y_3 + \cdots = 1.0$$
$$MW_m = x_1MW_1 + x_2MW_2 + x_3MW_3 + \cdot$$
$$C_m = y_1c_1 + y_2c_2 + y_3c_3 + \cdots$$

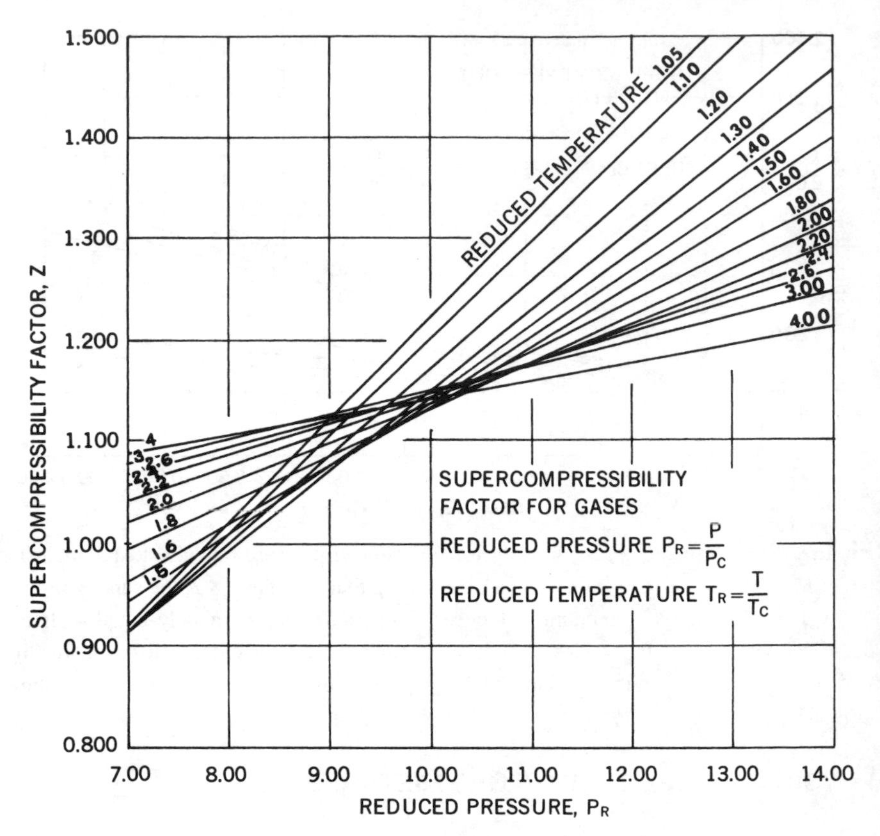

Figure 8.4 Supercompressibility factor for reduced pressure range, 7 to 14, inclusive.

where:

W_m, W_1, W_2, etc.	=	Weight of mixture and of constituents, respectively.
M_m, M_1, M_2, etc.	=	Number of mols of mixture and of constituents, respectively.
MW_m, MW_1, MW_2, etc	=	Molecular weight of mixture and of constituents, respectively.
C_m, C_1, C_2, etc.	=	Specific heats of mixture and of constituents, respectively.
x_1, x_2, x_3, etc.	=	Mol fraction of constituents in mixture.
y_1, y_2, y_3, etc.	=	Weight fraction of constituents in mixture.

Molal properties, such as molal specific heat, MC_m, are calculated on a molal basis.

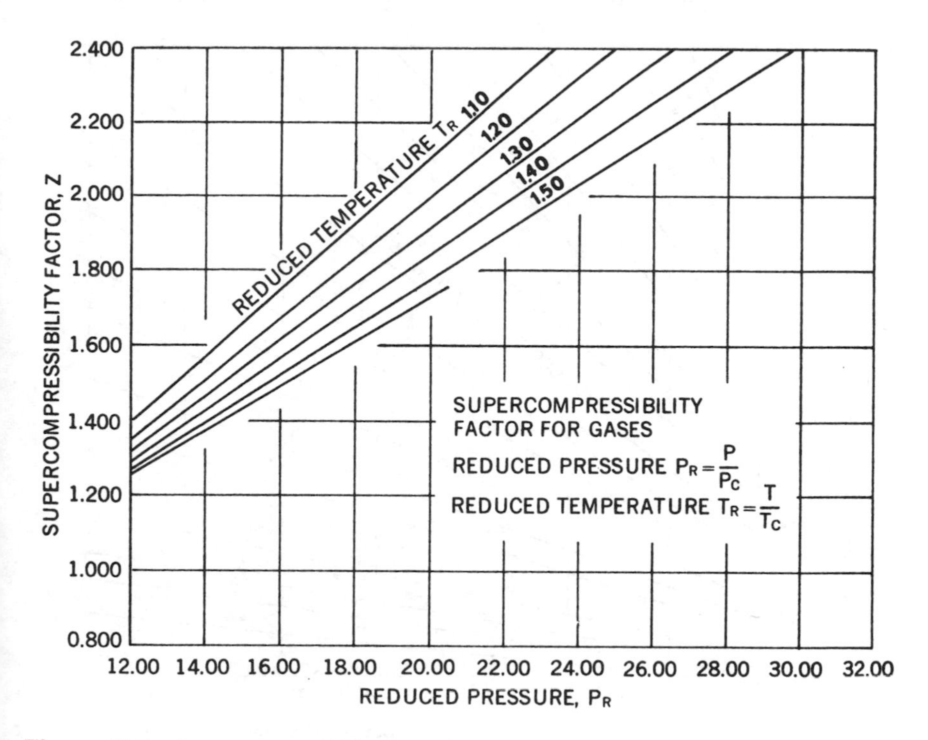

Figure 8.5 Supercompressibility factor for reduced pressure range, 12 to 32, inclusive.

An application of these relations is illustrated in the example, Table 8.1, which presents the computation of the physical characteristics of a typical natural gas, the composition of which is known on the volumetric basis. The molecular weight (MW_m) of this gas is found to be 19.75 and its specific gravity relative to air is

$$\text{sp.gr.} = \frac{19.75}{28.96} = 0.682$$

Example 8.2

Its other characteristics can be determined as follows:

$$R = \frac{1544}{MW} = \frac{1544}{19.75} = 78.18\text{, approximately}$$

$$c_v = c_p - \frac{R}{778} = 0.482 - \frac{78.78}{778} = 0.377$$

$$k = \frac{c_p}{c_v} = \frac{0.482}{0.377} = 1.279$$

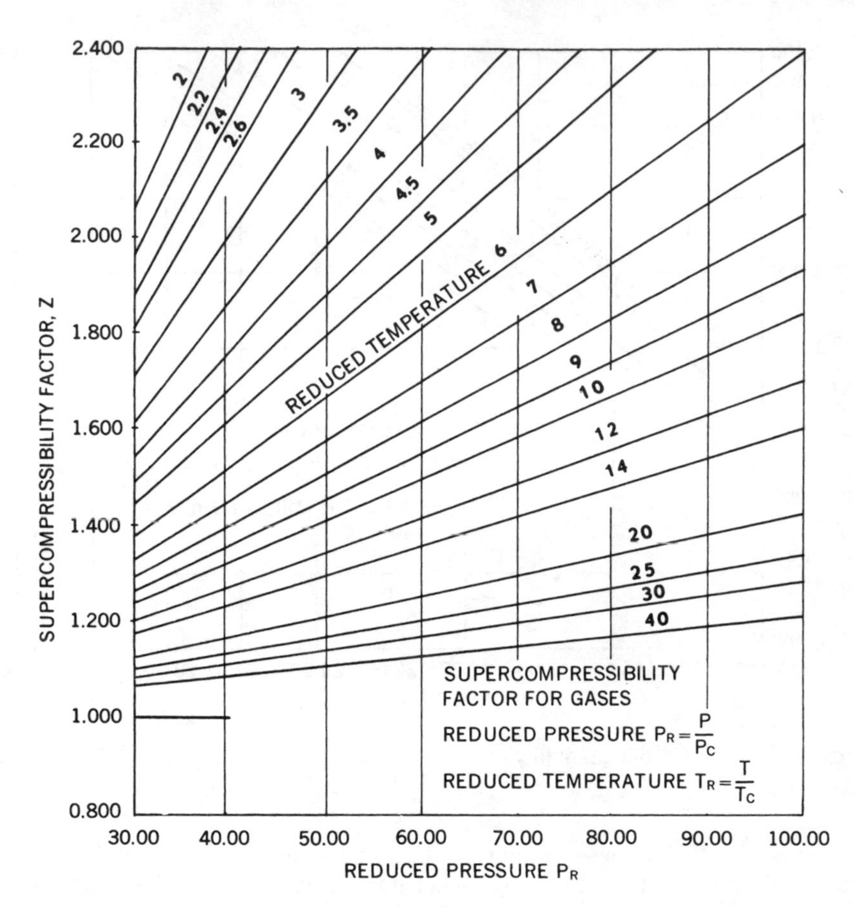

Figure 8.6 Compressibility factor for reduced pressure range, 30 to 100, inclusive.

Pseudo-critical Constants for Gas Mixtures

For a physical mixture of gases, not combined chemically, the usual method of determining a pseudo-critical constant presupposes a hypothetical gas having a critical constant equal to the sum of the products of the critical constants of the individual gases in the mixture times their respective mole fraction in the total mixture, i.e.,

$$p_c = p_{c1}x_1 + p_{c2}x_2 + p_{c3}x_3 \cdots + p_{cn}x_n$$

$$T_c + T_{c1}x_1 + T_{c2}x_2 + T_{c3}x_3 \cdots + T_{cn}x_n$$

where p_c, T_c are critical constants of the mixture, p_{c1}, p_{c2} and so on, and T_{c1}, T_{c2}, and so on, are critical constants of the constituents, and x_1, x_2, and so on, are mol fractions of each constituent in the mixture. This calculation is illustrated in Table 8.2.

Table 8.1 Computation of Characteristics of a Typical Natural Gas

Gas Component	Chemical Formula	Fraction by Volume	Molecular Weight	Vol. Fraction × Mol. Wt.	Fraction by Weight	c_p	Wt. Fraction × c_p	MC_p at 150°F	Vol. Fraction × MC_p
Methane	CH_4	0.832	16.04	13.35	0.675	0.526	0.355	8.97	7.46
Ethane	C_2H_6	0.085	30.07	2.56	0.130	0.409	0.053	13.78	1.17
Propane	C_3H_8	0.044	44.09	1.94	0.098	0.386	0.038	19.58	0.86
Butane	C_4H_{10}	0.027	58.12	1.57	0.080	0.397	0.032	26.16	0.71
Nitrogen	N_2	0.012	28.02	0.33	0.017	0.248	0.004	6.96	0.08
		1.000		M = 19.75	1.000		c_p = 0.482		MC_p = 10.28

Table 8.2 Computation of Pseudo-critical Temperature and Pressure of a Typical Natural Gas

Gas Component	Chemical Formula	Fraction by Volume	Critical Pressure	Mol Fraction × Crit Press.	Critical Temperature	Mol Fraction × Crit Temp.
Methane	CH_4	0.832	673.1	560	343.5	276
Ethane	C_2H_6	0.085	708.3	60	550.9	47
Propane	C_3H_8	0.044	617.4	27	666.26	29
Butane	C_4H_{10}	0.027	550.7	15	765.62	21
Nitrogen	N_2	0.012	492	6	227.2	3
		1.000		668 p_c of mixture		376 T_c of mixture

TEST PROCEDURE

Tests to determine the performance of air compressors and blowers or to establish compliance with performance guarantees can be of value only if they are conducted carefully and in strict conformity with accepted methods and standards. The Compressed Air and Gas Institute, therefore, endorses the ASME Test Code on Compressors and Exhausters (PTC 10) and for Displacement Compressors, Vacuum Pumps and Blowers (PTC 9), and recommends that all tests to establish performance he made according to the rules specified in these codes or according to the Institute's interpretation of these codes in this section. The Institute's endorsement of these codes includes acceptance of the ASME Code on General Instructions (PTC 1).

Copies of the ASME Test Codes may he purchased at a small cost from the American Society of Mechanical Engineers, New York.

The ASME Test Code on Compressors and Exhausters (PTC 10) and for Displacement Compressors (PTC 9) give complete instructions for testing compressors handling air. For compressors handling gases other than air, the codes are essentially complete, provided the gases conform to the perfect-gas laws or the generalized compressibility data are sufficiently accurate (see page 648). When a displacement-type air compressor is tested, the air must be discharged into the atmosphere; otherwise, reliable results cannot be obtained except under certain special circumstances specified in the code. When a centrifugal compressor is tested, air or gas may be discharged into the atmosphere or may be measured and retained within a closed system. Compressors handling gases for which none of the physical properties are known cannot be tested for capacity

Purpose

The first essential in any test is to establish its purpose. An air-compressor test is usually undertaken to determine the volume of air compressed and delivered in a given time under specified conditions or to determine the overall efficiency as a periodic check on operations, or for comparison with certain standards.

Capacity Measurement

The Compressed Air and Gas Institute has agreed to the methods described in ASME PTC-9 and PTC-10, also ISO 1217, as applicable. The form of the nozzle and all its associated dimensions for various sizes with throat diameters ranging from 1/8 to 24 in. are given in Figure. 8.7. The table in Figure. 8.7 gives approximate capacities for which each size of nozzle is suitable when discharging into or from the atmosphere. This table will be useful in selecting a nozzle for any particular test of a displacement-type compressor or of a centrifugal-type compressor discharging into the atmosphere. If the operating conditions are limited as specified on page 666, nozzles described in the ASME Code may be used. When the discharge pressure from a rotary displacement-type compressor is less than that required to meet the limitations given on page 678, the capacity may be determined by means of a slip test, page 678.

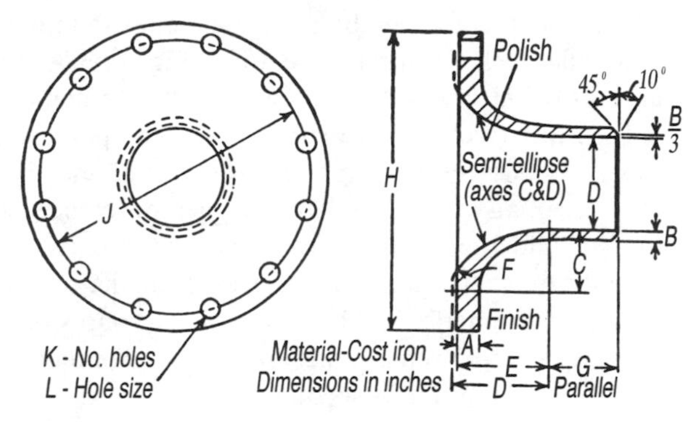

Long-radius low-ratio flange-type nozzle

D	A	B	C	E	F	G	H	J	K	L	Approx. flow rates, cfm	
											10" H_2O	10" H_2O
0.125	0.437	0.250	0.09	0.121	0.01	0.437	4.25	3.125	4	0.562	1	2
0.1875	0.437	0.250	0.13	0.181	0.01	0.468	4.25	3.125	4	0.562	2	4
0.250	0.437	0.250	0.17	0.242	0.01	0.500	4.25	3.125	4	0.562	4	8
0.375	0.625	0.250	0.25	0.363	0.02	0.562	7.50	6.00	4	0.750	9	18
0.500	0.625	0.250	0.34	0.484	0.03	0.625	7.50	6.00	4	0.750	16	32
0.750	0.625	0.250	0.50	0.726	0.04	0.750	7.50	6.00	4	0.750	36	71
1.000	0.937	0.250	0.67	0.699	0.05	0.875	9.00	7.50	8	0.750	62	127
1.375	1.000	0.250	0.92	1.332	0.07	1.063	11.00	9.50	8	0.875	119	239
2.000	1.000	0.313	1.33	1.938	0.10	1.500	11.00	9.50	8	0.875	253	506
2.500	1.000	0.375	1.67	2.422	0.13	1.875	11.00	9.50	8	0.875	397	790
3.000	1.000	0.375	2.00	2.906	0.15	2.500	11.00	9.50	8	0.875	565	1,127
4.000	1.125	0.438	2.67	3.875	0.20	3.000	13.50	11.75	8	0.875	1,010	2,020
5.000	1.188	0.500	3.33	4.844	0.25	3.750	16.00	14.25	12	1.000	1,590	3,160
6.000	1.250	0.500	4.00	5.812	0.30	4.500	19.00	17.00	12	1.000	2,260	4,510
8.000	1.438	0.625	5.33	7.750	0.40	6.000	23.50	21.25	16	1.125	4,050	8,100
10.000	1.688	0.625	6.67	9.688	0.50	7.500	27.50	25.00	20	1.250	6,350	12,600
12.000	1.875	0.750	8.00	11.625	0.60	9.000	32.00	29.50	20	1.375	9,100	18,200
18.000	2.375	1.000	12.00	17.438	0.90	13.500	46.00	42.75	32	1.625	18,000	36,000
24.000	2.750	1.125	16.00	23.250	1.20	18.000	59.50	56.00	44	1.625	39,500	78,000

Figure 8.7 Dimensions for standard nozzles.

Nozzle Coefficients

Coefficients for nozzles prescribed in the ASME Codes are nearly, but not quite, uniform for any particular nozzle diameter when used under limitations prescribed in the Codes. These coefficients vary with the differential pressure across the nozzle, the temperature of the air or gas flowing through the nozzle, and other factors. For arrangements A and B in Figure. 8.8 and for air, the coefficients are shown in Table 8.3 and must be selected with particular reference to the temperature and differential pressure across the nozzle as indicated by reference to the curves shown on Figure. 8.9, corresponding to the air temperature and differential pressure for any particular test. For arrangement C in Figure. 8.8, the nozzle coefficient may he taken as 0.993 to 0.995 when pressure before and after the nozzle may be maintained sensibly free from pressure and velocity pulsations.*

In general, test methods for compressors, whether of the displacement or centrifugal type, are essentially the same. The principal differences arise from the fact that in the former fluid flow is intermittent and pulsating while in the latter flow is steady and uniform, so that different rules covering the conduct of tests are required. As a matter of convenience and in order to avoid confusion, the two types of compressors are covered under separate headings in the Compressed Air and Gas Institute standards that are given in this section.

TESTS OF DISPLACEMENT COMPRESSORS, BLOWERS, AND VACUUM PUMPS

The paragraphs immediately following in this section constitute, in effect, a resume of the ASME Power Test Code (PTC 9), and afford an explanation of the methods used for air measurements in tests of displacement compressors, blowers, and vacuum pumps. Certain of the less important provisions of the code have been omitted in an effort to provide a simple exposition of the test methods employed. While the discussion outlined in this section has been directed primarily to two-stage air compressors, the same methods apply regardless of the number of stages of compression. Variations in setup and calculations for vacuum pump tests are included.

The ASME code applies to tests of complete compressor units when operated under conditions which permit discharging the gas compressed into the atmosphere or into pipe lines or reservoirs in which the pressure is maintained sensibly uniform and free from pressure or velocity pulsation. It is intended to cover the compressor only and is applicable for air compressors only when the unit is operated without an intake pipe or duct. The code provides that, when a compressor must be tested with an intake duct connected, an overall allowance must be made to compensate for the influence of the intake duct or pipe on the performance of the compressor.

*The exact value of the coefficient is stated as a function of the Reynolds number as follows:

Reynolds No.	Coefficient
200,000	0.987
250,000	0.990
300,000	0.992
350,000	0.993
400,000	0.994
450,000	0.995

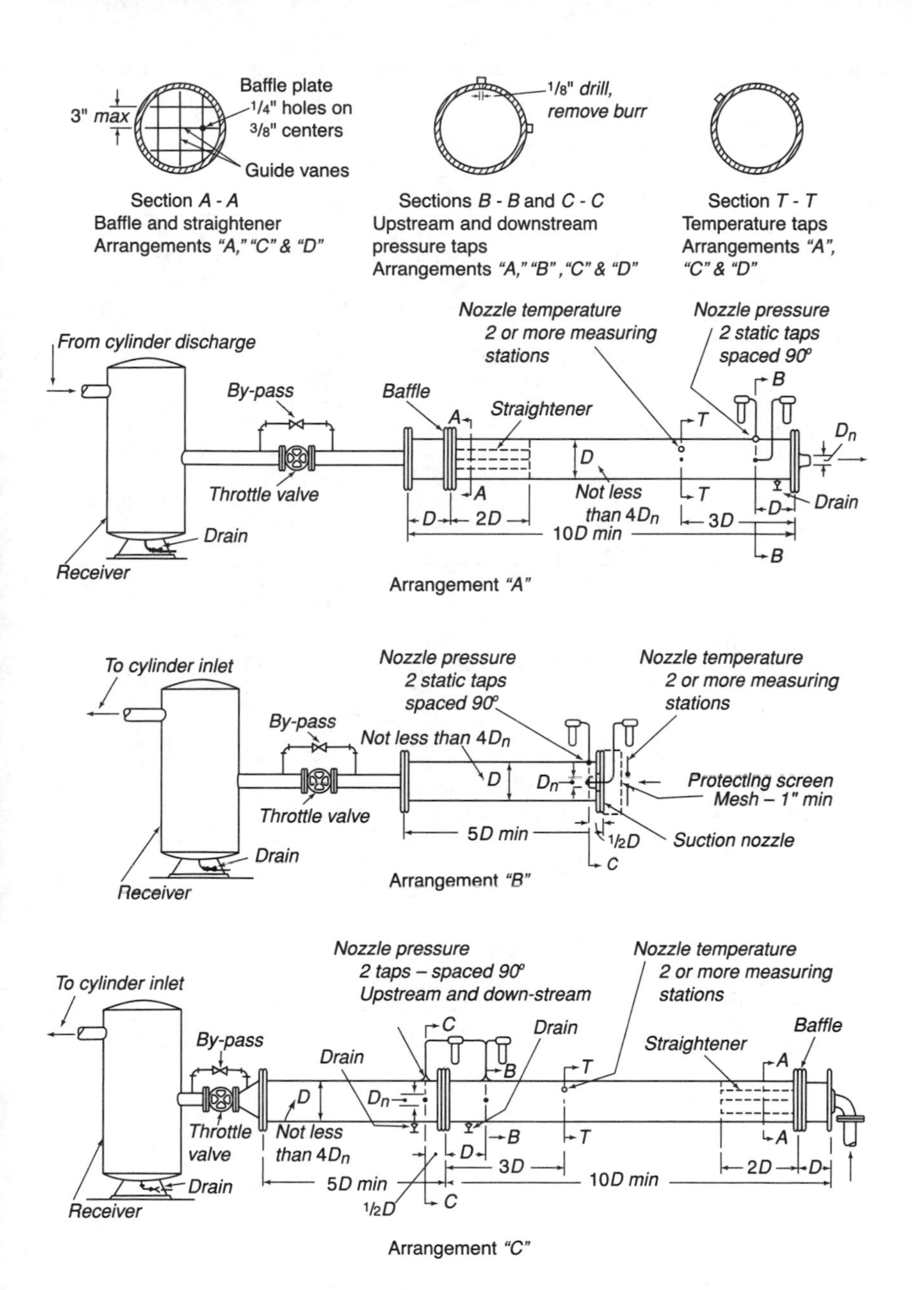

Figure 8.8 Coefficients for nozzles under various arrangements shown above are given in Table 8.3.

Table 8.3 Nozzle-flow Coefficients for air applicable to Arrangements A and B of Figure. 8.8.

	Nozzle Diameter, In.														
Curve	1/8	3/16	1/4	3/8	1/2	3/4	1	1 3/8	2	2 1/2	3	4	5	6	8
A	0.938	0.946	0.951	0.957	0.963	0.968	0.973	0.977	0.982	0.984	0.986	0.990	0.993	0.994	0.995
B	0.942	0.948	0.955	0.960	0.965	0.971	0.975	0.979	0.984	0.987	0.989	0.992	0.994	0.995	0.995
C	0.944	0.952	0.959	0.964	0.968	0.974	0.978	0.981	0.986	0.990	0.991	0.994	0.995	0.995	0.995
D	0.947	0.954	0.961	0.966	0.970	0.976	0.980	0.983	0.988	0.991	0.993	0.994	0.995	0.995	0.995
E	0.950	0.957	0.963	0.968	0.972	0.977	0.982	0.985	0.990	0.992	0.994	0.995	0.995	0.995	0.995
F	0.953	0.958	0.964	0.969	0.973	0.978	0.983	0.986	0.991	0.993	0.994	0.995	0.995	0.995	0.995
G	0.956	0.960	0.966	0.970	0.974	0.979	0.984	0.987	0.992	0.994	0.995	0.995	0.995	0.995	0.995
H	0.958	0.962	0.967	0.972	0.976	0.980	0.985	0.988	0.993	0.995	0.995	0.995	0.995	0.995	0.995
I	0.959	0.964	0.968	0.974	0.978	0.982	0.986	0.989	0.994	0.995	0.995	0.995	0.995	0.995	0.995
J	0.960	0.965	0.970	0.975	0.979	0.983	0.987	0.990	0.994	0.995	0.995	0.995	0.995	0.995	0.995
K	0.961	0.966	0.971	0.976	0.980	0.984	0.988	0.991	0.994	0.995	0.995	0.995	0.995	0.995	0.995
L	0.962	0.967	0.972	0.977	0.981	0.985	0.989	0.992	0.995	0.995	0.995	0.995	0.995	0.995	0.995
M	0.963	0.968	0.973	0.978	0.982	0.986	0.990	0.993	0.995	0.995	0.995	0.995	0.995	0.995	0.995
N	0.964	0.969	0.974	0.979	0.983	0.987	0.991	0.994	0.995	0.995	0.995	0.995	0.995	0.995	0.995

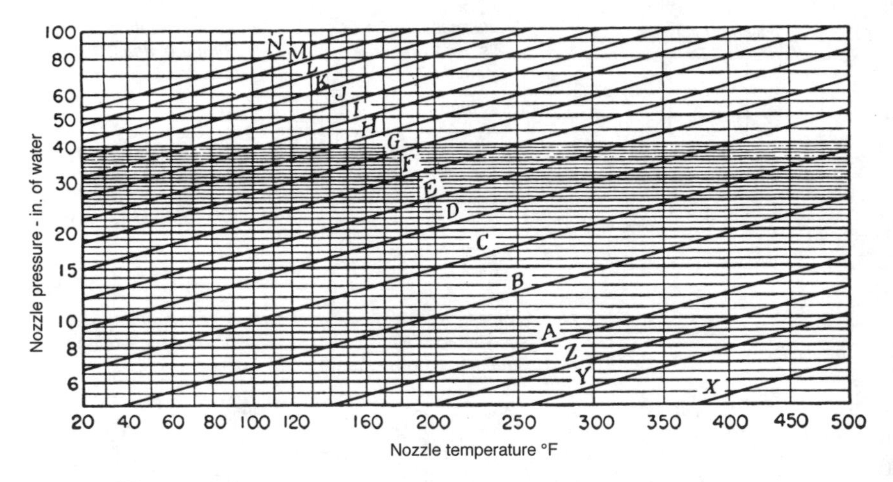

Figure 8.9 Curve for selecting nozzle coefficient from Table 8.3.

The power consumption of a multi-stage compressor depends not only on the operating conditions as to intake pressure, discharge pressure, speed, and so on, but also on the amount of heat removed in the intercoolers. Under winter conditions of operation, the cooling water temperature may be relatively high and the intake air temperature relatively low. Under this condition the degree of intercooling may be less than perfect. In the summertime the conditions may he reversed, and the degree of intercooling may he more than "perfect." The power consumption accordingly may vary 3 to 5 percent depending on the degree of intercooling obtained. Manufacturers' power-consumption statements are usually based on perfect intercooling; therefore, since cooling water conditions are extremely variable and it may be impossible to obtain perfect intercooling, a correction must he applied to the horsepower data to compensate for the variation. This correction applies only when testing a machine having two or more stages of compression with intercooling.

The barometric pressure is not subject to control by the test engineers. The discharge pressure is subject to control within certain limits, but it is usually difficult to hold it at exactly the desired test point.

If the test is made in humid summer weather or with the compressor intake in a location where warm, moist air is taken into the compressor, considerable moisture may be condensed out of the air between the compressor intake and the measuring nozzle. This may cause an error of as much as 1 or 2 percent in the capacity calculations.

To secure comparable results, the corrections for all of the preceding variables are discussed later.

Maximum Deviation from Specified Conditions

For each variable the maximum permissible deviation of the average operating conditions from conditions specified in the contract shall fall within the limits stated in column (2), Table 8.4.

Required Constancy of Test Operating Conditions

During any one test, no single value for any operating conditions shall fluctuate from the average for the test by any amount more than that shown in column (3), Table 8.4.

Method of Conducting the Test

Figure 8.10 shows diagrammatically the general scheme of the test setup for a two-stage displacement-type compressor. The compressor is isolated from the regular service line so that there can be no leakage into or out of the test system, and the entire output of the machine is throttled from the receiver to the low-pressure nozzle tank, for measurement. The actual delivered capacity of the compressor is calculated by a suitable formula using measured values of nozzle pressure, barometric pressure, air temperature at the nozzle, and air temperature at the compressor intake. The compressor is operated at constant speed, and the discharge pressure in the receiver is controlled by the rate of throttling into the nozzle tank.

Table 8.4 Maximum Allowable Variation in Operating Conditions

Variable	Deviation of Test From Value Specified	Fluctuation From Average During Any Test Run
Intake pressure	2 per cent of abs press	1 per cent
Compression ratio*	1 per cent	
Discharge pressure*	....	1 per cent
Intake temperature	....	1°F
Temperature difference, low-pressure in-take and exit air or gas from intercooler	15°F	
Speed	3 per cent	1 per cent
Cooling water inlet temperature	....	2°F
Cooling water flow rate	....	2 per cent
Nozzle temperature	....	3°F
Nozzle differential pressure	....	2 per cent
Voltage	5 per cent	2 per cent
Frequency	3 per cent	1 per cent
Power factor	1 per cent	1 per cent
Belt slip	3 per cent	None

*Discharge pressure shall be adjusted to maintain the compression ratio within the limits stated.

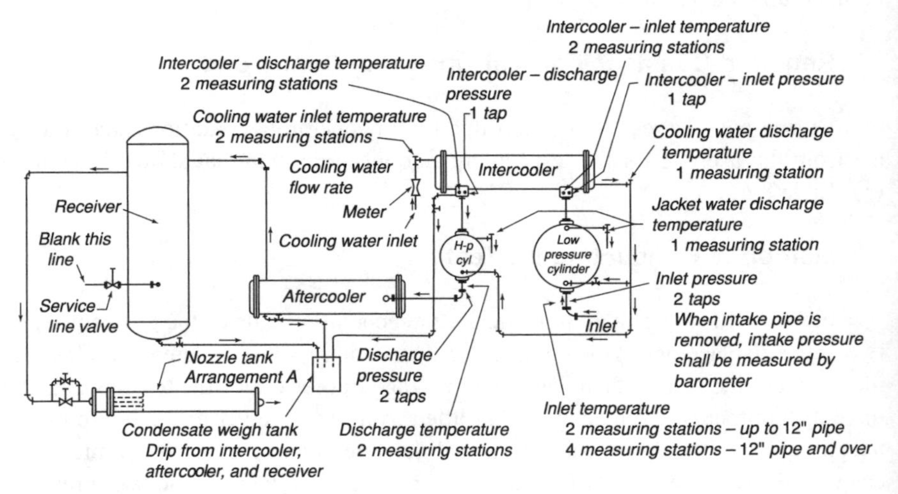

Figure 8.10 Arrangement of nozzle tank and other essential apparatus for test of multi-stage air compressor.

Pressure Pulsation

Prior to the test, the amplitude of pressure waves prevailing in the pipe system shall be measured at each of the stations described for inlet and discharged pressures. If the amplitude is found to exceed 10% of the average absolute pressure, methods for correction shall be mutually arranged.

For air compressors with atmospheric intake, removal of the intake pipe is mandatory, except by mutual agreement to the contrary. For all compressors where the intake pipe is used, an allowance must be agreed upon to compensate for pipe effects on capacity and horsepower.

Nozzle-tank Design

The accuracy normally expected from a low-pressure nozzle test is based on the assumption that the air stream approaching the nozzle flows straight and is free from turbulence, whirls, and spiral motions. It is practically impossible to measure the correct nozzle pressure without these conditions of flow. It is, therefore, extremely important to give detailed attention to the design and construction of the nozzle tank.

A conservative design of nozzle tank is shown in Figure. 8.11. The recommended minimum nozzle tank diameter is four times the maximum nozzle diameter, and the length of the tank should be at least 10 times its diameter or 40 times the maximum nozzle diameter.

The valve through which the air is throttled from the receiver to the nozzle tank frequently sets up a spiral flow, which necessitates the use of the baffle plate and guide vanes shown in Figure. 8.11. When the nozzle size is not more than 2 in., it is often satisfactory to use a long section of 8- or 10-in. pipe for the nozzle tank, and if the length of this pipe is made from two to three times that recommended in Figure. 8.11, the baffle plate and guide vanes are not necessary.

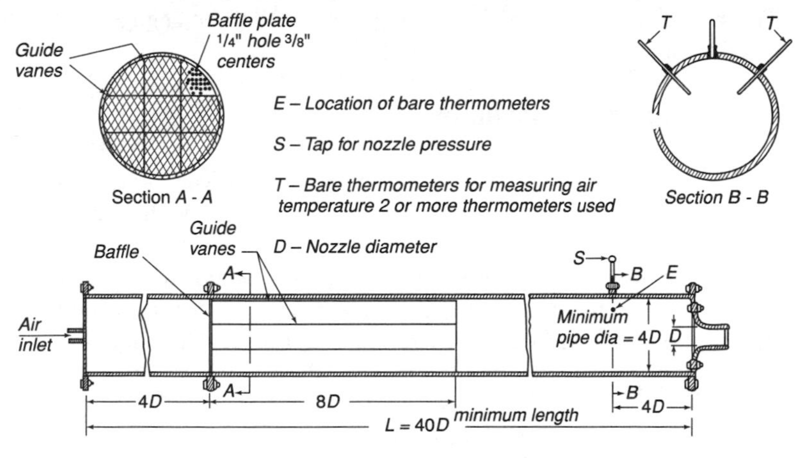

Figure 8.11 Details of nozzle tank for compressor test.

Nozzle Design and Selection

The ASME Test Code limits the use of nozzles for measuring the capacity of a displacement-type air compressor to those applications where the air may be discharged into the atmosphere from a reservoir or nozzle tank in which the pressure does not exceed 40 in. of water and is not less than 10 in. of water. These limiting conditions are such as to minimize the requirements for close tolerance as to nozzle contour. As measuring instruments under these conditions of operation, the accuracy depends primarily on (1) absolute roundness and uniformity of bore along the throat or straight portion of the nozzle and (2) smoothness and character of machine finish for the rounded entrance, with the throat of the nozzle tangent where it joins the rounded nozzle entrance.

Figure 8.7 gives dimensions for convenient nozzle sizes suitable for a large range of air capacity. One nozzle size can usually be selected to measure the output of a given compressor, at loads down to about 50% capacity.

In every case, the nozzle throat should he accurately measured in several directions to the nearest one-thousandth of an inch, and the average diameter used in the calculations.

Nozzle Pressure

When the ratio of gaging tank diameter to nozzle diameter is limited as specified in this code ($D_1/D_2 \geq 4$), the static pressure and total pressure in the gaging tank are sensibly identical. A manometer applied as shown in Figure. 8.11 shall be used for measuring this pressure. A tap connection for the manometer shall be located upstream one pipe diameter from the face of the nozzle. The inner bore of the manometer and the connecting tube shall never be less than 1/4 in., preferably 1/2 in. The inner bore shall not be contracted by projecting gaskets or other imperfections of manufacture or assembly. Nozzle pressures shall be measured in inches of water. The measuring scale used for the manometer shall be engine-divided with divisions of 0.1 in. or less, and accurate within 0.0100 in.

Manometer and Connections

A well-made manometer with an accurately graduated scale is an essential part of the apparatus for measuring nozzle pressure. Glass tubes with small bores are not suitable for manometers because of the error caused by capillarity. A glass tube 48 in. long, 5/8 in. outside diameter, and with a 3/8-in. bore is recommended.

Small air leaks in the manometer connections often cause serious errors, and after setup all joints should he carefully tested with a soap solution to detect small leaks.

Nozzle Temperature

In the calculation of air flow through the nozzle, the air temperature is of equal importance with the nozzle pressure. This temperature is measured on the upstream side of the nozzle with bare-bulb thermometers (no thermometer bulbs) located as shown in Figure. 8.11, and the test should not be started until the nozzle temperature has become approximately uniform.

Engraved stem thermometers of a good laboratory grade are held in the nozzle tank by small packing glands. The bulb of the thermometer should project into the tank as far as possible. To guard against accidental errors, two or more thermometers should be used, and they should be moved in and out of the tank to make sure that they are located to get the maximum temperature reading. The nozzle temperature should be determined to the nearest 1/2°F.

Discharge Pressure

The flow through the discharge pipe to the receiver or compressor piping system is intermittent and pulsating in character for all displacement-type compressors. If the natural period of the discharge pipe approaches resonance with the speed of the compressor, or if the discharge pipe is too small in diameter or too long, pressure waves of considerable amplitude are induced in the discharge pipe and the discharge receiver, and it becomes impossible to measure accurately the average discharge pressure by any suitable means for damping the gage. The code defines discharge pressures as the average shown by a pressure-time indicator diagram taken at a point on the discharge line immediately adjacent to the compressor cylinder during that period which corresponds to the delivery line on an indicator diagram taken from the compressor cylinder, that is, for the period during which the discharge valve is open (A to B in Figure. 8.12). This figure shows respectively a typical pipe indicator diagram and a cylinder indicator diagram with that portion of the former marked to show the average discharge pressure as defined in the Code. Upon agreement by the parties to the test, a Bourdon gage may be used to measure the discharge pressure.

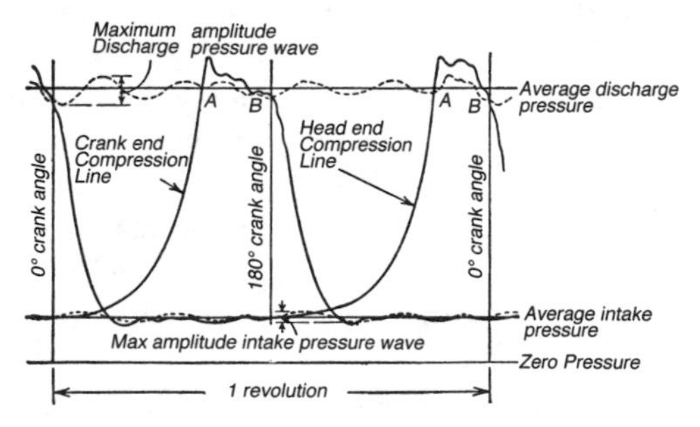

Figure 8.12 Pressure-time diagrams of double-acting compressor, illustrating cylinder pressures and pressure waves at measuring stations for inlet and discharge pressure.

For rotary-type displacement compressors, discharge pressure cannot be measured as previously specified. It shall be measured by determining the mean ordinate of a pressure curve obtained from a card drawn by an indicator operated with the drum rotated at approximately uniform speed (hand-pulled).

Intercooler pressure is not ordinarily an essential reading, but it should be taken as a check on the mechanical condition of the machine. The gage must be connected so that there is no needle vibration due to pulsations.

Intake Pressure

The performance of a compressor is extremely sensitive to variations in intake pressure, particularly to periodic fluctuations in pressure which approach resonance with the compression cycle or rotative speed of the unit (see pages 659-664). When intake pipes or ducts are connected, the intake pressure shall be regarded as the average pressure shown by a pressing-time indicator diagram taken at a point adjacent to the compressor intake during the intake portion of the stroke for the period during which the inlet valve is open (Figure. 8.12). The inlet pressure of an air compressor operating without an intake pipe shall be measured by the barometer.

Atmospheric pressure shall be measured with a Fortin-type mercurial barometer fitted with a vernier suitable for reading to the nearest 0.002 part of an inch. It shall have an attached thermometer for indicating the instrument temperature. It shall be located at the floor level of the compressor and supported on a structure free of mechanical vibrations.

Air Intake, Temperature, and Pressure

The actual delivery capacity of an air compressor must always be referred to the actual conditions of temperature and pressure at the compressor intake. The procedure to be followed in measuring intake temperature and pressure therefore is extremely important.

Air Temperature at Intercooler

Thermometer wells should be located at the inlet and outlet of the intercooler, and temperature readings should be taken at these points. The temperature of the air leaving the intercooler (entering the high-pressure cylinder) is particularly necessary. The cooling water should be adjusted to obtain as near perfect cooling as possible and should be adjusted to give approximately the same relation between the low-pressure inlet temperature and the high-pressure inlet temperature on all test runs.

At the same time it should be recognized that the cooling water required is a part of the compressor performance and is chargeable to the cost of operation.

Measurement of Cooling Water

Where a complete test is desired, the cooling water from the cylinder jackets and intercooler can best be measured by leading the water from the cooler outlet into a tank or tanks fitted with an orifice, preferably of the rounded-inlet type, and equipped with a gage glass and scale for actually measuring the head on the orifice. For most installations a suitable tank can be made from a section of 8- or 10-in. pipe about 4 ft. high, with a 3 or 4 in. coupling welded into the side. The orifice or nozzle used is exactly similar to that recommended for measuring the air. The coefficient of flow will be approximately 0.970, and the complete formula for calculation can be found in any standard engineers' handbook.

Air Temperature at Compressor Discharge

The temperature at the discharge of the high-pressure cylinder should be measured in a thermometer well. This reading is not used in calculations, but serves as an additional check on the mechanical condition of the compressor.

Speed

An accurate measurement of the total revolutions during the period of the test is required for calculating the revolutions per minute and the piston displacement of the unit. A mechanical counter geared to the compressor shaft or operated by a cam and indicating total revolutions is the best instrument to use. Tachometers and in-termittent counters should be avoided. Counter readings should be carefully controlled to eliminate changes in the discharge pressure and the nozzle-tank pressure, and also to eliminate inaccuracies in power-input readings.

Barometric Pressure

The barometric pressure must be known in order to make proper calculations of the air flow through the nozzle and for the determination of the absolute intake pressure. A barometer that has been carefully checked and compared with a standard barometer should he used, and if possible, it should be located in the near vicinity of the compressor. Simultaneous readings of the barometer and room temperature at the barometer should be taken at 1/2 -hour intervals throughout the test. If a reliable barometer is not available, an approximation may be obtained by using records of the nearest Weather Bureau station and correcting for the difference in altitude between the government station and the compressor.

Condensation

If the test is run during humid summer weather, or if the compressor intake is warm and moist, corrections to the capacity calculations will be required to compensate for the shrinkage in the volume of air due to condensation of moisture between the compressor intake and the measuring nozzle. It is necessary to drain all points, such as the intercooler, aftercooler, and receiver, and any low points in the piping between the compressor inlet and nozzle tank where moisture might collect. It is frequently difficult to obtain consistent values for the condensate, and it may be necessary to prolong the test to get dependable measurements. In cool weather or when the relative humidity of the atmosphere is below 30%, the correction for condensation is so small that it may be neglected. In summer weather, however, the correction may amount to as much as 2% of the total capacity.

Technique of Taking Readings

Readings should be taken as nearly simultaneously as possible and with sufficient frequency to ensure average results. This is important in obtaining the air-capacity data, but it is even more important when taking electrical power-input data.

CALCULATION OF TEST RESULTS: DISPLACEMENT COMPRESSORS

Actual Compressor Capacity

The ASME Code gives two sets of formulas for calculating the air quantity discharged through the nozzle. The first, known as adiabatic formulas, are complicated in form and rather difficult to use. The second set are simple approximations of the adiabatic formulas, and for conditions of flow permissible under the test code they give nearly identical results. The code specifies that the approximate formula shall be used for calculating the results of all acceptance tests. Two forms of this approximate formula are presented in these standards as follows:

$$Q_3 = 19.16 \frac{CD^2 T_3}{p_3} \sqrt{\frac{(p_1 - p_2)p_2}{T_1}}$$

$$Q_3 = 2.552 \frac{CD^2 T_3}{p_3} \sqrt{\frac{B_i}{T_1}}$$

where:

T_1 = Absolute Fahrenheit temperature, upstream side of nozzle.

T_3 = Absolute Fahrenheit temperature at which volume of flowing air is to be expressed; usually compressor intake temperature.

p_1 = Absolute static pressure, psi, upstream side of nozzle.

p_2 = Absolute total pressure, psi, downstream side of nozzle.

p_3 = Absolute pressure, psi, at which volume of flowing air is to be expressed; usually compressor intake.

B = Value p_2 expressed in in. Hg (32°F); in the usual case of a nozzle discharging into atmosphere, this is the barometer reading, corrected for the temperature of the mercury column, and the barometer calibration constant.

i = Differential pressure ($p_1 - p_2$), the pressure drop through the nozzle,expressed in inches of water column.

Q_3 = Cubic feet of air flowing per minute, expressed at pressure p_3 and temperature T_3, which in air-compressor testing are usually the conditions of the compressor intake.

D = Diameter of smallest part of nozzle throat, inches.

C = Coefficient of discharge.

Moisture Correction for Capacity

In the case of a compressor handling moist air, if some of the moisture is condensed and removed during the process of intercooling and aftercooling, the weight of the air and water vapor mixture which passes through the measuring nozzle will be less than that taken in by the compressor. In reducing the amount of air as shown by the nozzle measurements to terms of air at intake pressure and temperature, it is necessary to correct for the water thus removed, and this correction can most readily be made by reducing the capacity to terms of air at intake total pressure and temperature and discharge specific humidity. This correction may be made in either of two ways: through vapor pressure or through vapor density.

A simple method of approximating the correction is as follows. Having measured the condensation rate as discussed previously, the correction to be added to Q_3 is the product of the condensation rate and the specific volume for superheated steam at intake pressure and temperature. Equivalent volumes may be obtained from the curves given in Figure. 8.13.

Power Correction for Imperfect Intercooling

To correct for imperfect intercooling, there shall be added to the measured horsepower input calculated from the observed results of the test a horsepower correction as shown by the following formula:

$$\text{Horsepower correction} = \frac{1}{N}\frac{T_1 - T_1}{T_1} \text{ x } P$$

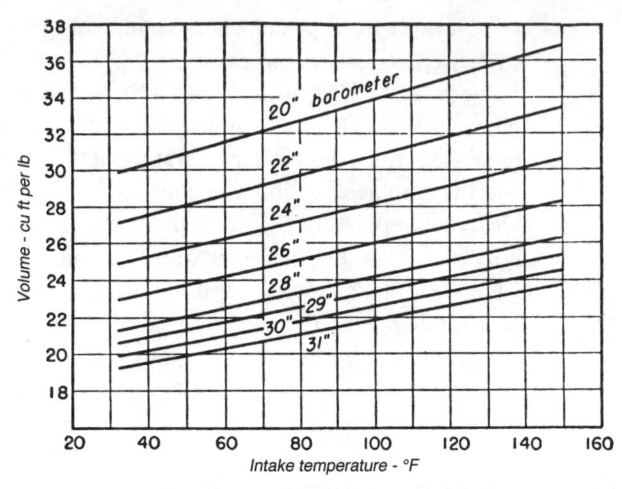

Figure 8.13 Equivalent volume condensate at atmospheric pressures and temperatures.

where:

P = Measured horsepower input.
N = Number of stages.
T_I = Absolute temperature of air or gas at compressor intake (Rankine).
T_t = Absolute temperature of air or gas leaving the intercooler in question (Rankine).

If the compressor in question is a multi-stage unit with more than one intercooler, a correction must be made for each intercooler, and in each case T_I will be the temperature of the air or gas at the compressor intake and T_I will be the temperature of the air or gas leaving the intercooler in question.

Power Correction for Variation in Intake Pressure

To correct for intake pressure, the horsepower correction as given by the following formula is added to the horsepower input calculated from the observed results of the test.

$$\frac{p_I - p_a}{P_a} \times P$$

where:

P = Measured horsepower input.
P_I = Absolute contract intake pressure.
P_a = Absolute intake pressure observed during test.

Power Correction for Variation In Compression Ratio

To correct for variations in compression ratio, there shall be added to the measured horsepower input calculated from the observed results of the test a horsepower correction as shown by the following formula:

$$\text{Horsepower correction} = \frac{\left(r_1^{(k-1)/Nk} - 1\right) - \left(r_t^{(k-1)/Nk} - 1\right)}{\left(r_t^{(k-1)/Nk} - 1\right)}$$

where:

P = Measured horsepower input.
N = Number of stages.
k = Ratio of specific heats (1.395 for air with some moisture as commonly employed in engineering).
r_1 = Ratio of compression under conditions guaranteed in contract.
r_t = Ratio of compression observed during test.

The formulas giving the correction for variation in compression ratio and also the correction for imperfect intercooling assume that the cylinder ratio of the compressor is such as to divide the work between the cylinders equally. These corrections will in all cases be small, so that, if for the compressor tested the division of work between cylinders is not exactly equal, the effect of such deviation on the correction will be negligible.

Power Correction for Speed

To correct for variations in speed, there shall be added to the measured horsepower input calculated from observed results of the test a horsepower correction as shown by the following formula:

$$\frac{N_c - N_t}{N_t} \times P_c$$

where:

P_c = Corrected horsepower input.
N_c = Contract speed, rpm.
N_t = Average speed during test, rpm.

Capacity Correction for Speed

To correct for variations in speed, a capacity correction as shown by the following formula shall be added to the measured capacity calculated from the observed results of the test.

$$\frac{N_c - N_t}{N_t} \times q$$

where:

q = Measured capacity calculated from observed results of test, cfm.

N_c = Contract speed, rpm.

N_t = Average speed during test, rpm.

Power Measurements

The following codes should be observed in determining the power input to the compressor:

Steam-engine drive: ASME Test Code for Reciprocating Steam Engines
Steam-turbine drive: ASME Test Code for Steam Turbines
Oil- or gas-engine drive: ASME Test Code for Internal Combustion Engines
Electric-motor drive:
    Direct-current motor: AIEE Standard No.5
    Induction motor: AIEE Standard No.9
    Synchronous motor: AIEE Standard No.7

It is, of course, essential that all power test apparatus be carefully calibrated and the readings taken with care corresponding to that used in obtaining the air-capacity results.

Example

The following is an example of the essential calculations and corrections for the test of a two-stage air compressor with electric-motor drive:

SPECIFIED CONDITIONS OF OPERATION:

1. Speed = 225 rpm.
2. Discharge pressure = 100 psig.
3. Intake pressure, atmospheric at sea-level elevation, normal barometric pressure = 14.7 psia.
4. Intercooling is perfect.

OBSERVED DATA:

5. Duration of test = 60 minutes.
6. Speed, average, by revolution counter = 222.5 rpm.
7. Intake pressure (barometer) = 14.41 psia.
8. Discharge pressure = 97.6 psig.
9. Barometer, corrected to 32°F = 29.348 in. Hg.
10. Gaging tank
    a. Nozzle diameter = 4.062 in.
    b. Nozzle pressure = 13.91 in. water
    c. Nozzle temperature = 83.0°F
11. Intake temperature = 84.0°F.
12. Psychrometric readings
    a. Wet bulb = 72.0°F
    b. Dry bulb = 84.0°F
13. Intercooler pressure = 25.2 psig.
14. Air temperature at discharge of intercooler = 98.0°F.
15. Total condensate collected from intercooler, aftercooler, and receiver = 45.8 lb.

MOTOR-INPUT METER READINGS, CORRECTED:

16. AC input to stator, wattmeter = 205.80 kW.
17. DC input to field = 8.48 kW.
18. Total motor losses from calibrations, including excitation = 18.05 kW.
19. Volume discharged through nozzle referred to intake pressure and temperature, using the Equation on page 766, nozzle coefficient C, Table 8.3 and Figure 8.8.

$$Q = \frac{2.552 \times 4.062^2 \times 544 \times 0.995}{14.41} \times \sqrt{\frac{29.348 \times 13.91}{543}} = 1371.6\,\text{cfm}$$

20. Correction for moisture condensed and removed
    a. Condensation rate = $\frac{45.8 \text{ lb}}{60 \text{ min}}$ = 0.76333 lb/min
    b. Equivalent volume at intake conditions (Figure 8.13) = 22.5 x 0.76333 = 17.2 cfm
21. Capacity as run = 1371.6 + 17.2 = 1388.8 cfm.
22. Bhp as run = [(205.80 + 8.48) – 18.05] ÷ 0.746 = 263.04.

TEST RESULTS CORRECTED TO SPECIFIED CONDITIONS:

23. Capacity correction for speed = $\dfrac{225\text{-}222.5}{222.5}$ x 1388.8 = 15.6 cfm

24. Horsepower corrections for

a. Intake pressure $\dfrac{14.7-14.41}{14.41}\times 263.04 = 5.29\text{ hp}$

b. Compression ratio = $\dfrac{\left(r_1^{(k-1)/Nk}-1\right)-\left(r_t^{(k-1)/Nk}-1\right)}{r_t^{(k-1)/Nk}-1}\times 263.04$

$r_1 = \dfrac{114.7}{14.7} = 7.8027$, $k = 1.3947$

$r_t\,\dfrac{112.3}{14.41} = 7.7731$, $\dfrac{(k-1)}{Nk} = 0.1415$

$$\frac{\left[(7.8027)^{0.1415}-1\right]-\left[(7.7731)^{0.1415}-1\right]}{\left(7.7731^{0.1415}-1\right)}\times 263.04$$

$$=\frac{(1.3374-1)-(1.3367-1)}{1.3367-1}\times 263.04 = 0.002079\times 263.04$$

$$= 0.05\text{hp}$$

c. Intercooling = $\left(\dfrac{1}{2}\times\dfrac{544-558}{544}\right)\times 263.04 = -3.38\text{ hp}$

d. Speed = $(263.04 + a + b + c)\times\dfrac{225-222.5}{222.5} = 2.98\text{ hp}$

25. Total bhp correction = $a + b + c + d$ = 5.44 hp.
26. Bhp corrected to contract conditions = 263.04 + 5.44 = 268.48 hp.
27. Capacity corrected to contract conditions = 1388.8 + 15.6 = 1404.4 cfm.
28. Bhp/100 cfm corrected to contract conditions.

$$= \frac{268.48\text{ x }100}{1404.4} = 19.117$$

Normally the air ihp/100 cfm is not the ultimate desired result. To show the complete performance, the overall cost or the overall efficiency to the basic source of power should be shown. This means that results should be expressed by kW/100, ehp/l00, pounds of steam/100, pounds of oil/100, or cubic feet of gas/100, depending on the source of power.

DISPLACEMENT-TYPE VACUUM PUMPS

For accurately measuring the capacity of a vacuum pump, the same general procedure should be followed as outlined for compressors. The apparatus, however, should be placed at the intake of the machine rather than at the discharge. The setup for a nozzle test of a dry vacuum pump is shown in Figure 8.14.

The nozzle tank used in Figure 8.15 is similar to that shown in Figure 8.12, but since the air enters as stream-line flow, the use of baffle plates and guide vanes is not essential.

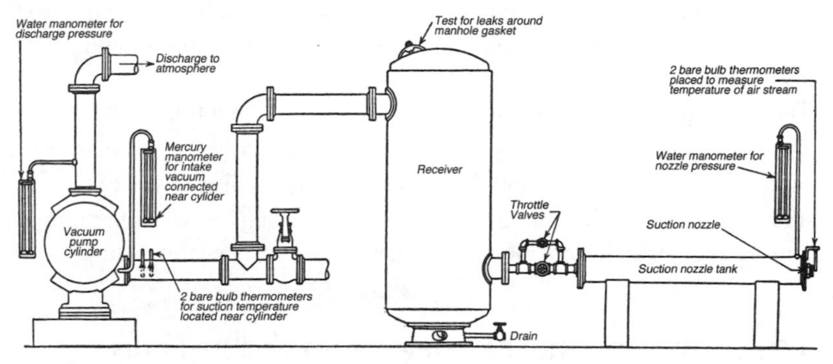

Figure 8.14 Setup for test of vacuum pump with nozzle in intake.

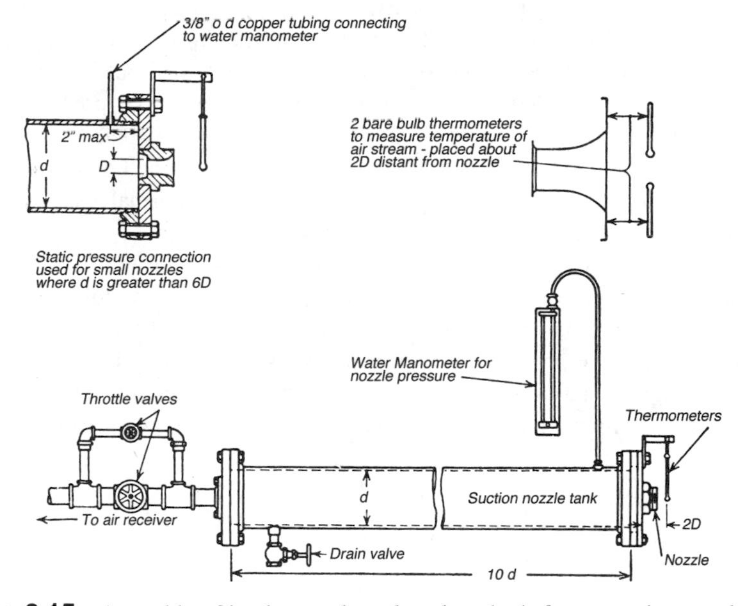

Figure 8.15 Assembly of intake nozzle tank and methods for measuring nozzle pressures.

The nozzle-tank diameter and length will be determined by the maximum nozzle size used, and a tank diameter at least four times the maximum nozzle diameter is satisfactory except that the smallest nozzle tank recommended is 6 in. in diameter by 10 ft. in length. It will be necessary to estimate the actual free air capacity of the vacuum pump at the desired test vacuum to choose the proper size of nozzle. This capacity can best be obtained from the manufacturer.

It is desirable to test a vacuum pump over a considerable range on either side of the test vacuum So that a curve will be procured from which the performance at the desired vacuum may be obtained.

It is essential that leakage into the system be absolutely eliminated. Extreme care is therefore necessary in making the setup. All pipe joints should be painted under vacuum.

The first equation for calculating the flow Q_3 through a nozzle is now applicable. The flow Q_3 is usually expressed at the temperature and pressure of the vacuum pump intake; p_1 will now be the barometric pressure, and p_2 the absolute pressure inside the nozzle tank. Since all the air entering the vacuum pump passes first through the nozzle, no moisture corrections are necessary for the flow or capacity calculations.

ROTARY COMPRESSORS, BLOWERS, AND VACUUM PUMPS

For displacement blowers and boosters of the type in which volumetric clearance is zero, when the discharge pressure is insufficient to provide throttling to the extent specified in the ASME Code, capacity may be determined by subtracting the leakage past the impellers from the gross displacement. This method does not lead to greater accuracy than nozzle measurements, but may be more convenient as a means of determining the approximate net capacity. The displacement may be determined from the measurements of the blower, and as defined on page 741, it is the product of the volume displaced per revolution and the normal speed in revolutions per minute. The leakage past the impellers is the product of the displacement per revolution and the number of revolutions per minute required to maintain the predetermined rated pressure with the discharge pipe from the blower or booster closed and the inlet pipe open to the atmosphere. The leakage test or tests may be conducted at the same pressure and temperature as the contract conditions, or the test may be run with a differential of 1 psi across the impellers and correction made for obtaining the slip at contract conditions by using the following formula.

$$S_c = S_t \sqrt{\Delta_c \frac{T_c G_t P_t}{T_t G_c p_c}}$$

where:

Δ_c = contract differential pressure, psi
S_t = slip at ~ differential, cfm
S_c = slip at contract conditions, cfm
T_t = Rankine temperature at test conditions
T_c = Rankine temperature at contract conditions
G_t = specific gravity at test conditions
G_c = specific gravity at contract conditions
p_t = discharge pressure at test conditions, psia
p_c = discharge pressure at contract conditions, psia

TESTS OF CENTRIFUGAL BLOWERS, COMPRESSORS AND EXHAUSTERS

The paragraphs that follow constitute a resume of the ASME Power Test Code (PTC 10). Certain of the less important provisions of the code have been omitted in order to provide a simple exposition of the test methods employed.

The ASME code provides standard directions for conducting and reporting tests on

centrifugal compressors or exhausters of the radial-flow, mixed-flow, or axial-flow types (hereafter inclusively covered by the term compressors) in which the gas specific weight change produced exceeds 7%. Apparatus of the centrifugal type for compressing or exhausting service in which the gas specific weight change is 7% or less shall be tested in accordance with the ASME Power Test Code for Fans (PTC11).

The capacity and power consumption of a centrifugal compressor depend on the composition of the gas, the density at intake, and the pressure rise. In the case of multi-stage group machines with intercooling between stage groups, the power also depends on the degree of intercooling. Manufacturers' statements of capacity and power are based on stipulated conditions of temperature, total pressure, and composition of gas prevailing at the compressor inlet, as well as on speed, pressure rise, and degree of intercooling (where intencoolers are used). Since these conditions cannot be subject to independent control by the test engineers, it is necessary to correct or adjust the test results to account for any deviation from specified conditions. To permit a direct comparison between test results and a manufacturer's guarantee, the correction for all of these variables is discussed later.

Where intercoolers are used between groups of stages, the conditions of heat removal and the drop in pressure in the intercooler and associated piping must be included as part of the operating conditions in any complete statement of performance guarantee.

In a multi-stage group compressor with intercooling and with a common driver, the contract performance must be determined for each stage group separately when the inlet conditions at the first stage and the degree of intercooling differ from the specified conditions. If the degree of intercooling is such that the deviations of density, pressure ratio, and q/n (ratio of capacity to revolutions per minute) at the inlets to both stage groups are within the limits stated in Table 8.5, corrections shall be applied to the performance of each stage group to reduce it to contract conditions as if each group were a single machine.

When testing a compressor, every effort should be made to have the operating conditions and composition of the gas as near as possible to those specified in the contract. A single contract condition or guarantee shall be established either by a complete characteristic curve or by not less than two test points that bracket the contract capacity by not more than ± 3 percent. The maximum deviation for which adjustment may be applied to any of the variables is given in Table 8.5. Under these conditions, the values of these variables, as calculated under the rules of this code, shall be accepted as indicating the performance under specified operating conditions.

Before starting a test, the machine shall be run for a sufficient length of time to assure steady conditions. The duration of a test shall be long enough to record sufficient data to demonstrate the uniformity of test operation, but in any event it shall be not less than 30-minute duration, but longer if the test code requirements of the driving element specify additional time. During the test, readings of each instrument having an important bearing on the calculation of results shall be taken at 5-minute intervals, the readings of each set being as nearly simultaneous as practicable. Throughout the test period, the machine shall be in continuous steady operation, the observed readings shall be consistent, and the maximum permissible fluctuation of any individual reading from the average shall be within the limits shown in column (3), Table 8.5.

Table 8.5 Maximum Allowable Variations in Operating Conditions, Centrifugal Compressors

Variable (1)	Deviation of Test From Value Specified (2)	Fluctuation From Average During Any One Test (3)
(a) Inlet pressure*	. .	2%
(b) Inlet temperature* (abs)	6%	4°F
(c) Intercooling, deg	30°F	4°F
(d) Discharge of pressure (abs)	. .	2%
(e) Capacity	4%	
(f) Molecular weight of gas*	10%	1%
(g) Ratio of specific heats	5%	
(h) Voltage	5%	2%
(i) Frequency	2%	1%
(j) Speed	5%	1%
(k) Power factor (synchronous motor)	1%	1%
(l) Nozzle pressure	. .	2%
(m) Nozzle temperature	. .	3°F
(n) Capacity speed ratio q/n	4%	
(o) Inlet specific weight*	10%	

*The combined effect of variables a, b, and f shall not produce a deviation greater than specified for inlet specific weight.

Test Arrangements

Four alternate test arrangements are provided in this code, and selection of the arrangement to be used for any particular test will depend on the type of compressor to be tested and upon the operating conditions.

Test arrangement 1 (Figure 8.16). Compressor with atmospheric inlet. The arrangement of the flow nozzle and the location of the instruments for measuring temperature, pressure, and so on, shall be as shown in arrangement A, Figure 8.8.

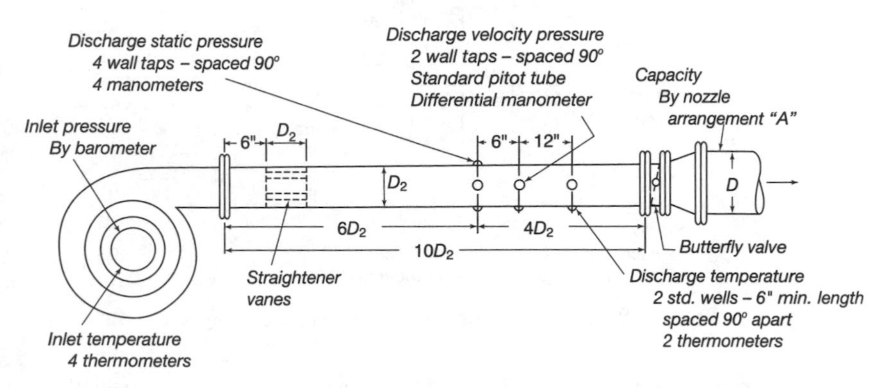

Figure 8.16 Test setup No. 1. volute-type compressor, atmospheric inlet.

Test arrangement 2 (Figure 8.17). Exhauster with atmospheric discharge. The arrangement of the flow nozzle and the location of instruments for measuring temperature, pressure, and so on, shall be as shown in arrangement B, Figure 8.8.

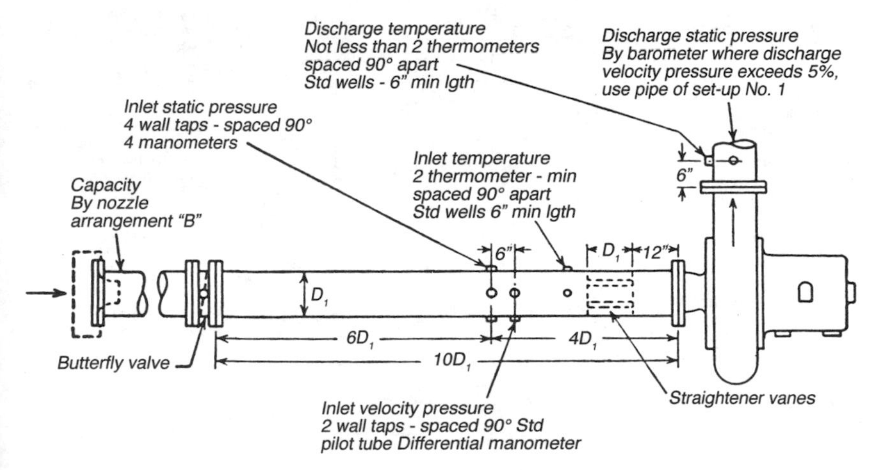

Figure 8.17 Test setup No.2. single-stage compressor, atmospheric discharge.

Test arrangement 3 (Figure 8.18). Compressor without intercooler. The arrangement of the flow nozzle and the location of instruments for measuring temperature, pressure, and so on, shall be as shown in arrangements A or C, Figure 8.8.

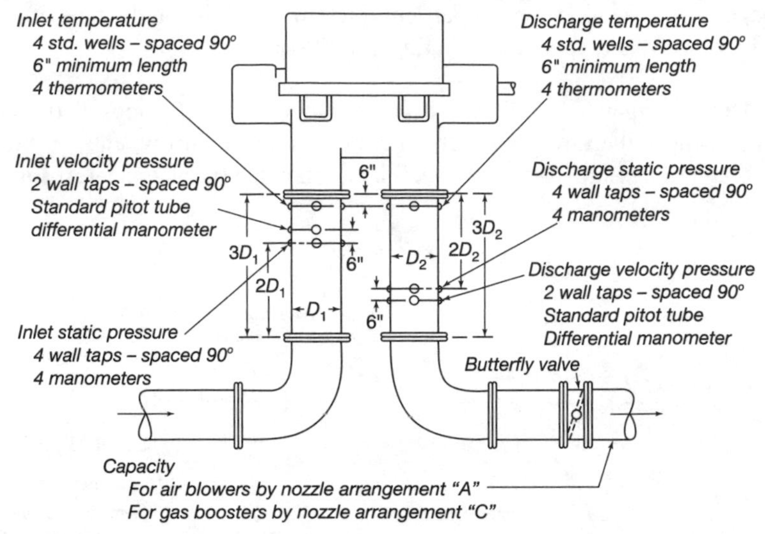

Figure 8.18 Test setup No.3. Multi-stage compressor.

Test arrangement 4 (Figure 8.19). Multi-stage group compressor with intercooler between stage groups. The arrangement of the flow nozzle and the location of instruments for measuring temperature, pressure, and so on, shall be as shown in arrangements A or C, Figure 8.8.

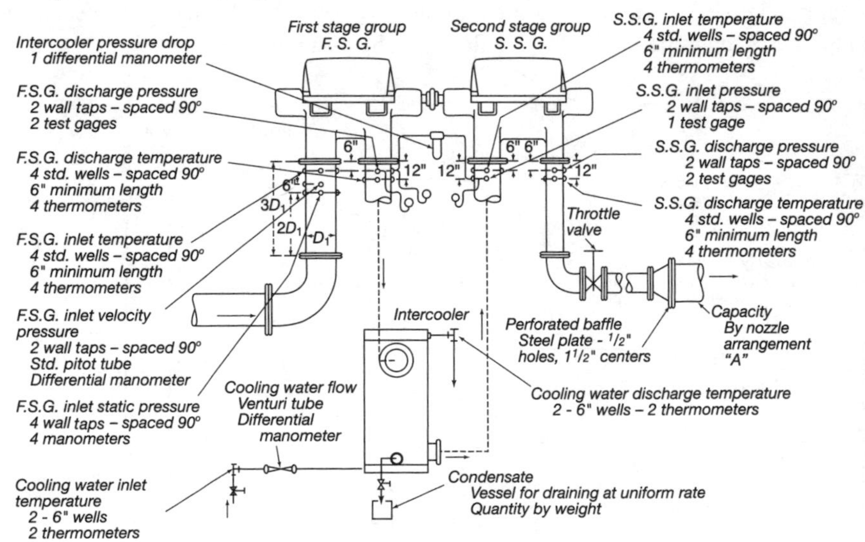

Figure 8.19 Test setup No. 4. Multi-stage groups with intercooler.

Barometric Pressure

The barometric pressure shall be read at intervals of 30 minutes, during the test from a mercury barometer of the type used by the U.S. Weather Bureau.

Inlet Pressure

When the inlet flange of the compressor is open to the atmosphere (Figure 8.16), the inlet pressure shall be taken as the barometric pressure adjacent to the compressor intake. When the inlet flange to the compressor is piped, the inlet pressure shall be the sum of the static and velocity pressure as computed from measurements made with instruments placed as shown in Figures 8.17, 8.18, or 8.19.

Discharge Pressure

When the discharge flange of the compressor is open to the atmosphere (Figure 8.17), the discharge pressure shall be taken as the sum of the barometric pressure and the velocity pressure at the plane of the discharge flange. If the velocity pressure is greater than 5 percent of the total pressure, the test shall be conducted with the outlet flange piped. When the outlet flange of the compressor is piped, the discharge pressure shall be the sum of the static and the velocity pressure as computed from measurements made with instruments placed as shown in Figures 8.16, 8.18, or 8.19.

Static Pressure

The static pressure shall be taken as the arithmetic average of the readings obtained by means of four wall taps, each connected to a separate manometer. The four taps shall be disposed at intervals of 90° around the circumference of the pipe. The diameter of the holes shall not be greater than one thirtieth of the pipe diameter (nor less than 1/8 in.), and they shall be drilled perpendicular to the pipe wall, with their inner edges free of burrs. Where the individual readings of the four wall taps differ from their mean by more than 1 percent, the cause shall be determined and corrected. If the cause is traceable to the flow pattern at the measuring section, and this cannot be corrected, a reliable test cannot be obtained. The pressure-measuring stations shall be located in a region where the flow is essentially parallel to the pipe wall. For the measurement of pressure or pressure differences in excess of 35 psig, dead-weight gages, or their equivalent, or calibrated gages shall be employed. For lower pressures or pressure differences, liquid manometers shall be used. Whichever of the above means of pressure measurement is employed, the instruments shall be so graduated that readings can be made within 1/2 percent of the absolute pressure.

Velocity Pressure

When the velocity pressure is not more than 5 percent of the total pressure, it shall be calculated on the basis of average velocity. The velocity shall be computed as the ratio of the quantity at the measuring section to the pipe area.

When the velocity pressure is more than 5 percent of the total pressure, it shall be determined by a pitot-tube traverse. The traverse shall consist of readings made at 10 traverse points across each of two diameter disposed at 90 degrees to each other. The traverse points shall be spaced at equal area positions. In a round pipe, the spacing shall be as defined in Figure 8.20. The pitot tube shall be of a type and design shown in Figure 8.21.

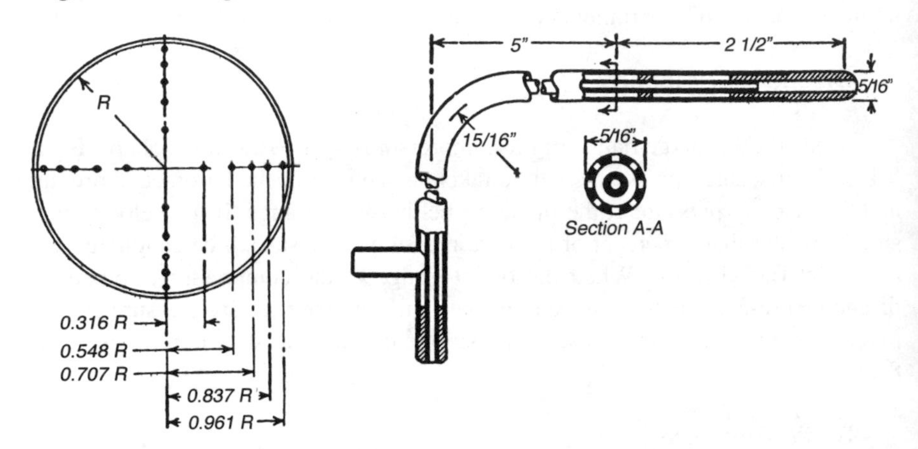

Figure 8.20 Traverse points in pipe.

Figure 8.21 Pitot tube.

Inlet Temperature

The inlet temperature shall be measured by four temperature-measuring devices. When the compressor has a piped inlet flange, the instruments shall be placed as shown in Figures 8.17, 8.18, and 8.19 and shall be disposed symmetrically and at 45° to the inlet-pressure-measuring location. For machines assembled for test with an atmospheric intake (Figure 8.16), the inlet total temperature is the atmospheric temperature measured in a region of substantially zero velocity (less than 125 fps) in the vicinity of the inlet flange. For machines assembled for test with an intake pipe, the intake temperature shall be the sum of the measured stream temperature and the velocity recovery effect. Thus the total temperature is

$$t_{1t} = t_1 + \frac{1}{Jc_p} \frac{V^2}{2g} (1 - \alpha)$$

where:

c_p = Specific heat at constant pressure.
g = Acceleration due to gravity (32.17 fps^2 at sea level and 45 degrees latitude).
t_1 = Measured temperature, °F.
V = Velocity at temperature-measuring station, fps.
α = Recovery factor of temperature-measuring device.
J = Mechanical equivalent of heat = 778 ft-lb/Btu.

For temperature-measuring devices, such as bare thermometer, wells, or thermocouples installed perpendicular to the stream flow, the recovery factor a equals 0.6. For thermocouples installed in such a fashion that their junction points essentially upstream, the recovery factor α equals 1.0.

Discharge Temperature

For an exhauster assembled for test with an open exhaust (Figure 8.17), the discharge temperature shall be the total temperature as measured at not less than four stations symmetrically disposed around the discharge flange. For a compressor or exhauster assembled for test with an exit pipe attached, four discharge temperatures shall be measured approximately in the plane of the discharge static pressure stations and disposed at 45° to them, as shown in Figures 8.16, 8.18, or 8.19. Velocity corrections to the measured discharge temperatures shall be made as explained on page 684.

Temperature Measurements

Depending on the operating conditions or on convenience, the temperatures may be measured by certified thermometers or calibrated thermocouples inserted into the pipe or into wells. The installation of the temperature-measuring device directly into the pipe without the addition of a well is desirable for temperatures below 300°F.

Whichever means is employed, the temperature device shall be so chosen that it can be read to within an accuracy of 0.2 percent of the absolute temperature. The average of the four readings at each measuring station shall be taken as the temperature of the fluid. If discrepancies between the individual readings and the average are greater than 0.2 percent of the absolute temperature, they shall be investigated and eliminated.

Capacity

The ASME Code provides for the measurements of capacity by means of a long-radius low-ratio nozzle, page 658, located (A) in the discharge pipe of the compressor and discharging to the atmosphere, (B) in the intake to the compressor and discharging into the compressor from the atmosphere, or (C) in either the intake or discharge pipe in a closed system (Figure 8.22). For arrangement (A) or (B), the nozzle diameter shall be such that the drop in pressure across the nozzle will not be less than 10 in. of water or greater than 100 in. of water. For arrangement (C) the nozzle diameter shall be such that the Reynolds number (Part 5, Chapter 4, equation 6, Par. 23, ASME Power Test Codes) will not be less than 300,000. For information regarding test nozzles and nozzle coefficients, see pages 658-659.

Nozzle Pressure

The nozzle pressure designates the differential pressure Δp across the nozzle used in the flow formulas (see page 690-691). In arrangements A and B, nozzle pressure is measured directly by the differential manometer when located as indicated in Figure 8.22, and must be used in Eq. (8.13) or (8.14). In arrangement C, Δp is measured by a differential manometer located as shown in Figure 8.22 and is to be used in Eq. (8.13) for calculating flow.

Differential Pressure

The differential pressure across the nozzle and the temperature ahead of the nozzle shall be measured with duplicate instruments independently connected at the locations shown in Figure 8.22. For arrangements B and C, the pressure alter the nozzle shall be read by two independent manometers connected to the downstream pressure taps as shown in Figure 8.22. For arrangement B (Figure 8.22), the upstream pressure is equal to the barometric pressure, while the downstream pressure is the difference between the barometric pressure and the differential pressure. For arrangement A (Figure 8.22), the downstream pressure is equal to the barometric pressure, while the upstream pressure is equal to the sum of the barometric and the differential pressures. For arrangement C (Figure 8.22), the upstream pressure is equal to the sum of the downstream pressure and the differential pressure. When arrangement A or C is employed, upstream straightening vanes shall be installed. The flow being measured must be sensibly steady, and the manometers must not show pulsations greater than 2 percent of the differential pressure. Any greater pulsation in the flow is to be corrected at its source; attempts to reduce the pulsations by damping the correcting piping to the manometers are not permissible.

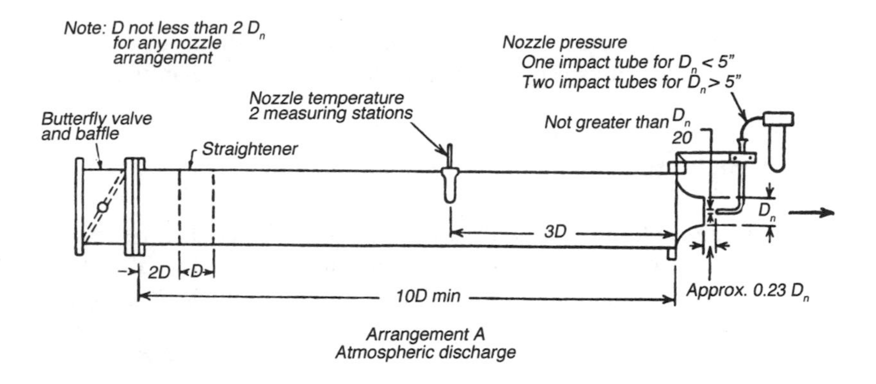

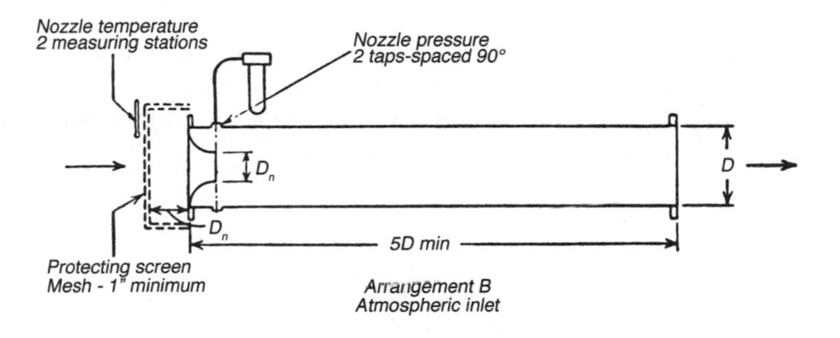

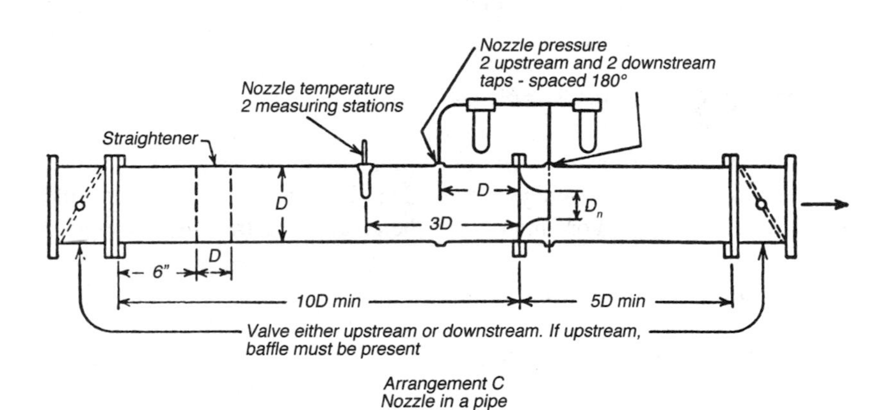

Figure 8.22 Various arrangements of flow nozzles for compressor tests.

Nozzle Temperature

The nozzle temperature shall be measured on the upstream side of the nozzle by instruments located as shown in Figure 8.22. No fewer than two instruments shall be used.

Cooling Water

When intercoolers are used, the flow of cooling water shall be measured by indicating-type meters accurate within 2 percent as shown by calibration under flow conditions corresponding to those obtaining during the test. The flow of cooling water through the intercooler may be adjusted to regulate the degree of intercooling.

Speed Measurement

An integrating revolution counter directly connected to a geared rotating shaft shall be used to record the total number of revolutions of the compressor during a test. The rate of speed shall be computed from the total number of revolutions during successive periods and the time of those periods.

Time Measurement

The date and time of day at which each individual test reading is taken and the time of day during which the test is conducted shall be recorded. The time of day may be determined by observation of timepieces by the individual observers, which timepieces have been compared with a master clock and are accurate to within 30 seconds per day.

Technique of Taking Readings

Readings should be taken as nearly simultaneously as possible and with sufficient frequency to ensure average results. This is important in obtaining the air-capacity data but is even more important when taking electrical power-input data.

Computation of Results

A complete presentation of the performance of a compressor must include a statement of the following significant quantities: capacity, pressure ratio, and power consumption. These quantities shall be stated for specific conditions of operation including pressure, inlet temperature, discharge pressure, rotative speed, and degrees of intercooling, including the temperature and quantity of the circulating water entering the intercoolers and the pressure drop across the intercoolers.

Before final calculations are undertaken, the observed data recorded during each test run shall be scrutinized. Readings at the beginning of any test run may be discarded provided the time interval covered by the acceptable data is not less than 30 minutes. If a sufficient number of consecutive test readings meeting the conditions on pages 679-682 do not cover the minimum time specified for the test, the test shall be repeated.

Velocity Pressure

When the velocity pressure is not more than 5 percent of the total pressure, the velocity shall be computed as the ratio of the quantity at the measuring section to the pipe area. The velocity pressure shall then be

$$\text{Velocity pressure, psi} = \frac{V_{av^2}\gamma}{2g \times 144} \quad (8.5)$$

When the velocity pressure (page 745) is greater than 5 percent of the total pressure, it shall be computed from a pitot-tube traverse in accordance with the following equations:

$$\text{Velocity pressure, psi} = \frac{\Sigma V_i^3 \gamma}{144 N 2 g V_{av}} \quad (8.6)$$

where:

V_i = Velocity at each traverse point (i) as determined by the pitot tube, fps.
Σ = Indicates that the sum is to be taken of the third powers of the velocity at each traverse point.
γ = Specific weight of the gas at the measuring section, lb/ft^3.
N = Number of traverse points (20).
g = Gravitational constant = 32.17 fps^2.
V_{av} = Average velocity of gas at the traverse section as determined by dividing the volume rate of flow at the section by the area of the section, fps.

Specific Weight

Computations of specific weight y shall be made from measured values of temperature, pressure, specific gravity, relative humidity, molecular weight, or chemical composition. Alternate formulas are given to facilitate the direct use of the measurable properties.

For any dry gas, where R is the gas constant and is defined as 1544/M:

For dry air:

$$\gamma = \frac{p_1 \times 144}{ZT_1R} \tag{8.7}$$

$$\gamma = \frac{p_1}{ZT_1} 2.70 \tag{8.8}$$

For any gas or air containing water vapor, where G is based on dry air:

$$\gamma = \frac{p_1}{ZT_1} G \times 2.70 \tag{8.8}$$

For any gas or air containing water vapor, where MW_o is the molecular weight of dry air:

$$\gamma = \frac{p_1 MW_1}{ZT_1 MW_o} 2.70 \tag{8.10}$$

For any gas or air containing water vapor, where RH is the relative humidity and Z= 1:

$$\gamma_m = \gamma_o \frac{1 - RHp_s}{p_1} + RH\gamma \tag{8.11}$$

For conversion of any known values of γ_x , to conditions p_1 and T_1, where Z = 1:

$$\gamma_m = \gamma_x \frac{p_1 T_x}{p_x T_1} \tag{8.12}$$

where:

Z	=	Supercompressibility factor as defined in the equation of state.
pv	=	ZRT; for a perfect gas and the common diatomic gases in the low-pressure range, Z is approximately 1; for air at pressures below 115 psia, Z shall be taken as 1.
R	=	Gas constant 1544/M, ft-lb/°F.
G	=	Specific gravity with respect to dry air.
MW	=	Molecular weight.
MW_o	=	Molecular weight of dry air = 28.96.
γ	=	Specific weight at point of measurement, lb/ft^3.
γ_o	=	Specific weight of dry gas at p_1 and T_1, lb/ift^3.
p_s	=	Saturation pressure of water vapor at T_1, psia.
γ_s	=	Specific weight of saturated water vapor at T_1, lb/ft^3.
RH	=	Relative humidity at measuring section.
p_1	=	Pressure at compressor inlet, psia.
T_1	=	Temperature at compressor inlet, °F abs.

For tests with air, relative humidity shall be determined from measurements of the wet- and dry-bulb temperatures and from the psychrometric tables (published by the Department of Agriculture, U.S. Weather Bureau). Table 8.6 gives values of specific gravity for moist air throughout the usual range of temperatures and degrees of saturation. The use of Table 8.6 shall be limited to a barometric pressure range of 28 to 30.5 in. Hg.

Table 8.6 Specific Gravity of Moist Air at Standard Sea-level Pressure

Temperature deg F	Relative Humidity, Per Cent									
	10	20	30	40	50	60	70	80	90	100
0	.9999	.9999	.9999	.9998	.9998	.9997	.9997	.9996	.9996	.9995
10	.9999	.9998	.9998	.9997	.9996	.9995	.9994	.9994	.9993	.9992
20	.9999	.9997	.9996	.9995	.9994	.9992	.9991	.9990	.9988	.9987
30	.9998	.9996	.9994	.9992	.9990	.9987	.9985	.9983	.9981	.9979
40	.9997	.9994	.9991	.9987	.9984	.9981	.9978	.9975	.9972	.9969
50	.9995	.9991	.9986	.9982	.9977	.9972	.9968	.9963	.9959	.9954
60	.9993	.9987	.9980	.9974	.9967	.9960	.9954	.9947	.9941	.9934
70	.9991	.9981	.9972	.9963	.9953	.9944	.9935	.9925	.9916	.9907
80	.9987	.9974	.9961	.9948	.9935	.9922	.9909	.9896	.9883	.9870
90	.9982	.9964	.9946	.9928	.9910	.9892	.9875	.9857	.9839	.9821
100	.9976	.9951	.9927	.9903	.9878	.9854	.9830	.9805	.9781	.9756
110	.9967	.9935	.9902	.9869	.9837	.9804	.9771	.9738	.9706	.9673
120	.9957	.9913	.9870	.9827	.9783	.9740	.9697	.9653	.9610	.9567
130	.9943	.9886	.9830	.9773	.9716	.9659	.9602	.9545	.9488	.9432

For gases other than air, where the chemical composition is likely to be variable, m shall be computed from values of G, measured directly by the gas balance or indirectly through chemical analysis.

Capacity

For tests with air, providing no condensation occurs and using the nozzle arrangement A or B of Figure 8.22 as described on pages 686-687, the capacity may be conveniently computed by the formula

$$Q_1 = \frac{36.0 C D_n^2 p_{2n} T_1 \sqrt{X(X+1)}}{p_1 \sqrt{T_n} \sqrt{G}} \tag{8.13}$$

For tests with gases (including air), providing no condensation occurs and using the nozzle arrangement C of Figure 8.22, the following formula shall be used [for nozzle arrangements A and B, either Eq. (8.13) or (8.14) may be used]:*

$$Q_1 \frac{31.5CD_n^2Y^1\sqrt{\gamma_n\Delta p}}{\gamma_1} \tag{8.14}$$

where:

Q_1 = Capacity, volume rate of flow at inlet conditions p_1 and T_1, cfm.
D_n = Nozzle-throat diameter, in.
C = Flow coefficient (Table 8.3).
T_n = Total temperature at upstream side of nozzle, °F abs.
D_1 = Diameter of nozzle pipe, in.
p_1 = Total pressure at compressor inlet, psia.
T_1 = Absolute temperature at compressor inlet, °F abs.
p_{2n} = Static pressure, downstream side of nozzle, psia.
p_{1n} = Total pressure, upstream side of nozzle, psia.
X = $(p_{1n}/p_{2n})^{(k-1)/k} - 1$; for standard air, $(k - 1)/k = 0.283$ (see Table 8.7).
r = p_{1n}/p_{2n}
γ_n = Specific weight of gas, upstream side of nozzle, lb/ft³.
γ_1 = Specific weight at compressor inlet, lb/ft³.
Δp = Differential pressure across nozzle $(p_{1n} - p_{2n})$, psi.

$$Y^1 = \left[\frac{k}{k-1}\left(\frac{p_{2n}}{p_{1n}}\right)^{2/k}\frac{1-p_{2n}/p_{1n}{}^{(k-1)/k}}{1-(p_{2n}/p_{1n})}\right]^{1/2} \div \left[1-\left(\frac{D_n}{D_1}\right)^4\left(\frac{p_{2n}}{p_{1n}}\right)^{2/k}\right]^{1/2}$$

*It shall be note that p_{1n}, when used in the Y^1 factors of Eq. (8.14), is static pressure and that the velocity of approach effect is accounted for by selecting values of Y^1 for the correct ratio of D_n/D_1, in Table 8.8.

However, when p_{1n} becomes total pressure, as in the case of nozzle arrangement A or B, the correct value of Y_1 is found under $D_n/D_1 = 0$.

To facilitate the use of Eq. (8.13), values of X have been computed for standard air, and are arranged in Table 8.7. In like manner, values of Y^1, for Eq. (8.14), are given in Table 8.8. Either of these tables may be used for air and gases in which the value of k lies between 1.39 and 1.40.

Values of the flow coefficient are given in Table 8.3. Selection of the values for C is made through the use of the curves of Figure 8.9, which serve to integrate the relation of nozzle pressure and nozzle temperature, and thereby avoid the necessity of computing Reynolds number.

When an intercooled compressor operates with moist air or gas, and the flow is measured on the discharge side of the compressor, a correction to the measured flow shall be made when any moisture is removed by the intercooler. This correction can be based on either the vapor pressure or the vapor density, and since the correction is small, any one of the generally used methods is acceptable (see page 671).

Theoretical Power for Compression

The theoretical power to compress the gas delivered by a compressor shall be computed for an isentropic compression. For a single-stage group and for diatomic gases, including air, where c_p is known, P_t shall be computed by:

$$P_t = \frac{wc_pT_{1t}}{42.42}\left[\left(\frac{p_2}{p_1}\right)^{(k-1)/k} - 1\right] \qquad (8.15)$$

When used for air or gases having a value of k between 1.39 and 1.40, Eq. (8.15) may be stated in terms of volume rate of flow and of X to permit the use of Table 8.7.

$$P_t = \frac{Q_1 p_1 X}{64.85} \qquad (8.16)$$

where:

c_p = Arithmetic average of specific heats at constant pressure between initial and final condition, Btu/lb/°F.

k = Arithmetic average of specific heat ration c_p/c_v between initial and final conditions.

p_1 = Inlet pressure of blower, psia.

p_2 = Discharge pressure of blower, psia.

Q_1 = Volume rate of flow at inlet conditions, cfm.

$$X = \left[\left(\frac{p_2}{p_1}\right)^{0.283} - 1\right] \quad \text{(values from Table 8.7)}$$

Table 8.7 Values of X for Standard Air and Perfect Diatomic Gases*

$$X = \left(\frac{p_1}{p_2}\right)^{0.283} - 1$$

r		0	1	2	3	4	5	6	7	8	9	Proportional Parts	
1.00	0.00	000	028	057	085	113	141	169	198	226	254	28	
1.01		282	310	338	366	394	422	450	478	506	534	1	2.8
1.02		562	590	618	646	673	701	729	757	785	812	2	5.6
1.03		840	868	895	923	951	978	006	034	061	089	3	8.4
1.04	0.01	116	144	171	199	226	253	281	308	336	363	4	11.2
1.05		390	418	445	472	500	527	554	581	608	636	5	14.0
1.06		663	690	717	744	771	798	825	852	879	906	6	16.8
1.07		933	960	987	014	041	068	095	122	148	175	7	19.6
1.08	0.02	202	229	255	282	309	336	362	389	416	442	8	22.4
1.09		469	495	522	549	575	602	628	655	681	708	9	25.2
1.10		734	760	787	813	840	866	892	919	945	971	27	
1.11		997	024	050	076	102	129	155	181	207	233	1	2.7
1.12	0.03	259	285	311	337	363	389	415	441	467	493	2	5.4
1.13		519	545	571	597	623	649	675	700	726	752	3	8.1
1.14		778	804	829	855	881	906	932	958	983	009	4	10.8
1.15	0.04	035	060	086	111	137	162	188	213	239	264	5	13.5
1.16		290	315	341	366	391	417	422	467	493	518	6	16.2
1.17		543	569	594	619	644	670	695	720	745	770	7	18.9
1.18		796	821	846	871	896	921	946	971	996	021	8	21.6
1.19	0.05	046	071	096	121	146	171	196	221	245	270	9	24.3
1.20		295	320	345	370	394	419	444	469	493	518	26	
1.21		543	567	592	617	641	666	691	715	740	764	1	2.6
1.22		789	813	838	862	887	911	936	960	985	009	2	5.2
1.23	0.06	034	058	082	107	131	155	180	204	228	253	3	7.8
1.24		277	301	325	350	374	398	422	446	470	495	4	10.4
1.25		519	543	567	591	615	639	663	687	711	735	5	13.0
1.26		759	783	807	831	855	879	903	927	951	974	6	15.6
1.27		998	022	046	070	094	117	141	165	189	212	7	18.2
1.28	0.07	236	260	283	307	331	354	378	402	425	449	8	20.8
1.29		472	496	520	543	567	590	614	637	661	684	9	23.4
1.30		708	731	754	778	801	825	848	871	895	918	25	
1.31		941	965	988	011	035	058	081	104	128	151	1	2.5
1.32	0.08	174	197	220	243	267	290	313	336	359	382	2	5.0
1.33		405	428	451	474	497	520	543	566	589	612	3	7.5
1.34		635	658	681	704	727	750	773	795	818	841	4	10.0
1.35		864	887	910	932	955	978	001	023	046	069	5	12.5
1.36	0.09	092	114	137	160	182	205	228	250	273	295	6	15.0
1.37		318	341	363	386	408	431	453	476	498	521	7	17.5
1.38		543	566	588	611	633	655	678	700	723	745	8	20.0
1.39		767	790	812	834	857	879	901	923	946	968	9	22.5
1.40		990	012	035	057	079	101	123	145	168	190	24	
1.41	0.10	212	234	256	278	300	322	344	366	389	411	1	2.4
1.42		433	455	477	499	521	542	564	586	608	630	2	4.8
1.43		652	674	696	718	740	761	783	805	827	849	3	7.2
1.44		871	892	914	936	958	979	001	023	045	066	4	9.6
1.45	0.11	088	110	131	153	175	196	218	239	261	283	5	12.0
1.46		304	326	347	369	390	412	433	455	476	498	6	14.4
1.47		520	541	562	584	605	627	648	669	691	712	7	16.8
1.48		734	755	776	798	819	840	862	883	904	925	8	19.2
1.49		947	968	989	010	032	053	074	095	116	138	9	21.6

Table 8.7 (continued)

r		0	1	2	3	4	5	6	7	8	9	Proportional Parts	
1.50	0.12	159	180	201	222	243	264	286	307	328	349		23
1.51		370	391	412	433	454	475	496	517	538	559	1	2.3
1.52		580	601	622	643	664	685	706	726	747	768	2	4.6
1.53		789	810	831	852	872	893	914	935	956	977	3	6.9
1.54		997	018	039	060	080	101	122	142	163	184	4	9.2
1.55	0.13	205	225	246	266	287	308	328	349	370	390	5	11.5
1.56		411	431	452	472	493	513	534	554	575	595	6	13.8
1.57		616	636	657	677	698	718	739	759	780	800	7	16.1
1.58		820	841	861	881	902	922	942	963	983	003	8	18.4
1.59	0.14	024	044	064	085	105	125	145	165	186	206	9	20.7
1.60		226	246	267	287	307	327	347	367	387	408		22
1.61		428	448	468	488	508	528	548	568	588	608	1	2.2
1.62		628	648	668	688	708	728	748	768	788	808	2	4.4
1.63		828	848	868	888	908	928	948	968	988	007	3	6.6
1.64	0.15	027	047	067	087	107	126	146	166	186	206	4	8.8
1.65		225	245	265	284	304	324	344	363	383	403	5	11.0
1.66		423	442	462	481	501	521	540	560	580	599	6	13.2
1.67		619	638	658	678	697	717	736	756	775	795	7	15.4
1.68		814	834	853	873	892	912	931	951	970	990	8	17.6
1.69	0.16	009	028	048	067	087	106	125	145	164	184	9	19.8
1.70		203	222	242	261	280	299	319	338	357	377		21
1.71		396	415	434	454	473	492	511	531	550	569	1	2.1
1.72		588	607	626	646	665	684	703	722	741	760	2	4.2
1.73		780	799	818	837	856	875	894	913	932	951	3	6.3
1.74		970	989	008	027	046	065	084	103	122	141	4	8.4
1.75	0.17	160	179	198	217	236	255	274	292	311	330	5	10.5
1.76		349	368	387	406	425	443	462	481	500	519	6	12.6
1.77		538	556	575	594	613	631	650	669	688	706	7	14.7
1.78		725	744	762	781	800	818	837	856	874	893	8	16.8
1.79		912	930	949	968	986	005	023	042	061	079	9	18.9
1.80	0.18	098	116	135	153	172	191	209	228	246	265		20
1.81		283	302	320	339	357	376	394	412	421	449	1	2.0
1.82		468	486	505	523	541	560	578	596	615	630	2	4.0
1.83		652	670	688	707	725	743	762	780	798	816	3	6.0
1.84		835	853	871	890	908	926	944	962	981	999	4	8.0
1.85	0.19	017	035	054	072	090	108	126	144	163	181	5	10.0
1.86		199	217	235	253	271	289	308	326	344	362	6	12.0
1.87		380	398	416	434	452	470	488	506	524	542	7	14.0
1.88		560	578	596	614	632	650	668	686	704	722	8	16.0
1.89		740	758	776	794	811	829	847	865	883	901	9	18.0
1.90		919	937	954	972	990	008	026	044	061	079		19
1.91		097	115	133	150	168	186	204	221	239	257	1	1.9
1.92		275	292	310	328	345	363	381	399	416	434	2	3.8
1.93		452	469	487	504	522	540	557	575	593	610	3	5.7
1.94		628	645	663	681	698	716	733	751	768	786	4	7.6
1.95		804	821	839	856	874	891	909	926	944	961	5	9.5
1.96		979	996	013	031	048	066	083	101	118	135	6	11.4
1.97	0.20	153	170	188	205	222	240	257	275	292	309	7	13.3
1.98		327	344	361	079	396	413	431	448	465	482	8	15.2
1.99		500	517	534	552	569	586	603	620	638	655	9	17.1
2.00	0.21	672	689	707	724	741	758	775	792	810	827		
2.01		844	861	878	895	913	930	947	964	981	998		
2.02	0.22	015	032	049	066	084	101	118	135	152	169		18
2.03		186	203	220	237	254	271	288	305	322	339	1	1.8
2.04		356	373	390	407	424	441	458	474	491	508	2	3.6
2.05		525	542	559	576	593	610	627	644	660	677	3	5.4
2.06		694	711	728	745	762	778	795	812	829	846	4	7.2
2.07		863	879	896	913	930	946	953	980	997	013	5	9.0

Table 8.7 (continued)

r		0	1	2	3	4	5	6	7	8	9	Proportional Parts	
2.08	0.23	030	047	064	080	097	114	130	147	164	181	6	10.8
2.09		197	214	231	247	264	281	297	314	331	347	7	12.6
												8	14.4
2.10		364	380	397	414	430	447	463	480	497	513	9	16.2
2.11		530	546	563	579	596	613	629	646	662	679		
2.12		695	712	728	745	761	778	794	811	827	844		
2.13		860	877	893	909	926	942	959	975	992	008		
2.14	0.24	024	041	057	074	090	106	123	139	155	172	17	
2.15		188	204	221	237	253	270	286	302	319	335	1	1.7
2.16		351	368	384	400	416	433	449	465	481	498	2	3.4
2.17		514	530	546	563	579	595	511	627	644	660	3	5.1
2.18		676	692	708	724	741	757	773	789	805	821	4	6.8
2.19		838	854	870	886	902	918	934	950	966	983	5	8.5
												6	10.2
2.20		999	015	031	047	063	079	095	111	127	143	7	11.9
2.21	0.25	159	175	191	207	223	239	255	271	287	303	8	13.6
2.22		319	335	351	367	383	399	415	431	447	463	9	15.3
2.23		479	495	511	526	542	558	574	590	606	622		
2.24		638	654	669	685	701	717	733	749	765	780		
2.25		796	812	828	844	859	875	891	907	923	938		
2.26		954	970	986	001	017	033	049	064	080	096	16	
2.27	0.26	112	127	143	159	175	190	206	222	237	253	1	1.6
2.28		269	284	300	316	331	347	363	378	394	409	2	3.2
2.29		425	441	456	472	488	503	519	534	550	566	3	4.8
												4	6.4
2.30		581	597	612	628	643	659	675	690	706	721	5	8.0
2.31		737	752	768	783	799	814	830	845	861	876	6	9.6
2.32		892	907	923	938	954	969	984	000	015	031	7	11.2
2.33	0.27	046	062	077	092	108	123	139	154	169	185	8	12.8
2.34		200	216	231	246	262	277	292	308	323	338	9	14.4
2.35		354	369	284	400	415	430	446	461	476	492		
2.36		507	522	538	553	568	583	599	614	629	644		
2.37		660	675	690	705	721	736	751	766	781	797		
2.38		812	827	842	857	873	888	903	918	933	948	15	
2.39		964	979	994	009	024	039	054	070	085	100	1	1.5
												2	3.0
2.40	0.28	115	130	145	160	175	190	205	220	236	251	3	4.5
2.41		266	281	296	311	326	341	356	371	386	401	4	6.0
2.42		416	431	446	461	476	491	506	521	536	551	5	7.5
2.43		566	581	596	611	626	641	656	671	686	701	6	9.0
2.44		716	730	745	760	775	790	805	820	835	850	7	10.5
2.45		865	879	894	909	924	939	954	969	984	993	8	12.0
2.46	0.29	013	028	043	058	073	087	102	117	132	147	9	13.0
2.47		162	176	191	206	221	235	250	265	280	295		
2.48		309	324	339	353	368	383	398	412	427	442		
2.49		457	471	486	501	515	530	545	559	574	589		
2.50	0.29	604	618	633	647	662	677	691	706	721	735		
2.51		750	765	779	794	808	823	838	852	867	881		
2.52		896	911	925	940	954	969	984	998	013	027		
2.53	0.30	042	056	071	085	100	114	129	144	158	173		
2.54		187	202	216	231	245	260	274	289	303	318		
2.55		332	346	361	375	390	404	419	433	448	462		
2.56		476	491	505	520	534	548	563	577	592	606		
2.57		620	635	649	663	678	692	707	721	735	750		
2.58		764	778	793	807	821	836	850	864	879	893		
2.59		907	921	936	950	964	979	993	007	021	036	14	
												1	1.4
2.60	0.31	050	064	079	093	107	121	136	150	164	178	2	2.8

Table 8.7 (continued)

r		0	1	2	3	4	5	6	7	8	9	Proportional Parts	
2.61		193	207	221	235	249	264	278	292	306	320	3	4.2
2.62		335	349	363	377	391	405	420	434	448	462	4	5.6
2.63		476	490	505	519	533	547	561	575	589	603	5	7.0
2.64		618	632	646	660	674	688	702	716	730	744	6	8.4
2.65		759	773	787	801	815	829	843	857	871	855	7	9.5
2.66		899	913	927	941	955	969	983	997	011	025	8	11.2
2.67	0.32	039	053	067	081	095	109	123	137	151	165	9	12.6
2.68		179	193	207	221	235	249	262	278	290	304		
2.69		318	332	346	360	374	388	402	416	429	443		
2.70		457	471	485	499	513	527	540	554	568	582	13	
2.71		596	610	624	637	651	665	679	693	707	720	1	1.3
2.72		734	748	762	776	789	803	817	831	845	858	2	2.6
2.73		872	886	900	913	927	941	955	968	982	996	3	3.9
2.74	.033	010	023	037	051	065	078	092	106	119	133	4	5.2
2.75		147	161	174	188	202	215	229	243	256	270	5	6.5
2.76		284	297	311	325	338	352	366	379	393	407	6	7.8
2.77		420	434	448	461	475	488	502	516	529	543	7	8.4
2.78		556	570	584	597	611	624	638	651	665	679	8	10.4
2.79		692	706	719	733	746	760	773	787	801	814	9	11.7
2.80		828	841	855	868	882	895	909	922	936	949		
2.81		963	976	990	003	017	030	044	057	070	084		
2.82	0.34	097	111	124	138	151	165	178	191	205	218	12	
2.83		232	245	259	272	285	299	312	326	339	352	1	1.2
2.84		366	379	393	406	419	433	446	459	473	486	2	2.4
2.85		500	513	526	540	553	566	580	593	606	620	3	3.6
2.86		633	645	660	673	686	700	713	726	739	753	4	4.8
2.87		766	779	793	806	819	832	846	859	872	886	5	6.0
2.88		899	912	925	939	952	965	978	991	005	018	6	7.2
2.89	0.35	031	044	058	071	084	097	110	124	137	150	7	8.4
												8	9.6
2.90		163	176	190	203	216	229	242	255	269	282	9	16.8
2.91		295	308	321	334	347	361	374	387	400	413		
2.92		426	439	452	466	479	492	505	518	531	544		
2.93		557	570	584	597	610	623	636	649	662	675		
2.94		688	701	714	727	740	753	767	780	793	806		
2.95		819	832	845	858	871	884	897	910	923	936		
2.96		949	962	975	988	001	014	027	040	053	066		
2.97	0.36	079	092	105	118	131	144	157	169	182	195		
2.98		208	221	234	247	260	273	286	299	312	324		
2.99		337	350	363	376	389	402	415	428	440	453		

r	0	1	2	3	4	5	6	7	8	9
3.0	0.3647	0.3659	0.3672	0.3685	0.3698	0.3711	0.3723	0.3736	0.3749	0.3761
3.1	0.3774	0.3786	0.3799	0.3811	0.3824	0.3836	0.3849	0.3861	0.3874	0.3886
3.2	0.3898	0.3911	0.3923	0.3935	0.3947	0.3959	0.3971	0.3984	0.3996	0.4008
3.3	0.4020	0.4032	0.4044	0.4056	0.4068	0.4080	0.4091	0.4103	0.4115	0.4127
3.4	0.4139	0.4150	0.4162	0.4174	0.4186	0.4197	0.4209	0.4220	0.4232	0.4244
3.5	0.4255	0.4267	0.4278	0.4290	0.4301	0.4313	0.4324	0.4335	0.4347	0.4358
3.6	0.4369	0.4380	0.4392	0.4403	0.4414	0.4425	0.4437	0.4448	0.4459	0.4470
3.7	0.4481	0.4492	0.4503	0.4514	0.4525	0.4536	0.4547	0.4558	0.4569	0.4580
3.8	0.4591	0.4602	0.4612	0.4623	0.4634	0.4645	0.4656	0.4666	0.4677	0.4688
3.9	0.4698	0.4709	0.4720	0.4730	0.4741	0.4752	0.4762	0.4773	0.4783	0.4794
4.0	0.4804	0.4815	0.4825	0.4835	0.4846	0.4856	0.4867	0.4877	0.4887	0.4898

For nozzles, $r = p_1/p_2$. For compressors and exhausters, $r = p_2 p_1$.

*Taken torn Moss and Smith, Engineering Computations for Air and Gases, Trans. ASME, vol.52, Paper APM-52-8.

For a multi-stage group with intercooling between N stage groups, and with the same restrictions specified for Eq. (8.15), the theoretical power for compression is:

$$P_t = \frac{Nwc_pT_{1t}}{42.42}\left[\left(\frac{p_2}{p_1}\right)^{(k-1)/Nk} - 1\right] \tag{8.17}$$

Table 8.8 Values for Y'

	D2/D1														
	k = 1.40					k = 1.35					k = 1.30				
P_2/P_1	0	0.2	0.3	0.4	0.5	0	0.2	0.3	0.4	0.5	0	0.2	0.3	0.4	0.5
1.00	1.000	1.001	1.004	1.013	1.033	1.000	1.001	1.004	1.013	1.033	1.000	1.001	1.004	1.013	1.033
0.99	0.995	0.995	0.999	1.007	1.027	0.994	0.995	0.999	1.007	1.027	0.994	0.995	0.998	1.007	1.026
0.98	0.989	0.990	0.993	1.002	1.021	0.989	0.990	0.993	1.001	1.020	0.988	0.989	0.992	1.001	1.020
0.97	0.984	0.985	0.988	0.996	1.015	0.983	0.984	0.987	0.995	1.014	0.983	0.983	0.986	0.995	1.013
0.96	0.978	0.979	0.982	0.990	1.009	0.978	0.978	0.981	0.990	1.008	0.977	0.977	0.980	0.989	1.007
0.95	0.973	0.974	0.977	0.985	1.002	0.972	0.973	0.976	0.984	1.001	0.971	0.972	0.974	0.982	1.000
0.94	0.967	0.968	0.971	0.979	0.996	0.966	0.967	0.970	0.978	0.995	0.965	0.966	0.968	0.976	0.993
0.93	0.962	0.963	0.965	0.973	0.990	0.961	0.961	0.964	0.972	0.989	0.959	0.960	0.962	0.970	0.987
0.92	0.956	0.957	0.960	0.967	0.984	0.955	0.955	0.958	0.966	0.982	0.953	0.954	0.956	0.964	0.980
0.91	0.951	0.951	0.954	0.961	0.978	0.949	0.950	0.952	0.960	0.976	0.947	0.948	0.950	0.957	0.973
0.90	0.945	0.946	0.948	0.956	0.971	0.943	0.944	0.946	0.953	0.969	0.941	0.942	0.944	0.951	0.966
0.89	0.939	0.940	0.943	0.950	0.965	0.937	0.938	0.940	0.947	0.963	0.935	0.935	0.938	0.945	0.959
0.88	0.934	0.934	0.937	0.944	0.959	0.931	0.932	0.934	0.941	0.956	0.929	0.929	0.932	0.938	0.953
0.87	0.928	0.928	0.931	0.938	0.953	0.925	0.926	0.926	0.935	0.950	0.922	0.923	0.926	0.932	0.946
0.86	0.922	0.923	0.925	0.932	0.946	0.919	0.920	0.922	0.929	0.943	0.916	0.917	0.919	0.926	0.939
0.85	0.916	0.917	0.919	0.926	0.940	0.913	0.914	0.916	0.923	0.936	0.910	0.911	0.913	0.923	0.932
0.84	0.910	0.911	0.913	0.920	0.933	0.907	0.908	0.910	0.916	0.930	0.904	0.904	0.907	0.919	0.925
0.83	0.904	0.905	0.907	0.913	0.927	0.901	0.902	0.904	0.910	0.923	0.897	0.898	0.900	0.916	0.918
0.82	0.898	0.899	0.901	0.907	0.920	0.895	0.895	0.898	0.904	0.917	0.891	0.891	0.894	0.900	0.911
0.81	0.892	0.893	0.895	0.901	0.914	0.889	0.889	0.891	0.897	0.910	0.885	0.885	0.887	0.893	0.904
0.80	0.886	0.887	0.889	0.895	0.907	0.883	0.883	0.885	0.891	0.903	0.878	0.879	0.880	0.886	0.897
0.79	0.880	0.881	0.883	0.889	0.901	0.876	0.877	0.879	0.884	0.896	0.872	0.872	0.874	0.880	0.890
0.78	0.874	0.875	0.877	0.882	0.894	0.870	0.870	0.872	0.878	0.889	0.865	0.865	0.868	0.873	0.883
0.77	0.868	0.869	0.871	0.876	0.887	0.864	0.864	0.866	0.871	0.882	0.859	0.859	0.861	0.866	0.876
0.76	0.862	0.862	0.864	0.869	0.881	0.857	0.858	0.859	0.865	0.876	0.852	0.852	0.854	0.859	0.869
0.75	0.856	0.856	0.858	0.863	0.874	0.851	0.851	0.853	0.858	0.869	0.845	0.846	0.848	0.852	0.862
0.74	0.849	0.850	0.852	0.857	0.867	1.844	0.845	0.846	0.851	0.862	0.839	0.839	0.841	0.845	08.55
0.73	0.843	0.844	0.845	0.850	0.860	0.838	0.838	0.840	0.845	0.855	0.832	0.832	0.834	0.838	0.848
0.72	0.837	0.837	0.839	0.844	0.854	0.831	0.831	0.833	0.838	0.848	0.825	0.825	0.827	0.831	0.841
0.71	0.830	0.831	0.832	0.837	0.847	0.825	0.825	0.827	0.831	0.840	0.818	0.819	0.820	0.824	0.834
0.70	0.824	0.824	0.826	0.830	0.840	0.818	0.818	0.820	0.824	0.833	0.811	0.812	0.813	0.817	0.826

If the velocity of approach is zero (as with a nozzle taking in air from the outside), D_1 is infinite, and D_2/D_1 is zero.

Shaft Power

Where compressors are driven by electric motors, shaft horsepower may be computed from measured values of the electrical input. For induction motors of the squirrel-cage type, the shaft horsepower shall be:

$$P_s = \frac{\text{electrical input x efficiency}}{0.746}$$

in which the efficiency has been determined by test.

For synchronous motors, shaft horsepower shall be computed as:

$$P_s = \frac{\text{electrical input - sum of the losses in kW}}{0.746}$$

where the losses are based on the prevailing voltage, armature current, and field current. The losses shall he established by test measurements of armature resistance, open-circuit core losses, short-circuit core losses, and the friction and windage losses. The complete loss shall be the sum of I^2R + core loss + load loss + excitation + friction and windage. (See PTC 10 for the definition of electrical input.)

The shaft horsepower output of a wound-rotor type of induction motor may be calculated in the same manner as outlined for the squirrel-cage type when the secondary winding is short-circuited and where measured efficiency data for this condition of operation are available. This Code does not provide for shaft horsepower determination with a wound-rotor type of motor when it is operated with external resistance in the secondary circuit.

Computations of shaft horsepower by the motor input method shall not be acceptable in the case of the induction-type motor when the output is less than one-half of the motor rating.

Compressor Efficiency

The compressor efficiency shall be computed as the ratio of the theoretical power to the shaft power.

$$\eta = \frac{P_t}{P_s}$$

Adjustment of Results to Specified Conditions

Tests that are made in accordance with this Code and that have deviations between test and specified conditions within the limits prescribed in column (2), Table 8.5, may be adjusted to the operating conditions specified by the following equations.

Adjustment of Capacity

When the speed at test conditions deviates from the specified speed by not more than 5 percent, capacity shall he adjusted by the equation

$$Q_{mc} = Q_m \frac{N_c}{N_m}$$

where:

N = speed, rpm
Q = capacity, cfm
subscript m refers to measured quantity
subscript c refers to specified quantity
subscript mc refers to adjusted value

Adjustment of Pressure Ratio

The adjustment of pressure ratio shall be in accordance with the relation

$$\left(r_{mc}^{(k_c-1)/k_c} - 1\right) = \left(\frac{N_c}{N_m}\right)^2 \frac{MW_c k_m (k_c - 1) T_{1m}}{MW_m k_c (k_m - 1) T_{1c}} \left(r_{mc}^{(k_m-1)/k_m} - 1\right)$$

For tests with air or with other gases in which the values of k and MW are the same for both test and contract conditions, and in which the value of k lies between 1.39 and 1.40, the following relation may be used to simplify the computations:

$$X_{mc} = X_m \left(\frac{N_c}{N_m}\right)^2 \frac{T_{1m} G_c}{T_{1c} G_m} \tag{8.18}$$

where:

X = $r^{0.283} - 1$ (see Table 8.7)
r = p_2/p_1
T_1 = inlet temperature, °F abs
MW = molecular weight
G = specific gravity

Having found the value of X_{mc} by Eq. (8.18), the corresponding value of r_{mc} is found from Table 8.7. The adjusted pressure is:

$$p_{2mc} = r_{mc} p_{1c}$$

Adjustment of Power

The shaft horsepower shall be adjusted by the equation:

$$P_{mc} = P_m \left(\frac{N_c}{N_m} \right)^2 \frac{T_{1m} p_{1c} M_c}{T_{1c} p_{1m} M_m}$$

For tests with air in which values of X have already been determined, the computations may be facilitated by the equivalent equation:

$$P_{mc} = P_m \frac{p_{1c} N_c X_{mc}}{p_{1m} N_m X_m}$$

When the compressor is driven by an electric motor, the adjusted kilowatt input shall be 0.746 P_{mc}/e, where e is the measured motor efficiency at the power output of P_{mc}. If the motor is of the induction type, the speed value of N_c used in the pressure and power correction formula shall be the actual speed at the power output P_{mc} as determined by plotting slip against kilowatt input.

If the compressor is driven by a steam turbine, the steam consumption shall be corrected in accordance with the Power Test Code for Steam Turbines. This Code prescribes limited correction for deviation in initial steam pressure, superheat, exhaust pressure, and load. In view of their complexity, the values for these corrections are preferably included as a part of the contract.

Examples

A complete presentation of observed data, computations, and adjustments is illustrated by Examples 8.3 and 8.4 for air compressors. In each case the test consists of two points which bracket the "guarantee" or specified point within the limits of capacity and speed given in Table 8.5. The adjusted results are compared with the specified pressure and power in the form of curves (Figures 8.23 and 8.24).

Example 8.3 illustrates the test of a multi-stage compressor driven by a direct-connected steam turbine of the straight condensing type. The test setup is arranged as shown in Figure 8.16, without pipe on the inlet, so that the inlet total pressure is measured by the barometer. The air output is measured by the nozzle setup in accordance with arrangement A of Figure 8.22. Steam is condensed and measured either by a pair of weigh tanks or by calibrated volume tanks. In accordance with the requirements of the Power Test Code for Steam Turbines, the correction values for deviation in operating conditions were established by an agreement in which the correction for superheat, initial steam pressure, and exhaust pressure was to be based on the ratio of available enthalpy, and in which the steam flow was to be directly proportioned to the load.

Table 8.9 shows the capacities, pressures, speeds, and steam flows for the two test points, "as run" and "after adjustment." The last column shows corresponding values for the specified point. The exact values of discharge pressure and steam flow to be compared with the guarantee are given in the curves of Figure 8.23.

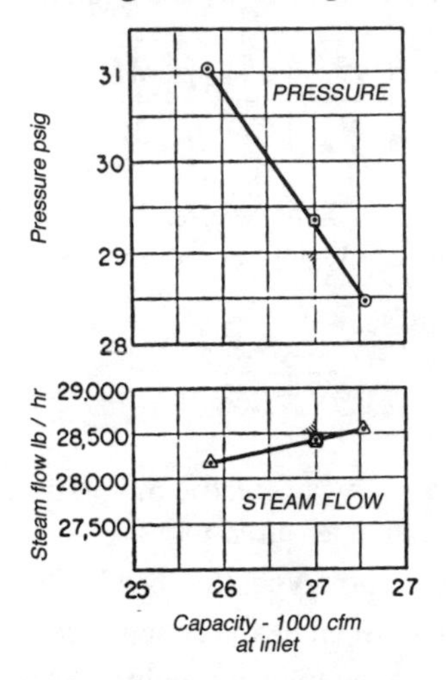

Figure 8.23 Values of discharge pressure and steam flow.

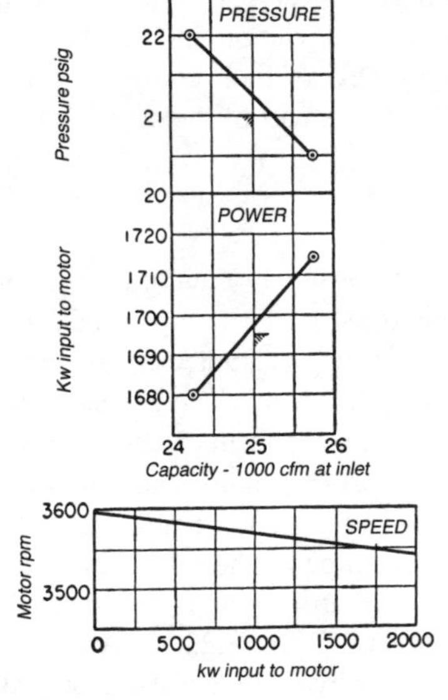

Figure 8.24 Corrected test values of pressure and power.

Example 8.3

Table 8.9 Summary of Computed Results

	Unit	As run (1)	As run (2)	Adjusted to specified conditions (1)	Adjusted to specified conditions (2)	Specified
Speed	rpm	4,763	4,746	4,700	4,700	4,700
Barometer	psia	14.50	14.47	14.7	14.7	14.7
Gas composition	....	Air	Air	Air	Air	Air
Specific Gravity of gas (dry air = 1)	....	0.9855	0.9853	0.9962	0.9962	0.9962
Inlet conditions:						
Pressure	psia	14.50	14.47	14.7	14.7	14.7
Temperature	deg F	90.0	89.4	70	70	70
Relative humidity	per cent	80.0	82	40	40	40
Capacity	cfm	26,340	27,970	25,990	27,700	27,000
Discharge pressure	psig	29,705	26,936	31.05	28.40	29.00
Turbine steam conditions:						
Steam pressure of throttle	psig	398.3	401.9	400	400	400
Total steam temperature	deg F	556.2	557.4	548	548	548
Exhaust pressure, lb	In. Hg	3.259	2.986	3.0	3.0	3.0
Total steam flow, including seal leakage	lb per hr	27,726	27,348	28,180	28,470	28,500

Average of Observed Readings

	Unit	Symbol	Test 1	Test 2
Speed	rpm	N_m	4,763	4,746
Inlet temperature of compressor	deg F	t_{1m}	90.0	89.4
Barometric pressure, corrected to 32F	in. Hg	....	29.52	29.461
Barometric pressure	psia	....	14.50	14.470
Pressure at compressor inlet flange	psia	p_{1m}	14.500	14.470
Psychrometer reading at inlet:				
Dry bulb	deg F	....	89.6	89.2
Wet bulb	deg F	....	84.3	84.5
Static pressure at compressor discharge	in. Hg	....	60.50	54.779
Temperature at compressor discharge	deg F	t_{2m}	357	344
Air measuring nozzle differential pressure measured by impact tube	in. H_20	....	26.22	29.34
Nozzle upstream temperature	deg F	t_n	221.3	215.8
Nozzle-throat diameter	in.	D_n	16.000	16.000
Room temperature at gage board	deg F	....	88.7	87.9
Compressor-discharge (circular section) diameter	in.	....	20	20
Compressor-inlet (circular section) diameter	in.	....	30	30
Steam pressure at turbine throttle valve, corrected for gage error and head of water in tubing	psig	....	398.3	401.9
Steam temperature at turbine throttle valve	deg F	....	556.2	557.4
Exhaust steam temperature	deg F	....	118.0	114.8
Vacuum at turbine exhaust flange	in. Hg	....	26.409	26.617
Vacuum corrected to 32F	in. Hg	....	26.264	26.475
Absolute pressure at turbine exhaust flange	in. Hg	....	3.259	2.986
Steam flow, condensate measurement	lb per hr	....	27,463	27,089
Turbine steam seal leakage	lb per hr	....	264	259

Table 8.9 continued

Calculation of Results	Unit	Symbol	Test 1	Test 2
Capacity (by equation 8.13)				
Density of water at room temperature	lb/cu in	....	0.035952	0.035958
Differential pressure across nozzle	psig	Δ_p	0.9425	1.0549
Pressure ratio across nozzle [(barometer + diff. press.) + barometer], p_{nl}/p_{n2}	....	r_m	1.0650	1.0729
X value from Table 8.7	....	....	0.01798	0.02011
Nozzle diameter squared	sq in.	D_n^2	256.000	256.000
Nozzle coefficient of discharge	....	C	0.995	0.995
Nozzle upstream temperature	deg F abs	T_{nm}	681.00	675.5
Temperature to which flow is referred	deg F abs	T_{lm}	549.7	549.1
Pressure to which flow is referred	psig	p_{lm}	14.500	14.470
$\text{Cap.} = \dfrac{36.0\text{x}0\times 0.995\times 256\times T_{lm}\sqrt{X(X+1)}}{\sqrt{T_n G}}$	cfm	Q_m	26,340	27,970
Inlet specific gravity:				
Relative humidity from psychrometer reading and psychrometric tables	per cent	....	80	82
Inlet temperature	deg F	t_{lm}	90	89.4
Specific gravity at inlet conditions from Table 8.6	....	G_{lm}	0.9855	0.9853
Total discharge pressure:				
Static pressure corrected to 32F	in. Hg	....	60.175	54.486
Static pressure (in Hg at 32F x 0.49115)	psig	....	29.555	26.716
Velocity-pressure calculations:				
Specific weight of air in discharge pipe:				
$(1)\ 0.0717\times\dfrac{29.555+14.500}{14.500}\times\dfrac{459.6+90.0}{459.6+357}$	lb per cu ft	....	0.1434	
$(2)\ 0.07009\times\dfrac{26.716+14.470}{14.47}\times\dfrac{459.6+89.4}{459.6+344}$	lb per cu ft	....	....	0.1368
Approximate flow rate in discharge pipe:				
$(1)\ \dfrac{26{,}340}{60}\times\dfrac{14.50}{44.1}\times\dfrac{816.7}{549.7}$	cfs	....	214.5	
$(2)\ \dfrac{27{,}810}{60}\times\dfrac{14.47}{41.3}\times\dfrac{803.7}{549.1}$	cfs	....	....	238.8
Area of discharge pipe	sq ft	....	2.182	2.182
Velocity in discharge pipe = cfs + area	fps	V_2	98.30	109.4
Velocity pressure:				
$(1)\ 0.1434\times\dfrac{(98.30)^2}{2\times 32.174}\times\dfrac{1}{144}$	psi	....	0.150	
$(2)\ 0.1363\times\dfrac{(109.4)^2}{2\times 32.174}\times\dfrac{1}{144}$	psi	....	....	0.176
Total discharge pressure	psig	....	29.703	26.892

Table 8.9 continued

	Unit	Symbol	Test 1	Test 2
Capacity adjustment:				
(1) $26{,}340 \times \frac{4{,}700}{4{,}763}$	cfm	Q_{mc}	25,990	
(2) $27{,}970 \times \frac{4{,}700}{4{,}746}$	cfm	Q_{mc}	....	27,700
Steam-flow adjustment:				
Enthalpy of steam at turbine throttle, as run	Btu per lb	....	1279.7	1280.0
Enthalpy of exhaust steam by isentropic expansion to exhaust pressure	Btu per lb	....	894.1	889.4
Enthalpy of steam at turbine throttle, at specified conditions	Btu per lb	....	1274.4	1274.4
Enthalpy of exhaust steam by isentropic expansion to specified exhaust pressure	Btu per lb	....	886.6	886.6
Theoretical enthalpy drop, as measured	Btu per lb	....	385.6	390.7
Theoretical enthalpy drop at specified conditions	Btu per lb	....	387.8	387.8
Steam flow adjusted to specified compressor intake conditions, speed and turbine steam conditions:				
(1) $27{,}462 \times \frac{14.7}{14.5} \times \frac{0.37886}{0.3709} \times \frac{4{,}700}{4{,}763} \times \frac{385.6}{387.8}$	lb per hr	....	27.917	
(2) $27{,}089 \times \frac{14.7}{14.47} \times \frac{0.35581}{0.34654} \times \frac{4{,}700}{4{,}746} \times \frac{390.7}{387.8}$	lb per hr	....	....	28,206
Turbine steam seal leakage	lb per hr	....	264	259
Total steam flow, adjusted to specified conditions	lb per hr	....	28,180	28,470

Adjustment of Test Results to Specified Inlet Conditions

	Unit	Symbol	Test 1	Test 2
Pressure adjustment:				
Pressure ratio as run:				
(1) (29.705 + 14.500)/14.500	....	r_m	3.0486	
(2) (26.892 + 14.470)/14.470	....	r_m	....	2.8585
X_m value from Table 8.7	....	X_m	0.3709	0.34613
X_{mc}:				
(1) $= 0.3709 \left(\frac{549.7}{529.7}\right)\left(\frac{0.9962}{0.9855}\right)\left(\frac{4{,}700}{4{,}763}\right)^2$	....	X_{mc}	0.37886	
(2) $= 0.34613 \left(\frac{549.1}{529.7}\right)\left(\frac{0.9962}{0.9853}\right)\left(\frac{4{,}700}{4{,}746}\right)^2$	....	X_{mc}	....	0.35581
Pressure ratio for X_{mc} from Table 8.7	....	r_{mc}	3.112	2.9318
Adjusted discharge pressure = $r_{mc} \times 14.696$	psia	P_{2mc}	45.746	43.097
Adjusted discharge pressure	psig	....	31.05	28.40

Example 8.4 illustrates the test of a multi-stage air compressor driven by a direct-connected induction motor. The test setup is given in Figure 8.18 except for the intake pipe. In this case no pipe is used, and the inlet total pressure is measured by the barometer.

The nozzle setup is in accordance with arrangement A of Figure 8.22. Motor output is based on the measured input and efficiency curves. The two test points, which bracket the specified capacity within the limits of Table 8.5, are corrected to the actual motor speed and not the specified speed, which is usually a nominal figure or that stamped on the motor name plate. The actual speed-load curve is established from the test measurements. The corrected test values of pressure and power to be compared with the specified values are shown in the curves of Figure 8.24.

Testing with Substitute Gases

The foregoing discussion of testing centrifugal compressors assumes that the test gas will be the same as the gas for which the machine is designed, or at least differ only to a minor degree. In factory tests, however, such duplication of the design gas is frequently impractical. State or local safety laws, insurance limitations or labor agreements, for example, often prevent the introduction of combustible or toxic gases onto the test floor of machinery manufacturers.

In such circumstances, tests must be conducted with substitute gases. Fortunately, the procedures for correlating the performance so obtained are well known, and quite accurate tests can be conducted despite this handicap. The essential step is to establish a test speed that will recreate dynamic similarity between test and design gases, within the compressor.

The test speed on the substitute gas is generally referred to as "equivalent speed" – it produces equivalent aerodynamics in the machines. Two factors must be considered in selecting this speed: (1) Mach number and (2) density ratio.

The equivalent test speed for a substitute gas at which the design Mach numbers obtain is determined from the ratio of the sonic velocity in the substitute gas to that in the design gas. It is usually sufficient to compute this ratio only at inlet conditions. The relation is:

$$N_{eq} = N_d \frac{\sqrt{(kgZRT_1)_s}}{\sqrt{(kgZRT_1)_d}}$$

The density ratio is the ratio of gas density at a given point in the compression to the density at inlet conditions. If equivalence of the density ratio (test versus design) is maintained throughout the compression cycle, the gas volume will be consistent with the intended use of the machine.

Example 8.4

Table 8.10 Summary of Computed Results

	Unit	As run (1)	As run (2)	Adjusted to specified conditions (1)	Adjusted to specified conditions (2)	Specified
Speed	rpm	3,554	3,551	3,551	3,549	3,550
Barometer	psia	14.580	14.620	14.7	14.7	14.7
Gas composition	....	Air	Air	Air	Air	Air
Specific Gravity of gas (dry air = 1)	....	0.9878	0.9881	0.9962	0.9962	0.9962
Inlet conditions:						
Pressure	psia	14.580	14.620	14.7	14.7	14.7
Temperature	deg F	87.4	88.5	70	70	70
Relative humidity	per cent	73	69	40	40	40
Capacity	cfm	24,363	25,869	24,340	25,850	25,000
Discharge pressure	psig	20.695	19.303	21.99	20.50	21.00
Electrical conditions:						
Line voltage	volts	2,215	2,208	....	....	2,200
Line current	amp	445.2	454.9			
Power input to motor	kw	1,605.5	1,638.7	1,680	1,715	1,695

Averages of Observed Readings

	Unit	Symbol	Test 1	Test 2
Speed	rpm	N_m	3,554	3,551
Inlet temperature of compressor	deg F	t_{1m}	87.4	88.5
Barometric pressure, corrected to 32F	in. Hg	....	29.685	29.767
Barometric pressure	psia	....	14.580	14.620
Inlet Pressure at compressor inlet flange	psia	p_{1m}	14.580	14.620
Psychrometer reading at inlet:				
Dry bulb	deg F	....	87.2	87.9
Wet bulb	deg F	....	80.0	79.4
Static pressure at compressor discharge	in. Hg	....	42.049	39.150
Temperature at compressor discharge	deg F	t_{2m}	249.6	287.8
Air measuring nozzle differential pressure measured by impact tube	in. H_2O	....	22.661	25.527
Nozzle upstream temperature	deg F	t_n	218.4	217.9
Nozzle-throat diameter	in.	D_n	16.000	16.000
Room temperature at gage board	deg F	....	86.7	87.2
Compressor-inlet (circular section) diameter	in.	....	30	30
Compressor-discharge (circular section) diameter	in.	....	20	20
Power input to motor	kw	P_m	1,605.5	1,638.7
Line voltage at motor terminal (average 3 phases)	volts	....	2,215	2,208
Line current (average 3 phases)	amp	....	445.2	454.9

Calculation of Results

	Unit	Symbol	Test 1	Test 2
Capacity (by equation 8.13)				
Density of water at room temperature	lb per cu in	....	0.035965	0.035962
Differential pressure across nozzle	psig	Δ_p	0.8150	0.9180
Pressure ratio across nozzle [(barometer + diff. press.) ÷ barometer]	....	r_n	1.0559	1.0628
X value from Table 8.7	....	....	0.01552	0.01739
Nozzle diameter squared	sq in.	D_n^2	256.0	256.0
Nozzle coefficient of discharge from Table 8.3	....	C	0.995	0.995
Nozzle upstream temperature	deg F abs	T_{nm}	678.1	677.6

Table 8.10 continued

	Unit	Symbol	Test 1	Test 2
Temperature to which flow is referred	deg F abs	T_{lm}	547.1	548.2
Pressure to which flow is referred	psig	p_{lm}	14.580	14.620
$\text{Cap.} = \dfrac{36.0 \times 0.995 \times 256 \times T_{lm}\sqrt{X(X+1)}}{\sqrt{T_n G}}$	cfm	Q_m	24,363	25,869
Inlet specific gravity:				
Relative humidity from psychrometer reading and psychrometric tables	per cent	....	73	69
Inlet temperature	deg F	....	87.4	88.5
Specific gravity at inlet conditions from Table 8.6	....	G_{lm}	0.9878	0.9881
Total discharge pressure:				
Static pressure corrected to 32 F	in. Hg	....	41.828	38.943
Static pressure (in Hg at 32F x 0.49115)	psig	....	20.544	19.127
Velocity-pressure calculations:				
Specific weight of air in discharge pipe:				
$(1)\ 0.07106 \times \dfrac{20.544 + 14.580}{14.580} \times \dfrac{547.1}{754.3}$	lb per cu ft	....	0.1243	....
$(2)\ 0.07113 \times \dfrac{19.127 + 14.620}{14.620} \times \dfrac{548.2}{747.5}$	lb per cu ft	....	....	0.1205
Approximate flow rate in discharge pipe:				
$(1)\ \dfrac{24{,}363}{60} \times \dfrac{14.580}{20.544 + 14.580} \times \dfrac{754.3}{547.1}$	cfs	....	231.3	....
$(2)\ \dfrac{25{,}869}{60} \times \dfrac{14.620}{19.127 + 14.620} \times \dfrac{747.5}{548.2}$	cfs	....	....	253.5
Area of discharge pipe	sq ft	....	2.182	2.182
Velocity in discharge pipe = cfs ÷ area	fps	V_2	106.5	116.8
Velocity pressure:				
$(1)\ 0.1243 \times \dfrac{(106.5)^2}{2 \times 32.174} \times \dfrac{1}{144}$	psi	....	0.152	....
$(2)\ 0.1205 \times \dfrac{(116.8)^2}{2 \times 32.174} \times \dfrac{1}{144}$	....	....	....	0.177
Total discharge pressure	psig	....	20.695	19.303

Adjustment of Test Results to Specified Inlet Conditions

	Unit	Symbol	Test 1	Test 2
Pressure adjustment:				
Pressure ratio as run:				
(1) (20.695 + 14.580)/14.580	....	r_m	2.4194	
(2) (19.303 + 14.620)/14.620	....	r_m	....	2.3203
X_m value from Table 8.7	....	X_m	0.28407	0.26898
Estimation of probable speed at specified inlet conditions, from motor rpm vs. kw curve. The kw will vary approximately as the inlet specific weight:				

Table 8.10 continued

	Unit	Symbol	Test 1	Test 2
(1) $1,605.5 \times \frac{0.07463}{0.07106} = 1686$ kW; corresp. rpm = rpm	....	....	3,551	
(2) $1,638.7 \times \frac{0.07470}{0.07113} = 1719$ kW; corresp. rpm = rpm	....	....	....	3,549
X_{mc}:				
(1) $= 0.28407 \frac{547.1}{529.7} \times \frac{0.9962}{0.9878} \times \left(\frac{3,551}{3,554}\right)^2$	....	X_{mc}	0.29544	
(2) $= 0.26898 \frac{548.2}{529.7} \times \frac{0.9962}{0.9881} \times \left(\frac{3,549}{3,551}\right)^2$	....	X_{mc}	....	0.28037
Pressure ratio for X_{mc} from Table 8.7	....	r_{mc}	2.4960	2.3949
Adjusted discharge pressure = $r_{mc} \times 14.7$	psia	p_{2mc}	36.691	35.205
Adjusted discharge pressure	psig	....	21.99	20.50
Capacity adjustment:				
(1) $24,363 \times \frac{3,551}{3,554}$	cfm	Q_{mc}	24,340	
(2) $24,869 \times \frac{3,549}{3,551}$	cfm	Q_{mc}	....	25,850
Power adjustment:				
Motor efficiency as run	per cent	....	95.3	95.4
Shp as run = kW × eff ÷ 746	hp	....	2,051.00	2,095.6
Shp adjusted to specified inlet conditions and probable speed:				
(1) $2051.0 \times \frac{14.7}{14.58} \times \frac{0.29540}{0.28407} \times \frac{3,551}{3,554}$	hp	....	2,148.8	
(2) $2095.6 \times \frac{14.7}{14.62} \times \frac{0.28032}{0.26896} \times \frac{3,549}{3,551}$	hp	....	....	2,196.0
Motor efficiency (data from motor manufacturer)	per cent	e	95.4	95.5
Power input to motor, adjusted to specified conditions. The speeds corresponding to the values of P_{mc} are sufficiently close to the estimated speeds for specified intake conditions.	kw	P_{mc}	1,680	1,715

The equivalent test speed for a substitute gas at which the design density ratio obtains is determined from the relations for polytropic compression. The computations are made with the temperatures and pressures at the inlet and discharge of the compressor. The relations are:

$$(r_s)^{1/n_s} = \left(\frac{\gamma_2}{\gamma_1}\right)_s = \left(\frac{\gamma_2}{\gamma_1}\right)_d = (r_d)^{1/n_d}, \quad r_s = (r_d)^{n_s/n_d}$$

$$N_{eq} = N_d \sqrt{\frac{\left[ZR^{n/(n-1)}T_2\left(r^{(n-1)/n} - 1\right)\right]_s}{\left[ZR^{n/(n-1)}T_1\left(r^{(n-1)/n} - 1\right)\right]_d}}$$

where:

N_{eq} = equivalent speed
N_d = design speed
r = compressor pressure ratio
γ_1 = inlet density
γ_2 = discharge density
Z = supercompressibility factor
g = acceleration due to gravity
k = ratio of specific heats
R = gas constant = 1544/mo1. wgt.
T_1 = inlet temperature, °R
n = polytropic compression exponent
s = substitute gas, subscript
d = design gas, subscript

For perfect dynamic similarity, both the Mach number and the density ratio with the substitute gas must be exactly the same as the design values, and the equivalent speeds calculated by relations (8.11) and (8.12) must be identical. In practice this is seldom possible and difference of a few percent between the two computed equivalent speeds is accepted.

The most desirable substitute gas for test purposes is, of course, air. For many light gas compressors both the design density ratio and the design Mach numbers can be very nearly obtained when compressing air at an equivalent speed below design speed. For gases heavier than air, the equivalent test speed is above the design speed and unsafe stresses may be encountered. In addition, properties of the heavier gases frequently preclude simulation of both the design density ratio and design Mach numbers at the same or nearly the same equivalent speeds. In these cases, a substitute gas with more suitable properties is selected, such as Freon.

To translate test points to design conditions, the following relations are applied:

$$Q_d = Q_s \frac{N_d}{N_s} \tag{8.22}$$

$$H_d = H_s \left(\frac{N_d}{N_s} \right)^2 \tag{8.23}$$

$$\eta_d = \eta_s \tag{8.24}$$

where:

Q	=	inlet volume
H	=	polytropic head
N_d	=	design speed
N_{eq}	=	equivalent test speed
η	=	polytropic efficiency
d	=	design gas, subscript
S	=	substitute gas, subscript

Although similar to the fan laws, these relations are valid only for translating data to the design speed from a proper equivalent speed where dynamic similarity was achieved; that is, the design density ratio and Mach numbers existed in the compressor. In addition, the relations apply only to the polytropic head and efficiency, and are not accurate for isentropic head and efficiency where the ratio of specific heats of the substitute gas differs from that of the design gas.

The pressure level at which compressor tests are run does not enter the dynamic similarity relations, and performance tests may ordinarily be run at a convenient pressure level. Where the test pressure is greatly different from the design pressure, however, some error will appear because of the difference in Reynolds numbers and friction factors within the compressor.

A compressor designed to operate at 1000 psi will show poorer performance at atmospheric pressure than at design pressure because the friction factors at the test conditions are higher than at design conditions. Where it is not practicable to test such compressors at the design pressure level, low pressure test results are corrected by the calculated ratio of the friction losses at test pressure to those at design pressure.

LUBRICATION

For satisfactory performance and freedom from wear, any machine or tool having moving parts with rubbing surfaces depends on adequate and efficient lubrication. Not only must the lubricant itself, whether grease, oil, or other liquid, be carefully chosen to meet the required conditions of service, but adequate means, including lubricators, oil ducts, and feeding mechanisms, must be provided to ensure dependable application wherever lubrication is needed. These are the two necessary requirements for good lubrication, but they are not sufficient without conscientious attention on the part of the operator, who must assume responsibility for maintaining the supply of lubricant, for guarding against contamination, and for adjusting the rate of feed when necessary. Even so-called "fully automatic" systems of lubrication require occasional attention on the part of the operator.

Table 8.11 Cost of Air Leaks

Size of Opening in.	Cu. ft. Air Wasted per Month at 100 Lb. Pressure, Based on an Orifice Coefficient of .65	Cost of Air Wasted per Month, Based on 13.677 cents per 1,000 Cu. ft.*
3/8	6,671,890	$912.52
1/4	2,290,840	$399.50
1/8	740,210	$101.22
1/16	182,272	$ 24.91
1/32	45,508	$ 6.24

*At 22 bhp/100 cfm (16.412 kW/100 cfm) and 5 cents/kWh.

Thus, it will be seen that good lubrication depends on three principal factors:

1. Type of lubricant.
2. Effective application.
3. Attention from the operator.

In the preceding chapters, specific information is given in regard to these items as applied to various types of compressors and compressed-air equipment.

It should be borne in mind that lubricants, and particularly greases and oils manufactured from petroleum, are so complex in nature and vary so widely as to physical properties, that detailed specifications describing any desired grade or quality cannot be written with any assurance that the specification will define exactly what is wanted. Manufacturers of compressed-air machinery limit their specifications as to lubrication requirements to cover only the more important physical characteristics, such as viscosity at one or more temperatures, fire point, pour point, etc. In each case they specify the particular kind of service intended, in what atmosphere the oil must operate, what are the minimum and maximum temperatures it must withstand, and they indicate and provide the means for applying the oil to bearings or rubbing surfaces. However, they must leave to the oil refiner and those who sell the lubricant the responsibility of furnishing an oil suitable in all other respects for the service intended. The necessity of following this practice as a matter of policy is obvious, since many important qualities of the lubricant depend entirely on the origin of the oil, on the processes used for refining it, and on other items over which only the oil supplier has control.

Oils and greases are usually known by their trade names, but as the oil-refining art progresses, improvements or variations in the quality of the oil may result, so that the complex qualities of any particular brand of oil may change from time to time without corresponding change in the brand name. For this and similar reasons, it is against the policy of the Compressed Air and Gas Institute to specify the kind and quality of lubricant required by brand or trade name. The Institute recommends that the user purchase oil and greases only from reputable oil companies and that he require the oil companies to guarantee the quality of their lubricant for the use intended.

LOSS OF AIR PRESSURE IN PIPING DUE TO FRICTION

All these data are based on nonpulsating flow and apply to clean and smooth pipe. The data included in Figure 8.25 and Tables 8.12 to 8.16 are calculated from the following formulas by E.G. Harris:*

**University of Missouri Bulletin*, vol. no. 4.

$$f = \frac{CLq^2}{rd^5}$$

$$C = \frac{0.1025}{d^{0.31}}$$

$$f = \frac{0.1025Lq^2}{rd^{0.31}d^5} = \frac{0.1025Lq^2}{rd^{(0.31+5)}} = \frac{0.1025Lq^2}{rd^{5.31}}$$

where:

f = pressure drop, psi
L = length of pipe, ft.
q = cubic feet of free air per second
r = ratio of compression (from free air) at entrance of pipe
d = actual internal diameter of pipe, in.
C = experimental coefficient

Figure 8.28 gives directly the pressure drop in pipes of up to 12-in. diameter, capacities up to 10,000 ft³ of free air per minute (atmospheric), for initial pressures up to 400 psig.

Tables 8.12 to 8.15 show directly the pressure drop in pipes of up to 12-in. diameter, for capacities up to 30,000 ft³ of free air per minute, and for initial pressures of 60, 80,100, and 125 lb.

Table 8.16 gives factors that can be used conveniently to determine the pressure drops in pipes of up to 12-in. diameter, for capacities up to 30,000 ft³ of free air per minute and for any initial pressure.

$$D = \frac{F}{P_m} \frac{L}{1000} \quad \text{(formula based on data of Fritzsche for steel pipe)}$$

where:

D = pressure drop in pipe line, psi
F = friction factor from Figure 8.25
P_m = mean pressure in pipe, psia
L = length of pipe, ft.
d = actual inside diameter of pipe, in.

Note: For first approximation, use known terminal pressures at either end of pipe. Add barometric pressure to gage pressure to get absolute pressure.

Table 8.17 shows the loss of pressure through screw pipe fittings, expressed in equivalent lengths of straight pipe.

Figure 8.26 will be found convenient in determining pressure drop due to pipe friction where comparatively large volumes are handled at low initial pressures, such as are encountered in centrifugal-blower applications.

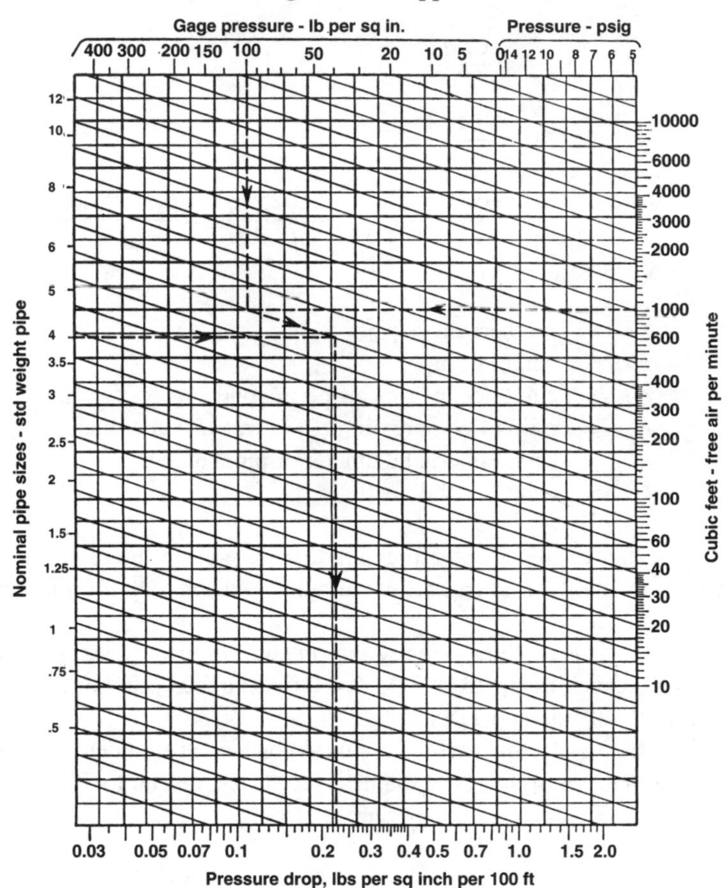

Figure 8.25 Loss of air pressure due to pipe friction for initial pressures up to 400 lb. *Problem:* 1,000 cfm free air (standard air) is to be transmitted at 100 psig pressure through a 4-in. standard weight pipe. What will be the pressure drop due to friction? *Solution by chart:* Enter the chart at the top, at the point representing 100 psig pressure, and proceed vertically downward to the intersection with a horizontal line representing 1,000 cfm, then parallel to the diagonal guide lines to the right (or left) to the intersection with a horizontal line representing a 4-in. pipe, then vertically downward to the pressure-loss scale at the bottom of the chart, where it is observed that the pressure loss would be 0.225 psi per 100 ft. of pipe. (Reprinted by permission from Walworth Co.)

Table 8.12 Loss of Air Pressure Due to Friction

Cu ft Free Air Per Min	Equivalent Cu ft Compressed Air Per Min	Nominal Diameter, In.											
		1/2	3/4	1	1 1/4	1 1/2	2	3	4	6	8	10	12
10	1.96	10.0	1.53	0.43	0.10								
20	3.94	39.7	5.99	1.71	0.39	0.18							
30	5.89	....	13.85	3.86	0.88	0.40							
40	7.86	....	24.7	6.85	1.59	0.71	0.19						
50	9.84	....	38.6	10.7	2.48	1.10	0.30						
60	11.81	....	55.5	15.4	3.58	1.57	0.43						
70	13.75	....	....	21.0	4.87	2.15	0.57						
80	15.72	....	....	27.4	6.37	2.82	0.75						
90	17.65	....	....	34.7	8.05	3.57	0.57	0.37					
100	19.60	....	....	42.8	9.95	4.40	1.18						
125	19.4	....	....	46.2	12.4	6.90	1.83	0.14					
150	29.45	....	....	....	22.4	9.90	2.64	0.32					
175	34.44	....	....	....	30.8	13.40	3.64	0.43					
200	39.40	....	....	....	39.7	17.60	4.71	0.57					
250	49.20	....	....	....	....	27.5	7.37	0.89	0.21				
300	58.90	....	....	....	....	39.6	10.55	1.30	0.31				
350	68.8	....	....	....	....	54.0	14.4	1.76	0.42				
400	78.8	....	....	....	....	....	18.6	2.30	0.53				
450	88.4	....	....	....	....	....	23.7	2.90	0.70				
500	98.4	....	....	....	....	....	29.7	3.60	0.85				
600	118.1	....	....	....	....	....	42.3	5.17	1.22				
700	137.5	....	....	....	....	....	57.8	7.00	1.67				
800	157.2	....	....	....	....	....	....	9.16	2.18				
900	176.5	....	....	....	....	....	....	11.6	2.76				
1,000	196.0	....	....	....	....	....	....	14.3	3.40				
1,500	294.5	....	....	....	....	....	....	32.3	7.6	0.87	0.29		
2,000	394.0	....	....	....	....	....	....	57.5	13.6	1.53	0.36		
2,500	492	....	....	....	....	....	....	....	21.3	2.42	0.57	0.17	
3,000	589	....	....	....	....	....	....	....	30.7	3.48	0.81	0.24	
3,500	688	....	....	....	....	....	....	....	41.7	4.68	1.07	0.33	
4,000	788	....	....	....	....	....	....	....	54.5	6.17	1.44	0.44	
4,500	884	....	....	....	....	....	....	....	....	7.8	1.83	0.55	0.21
5,000	984	....	....	....	....	....	....	....	....	9.7	2.26	0.67	0.27
6,000	1,181	....	....	....	....	....	....	....	....	13.9	3.25	0.98	0.38
7,000	1,375	....	....	....	....	....	....	....	....	18.7	4.43	1.34	0.51
8,000	1,572	....	....	....	....	....	....	....	....	24.7	5.80	1.73	0.71
9,000	1,765	....	....	....	....	....	....	....	....	31.3	7.33	2.20	0.87
10,000	1,960	....	....	....	....	....	....	....	....	38.6	9.05	2.72	1.06
11,000	2,165	....	....	....	....	....	....	....	....	46.7	10.9	3.29	1.28
12,000	2,362	....	....	....	....	....	....	....	....	55.5	13.0	3.90	1.51
13,000	2,560	....	....	....	....	....	....	....	....	....	15.2	4.58	1.77
14,000	2,750	....	....	....	....	....	....	....	....	....	17.7	5.32	2.07
15,000	2,945	....	....	....	....	....	....	....	....	....	20.3	6.10	2.36
16,000	3,144	....	....	....	....	....	....	....	....	....	23.1	6.95	2.70
18,000	3,530	....	....	....	....	....	....	....	....	....	29.2	8.80	3.42
20,000	3,940	....	....	....	....	....	....	....	....	....	36.2	10.8	4.22
22,000	4,330	....	....	....	....	....	....	....	....	....	43.7	13.2	5.12
24,000	4,724	....	....	....	....	....	....	....	....	....	51.9	15.6	5.92
26,000	5,120	....	....	....	....	....	....	....	....	....	....	18.3	7.15
28,000	5,500	....	....	....	....	....	....	....	....	....	....	21.3	8.3
30,000	5,890	....	....	....	....	....	....	....	....	....	....	24.4	9.5

In psi in 1000-ft of pipe, 60-lb gage initial pressure. For longer or shorter lengths of pipe the friction loss is proportional to the length, i.e., for 500 ft, one-half of the above; for 4,000 ft, four times the above, etc.

Table 8.13 Loss of Air Pressure Due to Friction

Cu ft Free Air Per Min	Equivalent Cu ft Compressed Air Per Min	Nominal Diameter, In.											
		1/2	3/4	1	1 1/4	1 1/2	2	3	4	6	8	10	12
10	1.55	7.90	1.21	0.34									
20	3.10	31.4	4.72	1.35	0.31								
30	4.65	70.8	10.9	3.04	0.69	0.31							
40	6.20	....	19.5	5.40	1.25	0.56							
50	7.74	....	30.5	8.45	1.96	0.87							
60	9.29	....	43.8	12.16	2.82	1.24	0.34						
70	10.82	....	59.8	16.6	3.84	1.70	0.45						
80	12.40	....	78.2	21.6	5.03	2.22	0.59						
90	13.95	....	....	27.4	6.35	2.82	0.75						
100	15.5	....	....	33.8	7.85	3.74	0.93						
125	19.4	....	....	46.2	12.4	5.45	1.44						
150	23.2	....	....	76.2	17.7	7.82	2.08						
175	27.2	....	....	....	24.8	10.6	2.87						
200	31.0	....	....	....	31.4	13.9	3.72	0.45					
250	38.7	....	....	....	49.0	21.7	5.82	0.70					
300	46.5	....	....	....	70.6	31.2	8.35	1.03					
350	54.2	....	....	....	....	42.5	11.4	1.39	0.33				
400	62.0	....	....	....	....	55.5	14.7	1.82	0.42				
450	69.7	....	....	....	....	....	18.7	2.29	0.55				
500	77.4	....	....	....	....	....	23.3	2.84	0.67				
600	92.9	....	....	....	....	....	33.4	4.08	0.96				
700	108.2	....	....	....	....	....	45.7	5.52	1.32				
800	124.0	....	....	....	....	....	59.3	7.15	1.72				
900	139.5	....	....	....	....	....	....	9.17	2.18				
1,000	155	....	....	....	....	....	....	11.3	2.68				
1,500	232	....	....	....	....	....	....	25.5	6.0	0.69			
2,000	310	....	....	....	....	....	....	45.3	10.7	1.21	0.29		
2,500	387	....	....	....	....	....	....	70.9	16.8	1.91	0.45		
3,000	465	....	....	....	....	....	....	....	24.2	2.74	0.64	0.19	
3,500	542	....	....	....	....	....	....	....	32.8	3.70	0.85	0.26	
4,000	620	....	....	....	....	....	....	....	43.0	4.87	1.14	0.34	
4,500	697	....	....	....	....	....	....	....	54.8	6.15	1.44	0.43	
5,000	774	....	....	....	....	....	....	....	67.4	7.65	1.78	0.53	0.21
6,000	929	....	....	....	....	....	....	....	....	11.0	2.57	0.77	0.29
7,000	1,082	....	....	....	....	....	....	....	....	14.8	3.40	1.06	0.40
8,000	1,240	....	....	....	....	....	....	....	....	19.5	4.57	1.36	0.54
9,000	1,395	....	....	....	....	....	....	....	....	24.7	5.78	1.74	0.69
10,000	1,550	....	....	....	....	....	....	....	....	30.5	7.15	2.14	0.84
11,000	1,710	....	....	....	....	....	....	....	....	36.8	8.61	2.60	1.01
12,000	1,860	....	....	....	....	....	....	....	....	43.8	10.3	3.08	1.19
13,000	2,020	....	....	....	....	....	....	....	....	51.7	12.0	3.62	1.40
14,000	2,170	....	....	....	....	....	....	....	....	60.2	14.0	4.20	1.63
15,000	2,320	....	....	....	....	....	....	....	....	68.5	16.0	4.82	1.84
16,000	2,480	....	....	....	....	....	....	....	....	78.2	18.2	5.48	2.13
18,000	2,790	....	....	....	....	....	....	....	....	....	23.0	6.95	2.70
20,000	3,100	....	....	....	....	....	....	....	....	....	28.6	8.55	3.33
22,000	3,410	....	....	....	....	....	....	....	....	....	34.5	10.4	4.04
24,000	3,720	....	....	....	....	....	....	....	....	....	41.0	12.3	4.69
26,000	4,030	....	....	....	....	....	....	....	....	....	48.2	14.4	5.6
28,000	4,350	....	....	....	....	....	....	....	....	....	55.9	16.8	6.3
30,000	4,650	....	....	....	....	....	....	....	....	....	64.2	19.3	7.5

In psi in 1000-ft of pipe, 80-lb gage initial pressure. For longer or shorter lengths of pipe the friction loss is proportional to the length, i.e., for 500 ft, one-half of the above; for 4,000 ft, four times the above, etc.

Table 8.14 Loss of Air Pressure Due to Friction

Cu ft Free Air Per Min	Equivalent Cu ft Compressed Air Per Min	Nominal Diameter, In.											
		1/2	3/4	1	1 1/4	1 1/2	2	3	4	6	8	10	12
10	1.28	6.50	.99	0.28									
20	2.56	25.9	3.90	1.11	0.25	0.11							
30	3.84	58.5	9.01	2.51	0.57	0.26							
40	5.12	....	16.0	4.45	1.03	0.46							
50	6.41	....	25.1	9.96	1.61	0.71	0.19						
60	7.68	....	36.2	10.0	2.32	1.02	0.28						
70	8.96	....	49.3	13.7	3.16	1.40	0.37						
80	10.24	....	64.5.	17.8	4.14	1.83	0.49						
90	11.52	....	82.8	22.6	5.23	2.32	0.62						
100	12.81	....	....	27.9	6.47	2.86	0.77						
125	15.82	....	....	48.6	10.2	4.49	1.19						
150	19.23	....	....	62.8	14.6	6.43	1.72	0.21					
175	22.40	....	....	....	19.8	8.72	2.36	0.28					
200	25.62	....	....	....	25.9	11.4	3.06	0.37					
250	31.64	....	....	....	40.4	17.9	4.78	0.58					
300	38.44	....	....	....	58.2	25.8	6.85	0.84	0.20				
350	44.80	....	....	....	....	35.1	9.36	1.14	0.27				
400	51.24	....	....	....	....	45.8	12.1	1.50	0.35				
450	57.65	....	....	....	....	58.0	15.4	1.89	0.46				
500	63.28	....	....	....	....	71.6	19.2	2.34	0.55				
600	76.88	....	....	....	....	....	27.6	3.36	0.79				
700	89.60	....	....	....	....	....	37.7	4.55	1.09				
800	102.5	....	....	....	....	....	49.0	5.89	1.42				
900	115.3	....	....	....	....	....	62.3	7.6	1.80				
1,000	128.1	....	....	....	....	....	76.9	9.3	2.21				
1,500	192.3	....	....	....	....	....	....	21.0	4.9	0.57			
2,000	256.2	....	....	....	....	....	....	37.4	8.8	0.99	0.24		
2,500	316.4	....	....	....	....	....	....	58.4	13.8	1.57	0.37		
3,000	384.6	....	....	....	....	....	....	84.1	20.0	2.26	0.53		
3,500	447.8	....	....	....	....	....	....	....	27.2	3.04	0.70	0.22	
4,000	512.4	....	....	....	....	....	....	....	35.5	4.01	0.94	0.28	
4,500	576.5	....	....	....	....	....	....	....	45.0	5.10	1.19	0.36	
5,000	632.8	....	....	....	....	....	....	....	55.6	6.3	1.47	0.44	0.17
6,000	768.8	....	....	....	....	....	....	....	80.0	9.1	2.11	0.64	0.24
7,000	896.0	....	....	....	....	....	....	....	....	12.2	2.88	0.87	0.33
8,000	1,025	....	....	....	....	....	....	....	....	16.1	3.77	1.12	0.46
9,000	1,153	....	....	....	....	....	....	....	....	20.4	4.77	1.43	0.57
10,000	1,280	....	....	....	....	....	....	....	....	25.1	5.88	1.77	0.69
11,000	1,410	....	....	....	....	....	....	....	....	30.4	7.10	2.14	0.83
12,000	1,540	....	....	....	....	....	....	....	....	36.2	8.5	2.54	0.98
13,000	1,668	....	....	....	....	....	....	....	....	42.6	9.8	2.98	1.15
14,000	1,795	....	....	....	....	....	....	....	....	49.2	11.5	3.46	1.35
15,000	1,923	....	....	....	....	....	....	....	....	56.6	13.2	3.97	1.53
16,000	2,050	....	....	....	....	....	....	....	....	64.5	15.0	4.52	1.75
18,000	2,310	....	....	....	....	....	....	....	....	81.5	19.0	5.72	2.22
20,000	2,560	....	....	....	....	....	....	....	....	....	23.6	7.0	2.74
22,000	2,820	....	....	....	....	....	....	....	....	....	28.5	8.5	3.33
24,000	3,080	....	....	....	....	....	....	....	....	....	33.8	10.0	3.85
26,000	3,338	....	....	....	....	....	....	....	....	....	39.7	11.9	4.65
28,000	3,590	....	....	....	....	....	....	....	....	....	46.2	13.8	5.40
30,000	3,850	....	....	....	....	....	....	....	....	....	53.0	15.9	6.17

In psi in 1000-ft of pipe, 100-lb gage initial pressure. For longer or shorter lengths of pipe the friction loss is proportional to the length, i.e., for 500 ft, one-half of the above; for 4,000 ft, four times the above, etc.

Table 8.15 Loss of Air Pressure Due to Friction

Cu ft Free Air Per Min	Equivalent Cu ft Compressed Air Per Min	Nominal Diameter, In.											
		1/2	3/4	1	1 1/4	1 1/2	2	3	4	6	8	10	12
10	1.05	5.35	0.82	0.23									
20	2.11	21.3	3.21	0.92	0.21								
30	3.16	48.0	7.42	2.07	0.47	0.21							
40	4.21	....	13.2	3.67	0.85	0.38							
50	5.26	....	20.6	5.72	1.33	0.59							
60	6.32	....	29.7	8.25	1.86	0.84	0.23						
70	7.38	....	40.5	11.2	2.61	1.15	0.31						
80	8.42	....	53.0	14.7	3.41	1.51	0.40						
90	9.47	....	68.0	18.6	4.30	1.91	0.51						
100	10.50	....	....	22.9	5.32	2.36	0.63						
125	13.15	....	....	39.9	8.4	3.70	0.98						
150	15.79	....	....	51.6	12.0	5.30	1.41	0.17					
175	18.41	....	....	....	16.3	7.2	1.95	0.24					
200	21.05	....	....	....	21.3	9.4	2.52	0.31					
250	26.30	....	....	....	33.2	14.7	3.94	0.48					
300	31.60	....	....	....	47.3	21.2	5.62	0.70					
350	36.80	....	....	....	....	28.8	7.7	0.94	0.22				
400	42.10	....	....	....	....	37.6	10.0	1.23	0.28				
450	47.30	....	....	....	....	47.7	12.7	1.55	0.37				
500	52.60	....	....	....	....	58.8	15.7	1.93	0.46				
600	63.20	....	....	....	....	....	22.6	2.76	0.65				
700	73.80	....	....	....	....	....	30.0	3.74	0.89				
800	84.20	....	....	....	....	....	40.2	4.85	1.17				
900	94.70	....	....	....	....	....	51.2	6.2	1.48				
1,000	105.1	....	....	....	....	....	63.2	7.7	1.82				
1,500	157.9	....	....	....	....	....	....	17.2	4.1	0.47			
2,000	210.5	....	....	....	....	....	....	30.7	7.3	0.82	0.19		
2,500	263.0	....	....	....	....	....	....	48.0	11.4	1.30	0.31		
3,000	316	....	....	....	....	....	....	69.2	16.4	1.86	0.43		
3,500	368	....	....	....	....	....	....	....	22.3	2.51	0.57	0.18	
4,000	421	....	....	....	....	....	....	....	29.2	3.30	0.77	0.23	
4,500	473	....	....	....	....	....	....	....	37.0	4.2	0.98	0.24	
5,000	526	....	....	....	....	....	....	....	45.7	5.2	1.21	0.36	
6,000	632	....	....	....	....	....	....	....	65.7	7.5	1.74	0.52	0.20
7,000	738	....	....	....	....	....	....	....	....	10.0	2.37	0.72	0.27
8,000	842	....	....	....	....	....	....	....	....	13.2	3.10	0.93	0.38
9,000	947	....	....	....	....	....	....	....	....	16.7	3.93	1.18	0.47
10,000	1,051	....	....	....	....	....	....	....	....	20.6	4.85	1.46	0.57
11,000	1,156	....	....	....	....	....	....	....	....	25.0	5.8	1.76	0.68
12,000	1,262	....	....	....	....	....	....	....	....	29.7	7.0	2.09	0.81
13,000	1,368	....	....	....	....	....	....	....	....	35.0	8.1	2.44	0.95
14,000	1,473	....	....	....	....	....	....	....	....	40.3	9.7	2.85	1.11
15,000	1,579	....	....	....	....	....	....	....	....	46.5	10.9	3.26	1.26
16,000	1,683	....	....	....	....	....	....	....	....	53.0	12.4	3.72	1.45
18,000	1,893	....	....	....	....	....	....	....	....	66.9	15.6	4.71	1.83
20,000	2,150	....	....	....	....	....	....	....	....	....	19.4	5.8	2.20
22,000	2,315	....	....	....	....	....	....	....	....	....	23.4	7.1	2.74
24,000	2,525	....	....	....	....	....	....	....	....	....	27.8	8.4	3.17
26,000	2,735	....	....	....	....	....	....	....	....	....	32.8	9.8	3.83
28,000	2,946	....	....	....	....	....	....	....	....	....	37.9	16.4	4.4
30,000	3,158	....	....	....	....	....	....	....	....	....	43.5	13.1	5.1

In psi in 1000-ft of pipe, 125-lb gage initial pressure. For longer or shorter lengths of pipe the friction loss is proportional to the length, i.e., for 500 ft, one-half of the above; for 4,000 ft, four times the above, etc.

Table 8.16 Factor for Calculating Loss of Air Pressure Due to Pipe Friction Applicable for any Initial Pressure*

Cu ft Free Air Per Min	Nominal Diameter, In.												
	1/2	3/4	1	1 1/4	1 1/2	1 3/4	2	3	4	6	8	10	12
5	12.7	1.2	0.5										
10	50.7	7.8	2.2	0.5									
15	114.1	17.6	4.9	1.1									
20	202	30.4	8.7	2.0	0.9								
25	316	50.0	13.6	3.2	1.4	0.7							
30	456	70.4	19.6	4.5	2.0	1.1							
35	811	95.9	26.2	6.2	2.7	1.4							
40	....	125.3	34.8	8.1	3.6	1.9							
45	....	159	44.0	10.2	4.5	2.4	1.2						
50	....	196	54.4	12.6	5.6	2.9	1.4						
60	....	282	78.3	18.2	8.0	4.2	2.2						
70	....	385	106.6	24.7	10.9	5.7	2.9						
80	....	503	139.2	32.3	14.3	7.5	3.8						
90	....	646	176.2	40.9	18.1	9.5	4.8						
100	....	785	217.4	50.5	22.3	11.7	6.0						
110	....	950	263	61.2	27.0	14.1	7.2						
120	....	....	318	72.7	32.2	16.8	8.6						
130	....	....	369	85.3	37.8	19.7	10.1	1.2					
140	....	....	426	98.9	43.8	22.9	11.7	1.4					
150	....	....	490	113.6	50.3	26.3	13.4	1.6					
160	....	....	570	129.3	57.2	29.9	15.3	1.9					
170	....	....	628	145.8	64.6	33.7	17.6	2.1					
180	....	....	705	163.3	72.6	37.9	19.4	2.4					
190	....	....	785	177	80.7	42.2	21.5	2.6					
200	....	....	870	202	89.4	46.7	23.9	2.9					
220	....	....	....	244	108.2	56.5	28.9	3.5					
240	....	....	....	291	128.7	67.3	34.4	4.2					
260	....	....	....	341	151	79.0	40.3	4.9					
280	....	....	....	395	175	91.6	46.8	5.7					
300	....	....	....	454	201	105.1	53.7	6.6					
320	....	....	....	....	....	....	61.1	7.5					
340	....	....	....	....	....	....	69.0	8.4	2.0				
360	....	....	....	....	....	....	77.3	9.5	2.2				
380	....	....	....	....	....	....	86.1	10.5	2.5				
400	....	....	....	....	....	....	94.7	11.7	2.7				
420	....	....	....	....	....	....	105.2	12.9	3.1				
440	....	....	....	....	....	....	115.5	14.1	3.4				
460	....	....	....	....	....	....	125.6	15.4	3.7				
480	....	....	....	....	....	....	137.6	16.8	4.0				
500	....	....	....	....	....	....	150.0	18.3	4.3				
525	....	....	....	....	....	....	165.0	20.2	4.8				
550	....	....	....	....	....	....	181.5	22.1	5.2				
575	....	....	....	....	....	....	197	24.2	5.7				
600	....	....	....	....	....	....	215	26.3	6.2				
625	....	....	....	....	....	....	233	28.5	6.8				
650	....	....	....	....	....	....	253	30.9	7.3				
675	....	....	....	....	....	....	272	33.3	7.9				
700	....	....	....	....	....	....	294	35.8	8.5				
750	....	....	....	....	....	....	337	41.4	9.7				
800	....	....	....	....	....	....	382	46.7	11.1				
850	....	....	....	....	....	....	433	52.8	12.5				
900	....	....	....	....	....	....	468	59.1	14.0				
950	....	....	....	....	....	....	541	65.9	15.7				
1,000	....	....	....	....	....	....	600	73.0	17.3	1.9			
1,050	....	....	....	....	....	....	658	80.5	19.1	2.1			

Table 8.16 (continued)

Cu Ft Free Air Per Min	Nominal Diameter, In.												
	1/2	3/4	1	1 1/4	1 1/2	1 3/4	2	3	4	6	8	10	12
1,100	....	....	....	....	....	....	723	88.4	21.0	2.4			
1,150	....	....	....	....	....	....	790	96.6	22.9	2.6			
1,200	....	....	....	....	....	....	850	105.2	25.0	2.8			
1,300	....	....	....	....	....	....	....	123.4	29.3	3.3			
1,400	....	....	....	....	....	....	....	....	33.9	3.8			
1,500	....	....	....	....	....	....	....	....	39.0	4.4			
1,600	....	....	....	....	....	....	....	....	44.3	5.1			
1,700	....	....	....	....	....	....	....	....	50.1	5.7			
1,800	....	....	....	....	....	....	....	....	56.1	6.4			
1,900	....	....	....	....	....	....	....	....	62.7	7.1	1.6		
2,000	....	....	....	....	....	....	....	....	69.3	7.8	1.8		
2,100	....	....	....	....	....	....	....	....	76.4	8.7	2.0		
2,200	....	....	....	....	....	....	....	....	83.6	9.5	2.2		
2,300	....	....	....	....	....	....	....	....	91.6	10.4	2.4		
2,400	....	....	....	....	....	....	....	....	99.8	11.3	2.6		
2,500	....	....	....	....	....	....	....	....	108.2	12.3	2.9		
2,600	....	....	....	....	....	....	....	....	117.2	13.3	3.1		
2,700	....	....	....	....	....	....	....	....	126	14.3	3.3		
2,800	....	....	....	....	....	....	....	....	136	15.4	3.6		
2,900	....	....	....	....	....	....	....	....	146	16.5	3.9		
3,000	....	....	....	....	....	....	....	....	156	17.7	4.1		
3,200	....	....	....	....	....	....	....	....	177	20.1	4.7		
3,400	....	....	....	....	....	....	....	....	200	22.7	5.3		
3,600	....	....	....	....	....	....	....	....	224	25.4	5.6	1.8	
3,800	....	....	....	....	....	....	....	....	250	28.4	6.6	2.0	
4,000	....	....	....	....	....	....	....	....	277	31.4	7.3	2.2	
4,200	....	....	....	....	....	....	....	....	305	34.6	8.1	2.4	
4,400	....	....	....	....	....	....	....	....	335	38.1	8.9	2.7	
4,600	....	....	....	....	....	....	....	....	366	41.5	9.7	2.9	
4,800	....	....	....	....	....	....	....	....	399	45.2	10.5	3.2	
5,000	....	....	....	....	....	....	....	....	433	49.1	11.5	3.4	
5,250	....	....	....	....	....	....	....	....	477	54.1	12.6	3.4	
5,500	....	....	....	....	....	....	....	....	524	59.4	13.9	4.2	1.6
5,750	....	....	....	....	....	....	....	....	....	64.9	15.2	4.6	1.8
6,000	....	....	....	....	....	....	....	....	....	70.7	16.5	5.0	1.9
6,500	....	....	....	....	....	....	....	....	....	82.9	19.8	5.9	2.3
7,000	....	....	....	....	....	....	....	....	....	96.2	22.5	6.8	2.6
7,500	....	....	....	....	....	....	....	....	....	110.5	25.8	7.8	3.0
8,000	....	....	....	....	....	....	....	....	....	125.7	29.4	8.8	3.6
9,000	....	....	....	....	....	....	....	....	....	159	37.2	10.2	4.4
10,000	....	....	....	....	....	....	....	....	....	196	45.9	13.8	5.4
11,000	....	....	....	....	....	....	....	....	....	237	55.5	16.7	6.5
12,000	....	....	....	....	....	....	....	....	....	282	66.1	19.8	7.7
13,000	....	....	....	....	....	....	....	....	....	332	77.5	23.3	9.0
14,000	....	....	....	....	....	....	....	....	....	387	89.9	27.0	10.5
15,000	....	....	....	....	....	....	....	....	....	442	103.2	31.0	12.0
16,000	....	....	....	....	....	....	....	....	....	503	117.7	35.3	13.7
18,000	....	....	....	....	....	....	....	....	....	636	148.7	44.6	17.4
20,000	....	....	....	....	....	....	....	....	....	....	184	55.0	21.4
22,000	....	....	....	....	....	....	....	....	....	....	222	66.9	26.0
24,000	....	....	....	....	....	....	....	....	....	....	264	79.3	30.1
26,000	....	....	....	....	....	....	....	....	....	....	310	93.3	36.3
28,000	....	....	....	....	....	....	....	....	....	....	360	108.0	42.1
30,000	....	....	....	....	....	....	....	....	....	....	413	123.9	48.2

*To determine the pressure drop in psi, the factor listed in the table for a given capacity and pipe diameter should be divided by the ratio of compression (from free air) at entrance of pipe, multiplied by the actual length of the pipe in feet, and divided by 1000.

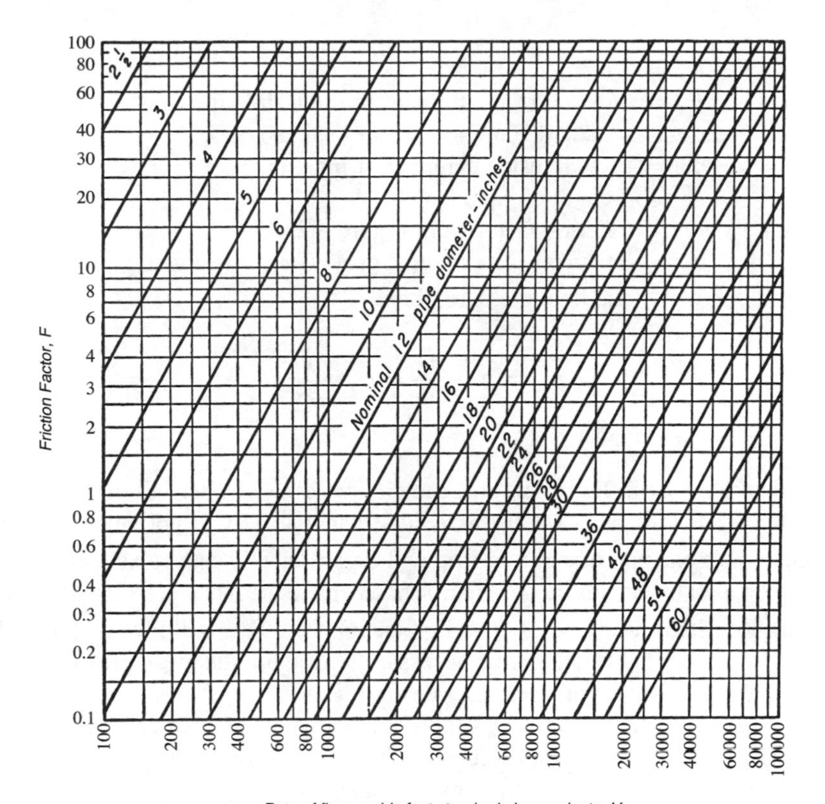

Figure 8.26 Loss of air pressure due to pipe friction measured at standard conditions of 14.7 psia and 60°F.

In all blower installations where a length of pipe is used to deliver air, either to the blower inlet or from the blower discharge or both, a certain amount of pressure is used up in forcing the air through these pipes. Ordinarily, when the combined length of the intake and discharge pipes is greater than ten pipe diameters, the drop in pressure is great enough to make a difference between the generated pressure and the pressure at the delivery end of the discharge pipe. This drop must be taken into consideration, especially if the pressure generated by the blower is to be very little in excess of that required to force the desired volume of air through the particular apparatus or system alone.

The formula for calculating this drop is:

$$D = 0.7 \frac{V^{1.85}}{d^5 p_m} \frac{L}{1000} = \frac{F}{p_m} \frac{L}{1000} \tag{8.25}$$

where:

D = Pressure drop, psi.
V = Volume of air flowing through the pipe in cfm measured at standard conditions (14.7 psia and 60°F).
d = Actual inside diameter of pipe, in.
L = Length of pipe, ft.*
F = Friction factor from Figure 8.25.
p_m = Mean pressure in pipe, psia. If p_1 is the initial pressure and p_2 the final pressure, then:

$$p_m = p_1 - \frac{D}{2} = p_2 + \frac{D}{2} \tag{8.26}$$

Substitute first in Eq. (8.25) the known pressure (whether initial or final) for p_m and solve for the drop D. If this is less than 1 lb, it will not be necessary to calculate the mean pressure as this drop will be sufficiently accurate for most work. If the drop as calculated is greater than 1 lb, calculate the mean pressure p_m by substituting Eq (8.26) the value of D as first calculated. This will give a close value for the drop D so that it will not be necessary to refigure the mean pressure again. Continued trial will give any accuracy desired.

The use of Eq. (8.25) has been simplified by Figure 8.26, which gives the value of the friction factor F to be substituted in the right-hand member of Eq. (8.25) for various rates of flow (cubic feet per minute standard air).

Equation (8.25) is based on a flowing air temperature in the pipe of 60°F. For other flowing temperatures, multiply D already found by $T/520$, where $T = 460 +$ actual "flowing" temperature (°F).

* Where there are bends in the pipe line, to derive L the linear length should be increased in accordance with the following data:

For each 90-degree bend with radius equal to:

(a) 1 pipe diameter, L should be increased 17.5 pipe diameters.
(b) 1 1/2 pipe diameters, L should be increased 10.4 pipe diameters.
(c) 2 pipe diameters, L should be increased 9.0 pipe diameters.
(d) 3 pipe diameters, L should be increased 8.2 pipe diameters.

TABLE 8.17 Loss of Pressure through Screw Pipe Fittings, Steam, Air, Gas*

Nominal Pipe Size, In.	Actual Inside Diameter, In.	Gate Valve	Long Radius Ell or On Run of Standard Tee	Standard Ell or On Run of Tee Reduced in Size 50 per cent	Angle Valve	Close Return Bend	Tee through Side Outlet	Globe Valve
1/2	0.622	0.36	0.62	1.55	8.65	3.47	3.10	17.3
3/4	0.824	0.48	0.82	2.06	11.4	4.60	4.12	22.9
1	1.049	0.61	1.05	2.62	14.6	5.82	5.24	29.1
1 1/4	1.380	0.81	1.38	3.45	19.1	7.66	6.90	38.3
1 1/2	1.610	0.94	1.61	4.02	22.4	8.95	8.04	44.7
2	2.067	1.21	2.07	5.17	28.7	11.5	10.3	57.4
2 1/2	2.469	1.44	2.47	6.16	34.3	13.7	12.3	68.5
3	3.068	1.79	3.07	6.16	42.6	17.1	15.3	85.2
4	4.026	2.35	4.03	7.67	56.0	22.4	20.2	112
5	5.047	2.94	5.05	10.1	70.0	28.0	25.2	140
6	6.065	3.54	6.07	15.2	84.1	33.8	30.4	168
8	7.981	4.65	7.98	20.0	111	44.6	40.0	222
10	10.020	5.85	10.00	25.0	139	55.7	50.0	278
12	11.940	6.96	11.0	29.8	166	66.3	59.6	332

* Adapted from Sabin Crocker, *Piping Handbook*, 4th ed., McGraw-Hill Book Company, Inc., New York, 1945. Given in equivalent lengths (feet) of straight pipe, schedule 40.

DATA, TABLES, FORMULAS

TABLE 8.18 Friction of Air in Hose, Pulsating Flow*

Size of Hose, Coupled Each End In.	Gage Pressure at Line, Lb	Cu ft. Air Per Min Passing through 50-ft Lengths of Hose													
		20	30	40	50	60	70	80	90	100	110	120	130	140	150
		Loss of Pressure (psi) in 50-ft Lengths of Hose													
	50	1.8	5.0	10.1	18.1										
	60	1.3	4.0	8.4	14.8	23.4									
1/2	70	1.0	3.4	7.0	12.4	20.0	28.4								
	80	0.9	2.8	6.0	10.8	17.4	25.2	34.6							
	90	0.8	2.4	5.4	9.5	14.8	22.0	30.5	41.0						
	100	0.7	2.3	4.8	8.4	13.3	19.3	27.2	36.6						
	110	0.6	2.0	4.3	7.6	12.0	17.6	24.6	33.3	44.5					
	50	0.4	0.8	1.5	2.4	3.5	4.4	6.5	8.5	11.4	14.2				
	60	0.3	0.6	1.2	1.9	2.8	3.8	5.2	6.8	8.6	11.2				
3/4	70	0.2	0.5	0.9	1.5	2.3	3.2	4.2	5.5	7.0	8.8	11.0			
	80	0.2	0.5	0.8	1.3	1.9	2.8	3.6	4.7	5.8	7.2	8.8	10.6		
	90	0.2	0.4	0.7	1.1	1.6	2.3	3.1	4.0	5.0	6.2	7.5	9.0		
	100	0.2	0.4	0.6	1.0	1.4	2.0	2.7	3.5	4.4	5.4	6.6	7.9	9.4	11.1
	110	0.1	0.3	0.5	0.9	1.3	1.8	2.4	3.1	3.9	4.9	5.9	7.1	8.4	9.9
	50	0.1	0.2	0.3	0.5	0.8	1.1	1.5	2.0	2.6	3.5	4.8	7.0		
	60	1.1	0.2	0.3	0.4	0.6	0.8	1.2	1.5	2.0	2.6	3.3	4.2	5.5	7.2
	70	.....	0.1	0.2	0.4	0.5	0.7	1.0	1.3	1.6	2.0	2.5	3.1	3.8	4.7
	80	.....	0.1	0.2	0.3	0.5	0.7	0.8	1.1	1.4	1.7	2.0	2.4	2.7	3.5
	90	.....	1.1	0.2	0.3	0.4	0.6	0.7	0.9	1.2	1.4	1.7	2.0	2.4	2.8
	100	.....	1.1	0.2	0.2	0.4	0.5	0.6	0.8	1.0	1.2	1.5	1.8	2.1	2.4
	110	.....	0.1	0.2	0.2	0.3	0.4	0.6	0.7	0.9	1.1	1.3	1.5	1.8	2.1
	50	.....	.....	0.1	0.2	0.2	0.3	0.4	0.5	0.7	1.1				
	60	.....	.....	.....	0.1	0.2	0.3	0.3	0.5	0.6	0.8	1.0	1.2	1.5	
	70	.....	.....	.....	0.1	0.2	0.2	0.3	0.4	0.4	0.5	0.7	0.8	1.0	1.3
1 1/4	80	.....	.....	.....	.....	0.1	0.2	0.2	0.3	0.4	0.5	0.6	0.7	0.8	1.0
	90	.....	.....	.....	.....	0.1	0.2	0.2	0.3	0.3	0.4	0.5	0.6	0.7	0.8
	100	.....	.....	.....	.....	.....	0.1	0.2	0.2	0.3	0.4	0.4	0.5	0.6	0.7
	110	.....	.....	.....	.....	.....	0.1	0.2	0.2	0.3	0.3	0.4	0.5	0.5	0.6
	50	.....	.....	.....	.....	.....	0.1	0.2	0.2	0.2	0.3	0.3	0.4	0.5	0.6
	60	.....	.....	.....	.....	.....	.....	0.1	0.2	0.2	0.2	0.3	0.3	0.4	0.5
1 1/2	70	.....	.....	.....	.....	.....	.....	.....	0.1	0.2	0.2	0.2	0.3	0.3	0.4
	80	.....	.....	.....	.....	.....	.....	.....	.....	0.1	0.2	0.2	0.2	0.3	0.4
	90	.....	.....	.....	.....	.....	.....	.....	.....	.....	0.1	0.2	0.2	0.2	0.3
	100	.....	.....	.....	.....	.....	.....	.....	.....	.....	.....	0.1	0.2	0.2	0.2
	110	.....	.....	.....	.....	.....	.....	.....	.....	.....	.....	0.1	0.2	0.2	0.2

*For longer or shorter lengths of hose the friction loss is proportional to the length, i.e., for 25 ft one-half of the above; for 150 ft, three times the above, etc.

TABLE 8.19 Effect of Altitude on Capacity of Single-stage Compressors

Altitude, ft.	25 psig		40 psig		60 psig		80 psig		90 psig		100 psig		125 psig	
	Compressor Ratio	Factor	Compressor Ratio	Factor	Compressor Ratio	Factor	Compressor Ratio	Factor	Compressor Ratio	Factor	Compressor Ratio	Factor	Compressor Ratio	Factor
Sea level	2.70	1.0	3.72	1.0	5.08	1.0	6.44	1.0	7.12	1.0	7.81	1.0	9.51	1.0
1,000	2.76	0.996	3.82	0.993	5.23	0.992	6.64	0.992	7.34	0.988	8.05	0.987	9.81	0.982
2,000	2.84	0.992	3.94	0.987	5.42	0.984	6.88	0.977	7.62	0.972	8.35	0.972	10.20	0.962
3,000	2.91	0.987	4.06	0.981	5.59	0.974	7.12	0.967	7.87	0.959	8.63	0.957	10.55	0.942
4,000	2.99	0.982	4.18	0.974	5.76	0.963	7.36	0.953	8.15	0.944	8.94	0.942	10.92	0.923
5,000	3.07	0.977	4.31	0.967	5.96	0.953	7.62	0.940	8.44	0.931	9.27	0.925	11.32	
6,000	3.14	0.972	4.42	0.961	6.13	0.945	7.84	0.928	8.69	0.917	9.55	0.908	11.69	
7,000	3.23	0.967	4.57	0.953	6.36	0.936	8.14	0.915	9.03	0.902	9.93	0.890	12.17	
8,000	3.32	0.962	4.71	0.945	6.56	0.925	8.42	0.900	9.33	0.886	10.26	0.873	12.58	
9,000	3.41	0.957	4.85	0.938	6.77	0.915	8.70	0.887	9.65	0.868	10.62	0.857	13.02	
10,000	3.50	0.951	5.00	0.931	7.00	0.902	9.00	0.872	10.00	0.853	11.00	0.340	13.50	
11,000	3.61	0.945	5.17	0.923	7.25	0.891	9.34	0.858	10.38	0.837	11.42	....	14.03	
12,000	3.72	0.938	5.35	0.914	7.53	0.878	9.70	0.839	10.79	0.818	11.88	....	14.60	
14,000	3.94	0.927	5.71	0.897	8.06	0.852	10.42	0.805	11.60	....	12.78	....	15.71	
15,000	4.09	0.918	5.94	0.887	8.42	0.836	10.88	0.784	12.12	....	13.36	....	16.43	

Factor for estimating, based on 7% cylinder clearance.

Note: To find the capacity of a compressor when it is used at an altitude, multiply the sea-level capacity of the unit by the factor corresponding to the altitude and the discharge pressure. The result will be the actual capacity of the unit at the altitude.

TABLE 8.20 Multipliers to Determine Air Consumption of Rock Drills at Altitudes and for Various Number of Drills*

Altitude, ft.	Number of Drills																
	1	2	3	4	5	6	7	8	9	10	12	15	20	25	30	40	50
	Multiplier- Assuming 90 psig Air Pressure																
0	1.000	2.0	3.0	4.0	5.0	6.0	6.3	7.2	8.1	9.0	10.8	12.0	16.0	20.0	22.5	30.0	37.5
1,000	1.032	2.1	3.1	4.1	5.2	6.2	6.5	7.4	8.4	9.3	11.1	12.4	16.5	20.6	23.2	31.0	38.7
2,000	1.065	2.1	3.2	4.3	5.3	6.4	6.7	7.7	8.6	9.6	11.5	12.8	17.0	21.3	24.0	31.9	40.0
3,000	1.100	2.2	3.3	4.4	5.5	6.6	6.9	7.9	8.9	9.9	11.9	13.2	17.6	22.0	24.8	33.0	41.3
4,000	1.136	2.3	3.4	4.5	5.7	6.8	7.1	8.2	9.2	10.2	12.3	13.7	18.2	22.7	25.6	34.1	42.6
5,000	1.174	2.3	3.5	4.7	5.9	7.0	7.4	8.5	9.5	10.5	12.7	14.1	18.8	23.5	26.4	35.2	44.0
6,000	1.213	2.4	3.6	4.9	6.1	7.3	7.6	8.7	9.8	10.9	13.0	14.6	19.4	24.2	27.4	36.4	45.5
7,000	1.255	2.5	3.8	5.0	6.3	7.5	7.9	9.0	10.2	11.3	13.6	15.1	20.0	25.1	28.2	37.6	47.1
8,000	1.298	2.6	3.9	5.2	6.4	7.8	8.2	9.3	10.5	11.7	14.0	15.6	20.8	26.0	29.2	38.9	48.7
9,000	1.343	2.7	4.0	5.4	6.7	8.1	8.5	9.7	10.9	12.1	14.5	16.1	21.5	26.9	30.3	40.3	50.4
10,000	1.391	2.8	4.2	5.6	7.0	8.3	8.8	10.0	11.3	12.5	15.0	16.7	22.3	27.8	31.3	41.7	52.2
12,500	1.520	3.0	4.6	6.1	7.6	9.1	9.6	10.9	12.3	13.7	16.4	18.2	24.3	30.4	34.2	45.6	57.0
15,000	1.665	3.3	5.0	6.7	8.3	10.0	10.5	11.9	13.5	15.0	18.0	20.0	26.6	33.3	37.5	49.9	62.4

*Air consumption for various number of drills based on fact that all drills will not operate at once. It will vary with rock, type of work, etc.

TABLE 8.21 Theoretical Horsepower Required at Altitude to Compress 100 Cubic Feet

Altitude, ft.	Isothermal Compression Single- and Two-stage Gage Pressure					Adiabatic Compression Single-stage Gage Pressure			Adiabatic Compression Two-stage Gage Pressure				
	60	80	100	125	150	60	80	100	60	80	100	125	150
0	10.4	11.9	13.2	14.4	15.5	13.4	15.9	18.1	11.8	13.7	15.4	17.1	18.7
1,000	10.2	11.7	12.9	14.1	15.1	13.2	15.6	17.8	11.6	13.5	15.1	16.8	18.3
2,000	10.0	11.4	12.6	13.8	14.8	13.0	15.4	17.5	11.4	13.2	14.8	16.4	17.9
3,000	9.8	11.2	12.3	13.5	14.4	12.8	15.2	17.2	11.2	13.0	14.5	16.1	17.5
4,000	9.6	11.0	12.1	13.2	14.1	12.6	14.9	16.9	11.0	12.7	14.2	15.7	17.1
5,000	9.4	10.7	11.8	12.8	13.7	12.4	14.7	16.5	10.8	12.5	13.9	15.4	16.7
6,000	9.2	10.5	11.5	12.5	13.4	12.2	14.4	16.2	10.6	12.2	13.6	15.1	16.4
7,000	9.0	10.3	11.2	12.2	13.0	12.0	14.2	16.0	10.4	12.0	13.4	14.8	16.0
8,000	8.9	10.0	11.0	11.9	12.7	11.8	14.0	15.7	10.2	11.8	13.1	14.5	15.6
9,000	8.7	9.8	10.7	11.6	12.4	11.6	13.7	15.4	10.0	11.6	12.8	14.1	15.3
10,000	8.5	9.6	10.4	11.4	12.1	11.5	13.5	15.1	9.8	11.3	12.6	13.8	15.0

Table 8.22 Approximate Brake Horsepower Required by Air Compressors

Altitude, ft.	Single-stage psig			Two-stage psig			
	60	80	100	60	80	100	125
0	16.3	19.5	22.1	14.7	17.1	19.1	21.3
1,000	16.1	19.2	21.7	14.5	16.8	18.7	20.9
2,000	15.9	18.9	21.3	14.3	16.5	18.4	20.5
3,000	15.7	18.6	20.9	14.0	16.1	18.0	20.0
4,000	15.4	18.2	20.6	13.8	15.8	17.7	19.6
5,000	15.2	17.9	20.3	13.5	15.5	17.3	19.2
6,000	15.0	17.6	20.0	13.3	15.2	17.0	18.8
7,000	14.7	17.3	19.6	13.0	14.9	16.6	18.4
8,000	14.5	17.1	19.3	12.7	14.6	16.2	18.0
9,000	14.3	16.8	18.9	12.5	14.3	15.9	17.6
10,000	14.1	16.5	18.6	12.3	14.1	15.6	17.2
12,000	13.6	15.9	17.9	11.8	13.5	15.0	16.5
14,000	13.1	15.2	17.2	11.3	12.9	14.3	15.7

Figures given are bhp per 100 ft^3 of free air per minute actually delivered. Bhp per 100 ft^3 of free air per minute will vary considerably with the size and type of compressor.

Table 8.23 Theoretical Horsepower Required to Compress Air from Atmospheric Pressure to Various Pressures, Mean Effective Pressures (mep)

Discharge Pressure			Isothermal Compression, Single or Multi-stage		Adiabatic Compression*					Per Cent of Power Saved by Two-stage over Single-stage Adiabatic Compression
					Single-stage		Two-stage			
psig	psig	Atm Abs	Mep	Theo-rectical hp Per 100 Cu. ft.	Mep	Theo-rectical hp Per 100 Cu. ft.	Mep, Psi Referred to Low-press. Air Cylinder	Theo-rectical hp Per 100 Cu. ft.	Theo-rectical Intercooler Gauge Pressure	
5	19.7	1.34	4.13	1.8	4.48	1.96				
10	24.7	1.68	7.57	3.3	8.21	3.58				
15	29.7	2.02	10.31	4.5	11.4	5.0				
20	34.7	2.36	12.62	5.5	14.3	6.2				
25	39.7	2.70	14.68	6.4	16.9	7.4				
30	44.7	3.04	16.30	7.1	19.2	8.4				
35	49.7	3.38	17.90	7.8	21.4	9.3				
40	54.7	3.72	19.28	8.4	23.4	10.2				
45	59.7	4.06	20.65	9.0	25.2	11.0				
50	64.7	4.40	21.80	9.5	27.0	11.8				
55	69.7	4.74	22.95	10.0	28.7	12.6				
60	74.7	5.08	23.90	10.4	30.3	13.3				
65	79.7	5.42	24.80	10.8	31.9	13.9				
70	84.7	5.76	25.70	11.2	33.3	14.6	29.2	12.8	20.6	12.3
75	89.7	6.10	26.62	11.6	34.7	15.2	30.2	13.3	21.6	12.5
80	94.7	6.44	27.52	12.0	36.0	15.7	31.3	13.7	22.7	12.7
85	99.7	6.78	28.21	12.3	37.3	16.3	32.3	14.1	23.6	13.5
90	104.7	7.12	28.93	12.6	38.6	16.9	33.2	14.5	24.5	14.2
95	109.7	7.46	29.60	12.9	39.8	17.4	34.2	14.9	25.5	14.4
100	114.7	7.80	30.30	13.2	40.9	17.9	35.0	15.3	26.3	14.5
110	124.7	8.48	31.42	13.7	43.2	18.9	36.7	16.1	28.1	14.8
120	134.7	9.16	32.60	14.2	45.2	19.8	38.3	16.8	29.8	15.1
130	144.7	9.84	33.75	14.7	47.2	20.7	39.6	17.3	31.5	16.4
140	154.7	10.52	34.67	15.1	49.2	21.5	40.8	17.9	32.9	15.7
150	164.7	11.20	35.59	15.5	51.0	22.3	42.3	18.5	34.5	17.1
160	174.7	11.88	36.30	15.8	.....	.....	43.6	19.0	36.1	
170	184.7	12.56	37.20	16.2	.....	.....	44.7	19.5	37.3	
180	194.7	13.24	38.10	16.6	.....	.....	45.8	20.0	38.8	
190	204.7	13.92	38.80	16.9	.....	.....	46.8	20.4	40.1	
200	214.7	14.60	39.50	17.2	.....	.....	47.8	20.9	41.4	
250	264.7	18.00	42.70	18.6	.....	.....	52.5	22.7	47.6	
300	314.7	21.40	45.30	19.7	.....	.....	56.5	24.5	53.4	
350	364.7	24.81	47.30	20.6	.....	.....	59.6	26.1	58.5	
400	414.7	28.21	49.20	21.4	.....	.....	62.7	27.4	63.3	
450	464.7	31.61	51.20	22.3	.....	.....	65.3	28.6	67.8	
500	514.7	35.01	52.70	22.9	.....	.....	67.8	29.6	71.2	
550	564.7	38.41	53.75	23.4	.....	.....	70.0	30.6	76.3	
600	614.7	41.81	54.85	23.9	.....	.....	72.3	31.3	80.5	

* Based on a value for *n* of 1.3947.

TABLE 8.24 Displacement of a Double-acting Piston at Various Piston Speeds (Piston Rods Not Deducted)

Diameter of cylinder, In.	Piston Speed, Fpm = 2 × Stroke (Ft) × Rpm (for Double-acting Cylinder)																		
	200	225	250	275	300	325	350	375	400	425	450	475	500	550	600	650	700	750	800
	Displacement, Cfm																		
3	9.8	11.1	12.3	13.5	14.7	16	17.2	18.5	19.6	21	22.2	23.4	24.6	27	29.4	31.8	34.4	36.8	39.2
4	17.4	19.7	21.8	24	26.2	28.5	30.7	32.9	34.8	37.2	39.4	41.5	43.6	48	52.4	56.7	61.4	65.2	69.5
5	27.2	30.8	34.2	37.5	41	44.5	48	51.4	54.4	58	61.6	65	68.4	75	82	88.7	96	102	109
6	39.2	44.4	49	54	59	64	69	74	78.4	84	88.8	93	98	108	118	128	138	148	157
7	53.4	60.5	67	74	80	87	94	100.6	106.8	114	121	127	134	148	160	174	188	201	214
8	70	79	87.5	96	105	114	123	132	140	149	158	166	175	192	210	227	246	262	280
9	88	100	110	122	133	144	155	166	176	188	200	210	220	244	266	287	310	332	352
10	109	123	136	150	164	178	192	206	218	233	246	260	272	300	328	354	384	409	436
11	132	149	165	182	198	215	232	250	264	282	298	315	330	364	396	430	464	496	528
12	156	177	196	216	236	256	276	296	312	335	354	374	392	432	472	514	552	585	625
13	184	208	230	255	276	300	325	345	368	392	416	440	460	510	552	601	650	690	736
14	212	240	266	295	320	348	375	400	424	455	480	510	532	590	640	695	750	795	850
15	244	276	305	337	368	400	432	460	488	521	552	580	610	674	736	798	864	915	976
16	278	315	350	385	420	455	490	525	556	595	630	660	700	790	840	910	980	1,010	1,110
17	314	355	395	435	472	512	550	591	628	672	710	750	790	870	944	1,020	1,100	1,180	1,260
18	352	400	440	485	530	575	620	660	704	755	800	840	880	970	1,060	1,150	1,240	1,320	1,410
19	392	445	490	540	590	640	690	740	784	840	890	930	980	1,080	1,180	1,280	1,380	1,470	1,570
20	436	490	545	600	656	710	765	820	872	930	980	1,040	1,090	1,200	1,312	1,420	1,530	1,640	1,750
22	530	595	660	725	790	860	930	990	1,060	1,120	1,190	1,260	1,320	1,450	1,580	1,720	1,860	1,990	2,040
24	628	710	790	865	940	1,020	1,100	1,180	1,256	1,340	1,420	1,500	1,580	1,730	1,880	2,040	2,200	2,360	2,520
26	738	830	920	1,020	1,100	1,200	1,300	1,390	1,476	1,570	1,660	1,750	1,840	2,040	2,200	2,400	2,600	2,770	2,970
28	854	900	1,070	1,180	1,280	1,390	1,500	1,600	1,780	1,820	1,920	2,030	2,140	2,360	2,560	2,780	3,000	3,200	3,420
30	980	1,110	1,230	1,350	1,480	1,600	1,720	1,810	1,960	2,100	2,220	2,340	2,460	2,700	2,960	3,200	3,440	3,680	3,920
32	1,120	1,260	1,400	1,540	1,670	1,820	1,960	2,100	2,240	2,380	2,520	3,660	2,800	2,080	3,340	3,630	3,920	4,200	4,480
34	1,260	1,420	1,570	1,740	1,890	2,050	2,220	2,360	2,520	2,680	2,840	3,000	3,140	3,480	3,780	4,090	4,440	4,720	5,040
36	1,420	1,600	1,770	1,950	2,120	2,300	2,500	2,650	2,840	3,000	3,200	3,360	3,540	4,240	4,240	4,620	5,000	5,320	5,680
38	1,560	1,780	1,960	2,170	2,360	2,560	2,750	2,950	3,120	3,350	3,560	3,750	3,920	4,720	4,720	5,110	5,500	5,850	6,250
40	1,740	1,970	2,180	2,400	2,620	2,850	3,060	3,280	3,480	3,720	3,940	4,150	4,360	5,240	5,240	5,680	6,120	6,520	6,960

A convenient formula for calculating the displacement of one double-acting cylinder is square of diameter (in.) x stroke (in.) x rpm x .0009 = piston displacement (cfm). This provides for a reasonable allowance for piston rod on all piston sizes from about 5 in. diameter to about 30 in. diameter. If there were no allowance for the piston rod, the multiplier would be .0009090.

TABLE 8.25 Discharge of Air Through an Orifice

Gage Pressure before Orifice, psi	Nominal Diameter, In.										
	1/64	1/32	1/16	1/8	1/4	3/8	1/2	5/8	3/4	7/8	1
	Discharge, Cu. ft. Free Air Per Min.										
1	.028	0.112	0.450	1.80	7.18	16.2	28.7	45.0	64.7	88.1	115
2	.040	0.158	0.633	2.53	10.1	22.8	40.5	63.3	91.2	124	162
3	.048	0.194	0.775	3.10	12.4	27.8	49.5	77.5	111	152	198
4	.056	0.223	0.892	3.56	14.3	32.1	57.0	89.2	128	175	228
5	.062	0.248	0.993	3.97	15.9	35.7	63.5	99.3	143	195	254
6	.068	0.272	1.09	4.34	17.4	39.1	69.5	109	156	213	278
7	.073	0.293	1.17	4.68	18.7	42.2	75.0	117	168	230	300
9	.083	0.331	1.32	5.30	21.1	47.7	84.7	132	191	260	339
12	.095	0.379	1.52	6.07	24.3	54.6	97.0	152	218	297	388
15	.105	0.420	1.68	6.72	26.9	60.5	108	168	242	329	430
20	.123	0.491	1.96	7.86	31.4	70.7	126	196	283	385	503
25	.140	0.562	2.25	8.98	35.9	80.9	144	225	323	440	575
30	.158	0.633	2.53	10.1	40.5	91.1	162	253	365	496	648
35	.176	0.703	2.81	11.3	45.0	101	180	281	405	551	720
40	.194	0.774	3.10	12.4	49.6	112	198	310	446	607	793
45	.211	0.845	3.38	13.5	54.1	122	216	338	487	662	865
50	.229	0.916	3.66	14.7	58.6	132	235	366	528	718	938
60	.264	1.06	4.23	16.9	67.6	152	271	423	609	828	1,082
70	.300	1.20	4.79	19.2	76.7	173	307	479	690	939	1,227
80	.335	1.34	5.36	21.4	85.7	193	343	536	771	1,050	1,371
90	.370	1.48	5.92	23.7	94.8	213	379	592	853	1,161	1,516
100	.406	1.62	6.49	26.0	104	234	415	649	934	1,272	1,661
110	.441	1.76	7.05	28.2	113	254	452	705	1,016	1,383	1,806
120	.476	1.91	7.62	30.5	122	274	488	762	1,097	1,494	1,951
125	.494	1.98	7.90	31.6	126	284	506	790	1,138	1,549	2,023

Based on 100% coefficient of flow. For well-rounded entrance multiply values by 0.97. For sharp-edged orifices a multiplier of 0.65 may be used.

This table will give approximate results only. For accurate measurements see ASME Power Test Code, Velocity Volume Flow Measurement.

Values for pressures from 1 to 15 psig calculated by standard adiabatic formula.

Values for pressures above 15 psig calculated by approximate formula proposed by S. A. Moss: $w = 0.5303\ aCp_1\ \ T_1$ where w = discharge in lb per sec, a = area of orifice in sq. in., C = coefficient of flow, p_1 = upstream total pressure in psia, and T_1 = upstream temperature in deg F abs.

Values used in calculating above table were $C = 1.0$, p_1 = gage pressure + 14.7 psi, $T_1 = 530$ F abs.

Weights (w) were converted to volumes using density factor of 0.07494 lb. per cu. ft. This is correct for dry air at 14.7 psia and 70°F.

Formula cannot be used where p_1 is less than two times the barometric pressure.

Table 8.26 Standard Weight of Welded and Seamless Steam, Air, Gas and Water Pipe

Nominal Pipe Size, In.	Diameter In.		Thickness, In.	Circumference, In.		Transverse Area, Sq In.			Length of Pipe, Ft Per Sq Ft		Length of Pipe Containing 1 Cu Ft
	External	Internal		External	Internal	External	Internal	Metal	External Surface	Internal Surface	
1/8	0.405	0.269	.068	1.272	0.845	0.129	0.057	0.072	9.431	14.199	2,533.775
¼	0.540	0.364	.088	1.696	1.144	0.229	0.104	0.125	7.073	10.493	1,383.789
3/8	0.675	0.493	.091	2.121	1.549	0.358	0.191	0.167	5.658	7.748	754.360
½	0.840	0.622	.109	2.639	1.954	0.554	0.304	0.250	4.547	6.141	473.906
¾	1.050	0.824	.113	3.299	2.589	0.866	0.533	0.333	3.637	4.635	270.034
1	1.315	1.049	.133	4.131	3.296	1.358	0.864	0.494	2.904	3.641	166.618
1 ¼	1.660	1.380	.140	5.215	4.335	2.164	1.495	0.669	2.301	2.768	96.275
1 ¼	1.900	1.610	.145	5.969	5.058	2.835	2.036	0.799	2.010	2.372	70.733
2	2.375	2.067	.154	7.461	6.494	4.430	3.355	1.075	1.608	1.847	42.913
2 ½	2.875	2.469	.203	9.032	7.757	6.492	4.788	1.704	1.328	1.547	30.077
3	3.500	3.068	.216	10.996	9.638	9.621	7.393	2.228	1.091	1.245	19.479
3 ½	4.000	3.548	.226	12.566	11.146	12.566	9.886	2.680	0.954	1.076	14.565
4	4.500	4.026	.237	14.137	12.648	15.904	12.730	3.174	0.848	0.948	11.312
5	5.563	5.047	.258	17.477	15.856	23.306	20.066	4.300	0.686	0.756	7.198
6	6.625	6.065	.280	20.813	19.054	34.472	28.891	5.581	0.576	0.629	4.984
8	8.625	8.071	.277	27.096	25.356	58.426	51.161	7.265	0.443	0.473	2.815
8	8.625	7.981	.322	27.096	25.073	58.426	50.027	8.399	0.443	0.478	2.878
10	10.750	10.020	.365	33.772	31.479	90.763	78.855	11.908	0.355	0.381	1.826
12	12.750	12.000	.375	40.055	37.699	127.676	113.097	14.579	0.299	0.318	1.273
14 OD	14.000	13.250	.375	43.982	41.626	153.938	137.886	16.052	0.272	0.288	1.044
15 OD	15.000	14.250	.375	47.124	44.768	176.715	159.485	17.230	0.254	0.268	0.903
16 OD	16.000	15.250	.375	50.265	47.909	201.062	182.654	18.408	0.238	0.250	0.788
17 OD	17.000	16.214	.393	53.407	50.938	226.980	206.476	20.504	0.224	0.235	0.697
18 OD	18.000	17.182	.409	56.549	53.979	254.469	231.866	22.603	0.212	0.222	0.621
20 OD	20.000	19.182	.409	62.832	60.262	314.159	288.986	25.173	0.191	0.199	0.498

Based on ASTM Standard Specification A53-33.

Table 8.27 Weight of Water Vapor in One Cubic Foot of Air and Various Temperatures and Percentages of Saturation

	Per cent of Saturation									
	10	20	30	40	50	60	70	80	90	100
Temperature deg F	Weight, Grains									
– 10	0.028	0.057	0.086	0.114	0.142	0.171	0.200	0.228	0.256	0.285
0	0.048	0.096	0.144	0.192	0.240	0.289	0.337	0.385	0.433	0.481
10	0.078	0.155	0.233	0.210	0.388	0.466	0.543	0.621	0.698	0.776
20	0.124	0.247	0.370	0.494	0.618	0.741	0.864	0.988	1.112	1.235
30	0.194	0.387	0.580	0.774	0.968	1.161	1.354	1.548	1.742	1.935
32	0.211	0.422	0.634	0.845	1.056	1.268	1.479	1.690	1.902	2.113
35	0.237	0.473	0.710	0.947	1.183	1.420	1.656	1.893	2.129	2.366
40	0.285	0.570	0.855	1.140	1.424	1.709	1.994	2.279	2.564	2.749
45	0.341	0.683	1.024	1.366	1.707	2.048	2.390	2.731	3.073	3.414
50	0.408	0.815	1.223	1.630	2.038	2.446	2.853	3.261	3.668	4.076
55	0.485	0.970	1.455	1.940	2.424	2.909	3.394	3.879	4.364	4.849
60	0.574	1.149	1.724	2.298	2.872	3.447	4.022	4.596	5.170	5.745
62	0.614	1.228	1.843	2.457	3.071	3.685	4.299	4.914	5.528	6.142
64	0.656	1.313	1.969	2.625	3.282	3.938	4.594	5.250	5.907	6.563
66	0.701	1.402	2.103	2.804	3.504	4.205	4.906	5.607	6.208	7.009
68	0.748	1.496	2.244	2.992	3.740	4.488	5.236	5.974	7.732	7.480
70	0.798	1.596	2.394	3.192	3.990	5.788	5.586	6.384	7.182	7.980
72	0.851	1.702	2.552	3.403	4.254	5.105	5.956	6.806	7.657	8.508
74	0.907	1.813	2.720	3.626	4.553	5.440	6.346	7.523	8.159	9.066
76	0.966	1.931	2.896	3.862	4.828	5.793	6.758	7.724	8.690	9.655
78	1.028	2.055	3.083	4.111	5.138	6.166	7.194	8.222	9.249	10.277
80	1.093	2.187	3.280	4.374	5.467	6.560	7.654	8.747	9.841	10.934
82	1.163	2.325	3.488	4.650	5.813	6.976	8.138	9.301	10.463	11.626
84	1.236	2.471	3.707	4.942	6.178	7.414	8.649	9.885	11.120	12.356
86	1.313	2.625	3.938	5.251	6.564	7.877	9.189	10.502	11.814	13.127
88	1.394	2.787	4.181	5.575	6.968	8.362	9.576	11.150	12.543	13.937
90	1.479	2.958	4.437	5.916	7.395	8.874	10.353	11.832	13.311	14.790
92	1.569	3.138	4.707	7.276	7.844	9.413	10.982	12.551	13.120	15.689
94	1.663	3.327	4.990	6.654	8.317	9.980	11.644	13.307	14.971	16.634
96	1.763	3.525	5.288	7.050	8.813	10.576	12.338	14.101	15.863	17.626
98	1.867	3.734	5.601	7.468	9.336	11.203	13.070	14.937	16.804	18.671
100	1.977	3.953	5.930	7.906	9.883	11.860	13.836	15.813	17.789	19.766

Table 8.28 Atmospheric Pressure and Barometer Readings at Different Altitudes

Altitude above Sea Level, ft	Atmospheric Pressure, Psi	Barometer Reading, In. Hg	Altitude above Sea Level, ft	Atmospheric Pressure, Psi	Barometer Reading, In. Hg
0	14.69	29.92	7,500	11.12	22.65
500	14.42	29.38	8,000	10.91	22.22
1,000	14.16	28.86	8,500	10.70	21.80
1,500	13.91	28.33	9,000	10.50	21.38
2,000	13.66	27.82	9,500	10.30	20.98
2,500	13.41	27.31	10,000	10.10	20.58
3,000	13.16	26.81	10,500	9.90	20.18
3,500	12.92	26.32	11,000	9.71	19.75
4,000	12.68	25.84	11,500	9.52	19.40
4,500	12.45	25.36	12,000	9.34	19.03
5,000	12.22	24.89	12,500	9.15	18.65
5,500	11.99	24.43	13,000	8.97	18.29
6,000	11.77	23.98	13,500	8.80	17.93
6,500	11.55	23.53	14,000	8.62	17.57
7,000	11.33	23.09	14,500	8.45	17.22
			15,000	8.28	16.88

TABLE 8.29 Average Gas Compositions Composition, Per Cent

		Composition, Per Cent							
Gas	By	H_2	CO	CO_2	CH_4	C_2H_6	O_2	N_2	Illuminating
Air (dry)	Volume	....	....	....	....	....	21.0	79.0	
	Weight	....	....	....	....	.. ...	23.2	76.8	
Natural gas	Volume	.....	....	....	85.0	14.0	....	1.00	
	Weight	....	....	....	75.20	23.25	....	1.55	
Blast-furnace gas	Volume	3.00	25.00	12.00	2.00	....	....	58.00	
	Weight	0.21	24.22	18.27	1.11	....	....	56.19	
Coke-oven gas	Volume	56.00	6.00	1.50	30.00	0.50	0.50	2.50	3.00
	Weight	11.27	14.90	6.64	48.32	1.51	1.61	4.28	11.47
Illuminating gas	Volume	35.00	25.00	5.50	12.50	2.50	0.50	5.00	14.00
	Weight	3.59	35.95	12.43	10.27	3.85	0.82	7.20	25.89

At 32°F and 29.92 in. Hg.

TABLE 8.30 K Value and Properties of Various Gases at 60°F and 14.7 Pounds Absolute*

Gas	Symbol	*Cp/Cv* = k	Sp. Gr. Air 1.00	Mol. Wt.	Lb Per Cu ft	Cu ft Per Lb	Boiling Point at Atmos. Press. Deg F	Crit. Temp., Deg F	Crit. Pressure, Lb Abs
Acetylene	C_2H_2	1.3	0.9073	26.0156	0.06880	14.534	- 118	96	910
Air	....	1.395	1.000	28.9752	0.07658	13.059	- 317	- 221	546
Ammonia	NH_3	1.317	0.5888	17.0314	0.04509	22,178	- 28	270	1,638
Argon	A	1.667	1.379	39.944	0.10565	0.467	- 302	- 187	705
Benzene	C_6H_6	1.08	2.6935	78.0468	0.20640	4.845	176	551	700
Butane	C_4H_{10}	1.11	2.067	58.078	0.15350	6.514	31	307	528
Butylene	C_4H_8	1.11	1.9353	56.0624	0.14826	6.7452	21	291	621
Carbon dioxide	CO_2	1.30	1.529	44.000	0.11637	8.593	- 109	88	1,072
Carbon disulfide	CS_2	1.20	2.6298	76.120	0.20139	4.965	115	523	1,116
Carbon monoxide	CO	1.403	0.9672	28.000	0.07407	13.503	- 313	- 218	514
Carbon tetrachloride	CCl_4	1.18	5.332	153.828	0.40650	2.4601	170	541	661
Carbureted water gas	....	1.35	0.4090						
Chlorine	Cl_2	1.33	2.486	70.914	0.18750	5.333	-30	291	118
Dichloromethane	CH_2Cl_2	1.18	3.005	84.9296	0.22450	4.458	105	421	1,490
Ethane	C_2H_6	1.22	1.049	30.0468	0.07940	12.594	- 127	90	717
Ethyl chloride	C_2H_5Cl	1.13	2.365	64.4960	0.17058	5.866	54	370	764
Ethylene	C_2H_4	1.22	0.9748	28.0312	0.07410	13.495	- 155	50	747
Flue gas	....	1.40							
Freon (F-12)	CCl_2F_2	1.13	4.520	120.9140	0.31960	3.129	- 21	233	580
Helium	He	1.66	0.1381	4.002	0.01058	94.510	- 452	- 450	33
Hexane	C_6H_{14}	1.08	2.7395	86.1092	0.22760	4.393	156	454	433
Hexylene	C_6H_{12}	....	2.9201	84.0936	0.22250	4.4951			
Hydrogen	H_2	1.41	0.06952	2.0156	0.00530	188.62	- 423	- 400	188
Hydrogen chloride	HCl	1.41	1.268	36.4648	0.09650	10.371	- 121	124	1,198
Hydrogen sulfide	H_2S	1.30	1.190	34.0756	0.09012	11.096	- 75	212	1,306
Isobutane	C_4H_{10}	1.11	2.0176	58.078	0.15365	6.5135	14	273	543
Isopentane	C_5H_{12}	....	2.5035	72.0936	0.19063	5.2451			
Methane	CH_4	1.316	0.5544	16.0312	0.04234	23.626	- 258	- 116	672
Methyl chloride	CH_3Cl	1.20	1.785	50.4804	0.13365	7.491	- 11	289	966
Naphthalene	$C_{10}H_8$	....	4.423	128.0624	0.33870	2.952			
Natural gas† (app. avg.)	....	1.269	0.6655	19.463	0.05140	19.451	....	- 80	670
Neon	Ne	1.642	0.6961	20.183	0.05332	18.784	- 410	- 380	389
Nitric oxide	NO	1.40	1.037	30.008	0.07935	12.605	- 240	- 137	954
Nitrogen	N_2	1.40	0.9672	28.016	0.07429	13.460	-320	- 232	492
Nitrous oxide	N_2O	1.311	1.530	44.016	0.11632	8.595	- 129	98	1,053
Oxygen	O_2	1.398	1.105	32.000	0.08463	11.816	- 297	- 182	730
Pentane	C_5H_{12}	1.06	2.471	72.0936	0.19055	5.248	97	387	485
Phenol	C_6J_5OH	....	3.2655	94.0486	0.24870	4.022	360	786	889
Propane	C_3H_8	1.15	1.562	44.0624	0.11645	8.587	- 48	204	632
Propylene	C_3H_6	....	1.4505	42.0468	0.11115	8.997	- 52	198	661
Refinery gas† (app. avg.)	....	1.20							
Sulfur oxide	SO_2	1.256	2.264	64.060	0.16945	5.901	14	315	1,141
Water vapor (steam)	H_2O	1.33‡	0.6217	18.0156	0.04761	21.004	212	706	3,206

* From "Plain Talks on Air and Gas Compression."
† To obtain exact characteristics of natural gas and refinery gas, the exact constituents must be known.
‡ This k value is given at 212ºF. All others are at 60ºF. Authorities differ slightly; hence above data are average results.

Table 8.31 Dynamic Viscosity of Gases at Atmospheric Pressure

	50F	100F	150F	200F	300F	400F	500F
Air	.37	.40	.42	.44	.49	.53	.56
Carbon Dioxide	.29	.31	.33	.36	.40	.45	.50
CO and N_2	.36	.38	.40	.43	.48	.52	.56
Ethane	.19	.20	.22	.23	.26	.29	.32
Hydrogen	.20	.20	.215	.23	.25	.27	.28
Methane	.23	.24	.25	.27	.29	.32	.35
Oxygen	.41	.44	.46	.49	.53	.58	.63

Slugs per ft. sec. x 10^6.

Table 8.32 Approximate Coefficient of Friction for Clean Commercial Iron and Steel Pipe, Steady Flow

Reynold's Number	Coefficient of Friction
20,000	.0077
50,000	.0063
100,000	.0055
200,000	.0048
500,000	.0041
1,000,000	.0036
5,000,000	.0032
10,000,000	.0026

TABLE 8.33 Number of Tools That Can Be Operated by One Compressor

	Compressor Capacity, Cu Ft Per Min																	
	75		85		125		150		250		365		600		900		1200	
	Air Pressure, Psi																	
	70	90	70	90	70	90	70	90	70	90	70	90	70	90	70	90	70	90
Rock drills:																		
Very light, wet or dry	2	2	3	2	4	3	5	4	8	7	11	10	19	16	30	24	38	32
Very light, blower style	2	1	2	1	3	2	4	3	6	5	9	8	16	14	24	21	32	38
Light, wet or dry	1	1	1	1	2	1	3	2	4	3	6	5	10	8	15	12	20	16
Light, blower style	1	1	1	1	1	1	2	1	3	2	5	4	9	7	13	11	18	14
Medium, wet or dry	1	..	1	..	1	1	2	1	3	2	5	3	7	6	10	9	14	12
Medium, blower style	1	..	1	..	1	1	2	1	3	2	4	3	7	6	10	9	14	12
Heavy, blower	..	..	1	..	1	1	1	1	2	2	4	3	6	4	9	7	12	9
Wagon & crawler drills:																		
Lightweight	..	..	..	..	..	..	1	..	1	1	2	1	3	2	4	3	6	4
Medium-weight	..	..	..	..	..	..	..	..	1	..	1	1	2	1	3	2	4	3
Heavyweight	..	..	..	..	..	..	..	..	..	..	1	..	1	1	2	1	2	2
Paving breaker:																		
Light, horizontal and light work	2	1	2	1	3	2	4	3	7	5	10	8	15	13	22	20	30	26
Medium, light and general work	1	1	1	1	2	2	3	2	5	4	8	6	14	10	21	15	27	20
Heavy, general work	1	1	1	1	2	1	2	2	4	3	6	5	10	7	15	11	20	16
Heavy, concrete breaking	1	1	1	1	2	1	2	2	4	3	6	5	10	8	15	12	20	16
Heavy, all-purpose	1	1	1	1	2	1	2	2	4	3	6	5	10	8	15	12	20	16
Sheeting drivers:																		
Light, gravel	1	1	2	1	2	1	4	2	6	4	10	6	14	12	21	18	28	24
Light, general	1	1	2	1	2	1	4	2	6	4	10	6	14	12	21	18	28	24
Light, stiff ground	1	1	2	2	3	2	4	3	7	5	13	8	15	14	22	21	30	28
Drill-steel cutter	..	..	1	1	1	1	2	2	3	2	4	3	*	*	*	*	*	*
Sharpeners:																		
Lightweight	1	1	2	2	3	3	4	4	6	6	*	*	*	*	*	*	*	*
Medium-weight	1	1	1	1	1	1	2	2	3	3	*	*	*	*	*	*	*	*
Heavyweight, high production	..	..	..	..	..	..	1	1	1	1	2	2	4	4	6	6	8	8
Heavyweight, greater production	..	..	..	..	..	..	1	1	1	1	2	2	3	3	4	4	6	6
Furnaces:																		
Light	3	3	5	5	6	6	10	10	*	*	*	*	*	*	*	*	*	*
Heavy	2	2	4	4	5	5	8	8	*	*	*	*	*	*	*	*	*	*
Heating bits, rods and steels	2	2	4	4	5	5	8	8	*	*	*	*	*	*	*	*	*	*
Grinder for detachable rock-drill bits	1	1	1	1	2	2	3	2	5	4	7	6	12	10	18	15	24	20
Diggers:																		
Light, medium-duty and trench digger	2	1	3	2	4	3	5	4	8	5	10	9	16	13	24	20	32	27
Medium- and heavy-duty	2	1	2	1	3	2	4	3	7	5	10	8	16	13	24	20	32	27
Heavy-duty	1	1	1	1	2	2	3	3	5	4	7	7	12	11	17	15	27	25
Riveting hammers:																		
4-in. stroke, ⅜-in. capacity	3	2	4	3	6	5	9	7	13	11	24	18	*	*	*	*	*	*
5-in. stroke, ¾ in. capacity	3	2	4	3	6	5	8	6	12	10	20	15	*	*	*	*	*	*
6-in. stroke, ⅞-in. capacity	2	2	3	2	5	4	8	5	11	9	19	14	*	*	*	*	*	*
8-in. stroke, 1⅛-in. capacity	2	2	3	2	5	4	7	5	10	8	18	14	*	*	*	*	*	*
9-in. stroke, 1¼-in. capacity	2	1	3	2	4	3	7	5	9	7	17	13	*	*	*	*	*	*

TABLE 8.33 continued

	Compressor Capacity, Cu Ft Per Min																	
	75		85		125		150		250		365		600		900		1200	
	Air Pressure, Psi																	
	70	90	70	90	70	90	70	90	70	90	70	90	70	90	70	90	70	90
Jam rivets, ⅞-in. capacity	1	1	1	1	2	1	3	2	4	3	7	5	*	*	*	*	*	*
Short jam rivets, 1⅛-in. capacity	1	1	1	1	2	1	3	2	4	3	7	5	*	*	*	*	*	*
Comp. jam rivets, ¾ in.	4	2	4	3	6	5	9	6	14	10	18	12	*	*	*	*	*	*
Rivet buster up to ⅞-in. cap.	2	1	3	2	4	4	6	5	8	7	14	11	*	*	*	*	*	*
Caulking and chipping hammers:																		
1-in. stroke, medium chipping	5	3	6	4	9	6	10	7	18	11	26	17	42	28	64	47	85	57
2-in. stroke, medium and heavy chipping	5	3	6	4	9	6	10	7	17	11	26	17	42	28	64	47	85	57
3-in. stroke, heavy chipping	5	3	6	4	9	6	10	7	17	11	26	17	42	28	64	47	85	57
3-in. stroke, general chipping	5	3	6	4	9	6	10	7	17	11	26	17	42	28	64	47	85	57
Light caulking and scaling	5	3	6	4	9	6	10	7	17	11	26	17	42	28	64	47	85	57
Light and medium caulking and scaling	5	3	6	4	9	6	10	7	17	11	26	17	42	28	64	47	85	57
Scaling tools:																		
Light-duty and boiler scaling tool	11	9	13	10	19	15	23	18	38	31	56	45	*	*	*	*	*	*
Heavy-duty scaling hammer	8	7	10	8	15	12	18	15	31	25	45	36	*	*	*	*	*	*
Grinders:																		
Die grinder	6	5	7	5	10	8	12	10	20	16	30	24	*	*	*	*	*	*
Grinder 2-in. diameter wheel cap	2	2	2	2	4	3	5	4	8	6	12	10	20	16	*	*	*	*
Grinder 6-in. diameter wheel cap	1	1	2	1	3	2	3	3	6	5	9	7	15	12	*	*	*	*
Grinder 8-in. diameter wheel cap	1	1	1	1	2	1	2	2	4	3	6	5	11	9	17	13	*	*
Wire brushing machine, 8-in. radial brush cap	1	1	2	1	3	2	3	3	6	5	9	7	15	12	*	*	*	*
For squaring shanks on rock-drill steels	2	2	3	2	4	3	4	4	9	7	15	11	*	*	*	*	*	*
Grinders:																		
5-in. cup wheel or 7-in. sanding pad cap	1	1	2	1	3	2	3	3	6	5	9	7	15	12	*	*	*	*
6-in. cup wheel or 9-in. sanding pad cap	1	1	1	1	2	1	2	2	4	3	6	5	11	8	*	*	*	*
Drills, drilling, reaming, tapping:																		
Heavy-duty drilling machine up to ⅜-in. cap	3	2	3	3	5	4	6	5	11	8	16	13	*	*	*	*	*	*
Drilling and reaming up to ⅞-in.	1	1	2	1	3	2	3	3	6	5	9	7	*	*	*	*	*	*
Close-quarter drilling and reaming machine up to 1 1/16-in.	2	2	3	2	5	4	7	5	11	9	18	14	*	*	*	*	*	*
Machines up to 1 ¼ in.	..	..	..	..	..	..	..	..	6	5	8	7	*	*	*	*	*	*
Machines up to 2 in.	..	..	..	..	..	..	..	..	3	2	5	3	*	*	*	*	*	*
Drilling and reaming machines up to 1 ¼ in. cap.	1	1	1	1	2	2	3	2	4	3	8	5	*	*	*	*	*	*
Machines up to 2-in. cap.	..	..	..	..	1	1	2	1	3	2	5	4	*	*	*	*	*	*

TABLE 8.33 continued

	Compressor Capacity, Cu Ft Per Min																	
	75		85		125		150		250		365		600		900		1200	
	Air Pressure, Psi																	
	70	90	70	90	70	90	70	90	70	90	70	90	70	90	70	90	70	90
Wood borers:																		
Wood-boring machines up to 1-in. cap.	1	1	2	1	3	2	3	3	6	5	9	7	15	12	*	*	*	*
Up to 4-in cap.	2	2	3	2	5	4	7	5	10	8	16	13	*	*	*	*	*	*
Torque wrenches:																		
Right-angle nut running machines up to ¾ in. bolt cap. torque type	1	1	2	1	3	2	3	3	6	5	9	7	15	12	22	18	30	24
Reversible nut running machine, impact type:																		
Capacity up to ⅜-in, bolt size nuts	5	4	5	4	8	6	10	8	17	13	25	20	41	33	62	50	83	66
Up to ⅝ in.	3	2	3	2	5	4	6	5	10	8	15	12	25	20	37	30	50	40
Up to ¾ in.	2	2	3	2	4	3	5	4	9	7	13	10	21	17	32	26	43	35
Up to 1 ¼ in.	2	1	2	2	4	3	4	3	8	6	11	9	19	15	29	23	38	31
Up to 1 ¾ in.	..	..	1	1	1	1	3	2	3	3	6	4	8	6	12	9	18	14
Port. circ. saw, 12-in. blade	2	1	2	2	4	3	6	4	9	7	16	12	*	*	*	*	*	*
Concrete vibrators:																		
Light-duty concrete vibrator	2	2	3	2	6	4	9	7	14	10	22	17	*	*	*	*	*	*
Medium, heavy-duty	1	..	1	1	2	1	3	2	5	3	8	6	*	*	*	*	*	*
Port. sing. drum hoists:																		
Capacity up to 2,000 lb pull, slow rope speed on single line	..	..	..	1†	..	1†	..	2†	..	3†	..	6†	..	7†	..	9†	..	14†
High rope speed	..	..	..	..	..	..	..	..	..	1†	..	2†	..	2†	..	3†	..	4†
Up to 3,500 lb	..	..	..	..	..	..	..	..	..	..	..	1†	..	1†	..	2†	..	3†
Heavy-duty backfill-tamping machine	1	1	3	2	5	3	5	4	9	7	11	9	14	11	21	16	28	22
Sump pump:																		
Portable centrifugal-type, low heads	1	..	1	..	2	1	2	2	3	2	5	4	9	7	14	11	19	15
Two tandem connected pumps, higher heads	..	..	..	..	..	..	1	1	2	1	2	1	3	3	5	4	8	6
Portable centrifugal-type, high heads	..	..	..	..	..	..	1	1	1	1	1	1	2	2	3	3	4	4

* Compressor capacity more than that needed for usual number of tools.

† Hoist figures based on 80-lb pressure.

This table is based upon a factor of intermittent use, that is, that all tools are not operated simultaneously. Many conditions may develop in which more or fewer tools may be operated than shown above.

GLOSSARY

Absolute Pressure – Total pressure measured from zero.

Absolute Temperature – See **Temperature, Absolute.**

Absorption – The chemical process by which a hygroscopic desiccant, having a high affinity with water, melts and becomes a liquid by absorbing the condensed moisture.

Actual Capacity – Quantity of gas actually compressed and delivered to the discharge system at rated speed and under rated conditions. Also called Free Air Delivered (FAD).

Adiabatic Compression – See **Compression, Adiabatic.**

Adsorption – The process by which a desiccant with a highly porous surface attracts and removes the moisture from compressed air. The desiccant is capable of being regenerated.

Air Receiver – See **Receiver.**

Air Bearings – See **Gas Bearings.**

Aftercooler – A heat exchanger used for cooling air discharged from a compressor. Resulting condensate may be removed by a moisture separator following the aftercooler.

Atmospheric Pressure – The measured ambient pressure for a specific location and altitude.

Automatic Sequencer – A device which operates compressors in sequence according to a programmed schedule.

Brake Horsepower (bhp) – See **Horsepower, Brake.**

Capacity – The amount of air flow delivered under specific conditions, usually expressed in cubic feet per minute (cfm).

Capacity, Actual – The actual volume flow rate of air or gas compressed and delivered from a compressor running at its rated operating conditions of speed, pressures, and temperatures. Actual capacity is generally expressed in actual cubic feet per minute (acfm) at conditions prevailing at the compressor inlet.

Capacity Gauge – A gauge that measures air flow as a percentage of capacity, used in rotary screw compressors.

Check Valve – A valve which permits flow in only one direction.

Clearance – The maximum cylinder volume on the working side of the piston minus the displacement volume per stroke. Normally it is expressed as a percentage of the displacement volume.

Clearance Pocket – An auxiliary volume that may be opened to the clearance space, to increase the clearance, usually temporarily, to reduce the volumetric efficiency of a reciprocating compressor.

Compressibility – A factor expressing the deviation of a gas from the laws of thermodynamics. (See also **Supercompressibility**)

Compression, Adiabatic – Compression in which no heat is transferred to or from the gas during the compression process.

Compression, Isothermal – Compression in which the temperature of the gas remains constant.

Compression, Polytropic – Compression in which the relationship between the pressure and the volume is expressed by the equation PV^n is a constant.

Compression Ratio – The ratio of the absolute discharge pressure to the absolute inlet pressure.

Constant Speed Control – A system in which the compressor is run continuously and matches air supply to air demand by varying compressor load.

Critical Pressure – The limiting value of saturation pressure as the saturation temperature approaches the critical temperature.

Critical Temperature – The highest temperature at which well-defined liquid and vapor states exist. Sometimes it is defined as the highest temperature at which it is possible to liquify a gas by pressure alone.

Cubic Feet Per Minute (cfm) – Volumetric air flow rate.

cfm, Free Air – cfm of air delivered to a certain point at a certain condition, converted back to ambient conditions.

Actual cfm (acfm) – Flow rate of air at a certain point at a certain condition at that point.

Inlet cfm (icfm) – Cfm flowing through the compressor inlet filter or inlet valve under rated conditions.

Standard cfm – Flow of free air measured and converted to a standard set of reference conditions (14.5 psia, 68°F, and 0% relative humidity).

Cut-In/Cut-Out Pressure – Respectively, the minimum and maximum discharge pressures at which the compressor will switch from unload to load operation (cut in) or from load to unload (cut out).

Cycle – The series of steps that a compressor with unloading performs; 1) fully loaded, 2) modulating (for compressors with modulating control), 3) unloaded, 4) idle.

Cycle Time – Amount of time for a compressor to complete one cycle.

Degree of Intercooling – The difference in air or gas temperature between the outlet of the intercooler and the inlet of the compressor.

Deliquescent – Melting and becoming a liquid by absorbing moisture.

Desiccant – A material having a large proportion of surface pores, capable of attracting and removing water vapor from the air.

Dew Point – The temperature at which moisture in the air will begin to condense if the air is cooled at constant pressure. At this point the relative humidity is 100%.

Demand – Flow of air at specific conditions required at a point or by the overall facility.

Diaphragm – A stationary element between the stages of a multi-stage centrifugal compressor. It may include guide vanes for directing the flowing medium to the impeller of the succeeding stage. In conjunction with an adjacent diaphragm, it forms the diffuser surrounding the impeller.

Diaphragm cooling – A method of removing heat from the flowing medium by circulation of a coolant in passages built into the diaphragm.

Diffuser – A stationary passage surrounding an impeller, in which velocity pressure imparted to the flowing medium by the impeller is converted into static pressure.

Digital Controls – See **Logic Controls.**

Discharge Pressure – Air pressure produced at a particular point in the system under specific conditions.

Discharge Temperature – The temperature at the discharge flange of the compressor.

Displacement – The volume swept out by the piston or rotor(s) per unit of time, normally expressed in cubic feet per minute.

Droop – The drop in pressure at the outlet of a pressure regulator, when a demand for air occurs.

Dynamic Type Compressors – Compressors in which air or gas is compressed by the mechanical action of rotating impellers imparting velocity and pressure to a continuously flowing medium. (Can be centrifugal or axial design.)

Efficiency – Any reference to efficiency must be accompanied by a qualifying statement which identifies the efficiency under consideration, as in the following definitions of efficiency:

> **Efficiency, Compression** – Ratio of theoretical power to power actually imparted to the air or gas delivered by the compressor.
>
> **Efficiency, Isothermal** – Ratio of the theoretical work (as calculated on a isothermal basis) to the actual work transferred to a gas during compression.
>
> **Efficiency, Mechanical** – Ratio of power imparted to the air or gas to brake horsepower (bhp).
>
> **Efficiency, Polytropic** – Ratio of the polytropic compression energy transferred to the gas, to the actual energy transferred to the gas.
>
> **Efficiency, Volumetric** – Ratio of actual capacity to piston displacement.

Exhauster – A term sometimes applied to a compressor in which the inlet pressure is less than atmospheric pressure.

Expanders – Turbines or engines in which a gas expands, doing work, and undergoing a drop in temperature. Use of the term usually implies that the drop in temperature is the principle objective. The orifice in a refrigeration system also performs this function, but the expander performs it more nearly isentropically, and thus is more effective in cryogenic systems.

Filters – Devices for separating and removing particulate matter, moisture or entrained lubricant from air.

Flange connection – The means of connecting a compressor inlet or discharge connection to piping by means of bolted rims (flanges).

Fluidics – The general subject of instruments and controls dependent upon low rate of flow of air or gas at low pressure as the operating medium. These usually have no moving parts.

Free Air – Air at atmospheric conditions at any specified location, unaffected by the compressor.

Full-Load – Air compressor operation at full speed with a fully open inlet and discharge delivering maximum air flow.

Gas – One of the three basic phases of matter. While air is a gas, in pneumatics the term gas normally is applied to gases other than air.

Gas Bearings – Load carrying machine elements permitting some degree of motion in which the lubricant is air or some other gas.

Gauge Pressure – The pressure determined by most instruments and gauges, usually expressed in psig. Barometric pressure must be considered to obtain true or absolute pressure.

Guide Vane – A stationary element that may be adjustable and which directs the flowing medium approaching the inlet of an impeller.

Head, Adiabatic – The energy, in foot pounds, required to compress adiabatically to deliver one pound of a given gas from one pressure level to another.

Head, Polytropic – The energy, in foot pounds, required to compress polytropically to deliver one pound of a given gas from one pressure level to another.

Horsepower, Brake – Horsepower delivered to the output shaft of a motor or engine, or the horsepower required at the compressor shaft to perform work.

Horsepower, Indicated – The horsepower calculated from compressor indicator diagrams. The term applies only to displacement type compressors.

Horsepower, Theoretical or Ideal – The horsepower required to isothermally compress the air or gas delivered by the compressor at specified conditions.

Humidity, Relative – The relative humidity of a gas (or air) vapor mixture is the ratio of the partial pressure of the vapor to the vapor saturation pressure at the dry bulb temperature of the mixture.

Humidity, Specific – The weight of water vapor in an air vapor mixture per pound of dry air.

Hysteresis – The time lag in responding to a demand for air from a pressure regulator.

Impeller – The part of the rotating element of a dynamic compressor which imparts energy to the flowing medium by means of centrifugal force. It consists of a number of blades which rotate with the shaft.

Indicated Power – Power as calculated from compressor-indicator diagrams.

Indicator Card – A pressure – volume diagram for a compressor or engine cylinder, produced by direct measurement by a device called an indicator.

Inducer – A curved inlet section of an impeller.

Inlet Pressure – The actual pressure at the inlet flange of the compressor.

Intercooling – The removal of heat from air or gas between compressor stages.

Intercooling, Degree of – The difference in air or gas temperatures between the inlet of the compressor and the outlet of the intercooler.

Intercooling, Perfect – When the temperature of the air or gas leaving the intercooler is equal to the temperature of the air or gas entering the inlet of the compressor.

Isentropic Compression – See **Compression, Isentropic.**

Isothermal Compression – See **Compression, Isothermal.**

Leak – An unintended loss of compressed air to ambient conditions.

Liquid Piston Compressor – A compressor in which a vaned rotor revolves in an elliptical stator, with the spaces between the rotor and stator sealed by a ring of liquid rotating with the impeller.

Load Factor – Ratio of average compressor load to the maximum rated compressor load over a given period of time.

Load Time – Time period from when a compressor loads until it unloads.

Load/Unload Control – Control method that allows the compressor to run at full-load or at no load while the driver remains at a constant speed.

Modulating Control – System which adapts to varying demand by throttling the compressor inlet proportionally to the demand.

Multi-casing Compressor – Two or more compressors, each with a separate casing, driven by a single driver, forming a single unit.

Multi-stage Axial Compressor – A dynamic compressor having two or more rows of rotating elements operating in series on a single rotor and in a single casing.

Multi-stage Centrifugal Compressor – A dynamic compressor having two or more impellers operating in series in a single casing.

Multi-stage Compressors – Compressors having two or more stages operating in series.

Perfect Intercooling – The condition when the temperature of air leaving the intercooler equals the of air at the compressor intake.

Performance Curve – Usually a plot of discharge pressure versus inlet capacity and shaft horsepower versus inlet capacity.

Piston Displacement – The volume swept by the piston; for multi-stage compressors, the piston displacement of the first stage is the overall piston displacement of the entire unit.

Pneumatic Tools – Tools that operate by air pressure.

Polytropic Compression – See **Compression, Polytropic.**

Polytropic Head – See **Head, Polytropic.**

Positive Displacement Compressors – Compressors in which successive volumes of air or gas are confined within a closed space, and the space mechanically reduced, results in compression. These may be reciprocating or rotating.

Power, theoretical (polytropic) – The mechanical power required to compress polytropically and to deliver, through the specified range of pressures, the gas delivered by the compressor.

Pressure – Force per unit area, measured in pounds per square inch (psi).

Pressure, Absolute – The total pressure measured from absolute zero (i.e. from an absolute vacuum).

Pressure, Critical – See **Critical Pressure.**

Pressure Dew Point – For a given pressure, the temperature at which water will begin to condense out of air.

Pressure, Discharge – The pressure at the discharge connection of a compressor. (In the case of compressor packages, this should be at the discharge connection of the package)

Pressure Drop – Loss of pressure in a compressed air system or component due to friction or restriction.

Pressure, Intake – The absolute total pressure at the inlet connection of a compressor.

Pressure Range – Difference between minimum and maximum pressures for an air compressor. Also called cut in-cut out or load-no load pressure range.

Pressure Ratio – See **Compression Ratio.**

Pressure Rise – The difference between discharge pressure and intake pressure.

Pressure, Static – The pressure measured in a flowing stream in such a manner that the velocity of the stream has no effect on the measurement.

Pressure, Total – The pressure that would be produced by stopping a moving stream of liquid or gas. It is the pressure measured by an impact tube.

Pressure, Velocity – The total pressure minus the static pressure in an air or gas stream.

Rated Capacity – Volume rate of air flow at rated pressure at a specific point.

Rated Pressure – The operating pressure at which compressor performance is measured.

Required Capacity – Cubic feet per minute (cfm) of air required at the inlet to the distribution system.

Receiver – A vessel or tank used for storage of gas under pressure. In a large compressed air system there may be primary and secondary receivers.

Reciprocating Compressor – Compressor in which the compressing element is a piston having a reciprocating motion in a cylinder.

Relative Humidity – The ratio of the partial pressure of a vapor to the vapor saturation pressure at the dry bulb temperature of a mixture.

Reynolds Number – A dimensionless flow parameter ($\vartheta v \rho / \mu$), in which ϑ is a significant dimension, often a diameter, v is the fluid velocity, ρ is the mass density, and μ is the dynamic viscosity, all in consistent units.

Rotor – The rotating element of a compressor. In a dynamic compressor, it is composed of the impeller(s) and shaft, and may include shaft sleeves and a thrust balancing device.

Seals – Devices used to separate and minimize leakage between areas of unequal pressure.

Sequence – The order in which compressors are brought online.

Shaft – The part by which energy is transmitted from the prime mover through the elements mounted on it, to the air or gas being compressed.

Sole Plate – A pad, usually metallic and embedded in concrete, on which the compressor and driver are mounted.

Specific Gravity – The ratio of the specific weight of air or gas to that of dry air at the same pressure and temperature.

Specific Humidity – The weight of water vapor in an air-vapor mixture per pound of dry air.

Specific Power – A measure of air compressor efficiency, usually in the form of bhp/100 acfm.

Specific Weight – Weight of air or gas per unit volume.

Speed – The speed of a compressor refers to the number of revolutions per minute (rpm) of the compressor drive shaft or rotor shaft.

Stages – A series of steps in the compression of air or a gas.

Standard Air – The Compressed Air & Gas Institute and PNEUROP have adopted the definition used in ISO standards. This is air at 14.5 psia (1 bar); 68°F (20°C) and dry (0% relative humidity).

Start/Stop Control – A system in which air supply is matched to demand by the starting and stopping of the unit.

Supercompressibility – See **Compressibility.**

Surge – A phenomenon in centrifugal compressors where a reduced flow rate results in a flow reversal and unstable operation.

Surge Limit – The capacity in a dynamic compressor below which operation becomes unstable.

Temperature, Absolute – The temperature of air or gas measured from absolute zero. It is the Fahrenheit temperature plus 459.6 and is known as the Rankine temperature. In the metric system, the absolute temperature is the Centigrade temperature plus 273 and is known as the Kelvin temperature.

Temperature, Critical – See **Critical Temperature.**

Temperature, Discharge – The total temperature at the discharge connection of the compressor.

Temperature, Inlet – The total temperature at the inlet connection of the compressor.

Temperature Rise Ratio – The ratio of the computed isentropic temperature rise to the measured total temperature rise during compression. For a perfect gas, this is equal to the ratio of the isentropic enthalpy rise to the actual enthalpy rise.

Temperature, Static – The actual temperature of a moving gas stream. It is the temperature indicated by a thermometer moving in the stream and at the same velocity.

Temperature, Total – The temperature which would be measured at the stagnation point if a gas stream were stopped, with adiabatic compression from the flow condition to the stagnation pressure.

Theoretical Power – The power required to compress a gas isothermally through a specified range of pressures.

Torque – A torsional moment or couple. This term typically refers to the driving couple of a machine or motor.

Total Package Input Power – The total electrical power input to a compressor, including drive motor, belt losses, cooling fan motors, VSD or other controls, etc.

Unit Type Compressors – Compressors of 30 bhp or less, generally combined with all components required for operation.

Unload – (No load) Compressor operation in which no air is delivered due to the intake being closed or modified not to allow inlet air to be trapped.

Vacuum Pumps – Compressors which operate with an intake pressure below atmospheric pressure and which discharge to atmospheric pressure or slightly higher.

Valves – Devices with passages for directing flow into alternate paths or to prevent flow.

Volute – A stationary, spiral shaped passage which converts velocity head to pressure in a flowing stream of air or gas.

Water-cooled Compressor – Compressors cooled by water circulated through jackets surrounding cylinders or casings and/or heat exchangers between and after stages.

METRICATION

TABLE 8.34 Preferred SI (Metric) Units

Quantity	Unit	Symbol	Metric-U.S. Customary Unit Equivalents	Remarks
LENGTH	meter	m	1 m = 1000 mm = 39.37 in. = 3.281 ft.	Use mm for dimensions on product engineering drawings, μm for surface finish, clearance & vibration amplitude.
	millimeter	mm	25.4 mm = 1 inch	
	micrometer	μm	1 μm = 10^{-6} m	
			25.4 μm = 1 mil = .001 inch	
AREA	square meter	m^2	1 m^2 = 10.764 $ft.^2$	
	square millimeter	mm^2	645.16 mm^2 = 1 $inch^2$	
			100 m^2 = 119.6 yd. $inch^2$	
			10,000 m^2 = 2.47 acres	
MASS	kilogram	kg	1 kg =2.205 lb_m	
VOLUME FLOW RATE-GASES	cubic meter per second	m^3/s	1 m^3/s = 2118.9 ft^3/min	Allow time to vary to provide suitable numbers.
	cubic meter per minute	m^3/min	1 m^3/min = 35.315 ft^3/min	
VOLUME FLOW RATE-LIQUIDS	liter per second	L/s	1 L/sec = 15.85 gpm	Allow time to vary to provide suitable numbers.
	liter per minute	L/min	1 L/min = .2642 gpm	
PRESSURE	bar	bar	1 bar = 14.5 psi (lb_f/lb^2 = 100 kPa)	bar more commonly used in industry. kPa used in Academia and technical publications.
	kilopascal	kPa	1 kPa = 1 kN/m^2 = 145 psi	
STRESS	megapascal	MPa	1 Mpa = MN/m^2 = 10^6 Pa = 145 psi (lb_f/in^2)	
POWER	watt	W	1 W = 1 J/s = 1 N · m/s = 44.25 ft-lb_f/min	
	kilowatt	kW	1 kW = 1 kJ/s = 1.34 hp (horsepower)	
VOLUME SPECIFIC ENERGY-GASES	kilojoule per cubic meter	kJ/m^3		1 kJ/m^3 = 1 J/L
VOLUME SPECIFIC ENERGY-LIQUIDS	joule per liter	J/L		
ENERGY, WORK, QUANTITY OF HEAT	joule	J	1 J = n · m = .7376 ft-lb_f 1 J = .948 x 10^{-3} Btu 1 kJ = .948 Btu	

METRICATION

TABLE 8.34 Preferred SI (Metric) Units (continued)

ROTATIONAL SPEED	revolution per second	r/s		
	revolution per minute	r/min		
VOLUME-GASES	cubic meter	m^3	1 m^3 = 35.315 ft^3	
VOLUME-LIQUIDS	liter	L	1 liter = .2642 gallon	
DENSITY	kilogram per cubic meter	kg/m^3	1 kg/m^3 = .0624 lb_m/ft^3	
VELOCITY	meter per second	m/s	1 m/s = 3.281 ft/sec	
VELOCITY-VEHICLE	kilometer per hour	km/h	1 km/h = .6214 miles/hr	
TEMPERATURE	degrees Celsius	°C	$t_{°C} = (t_{°F} - 32)\ 5/9$	Use °C for Celsius temperature and K for absolute temperature (thermodynamic)
	Kelvin	K	$T_k = T_{°C} + 273.15$	
SOUND PRESSURE LEVEL	decibel	dB		Reference level is 20 μ Pa = .0002 μ bar. Therefore unit remains the same.
VISCOSITY (DYNAMIC)	millipascal-second	mPa · s	1 mPa · s = 1 cP (centipoise)	
VISCOSITY (KINEMATIC)	square millimeter per second	mm^2/s	1 mm^2/s = 1 cSt (centistoke)	
MOISTURE CONTENT	kilogram per cubic feet	kg/m^3	1 kg/m^3 = .0624 lb_m/ft^3	
FORCE	Newton	N	1 N = .2248 lb_f	
	kilonewton	kN	1 kN = 224.8 lb_f	
MOMENT OF FORCE (TORQUE)	newton-meter	N · m	1 N · m = .7376 lb_f-ft	
MOMENT OF INERTIA	kilogram-meter squared	kg · m^2	1 kg · m^2 = 23.73 lb_m-ft^2	
FREQUENCY	hertz	Hz	1 Hz = 1 cycle per second	
GAS CONSTANT	joule per kilogram-kelvin	J(kg · K)	1 J/(kg · K) = .1859 ft-lb_f/lb_m-°R	
SPECIFIC HEAT	joule per kilogram-kelvin	J/(kg · K)	1 J/(kg · K) = .2389 x 10^{-3} Btu/lb_m-°R	

Where there is a choice of SI units depending on quantity, the reference number has been put against the unit likely to be most frequently used.

1. The three units based on cm, dm and m, respectively, roughly correspond to use with fluidics, pneumatic controls, tools (consumption),up to medium-sized compressors, and large compressors. The alternatives of l/s and ml/s were rejected not only because the liter tends to be associated with liquids, but also because of the danger of confusion with l/min., widely used in Europe. One dm^3/s = approximately 2.1 cfm; that is, halving existing cfm tables is accurate within 5 percent and, in the case of consumption, cautious from the user's point of view.
2. This is the consistent unit but the long established use of rpm may call for the continued use of this alternative for some time, but this practice is not to be encouraged.
3. Weights of compressors, air tools, pneumatic equipment, and so on, will normally be described in these units.
4. Standard reference atmospheric conditions are as contained in ISO 1217 [i.e., 1 bar (14.5 psia); 20°C (68°F); 0 percent relative humidity (dry)].
5. The smaller unit (1 millibar = 100 N/m^2) will be used with fluidics and very low pressures. The high vacuum industry may use N/m^2 or rather the internationally and U-K preferred Pascal (Pa); 1 Pa = 1 N/m^2). As with pressure units hitherto in use, "absolute" or "gage" have to be stated where doubt could arise.
   At least one point in any document mentioning *bar*, the conversion 1 bar = 100 kPa should be stated as shown. Submultiples and multiples of Pa are used as with N/m (e.g., mPa, kPa, MPa).
   Designers of air receivers relating the pressure in bars to the MPa stress in the shell in one formula must not forget to include a factor of 10 (10 bars = 1 MPa).
   Users of low pressures and the fluidics industry have come across the use of inches water gage and mm of H_2O. 1 mm H_2O = 0.0985 m bar = 9.85 Pa approximately. Use of the w.g. will continue.
6. See also Note 5 for the explanation of MPa and the reason why this will replace the more cumbersome fraction, NM/m^2, preferable to N/mm^2. 1 ton/in.^{2} = 15.44 MPa.
7. $J = N \cdot m = W \cdot s$, for $W = N \cdot m/s$. 746 W = 1 hp.
8. We are advised by BICEMA that the term *brake kilowatts* is likely to be used as standard practice in describing power outputs previously quoted in bhp (e.g., for prime moves such as diesel engines of portable compressors).

The following is a list of abbreviations of Metric SI Units in the order of their appearance in the last column of Table 8.34:

mm	millimeter (1 m = 1000 mm = 39.37 in. = 3.281 ft.)
m	meter
dm	decimeter (10 dm = 1 m)
cm	centimeter (100 cm = 1 m)
l	liter (originally 1 kg of water). In 1964 the liter was redefined as to be equal to 10^{-3} m^3 = 1 dm^3.
km	kilometer (1000 m)
h	hour
s	second
ml	milliliter (1000 ml = 1 l) = 1 cm^3 (do not write ccm, cc, or ccs)
Hz	hertz (1 Hz = 1 cycle per second)
g	gram
kg	kilogram (= 1000 g)
t	ton (= 1000 kg). The abbreviation is not so widely used as, for instance g and kg, hence the unit is named full in the table.
N	newton. The force that will accelerate a freely movable mass of 1 kg by 1 m/s^2.
kN	kilonewton = $N \times 10^3$
MN	meganewton = $n \times 10^6$
J	Joule (see note 7)
W	watt
kW	kilowatt (= 1000 W)
C	Celsius = centigrade. The use of the word centigrade is deprecated.
K	Kelvin. Note that the ° sign is not used when quoting temperatures in kelvins.
cSt	centistokes.

TABLE 8.35 Metric Conversion Factors

To convert:	Into:	Multiply by:
Atmospheres	Dynes per cm^2	1.0132×10^6
Atmospheres	Kilograms per square meter	1.0332×10^4
Amtospheres	Millimeters of mercury at 0°C	760
Atmospheres	Newtons per square meter	1.0133×10^5
British thermal units (Btu)	Kilogram-calories	0.2520
Centimeters	Feet	3.281×10^{-2}
Centimeters	Inches	0.3937
Centimeters	Mils (10^{-3} in.)	393.7
Centimeters per second	Feet per minute	1.969
Centimeters per second per second	Feet per second per second	3.281×10^{-2}
Circular mils	Square centimeters	5.067×10^{-6}
Cubic inches	Cubic centimeters	16.39
Cubic inches	Cubic meters	1.639×10^{-5}
Cubic inches	Liters	1.639×10^{-2}
Degrees Fahrenheit	Degrees centigrade	°C = 5/9 (°F – 32)
Dynes	Pounds	2.248×10^{-6}
Dyne-centimeters	Pounds-feet	7.376×10^{-8}
Grams	Ounces (avoir.)	3.527×10^{-2}
Grams per cm^3	Pounds per ft^3	62.43
Gram-cm^2	Pound-ft^2	2.37285×10^{-6}
Gram-cm^2	Slug-ft^2	7.37507×10^{-8}
Inches	Centimeters	2.540
Joules (int.)	Foot-pounds	0.7376
Kilograms	Pounds	2.205
Kilogram-calories	Foot-pounds	3.088
Kilometers	Feet	3.281
Liters	Gallons (U.S. liquid)	0.2642
Meters	Yards	1.094
Meters per second	Feet per second	3.28
Newton meters	Pound-feet	0.7376
Ounces (avoir.)	Grams	28.35
Pints (liquid)	Liters	0.4732
Pounds (avoir.)	Grams	453.6
Square Centimeters	Square feet	1.076×10^{-3}
Square Centimeters	Square inches	0.1550
Square feet	Square meters	0.09290

STANDARDS*

We suggest you reference the latest edition of the standards listed below.

Standards Organizations

AGMA = American Gear Manufacturers Association
ANSI = American National Standards Institute
API = American Petroleum Institute
ASME = American Society of Mechanical Engineers
CAGI = Compressed Air & Gas Institute
ISA = Instrument Society of America
ISO = International Standards Organization
NFPA = National Fluid Power Association
OSHA = Occupational Safety and Health Act
PNEUROP = European Committee of Manufacturers of Compressors, Vacuum Pumps and Pneumatic Tools

Note: ANSI and ISO Standards are available through:
ANSI, 25 West 43rd Street, 4th Floor, New York, NY 10036
Telephone: 1-212-642-4900 Fax: 1-212-398-0023 – www.ansi.org

Standards – Compressors

PN2CPTC1*	Acceptance Test Code for Bare Displacement Air Compressors
PN2CPTC2*	Acceptance Test Code for Electrically Driven Packaged Displacement Type Air Compressors
PN2CPTC3*	Acceptance Test Code for I.C. Engine Driven Packaged Displacement Type Air Compressors

* The standards have been incorporated in an Appendix to the latest edition of ISO 1217.

ISO 1217	Displacement Compressors - Acceptance Tests
ISO 5388	Stationary Air Compressors - Safety Rules and Code of Practice
ISO 5389	Turbocompressors - Performance Test Code
ISO 5390	Compressors - Classification
ISO 5941	Compressors, Pneumatic Tools and Machines-Preferred Pressures
ISO 6798	Reciprocating Internal Combustion Engines - Measurement of Airborne Noise
PN8NTC2.3	Measurement of Noise Emitted by Compressors. (Available from PNEUROP)
ISO 2151	Acoustics - Noise Test Code for Compressors and Vacuum Pumps Engineering Method (Grade 2)
API 617	Centrifugal Compressors for Petroleum, Chemical and Gas Industry Services
API 618	Reciprocating Compressors for Petroleum, Chemical and Gas Industry Services

API 619	Rotary Compressors for Petroleum, Chemical and Gas Industry Services
API 672	Packaged Integral Geared Centrifugal Air Compressors
API 681	Liquid Ring Vacuum Pumps and Compressors
ASME B19.1	Safety Standard for Air Compressor Systems
ASME B19.3	Safety Standard for Compressors for Process Industries

Standards – Compressed Air Dryers

CAGI ADF100	Refrigerated Compressed Air Dryers - Methods for Testing and Rating
ANSI/CAGI ADF200	Dual Tower Regenerative Desiccant Compressed Air Dryers - Methods for Testing and Rating
ANSI/CAGI ADF300	Standard for Single Tower (Deliquescent) Compressed Air Dryers - Methods for Testing and Rating
ISO 7183-1	Compressed Air Dryers - Specifications and Testing
ISO 7183-2	Compressed Air Dryers - Part 2: Performance Ratings

Standards – Compressed Air Filters

ANSI/CAGI ADF400	Standards for Testing and Rating Coalescing Filters
ANSI/CAGI ADF500	Standard for Measuring the Adsorption Capacity of Oil Vapor Removal Adsorbent Filters
ANSI/CAGI ADF600	Standards for Particulate Filters
ANSI/CAGI ADF700	Standards for Membrane Compressed Air Dryers
ISO 8573-1	Compressed Air for General Use - Part 1: Contaminants and Quality Classes
ISO 8573-2	Compressed Air for General Use - Part 2: Test Methods for Aerosol Oil Content
ISO 8573-3	Compressed Air for General Use - Part 3: Test Methods for Humidity
ISO 8573-4	Compressed Air for General Use - Part 4: Test Methods for Solid Particle Content
ISO 8573-5	Compressed Air for General Use - Part 5: Test Methods for Oil Vapor and Organic Solvent Content
ISO 8573-6	Compressed Air for General Use - Part 6: Test Methods for Gaseous Contaminant Content
ISO 8573-7	Compressed Air for General Use - Part 7: Test Methods for Viable Microbiological Contaminant Content
ISO 8573-8	Compressed Air for General Use - Part 8: Contaminants and Purity Classes (by Mass Concentration of Solid Particles)
ISO 8573-9	Compressed Air for General Use - Part 9: Test Methods for Liquid Water Content
ISO 8573-10	Compressed Air for General Use - Part 10: Test Methods for Mass Concentration of Solid Particle Content

Standards – Pneumatic Tools

CAGI B186.1	Safety Code for Portable Air Tools
ISO 2787	Rotary and Percussive Tools - Performance Tests
ISO 5391	Pneumatic Tools and Machines - Vocabulary
ISO 5393	Rotary Pneumatic Tools for Threaded Fasteners - Performance Tests
ISO 6544	Hand-held Pneumatic Assembly Tools for Installing Threaded Fasteners - Reaction Torque Impulse Measurements
ISO 15744	Acoustics - Noise Test Code for Hand-held Non-Electric Power Tools - Engineering Method
PN8NTC1.2	Measurement of Noise Emitted by Hand-held Pneumatic Tools (Available from PNEUROP)
ANSI B7.1	The Use, Care, and Protection of Abrasive Wheels
ANSI B7.5	Safety Code for the Construction, Use, and Care of Gasoline-Powered, Hand-held, Portable, Abrasive Cutting off Machines
ANSI B30.16	Overhead Hoists (Underhung)
ANSI B74.18	American National Standard for Grading of Certain Abrasive Grain on Coated Abrasive Material
ISO 8662-1	Measurement of Vibration in Hand-held Power Tools - Part 1: General
ISO 8662-2	Measurement of Vibrations in Hand-held Power Tools - Part 2: Chipping Hammers, Riveting Hammers
ISO 8662-3	Measurement of Vibrations in Hand-held Power Tools - Part: 3: Rotary Hammers and Rock Drills
ISO 8662-4	Measurement of Vibrations in Hand-held Power Tools - Part 4: Grinders
ISO 8662-5	Measurement of Vibrations in Hand-held Power Tools - Part 5: Breakers and Hammers for Construction
ISO 8662-6	Measurement of Vibrations in Hand-held Power Tools - Part 6: Impact Drills
ISO 8662-7	Measurement of Vibrations in Hand-held Power Tools - Part 7: Wrenches, Screwdrivers and Nutrunners with Impact, Impulse or Ratcheting Action
ISO 8662-8	Measurement of Vibrations in Hand-held Power Tools - Part 8: Polishers and Rotary, Orbital and Random Orbital Sanders
ISO 8662-9	Measurement of Vibrations in Hand-held Power Tools - Part 9: Rammers
ISO 8662-10	Measurement of Vibrations in Hand-held Power Tools - Part 10: Nibblers and Shears
ISO 8662-11	Measurement of Vibrations in Hand-held Power Tools - Part 11: Fastener Driving Tools

ISO 8662-12	Measurement of Vibrations in Hand-held Power Tools - Part 12: Saws and Files with Reciprocating Action and Saws with Oscillating or Rotating Action
ISO 8662-13	Measurement of Vibrations in Hand-held Power Tools - Part 13: Die Grinders
ISO 8662-14	Measurement of Vibrations in Hand-held Power Tools - Part 14: Stone Working Tools and Needle Scalers
ISO 8662-16	Measurement of Vibrations in Hand-held Power Tools - Part 16: Screw Drivers (rapping)
ISO 5349-1	Mechanical Vibration - Measurement and Evaluation of Human Exposure to Hand-Transmitted Vibration - Part 1: General Requirements
ISO 5349-2	Measurement and Evaluation of Human Exposure to Hand-Transmitted Vibration - Part 2: Practical Guidance for Measurement at the Work Place

General

ANSI/ISA S7.0.01	Quality Standard for Instrument Air
ANSI S12.12	Engineering Method for Determination of Sound Power Levels of Noise Sources Using Sound Intensity
ASME B31.1	Power Piping
ASME BPVC Section VIII	Rules for Construction of Pressure Vessels
ISO 2398	Rubber Hose, Textile-Reinforced, for Compressed Air -Specification
ISO 3746	Acoustics – Determination of Sound Power Levels of Noise Sources Using Sound Pressure - Survey Method Using an Enveloping Measurement Surface Over a Reflecting Plane
ISO 3747	Acoustics – Determination of Sound Power Levels of Noise Sources Using Sound Pressure - Survey Method Using a Reference Sound Source
ISO 3857-1	Compressors, Pneumatic Tools and Machines - Vocabulary - Part 1: General
ISO 3857-2	Compressors, Pneumatic Tools and Machines - Vocabulary - Part 2: Compressors
ISO 3857-3	Compressors, Pneumatic Tools and Machines - Vocabulary - Part 3: Pneumatic Tools and Machines.
NFPA 99	Health Care Facilities
ANSI Z87.1	Occupational and Educational Eye and Face Protection
ANSI S12.6	Hearing Protectors

OSHA Regulations, 29 CFR, 1926.302 – Safety Equipment
OSHA Regulations, 29 CFR, 1910.133 – Eye and Face Protection
OSHA Regulations, 29 CFR, Section 1910.95 – Occupational Noise Exposure
OSHA Appendix F – Table 1 – Breathing Air Systems for use with Pressure-Demand Supplied Air Respirators in Asbestos Abatement

INDEX

A

B

C

D

H

I

K

L

M

N

O

P

Q

R

S

T

U

V

W

X